Fundamentals of ANALYTICAL
CHEMISTRY

Fundamentals

of ANALYTICAL CHEMISTRY

Douglas A. Skoog
Stanford University

Donald M. West
San Jose State College

HOLT, RINEHART AND WINSTON

New York
Chicago
San Francisco
Toronto
London

preface

Our principal aim in the preparation of this text has been to provide the reader with an appreciation for the over-all analytical process. It is our belief that this end is best served through consideration of the theoretical backgrounds of chemical analysis. We in no way deny the need for development of sound laboratory techniques; we feel, rather, that these become obvious to the person who understands thoroughly the basis for the experiment he performs. In the same vein we believe that, left to his own devices, the person who recognizes the total problems associated with a quantitative analysis is most likely to make a prudent choice among the methods available to him.

In pursuit of this aim we have sought:

To stress that the final operation in a quantitative analysis represents but one aspect of a much larger problem.

To minimize the rather arbitrary distinction between the so-called classical and instrumental methods through emphasis upon the similarities rather than the differences between them.

To present, in a first course, a representative sampling of the common methods for completion of an analysis.

To provide the reader with rational methods of approach to the mathematical aspects of analytical chemistry, with particular emphasis upon the calculations associated with chemical equilibrium.

In keeping with our belief that a good text should also serve a reference function, we have included material that is not ordinarily a part of an elementary course. Numerous and frequent references to the chemical literature are included as guides to those who wish further information upon selected topics.

Specific directions are supplied for the common laboratory operations, and for a number of representative methods of analysis as well.

We have divided the material into six sections. Part 1 contains an introductory chapter, a review of simple stoichiometric relationships, and an introduction to the problems associated with the evaluation of experimental data. Part 2 deals with the general topic of gravimetric analysis, and Part 3 with volumetric analysis. Part 4 is concerned with the field of electroanalytical chemistry and Part 5 with optical methods for completion of the analysis. Part 6 covers aspects of chemical analysis other than those associated with the final measurement.

A fair degree of flexibility should result from this arrangement. Chapter 2 will represent a review for most students and needs not be formally assigned. Chapters 5 and 11 deal with practical matters and are thus the province of the laboratory portion of the course. If desired, volumetric analysis can be considered before gravimetric analysis. Parts 4 and 5 may become the basis for a second semester course after the student has had some training in organic and physical chemistry, although the attempt has been made to present this material in a way that does not require these courses as prerequisites. The subject matter of Part 6 can be covered in the introductory course or postponed to the second semester, as desired.

We wish to thank Professors Joseph Jordan and Richard Bowers who read the entire manuscript and offered many helpful criticisms and suggestions.

D.A.S.
D.M.W.

November, 1962
Stanford, California
San Jose, California

contents

part 2. Gravimetric Analysis

part 3. Volumetric Analysis

part 4. Electroanalytical Methods

part 5. Optical Methods of Analysis

part 6. The Complete Analysis

Fundamentals of ANALYTICAL
CHEMISTRY

part 1. *Introduction*

chapter 1. *The Scope of Analytical Chemistry*

Analytical chemistry comprises the techniques and methods that provide answers to the questions "What?" and "How Much?" with respect to the chemical composition of a sample of matter. The former is the province of qualitative analysis, with which most readers have had an introduction. Quantitative analysis is concerned with the problems attending the determination of the amount of species present in a given sample. Although our present concern is principally with the latter, it is instructive to consider the similarities and dissimilarities between these two broad types of analysis. Both make use of whatever physical or chemical properties are convenient to the purpose at hand, be it detection or estimation; in some instances, the same property is useful for both purposes. Both generally require preliminary steps to isolate the species in question from those others that will interfere with the analytical process. It is noteworthy, however, that partial loss of the species of interest can be tolerated during the separations preceding a qualitative test; it is necessary only that sufficient remains to give a test. In contrast, these losses must be kept to an absolute minimum during the separations preliminary to a quantitative determination. A qualitative test can be based upon a chemical reaction that does not proceed to completion; it may actually consist of several reactions occurring simultaneously to give a number of products. On the other hand, the quantitative estimation of some species requires that the observed effect be related to its concentration in some reproducible manner; in general, therefore, the reactions useful for quantitative analysis are restricted to those that proceed to essential completion with the formation of a single set of products. Furthermore, when a quantitative reaction is complete, it must be possible to measure some property of the system that is reproducibly related to the amount of species being analyzed. These additional limitations restrict the number of reactions that are suitable for quantitative analysis.

The Importance of Analytical Chemistry

Historically, analytical chemistry has always occupied a vital position in the development of chemistry. The successful elucidation of the process of combustion by Lavoisier was due mainly to his employment of a balance in his investigations; he was among the first to recognize the immense power of quantitative measurements in chemical research. The atomic concept of matter dates back at least to ancient Greece, and certainly was not original with John Dalton. Dalton's contribution, above all, was to introduce a quantitative aspect to this notion—an aspect that was verifiable by actual experiment. In a very real sense, then, chemical analysis provided the support necessary to convert the atomic theory from a philosophical abstraction into something of physical significance.

Early chemistry was principally analytical in nature. Only as the body of experimental fact increased did it become possible for the chemist to specialize—according to his interests—in other fields. Irrespective of choice, however, he continued to rely heavily upon analytical methods and techniques to provide him with experimental information. Analytical chemistry thus assumed the supporting role of an indispensible tool in advancing the state of knowledge in the fields of inorganic, organic, and physical chemistry.

This situation is as applicable to the chemistry of today as to that of the past; every experimental investigation relies, to an extent, upon the results of analytical measurements. A thorough background in analytical chemistry is thus a vital necessity for all who aspire to be chemists, regardless of their field of specialization. Nor need these remarks be limited to prospective chemists. Investigators in virtually all of the physical and biological sciences are obliged to make use of analytical data in the course of their work. The physician relies heavily upon the results of the analysis of body fluids in making his diagnoses. Analytical techniques are indispensible in the biochemist's study of living matter and its metabolic processes. The classification of a mineral is incomplete without knowledge of its chemical composition. Analytical techniques are employed by the physicist in identifying the products of high-energy bombardments. A catalogue such as this can be extended virtually without limit.

The employment of analytical chemistry in modern industry is of inestimable importance. It is difficult to imagine an article of present-day commerce whose raw materials have not, at some stage, been subjected to analytical control. The uniform quality of the paper upon which these words are printed is due in part to careful analysis during the various phases of its production; hundreds of analyses are performed upon the materials that go into as complex a commodity as an automobile.

Finally, aside from these highly practical considerations, a study of quanti-

Table 1-1

CLASSIFICATION OF COMMON ANALYTICAL METHODS

General Classification	Subclassification	Quantity Measured
Gravimetric	direct method	weight of compound containing species
	indirect method	loss of weight due to volatilization of species
Volumetric	titration methods	volume of solution that is chemically equivalent to the wanted species
	gas analysis	volume of a gaseous species produced or consumed
Optical	emission spectroscopy	radiation emitted by species
	absorption spectroscopy including colorimetry	radiation absorbed by a species
	polarimetry	rotation of polarized light by species
	refractometry	refractive index of solution of species
	turbidimetry, nephelometry	scattering of radiation by species
Electro-analytical	potentiometry	potential of an electrode in equilibrium with the species
	conductimetry	conductance of a solution of the species
	coulometry	quantity of current equivalent to the species
	polarography	current associated with reaction at a polarizable electrode
	high-frequency methods	capacitance of solution of species
Miscellaneous	mass spectroscopy	charge to mass ratio of decomposition products of species
	radiochemical methods	radioactive decay of species
	thermal-conductivity methods	thermal conductance of species
	enthalpy titrations	heat of reaction of species

tative analysis is of benefit in that it places the highest premium upon careful, orderly work and intellectually honest observation; regardless of one's ultimate field of endeavor, these are traits worthy of cultivation.

Classification of the Methods of Quantitative Analysis

The ultimate aim of a quantitative analysis is to ascertain how much of a given species is present in a sample of matter; depending upon the procedure employed, this may be accomplished directly or very indirectly. Regardless of how it is done, however, a final measurement of some sort is inherent in every determination and from this, the quantity of the species in question is derived. It is convenient to classify the methods of quantitative analysis according to the nature of this final measurement. Thus if this consists of securing the weight of a solid, the method is classed as a *gravimetric* analysis; where the final measurement involves determination of a volume, the method is called a *volumetric* analysis; if the absorption of light is measured, the procedure is sometimes termed a *colorimetric* analysis; and where an electrical property is determined, the method can be classified as *electroanalytical*. As may be seen from Table 1-1, the variety of measurements that are useful in analysis are many.

From a historical standpoint, the majority of early analytical methods were either gravimetric or volumetric procedures; for this reason these are sometimes known as *classical* methods of analysis. Procedures based on the measurement of optical, electrical, thermal, and other properties were developed later, and these as a group are sometimes called *instrumental* methods. In many respects this dual classification is unfortunate, for it implies that classical and instrumental methods are basically different. In reality, fundamental differences between the two categories do not exist; both are based upon the correlation of a physical measurement with concentration; both employ an instrument for this measurement; to the extent that neither is specific, separations often precede both types of analysis. Finally, the treatment of the sample preliminary to the analysis is basically the same for both. The classification of methods as classical or instrumental is thus founded largely on chronological development.

Steps Common to All Quantitative Analytical Methods

A number of preliminary operations are likely to be a part of all analytical procedures. Although detailed treatment of these must be deferred to subsequent chapters, it is worthwhile to indicate their nature at this point.

1. Decision regarding the method to be used. Careful consideration must be given to the choice of method employed in the analysis of a substance. In making his choice the chemist must take several factors into account. Principal among these is the reliability with which the data must be known and the time allowable for obtaining the results. Quite generally the time required for an analysis increases tremendously as the accuracy requirement becomes more strict. Often, then, the chemist is faced with the difficult task of trying to find an analytical method that represents a satisfactory compromise between the demands of accuracy and time expended.

Another consideration will be the number of samples to be analyzed. When these are many, the chemist can afford to choose a method that requires such preliminary preparations as the assembling of instrumental equipment, preparation of solutions, or calibrations. On the other hand, with only a single sample or a few samples he may find it more expedient to choose a procedure that avoids such preliminary steps. Finally, the method chosen will always be governed by the complexity of the substance being analyzed and whether the analysis is for a single constituent or for all components in the sample.

The choice of method is often a very difficult step in an analysis, and one that requires a good deal of experience and chemical intuition.

2. Sampling. To produce meaningful results the analysis must be performed on a sample whose composition faithfully reflects that of the bulk of material from which it was taken. The problem of procuring a representative sample is frequently an imposing one. Consider, for example, a railroad car containing 25 tons of silver ore. Buyer and seller must come to agreement regarding the value of the shipment; this will be based primarily upon its silver content. The ore itself is inherently heterogeneous, consisting of lumps of varying size and of varying silver content. The actual assay of this shipment is performed upon a sample that weighs perhaps 1 gram; its composition must be representative of the 25 tons—or approximately 22,700,000 grams—of ore in the shipment. It is clear that the selection of a small sample for this analysis cannot be a simple one-step operation; in short, a systematic preliminary manipulation of the bulk of material will be required before it becomes possible to select 1 gram and have any confidence that its composition is typical of the nearly 23,000,000 grams from which it was taken.

3. Preparation of the sample for analysis. Once a sample is selected, it must be converted into a form that is suitable for analysis. With solid materials this may involve grinding to reduce the particle size and then mixing to insure homogeneity of the solid. The former operation is particularly important with difficultly soluble substances since the rate of solution is increased greatly by reduction in the size of particles; the sampling operation itself frequently includes preliminary crushing and grinding steps.

Another preparatory operation of considerable importance with most solids is that of removal of adsorbed moisture. Often the percentage composition of a substance will depend upon the humidity of its surroundings at the time of the analysis owing to the adsorption or desorption of water. In order to avoid the problems arising from such variations it is common practice to attempt the removal of all such adsorbed moisture and base the analysis on a dry sample. Frequently this will involve heating of the material to about 100° C, or exposing it to a dry atmosphere for suitable periods of time.

4. Procurement of a measured quantity of the sample. Quantitative analytical results are usually reported in relative terms—that is, in some way that expresses the quantity of the desired component present per unit weight or volume of sample. Percentage composition is probably the most commonly employed of these. In order to express the results in a meaningful manner it is therefore necessary to know the weight of the sample upon which the analysis is performed. The purpose of all subsequent steps is the evaluation of the weight of the desired species present in this weight of sample.

5. Solution of the sample. Many, but by no means all, methods of analysis are performed upon a solution of the substance being analyzed. In these cases, the choice of solvent and method of solution becomes an important preliminary to the analysis.

The choice of solvent is limited by several considerations. Generally, it is desirable that the solvent be effective in dissolving all of the components of the sample, not just the species being determined. The solution process should take place in a reasonable period of time. The chemical composition of the solvent should be such as to offer no interference in the subsequent steps of the analysis; lacking this quality, such interferences it does produce should be easy to eliminate.

In general, it is advisable to employ the mildest solvent and dissolving temperature that is consonant with the expeditious solution of the sample. The needless employment of concentrated reagents can produce a reaction that is sufficiently violent to cause the physical loss of a portion of the finely divided sample. A considerable number of inorganic substances (particularly chlorides) are appreciably volatile at moderate temperatures, and the indiscriminate use of heat to hasten the dissolving process often leads to serious losses of the substance of interest.

6. Separation of interfering substances. There are few, if any, chemical or physical properties of importance in analysis that are entirely unique to a single chemical species; rather, the reactions used and the properties measured are generally characteristic of a number of elements or compounds. This lack of truly specific reactions and properties adds greatly to the difficulties faced by the chemist undertaking an analysis; it means that a scheme must be devised for isolating the species of interest from all others present in the original material that produce an effect upon the final measurement. Compounds or elements that thus prevent the direct measurement of the species being determined are called *interferences*, and their separation prior to the final measurement constitutes an important step in most analyses. No hard and fast rules can be given for the elimination of interferences; this problem is often the most imposing aspect of the analysis.

7. The completion of the analysis. All of the preliminary steps in an analysis are undertaken in order to make the final measurement a true gauge of the quantity of the species being determined. An indication of diversity that characterizes final measurements can be gained from inspection of Table 1-1. It is with this aspect that an introductory course in quantitative analysis is chiefly concerned. The student should constantly bear in mind, however, that in most instances the final measurement is the least difficult aspect of the analysis.

8. Calculation and interpretation of results. The numerical values of the experimental data comprising the final measurement bear a direct relationship to the quantity of species present in the sample. Since the information desired from the analysis is usually extensive in nature, it is necessary to relate mathematically the value of the final measurement to the weight of the sample employed in its procurement.

Finally, the report of a quantitative analysis is incomplete without some indication being given regarding the reliability of the results. Such an evaluation may be based upon previous experience with the method in question, upon an estimate of the reliability that is likely from the method employed, or from the analytical data themselves.

While the foregoing can be considered to represent the steps common to all quantitative analyses, it must be recognized that no inclusive statement can be made regarding their relative difficulty. Each analysis undertaken is unique to the extent of the importance of the problems arising in the various steps; that which represents the most formidable aspect of one determination may actually be of negligible importance in another.

It should further be pointed out that the mechanical skills necessary to the completion of a successful analysis can readily be taught to persons having little or no scientific background; indeed, a routine analysis may be performed better by the technician than by his more highly trained supervisor. However, an intelligent appraisal of the factors of importance (and those of no import) in the several steps comprising a quantitative analysis requires more chemical knowledge than that possessed by the technician. It is this added knowledge with which this text is concerned.

chapter 2. *Some Elementary Concepts Important to Quantitative Analysis*

A few fundamental concepts are of particular significance to analytical chemistry. These have to do with the chemical composition of aqueous solutions of solutes, the methods for expressing the concentrations of solutions, and the extent to which chemical reactions proceed.

THE CHEMICAL COMPOSITION OF SOLUTIONS

Strong and Weak Electrolytes

Most of the compounds with which we shall be concerned here are electrolytes; that is, in aqueous solutions they are dissociated to a greater or lesser degree into ions. A convenient method of classifying compounds is on the basis of the extent to which they do dissociate. Strong electrolytes are those that are completely or nearly completely dissociated; weak electrolytes are those that are only partially ionized.

It will frequently be important for our purposes to recognize whether we are dealing with solutions of weak or strong electrolytes. Table 2-1 will serve as an aid in making this distinction.

Table 2-1

CLASSIFICATION OF ELECTROLYTES

Strong Electrolytes	Weak Electrolytes
1. The inorganic acids HNO_3, $HClO_4$, H_2SO_4,[1] HCl, HI, HBr, $HClO_3$, $HBrO_3$	1. Many inorganic acids such as H_2CO_3, H_3BO_3, H_3PO_4, H_2S, H_2SO_3, etc.
2. Alkali and alkaline-earth hydroxides as well as some of the heavy metal hydroxides	2. Most organic acids
3. Most salts	3. Ammonia and most organic bases
	4. Halides, cyanides, and thiocyanates of Hg, Zn, and Cd

[1] H_2SO_4 is completely dissociated into HSO_4^- and H^+ ions and for this reason is called a strong electrolye. However, it should be noted that the HSO_4^- ion is a weak electrolyte, being only partially dissociated.

Acid-base Concepts

Historically speaking, the classification of certain substances as acids or bases was founded upon several characteristic properties that these compounds impart to an aqueous solution. For example, an acid causes litmus to turn red, gives a sharp taste to water, and reacts with a base to form a salt. A base, on the other hand, gives a bitter taste and a slippery feel to water, turns litmus blue, and produces a salt with an acid.

In the late nineteenth century, Arrhenius proposed a more sophisticated basis of classification. He defined acids as substances that dissociate into hydrogen ions and anions when dissolved in water; bases, on the other hand, give hydroxyl ions and cations upon the same treatment. This proposal was a fruitful one, indeed, for it led to a quantitative treatment of the idea of acidity and basicity. Thus, by measuring the degree of dissociation of an acid or a base in aqueous solution, it became possible to compare the relative strengths of these compounds. In addition, the theory provided a foundation for the mathematical treatment of the equilibria established when acids and bases react with one another.

A serious limitation to the Arrhenius theory is that it fails to recognize the role played by the solvent in the dissociation process. It remained for Brønsted and Lowry, in 1923, to propose independently a more generalized concept of acids and bases. They defined an acid as any substance that donates a proton and a base as any substance that accepts a proton. According to this definition a base is formed whenever an acid loses a proton; that is,

$$\text{acid} \rightleftharpoons H^+ + \text{base}$$

The base produced by such a process is called the *conjugate base* of the acid; it may be an anion, a cation, or a neutral particle. The acid in this instance can also be considered to be the *conjugate acid* of the base taking part in the reaction.

An important aspect of the Brønsted concept is the involvement of the solvent in the "dissociation" of an acid or a base. For example, when gaseous hydrogen chloride is dissolved in water, an acid-base reaction occurs in which the water acts as a base (proton acceptor).

$$\begin{array}{cccc} HCl + & H_2O & \rightleftharpoons H_3O^+ + & Cl^- \\ \text{acid}_1 & \text{base}_1 & \text{acid}_2 & \text{base}_2 \end{array}$$

Here, chloride ion is the conjugate base of the acid HCl, and hydronium ion is the conjugate acid of the base water. On the other hand, when a base such as ammonia is dissolved in water, the solvent behaves as an acid (proton donor).

$$\begin{array}{cccc} NH_3 + & H_2O & \rightleftharpoons NH_4^+ + & OH^- \\ \text{base}_1 & \text{acid}_1 & \text{acid}_2 & \text{base}_2 \end{array}$$

The dissociation of water, in this view, is simply an acid-base reaction.

$$\begin{array}{cccc} H_2O + & H_2O & \rightleftharpoons H_3O^+ + & OH^- \\ \text{acid}_1 & \text{base}_1 & \text{acid}_2 & \text{base}_2 \end{array}$$

Reactions such as these are not confined exclusively to water but are found also in other solvents that have acidic or basic properties. Thus, when HCl is dissolved in ethyl alcohol or in glacial acetic acid, we may write

$$HCl + C_2H_5OH \rightleftharpoons C_2H_5OH_2^+ + Cl^-$$
$$HCl + HC_2H_3O_2 \rightleftharpoons H_2C_2H_3O_2^+ + Cl^-$$

The extent to which these reactions proceed depends upon the basic strength of the solvent. For example, water is a sufficiently strong base to cause the complete dissociation of hydrochloric acid, and undissociated hydrochloric acid molecules do not exist as such in aqueous solution. On the other hand, glacial acetic acid causes only partial dissociation of the hydrogen chloride molecules because it is a much weaker base; in this solvent, then, hydrogen chloride is no longer a "strong" acid.

An important consequence of the Brønsted concept is that the classification of acids or bases as strong or weak now becomes dependent upon the solvent system. For example, perchloric, hydrochloric, and hydrobromic acids are all completely dissociated and thus strong acids in aqueous solution. Differences in strength are observed, however, if a less basic solvent is used. Thus, in glacial acetic acid only perchloric acid dissociates completely; of the other two, hydro-bromic appears to be the stronger acid. We may, therefore, say that perchloric acid is inherently a stronger acid than either of the other two and that hydro-bromic acid is stronger than hydrochloric acid. Water has a leveling effect on the strengths of the three acids by converting each completely to the hydronium ion and the corresponding anion. Actually then, the hydronium ion is the strongest acid that can exist in an aqueous solution. Similarly, the ion $H_2C_2H_3O_2^+$ is the strongest acid that can exist in glacial acetic acid solutions, any stronger acid being converted completely to this species by an acid-base reaction with the solvent. This species is produced by a reaction analogous in every way with that for the formation of hydronium ion

$$HC_2H_3O_2 \; + \; HC_2H_3O_2 \; \rightleftharpoons \; H_2C_2H_3O_2^+ \; + \; C_2H_3O_2^-$$
$$\text{acid}_1 \qquad\quad \text{base}_1 \qquad\quad \text{acid}_2 \qquad\quad \text{base}_2$$

Here, the acetate ion corresponds to the hydroxyl ion in water.

There are a number of inert solvents, such as benzene and carbon tetra-chloride, that have no appreciable acidic or basic properties; when an acid is dissolved in these, no dissociation can occur. The acidic property of the solute becomes apparent, however, when a soluble base is added to the solution; then an equilibrium between the acid and base is set up.

$$HA \; + \; B \; \rightleftharpoons \; HB \; + \; A$$
$$\text{acid}_1 \quad\; \text{base}_1 \quad\; \text{acid}_2 \quad\; \text{base}_2$$

The position of this equilibrium is determined by the acidic and basic strengths of the reactants and products.

From these rather brief remarks we see that the Brønsted concept of acids and bases is particularly useful when comparing the behavior of substances in several solvents. In considering the reaction of solutes in a single solvent, the classical Arrhenius approach is convenient and adequate for most purposes; it is the one that we shall use predominately in this text since the majority of the reactions with which we will be concerned are carried out in water as a solvent. Thus, we will use the symbol H^+ to represent the solvated proton and speak of the ammonium ion as the component of a salt rather than as an acid. In a few instances, we will need to employ Brønsted's ideas, however, particular-ly when dealing with acid-base titrations in nonaqueous solvents (Chap. 16). For a complete discussion of the Brønsted concept and its implications in ana-lytical chemistry, a number of excellent references are available.[2]

Dissociation of Water

Water is an extremely weak electrolyte, being dissociated to the extent of only a few parts per billion. This ionization, minute as it is, assumes considerable

[2] I. M. Kolthoff and P. J. Elving, *Treatise on Analytical Chemistry*, Part I, **1**, Chaps. 11 and 13. New York: The Interscience Encylopedia, Inc., 1959.

importance because of the frequency with which its products influence chemical reactions in aqueous solutions.

The dissociation of water is most accurately represented by the equation

$$2H_2O \rightleftharpoons H_3O^+ + OH^-$$

The positive ion so formed is called the *hydronium ion*, the proton being bonded to the water molecule via a covalent bond involving one of the unshared pair of electrons of the oxygen. This hydrate is so remarkably stable that there are essentially no simple protons, as such, in aqueous solution. Higher hydrates of the proton are also undoubtedly present, but these are nowhere near as stable as the hydronium ion itself.

In order to emphasize the unique character of the singly hydrated proton, many chemists choose to use the notation H_3O^+ in writing equations for reactions in which the proton is a participant. For the purpose of describing the stoichiometry of a reaction in aqueous solution, however, the formula H^+ is equally satisfactory, and this notation does possess the distinct advantage of simplifying the process of writing and balancing equations. For this reason, then, we shall follow the common convention of representing the hydrogen ion as H^+ unless a special need arises for conveying a more complete picture of this ion in solution. The reader should recognize, however, that hydration of the proton in aqueous solution is very extensive.

SOME WEIGHT AND CONCENTRATION UNITS

Units of Weight

In the laboratory the chemist generally measures and records the masses of objects in units of grams, milligrams, or micrograms (the microgram is 10^{-6} gram and is sometimes called a *gamma*). In carrying out chemical calculations, however, he finds it more convenient to convert these into units that express the weight relationships among substances in terms of small whole numbers. The gram formula weight, the gram molecular weight, and the gram equivalent weight are such units, and they find wide application in the computations of analytical chemistry; these terms are commonly referred to as the *formula weight*, the *molecular weight*, or the *equivalent weight*, respectively.

We shall defer a definition of the gram equivalent weight until a later time; but it will be worthwhile to consider the other two units here and to differentiate rather carefully between them. Before doing this, however, we will discuss briefly the significance of the chemical formula.

The empirical formula for a substance is the simplest combination of the number and species of atoms present in that substance. It is also the chemical formula unless experimental evidence indicates that the fundamental aggregate of atoms is actually some multiple of the empirical formula. For example, we use H_2 as the chemical formula for hydrogen gas because there is abundant evidence that this substance is ordinarily found in the form of a diatomic molecule. On the other hand Ne serves adequately to describe the composition of neon gas which is monatomic in nature.

The entity expressed by the chemical formula, however, may or may not be real. For example, there is no evidence for the existence of molecules of sodium chloride as such in either the pure solid or in solution. Rather, this substance is made up of sodium ions and chloride ions, no one of which can be said to be in simple combination with any other single ion. Despite this it is convenient to assign the formula NaCl to sodium chloride. Also the chemical formula of a substance is often the formula of the predominant species only. Thus, liquid water contains H_2O, H_4O_2, H_3O^+, OH^-, and others. The formula given to water is that of the predominant species, H_2O, and as such is only an approximation of the composition of the real substance.

The *gram formula weight (GFW)* of a substance is the summation of the atomic weights, in grams, of all the atoms appearing in the chemical formula for that substance. Thus, the formula weight for hydrogen is 2.016 grams and that for sodium chloride is 58.446 grams. The definition for the formula weight carries no inference regarding the existence or nonexistence of the substance for which it is calculated.

We shall define the *gram molecular weight* as the formula weight of a single, real chemical species. According to this definition, then, the molecular weight of H_2 is 2.016 grams; no molecular weight can be assigned to the species NaCl since there is no evidence for its existence. We can, however, properly assign weights to Na^+ (22.991 grams) and Cl^- (35.457 grams) since these are real chemical species; these might more properly be called "ionic weights" rather than "molecular weights."

Since we have restricted its use to real, chemical species, 1 gram molecular weight of any substance has enumerative significance, being the weight of 6.02×10^{23} particles of that species; this number of particles is commonly referred to as the *mol*. Thus, 100 grams of H_2 contains $100/2.016 = 49.60$ mols of this substance, or $49.60 \times 6.02 \times 10^{23}$ molecules. A like weight of sodium chloride contains $100/58.448 = 1.711$ formula weights of NaCl, 1.711 mols of Na^+, and 1.711 mols of Cl^-.

Let us further distinguish between the formula and molecular weight by again considering a quantity of water that weighs 18.016 grams. This sample clearly contains 1 gram formula weight of water; it contains slightly less than 1 mol of water, however, because some is in the form of H_4O_2, H_3O^+, OH^-, and such other species as may be present.

Not all chemists make this differentiation between formula weight and molecular weight, and these terms are frequently employed synonymously. We have chosen to use the more restricted definition for molecular weights because we believe this leads to less ambiguity insofar as the expression of concentration of solutions is concerned.

The terms *milliformula weight* and *millimolecular weight* are also encountered; these are simply 1/1000 of a formula weight and molecular weight, respectively.

Methods for the Expression of Concentration

Formality or formal concentration. The *formality*, *F*, of a solution expresses the number of gram formula weights of a solute that are present in 1 liter of solution. It follows that this also gives the number of milliformula weights of solute in 1 ml of the solution.

Molarity or molar concentration. The *molarity*, *M*, of a solution defines the number of gram molecular weights (or mols) of a species in 1 liter of solution, or the number of millimolecular weights in 1 ml of solution.

The molar concentration and the formal concentration of a given solution are usually not the same. It is worthwhile illustrating this. A solution prepared by dissolving 1 gram formula weight of oxalic acid (90.04 grams of $H_2C_2O_4$) in water and diluting to exactly 1 liter is, by definition, 1 formal. The molar concentration, however, will be less than one because oxalic acid undergoes dissociation to the extent of about 22 percent. As a consequence the concentration of the *species* $H_2C_2O_4$ in this solution will be approximately 0.78 molar; in contrast the concentration of the *substance* oxalic acid will be 1.00 formal.

Another example would be a solution prepared by dissolving 1 formula weight of sodium chloride (58.45 grams) in sufficient water to give 1 liter of solution. This will be 1.00 formal with respect to NaCl, by definition; the concentration of both Na^+ and Cl^- will be 1.00 molar. The molar concentration of the species NaCl will be zero. This is the general situation for solutions of all strong electrolytes.

From these examples we may conclude that quantitative information is needed with regard to the behavior of the solute in the solution before its molar concentration can be specified. On the other hand the formal concentration may be calculated from the recipe used for preparing the solution and the formula weight of the solute.

Normality or normal concentration. This method of expressing concentration will be defined in Chapter 10.

Percentage concentration. Chemists frequently denote the concentration of solutions in terms of percentages. This unfortunate practice often leads to ambiguity, inasmuch as the percentage composition of a solution can be expressed in any of several ways. Three of the more common methods are defined as follows:

$$\text{weight percent} = \frac{\text{weight of solute}}{\text{weight of solution}} \times 100$$

$$\text{volume percent} = \frac{\text{volume of solute}}{\text{volume of solution}} \times 100$$

$$\text{weight-volume percent} = \frac{\text{weight of solute, grams}}{\text{volume of solution, ml}} \times 100$$

The reader should note that the denominator in each case refers to the solution and not to the solvent alone. Furthermore the first two expressions are independent of the units of volume or weight used in their definition whereas the third is not. Weight percent has the great advantage of being independent of temperature while the other percentages are not; for this reason it is widely used. For example, the data found on the labels of commercial aqueous reagents, such as the mineral acids or ammonia, are of this form; that is, concentrated nitric acid is stated to be 70 percent HNO_3, which means that the reagent contains 70 grams of HNO_3 per 100 grams of the concentrated solution.

Weight-volume percent is frequently used to give the concentration of dilute aqueous solutions of solid reagents; thus, a 5-percent silver nitrate solution is one prepared by dissolving 5 grams of silver nitrate in water and diluting to 100 ml.

To avoid ambiguity, the use of percent composition as a means of expressing concentration requires that the type of percentage employed be explicitly stated. This information is often omitted, which makes it necessary for the reader to decide intuitively which of the several types has been employed by the writer. Large uncertainties in the interpretation of experimental procedures or data can result from this omission. In the absence of specific information one commonly assumes that a volume percent is being used if the solution involves a liquid solute and a weight-volume percent if the solute is a solid.

Parts per million. In dealing with concentrations of very dilute solutions, percentage compositions become awkward to use because of the number of zeros needed to place the decimal point. Under these circumstances the concentration can be conveniently expressed in terms of parts per million, ppm. This is defined by the equation

$$ppm = \frac{\text{weight of solute}}{\text{weight of solution}} \times 1,000,000$$

Thus a solution containing 0.0003 percent nickel by weight contains 3 ppm of nickel.

CHEMICAL EQUILIBRIUM

Most of the chemical reactions with which we shall be dealing yield products that react with one another to reform appreciable quantities of the starting substances. As a consequence these reactions are never entirely complete, but instead achieve a state in which the two opposing reactions occur at the same rate. For example, when iodide ion is added to a solution containing ferric ion, rapid formation of the orange-red triiodide ion occurs.

$$2Fe^{3+} + 3I^- \rightleftharpoons 2Fe^{2+} + I_3^-$$

The triiodide ions, however, are capable of reacting with the ferrous ions to produce the starting species. The concentration of the triiodide ion could be readily measured by the color it imparts to this system. Such a measurement would reveal that the quantity of this substance very rapidly reaches a constant level and remains there so long as conditions are not altered. If, however, we were able to examine this reaction on a molecular level, we would find that the two opposing reactions were occurring continuously, and the observed constant concentration of triiodide ion was the result of the equality in the reaction rates. Under these circumstances, an equilibrium state has been reached.

The relative concentration of reactants and products at equilibrium (that is, the position of the equilibrium) is greatly dependent upon the inherent nature of the substances involved. Thus if the species on the left side of a chemical equation have a much greater tendency to react than those on the right, the equilibrium will tend to lie to the right.

The position of an equilibrium can be altered by the application of stress to the system; this can take the form of an alteration in temperature, pressure, or concentration of one of the reactants. The effects of these variables can be predicted from the well-known *principle of LeChatelier* which states that a chemical equilibrium will always shift in such a direction as to counteract the effects of an added stress. Thus an increase in temperature will cause an equilibrium to shift in a direction that absorbs heat. Similarly an increase in pressure will result in a shift favoring production of those participants that occupy the smallest volume. The alteration of an equilibrium through the addition of a participating species is of special interest to the analytical chemist. We shall find it necessary to examine the quantitative aspects of this effect in considerable detail in later chapters.

Equilibrium Constants

Formulation of equilibrium constants. We have thus far considered the effects of temperature, pressure, and concentration on a chemical equilibrium in a qualitative way; it is quite possible to describe the influence of the latter two variables in quantitative terms by means of mathematical equations which are called *equilibrium-constant expressions*. These are of great importance in analytical chemistry; large portions of this text will be devoted to their application to chemical problems. The following paragraphs will serve as a brief introduction to the subject.

We will first consider a generalized equation for a chemical equilibrium

$$lL + mM + \cdots \rightleftharpoons pP + qQ + \cdots \tag{2-1}$$

where the capital letters represent the formulae of the chemical species participating in the reaction and the small letters are the integers required to balance the chemical equation. Thus this equation states that l mols of L reacts with m mols of M, etc., to yield p mols of P and q mols of Q, etc. The equilibrium-constant expression for this reaction is

$$K = \frac{[P]^p [Q]^q \cdots}{[L]^l [M]^m \cdots} \tag{2-2}$$

where the letters in brackets represent the molar concentrations of dissolved solute species or partial pressures (in atmospheres) if the reacting species are gases.[3] These concentrations are raised to powers that correspond numerically to the molar proportions required in the balanced equation for the process. The letter K represents a numerical constant that is called the *equilibrium constant* for the reaction. By convention, the products of the reaction *as written* are always placed in the numerator and the reactants in the denominator.

Equation (2-2) applies to all chemical equilibria and represents a mathematical expression of the *law of chemical equilibrium* or the *mass law*. It has a sound theoretical basis and can be derived from fundamental concepts. Furthermore, the expression has been found experimentally to apply to a wide variety of chemical reactions. It is certainly one of the most important generalizations of chemistry.

Equilibrium constants have been measured for a large number of chemical reactions; their values vary over a range of 10^{100} or more. A large value of K for an equilibrium clearly indicates that the reaction proceeds far to the right, while a value much smaller than unity indicates the position of equilibrium lies to the left. When the concentration of one of the participants in a reaction at equilibrium is changed independently, the concentrations of the other species readjust in such a way as to maintain K constant. When this occurs, the equilibrium is said to have been shifted. For example, the addition of substance L to the equilibrium represented by equation (2-1) would require a decrease in the concentration of M and a corresponding increase in the concentrations of P and Q until equilibrium was again achieved. At this new equilibrium the new concentrations of L, M, P, and Q would be such as to again give the constant K when substituted into (2-2).

Effect of temperature on equilibrium constants. The numerical value of an equilibrium constant is independent of concentration and pressure but dependent on the temperature at which the reaction takes place. As we might expect from our previous discussion the numerical values of equilibrium constants for some reactions increase with temperature increases, while others decrease and some change very little at all. For example, the equilibrium constant for the endothermic reaction

$$H_2O \rightleftharpoons H^+ + OH^-$$

increases by a factor of almost 100 as the temperature is raised from 0 to 100° C. On the other hand the constant for the exothermic reaction

$$H_2 + I_2 \rightleftharpoons 2HI$$

decreases with increasing temperature and is only 0.6 as great at 400° C as it is at 300° C. Thus a certain amount of care is required in the use of equilibrium constants to be sure that they are applied only at the proper temperature.

---

[3] The use of molar concentrations and partial pressures in this equation represents an approximation. More accurately the bracketed letters represent what are called *activities*. This is discussed further on page 23.

Some Common Types of Equilibrium Constants

The analytical chemist is vitally concerned with chemical equilibria in aqueous solutions. Six types are commonly encountered. These are listed in Table 2-2. Each will be considered in detail in later chapters.

Role of water. A concentration term for water does not appear in the equilibrium expressions in Table 2-2, even in those instances where water is a reactant or product. This is because these expressions apply only to relatively dilute systems where the concentration of water is enormous with respect to the other substances involved. As a consequence, its concentration remains essentially constant despite shifts in the equilibria, and the term for this substance can thus be considered to be part of the constant, K.

Role of a solid. In a heterogeneous equilibrium no concentration term is found for the solid in the equilibrium-constant expression. In this instance, the expression applies only when an excess of the solid phase is present in contact with the aqueous solution. The position of the equilibrium is determined by the concentration of the reactant *in the solid phase*. This, however, is *constant* and *independent* of the amount of solid and thus can be included in the constant, K_{sp}.

Stepwise equilibria. A number of weak electrolytes are capable of dissociation in a stepwise manner and dissociation-constant expressions can be written for each step. For example, oxalic acid dissociates to give both bioxalate and oxalate ions; thus we may write

$$H_2C_2O_4 \rightleftharpoons H^+ + HC_2O_4^- \qquad K_1 = \frac{[H^+][HC_2O_4^-]}{[H_2C_2O_4]}$$

and

$$HC_2O_4^- \rightleftharpoons H^+ + C_2O_4^{2-} \qquad K_2 = \frac{[H^+][C_2O_4^{2-}]}{[HC_2O_4^-]}$$

The subscripts on the constants are used to indicate the dissociation step to which K refers. The numerical value of K_2 will be less than K_1.

Numerical values for hydrolysis constants. The value for the hydrolysis constant for the salt of a weak acid or base is readily related to the ion-product constant for water and the dissociation constant for the acid or base. For example, the hydrolysis reaction for acetate ion is

$$OAc^- + H_2O \rightleftharpoons HOAc + OH^- \qquad K_h = \frac{[OH^-][HOAc]}{[OAc^-]}$$

In this instance, it is easily shown that

$$K_h = \frac{K_w}{K_a}$$

where K_a is the dissociation constant for acetic acid. Thus if we make appropriate substitutions for K_w and K_a, we find

$$K_h = \frac{[H^+][OH^-]}{[H^+][OAc^-]/[HOAc]}$$

Table 2-2

EQUILIBRIA AND EQUILIBRIUM CONSTANTS OF IMPORTANCE TO ANALYTICAL CHEMISTRY

Type of Equilibria	Name and Symbol of Equilibrium Constants	Typical Example	Equilibrium-constant Expressions
Dissociation of water	ion-product constant, K_w	$H_2O \rightleftharpoons H^+ + OH^-$	$K_w = [H^+][OH^-]$
Heterogeneous equilibrium between a slightly soluble substance and its ions in a saturated solution	solubility product, K_{sp}	$Ag_2CrO_4 \rightleftharpoons 2Ag^+ + CrO_4^{2-}$	$K_{sp} = [Ag^+]^2[CrO_4^{2-}]$
Dissociation of a weak acid or base	dissociation constant, K_d, K_a or K_b	$HCN \rightleftharpoons H^+ + CN^-$	$K_d = \dfrac{[H^+][CN^-]}{[HCN]}$
Hydrolysis of a salt	hydrolysis constant, K_h	$CN^- + H_2O \rightleftharpoons HCN + OH^-$	$K_h = \dfrac{[HCN][OH^-]}{[CN^-]}$
Dissociation of a complex ion	instability constant, K_{inst}	$Ni(CN)_4^{2-} \rightleftharpoons Ni^{2+} + 4CN^-$	$K_{inst} = \dfrac{[Ni^{2+}][CN^-]^4}{[Ni(CN)_4^{2-}]}$
Oxidation-reduction equilibrium	K	$MnO_4^- + 5Fe^{2+} + 8H^+ \rightleftharpoons Mn^{2+} + 5Fe^{3+} + 4H_2O$	$K = \dfrac{[Mn^{2+}][Fe^{3+}]^5}{[MnO_4^-][Fe^{2+}]^5[H^+]^8}$

which rearranges to the expression for the hydrolysis constant

$$K_h = \frac{[\text{HOAc}]\,[\text{OH}^-]}{[\text{OAc}^-]}$$

Few tables contain numerical values for hydrolysis constants since these may be readily calculated from dissociation constants.

Effect of Electrolyte Concentration on Chemical Equilibria

The equilibrium law, as we have presented it, is what is called a *limiting law* in the sense that it applies exactly only in very dilute solutions in which the electrolyte concentration is very small. At intermediate and high concentrations, marked variations in the numerical values of equilibrium constants based on molar concentrations are observed. This is demonstrated by the data in Table 2-3 which gives some values for the dissociation expression of acetic acid when the concentration terms are expressed in mols per liter. It is clear from these data that the degree of dissociation of acetic acid is appreciably greater in a sodium chloride solution than in pure water. Similar effects are encountered in the presence of other electrolytes; furthermore, other types of equilibria are affected in an analogous fashion. Thus the solubility of barium sulfate is greater in a solution of potassium nitrate than in water (see p. 150).

Table 2-3

DISSOCIATION CONSTANTS FOR ACETIC ACID IN SOLUTIONS OF SODIUM CHLORIDE AT 25° C [4]

Concentration of NaCl, Formality	Apparent K_d
0.00	1.75×10^{-5}
0.02	2.29×10^{-5}
0.11	2.85×10^{-5}
0.51	3.31×10^{-5}
1.01	3.16×10^{-5}

A large number of empirical observations on the effects of electrolyte concentration on chemical equilibria have revealed several important facts. One of these is that the magnitude of the effect is very much dependent upon the charges of the participants in the reaction. Where all are neutral particles, little variation in the equilibrium constant is observed; on the other hand, the

[4] H. S. Harned and B. B. Owen, *The Physical Chemistry of Electrolyte Solutions*, 3d ed., 676. New York: Reinhold Publishing Corporation, 1958.

effects become greater as the charges on the reactants or products increase. Thus, for example, of the following equilibria :

$$AgCl \rightleftharpoons Ag^+ + Cl^-$$

$$BaSO_4 \rightleftharpoons Ba^{2+} + SO_4^{2-}$$

the second is shifted further to the right in the presence of moderate amounts of potassium nitrate than is the first.

A second pertinent generality is that the observed effects are essentially independent of the kind of electrolyte and dependent only upon an electrolyte concentration parameter of the solution called the *ionic strength*. This is defined by the equation

$$\text{ionic strength} = \mu = \tfrac{1}{2} \left(m_1 Z_1^2 + m_2 Z_2^2 + m_3 Z_3^2 + \cdots \right) \tag{2-3}$$

where m_1, m_2, m_3, etc., are molar concentrations of the various ions in the solution and Z_1, Z_2, Z_3, etc., are their charges. The method of calculating the ionic strength of a solution is shown by the following examples.

Example. What is the ionic strength of a 0.1 F solution of KNO_3 and of a 0.1 F solution of Na_2SO_4?
For the KNO_3 solution m_{K^+} and $m_{NO_3^-}$ is 0.1 and

$$\mu = \tfrac{1}{2} (0.1 \times 1^2 + 0.1 \times 1^2) = 0.1$$

For the Na_2SO_4 solution $m_{Na^+} = 0.2$ and $m_{SO_4^{2-}} = 0.1$. Therefore

$$\mu = \tfrac{1}{2} (0.2 \times 1^2 + 0.1 \times 2^2) = 0.3$$

Example. What is the ionic strength of a solution that is 0.05 F in KNO_3 and 0.1 F in Na_2SO_4?

$$\mu = \tfrac{1}{2} (0.05 \times 1^2 + 0.05 \times 1^2 + 0.2 \times 1^2 + 0.1 \times 2^2)$$
$$= 0.35$$

From these examples it is apparent that the ionic strength of an electrolyte solution made up solely of singly charged ions is simply the sum of the formal concentrations of the various salts present. Where multiply charged species are involved, however, the ionic strength is greater than the formal concentration.

The ionic strength is a most important parameter, for it is found that the effects in which we are interested are independent of the kind of electrolyte so long as the ionic strength is held constant. Thus the degree of dissociation of acetic acid is the same in the presence of sodium chloride, potassium nitrate, barium iodate, or aluminum nitrate provided that the concentrations of these species are such that the ionic strength is fixed.

Activity and activity coefficients. The effects we have just described can be attributed to the influence of a charged environment on the behavior of the anions and cations participating in an equilibrium reaction. At moderate ionic strengths, this environment has the consequence of making an ion less effective in influencing the position of a chemical equilibrium. Thus in a solution of acetic acid containing sodium chloride, the acetate ions and hydronium ions are each surrounded by particles of opposite charge; this charged atmosphere makes less probable the recombination of the two species to form the undissociated acid molecule. The result of this is a greater degree of dissociation.

In order to describe, in quantitative terms, the effect of ionic strength on equilibria, chemists have chosen to invent a concentration parameter called the *activity* of a substance; this is defined as follows :

$$a_A = [A]f_A \tag{2-4}$$

where a_A is the activity of the species A, $[A]$ is its molar concentration, and f_A is a dimensionless number called the *activity coefficient*. Now the activity coefficient (and thus the activity) of A varies with ionic strength in such a way that when a_A is employed in equilibrium-constant expressions, the numerical value of the equilibrium constant is independent of the ionic strength. For example, we may write for the dissociation of acetic acid

$$K_d = \frac{a_{H^+} \cdot a_{OAc^-}}{a_{HOAc}} = \frac{[H^+][OAc^-]}{HOAc} \times \frac{f_{H^+} \cdot f_{OAc^-}}{f_{HOAc}}$$

where f_{H^+}, f_{OAc^-}, and f_{HOAc} vary with ionic strength in such a manner that K_d is constant over a wide range of ionic strengths (in contrast to the *apparent* K_d shown in Table 2-3).

Properties of activity coefficients.   It is necessary that we examine briefly the important properties of activity coefficients.

1. The activity coefficient of a species can be thought of as being a measure of the effectiveness of that ion in influencing an equilibrium in which it plays a part. In very dilute solutions of minimal ionic strength, this effectiveness becomes constant and the activity coefficient acquires a value of unity; under these circumstances the activity and molar concentration become identical. As the ionic strength increases, however, the ion becomes less effective in its behavior and the activity coefficient decreases. We may summarize this behavior in terms of equation (2-4). At moderate ionic strengths $f_A < 1$; as the solution approaches infinite dilution, however, $f_A \rightarrow 1$ and thus $a_A \rightarrow [A]$.

At high ionic strengths, the activity coefficients of some species increase and may even become greater than one. Interpretation of the behavior of solutions in this region is difficult; we shall confine our discussion to regions of low or moderate ionic strengths (that is, where $\mu < 0.1$).

The behavior of typical activity coefficients as a function of ionic strength is shown in Figure 2.1.

2. In solutions that are not too concentrated, the activity coefficient of a given species is independent of the specific nature of the electrolyte and dependent only upon the ionic strength.

3. For a given ionic strength, the activity coefficient of a substance departs further from unity as the charge carried by the substance increases. This is

illustrated in Figure 2.1 where curves for the activity coefficient of singly, doubly, and triply charged ions are shown. The activity coefficient of an uncharged molecule is approximately one regardless of ionic strength.

4. For ions of the same charge, activity coefficients are approximately the same at any given ionic strength. Small variations are found, however, which can be correlated with the effective diameter of the hydrated ions.

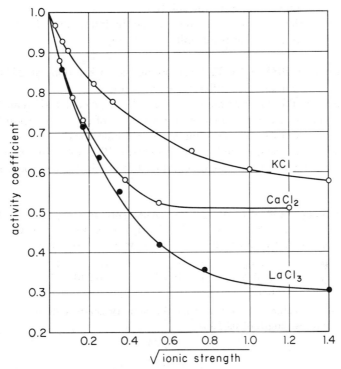

Fig. 2.1 Effect of Ionic Strength on Activity Coefficients.

5. The activity coefficient of a given ion is useful in describing the effective behavior of that ion in all equilibria in which it participates. For example, at a given ionic strength a single activity coefficient for cyanide ion describes the influence of that species upon any of the following equilibria:

$$HCN \rightleftharpoons H^+ + CN^-$$

$$AgCN \rightleftharpoons Ag^+ + CN^-$$

$$Ni(CN)_4^{2-} \rightleftharpoons Ni^{2+} + 4CN^-$$

Evaluation of activity coefficients. In 1923 P. Debye and E. Hückel derived a theoretical expression for the calculation of activity coefficients of ions.[5] The results of their derivation is known as the Debye-Hückel equation, which may be written as follows:

$$-\log f_A = \frac{0.5085\, Z_A^2\, \sqrt{\mu}}{1 + 0.3281\alpha_A\sqrt{\mu}} \tag{2-5}$$

where

f_A = activity coefficient of the species A
Z = charge on the species A
μ = ionic strength of the solution
α_A = the effective diameter of the hydrated ion A in Ångstrom units (1 Ångstrom unit = 10^{-8} cm)

The constants 0.5085 and 0.3281 are applicable to solutions at 25° C. Other values must be employed at different temperatures.

Values for the size parameter α_A have been given by Kielland[6] so that calculation of the activity coefficients for a variety of species is possible. Table 2-4 lists the activity coefficients for some common ions calculated from the Debye-Hückel expression and these diameters. The following example will demonstrate how this table may be employed to obtain activity coefficients and activities.

Example. Calculate the activity coefficients and activities of Ba^{2+} and Cl^- in a solution that is 0.0036 F in $BaCl_2$ and 0.0400 F in NaCl.
Let us first calculate the ionic strength of the solution

$$\mu = \tfrac{1}{2}(m_{Ba^{2+}} \times 2^2 + m_{Cl^-} \times 1^2 + m_{Na^+} \times 1^2)$$
$$= \tfrac{1}{2}(0.0036 \times 4 + 0.0472 \times 1 + 0.040 \times 1)$$
$$= 0.051$$

From Table 2-4 we find that the activity coefficients for the two ions at an ionic strength of 0.050 is

$$f_{Ba^{2+}} = 0.46$$
$$f_{Cl^-} = 0.80$$

Had the value for μ been further away from the value for which there was an entry in the table, we could have obtained the activity coefficients by interpolation.
The activity of the two ions could be obtained as follows:

$$a_{Ba^{2+}} = [Ba^{2+}] \cdot f_{Ba^{2+}} = 0.0036 \times 0.46$$
$$= 0.00166 \text{ mol/liter}$$
$$a_{Cl^-} = [Cl^-] \cdot f_{Cl^-} = 0.0472 \times 0.80$$
$$= 0.0378 \text{ mol/liter}$$

Experimental verification of individual activity coefficients such as those shown in Table 2-4 is, unfortunately, impossible; all experimental methods

[5] P. Debye and E. Hückel, *Physik. Z.*, **24**, 185 (1923).
[6] J. Kielland, *J. Am. Chem. Soc.*, **59**, 1675 (1937).

Table 2-4

ACTIVITY COEFFICIENT FOR IONS AT 25° C [7]

Ion	Ion Size, α_A, A	Activity Coefficient at Indicated Ionic Strengths				
		0.001	0.005	0.01	0.05	0.1
H^+	9	0.967	0.933	0.914	0.86	0.83
Li^+, $C_6H_5COO^-$	6	0.965	0.929	0.907	0.84	0.80
Na^+, IO_3^-, HSO_3^-, HCO_3^-, $H_2PO_4^-$, $H_2AsO_4^-$, OAc^-	4-4.5	0.964	0.928	0.902	0.82	0.78
OH^-, F^-, SCN^-, HS^-, ClO_3^-, ClO_4^-, BrO_3^-, IO_4^-, MnO_4^-	3.5	0.964	0.926	0.900	0.81	0.76
K^+, Cl^-, Br^-, I^-, CN^-, NO_2^-, NO_3^-	3	0.964	0.925	0.899	0.80	0.76
Rb^+, Cs^+, Tl^+, Ag^+, NH_4^+	2.5	0.964	0.924	0.898	0.80	0.75
Mg^{2+}, Be^{2+}	8	0.872	0.755	0.69	0.52	0.45
Ca^{2+}, Cu^{2+}, Zn^{2+}, Sn^{2+}, Mn^{2+}, Fe^{2+}, Ni^{2+}, Co^{2+}, $Phthalate^{2-}$	6	0.870	0.749	0.675	0.48	0.40
Sr^{2+}, Ba^{2+}, Cd^{2+}, Hg^{2+}, S^{2-}	5	0.868	0.744	0.67	0.46	0.38
Pb^{2+}, CO_3^{2-}, SO_3^{2-}, $C_2O_4^{2-}$	4.5	0.868	0.742	0.665	0.46	0.37
Hg_2^{2+}, SO_4^{2-}, $S_2O_3^{2-}$, CrO_4^{2-}, HPO_4^{2-}	4.0	0.867	0.740	0.660	0.44	0.36
Al^{3+}, Fe^{3+}, Cr^{3+}, La^{3+}, Ce^{3+}	9	0.738	0.54	0.44	0.24	0.18
PO_4^{3-}, $Fe(CN)_6^{3-}$	4	0.725	0.50	0.40	0.16	0.095
Th^{4+}, Zr^{4+}, Ce^{4+}, Sn^{4+}	11	0.588	0.35	0.255	0.10	0.065
$Fe(CN)_6^{4-}$	5	0.57	0.31	0.20	0.048	0.021

give only a mean activity coefficient for the positively and negatively charged ions in a solution. It should be pointed out, however, that *mean* activity coefficients calculated from the data in Table 2-4 agree satisfactorily with the experimental values.

While the Debye-Hückel relationship is a useful one for calculating activity coefficients at moderate electrolyte concentration, it is not very applicable to solutions having ionic strengths greater than 0.1.

Equilibrium calculations employing activity coefficients. More accurate information can be gained when activities rather than molar concentrations are employed in conjunction with equilibrium constants. The following example will demonstrate how the data in Table 2-4 can be used for this purpose.

Example. Calculate the solubility of AgCl in a solution which is 0.10 F in KNO_3 employing (a) molar concentrations and (b) activities.

(a) The equilibrium in this instance is

$$AgCl \rightleftharpoons Ag^+ + Cl^-$$

[7] Data taken from J. Kielland, *J. Am. Chem. Soc.*, **59**, 1675 (1937).

and we may write

$$K_{sp} = 1.82 \times 10^{-10} = [Ag^+].[Cl^-]$$

If we let x be the solubility in formula weights per liter, then since all of the dissolved AgCl is present as Ag^+ and Cl^-, we may say

$$x = [Ag^+] = [Cl^-]$$

and substituting we obtain

$$x^2 = 1.82 \times 10^{-10}$$
$$x = 1.35 \times 10^{-5} \text{ mol/liter}$$

(b) In this case we will write

$$K_{sp} = [Ag^+][Cl^-] \cdot f_{Ag^+} \cdot f_{Cl^-}$$

Again the solubility of the AgCl will be equal to the molar concentration of Ag^+ and Cl^- and

$$x = [Ag^+] = [Cl^-]$$

Substituting gives

$$x^2 \cdot f_{Ag^+} \cdot f_{Cl^-} = 1.82 \times 10^{-10}$$

In order to evaluate the activity coefficients we need to know the ionic strength which is

$$\mu = \tfrac{1}{2}(0.1 \times 1^2 + 0.1 \times 1^2) = 0.1$$

Referring to Table 2-4 we find at $\mu = 0.1$

$$f_{Ag^+} = 0.75$$
$$f_{Cl^-} = 0.76$$

Thus we may write

$$x^2 = \frac{1.82 \times 10^{-10}}{0.76 \times 0.75}$$

$$x = 1.79 \times 10^{-5} \text{ mol/liter}$$

This is undoubtedly a better figure for the molar solubility of AgCl than the 1.35×10^{-5} mol/liter obtained in (a).

Omission of activity coefficients in equilibrium calculations. We shall nearly always neglect activity coefficients and simply use molar concentrations in all applications of the equilibrium law. This simplifies the calculations and greatly reduces the amount of data needed. For most purposes the errors introduced by the assumption of unity for the activity coefficient will not be large enough to lead to false conclusions. The reader should remember that the results so obtained are approximations. This is apparent from the example just considered. In the typical case, errors of from 1 to 10 percent relative are to be expected as a result of disregarding activity coefficients; not infrequently these may become as great or greater than 100 percent relative.

The student should also recognize the conditions under which the approximation of concentration for activity is likely to lead to the largest errors. These will occur when the ionic strength is great (0.1 or larger) and the ions involved have multiple charges (see Table 2-4). In dealing with dilute solutions (ionic strength < 0.01) of nonelectrolytes or singly charged ions, the use of concentrations in mass-law calculations often leads to rather accurate results. We shall have occasion to make further reference to the accuracy of equilibrium calculations in later chapters dealing with specific applications.

problems

1. How many formula weights and how many milliformula weights are present in the following?
   (a) 14.1 grams $AgNO_3$ ans. 8.3×10^{-2} formula wt; 83 milliformula wt
   (b) 137 mg $BaSO_4$ ans. 5.88×10^{-4} formula wt
   (c) 20.0 ml of 0.100 F KCN   ans. 2.00×10^{-3} formula wt
   (d) 20.0 liters of 0.200 F Na_2CO_3 ans. 4.00 formula wt
   (e) 1 liter of pure water at 25° C

2. How many milliformula weights are represented by the following?
   (a) 1.73 grams $K_4Fe(CN)_6$
   (b) 3.10 grams NaCl
   (c) 937 ml of 0.100 F NaCl
   (d) 1.94 liters of 0.200 F KCl

3. Calculate the following weights in grams :
   (a) 2.00 milliformula weights NaCl   ans. 0.117 gram
   (b) 0.500 formula weight KCl   ans. 37.2 grams
   (c) 0.100 mol K^+

4. What weight in grams of Na_2SO_4 is present in the following?
   (a) 1.00 liter of 0.100 F Na_2SO_4
   (b) 13.6 ml of 0.200 F Na_2SO_4
   (c) 16.0 milliformula weights Na_2SO_4
   (d) 15.0 ml of a solution which is 3 percent Na_2SO_4 (weight-volume)

5. A solution was prepared by dissolving 1.30 grams $AgNO_3$ in water and diluting to 50.0 ml. Calculate the following :
   (a) The weight-volume percent $AgNO_3$ ans. 2.60 percent
   (b) The number of milliformula weights $AgNO_3$ taken
   (c) The formal concentration of the solution
   (d) The number of milliformula weights Ag^+ present

6. Calculate the formal concentration of the following :
   (a) A solution prepared by dissolving 1.00 milliformula weight $AgNO_3$ in water and diluting to 1.00 ml
   (b) A solution prepared by dissolving 4.70 milliformula weights Na_2SO_4 in water and diluting to 1327 ml   ans. 3.54×10^{-3} F
   (c) A solution containing 18.3 mg NaCl in 123 ml of water
   ans. 2.55×10^{-3} F
   (d) A solution containing 196 grams KCN in 2.00 liters

7. Calculate the formal concentration of the following :
   (a) A solution containing 63 formula weights $KMnO_4$ in 100 liters of solution
   (b) A 4.70 percent (weight-volume) $BaCl_2$ solution

8. What is the formal concentration of a solution that is 3 percent KNO_3 by weight? Assume a specific gravity of 1.01 for the solution.   ans. 0.30 F

9. What is the formal concentration of a solution prepared :
    (a) By dissolving 71.7 grams alcohol (C_2H_5OH) in water and diluting to 647 ml
    (b) By dissolving 1.71 milliformula weights alcohol in water and diluting to 10.0 ml

10. What is the formal concentration of the following :
    (a) A mixture of 13.0 ml of 0.131 F $AgNO_3$ and 87 ml of water   ans. 0.0170 F
    (b) A mixture of 67 ml of 0.100 F NaCl diluted to 450 ml   ans. 0.0149 F
    (c) A mixture prepared by diluting 63 liters of 1.00 F HCl to 1000 liters
                                                              ans. 0.063 F
    (d) A mixture of 8000 gallons of 1.00 F NaOH and enough water to make 20,000 gallons

11. Water allowed to equilibrate with the atmosphere is 1.5×10^{-5} F in CO_2.
    (a) What weight of CO_2 is present in 100 ml of water ?
    (b) What volume of CO_2 (measured at S.T.P.) will dissolve in 100 liters of water ?   ans. 33.6 ml

12. How would you prepare the following ?
    (a) 365 ml of 0.170 F $AgNO_3$ from the solid salt
                         ans. Dissolve 10.5 grams $AgNO_3$ and dilute to 365 ml
    (b) 2.00 liters of 0.100 F $AgNO_3$ from 0.600 F $AgNO_3$
    (c) 10.00 liters of exactly 0.1000 F KSCN from the salt
                         ans. Dissolve 97.2 grams KSCN and dilute to 10.00 liters
    (d) 434.0 ml of 0.100 F KSCN from 0.157 F KSCN

13. How would you prepare the following ?
    (a) A 5 percent (weight-volume) solution of NaCl
                         ans. Dissolve 5 grams NaCl in H_2O and dilute to 100 ml
    (b) A 5 weight-percent solution of NaCl
                         ans. Dissolve 5 grams NaCl in 95 ml of water

14. Describe a method for the preparation of the following :
    (a) 250 ml of approximately 0.2 F $KMnO_4$ from the solid
    (b) 500 ml of 0.71 F $BaCl_2$ from solid $BaCl_2 \cdot 2H_2O$
    (c) A liter of solution which is 0.100 M in Na^+; starting with NaCl
    (d) A liter of solution which is 0.100 M in Na^+; starting with Na_2CO_3

15. How would you prepare the following :
    (a) 3700 ml of 0.100 F $NaHCO_3$ from a 1.00 F solution
                         ans. Dilute 370 ml of the 1 F solution to 3700 ml
    (b) 220 ml of 0.0050 F NaCl from a 5 percent (weight-volume) solution

16. A bottle of concentrated HCl has the following information on its label : specific gravity 1.185; percent HCl 36.5.
    (a) How many grams HCl are contained in each ml ?
                                                      ans. 0.433 gram/ml
    (b) What is the formal concentration of HCl ?   ans. 11.9 F
    (c) How would you prepare 200 ml of 3.0 F HCl from the concentrated reagent ?   ans. Dilute 50 ml of the concentrated acid to 200 ml

17. What volume in milliliters (at standard conditions) of dry HCl gas is contained in one liter of concentrated HCl ?

18. Concentrated HNO_3 has the following properties : specific gravity 1.42; percent HNO_3 69.

(a) How would you prepare a liter of 0.5 F HNO_3 from the concentrated reagent?

(b) What volume of the concentrated acid is equivalent to 37 grams of HNO_3?

(c) What is the formal concentration of the concentrated reagent?

19. Lead ion reacts with IO_3^- to form an insoluble precipitate.
$$Pb^{2+} + 2IO_3^- \rightleftharpoons Pb(IO_3)_2$$

(a) What weight of KIO_3 will react with 0.100 formula weight of Pb^{2+}?
ans. 42.8 grams

(b) How many grams $Pb(IO_3)_2$ could be formed from 7.00 milliformula weights Pb^{2+}?
ans. 3.90 grams

(c) How many grams KIO_3 are required to react completely with 2.00 grams $Pb(NO_3)_2$?
ans. 2.58 grams

(d) How many grams $Pb(IO_3)_2$ are formed upon mixing 4.00 grams $Pb(NO_3)_2$ and 4.00 grams KIO_3?
ans. 5.21 grams

(e) What volume in milliliters of 0.300 F KIO_3 is required to precipitate completely 1.00 milliformula weight Pb^{2+}?
ans. 6.67 ml

20. How much of the substance in the second column is needed to react completely with the indicated amount of the substance in the first column?

(a) 13.1 milliformula weights HCl   (a) grams NaOH

(b) 14.7 grams H_2SO_4 (b) grams NaOH

(c) 100 milliformula weights $BaCl_2$ (c) ml of 0.25 F $AgNO_3$

(d) 30 ml of 0.100 F $BaCl_2 \cdot 2H_2O$ (d) ml of 0.25 F $AgNO_3$

(e) 18.1 mg $BaCl_2 \cdot 2H_2O$ (e) ml of 0.25 F $AgNO_3$

(f) 100 grams $AlCl_3$ (f) grams $AgNO_3$

(g) 50 grams NaOH   (g) ml of concentrated HCl specific gravity 1.18, percent HCl 36

(h) 500 ml of 0.200 F $AgNO_3$ (h) ml of 0.300 F K_2CrO_4

(i) 1.00 ml of 0.2 F $AgNO_3$ (i) mg K_2CrO_4

(j) 1 liter of 0.050 F $Ba(OH)_2$ (j) ml of HCl of density 1.12 and containing 24.0 percent HCl

21. How many grams of 90.0 percent pure $Fe(NH_4)_2(SO_4)_2 \cdot 6H_2O$ are required to prepare 1.00 liter of 0.200 F solution of Fe^{2+}?

22. How many milliliters of H_2SO_4 of specific gravity 1.84 and containing 98 percent H_2SO_4 should be taken to prepare 5 liters of 0.10 F acid?

23. How many grams of $FeSO_4$ will react with 0.600 gram $KMnO_4$?
$$5Fe^{2+} + MnO_4^- + 8H^+ \rightarrow 5Fe^{3+} + Mn^{2+} + 4H_2O$$

24. How many grams of $H_2C_2O_4 \cdot 2H_2O$ are required to precipitate 37.0 mg Ca^{2+} as CaC_2O_4?

25. How many milliliters of 0.100 F $Na_2C_2O_4$ are required to precipitate 37.0 mg Ca^{2+}?

26. What volume of H_2S gas (measured at S.T.P.) would just precipitate 1.00 gram Zn^{2+}?

27. When $NaHCO_3$ is ignited, the following reaction occurs quantitatively:
$$2NaHCO_3 \rightarrow Na_2CO_3 + H_2O + CO_2$$

(a) What would be the loss in weight when 1.00 gram $NaHCO_3$ is ignited?

(b) How many formula weights of CO_2 are formed when 2.00 grams $NaHCO_3$ is ignited?

(c) What volume of CO_2 (measured at S.T.P.) would be formed from 2.00 grams $NaHCO_3$?

28. What weight of AgCl would be obtained by decomposing 0.312 gram of an inorganic compound having the formula C_6H_5Cl and precipitating all of the chloride as AgCl?   ans. 0.397 gram

29. Exactly 22.1 ml of an HCl solution were required to react completely with 0.176 gram of pure $CaCO_3$ by the reaction
$$CO_3^{2-} + 2H^+ \rightarrow H_2O + CO_2$$
Calculate the formal concentration of HCl.

30. It was found that 27.6 ml of a $AgNO_3$ solution would just precipitate all Cl^- in 88.1 mg of $BaCl_2 \cdot 2H_2O$. What was the formal concentration of the $AgNO_3$?

31. Calculate the ionic strengths of the following solutions:
    (a) 0.0500 F LiCl   ans. 0.0500
    (b) 0.0500 F $Ba(NO_3)_2$ ans. 0.15
    (c) 0.0500 F $MgSO_4$
    (d) 0.0500 F $Al_2(SO_4)_3$
    (e) A solution 0.0500 F in NaCl and 0.0500 F in KNO_3 ans. 0.100
    (f) A solution 0.0600 F in KNO_3 and 0.100 F in $Al(NO_3)_3$ ans. 0.66
    (g) A solution 0.0300 F in $BaCl_2$ and 0.0800 F in Na_2SO_4
    (h) A solution which was 0.100 F each in $NaNO_3$, $Cd(NO_3)_2$, and $Al_2(SO_4)_3$

32. From the data in Table 2-4 calculate the activity coefficients of each of the following ions:
    (a) Zn^{2+} in a solution that was 1×10^{-6} F in $Zn(NO_3)_2$ and 0.01 F in KNO_3
                                                                         ans. 0.675
    (b) Zn^{2+} in a solution that was 0.02 F in KNO_3 and 0.01 F in $Zn(NO_3)_2$
                                                                         ans. 0.48
    (c) Na^+ in a 0.050 F solution of NaCl
    (d) Al^{3+} in a solution that was 1×10^{-6} F in $Al(NO_3)_3$ and 0.0500 F in KNO_3
    (e) Ba^{2+} and SO_4^{2-} ions in a solution that was 0.02 F in $BaCl_2$, 0.04 F in KNO_3, and saturated with $BaSO_4$
    (f) H^+ and OAc^- in a solution that was 0.1 F in KCl and 0.001 F in HOAc
    (g) HPO_4^{2-} in a solution that was 1×10^{-3} F in Na_2HPO_4 and 2×10^{-3} F in NaCl

33. What is the activity of Ba^{2+} and Cl^- in a solution that is 0.005 F in $BaCl_2$ and 0.035 F in NaCl?   ans. $a_{Ba^{2+}} = 0.0023$
                                                                             $a_{Cl^-} = 0.036$

34. Calculate the activity of Al^{3+} in a solution that is 1×10^{-5} F in $Al_2(SO_4)_3$ and 0.050 F in KNO_3.

35. Calculate the activity of H^+ in water that is 0.100 F in KNO_3.

36. What is the activity of SO_4^{2-} and Na^+ ions in a solution that is 0.03 F in Na_2SO_4 and 0.01 F in NaCl?

chapter 3. *The Evaluation of Analytical Data*

If a physical measurement is to be of much worth to its user, some idea regarding the reliability of its value is essential. Unfortunately the measurement of any physical quantity is subject to a degree of uncertainty; the experimenter can only strive to achieve an acceptably close approach to the actual value for the quantity. The estimation of the uncertainties affecting measurements, while often difficult, is thus an essential task of the experimental scientist. This chapter is devoted to a consideration of some of the problems associated with these estimations.

The accuracy to be expected from a measurement is directly related to the time and effort expended in its attainment. The addition of another decimal place to the value for a physical quantity often involves days, months, or even years of labor. The experimenter is thus unavoidably faced with the necessity of making a compromise between high accuracy and expenditure of effort; the reader should recognize that this dichotomy exists in all scientific work. Because of it, the wise investigator always gives serious consideration to the uses to which his data are to be put before undertaking any measurements; this way he avoids the waste of large amounts of time and effort necessary for the attainment of unneeded accuracies. The authors feel that this is a point that cannot be too strongly stressed; the scientist's time is far too valuable to be wasted in the indiscriminate pursuit of the last possible decimal place where such is unnecessary.

In an introductory course in quantitative analysis, the problems associated with the estimation of the required accuracy and the attainment thereof are, of necessity, generally less formidable than those encountered by the working chemist. The student is provided with homogeneous samples and established methods for their analysis; while his time in the laboratory is always at a premium, it is often secondary to the attainment of the best possible set of results. Notwithstanding these advantages, his experimental data occasionally appear to contain serious uncertainties; he is then confronted with the problem of evaluating his results and deciding which of these should be included in his final report. As the student progresses into more advanced work he will be faced with the problem of analyzing more complex substances for which no well-established methods and techniques are available. The evaluation of reliability in such cases becomes more perplexing; yet without some measure of this, the experimental data obtained are of little use. An understanding of the problems inherent in the appraisal of the accuracy of measurements is thus of the greatest importance to the chemist.

Definition of Terms

Each of the measurements in an analysis is subject to uncertainty. For the present we need only recognize that these uncertainties are responsible for the variations observed among measurements of the same quantity. In practice the chemist generally makes replicate measurements, his reasoning being that the confidence in which his results can be held is increased by demonstrating their reproducibility. Having obtained several values for some quantity, he is then confronted with the problem of arriving at the best value for the measurement. Two quantities, the *mean* and the *median*, are available to him for this purpose.

The Mean

The *mean*, *arithmetic mean*, and *average* are synonymous terms that refer to the numerical value obtained by dividing the sum of a set of replicate measurements by the number of individual results in the set.

The Median

The *median* of a set is that value about which all others are equally distributed, half being numerically greater and half being numerically smaller. If the set consists of an odd number of measurements, selection of the median may be made directly; for a set containing an even number of measurements, the average value of the central pair is taken as the median.

> **Example.** Calculate the mean and median for 10.06, 10.20, 10.08, 10.10.

$$\text{mean} = \frac{10.06 + 10.20 + 10.08 + 10.10}{4} = 10.11$$

> Since there are an even number of measurements in the set, the median is given by the average of the middle pair of results, which is

$$\frac{10.08 + 10.10}{2} = 10.09$$

We shall see that in the ideal case the mean and median are numerically equal; this fails to be true more often than not, however, when only a small set of measurements has been taken.

Precision

The term *precision* is frequently used to describe the reproducibility of results. It can be defined as the agreement between the numerical values of two or more measurements that have been made *in an identical fashion.*

Absolute deviation   There are several ways in which the precision of a result can be expressed. Perhaps the most commonly employed of these is in terms of absolute deviation, which is simply the difference between an experimental value and that which is taken as the best for the set. The arithmetic mean is usually employed by chemists for the latter although there are instances where the median is preferable. This relationship may be expressed by

$$D = O - M \tag{3-1}$$

where D is the absolute deviation of the observed value O from the arithmetic mean (or median) M of several identical measurements.

Inspection of a typical set of analytical data will serve to illustrate this concept. Suppose that a series of analyses has revealed the percentage of chloride in a sample to be 24.39, 24.20 and 24.28 percent, respectively. We observe at the outset that there is a *spread* (or range) of 0.19 percent separating the high and low values of the set. It should be noted in passing that this information is, of itself, a measure of the precision of the analysis. In the absence of any qualifying information, we have no way of judging which of the three determinations is the most reliable; by taking the average of 24.29 percent for this, we are inferring that there was an equal likelihood for the individual determinations to be high and low. Having decided upon the most likely value to report, we are able to calculate the deviation of the individual results from this; the average of these deviations is a measure of the precision of the analysis. In calculating the average deviation, the sign of the individual deviation is not taken into account; we are concerned only with its absolute value, not whether it is high or low. In the present instance

Sample	Percent Chloride	Deviation from Average
1	24.39	0.10
2	24.20	0.09
3	24.28	0.01
Average	24.29	0.07

We may thus report the result as 24.29 ± 0.07 percent chloride.

Relative deviation. Precision may also be expressed in relative terms as *percentage average deviation* or as *average deviation in parts per thousand*. Thus, for the above data the average value is 24.29, and

$$\text{relative deviation} = \frac{0.07 \times 100}{24.29} = 0.29 \cong 0.3 \text{ percent}$$

$$\text{relative deviation} = \frac{0.07 \times 1000}{24.29} = 2.9 \cong 3 \text{ parts per thousand}$$

Chemists frequently express the precision of their results in terms of absolute or relative deviation because these are readily computed quantities. Of greater value is the standard deviation, which will be discussed in a later section.

Accuracy

The term *accuracy* denotes the nearness of a measurement to its accepted value and is expressed in terms of *error*. The reader should note the fundamental difference between this term and precision. Accuracy involves a comparison with respect to a true, or accepted value; in contrast, precision compares a result with the best value of several measurements made in the same way.

Absolute error. The absolute error can be defined as

$$E = O - A \tag{3-2}$$

where the error E is the difference between the observed value O and the accepted value A. In many instances the accepted value of A is itself subject to considerable uncertainty. As a consequence it is frequently difficult to arrive at a realistic estimate for the error of a measurement.

Returning to the example given above, suppose that the true or accepted value for the percentage of chloride in the sample is 24.34 percent. The absolute error is thus $24.29 - 24.34 = -0.05$ percent; here we ordinarily retain the sign of the error to indicate whether the result is high or low.

Relative error. Equation (3-2) defines error in absolute terms; a more useful quantity is the *relative error*. As before, this is commonly expressed as a percentage or in parts per thousand. The calculation is also analogous; however, the accepted value is used as the basis for comparison in this case. For example, we find

$$\text{relative error} = -\frac{0.05 \times 100}{24.34} = -0.21 \cong -0.2 \text{ percent}$$

$$\text{relative error} = -\frac{0.05 \times 1000}{24.34} = -2.1 \cong 2 \text{ parts per thousand } low$$

In summary, then, the accuracy of a measurement can be ascertained only if the true, or accepted, value of that measurement is available; given a set of replicate measurements, the precision can always be expressed. In the absence of further information, the attainment of good precision is the only indication of a successful analysis; we shall see, however, that this is not necessarily a valid criterion. Judgments based upon precision alone must always be considered guardedly.

Types of Error in Analytical Results

The errors that accompany the performance of an analysis may be classified into two broad categories depending upon their origin. *Determinate errors* are those for which we can account, in principle if not in actuality. *Indeterminate errors* are due to random errors that arise from extending a system of measurement to its maximum; the effect of indeterminate error upon a set of results can often be reduced to acceptable limits but can never be entirely avoided.

DETERMINATE ERRORS

Sources

A complete listing of conceivable determinate errors is a patent impossibility; we can, however, recognize that they may be traced to the *personal* errors of the experimenter, the *instrumental* errors of his measuring devices, the errors that repose in the *method* of analysis he employs, or a combination of these.

Personal errors.   In a preponderance of cases where a gross error exists in the results obtained by an established method of analysis, the cause lies with the experimenter. He may have failed to take down the correct value for a weighing, having either misread his weights or transposed the correct numbers. He may have erred in the reading of a volume. The beginning student in quantitative analysis may incur errors of a manipulative nature, failing to transfer materials completely from one container to another or allowing a sample to become contaminated in some way. Failure to introduce the proper amount of a reagent in the course of an analysis must also be classed as a personal error.

These very common errors can, of course, be easily avoided by the exercise of care. For example, conscientious verification of every experimental reading is excellent practice; the time required for this is modest indeed and will serve to eliminate errors of this sort. In the event a manipulative error is suspected, a brief notation to this effect should be entered in the notebook. This entry, *made at the time the accident is detected*, should include the nature of the suspected lapse and the sample that might be affected. If the final result for this sample does in fact diverge seriously from the others in the set, this notation may serve as justification for omitting it. A thorough understanding of the principles underlying an analysis is the best insurance against the addition of improper quantities of the various reagents that are required.

One personal error that should always be guarded against is *prejudice*. Where an estimation is involved, there is a natural tendency for the experimenter to choose that value which is most favorable to him. Intellectual honesty is the keystone of all science; it is essential for the attainment of significant analytical results.

Instrumental errors. Instrumental errors are those that are attributable to imperfections in the tools with which the analyst works. The tolerances of the chemist's weights may be the potential source of an instrumental error; mishandling of analytical weights will often cause them to differ significantly from their nominal values. Volumetric equipment—burets, pipets, and volumetric flasks—frequently deliver or contain volumes slightly different from those indicated by their graduations; for analytical work of the highest order, these variations must be measured by calibration and taken into account; otherwise they become a source of instrumental error.

Method errors. Analytical procedures are also subject to limitations; these give rise to errors that can be traced to the method. For example, even the least soluble of substances has a finite solubility. In gravimetric analysis the chemist is confronted with the problem of isolating the element to be determined in the form of a precipitate of the greatest possible purity. If he fails to wash it sufficiently, the precipitate will be contaminated with foreign substances and have a spuriously high weight. On the other hand, excessive washing may lead to the loss of weighable quantities of the precipitate; this will have the opposite effect upon the results. It should be clear that an error will result in either case, and that there is a limit of accuracy that can be attained by such an analysis.

A method error that is frequently encountered in volumetric analysis arises from the fact that a volume of reagent in excess of the theoretical is required to cause the change of color that signifies completion of the reaction. As a consequence the ultimate accuracy of an analysis is often limited by the very phenomenon that makes the determination possible.

The undiscovered presence of interfering contaminants in either sample or reagents can lead to serious determinate errors in the results yielded by an analysis.

Errors inherent in a method are probably the most serious of determinate errors in that they are the most likely to remain undetected.

Effect of Determinate Error on the Results of an Analysis

Omitting those personal errors due to carelessness that influence only an individual result, it is possible to discern two types of determinate errors. *Constant errors* are those whose magnitude is independent of the size of the quantity measured; *proportional errors*, on the other hand, depend in absolute magnitude upon the quantity measured and increase or decrease in proportion to its size.

Constant errors. For a given analytical procedure, a constant error will be more serious as the size of the quantity measured decreases. For the purpose of illustration we will suppose that the washing of a slightly soluble precipitate with 100 ml of water is called for, and that 0.5 mg are lost during this process. If 500 mg of precipitate are so treated, the relative error due to this cause will be — (0.5 × 100/500) or — 0.1 percent, which for most purposes is tolerable. The loss of this quantity from 50 mg of precipitate will be responsible for a relative error of — (0.5 × 100/50) or 1 percent, which may no longer be acceptable.

The amount of reagent required to bring about the color change in a volumetric analysis is also an example of constant error. This volume, usually small, remains the same regardless of the total volume of reagent required. Again, the relative error will be more serious as the total volume decreases. Clearly, one way of minimizing the effect of constant errors is to use as large a sample as is consistent with the method at hand.

Proportional errors. The presence of interfering contaminants, if not eliminated in some manner, will lead to an error of the proportional variety. For example, a method widely employed for the analysis of copper involves reaction of the cupric ion with potassium iodide; the quantity of iodine produced in the reaction is then measured. Ferric iron, if present, will also liberate iodine from potassium iodide. Unless steps are taken to prevent this interference, the analysis will yield erroneously high results for the percentage of copper since the iodine produced is a measure of the sum of the copper and iron in the sample. Thus the magnitude of this error is fixed by the extent of iron contamination, and will produce the same relative effect regardless of the size of sample taken for analysis. If the sample size is doubled, for example, the amount of iodine liberated by both the copper and the iron contaminant will be doubled; and while the absolute error will also undergo a twofold increase the *relative* error will remain unchanged.

Every analytical procedure is potentially subject to determinate errors, both constant and proportional. Many of the precautions that must be followed in performing such procedures are necessary to minimize or eliminate these sources of uncertainty; the serious student will profit by keeping in mind the potential sources for error in his work and learning to recognize the effect these might have upon his results.

The Detection of Determinate Errors

From the standpoint of magnitude, determinate errors are the more important in an analysis, and in some instances are difficult to detect. They can affect a single result, a series of results, or an entire method of analysis depending upon their nature and the phase of the analysis in which they occur. A personal error, such as the misreading of a weight while weighing a sample, would result in a spurious value for that sample alone; the same error incurred during the preparation of a reagent might cause the propagation of a proportional error throughout all analyses undertaken with that solution. In the first instance the divergence of the result from the others in the series would possibly provide a clue that the error had been incurred. In the absence of knowledge of the approximate value to be expected, the error in the latter case could well pass undetected. An error of this sort would be revealed by repetition of the analysis with another reagent or by a check analysis with the original solution against a sample of established purity.

Instrumental errors attributable to the discrepancies from the nominal values in analytical weights and the volumes contained by volumetric flasks or delivered by pipets and burets are generally quite small in new equipment of modern manufacture; with heavy use or mishandling, however, they can acquire serious magnitude. To eliminate determinate error from these sources, the periodic calibration of such equipment is a necessity.

Errors of method are the most serious, since they affect all results obtained by that method. The detection of errors inherent in an analytical scheme may take any of several courses; some of those that are generally applicable merit discussion.

Analysis of standard samples. A method may be tested by employing it for the analysis of synthetic samples whose over-all composition approximates that of the substance for which the analysis is being considered. Great care is exercised in the preparation of these standard samples to insure that the concentration of the constituent to be determined is known with great accuracy.

Proportional as well as constant errors in a method may be detected through the use of standard samples. Studies aimed at the development or improvement of analytical procedures often utilize this approach; but the preparation of samples whose composition resembles that of a complex natural substance frequently is difficult. Furthermore, the difficulties are multiplied by the requirement that the exact concentration of one of the constituents be known as a result of the method of preparation. In some cases these problems can be so imposing they prevent the application of this procedure.

Independent analysis. Closely associated with the foregoing is the parallel analysis of a sample by a method of established reliability that is independent of the one under investigation. This is of particular value where samples of known purity are not available; it also has the effect of supplying a frame of reference by which the utility of a new method may be judged. In general, the more widely the independent method differs from the one under study, the more suitable it is for this purpose; this tends to eliminate the possibility that some common factor in the sample will affect both equally and thus remain undetected.

The National Bureau of Standards should be mentioned in this connection since it has available for purchase a fairly large number of common substances that have been carefully analyzed for one or more constituents. These are very valuable for the testing of analytical procedures for accuracy.

Blank determinations. Constant errors affecting physical measurements can frequently be evaluated with a blank determination; this consists of performing all steps of the analysis in the absence of a sample. The result is then applied as a correction to the actual measurements.

Blank determinations are of particular value in exposing constant errors that are due to causes such as the presence of interfering contaminants in the reagents employed in the analysis and the slight deficiencies often encountered in the indication of titration end points; the latter are referred to as *indicator corrections*.

Variation in sample size. In detecting errors in an analysis, applying the procedure to different-sized samples of the same material is sometimes helpful. The presence of constant errors will become obvious in the discrepancies appearing among the results of the several analyses. For example, the data in Table 3-1 were obtained from various sample sizes of a silver alloy containing exactly 20 percent silver. In each of these analyses, there was an error of 2 mg in the weight of silver found. This resulted in low values for the percentage of silver. The differences in the percent silver found between samples, however, becomes less as the sample size is increased, and the data appear to be approaching a constant value with the larger samples. This is shown graphically in Figure 3.1 where the presence of constant error becomes very obvious.

Table 3-1

EFFECT OF A CONSTANT ERROR OF 2 MG ON THE ANALYSIS OF A
SILVER ALLOY

Weight of Sample, grams	Weight of Silver Found, grams	Silver Found, percent
0.2000	0.0378	18.90
0.5000	0.0979	19.58
1.0000	0.1981	19.81
2.0000	0.3982	19.91
5.0000	0.9980	19.96

By way of contrast, proportional error goes undetected by this procedure. In the same figure are plotted data from the analysis of various samples of the same material by a method containing a proportional error of 5 parts per thousand. The data from which this graph was drawn are given in Table 3-2. The plot in this case is a straight line; clearly, without knowledge of the true percentage silver in the alloy, the existence of such an error would go unnoticed.

Table 3-2

EFFECT OF A PROPORTIONAL ERROR OF 5 PARTS PER THOUSAND ON
THE ANALYSIS OF A SILVER ALLOY

Weight of Sample, grams	Weight of Silver Found, grams	Silver Found, percent
0.2000	0.0402	20.10
0.5000	0.1006	20.12
1.0000	0.2009	20.09
2.0000	0.4021	20.11
5.0000	1.0051	20.10

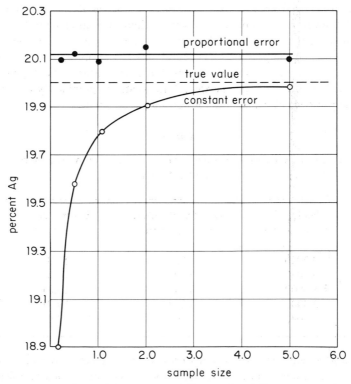

Fig. 3.1 Effect of Constant and Proportional Errors upon the
Results of a Silver Analysis. (See Tables 3-1 and 3-2)

INDETERMINATE ERROR

Sources

The necessity for making estimations is inherent in the process of collecting data for the measurement of any quantity. Common among these in quantitative analysis are the interpolations involved in the observation of balance swings and in judging the liquid level in volumetric apparatus. It follows that the final value of any measurement will be uncertain in an amount that is dependent upon the relative magnitude of the estimations involved in its evaluation. Appropriate design of an experiment will serve to reduce this uncertainty to a tolerable limit, but there is no way in which it can be entirely eliminated. It must be accepted that all physical measurements are subject to a degree of uncertainty, or indeterminate error.

Effect of Indeterminate Error on Results

Let us consider the effect of the phenomenon of indeterminate error upon the results of the relatively simple process of calibrating a pipet. The details for this and other calibrations are given in Chapter 11; here it is sufficient to recognize that the procedure is based upon determining the weight of liquid of known density (commonly water) which the pipet is found to deliver. A set of data pertaining to a 10-ml volumetric pipet is given in Table 3-3, and indicates

Table 3-3

CALIBRATION OF A 10-ML VOLUMETRIC PIPET

Trial	Volume of Water Delivered, ml	Trial	Volume of Water Delivered, ml	Trial	Volume of Water Delivered, ml
1	9.975	5	9.980	9	9.973
2	9.981	6	9.967	10	9.972
3	9.982	7	9.971	11	9.989
4	9.977	8	9.968	12	9.975
				Average	9.976

that a spread of 0.022 ml separates the high and low values of the set even though the measurements were performed with scrupulous care; we attribute this scatter in the results to the effect of indeterminate error. Examination of the calibration process shows that the necessity for the exercise of judgment arises four times in each determination; the uncertainty of each will contribute to the indeterminate error in the results.

1. Since density is a temperature-dependent property of a substance, the reading of a thermometer is necessary. A slight uncertainty is propagated into the final result from this source.

2. In obtaining the weight of water delivered by the pipet, two weighings are required; the estimation of balance swings is thus a potential source of uncertainty. However, since the density of water does not differ greatly from one, the weight delivered by this pipet must have been close to 10 grams. The error involved in weighing is clearly going to be negligible; an error of even 1 mg in a 10-gram sample represents a relative uncertainty of only 1 part in 10,000.

3. The water level must be made to coincide with the mark etched on the pipet.

4. The volume of water delivered by the pipet is dependent upon the time allowed for drainage.

The uncertainty in the calibration is due principally to these latter two sources. We are unable, from the data at hand, to define the contribution of each to the over-all uncertainty in the results; but since the two steps in question are independent of each other, we can expect that in some trials these uncertainties tended to cancel themselves while in others they reinforced each other. The indeterminate error of the process is thus the resultant of the individual uncertainties in the various measurements. In light of the foregoing, the fact that there is only one pair of duplicate results in the set is not surprising.

The Normal-error Curve

Two very important assumptions form the basis for the application of statistics to the problem of indeterminate error; experimental evidence indicates that we are justified in making them. First, results will be high or low with equal probability owing to the effect of indeterminate error. This is another way of stating that an average value is likely to be more reliable than any individual member of a set. Notice that this behavior is altogether different from that of determinate error. Second, indeterminate error will cause small deviations to occur more often than large deviations. It follows that the likelihood of indeterminate error being responsible for a result that deviates seriously from the average, while not impossible, is remote indeed.

The consequences of these assumptions may be envisioned by considering a very large number of measurements of some quantity that have been made in the same manner. If we plot the deviations of the individual measurements from the arithmetic mean against the number of times each such deviation occurs, we obtain curves having the shape of those in Figures 3.2 and 3.3. Such curves are called *normal-error curves*; they are of great importance because they describe graphically the distribution of indeterminate error in a typical physical measurement.

Curves having identical shapes will result from a plot of the numerical values of the measurements against the number of times each is observed. In this event the average value of the measurement will exhibit the greatest frequency of occurrence, which is consistent with the finding that zero deviation from the average is the most frequently observed.

In Figure 3.2 the solid line is the normal-error curve for a measurement that is entirely free of determinate error. We see that the deviations from the average distribute themselves symmetrically about zero, measurements possessing

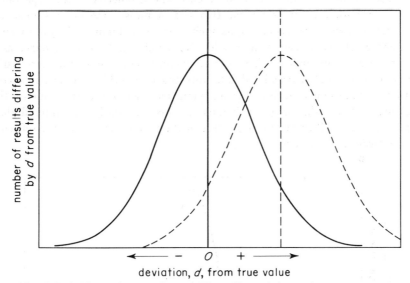

deviation, *d*, from true value

Fig. 3.2 A Normal-error Curve. The effect of determinate error upon a normal distribution is illustrated by the broken line.

zero deviation occurring with the highest frequency. Under these circumstances the mean or average value for the measurement is the true value; furthermore the median value corresponds numerically to the average since positive and negative deviations occur with equal frequency.

The effect of an undetected, uniform *determinate* error on the normal-error curve is illustrated by the dotted curve in Figure 3.2. While the shape is unaltered by the presence of such an error, the curve is displaced so that the arithmetic average of the measurements is no longer equal to the true value. The extent to which the curve is displaced is indicative of the size of the determinate error.

It cannot be too strongly emphasized that the preceding paragraphs describe the operation of *indeterminate* error for any physical measurement. This is why the distribution of replicate measurements caused by indeterminate error is predictable to a degree; we can, in short, make statements concerning the *probability* of occurrence of an indeterminate error of a given magnitude.

The properties of the normal-error curve will now be examined in greater detail. If we measure some quantity by two entirely different methods, there is no reason that the normal-error curves for the two processes be identical; the magnitude of the indeterminate error, after all, depends upon the method rather than the nature of the measurement. For example, we wish to measure the diameter of a piece of glass tubing. We perform this first with a ruler and then with a micrometer caliper, taking a very large number of data with each device. Each set of measurements will be affected by indeterminate error; the magnitude of this error, however, will certainly be smaller for the micrometer caliper data because of the greater reproducibility afforded by this method of measurement. The effect of this difference will be reflected in the appearance of the normal-error curves for the two sets of measurements. If sufficiently large sets of data were taken, graphs such as those in Figure 3.3 would result. The more refined measurement would give a taller and narrower curve as a result of the smaller indeterminate error, whereas the curve for the more crude measurement would be broad and flat. Notwithstanding this difference in shape, both curves are similar in that they are symmetrical about the true value for the measurement and that large deviations are less frequently encountered than small ones. The difference in magnitude of the indeterminate error is reflected in the breadth of the curve.

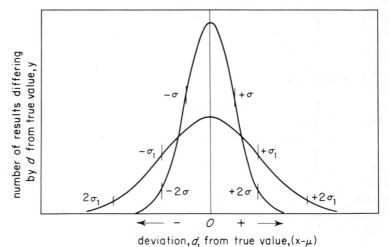

Fig. 3.3 Normal-error Curves. Determination of the same quantity by two methods of inherently different reliability.

The two curves in Figure 3.3 and all other normal-error curves are described by the following mathematical expression:

$$y = \frac{e^{-(x-\mu)^2/2\sigma^2}}{\sigma\sqrt{2\pi}} \tag{3-3}$$

In this equation, x is the value of an individual measurement and μ is the arithmetic mean of a very large number of values for this measurement. The quantity $(x-\mu)$ is thus the deviation of x from the mean; this is the quantity that is plotted as the abscissa of the normal-error curve. The term y is the frequency with which any particular deviation occurs and is the ordinate of the normal-error curve. The symbol π has its usual meaning; e represents the base for Naperian logarithms, 2.718.... The term σ is called the *standard deviation* and is a constant having a unique value for any set consisting of a very large number of measurements. The breadth of the normal-error curve is fixed by the numerical value of σ; σ is thus a very useful measure of the dispersion to be expected owing to the operation of indeterminate error. We therefore need to examine this constant further.

The Standard Deviation

Inspection of equation (3-3) has shown that a unique curve exists for each value of the standard deviation. This constant, then, provides the means of relating the magnitude of a given indeterminate error to its probability of occurrence. Thus, it may be shown mathematically that in a normally distributed population 68.3 percent of the results will differ from the average by less than one standard deviation ($\pm 1 \times \sigma$), 95.5 percent by less than ($\pm 2 \times \sigma$), 99.7 percent by less than ($\pm 3 \times \sigma$), and so forth (see Fig. 3.3). In practice, if an individual result deviates by, say, 3σ from the mean, we are able to conclude that the likelihood is very remote that this deviation is due to the operation of indeterminate error.

It is possible to compute the standard deviation for a set of experimental data by summing the squares of the individual deviations from the mean (μ), dividing by the number of measurements in the set (N), and extracting the square root of this quotient.

$$\sigma = \sqrt{\frac{\sum_{i=1}^{N} (x_i - \mu)^2}{N}} \qquad (3\text{-}4)$$

This equation is of limited utility since it is strictly applicable only to a set of data containing an infinite number of measurements (that is, where $N \to \infty$). In any actual situation, we must calculate the deviations of the individual measurements from the mean, m, of a limited number of measurements rather than from μ, the mean of an infinite set; we have no assurance that the numerical values of these two are identical. It can be shown, however, that this defect is largely compensated for by using ($N - 1$) as the divisor rather than N. Equation (3-4) then takes the form

$$s = \sqrt{\frac{\sum_{i=1}^{N} (x_i - m)^2}{N - 1}} \qquad (3\text{-}5)$$

The quantity s is based upon a finite set of experimental data and is thus an estimate of the true standard deviation (σ). We see that s approaches σ more and more closely as the number of measurements in the set (N) becomes large.

The difference $(x_i - m)$ is the absolute deviation of the individual result from the mean; the standard deviation is thus obtained in absolute terms. It is also proper to employ the relative deviation $(x_i - m)/m$ in place of the former quantity; in this event equation (3-5) yields an estimate of the relative standard deviation, s_r.

Example. Estimate the standard deviation s for the following series of data :

$$4.28,\ 4.21,\ 4.30,\ 4.36,\ 4.26,\ 4.33$$

The arithmetic mean (m) for this set is 4.29. The individual deviations may then be computed and squared.

	d	d^2
4.28	0.01	0.0001
4.21	0.08	0.0064
4.30	0.01	0.0001
4.36	0.07	0.0049
4.26	0.03	0.0009
4.33	0.04	0.0016

$$0.0140 = \Sigma(x_i - m)^2$$

The denominator $(N - 1)$ is equal to 5.
Substituting these into equation (3-5) we find

$$s = \sqrt{\frac{0.0140}{5}}$$

$$= \sqrt{0.0028} = 0.053$$

Example. Estimate the relative standard deviation for this set of data. We proceed as before, but compute the relative deviation. The terms m and $(N - 1)$ are unchanged.

	d	d_r	d_r^2
4.28	0.01	0.002	0.000004
4.21	0.08	0.019	0.000361
4.30	0.01	0.002	0.000004
4.36	0.07	0.016	0.000256
4.26	0.03	0.007	0.000049
4.33	0.04	0.009	0.000081

$$0.000755 = \Sigma(x_i - m)^2/m^2$$

Therefore

$$s_r = \sqrt{\frac{0.000755}{5}}$$

$$= \sqrt{0.000151} \cong 0.012$$

Notice that the same result is obtained by dividing the estimate of the absolute standard deviation by the mean value for the measurement

$$s_r = \frac{0.053}{4.29} \cong 0.012$$

The relative standard deviation may be expressed in terms of percentage or of parts per thousand by multiplying the foregoing estimate by 100 or 1000.

In summary, when reference is made to the standard deviation of some experimental quantity, it is to be understood that this is a prediction of the interval about the mean within which 68 percent of an infinite number of replicate measurements of this quantity would be expected to lie.

The Variance

The quantity s^2 is called the *variance* by statisticians and is an additive property. We shall see that advantage may be taken of this in estimating the precision of a multistep process; the over-all variance is simply the sum of the variances of the individual steps. Extraction of the square root, as before, yields the standard deviation for the over-all process. While the variance possesses this important attribute, the less fundamental standard deviation is preferred by most chemists as a means of expressing precision because it carries the same dimensions and has a numerical magnitude that is appropriate to the data to which it pertains.

THE EFFECT OF DETERMINATE AND INDETERMINATE ERROR ON PRECISION

While it is convenient to treat the two major types of error individually, separating the contribution of each to the precision, or lack thereof, in a practical situation is not always easy. A part of the spread existing in a set of replicate measurements is inevitably due to the indeterminate error inherent to the process. Also any determinate error that does not produce a proportional effect upon the results will also be reflected in the degree of precision attained. Precision is thus a measure of indeterminate error provided we can establish that the results are free of constant errors or those gross errors that affect at most only a few of the results in the set.

As a corollary to the foregoing it may be stated that the presence of determinate error may or may not be reflected in the precision of a set.

The Statistical Treatment of Small Sets of Data

When confronted with a set of data comprising a small number of measurements, we are dealing with a situation that is far removed from the infinite set of the previous section; analytical results are most commonly encountered in sets that contain from two to perhaps six measurements only. Furthermore, one value in a set often diverges considerably from the rest. In any actual situation we do not know whether this outlying result is due to an undetected gross error or to an infrequently occurring indeterminate error of large magnitude; in the former instance the result clearly should be rejected while in the latter it represents a statistically significant part of the data. The application of statistical methods can aid in this decision of retention or rejection by providing a knowledge of the *probability* that the datum is or is not a legitimate member of the set. We are only justified, positively, in dropping a value *when it is known to have been affected by a gross error*; this is an excellent argument for the conscientious maintenance of all observations in a laboratory notebook.

Statistical tests are employed by the analyst in two ways. They may be applied phenomenologically to provide an estimate of the precision based upon the experimental data that have been collected. The average or median is taken as the best value for the set, the average deviation, standard deviation, or the spread being employed to express the precision. In addition, statistics may be used to estimate the probable limit of precision inherent to a given process. This, in turn, provides an estimate of the indeterminate error to be expected in an analysis.

The Evaluation of Experimental Data

Table 3-4 gives the results of a series of replicate analyses for the percentage of calcium oxide in a sample of calcite ($CaCO_3$). There are various ways to express the reproducibility of these data, and they are arranged in order of increasing magnitude rather than the order in which they were obtained; this is an aid in selecting the median.

Average deviation.   As described previously, the average deviation is obtained by summing the deviations of the individual measurements from the average value (56.06 percent) and dividing this sum (0.38 percent) by the number of trials (5) in the set. We could thus report these results as 56.06 $\pm$ 0.08 percent CaO; alternatively, the precision could be expressed in the relative terms of percentage deviation or parts per thousand.

These data show that the result from the fifth trial deviates rather considerably from the others; intuitively we question its validity. By arbitrarily discarding this measurement, however, we are adjusting the data to conform with a subjective notion of how they should occur; statisticians refer to this as the introduction of *bias*. Retention of the measurement has its disadvantages also; a widely diverging result has a disproportionate effect upon the average value of

Table 3-4

ANALYSIS FOR THE PERCENTAGE OF CALCIUM OXIDE IN A SAMPLE OF CALCITE

Trial	CaO Found, percent	Deviation from Average (56.06 percent) d	d^2	Deviation from Median (56.04 percent) d_m
3	55.95	0.11	0.0121	0.09
4	56.00	0.06	0.0036	0.04
2	56.04 (median)	0.02	0.0004	0.00
1	56.08	0.02	0.0004	0.04
5	56.23	0.17	0.0289	0.19
Average	56.06	0.08		0.07

a set of measurements. We can avoid the necessity of making this choice by selecting the median as being the most reliable estimate available for the true value. In the present instance the median is 56.04 percent. The average deviation is calculated as before and the results would now be reported as 56.04 ± 0.07 percent CaO.

It is advantageous to select the median, rather than the average, as the best value for a small set of measurements. We have seen that its employment makes possible the inclusion of all the data in a set without undue influence from an outlying value. Also the average deviation of a set of measurements from the median will never exceed that calculated from the average; in the present instance, a slight improvement was achieved. Finally, for a set consisting of three measurements free from gross error, it has been shown that the median value of the three is more reliable than the average of the two closest measurements.[1]

Use of the average deviation as a criterion for precision suffers from the disadvantage of lacking reproducibility; more often than not the average deviation of a large set will be greater than that found for subsets made up from the same data.[2] Since most of the data in quantitative analysis are obtained in small sets, the average deviation of these will supply a misleading indication of the attainable precision of the method employed in procuring them.

The standard deviation. The data of Table 3-4 can be considered as being five of an infinite number of values we might have obtained; as such they should fit under the normal-error curve in some fashion. The standard deviation s calculated for the set should thus be indicative of the distribution to be expected from this method of measurement; it has the further advantage of showing less

[1] National Bureau of Standards, *Technical News Bulletin*, July, 1949; *J. Chem. Ed.*, **26**, 673 (1949).

[2] W. J. Youden, *Statistical Methods for Chemists*, 8. New York: John Wiley and Sons, Inc., 1951.

variation as the size of the set is changed than does the average deviation. Computation may be accomplished by use of equation (3-5). The mean is used exclusively since the smallest standard deviation results from its employment. Taking the values for d^2 from Table 3-4, we calculate the standard deviation to be

$$s = \sqrt{\frac{0.0121 + 0.0036 + 0.0004 + 0.0004 + 0.0289}{5 - 1}}$$

$$= 0.106 \text{ or } 0.11 \text{ percent.}$$

Accordingly we should expect approximately two thirds of the results of subsequent analyses to fall in the range of plus or minus one standard deviation (0.11 percent) about the average value of 56.06 percent, and nineteen out of twenty within twice this range (2×0.11 percent). These estimates are, of course, provisional upon the reliability of s. It is fortunate that a very rapid increase in reliability attends an increase in the number of individuals in the set.

There is another method available for estimating the standard deviation for a set of experimental data. This involves multiplication of the spread between the highest and lowest values of the set by a statistical factor, K_M. Numerical values for this factor have been determined from statistical theory; as may be seen in Table 3-5, an increase in the number of measurements in the set results in a decrease in the value of K_M. Applied to the present example, we see that K_M for five measurements is 0.43, and that the maximum spread in the data is 0.28 percent. The product of these yields an estimate of 0.12 percent for the standard deviation, which is in good agreement with that calculated in the previous section.

Table 3-5 [3]

VALUES OF K FOR ESTIMATION OF STANDARD DEVIATION FROM THE RANGE

Number of Observations	K_M
2	0.89
3	0.59
4	0.49
5	0.43
6	0.40
7	0.37
8	0.35
9	0.34
10	0.33

[3] Reproduced from R. B. Dean and W. J. Dixon, *Anal. Chem.*, **23**, 636 (1951), by permission.

Confidence Limits

Calculation of the standard deviation for a set of data is valuable in that it provides an indication of the precision inherent to the particular method of measurement. It does not, however, of itself allow us to predict how closely the true mean, μ, has been approached by an experimentally derived mean, m, because it is practical to secure analytical data in relatively small sets only. Unfortunately the average (m) of such a small number of measurements may deviate considerably from the true mean (μ). The probability of such a difference being large becomes less as the number of measurements (N) in the set increases and if the precision is high.

The true mean value (μ) of a measurement is a constant that must always remain unknown. With the aid of statistical theory, however, limits may be set about the experimentally determined mean (m) within which we might expect to find the true mean with some given degree of probability. The limits obtained in this manner are called *confidence limits*; the range defined by these limits is known as the *confidence interval*.

It is worthwhile to consider qualitatively some of the properties of the confidence interval. For a given set of data the size of the interval depends in part upon the odds for correctness desired. Clearly, if a prediction is to be absolutely correct, we will have to choose an interval about the mean that is large enough to include all conceivable values (μ) might take; this, of course, is of no value as a prediction. On the other hand, the interval does not need to be this large if we are willing to accept the probability of being correct 99 times in 100; it will be even smaller if 95 percent correctness is acceptable. In short, as the probability for making a correct prediction becomes less favorable, the interval included by the confidence limits becomes smaller. These limits can be brought so close together that predictions based upon them stand to be incorrect a prohibitively large portion of the time; this condition, too, is of marginal utility.

It can be shown that the confidence limit may be calculated from the following expression:

$$\text{confidence limit} = \frac{ts}{\sqrt{N}} \tag{3-6}$$

where s is the estimated standard deviation determined as described previously, and N is the number of measurements included in the set. The term t is dependent upon the number of measurements as well as the degree of probability for correctness that is desired. Statisticians have compiled tables for numerical values of t for various confidence levels. Representative values are given in Table 3-6.

Let us calculate the 95-percent confidence interval for the calcium oxide data. The standard deviation has been estimated to be 0.11 (p. 52). The value of t for the 95-percent confidence level and five measurements is 2.8. Thus

$$\text{95-percent confidence limit} = \pm \frac{(2.8)(0.11)}{\sqrt{5}} = \pm 0.14 \text{ percent}$$

Table 3-6

VALUES OF t FOR VARIOUS LEVELS OF PROBABILITY

Number of Observations (N)	Factor for Confidence Interval of 80%	90%	95%	99%
2	3.08	6.31	12.7	63.7
3	1.89	2.92	4.30	9.92
4	1.64	2.35	3.18	5.84
5	1.53	2.13	2.78	4.60
6	1.48	2.02	2.57	4.03
7	1.44	1.94	2.45	3.71
8	1.42	1.90	2.36	3.50
9	1.40	1.86	2.31	3.36
10	1.38	1.83	2.26	3.25
11	1.37	1.81	2.23	3.17
12	1.36	1.80	2.20	3.11
13	1.36	1.78	2.18	3.06
14	1.35	1.77	2.16	3.01
15	1.34	1.76	2.14	2.98

We may therefore state that there are 95 chances out of 100 that the true average for this analysis lies within ± 0.14 percent of 56.06—that is, that μ is included in the interval between 55.92 and 56.20 percent. An analogous calculation indicates that the limits at the 99-percent confidence level are ± 0.23 percent.

Prediction of the Reproducibility of a Process

Evaluation of the reproducibility that might reasonably be expected from a given method is often useful to the analyst; this provides him with an estimate of the maximum precision limits of the method in question and allows him to decide in advance whether it is capable of giving the reproducibility he requires.

An analysis generally requires the performance of several measurements; each of these is subject to indeterminate error. We have already seen that the probability of observing a high value is equal to that for a low one, and that small deviations are more likely than large ones. It follows, then, that the individual errors accumulated in the course of combining several measurements can either be additive or canceling to some degree. The cumulative effect of these upon the final result is what we wish to estimate; in so doing, we must take into account this variation in behavior of the individual errors.

The manner in which individual deviations affect the final result is dependent upon the nature of the calculation. Where the precision is being estimated for a result that is the sum of or difference between measurements, the *absolute* deviation of each is employed; where the result in question is a product or quotient, the appropriate *relative* deviations must be employed.

Table 3-7 gives estimates of precision to be expected in common analytical measurements. If we consider these estimates to be approximations of the standard deviation, by squaring each we will obtain the respective variances (see p. 49). Recalling that these latter are additive, it follows that the square root of the sum of the individual variances should be an approximation of the standard deviation for the over-all process. Thus, designating the individual measurements as $A, B, C, ...$, the absolute deviations associated with these as $a, b, c, ...$, and the respective relative deviations as $a_r, b_r, c_r, ...$, the absolute deviation x and relative deviation x_r for the resultant X obtained by combining these will be

$$x = \sqrt{a^2 + b^2 + c^2 + \cdots} \qquad (3\text{-}7)$$

if X is a sum or difference of the individual measurements, and

$$x_r = \sqrt{a_r^2 + b_r^2 + c_r^2 + \cdots} \qquad (3\text{-}8)$$

if X is a product or quotient.

Table 3-7

ESTIMATED PRECISION OF COMMON LABORATORY APPARATUS

Burets

+ 0.02 ml for each reading

Pipets[4]

5 ml	$\pm$ 0.01 ml
10	0.02
15	0.03
20	0.03
25	0.03
50	0.05
100	0.08

Volumetric flasks,[4] "to contain" basis

25 ml	$\pm$ 0.03 ml
50	0.05
100	0.08
250	0.12
500	0.15
1000	0.30
2000	0.50

A weighing on an analytical balance is reproducible to $\pm$ 0.0001 gram.

[4] Based upon allowed tolerances, National Bureau of Standards; Cf. Circular 602 "Testing of Glass Volumetric Apparatus." The reproducibility of measurement should not exceed the tolerance; these, then, serve as conservative estimates of precision.

An analytical calculation generally takes the form of a quotient; further, the numerator as well as the denominator may be the sum or difference of experimental observations. In an instance such as this the deviations due to the component sums and differences are estimated before considering the products and quotients. To illustrate, we shall estimate the precision to be expected in determining the concentration of a solution of hydrochloric acid. A quantity of sodium carbonate is weighed carefully and dissolved in water. The volume of acid required to neutralize this is measured, an indicator being employed for the detection of the end point; the concentration of the hydrochloric acid solution is calculated (in terms of normality) from the following equation :

$$\text{concentration of HCl} = \frac{\text{weight of } Na_2CO_3 \text{ taken, grams}}{k \times \text{ml HCl required}}$$

For our purposes we can assume that a highly reliable value for the constant k is available.

Consider the following experimental data:

weight of bottle $+ Na_2CO_3$	31.4057 ± 0.0001 grams
weight of bottle less sample	31.1536 ± 0.0001
weight of Na_2CO_3 taken	0.2521 gram
final volume of HCl	40.19 ± 0.02 ml
initial volume of HCl	0.15 ± 0.02
volume of HCl required	40.04 ml

Both the volume of hydrochloric acid and the weight of sodium carbonate are obtained by taking the difference between two experimental measurements, each of which is subject to uncertainty on the order of that given in Table 3-7. An additional uncertainty of, say, 0.03 ml, associated with the judgment of the end-point color of the indicator will also affect the volume term and must be taken into account.

The estimated uncertainty in the weight of sodium carbonate is given by

$$s = \pm \sqrt{(0.0001)^2 + (0.0001)^2}$$
$$= \pm 0.00014 \text{ gram}$$

The relative deviation thus will be

$$\pm \frac{0.00014}{0.2521} = \pm 0.00055, \text{ or about 0.6 part per thousand}$$

The estimated uncertainty in the volume of hydrochloric acid is calculated similarly

$$s = \pm \sqrt{(0.02)^2 + (0.02)^2 + (0.03)^2}$$
$$= \pm 0.041 \text{ ml}$$

This will yield a relative deviation of

$$\frac{0.041}{40.04} \cong 0.001, \text{ or about 1 part per thousand}$$

Since the value of k is known with great certainty, we assume that its contribution to the uncertainty of the over-all process is negligible.

Hence the estimated precision for the concentration of hydrochloric acid will be

$$s_r = \pm \sqrt{(0.6)^2 + (1.0)^2}$$

$$= 1.2 \text{ parts per thousand}$$

This is an approximation, in terms of standard deviation, of the ultimate precision to be expected of this method for weights and volumes of reactants that approach those of the example. However, while the absolute uncertainties of the measurements are invariant, the precision of the derived result is dependent upon the respective relative deviations; these, of course, are functions of the quantity of material involved. Thus, were we to reduce the weight of sodium carbonate taken by a factor of ten, the relative deviation from uncertainties in weighing would increase by a like amount. This sample would require only one tenth the volume of hydrochloric acid; a tenfold increase in the relative deviation of the volume term would also occur. Under these circumstances the estimated relative deviation in the final result would be in excess of ten parts per thousand.

THE OUTLYING RESULT

When a set of data contains an outlying result that appears to deviate excessively from the average or median, the decision must be made to retain or disregard it. The choice of criteria for the rejection of a suspected result has its perils. If we demand overwhelming odds in favor of rejecting a questionable measurement and thereby render this difficult, we run the risk of retaining results that are spurious and that have an inordinate effect on the average of the data. On the other hand, if we set lenient limits on precision and make easy the rejection of a result, we stand to discard measurements that rightfully belong to the set. It is an unfortunate fact that there is no universal rule that can be invoked to settle this question.

Three statistical tests widely employed as guides in making this decision are presented for consideration and discussion; each is applied to the calcium oxide data from Table 3-4.

The 4d Rule

Scope. Requires four or more results in the set.

Method. Disregarding the questionable result, compute the average deviation for the remaining members of the set from the new average.

Criterion. If the questionable result differs from the new average by more than four times the average deviation of the remaining members, it is to be rejected.

Applying the $4d$ rule to the calcium oxide data

Average of trials 3, 4, 2, 1 = 56.02 percent

Average deviation of trials 3, 4, 2, 1 from 56.02 percent = 0.04 percent

Deviation of outlying result from 56.02 percent = 56.23 — 56.02 = 0.21 percent

Since 0.21 > (4 × 0.04), reject trial 5

The 2.5d Rule

Scope and **method**   are the same as for the $4d$ rule.

Criterion. If the questionable result differs from the new average by an amount greater than 2.5 times the average deviation of the remaining members, it is to be rejected.

Again, using the calcium oxide data for illustration

$$\text{Since } 0.21 > (2.5 \times 0.04), \text{ reject trial 5}$$

The Q Test

Scope.   Requires three or more measurements.

Table 3-8[5]

CRITICAL VALUES FOR REJECTION QUOTIENT Q

Number of Observations	Q
2	—
3	0.94
4	0.76
5	0.64
6	0.56
7	0.51
8	0.47
9	0.44
10	0.41

[5] Reproduced from R. B. Dean and W. J. Dixon, *Anal. Chem.*, **23**, 636 (1951), by permission.

Method. Compute the spread separating the highest and lowest result in the set. Divide this into the spread between the questionable result and its nearest neighbor to obtain the quotient Q. Compare the value of Q with the appropriate figure in Table 3-8.

Criterion. The questionable result may be rejected with 90 percent confidence if Q exceeds that in the table.

In the case of the calcium oxide analysis:

Spread (w) between the highest and lowest values $= 0.28$ percent

Spread is $56.23 - 56.08 = 0.15$ percent between questionable result and nearest neighbor.

Therefore Q is equal to $0.15/0.28 = 0.54$

Tabulated value for Q (5 measurements) $= 0.64$; hence retain trial 5.

Application of these tests to the same set of data reveals that they do not necessarily yield consistent answers regarding the fate of a questionable result. The difficulty lies, for the most part, in the fact that the $4d$ and $2.5d$ tests require larger sets of data for validity; employment for the evaluation of small sets is open to serious objections. For example, provided the set is sufficiently large, the probability that an outlying value rejected by the $4d$ test is indeed spurious is on the order of 99 to 1; this criterion is strict to the point that erroneous measurements are likely to be retained. This is why the $2.5d$ test, which will correctly recommend rejection 19 times out of 20, was proposed. It has been found, however, that in sets consisting of only four normally distributed values, the outlying measurement will differ by $4d$ or more from the average of the other three nearly 60 percent of the time[6]; this frequency of occurrence is, of course, even greater if $2.5d$ is chosen as the criterion. That the rejection of a statistically significant observation from a small set is justified on the grounds of its disproportionate effect upon the average is, of course, open to argument. An objection to employment of the Q test[7] is that for small sets of data the rejection quotient is so large that erroneous data are likely to be retained. Its employment, nonetheless, is to be preferred to the other two.

Clearly, the blind application of statistics in deciding upon retention or rejection of suspect measurements in small sets of data is not likely to be any more fruitful than an arbitrary decision in this matter. The application of good judgment is also necessary for a meaningful statistical evaluation of a small set of data. One of the best aids in this regard is the estimate of precision to be expected from a method. This can be based upon past experience with the analysis or alternatively upon a calculation of the estimated reproducibility of the process.

Good precision in results is the mark of a careful analyst; the natural tendency is to reject data in order to improve the appearance of the set. Conservatism in this matter must be assiduously cultivated.

[6] W. J. Blaedel, V. W. Meloche, and J. A. Ramsay, *J. Chem. Ed.*, **28**, 643 (1951).

[7] R. B. Dean and W. J. Dixon, *Anal. Chem.*, **23**, 636 (1951).

In light of the foregoing, a number of recommendations suggest themselves for the treatment of a set of results that contains a suspect value.

(1) Estimate the precision that can reasonably be expected from the method as it was employed. Be certain, in short, that the outlying result actually is questionable.

(2) Re-examine carefully all data relating to the questionable result to see if a gross error has affected its value. Remember that the certain knowledge of a gross error is the only sure justification for disregarding a doubtful measurement.

(3) Repeat the analysis, if sufficient sample and time are available. Agreement of the newly acquired data with those that appear to be valid will lend weight to the notion that the outlying result should be rejected.

(4) If further data cannot be secured, apply the Q test to the existing set to see if the doubtful result can be rejected on these grounds.

(5) In the event that retention is necessary from the standpoint of the Q test, give consideration to the reporting of the median, rather than the average value of the set; as pointed out previously, this will tend to minimize the influence of the outlying value.

SIGNIFICANT FIGURES

The report of a set of results should include not only the "best" value for the set, be it average or median, but also give an indication of the precision observed between the individual members; we have seen that the latter may be expressed in terms of the average deviation or the standard deviation. Either of these serves to reveal the number of digits that we are entitled to include in an answer; common practice dictates that all numbers known with certainty shall be reported, and that in addition the first uncertain one shall also be included.

For example, the average of 61.64, 61.41, 61.55, and 61.62 is 61.555. The average deviation of the individuals from the mean is ± 0.075. Clearly, the number in the second decimal place is subject to uncertainty; such being the case, all numbers in succeeding decimal places are without meaning and we are forced to round the average value accordingly. The question of taking 61.55 or 61.56 must next be considered, 61.555 being equally spaced between them. A good guide here is always to round to the nearest even number; in this way any tendency to round in a set direction is eliminated, there being an equal likelihood that the nearest even number will be the higher or the lower in any given instance. Thus, we report the foregoing results as 61.56 ± 0.08.

In the absence of qualifying information of this sort, it is assumed that the last and uncertain digit is known to within plus or minus one unit. Thus the ratio of the circumference of a circle to its diameter (π) has been calculated to many decimal places. When using 3.1416 for π, we are inferring that this is the best value for the constant to four decimal places, but that its true value lies somewhere between 3.1415 and 3.1417. A further complication arises when we wish to express very large numbers whose values are subject to considerable uncertainty; a case in point is Avogadro's number. We know with certainty that the first three digits are 6, 0, and 2, and that the next is uncertain but probably is 3. Since the digits that follow are not known, and since it is necessary to indicate the position of the decimal point relative to 6023, we use zeros in place of the unknown digits. Here, the zeros indicate the order of magnitude of the number only, and have no other meaning. It is clear that we must make a distinction between those figures that have physical significance—that is, significant figures—and those that are either unknown or meaningless owing to the inadequacies of measurement. The following rules will be found useful in this regard.

1. *The significant figures in a number comprise all those digits whose values are known with certainty plus the first digit whose value is uncertain.* The position of the decimal point is irrelevant. Thus, 0.12345, 1.2345, 123.45, and 12,345 all contain five significant figures.

2. *Zeros are significant when part of the number; they are not significant when employed to indicate order of magnitude.* Zeros bounded left and right by digits other than zero are always significant. Thus 21.03 contains four significant figures; so does 20.03.

Zeros bounded only on the right by digits are never significant, for here they are being used to indicate the position of the decimal point. Thus 0.123, 0.000123, and 0.000000123 all contain three significant figures.

Zeros bounded by digits only on the left present a problem in that they may or may not be significant. If the zeros are a part of the number, as well they may be, they are significant. Thus, the weight of a 20-mg weight that carries no correction (to a tenth of a milligram) is known to three significant figures, 20.0 mg. In expressing this as 0.0200 gram, the number of significant figures has not changed. If, on the other hand, we wish to express the volume of a 2-liter beaker as 2000 ml, the latter number contains but one significant figure; the zeros simply indicate the order of magnitude. It can, of course, happen that the beaker in question has been found by experiment to contain 2.0 liters; here the zero following the decimal point infers that the volume is known to plus or minus 0.1 liter and hence is significant. When this volume is expressed in milliliters, the zero following the 2 is still significant in this case but the other two zeros are not. The confusion that inevitably attends this dual employment of zeros is easily eliminated by the use of exponential notation; that is, in the present instance we could indicate the volume as 2.0×10^3 ml.

Significant Figures in a Derived Result

The calculations of analytical chemistry often involve the manipulation of numerical quantities that contain widely varying numbers of significant figures. It is thus necessary to consider how many significant figures there are in a derived result.

The manner in which the uncertainty of an individual quantity is propagated into the final result depends upon the nature of the mathematical operation.

By way of example, we shall carry out the processes of addition, subtraction, multiplication, and division with the numbers 142.7 and 0.081. That the last digit in each number is uncertain to the extent of one unit must be assumed, since no amplifying information in this regard has been supplied. Thus

| | Uncertainty | |
Number	Absolute	Relative, parts per thousand
142.7	0.1	0.7
0.081	0.001	12

Addition.   Upon adding these numbers, we obtain the sum

$$142.7$$
$$0.081$$
$$\overline{142.781}$$

Recalling that the significant figures in a number include all those that are known with certainty plus the first uncertain one, we see that this sum will have to be rounded off to 142.8 since there is an uncertainty in the first decimal place. We are thus entitled to report the answer to four significant figures. An entirely too pessimistic estimate $[142.7 \pm (0.012 \times 142.7) = 142 \pm 2]$ would have resulted had we employed the relative, rather than the absolute uncertainties of the two numbers.

Subtraction.   The arguments advanced for addition may be applied with equal force when taking the difference between two numbers.

Multiplication.   Upon multiplying 142.7 by 0.081, we find that the product is equal to 11.5587. Here the number of decimal places that we are entitled to report is governed by the relative uncertainties in the two numbers. This may be roughly demonstrated by considering the range of answers that could result from the assumed uncertainties in these numbers—that is, in multiplying 0.081 ± 0.001 by 142.7 ± 0.1, the maximum estimated value is given by

$$0.082 \times 142.8 = 11.7096$$

and the minimum by

$$0.080 \times 142.6 = 11.4080$$

The difference between these is $(11.7096 - 11.4080) = 0.3016$. Thus, the answer is uncertain by about

$$\pm \frac{0.1508 \times 1000}{11.5587} = \pm 13 \text{ parts per thousand}$$

This closely approximates the relative uncertainty inherent in 0.081; thus the uncertainty in a product may be estimated from the relative deviation of the least certain quantity involved.

In the present instance

$$0.012 \times 11.5587 = 0.138$$

Hence, uncertainty appears in the first decimal place; the answer must be rounded to 11.6.

Division. The same considerations hold true in the case of a division as in a multiplication; again it is the relative uncertainty of the least certain quantity that dictates the number of significant figures in the quotient.

$$\frac{142.7}{0.081} = 1761.728 \ldots$$

The relative uncertainty in this quotient will be about twelve parts per thousand; the absolute uncertainty will be

$$0.012 \times 1761.728 = 21$$

and the answer must be rounded off. To report it as 1760 involves ambiguity since this notation is incapable of indicating that the zero is not significant. It is preferable to employ the exponential notation of 1.76×10^3.

These examples illustrate three important rules with regard to the number of significant figures in a derived result :

(1) Where the mathematical operation is an addition or subtraction, it is the *absolute* uncertainty of the component quantities that determines the number of significant figures in the sum or difference.

(2) Where the mathematical operation is a multiplication or division, it is the *relative* uncertainty of the component quantities that governs the number of significant figures in the product or quotient.

(3) Where the quantities being manipulated have widely differing degrees of uncertainty, the number of significant figures in the derived result will be limited by the least certain of the component quantities; whether the absolute or relative uncertainty should be considered depends upon the nature of the calculation.

Evaluation of the number of significant figures to be shown in the result of a multistep calculation is accomplished by first considering any sums or differences, and then the products or quotients. This is best illustrated by an example. We wish to know the number of significant figures that should appear in the answer to

$$\frac{(41.27 - 0.414)\,(0.0521)\,(7.090)}{(0.5135 + 0.0009)} = 29.3385 \ldots$$

Since no further information is given, each quantity in this expression is considered to be uncertain to the extent of one unit in the last decimal place.

We first consider the sums and differences, recalling that the absolute uncertainty is important for these. The difference in the numerator turns out to be 40.856; since an uncertainty exists in the second decimal place, however, this should be rounded to 40.86. Notwithstanding the fact that 0.0009 contains but one significant figure, the sum in the denominator will still contain the four figures of the larger number since the uncertainty in its value continues to reside in the fourth decimal place; it is thus taken as 0.5144.

We shall now compute the relative uncertainty of the individual quantities in the simplified equation

$$\frac{(40.86)\ (0.0521)\ (7.090)}{(0.5144)} = 29.3414\ldots$$

Number	Relative uncertainty, parts per thousand
40.86 ± 0.01	0.2
0.0521 ± 0.0001	1.9
7.090 ± 0.001	0.1
0.5144 ± 0.0001	0.2

The uncertainty in 0.0521 is approximately ten times that of any of the other three quantities; we are justified in stating that this figure will govern the number of significant figures in the answer.
Thus

$$29.3414\ldots \times 0.0019 = 0.0557\ldots = 0.06$$

Uncertainty will appear in the second decimal place in the answer; it should therefore be rounded off to 29.34.

Much time can be saved in the performance of calculations if an estimate of uncertainty is made before the actual computation is begun; with practice this may be done by inspection. It is generally advisable, in computation, to carry one figure in excess of the number estimated, and then round off the answer. The numerical value may otherwise suffer somewhat owing to the rounding-off process. Had the components of the foregoing example been rounded before calculation, an answer of 29.36 would have resulted.

Caution must be exercised in evaluating the number of significant figures in a derived result that is numerically close to some power of ten. For example, the relative error in 999 (three significant figures) is substantially identical to that in 1001 (four significant figures). Thus a result that is somewhat less than some power of ten possesses a smaller relative error than the number of significant figures indicate; this must be taken into account when rounding off such an answer.

The Estimated Uncertainty in a Derived Result

The foregoing procedures serve only to indicate the number of significant figures in the derived result; however, they do not give a realistic numerical estimate of the probable uncertainty in this quantity when all of the components in a calculation possess the same order of uncertainty. This is best illustrated by an example. In the following ratio

$$\frac{0.210}{0.197} = 1.06598\ldots$$

both divisor and dividend have essentially the same uncertainty (0.5 percent). This will cause the quotient to be uncertain in the third decimal place (1.06... × 0.005). According to one convention, however, expression of this quotient as 1.066 infers a reliability of one part per thousand which simply is not justified in this case. We are thus forced to round off to 1.07, this being a better representation for the quotient.

If we wish to retain four figures in this result, an estimate of the probable uncertainty must be made and included as part of the report. The relative uncertainty could possibly be as great as 1 percent; this may be seen, as before, by evaluating the quotients (0.211/0.196) and (0.209/0.198) which approximate the extremes to which the result is uncertain. It is equally possible that the two component uncertainties could exactly cancel one another. Neither of these is likely, however. More probably the individual uncertainties will affect the quotient to some intermediate extent. An evaluation of this, based upon the square root of the sum of the squares of the component uncertainties will yield a more realistic estimate of the uncertainty to be expected in the final answer. In the present instance this is found to be 0.7 percent. With this information we would be entitled to report the quotient as 1.066 ± 0.007. If we do not wish to include the numerical estimate of uncertainty, it would be necessary to round the answer to 1.07, in agreement with the treatment based upon significant figures.

In summary, then, there are two methods available for ascertaining the proper number of figures to which a derived result should be reported. Whenever there is an individual uncertainty that is much greater than the others in the computation, it may safely be used as the basis for decision. In the event of doubt regarding the propriety of this approach, the analyst should resort to the more laborious estimation of probable error and base his decision upon the result of this calculation.

suggested references

In an introductory text, it is only possible to indicate a few of the many applications of statistical methods that are of value to the analytical chemist. The following references are suggested for those who wish to pursue further this important topic.

1. C. A. Bennett and N. L. Franklin, *Statistical Analysis in Chemistry and the Chemical Industry*, New York: John Wiley and Sons, 1954.
2. W. J. Blaedel, V. W. Meloche, and J. A. Ramsay. "A Comparison of Criteria for the Rejection of Measurements," *J. Chem. Ed.*, **28**, 643 (1951).
3. K. A. Brownlee, *Industrial Experimentation*, 4th Amer. ed. New York: Chemical Publishing Co., 1953.
4. R. B. Dean and W. J. Dixon, "Simplified Statistics for Small Numbers of Observations," *Anal. Chem.*, **23**, 636 (1951).

5. W. J. Dixon and F. J. Massey, *Introduction to Statistical Analysis*, 2 ed. New York: McGraw-Hill Book Co., 1957.

6. R. A. Johnson, "Indeterminate Error Estimates from Small Groups of Replicates," *J. Chem. Ed.*, **31**, 465 (1954).

7. National Bureau of Standards, *Technical News Bulletin*, July 1949; "The Best Two out of Three ?" *J. Chem. Ed.*, **26**, 673 (1949).

8. W. J. Youden, *Statistical Methods for Chemists*, New York: John Wiley and Sons, 1951.

problems

1. Replicate analyses of a sample of blood meal has produced the following percentages for nitrogen content: 4.16, 4.21, 4.18, 4.12.
   For these data, calculate the following :
   (a) The average value for percent nitrogen.
   (b) The median value for percent nitrogen.   ans. 4.17 percent
   (c) The average deviation of these results in absolute terms.
   (d) The average deviation of these results in parts per thousand.
            ans. ± 7 parts per thousand

2. Calculate the average deviations for each of the following sets of data in absolute and in relative terms.

A	*B*
93.6 percent	10.02 percent
93.7	10.01
93.5	10.09

   (a) Which set possesses the larger absolute deviation?   ans. *A*
   (b) Which set possesses the larger relative deviation?   ans. *B*

3. If

$$s = \sqrt{\frac{(x_1 - m)^2 + (x_2 - m)^2 + (x_3 - m)^2 + (x_n - m)^2}{n - 1}}$$

   show that

$$(n - 1)s^2 = (x_1^2 + x_2^2 + x_3^2 + x_n^2) - \frac{(x_1 + x_2 + x_3 + x_n)^2}{n}$$

   NOTE : This is simply an algebraic transformation of equation 3-5 into an alternate form. If a table of squares is available, this alternate form is a more convenient route to the estimation of the standard deviation for a set of data. As a hint in performing this transformation, recall that

$$m = \frac{x_1 + x_2 + x_3 + x_n}{n}$$

4. The estimated accuracy for an iron analysis is 0.3 percent, and that for an analysis of manganese is 10 percent. A series of iron-manganse alloys is to be analyzed by experimentally determining the percentage of one component and obtaining the percentage of the other by difference. What is the maximum

number of significant figures that can appear in the result if the iron analysis is employed for a sample that contains approximately the following :

(a) 99.5 percent iron?

(b) 50 percent iron?

(c) 1 percent iron?

Repeat the calculation for the situation where the manganese analysis is employed.

5. How many significant figures are there in the following :

(a) 0.062005	(g) 60.025	(m) 0.0003014	(s) 6.111
(b) 31.4	(h) 3.14×10^{-2}	(n) 35.458	(t) 0.002605
(c) 0.00625	(i) 4.2	(o) 91.22	(u) 2.6528
(d) 2.81	(j) 620.1	(p) 0.0011	(v) 0.0314
(e) 0.60025	(k) 96.494	(q) 0.014334	(w) 0.0101
(f) 41.3798	(l) 44.21	(r) 1.008	(x) 0.01922

6. Express each of the numbers in Problem **5** to three significant figures.

7. Calculate the standard deviation for the calibration data in Table 3-3.

ans. 0.006 ml

8. Divide the data of Table 3-3 into two sets, the one consisting of trials 1-6, the other trials 7-12. For each of these subsets calculate and compare the following :

(a) The average.

(b) The median.

(c) The average deviation, calculated from the medians.

(d) The standard deviation (1) from equation 3-5 and (2) from K in Table 3-5.

(e) The 95-percent confidence interval.

9. Express the result of each of the following calculations to the proper number of significant figures :

(a) $4.1374 + 2.81 + 0.0603 = 7.0077$

(b) $4.1374 - 0.0603 = 4.0771$

(c) $4.1374 - 2.81 = 1.3274$

(d) $2.81 - 0.0603 = 2.7497$

(e) $4.1374 - (2.81 + 0.0603) = 1.2671$

10. Express the result of each of the following calculations to the proper number of significant figures :

(a) $14.37 \times 6.44 = 92.5428$

(b) $0.0613 \times 0.4044 = 0.02478972$

(c) $0.0613 \div 0.4044 = 0.151582$

(d) $0.841 \div 297.2 = 0.00282974$

(e) $4.1374 \times \dfrac{0.841}{297.2} = 0.0117077$

11. Express the result of each of the following calculations to the proper number of significant figures :

(a) $\dfrac{4.178 + 0.0037}{60.4} = 0.0692334 \ldots$ \qquad ans. 0.0692

(b) $\dfrac{4.178 \times 0.0037}{60.4} = 0.000255937 \ldots$

(c) $\dfrac{4.178 - 4.032}{1.217} = 0.119967 \ldots$

(d) $\dfrac{4.178 + 4.032}{1.217} = 6.74609 \ldots$

(e) $\dfrac{(6.3194 - 4.1387)\,(204.2)}{0.2148} = 2{,}073.08 \ldots$

12. Evaluate the probable uncertainty in the result for each of the calculations in Problem 11.

13. When a method of analysis was tested upon samples of varying size the following data were obtained:

Sample size, gram	Weight of component found, gram
1.0021	0.4171
0.7997	0.3329
0.6014	0.2506
0.4011	0.1674
0.1995	0.0837
0.1008	0.0426

What, if anything, do these data reveal relative to the existence of determinate error in this analysis?

14. For each of the following sets of data:

7.031	31.41	63.74	90.91
7.039	30.64	63.62	90.42
7.126	31.52	63.93	90.31
7.027	31.18	63.68	90.24

(a) Apply the Q test to see if rejection of the outlying result is justified.
(b) Report the best value for each set and defend your choice.
(c) Calculate the 95-percent confidence interval for each set.

15. What is the minimum volume that can be delivered from a 50-ml buret such that the relative error of the measurement is
(a) 5 percent
(b) 1 percent   ans. 4 ml
(c) 0.1 percent

part 2. *Gravimetric Analysis*

chapter 4. *An Introduction to Gravimetric Methods*

CLASSIFICATION OF GRAVIMETRIC METHODS

The term *gravimetric* pertains to a weight measurement, and a gravimetric method is one in which the analysis is completed by a weighing operation. Two types of gravimetric analyses may be distinguished. In the first of these the substance to be determined is isolated from the other constituents in the sample by formation of an insoluble precipitate; the analysis is completed by determining the weight of this precipitate, or of some substance formed from it, by suitable treatment. The second general type of gravimetric analysis takes advantage of the property of volatility; here the substance to be determined is isolated by distillation. The product may either be collected and weighed, or the weight loss in the sample as a result of the distillation may be measured. Of the two, precipitation methods are the more widely used.

71

Precipitation Methods

Not all insoluble precipitates are well suited for gravimetric analysis. Several of those that are of importance in the qualitative analysis scheme, for example, are not used for various reasons. It is worthwhile here to consider what properties are required in order that a precipitate be applicable for a quantitative precipitation method.

Solubility.   Clearly a precipitate for gravimetric work must be sufficiently insoluble so that the amount lost does not seriously affect the outcome of the analysis. Where the quantity of substance being determined is low and where the demands for accuracy are high, solubility losses may be of real concern.

Purity.   The physical properties of a gravimetric precipitate should be such that it can be readily freed from normally soluble contaminants by fairly simple treatment. During formation, all precipitates carry down some soluble constituents from the solution. The applicability of a precipitate to gravimetric analysis will depend upon the quantity of these contaminants and the ease with which they can be removed.

Filterability.   It must be possible to isolate quantitatively the solid precipitate from the liquid phase by reasonably simple and rapid filtration methods. Whether or not a precipitate meets this requirement will depend upon its particle size. If this is too small, filtration may become difficult indeed.

Chemical composition.   A gravimetric precipitate must be either of known chemical composition or readily converted to a compound of known composition. Only when this is so can the precipitate weight be used for the calculation of the percent composition of the original sample.

Other desirable properties.   Preferably the weighing form of a precipitate should be nonhygroscopic and in other ways nonreactive with the atmosphere in order to simplify the final weighing process. Furthermore, there is some advantage to be gained if the weighing form has a high formula weight. This tends to minimize weighing errors and errors resulting from reaction of the precipitate with the atmosphere.

Another important consideration in gravimetric analysis is the specificity of the reagent used for the formation of the precipitate. Ideally, a single element in the periodic table would precipitate with a given reagent under a given set of conditions. Unfortunately, few of the reactions available to the chemist even approach such specific behavior and most reagents, at best, can only be classified as group reagents that will precipitate several ions. As a result, preliminary separation must often precede the final gravimetric precipitation.

Few, if any, of the precipitates commonly used for gravimetric analysis have all of the desirable properties mentioned here and the chemist must "make do" with what is available to him. In the next chapters will be found more detailed discussions concerning some of these important properties and the variables that affect them.

Volatilization Methods

Gravimetric methods based on the volatilization of a compound containing the element to be determined fall into two categories. First, the evolved substance is somehow collected and weighed; second, the determination is based upon the loss of weight suffered by the original material. The former is to be preferred, being a good deal more specific and less subject to uncertainty.

Perhaps the two most common examples of the first method are the determination of water and the analysis of carbonates. Water can be separated from most inorganic compounds by ignition; the evolved water can then be absorbed on any one of several solid desiccants. The weight of water evolved may be calculated from the gain in weight of the absorbent.

Carbonates are readily decomposed by acids, the evolved carbon dioxide being removed from solution by distillation. As in the case of water there are solid absorbents that will remove carbon dioxide from a gas stream; from the increase in weight of such an absorbent the chemist can calculate the weight of carbon dioxide evolved from the sample.

Water is also frequently determined by the indirect method. In this case the weighed sample containing the water is simply ignited and then reweighed. The loss in weight is assumed to be due to the distillation of the moisture from the sample. Unfortunately this assumption is often not a good one; ignition of many compounds results in their decomposition and consequent change in weight irrespective of the presence of water.

CALCULATIONS OF RESULTS FROM GRAVIMETRIC ANALYSES

A gravimetric analysis is based upon the experimental measurement of two quantities, the weight of sample taken and the weight of a compound of known composition derived from that sample. The conversion of these data to a number expressing the concentration of the substance sought is a relatively simple process.

Generally, we shall want to express our gravimetric results in terms of percentage composition.

$$\text{percent } A = \frac{\text{weight of } A}{\text{weight of sample}} \times 100$$

In the usual case, however, the experimental data will not include the weight of A directly, but rather the weight of some compound containing A or chemically equivalent to A. To convert this to the desired quantity involves multiplication of the experimental datum by a constant called the *chemical factor* or the *gravimetric factor*. The nature of this constant can be readily seen by considering a few examples.

Example. A precipitate of AgCl was found to weigh 0.204 gram.

(a) To what weight of Cl would this correspond? From the formula for AgCl, we conclude that

number of formula weights Cl = number of formula weights AgCl

Since

$$\text{number of formula weights AgCl} = \frac{0.204}{\text{GFW AgCl}}$$

and

weight Cl = number of formula weights Cl × GFW Cl

Then

$$\text{weight Cl} = \frac{0.204 \times \text{GFW Cl}}{\text{GFW AgCl}}$$

$$= 0.204 \times \frac{35.5}{143.3} = 0.0505 \text{ gram}$$

(b) To what weight of $AlCl_3$ would this correspond? We know that one $AlCl_3$ would yield three AgCl. Therefore

number of formula weights $AlCl_3$ = ⅓ number of formula weights AgCl

By the above arguments

$$\text{weight AlCl}_3 = 0.204 \times \frac{\text{GFW AlCl}_3}{3 \times \text{GFW AgCl}}$$

$$= 0.204 \times \frac{133.3}{3 \times 143.3} = 0.0633 \text{ gram}$$

We see that both calculations are similar in that we multiply the known weight of the substance by a ratio of gram formula weights. This ratio is the gravimetric factor and in the cases above it is

(a) $\dfrac{\text{GFW Cl}}{\text{GFW AgCl}}$ (b) $\dfrac{\text{GFW AlCl}_3}{3 \times \text{GFW AgCl}}$

In the latter case, it was necessary to multiply the formula weight of AgCl by three in order to balance the number of chlorides in the numerator and denominator.

Example. What weight of Fe_2O_3 can be obtained from 1.00 gram of Fe_3O_4? What is the gravimetric factor for this conversion?

Here it is necessary to assume that all of the Fe in the Fe_3O_4 is transformed into Fe_2O_3 and that extra oxygen is available to accomplish this change. That is,

$$2Fe_3O_4 + [O] = 3Fe_2O_3$$

We see from this equation that 3/2 formula weights of Fe_2O_3 are obtained from 1 formula weight of Fe_3O_4. Thus the number of formula weights of Fe_2O_3 is greater than the number of formula weights of Fe_3O_4 by a factor of 3/2 or number of formula weights $Fe_2O_3 = 3/2 \times$ number of formula weights Fe_3O_4

$$= \frac{3}{2} \times \frac{\text{weight } Fe_3O_4}{\text{GFW } Fe_3O_4}$$

and

$$\text{weight of } Fe_2O_3 = \frac{3}{2} \times \frac{\text{weight } Fe_3O_4}{\text{GFW } Fe_3O_4} \times \text{GFW } Fe_2O_3$$

or rearranging

$$\text{weight of } Fe_2O_3 = \text{weight } Fe_3O_4 \times \frac{3 \times \text{GFW } Fe_2O_3}{2 \times \text{GFW } Fe_3O_4}$$

and substituting numerical values

$$\text{weight } Fe_2O_3 = 1.00 \times \frac{3 \times 160}{2 \times 232} = 1.03 \text{ grams}$$

$$\text{gravimetric factor} = \frac{3 \times \text{GFW } Fe_2O_3}{2 \times \text{GFW } Fe_3O_4}$$

We can generalize by defining the gravimetric factor as follows:

$$\text{gravimetric factor} = \frac{\text{GFW of the substance sought}}{\text{GFW of the substance weighed}} \times \frac{a}{b}$$

where a and b are small integers that take such values as are necessary to make the formula weights in the numerator and denominator *chemically equivalent*. We have shown, further, that

$$\begin{matrix} \text{weight of substance} \\ \text{sought, grams} \end{matrix} = \begin{matrix} \text{weight of substance} \\ \text{weighed, grams} \end{matrix} \times \begin{matrix} \text{gravimetric} \\ \text{factor} \end{matrix}$$

Tables of gravimetric factors and their logarithms are to be found in chemical handbooks. A few are listed in Table 4-1.

Table 4-1

TYPICAL GRAVIMETRIC FACTORS

Substance Sought	Substance Weighed	Gravimetric Factor
$BiCl_3$	Bi_2O_3	$\dfrac{2 \text{ GFW } BiCl_3}{\text{GFW } Bi_2O_3}$
KNO_3	K_2PtCl_6	$\dfrac{2 \text{ GFW } KNO_3}{\text{GFW } K_2PtCl_6}$
K_3PO_4	K_2PtCl_6	$\dfrac{2 \text{ GFW } K_3PO_4}{3 \text{ GFW } K_2PtCl_6}$
P_2O_5	$Mg_2P_2O_7$	$\dfrac{\text{GFW } P_2O_5}{\text{GFW } Mg_2P_2O_7}$

In all the gravimetric factors considered so far, establishment of the chemical equivalence of the numerator and denominator could be accomplished merely by balancing the number of atoms of the element other than oxygen that was common to both. There are, however, instances where this procedure is not feasible. For example, suppose that the sulfate in ferric sulfate were precipitated and weighed as barium sulfate and we wished to know the weight of iron in the original compound. Each ferric sulfate is equivalent to three barium sulfates; two irons, then, are equivalent to three barium sulfates. Therefore the gravimetric factor for calculation of the weight of iron would be

$$\frac{2 \times \text{GFW Fe}}{3 \times \text{GFW BaSO}_4}$$

In this case, the substances in the gravimetric factor are not directly related by a common element, but by knowing the chemical reactions relating them we can determine their chemical equivalence.

The example given below demonstrates the use of the gravimetric factor in the calculation of results from an analysis.

Example.   A 0.4000-gram sample of a mixture containing only K_2SO_4 and Na_2SO_4 was dissolved and the sulfate precipitated as $BaSO_4$. After suitable treatment the weight of the $BaSO_4$ was found to be 0.5760 gram. Calculate the percent Na_2SO_4 and K_2SO_4 in the sample.

Since we have two unknowns we must be able to write two independent equations. We know that

$$\text{weight } Na_2SO_4 + \text{weight } K_2SO_4 = 0.4000 \text{ gram}$$

and that

weight $BaSO_4$ from the Na_2SO_4 + weight $BaSO_4$ from the $K_2SO_4 = 0.5760$ gram

Rewriting

$$\text{weight } Na_2SO_4 \times \frac{\text{GFW BaSO}_4}{\text{GFW Na}_2SO_4} + \text{weight } K_2SO_4 \times \frac{\text{GFW BaSO}_4}{\text{GFW K}_2SO_4}$$
$$= 0.5760 \text{ gram}$$

or

$$\text{weight } Na_2SO_4 \times \frac{233.4}{142.0} + \text{weight } K_2SO_4 \times \frac{233.4}{174.2} = 0.5760 \text{ gram}$$

Substituting the first equation we obtain

$$(0.4000 - \text{weight } K_2SO_4) \times \frac{233.4}{142.0} + \text{weight } K_2SO_4 \times \frac{233.4}{174.2}$$
$$= 0.5760 \text{ gram}$$

Solving for weight K_2SO_4

$$\text{weight } K_2SO_4 = 0.268 \text{ gram}$$

Therefore

$$\text{percent } K_2SO_4 = \frac{0.268}{0.400} \times 100 = 67$$

and

$$\text{percent } Na_2SO_4 = 33$$

problems

1. Write gravimetric factors for each of the following :

Substance sought	Substance weighed
Mn_3O_4	MnO_2
$K_2Cr_2O_7$	$PbCrO_4$
$FeSO_4(NH_4)_2SO_4 \cdot 6H_2O$	Fe_2O_3
CCl_4	$AgCl$
$2K_2CO_3 \cdot UO_2CO_3$	CO_2

2. A 0.396-gram sample containing $BaCl_2 \cdot 2H_2O$ was analyzed by precipitation of the Cl^- with $AgNO_3$. This yielded 0.328 gram of $AgCl$.

(a) Calculate the percent $BaCl_2 \cdot 2H_2O$ in the sample. ans. 70.6 percent

(b) What weight of H_2O is associated with the $BaCl_2$ in the sample ?
ans. 0.0412 gram

3. A 0.600-gram sample containing $Al_2(SO_4)_3$ was dissolved and the aluminum precipitated as the basic oxide, $Al_2O_3 \cdot xH_2O$. This was ignited and gave a 0.163-gram residue of Al_2O_3. Calculate the percent $Al_2(SO_4)_3$ in the sample.

4. How many pounds of cobalt can be obtained from 30.0 lb of Co_3S_4 ?

5. A 0.886-gram sample containing CaC_2O_4 was ignited at high temperature. After ignition the weight was 0.614 gram. Calculate the percent CaC_2O_4. (Assume loss in weight due to $CaC_2O_4 \rightarrow CaO + CO_2 + CO$) ans. 54.6 percent

6. How many tons of H_2SO_4 can theoretically be produced from 1 ton of pyrite, FeS_2 ?

7. How many grams of $Pb(NO_3)_2$ are required to precipitate all of the chromium in 1.50 grams $K_2Cr_2O_7$ if the precipitate composition is $PbCrO_4$?
ans. 3.38 grams

8. A 0.843-gram sample of brass after suitable treatment yielded 0.0162 gram of SnO_2, 0.0228 gram of $PbSO_4$, 1.295 grams of $CuSCN$, and 0.316 gram of $Zn_2P_2O_7$. Calculate the percentage of each element in the alloy.

9. The carbon in a 0.710-gram sample of steel was determined by ignition in air which converted the C to CO_2. The CO_2 was absorbed on Ascarite and found to weigh 0.0175 gram. What is the percent C in the steel ?

10. A mixture of salts was dissolved and the chloride and iodide present precipitated as the silver salts. A 0.8000-gram sample gave a silver halide precipitate weighing 0.5941 gram. The precipitate was then ignited in a stream of Cl_2 gas which converted the AgI in the mixture to $AgCl$. The weight of the precipitate was 0.4592 gram. Calculate the percent of Cl and I in the sample.
ans. 23.4 percent I and 7.67 percent Cl

11. A 0.4000-gram mixture containing only $NaCl$ and KCl yielded a precipitate of $AgCl$ that weighed 0.9205 gram. Calculate the percentage of each salt present.

12. A 0.510-gram sample containing $Ca(ClO_3)_2$ was dissolved and the ClO_3^- reduced to Cl^- which was precipitated as $AgCl$. The precipitate weighed 0.330 gram. Calculate the percent Ca in the sample.

13. What sample weight must be taken so that each milligram of ignited $Mg_2P_2O_7$ will represent 0.05 percent MgO.   ans. 0.7243 gram

14. A 0.500-gram sample containing 5.0 percent Fe, 6.2 percent Al, and 1.8 percent Ti was dissolved and the metals precipitated as the basic oxides. What was the weight of the mixture of Fe_2O_3, Al_2O_3 and TiO_2?

15. A 1.42-gram sample containing phosphorus was dissolved and the P precipitated as $(NH_4)_3PO_4 \cdot 12MoO_3$. The precipitate was redissolved and the molybdate precipitated as $PbMoO_4$. The precipitate weight was 0.0820 gram. What was the percent P_2O_5 in the original sample? What is the gravimetric factor in this calculation?

ans. 0.093 percent P_2O_5

Gravimetric factor $= \dfrac{P_2O_5}{24PbMoO_4}$

16. What weight of $Ca_3(PO_4)_2$ will yield 1.00 gram of CaO? What weight of H_3PO_4 could be obtained from that amount of $Ca_3(PO_4)_2$?

17. A sample was made up of pure $BaCO_3$ and Na_2CO_3 in a formal ratio of 1:1. How many grams of CO_2 could be obtained from 1.50 grams of the mixture?

18. An iron ore sample contained 1.31 percent H_2O and 35.11 percent Fe. What would be the percent Fe if the sample were dried before analysis?

19. A 0.812-gram sample was dissolved and the Fe and Al present precipitated and ignited to a 0.140-gram mixture of Fe_2O_3 and Al_2O_3. This residue was further ignited in the presence of H_2 which left the Al as Al_2O_3 but converted the Fe_2O_3 to Fe. This was found to weigh 0.120 gram. Calculate the percent Fe and Al in the original sample.

chapter 5. *The Techniques and Tools of Gravimetric Analysis*

A gravimetric analysis involves determination of the weight of a substance produced from a given weight of sample. The common operations of gravimetric analysis thus include weighing, filtration, and drying or ignition of the isolated substance. We shall now consider the various implements employed by the analytical chemist in the performance of these operations.

THE MEASUREMENT OF MASS

Accurate weighing data are a requisite for virtually all chemical methods of analysis. In addition, the approximate weight of substances is frequently required in analytical work. Thus we need to consider the various types of weighing instruments and the techniques employed in their use.

The Distinction between Mass and Weight

The reader should clearly recognize that there is a difference between the concepts of mass and weight. The more fundamental of these is *mass*—an invariant measure of the quantity of matter in an object. The *weight* of an object, on the other hand, is the force of attraction exerted between it and its surroundings, principally the earth. Since this gravitational attraction is subject to slight geographical variation with altitude and latitude, the weight of an object is likewise a somewhat variable quantity. For example, the weight of a crucible would be less in Denver than in Atlantic City since the attractive force between it and the earth is less at the greater altitude. Similarly, it would weigh more in Seattle than in Panama; since the earth is somewhat flattened at the poles, the force of attraction increases appreciably with latitude. The mass of this crucible, on the other hand, remains constant regardless of the location in which it is measured.

Weight and mass are simply related to one another through the familiar expression

$$W = Mg$$

where the weight, W, is given by the product of the mass, M, of the object and the acceleration due to gravity, g.

In chemical analysis we are invariably interested in the determination of mass, since we do not wish our results to be dependent upon the locality in which the experiment was performed. This is readily accomplished by comparing the mass of an unknown to that of objects of known mass through the use of a balance; since the quantity g affects both known and unknown to exactly the same extent, a fair measure of masses will result.

The distinction between weight and mass is not always observed; in common usage the operation of comparing masses is called *weighing*, and the objects of known mass as well as the results of the process are called *weights*. While the two terms are hereafter used synonymously, strictly speaking it is *mass* to which we refer.

Equal-arm Analytical Balances

Construction. The principal moving part of a balance is the *beam*. In an equal arm design the point of pivot is a central agate knife edge that bears upon an agate surface mounted at the top of a fixed pillar. Two outer knife edges are located equidistant from the fulcrum; these support the *stirrups* which serve as the links through which the balance pans are connected to the beam. Movement of the beam is gauged by observing the motion of a vertical pointer with respect to a deflection scale at the base of the pillar.

The beam of the typical equal-arm balance is calibrated at intervals to accommodate a small metal rider. Movement of the rider along the beam has the effect of altering the loading on the balance; its position is adjusted with a rodlike control that extends through the balance case. The use of a rider eliminates the necessity for handling large numbers of small weights.

Analytical balances are adjusted with their riders in place over the zero marking on the beam; the rider should be so placed when determining the no-load equilibrium point.

The three knife edges of a balance are precision ground to reduce to a minimum the area of contact with the bearing surface. As a consequence, loading at the point of contact between knife edge and bearing is severe; for the central knife edge this is on the order of hundreds or thousands of kilograms per square centimeter. Sudden shocks can result in extensive damage to either of these parts. In order to prevent this, an analytical balance is equipped with a *beam support* system which isolates the knife edges from their bearing surfaces by lifting the beam slightly and supporting its weight. This safeguard, unfortunately, is effective only when properly used; the beam support must always be engaged when the balance is not in use and when a change is being made in the loading on the pans. The beam support is operated by a knob on the front of the balance case, usually in the center. *Pan arrests*, also actuated by a control on the front of the case, tend to minimize the independent oscillation of the pans owing to poor arrangement of loads and also provide a means of releasing the beam in a reproducible manner; these too should always be engaged under the same circumstances as described for the beam support.

Theory of operation. In essence, an equal-arm analytical balance acts as a first-class lever. The addition of weight to one side of such a lever at rest will cause it to seek a new rest position. The force moment tending to change its position is given by the product of the mass involved and the horizontal distance from the fulcrum through which it is acting. When the lever again achieves a position of equilibrium, the force moment tending to give it motion in one direction is exactly balanced by that which tends to impart motion in the opposite sense. Thus, we may write for the equilibrium state that

$$F_1 = F_2$$

where F_1 and F_2 represent these opposing forces. We may also express this relationship as

$$M_1L_1 = M_2L_2 \tag{5-1}$$

where M_1 is the mass located a distance L_1 to the left of the fulcrum, and M_2 and L_2 stand for the analogous mass and lever arm to the right of the fulcrum. An equal-arm lever will assume an equilibrium position that is horizontal when the force moment is identical its left and right sides; provided the two arms have equal masses and lengths, the loads they carry will also be identical. The weighing operation thus consists of duplicating, under load, the equilibrium position of the unloaded balance.

Stability of a balance. When set in motion, the beam of an analytical balance oscillates slowly about some equilibrium point at which it will eventually come to rest. In order for this to occur it is necessary that the center of gravity of the beam and pans be *below* the central knife edge. Under these conditions, the weight of the beam tends to act as a restoring force that will return the deflected beam to a stable equilibrium position.

Sensitivity of the analytical balance. The *sensitivity* of a balance describes the magnitude of the change in equilibrium position of the beam resulting from the addition of a given weight to one of the pans. Since this property is related to the attainable accuracy of weighing, the factors affecting it warrant examination.

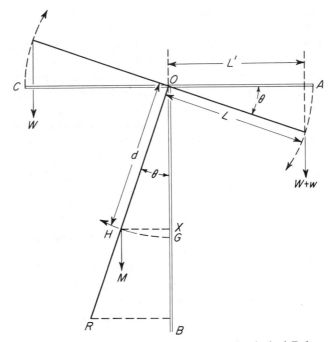

Fig. 5.1 Schematic Diagram, Equal-arm Analytical Balance.

An equal-arm balance is shown schematically in Figure 5.1. The rest position of the balance is indicated by *ABC* in this diagram. We have seen that this position will also be assumed when a load, *W*, is placed on each pan; the force moments will have been increased over those of the unloaded balance, but they will still exactly oppose one another. Let the mass of the beam and its appurtenances be *M* and let the center of gravity of this mass be located at point *G*, which is a distance *d* below the fulcrum.[1]

Now we shall introduce the additional small weight *w* to the right-hand pan. The mass tending to impart clockwise motion to the beam will now be given by $(W + w)$; when the beam achieves its new equilibrium position this

---

[1] Figure 5.1 is considerably distorted for the purposes of clarity; *G* is actually located a very short distance from the fulcrum, *d* being on the order of 0.1 to 0.2 mm.

will be acting at a distance L' from the fulcrum. The force moment will then be

$$F_2 = (W + w) L \cos \theta$$

since

$$L' = L \cos \theta$$

Owing to the shifting of the beam from the horizontal, its center of mass will undergo a shift to the left to a new position H. The mass of the beam M now effectively acts at the distance $\overline{HX}$ in opposing the imbalance of loading on the pans; additionally, of course, the load W on the left pan makes a contribution in the same sense. Thus the force moments tending to shift the beam in a counter-clockwise direction is the sum of these.

$$F_1 = M(\overline{HX}) + WL'$$

The distances $\overline{HX}$ and L' may be expressed in terms of fixed dimensions of the balance. This expression then becomes

$$F_1 = Md \sin \theta + WL \cos \theta$$

Since the condition of equilibrium entails equality between the opposing force moments, we may write

$$(W + w) L \cos \theta = Md \sin \theta + WL \cos \theta$$

After factoring out $WL \cos \theta$ and rearranging, this equation becomes

$$\frac{\sin \theta}{\cos \theta} = \tan \theta = \frac{wL}{Md} \tag{5-2}$$

Tangent θ, then, is a measure of the *sensitivity* of the balance, θ being the angular deflection induced by a given inequality, $\underline{w}$, in loading between the pans. In linear units, the sensitivity is the deflection $\overline{RB}$ of the pointer whose length is $\overline{OB}$. Thus, we may rewrite equation (5-2) as

$$\overline{RB} = \frac{wL \, \overline{OB}}{Md} \tag{5-3}$$

This allows us *to express the sensitivity in the more convenient terms of scale divisions deflection ($\overline{RB}$) per milligram (w).*

The sensitivity of a balance may also be expressed as the weight difference in milligrams required to cause a deflection of one scale division. This is sometimes referred to as the *sensibility*, or reciprocal sensitivity.

Factors affecting sensitivity. Equation (5-2) suggests that high sensitivity can be attained by employing a long beam whose mass is small and whose center of mass is located very close to the fulcrum. A number of practical aspects, however, sharply limit incorporation of extremes in balance design. For example, to have a long beam and still retain a rigid structure requires a prohibitive increase in mass M. Further, the time required for the completion of a cycle of swings, ideally between 10 and 20 seconds, increases with increasing beam length and decreasing distance, d, between the central knife edge and the center of mass.[2] Thus, the design of any analytical balance requires a compromise between these practical limitations and the desirable quality of rather high sensitivity.

The sensitivity of most analytical balances can be varied somewhat by changing d by vertical movement of a small weight that is attached to the pointer.

Variation of sensitivity with load. According to equation (5-3) the sensitivity of a balance should be independent of loading, since L, M, d, and $\overline{OB}$ are all constant for a given balance. In actual practice, however, the sensitivity may vary appreciably according to the load on the pans. Experiment has shown that this effect cannot be attributed to bending of the beam and that it results, rather, from failure of the terminal knife edges to lie in the same plane as the central knife edge.[3] With increased loadings, also, mechanical imperfections in the knife edges and their bearing surfaces give rise to frictional effects that are probably of comparable magnitude.

Auxiliary Devices

The weighing operation can be performed more rapidly on a balance equipped with any of several supplementary devices; these function either to expedite achievement of the equilibrium position of the beam or to reduce the number of weights that need be manipulated by hand.

Dampers. Dampers act to shorten the time required for the beam to come to rest. This is accomplished by either of two methods. An *air damper* consists of a piston that moves within a concentric cylinder. The former is attached to the beam while the latter is firmly mounted to the balance case. When the beam is set in motion, the enclosed air suffers slight expansions and contractions because of the close spacing between piston and cylinder. The beam comes rapidly to equilibrium as a consequence of this opposition to its movement. Air dampers suffer from being somewhat bulky; considerable attention must also be paid to positioning of the loads on the pans to ensure that piston and cylinder do not touch. These are not serious limitations for single pan (unequal-arm) balances; it is on these that air dampers are most widely employed. Of more common use is the *magnetic damper*. A metal (generally aluminum) plate secured to the end of the beam is positioned between the poles of a fixed permanent magnet. When the beam is set in motion and the plate moves through the field of the magnet, the currents induced tend to oppose oscillation and to bring the beam rapidly to its equilibrium position.

---

[2] The time, t, required for the completion of a cycle of swings is given by

$$t = 2\pi L \sqrt{2W/Mgd}$$

The symbols in this equation have their previously assigned meanings.

[3] K. J. Mysels, *J. Chem. Ed.*, **32**, 518 (1955).

Chain balances. As illustrated in Figure 5.2, a balance of this type is fitted with a small gold chain that is attached some distance to the right of the fulcrum and also to an externally operated crank. The amount of weight supported by the beam can be varied by lengthening or shortening the chain. A scale arrangement and vernier makes possible the direct reading of weight so introduced. Since the chain generally allows introduction of any weight within the range of 0 to 100 mg, the beam on such a balance will be notched for a rider that allows addition of 100-mg increments from 0 to 1 gram. Thus no weight smaller than 1 gram need be manually introduced.

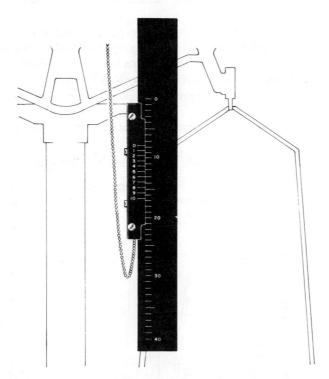

Fig. 5.2 The Chain Balance. Note arrangement of chain and vernier. Loading indicated is 7.6 mg.

A chain balance fitted with magnetic damping constitutes a weighing device capable of yielding accurate results for a minimum expenditure of time.

Keyboard balances. Keyboard balances represent a further refinement with respect to the expeditious handling of weights. Knobs or keys on the case of the balance allow the rapid addition of weights up to several grams in size. Balances of this type are often arranged so that any weight smaller than 100 mg is read directly from a projected image. Both single- and double-pan balances are available which incorporate these features.

Unequal-arm Balances

Figure 5.3 shows the essential features of an unequal-arm balance. It is seen to consist of a single pan (1) suspended from the end of the shorter arm of the beam. Also on this side of the fulcrum is a full complement of weights (2). At the longer end of the beam is a counterweighted damper (3) designed to impart equilibrium to the system. Placement of an object on the pan results in an imbalance that is corrected by *removal* of weights from the beam; reattainment of balance requires that the mass of the weights removed be equal to the mass of the object that replaced them. An unequal-arm balance thus operates under conditions of constant load and has the advantage of possessing a truly constant sensitivity.

Speed of operation is provided by the damper, the mechanical handling of weights greater than 100 mg (4), and an optical system (5) that projects a direct reading of weights smaller than this amount on a frosted glass window. Weighing with an unequal-arm balance reduces simply to manipulation of several dials and reading of the fractional weight projected upon the window.

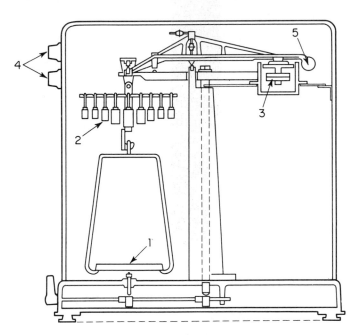

Fig. 5.3 Schematic Diagram, Unequal-arm Balance. (Diagram courtesy of Mettler Instrument Corporation, Hightstown, New Jersey.)

Other Types of Balances

In a ranking with respect either to sensitivity or permissible loading the ordinary analytical balance occupies an intermediate position. Typically, the maximum allowable load on each pan is 200 grams; the sensitivity commonly runs between 1.5 and 4 divisions per milligram. By way of contrast, *semimicro-balances* usually have capacities no greater than 100 grams and reciprocal sensitivities on the order of 0.05 mg per division. *Microbalances* accommodate only 20 to 50 grams on each pan but have reciprocal sensitivities in the range 0.01 to 0.001 mg per division. Yet another order of diminution is possible through the use of quartz fibers, either in the form of beams or helices. Such weighing devices have very low capacities but are sensitive to as little as 0.005 microgram (1 microgram $= 1 \times 10^{-6}$ gram).

Since approximate weights are frequently needed in the analytical laboratory, balances that are less sensitive but more rugged than the analytical balance are also a requirement.

Weights

For each balance there should be a set of weights. These will be of such number and denomination as to allow, in conjunction with the riders and chains built into the balance, addition of any weight within the capacity of the balance. Weights are manufactured from materials that render them resistant to changes in mass. Brass plated with platinum, gold, or rhodium is preferred for denominations larger than 1 gram, although highly polished, nonmagnetic stainless steel and lacquered brass weights also give good service. These integral weights are either carefully machined from a single piece of metal or are built in two pieces; with the latter, final adjustment of mass has been accomplished by inserting small compensating weights in the space between the two parts. Fractional weights— that is, those smaller than 1 gram—are fabricated from sheets of platinum, tantalum, or aluminum.

Classes of weights. Notwithstanding painstaking care in manufacture, some variation must be expected between the nominal and actual value of a weight; such variation can be determined through calibration of the weight against a known standard. Clearly, the cost of producing a set of weights increases rapidly as the acceptable tolerances are made smaller. As a corollary to this, it should also be clear that in most instances the use of extremely precise weights is unnecessary; a well-calibrated set will serve as well.

The National Bureau of Standards has established *classes* of weights based upon the type of work for which they are intended. The Bureau has also set acceptable tolerance limits for weights within each class; a partial listing of these is given in Table 5-1. Finally, specifications have been set down with respect to the materials and construction that are acceptable for each class. The interested reader should consult the *National Bureau of Standards Circular 547*, section 1 (1954) for further information on this subject.

Calibration. Probably the simplest method of calibration involves a direct comparison of each weight in a set with one whose value is known with certainty. It is also advisable to check individual calibrations by comparing groups of weights with the heavier pieces in the standard set. Sets of weights

Table 5-1

ACCEPTANCE TOLERANCES, NATIONAL BUREAU OF STANDARDS[4]

Nominal Weight, grams	Tolerance for Indicated Class[5], mg					
	M	S	S-1[7]	P	Q	T
	group[6]	group[6]				
100	0.50	0.25	1.0	2.0	9.0	100
50	0.20	0.12	0.60	1.2	5.6	62
30	0.15	0.074 ⎫	0.45	0.90	4.0	44
20	0.10	0.074 ⎬ 0.154	0.35	0.70	3.0	33
10	0.050	0.074 ⎭	0.25	0.50	2.0	21
5	0.034 ⎫	0.054 ⎫	0.18	0.36	1.3	13
3	0.034 ⎬ 0.065	0.054 ⎬ 0.105	0.15	0.30	0.95	9.4
2	0.034 ⎮	0.054 ⎮	0.13	0.26	0.75	7.0
1	0.034 ⎭	0.054 ⎭	0.10	0.20	0.50	4.5
0.5000	0.0054 ⎫	0.025 ⎫	0.080	0.16	0.38	3.0
0.3000	0.0054 ⎬ 0.0105	0.025 ⎬ 0.055	0.070	0.14	0.30	2.2
0.2000	0.0054 ⎮	0.025 ⎮	0.060	0.12	0.26	1.8
0.1000	0.0054 ⎭	0.025 ⎭	0.050	0.10	0.20	1.2
0.0500	0.0054 ⎫	0.014 ⎫	0.042	0.085	0.16	0.88
0.0300	0.0054 ⎬ 0.0105	0.014 ⎬ 0.034	0.038	0.075	0.14	0.68
0.0200	0.0054 ⎮	0.014 ⎮	0.035	0.070	0.12	0.56
0.0100	0.0054 ⎭	0.014 ⎭	0.030	0.060	0.10	0.40
0.0050	0.0054 ⎫	0.014 ⎫	0.028	0.055	0.080	——
0.0030	0.0054 ⎬ 0.0105	0.014 ⎬ 0.034	0.026	0.052	0.070	——
0.0020	0.0054 ⎮	0.014 ⎮	0.025	0.050	0.060	——
0.0010	0.0054 ⎭	0.014 ⎭	0.025	0.050	0.050	——

[4] From the *National Bureau of Standards Circular* 547, section 1, p. 2, 1954.

[5] Acceptance tolerances for Class J (microweight standards, 50-0.05 mg): 0.003 mg for readily compared weights, and not less than the tolerance for the nearest Class M weight for those of noncomparable demoninations.

[6] Additional restriction: no combination of weights shall have a total correction exceeding that of "group".

[7] Additional restriction: not more than one third of the weights of a set of new or newly adjusted weights may be in error by more than one half of these tolerances, and all weights shall be correct within these tolerances.

can be obtained that have been calibrated and certified by the National Bureau of Standards.[8] These may be employed as standards in calibrating the working set for the balance.

[8] Details concerning this service are to be found in *National Bureau of Standards Circular* 547, section 1, 1954.

Although it is unquestionably adequate for most purposes, this direct method is not the only, nor is it necessarily the best, calibration procedure. Where the weights will serve as the basis for very refined work, the Richards method[9] or one of its modifications[10] should be employed; this involves comparing the masses of all the weights in a set with respect to a single member. The relative masses can then be corrected in absolute terms if the absolute mass of the reference piece is available. This method is slightly more involved than direct comparison of individual weights, but is considered to be of somewhat greater reliability. This technique is described in detail on page 96.

Each individual weight correction is most conveniently expressed as that quantity which must be applied to the nominal value of the weight; thus, a 5-gram weight carrying a correction of — 0.4 mg has an actual weight of 4.9996 grams.

Weighing with an Equal-arm Balance

A number of methods exist for determining the weight of an object with an equal-arm analytical balance. Regardless of the method, the pointer on the beam and the deflection scale at the base of the central pillar are used to determine the state of equality of loading between the pans. In this connection two terms need defining. The *zero point* refers to the equilibrium position of the pointer for the empty balance. A *rest point* is the equilibrium position assumed by the pointer when a load has been imposed on the balance. In these terms, then, the weighing operation consists of determining the weight necessary to produce a rest point that corresponds with the zero point. Owing to the prohibitive length of time necessary for an undamped beam to come to rest, its equilibrium position is estimated from an observation of the magnitude of the swings of its pointer.

The deflection scale. The deflection scale of the typical analytical balance does not have numerical values assigned to the division lines. All that is required is a consistent numbering system, and this is left to the discretion of the operator. We shall employ the notation that assigns a value of 10 to the midline of the scale. The lines to the left, then, represent integers less than 10 while those to the right have values greater than this number.

We shall restrict discussion to the method of long swings and to that of short swings. The former is included because of its established reliability for the most exacting work; the latter, being the quicker method, is recommended for all routine weighing. The principal difference between these schemes is the number of observations made upon the pointer; the long swing method involves an odd number of these observations (either 5 or 3), while the short swing method uses but two observations under somewhat restricted conditions.

[9] T. W. Richards, *J. Am. Chem. Soc.*, **22**, 144 (1900).

[10] W. M. MacNevin, *The Analytical Balance*, 46. Sandusky, Ohio: Handbook Publishers, Inc., 1951.

Figure 5.4 illustrates why an odd number of observations is preferred. This indicates the position the pointer takes in making two and one-half full cycles, or five "swings." In actuality these would all be superimposed on a single arc. Each swing is less than the previous one by an amount, d, owing to the operation of damping and frictional forces. This is an essentially constant decrement for any small set of swings (d in Fig. 5.4 is somewhat exaggerated for the purposes of clarity). Now, if the maximum position reached by the pointer on the first swing we choose to observe is a distance A from the as yet unknown rest point Z, our observed reading for this swing will be $(Z + A)$. On the return swing, the pointer will fail to reach $- A$ by the amount of the decrement d. Similarly, the following swing will be short of $(Z + A)$ by the amount of $2d$. The course of subsequent swings can be predicted in the same manner. Using the data of Figure 5.4, we see that the average of the two swings observed to the left is $Z - (A - 2d)$, while that for those to the right is $Z + A - 2d$. Clearly, the average of these two will yield the value of the rest point. As long as an odd number of consecutive swings are observed, the average of the mean swing yields an estimate of the rest point; on the other hand, if an even number of consecutive readings are averaged, values other than Z are consistently obtained.

The method of long swings. To obtain an estimate of the zero point of the balance, lower the beam support gently, and then release by disengaging the pan arrests (this is the order always followed in freeing the beam; the opposite order is used when arresting the beam); allow the beam to swing freely through several cycles. Motion can be imparted to the beam, if necessary, by temporarily displacing the rider (or chain) from its zero position. The balance case is kept closed throughout this process. Next observe and record the maximum position attained by the pointer during a right-hand swing, estimating this to the nearest tenth of a division. Continue until observations of three or five consecutive maxima have been made. Calculate the average value for the swings in each direction; take the mean of these two quantities as the zero point of the balance. Since uncertainty resides in the first decimal place, the calculations should not be carried further than this.

Example. Calculate the zero point for a balance if the following consecutive readings were observed: 12.4, 6.9, 12.2, 7.1, 11.9.

readings	6.9	12.4
	7.1	12.2
	——	11.9
	14.0	——
		36.5
averages	7.0	12.2

$$\text{zero point} = \frac{7.0 + 12.2}{2} = 9.6$$

It should be noted parenthetically that the zero point will undergo slight changes in position with time owing to changes in temperature and humidity;

as a consequence, its value should be determined for each series of weighings. Pronounced shifts in the zero point are indicative of trouble; in such event the instructor should be consulted.

After ascertaining that the pan arrests and beam supports have been re-engaged, place the object to be weighed on the left pan of the balance. Make a rough estimate of its mass and, using forceps, place on the right pan the single weight that is believed to have a slightly greater mass. Gently and partially disengage the beam support, and note the behavior of the pointer; it will swing away from the pan that has the heavier load. If no oscillation occurs, the loadings are within 1 to 2 grams of each other; under such circumstances, cautious disengagement of the pan arrests will allow determination of which pan has the heavier load. Continue a systematic trial-and-error loading of weights until the closest approach to balance has been achieved where the object is still heavy. Then use the rider in a similar manner; the balance case should be closed henceforth. When balance has apparently been obtained, take a rest point in the manner described for determining the zero point and compare this with the zero point. Subsequent corrections with the rider can be checked in the same fashion until balance is achieved.

A significant saving in time can be made by employing the balance sensitivity to calculate the number in the final decimal place. Here, the weight of the object is determined to the nearest whole milligram and a rest point (RP_1) is obtained. Then 1 mg is either added or subtracted from the load on the balance and another rest point (RP_2) is taken. The difference between these yields the sensitivity of the balance for the particular loading involved. With this the correction needed to shift either rest point into coincidence with the zero point can be readily calculated.

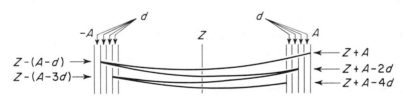

Fig. 5.4 Schematic Diagram Illustrating the Damping of the Pointer of an Analytical Balance.

Example. Zero point (ZP)   10.2
          Rest point with 10.422 grams (RP_1) 11.1
          Rest point with 10.423 grams (RP_2) 8.7
          Sensitivity (11.1 — 8.7)   2.4 divisions per mg
          Difference between RP_1 and ZP   0.9 division
          Since

$$\frac{1 \text{ mg}}{2.4 \text{ div}} = \frac{x \text{ mg}}{0.9 \text{ div}}$$

    then

$$x = 0.4 \text{ mg}$$

This correction should be *added* to 10.422 grams, because the rest point corresponding to this loading lies to the *right* of the zero point. Thus the weight of the object is $10.422 + 0.004 = 10.4224$ grams. Note that the same result is obtained from RP_2 data

Difference between ZP and RP_2 1.5 divisions

Since

$$\frac{1 \text{ mg}}{2.4 \text{ div}} = \frac{x \text{ mg}}{1.5 \text{ div}}$$

then

$$x = 0.6 \text{ mg}$$

This correction should be subtracted from 10.423 grams, because the rest point corresponding to this loading lies to the *left* of the zero point. Thus, the weight is again $10.423 - 0.0006 = 10.4224$ grams.

The method of short swings.   The decrement d becomes vanishingly small when the amplitude of the swings is small (from 1 to 3 divisions on either side of the mean); under these circumstances the arithmetic mean of a consecutive left and right swing of the pointer gives a sufficiently close approximation of the equilibrium point. Thus, only two observations are made and the average is taken for use; this represents a considerable saving of time. Whether the weight is determined to a tenth of a milligram directly or is calculated from the sensitivity is a matter of personal preference; with a knowledge of the approximate sensitivity of the balance, the direct method is quite rapid.

Weighing with a damped balance.   A damped balance, particularly if also equipped with a chain, is well adapted for the direct determination of the weight of an object. Here the zero point is determined by releasing the beam support and pan arrest and allowing time for the pointer to cease moving. Weighing is accomplished by placing the object on the left pan and weights on the right until a rest point equal to the zero point is obtained. It is preferable to adjust the rider and chain to exact balance.

Weighing with a Single-pan Balance

Obtaining the weight of an object with a modern single-beam balance is a very simple operation.

To adjust the empty balance, rotate the arrest knob. Then manipulate the zero adjusting control until the illuminated scale indicates a reading of zero. Next, arrest the balance and place the object to be weighed on the pan. Again turn the arrest knob, this time to its clockwise position. Rotate the dial controlling the heaviest likely weight for the object until the illuminated scale changes position or the notation "remove weight" appears; then turn back the knob one stop. Repeat this procedure with the other dials, working successively through the lighter weights. Then turn the arrest knob to its counterclockwise position and allow the balance to achieve equilibrium. The weight of the object is found by taking the sum of the weights that have been introduced with the dials and that which appears on the illuminated scale. The vernier is helpful in reading this scale to the nearest tenth of a milligram.

Errors in the Weighing Operation

In common with all physical measurements, weighing data are subject to error. The reader should bear in mind, however, that weighing is generally the most reliable step in a chemical analysis. Errors arising from this source amount typically to fractions of a milligram. Only where the method is sufficiently reliable in its other aspects, or where a very small sample must be taken, is it necessary to take into account the potential sources of error described in the following paragraphs.

Buoyancy. Comparison of the masses of objects and weights is potentially subject to error owing to the buoyant force of the medium (air) in which the comparison is performed. An object completely surrounded by a medium will have displaced a volume of the medium equal to its own. As a consequence, it is the *apparent* mass we obtain when we weigh an object in air; this is equal to the difference between its true mass and that of the air it has displaced. The same consideration holds true with respect to the weights used in the process. Thus, the condition of balance is characterized by the equilibrium

$$(W_1 - w_1) L_1 = (W_2 - w_2) L_2 \qquad (5\text{-}4)$$

where w_1 and w_2 represent the weights of air displaced, respectively, by the object W_1 and the weights W_2. If we assume the lever arms L_1 and L_2 to be equal, equation (5-4) may be rewritten as

$$W_1 = (w_1 - w_2) + W_2 \qquad (5\text{-}5)$$

Now, since the density d is equal to the ratio of mass to volume ($d = w/v$) and, further, since the volumes of air displaced are equal to the volumes of the respective masses (V_1 and V_2), equation (5-5) becomes

$$W_1 = d_{\text{air}} (V_1 - V_2) + W_2$$

But

$$V_1 = \frac{W_1}{d_1} \qquad \text{and} \qquad V_2 = \frac{W_2}{d_2}$$

where d_1 and d_2 are the densities of the object and the weights. Hence

$$W_1 = d_{\text{air}} \left(\frac{W_1}{d_1} - \frac{W_2}{d_2} \right) + W_2$$

Finally, since for all practical purposes at balance $W_1 \cong W_2$

$$W_1 = W_2 + W_2 \left(\frac{d_{\text{air}}}{d_1} - \frac{d_{\text{air}}}{d_2} \right) \qquad (5\text{-}6)$$

The densities of materials commonly employed in the construction of analytical weights are compiled in Table 5-2. The density of air can be taken as 0.0012 gram per ml.

Table 5-2

DENSITY OF METALS USED IN THE MANUFACTURE OF WEIGHTS

Metal	Density, grams/ml
Aluminum	2.7
Brass	8.0
Gold	19.3
Platinum	21.4
Stainless steel	7.8
Tantalum	16.6

Equation (5-6) indicates that when the object is less dense than the weights, its apparent weight W_2 is less than is actually the case; similarly, an object that is more dense than the weights will have an apparent weight that is greater than its actual weight. Finally, when object and weights have the same density, the buoyancy correction becomes zero.

Figure 5.5 correlates the effect of buoyancy with the density of the weighed object when brass (or stainless steel) weights are employed. The relative error due to this effect is most serious when objects of low density are being weighed; even so, its magnitude is rather low. In general, then, buoyancy is an effect that must be taken into account when the highest accuracy is desired, particularly

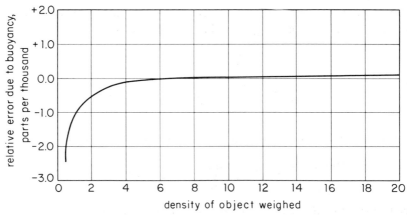

Fig. 5.5 Effect of Bouyancy upon Weighing Data. The relative error (with brass weights) incurred is shown as a function of the density of the object weighed.

when objects of low density are involved; for routine analytical work a correction for it is seldom necessary.

Errors in the weights used. Probably the most common source of weighing error resides in the weights themselves. Although fabricated from corrosion-resistant metals, weights can be significantly in error owing to hard use, or more seriously, to misuse. Periodic calibration against a standard set is recommended.

Other sources of error. A considerable number of weighing errors can be attributed to the analyst. The misreading of weights is a particularly common source of serious error. Attempts to obtain the weight of an object while it is still hot will result in error due to convection currents within the balance case. The placement of the balance should be such that it is protected from direct sunlight and other heat sources such as radiators, steam pipes, and the like; the length of the beam can be appreciably altered owing to unequal heating.

Occasionally a static charge will develop on porcelain or glass objects sufficient to cause erratic performance of the balance; this is a particularly serious problem where the humidity is low. Spontaneous discharge often occurs after a short period. The use of a faintly damp chamois to wipe the object is a recommended preventative.

A minor source of error in weighing with the two-pan balance arises from inequality in the lengths of the balance arms. With a good quality balance, the magnitude of this error will be vanishingly small.

Special Weighing Techniques

For calibrations and for work of the utmost refinement the method of *Gauss* or of *Borda* may be employed to eliminate error arising from inequalities in the lengths of the balance arms. In the Gauss method the object is weighed first on one pan and then on the other. The square root of the product of the two weighings yields a value for the mass of the object that is independent of uncertainty owing to unequal balance arms. The technique is also known as double weighing, or transposition weighing. Borda's method, or the method of substitution, involves the achievement of approximate balance between the object on the *right* pan and a set of counter weights on the left. Analytical weights are then substituted for the object, the same point of balance being used. As before, use of this method eliminates the error due to unequal arm lengths since both the object and the weights are compared to a common mass through an identical lever arm arrangement.

Summary of Rules Governing the Use of an Analytical Balance

Continued good performance from an analytical balance is highly dependent upon the treatment accorded it. Similarly, reliable data are obtained only when careful attention is paid to the details of the weighing operation. Since weighing data are required for virtually every quantitative analysis, we shall summarize the rules and precautions that relate to their acquisition.

To avoid damage or minimize wear to the balance and weights:

1. Be certain that the beam support and pan arrests are engaged whenever loading on the balance is being changed and when the balance is not in use.

2. Insofar as possible, center the loads on the pans.

3. When freeing the beam, first release the beam support slowly and then the pan arrests. When arresting the beam, the opposite order should be followed; the pan arrests should be engaged as the pointer passes through the center of the deflection scale.

4. Protect the balance from corrosion. Only vitreous materials and non-reactive metal or plastic objects should be placed directly on the pans. Weighing of volatile materials requires special precautions.

5. Do not attempt to adjust the balance without the prior consent of the instructor.

6. Handle weights gently and with special forceps only. Weights should never be touched, as perspiration from one's hands can initiate corrosion. Keep weights in their closed box except during use.

7. If possible, avoid bringing weights into the laboratory.

8. Balance and case should be kept scrupulously clean. A camel's hair brush is useful for removal of spilled material or dust.

To obtain reliable weighing data:

9. Place the object on the left pan, the weights on the right.

10. Do not attempt to weigh an object until it has returned to room temperature.

11. Do not touch a dried object with bare hands; handle it with tongs or use finger pads to protect against the uptake of moisture.

12. If a weight is accidentally mishandled, it should be recalibrated before being used again.

13. At the completion of the weighing, sum the weights at least twice to avoid arithmetical and reading errors. Some analysts do this by counting first the weights on the pan and then the empty spots in the weight box.

The Calibration of Weights

The weighings associated with a quantitative analysis are generally employed in the form of a ratio. As a result, the minimum requirement for a set of weights is that the masses of the individuals be known with reliability *relative to one another*. This is the principal aim of a calibration; although unquestionably useful, absolute values are actually of secondary interest. The directions that follow represent a modification of the *Richards* method (page 89). The mass of each piece is first determined with respect to a provisional standard whose nominal weight is the same as the rider. Weighing by substitution is used to eliminate any error due to unequal balance arms; an auxiliary box of weights is a useful source for the required tares. Any error in the provisional standard causes the relative weights for the heavier pieces to differ markedly from their

Table 5-3

CALIBRATION OF WEIGHTS

Weights on Left Pan	Rest Point	Sensitivity div/mg	Correction div	Correction mg	Weight Relative to Standard, grams		Aliquot Weight, grams	Correction†, mg
std*	9.6				0.0100			
0.01 gram	10.0		+ 0.4	+ 0.2		0.0102	0.0101	+ 0.1
0.01 + 1 mg	7.4	2.6						
std + 0.01	9.8				0.0202			
0.02′	9.9		+ 0.1	0.0		0.0202	0.0202	0.0
0.02′ + 1 mg	7.1	2.8						
0.02″	9.6		− 0.2	− 0.1		0.0201	0.0202	− 0.1
std + 0.02′ + 0.02″	9.4				0.0503			
0.05	9.6		+ 0.2	+ 0.1		0.0504	0.0504	0.0
0.05 + 1 mg	6.8	2.8						
std + 0.02′ + 0.02″ + 0.05	9.5				0.1007			
0.10	9.6		+ 0.1	0.0		0.1008	0.1008	0.0
0.10 + 1 mg	7.1	2.5						
1.0 + 2.0′ + 2.0″ + 5.0	8.7				10.0815			
10′ (standard)	8.9		+ 0.2	+ 0.1		10.0816	10.0816	0.0
10′ + 1 mg	6.3	2.6						
10″	10.7		+ 2.0	+ 0.8		10.0823	10.0816	+ 0.7
10′ + 10″	12.0				20.1639			
20	10.3		− 1.7	− 0.7		20.1632	20.1632	0.0
20 + 1 mg	7.8	2.5						
1 + 2′ + 2″ + 5 + 10′ + 10″ + 20	9.5				50.4086			
50	10.3		+ 0.8	+ 0.3		50.4089	50.4080	+ 0.9
50 + 1 mg	7.8	2.5						

* The provisional standard (std) used is a 0.01 gram weight.
† Correction with 10′ gram weight as standard.

nominal values. It is customary, therefore, to complete the calibration by adjusting these relative data to a new standard, usually a 10-gram weight; absolute corrections can be evaluated if the mass of this heavier standard is known.

These directions refer specifically to a balance equipped with a 10-mg rider; they can readily be adapted to other balances as well.

Procedure.   1. Place the provisional standard 10-mg weight on the left pan, a tare (for example, another 10-mg weight) on the right pan, and determine the rest point. Replace the provisional standard with the 10-mg weight from the set and again determine the rest point. Evaluate the sensitivity of the balance. Use this and the rest-point data to compare the mass of the weight with that of the provisional standard.

2. Determine a rest point for the provisional standard plus the 10-mg weight with respect to a new tare. Repeat, using a 20-mg weight from the set. Treat the data as before.

3. Determine the relative mass for each weight in the set with respect to the provisional standard and calibrated pieces by means of a systematic comparison against suitable tares.

4. Select a 10-gram weight from the set as standard, and determine the aliquot weight of every other piece in terms of its nominal value as compared to the relative mass of this new standard. Thus, the aliquot weight of a 2-gram piece will be exactly one fifth of the standard's relative weight, that for a 0.1-gram piece will be given by one hundredth of this weight, etc.

5. Determine the correction for each piece by subtracting its aliquot weight from its relative weight.

NOTES:

1. With a uniform weight (that is, the tare) on the right pan, smaller rest points correspond to lighter weights.

2. Because the value for each relative weight depends upon those that have already been determined, great care must be exercised to avoid computational errors while working up the calibration data.

Table 5-3 consists of a set of typical calibration data.

Procedure for the calibration of a notched beam.   Place a small tare on the left pan. Adjust the chain to the integral milligram mark that affords a convenient rest point. Use this as the reference point for subsequent calibrations. For each rider position determine the rest point with respect to calibrated weights on the left pan. Evaluate the sensitivity, and compute the correction on the basis of the known loading and the difference between the rest point and that with the tare only.

NOTE:

Integral weights may readily be calibrated relative to a 1-gram rider. See Problem 7 at the end of this chapter.

Procedure for the calibration of a chain.   Compare the chain at 10-mg intervals against fractional weights of known value. Calculate the correction to be applied from the difference between the rest and zero points, and the sensitivity. If the balance is equipped with a damper, it may be more convenient to determine the chain reading required to produce a rest point that is identical with the zero point. Plot the correction to be applied against the corresponding chain loading.

THE TOOLS OF GRAVIMETRIC ANALYSIS

Composition of Chemical Equipment

Table 5-4 gives some indication of the variety of substances used in the manufacture of chemical apparatus. It is presented as a guide only, because virtually every material tabulated possesses further, rather specialized limitations which may loom large in specific applications.

Table 5-4

CHARACTERISTICS OF MATERIALS FROM WHICH CHEMICAL EQUIPMENT
IS MANUFACTURED

Vitreous Materials:	Transparent or translucent, brittle, good electrical insulation, poor thermal conduction; attacked by HF and fluorides, fused alkalis, many basic oxides and salts.
Fused quartz	Excellent resistance to thermal shock, inert toward halogens, most acids, oxidizing or reducing atmosphere. Maximum working temperature, 1050° C. Trade mark: Vitreosil[11]
High silica glass	Similar in properties to fused quartz, almost as resistant to abrupt temperature change, more resistant toward alkalis than borosilicate glasses. Maximum working temperature, 1000° C. Trade mark: Vycor[12]
Borosilicate glass	Resistant to temperature changes on the order of 150° C, appreciably attacked by aqueous alkalis on heating or long standing. Practical working temperature, about 200° C. Most widely used variety in analytical work. Trade marks: Pyrex[12], KIMAX Resistant[13]
Alkali resistant glass	More sensitive to thermal shock, virtually boron free. Trade mark: Corning[12]
Soft, or lime glass	Of variable composition, very sensitive to thermal shock, readily attacked by aqueous alkalis.

[11] (R) Thermal American Fused Quartz Company, 18-20 Salem Street, Dover, New Jersey.
[12] (R) Corning Glass Works, Corning, New York.
[13] (R) Kimble Glass Co., Toledo 1, Ohio.

Table 5-4 (Continued)

Ceramic Materials:	Opaque, brittle, excellent electrical insulation; attacked by HF, H_3PO_4, concentrated solutions of, and fused, caustic alkalis. Insensitive to abrupt changes in temperature.
Aluminum oxide	Maximum working temperature, 1450° C. Inert with respect to most acids, dilute alkaline solutions, oxidizing atmosphere. Trade mark: ALUNDUM[14].
Porcelain	Maximum working temperatures, glazed ware, 1100° C, unglazed ware, 1400° C.
Metals:	Very wide variation in properties and reactivities. For the most part, ductile, good conductors of heat and electric current, quite insensitive to thermal shock.
Platinum	Generally alloyed with iridium or rhodium for increased hardness. Melting point 1770° C. See also section in text.
Gold	Can be used in place of platinum for low-temperature applications. Maximum working temperature, 950° C. Preferable to platinum for alkali fusions.
Silver	Working temperature, 900° C. Principal application; dishes for evaporation of alkaline solutions.
Stainless steel	Maximum working temperatures, 400° to 500° C where constant weight is required. Unaffected by hot, cold alkalis, and acids with the exception of HCl, dilute H_2SO_4, and boiling HNO_3.
Nickel, Iron	Preferred containers for peroxide fusions; fused sample will be contaminated with metal of the container, owing to attack by the flux.
Plastic Materials:	Opaque, not brittle, good electrical insulation; not attacked by many inorganic acids and bases; cannot be employed at elevated temperatures.
Polystyrene	Maximum tolerable temperature, $\cong 70°$ C. Dimensionally stable, somewhat brittle. Not attacked by HF, fluorides. Unsuitable for use with many organic solvents.
Polyethylene	Maximum tolerable temperature, $\cong 115°$ C. Flexible, not affected by alkali solutions, HF or its salts. Attacked by many organic solvents.

---

[14] ® Norton Company, Worcester 6, Massachusetts.

Platinum.   The chemically valuable properties of this soft, dense metal include its resistance to attack by most mineral acids, including hydrofluoric acid; its inertness with respect to many molten salts; its resistance to oxidation, even at elevated temperatures; and its very high melting point.

With respect to limitations, platinum is readily dissolved on contact with aqua regia and with mixtures of chlorides and oxidizing agents generally. At elevated temperatures, it is also dissolved by fused alkali oxides, peroxides, and to some extent hydroxides. When heated strongly it readily alloys with such metals as gold, silver, copper, bismuth, lead, and zinc; because of this predilection toward alloy formation, contact between heated platinum and other metals or their readily reduced oxides must be avoided. Slow solution of platinum accompanies contact with fused nitrates, cyanides, alkali, and alkaline-earth chlorides at temperatures above 1000° C; bisulfates attack the metal slightly at temperatures above 700° C. Surface changes result from contact with ammonia, chlorine and volatile chlorides, sulfur dioxide, and gases possessing a high percentage of carbon. At red heat, platinum is readily attacked by arsenic, antimony, and phosphorous, the metal being embrittled as a consequence. A similar effect occurs upon high-temperature contact with selenium, tellurium, and, to a lesser extent, sulfur and carbon. Finally, when heated in air for prolonged periods at temperatures greater than 1500° C, a significant loss in weight due to volatilization of the metal must be expected.

Rules Governing the Use of Platinum Ware

1. Use platinum equipment only in those applications that will not affect the metal. Where the nature of the system is in doubt, demonstrate the absence of potentially damaging components before committing platinum ware to use.

2. Avoid violent changes in temperature; deformation of a platinum container can result if its contents expand upon cooling.

3. Supports made of clean, unglazed ceramic materials, fused silica, or platinum itself may be safely used in contact with incandescent platinum; tongs of nichrome or stainless steel may be employed only after the platinum has cooled below the point of incandescence.

4. Clean platinum ware with an appropriate chemical agent immediately following use; recommended cleaning agents are hot chromic acid solution for removal of organic materials, boiling hydrochloric acid for removal of carbonates and basic oxides, and fused potassium bisulfate for the removal of silica, metals, and their oxides. A bright surface should be maintained by burnishing with sea sand.

5. Avoid heating platinum under reducing conditions, particularly in the presence of carbon. Specifically, (a) do not allow the reducing portion of a burner flame to contact a platinum surface, and (b) char filter papers under the mildest possible heating conditions and with free access of air.

Equipment Associated with the Weighing Operation

Many solid substances absorb atmospheric moisture and as a consequence change in composition. This effect assumes appreciable proportions when a large surface area is exposed, as in the case of a sample that has been finely ground. Thus, wherever possible, analyses are based upon dried samples in order to free the results from dependence upon the humidity of the surrounding atmosphere.

Weighing bottles.   Samples are conveniently dried and stored in weighing bottles, two common varieties of which are illustrated in Figure 5.6. Ground

glass contacting surfaces ensure a snug fit between container and lid. In the newer design, the lid acts as a cap; with this style of weighing bottle there is less possibility of the sample being entrained on and subsequently lost from a ground glass surface. Weighing bottles usually have numbers from 1 to 100 etched on their sides for purposes of identification.

Fig. 5.6 Typical Weighing Bottles.

Polyethylene weighing bottles are commercially available; ruggedness is the principal advantage possessed by these over their glass counterparts.

Desiccators, desiccants.   Oven drying is by all odds the most convenient method for the evolution of adsorbed moisture from a sample. The technique cannot, of course, be employed where the sample undergoes decomposition at the ambient temperature of the oven. Furthermore, with some types of solids, the temperatures attainable in ordinary drying ovens are insufficient to effect complete removal of the bound water. This is discussed in some detail in Chapter 31.

While they cool, dried materials are stored in desiccators; these provide a measure of protection from the up-take of moisture. As illustrated in Figure 5.7, the base section of a desiccator contains a quantity of chemical drying agent. Samples are placed on a perforated plate that is supported by a constriction in the wall. Lightly greased ground glass surfaces ensure a tight seal between lid and base.

Table 5-5 gives a partial listing of substances that find application as desiccants; they are arranged according to their effectiveness in removing moisture from a gas stream.[15] This order is not necessarily preserved when these drying agents are employed for the filling of desiccators, and other

Fig. 5.7 A Typical Desiccator.

[15] J. H. Bower, *J. Research Nat. Bur. Standards*, **12**, 241 (1934); **33**, 199 (1944).

factors may be of greater importance in the latter application.[16] For example, a chemical drying agent typically requires an hour or more to abstract moisture from the air contained in a desiccator; absorption is initially rapid and becomes less so with time.[17] Since objects are often stored for short periods only and since some displacement of dried air attends the placement or removal of objects from the desiccator, the atmosphere within this device may not be nearly as anhydrous as the data in Table 5-5 indicate. In addition, anhydrous substances often take up appreciable amounts of moisture very rapidly and then lose it reluctantly, if at all, when stored over a desiccant. Thus, the processes by which water is partitioned between hygroscopic substances and between these substances and the surrounding atmosphere may be fast or slow; the rates at which these equilibria are established are frequently of more importance in determining the degree of protection against the uptake of atmospheric moisture than the relative effectiveness of the particular desiccant employed.

Notwithstanding the foregoing limitations, the atmosphere within a desiccator is likely to be more anhydrous than that of the laboratory and is therefore a good repository for dried samples. Principally because of its low cost, calcium chloride is probably the most widely used drying agent for the filling of desiccators. Inspection of Table 5-5 suggests that its effectiveness in protecting, for example, a freshly ignited precipitate of aluminum oxide is questionable, and for this reason special precautions are required in such instances; instructions for the manipulation of very hygroscopic substances are to be found in a later section of this chapter.

Crucibles

Simple crucibles. Simple crucibles serve as containers only. Of the two varieties available the more common comprises those whose weight remains constant within the limits of experimental error while in use. They are generally made of porcelain, aluminum oxide, silica, or platinum, and are employed to convert precipitates into suitable weighing forms. In addition, there are other crucibles made of nickel, iron, silver, or gold that serve as containers for the high-temperature fusion of difficultly soluble samples. These are appreciably attacked by their contents; the fused mass will thus hold contaminants derived from the crucible. For this type of use the analyst chooses the crucible whose components will offer the least interference in subsequent phases of the analysis.

Filtering crucibles. Filtering crucibles may also be subdivided into two groups. There are those in which the filtering medium is an integral part of the crucible. In addition, there is the *Gooch crucible*, in which a filter mat (usually, but not always, asbestos) is supported upon a perforated bottom. A description of these is given in the next section.

The great saving in time often afforded by filtering crucibles is due to the

---

[16] R. Belcher, *Analyst*, **71**, 236 (1946).
[17] H. S. Booth and L. McIntyre, *Ind. Eng. Chem., Anal. Ed.*, **8**, 148 (1936).

Table 5-5

COMMON DRYING AGENTS

Agent	Capacity	Deliquescent	Condition for Regeneration	Nature of Reaction with Water	Mg Water Remaining in 1 Liter of Dried Air[18]
Phosphorous pentoxide	low	yes	difficult	acidic	2.6×10^{-4}
Barium oxide	moderate	no	difficult	alkaline	6.5×10^{-4}
Aluminum oxide	low	no	175° C	neutral	1×10^{-3}
Anhydrous magnesium perchlorate (Anhydrone[19] or Dehydrite[20])	high	yes	240° C vacuum	neutral	2×10^{-3}
Calcium oxide	moderate	no	500° C	alkaline	3×10^{-3}
Calcium sulfate (Drierite[21])	moderate	no	275° C	neutral	5×10^{-3}
Silica gel	low	no	120° C	neutral	6×10^{-3}
Potassium hydroxide, stick	moderate	yes	difficult	alkaline	1.4×10^{-2}
Anhydrous calcium chloride	high	yes	difficult	neutral	1.4

[18] J. H. Bower, *J. Research Nat. Bur. Standards*, **12**, 241 (1934); **33**, 199 (1944).
[19] (R) J. T. Baker Chemical Company.
[20] (R) Arthur H. Thomas Company.
[21] (R) W. A. Hammond Drierite Co.

use of a partial vacuum to speed the process. Heavy walled filter flasks equipped with adapters to accommodate the crucibles are employed for this purpose. Adapters for filter crucibles are encountered in considerable variety, two of which are illustrated in Figure 5.8; a diagram of a complete suction filtration train is shown in Figure 5.17.

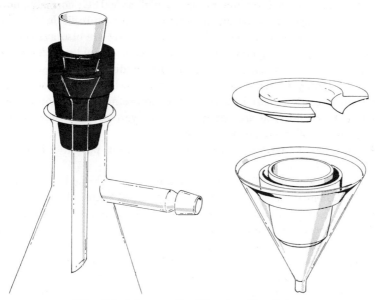

Fig. 5.8 Adapters for Filtering Crucibles.

Equipment Employed in the Filtration Process

An extremely common unit operation of gravimetric analysis involves the separation of a solid phase from the liquid in which it was formed; this requires the use of filtering media that will retain all of the former while offering little or no resistance to the flow of the latter.

Filtering media. Asbestos is commonly used as the filtering medium in a Gooch crucible, being formed as a mat by suction of an aqueous suspension of the material through the small holes in the bottom of the crucible. Asbestos is a mineral of somewhat variable composition and structure. Since some forms are appreciably water soluble, use should be limited to asbestos that has been especially selected and prepared for filtration purposes.

When using asbestos, care must be taken to bring the crucible and mat to constant weight under the same conditions as will be employed in the analysis; since the material is appreciably hygroscopic, uniform treatment is a necessity for satisfactory performance.

Small circles of glass matting are available commercially, and these can be employed in the bottom of a Gooch crucible in place of the asbestos mat; these are used in pairs to protect against accidental disintegration while adding liquid.

They can tolerate temperatures in excess of 500° C and are a good deal less hygroscopic than asbestos.

Glass also finds application as a filtering medium in the form of fritted disks sealed permanently into filtering crucibles. These are known as sintered glass crucibles and are available in a variety of porosities. With extreme care, their use can be extended to temperatures as high as 500° C; normally, however, 150° to 200° C represents the upper practical temperature limit.

Filtering crucibles made entirely of fused quartz are also available; these may be taken to high temperatures and can be cooled rapidly without damage.

Unglazed porcelain can serve as a filtering medium. Crucibles of this type have the versatility of temperature range possessed by the Gooch crucible and are not as costly as fused quartz. They require no preparation comparable to the Gooch crucible and are not appreciably hygroscopic. Crucibles made of aluminum oxide offer similar advantages.

A tabulation of filtering crucibles and their nominal porosities is to be found in the Appendix.

A crucible in which the filtering medium is a burnished platinum mat is known as a *Monroe crucible*; it has the advantages of chemical inertness and the ability to withstand elevated temperatures. The original literature should be consulted for the details of preparation of a Monroe crucible; the mat is produced by the thermal decomposition of ammonium chloroplatinate.[22]

Paper is an important filtering medium for analysis; this material, however, is noticeably hygroscopic, and it is quite impractical to weigh a solid on a large disk of paper because the weight of the paper cannot be reproduced. Thus it is necessary to remove the paper before weighing is attempted; this is done by ignition in air.

Ashless filter paper, so-called because of the extremely small inorganic residue that remains after its ignition, is an important filtering medium for analysis. The fibers that go into the manufacture of this paper have been washed with hydrochloric and hydrofluoric acids; this ultimately results in a paper that is virtually free of inorganic matter. Thus, typically, 9 or 11 cm circles of such paper will leave an ash weighing less than 0.1 mg, an amount that is ordinarily negligible. Ashless paper is manufactured in various grades of porosity. Characteristics of papers from different sources are tabulated in the Appendix.

Gelatinous precipitates, such as hydrous ferric oxide, present special problems owing to their tendency to clog the pores of the paper upon which they are being retained. This problem can be minimized by mixing a dispersion of ashless filter paper pulp with the precipitate prior to filtration. The pulp may be prepared by briefly treating a piece of ashless paper with concentrated hydrochloric acid and washing the disintegrated mass free of acid; tablets of pulp are also commercially available.

Table 5-6 summarizes the characteristics of common filtering media.

[22] W. O. Snelling, *J. Am. Chem. Soc.*, **31**, 456 (1909).

Table 5-6

COMPARISON OF FILTERING MEDIA FOR GRAVIMETRIC ANALYSIS

	Paper	Asbestos (Gooch)	Glass Crucible	Porcelain Crucible	Aluminum Oxide Crucible
Speed of filtration	slow	rapid	rapid	rapid	rapid
Convenience and ease of preparation	somewhat troublesome and inconvenient	somewhat troublesome and inconvenient	convenient	convenient	convenient
Maximum ignition temperature	none	$1200°$ C	$200-500°$ C	$1100°$ C	$1450°$ C
Chemical reactivity	carbon from paper has reducing properties	quite inert	quite inert	quite inert	quite inert
Control of porosity	wide variety of porosities available	little control possible	several porosities available	several porosities available	several porosities available
Ability to handle gelatinous precipitates	satisfactory	unsuitable; filter tends to clog	unsuitable; filter tends to clog	unsuitable; filter tends to clog	unsuitable; filter tends to clog
Cost	low	intermediate	high	high	high

Equipment Associated with the Heat Treatment of Precipitates

Ovens. Many precipitates may be weighed directly after the low-temperature removal of moisture; drying ovens are most advantageously employed. These are heated electrically and are capable of maintaining uniform temperatures to within a degree or less. The maximum attainable temperature will range from 140° to 260° C depending upon make and model; for many precipitates, 110° C is a satisfactory drying temperature.

The efficiency of a drying oven is increased significantly when air circulation within it is augmented by means of a pump. Further refinement in the drying process is achieved by predrying the air to be circulated and by the use of vacuum ovens through which a small flow of predried air is maintained.

Heat lamps. Commercially available heat lamps are useful for laboratory drying and provide temperatures capable of charring filter paper. A very convenient method of treating precipitates contained on paper involves the use of heat lamps for initial drying and charring followed by ignition at elevated temperatures in a muffle furnace.

Burners. A wide variety of burners is to be found in the laboratory. The attainable temperatures depend largely upon the combustion properties of the gas employed and upon the design of the burner itself. Three varieties common to most analytical laboratories are illustrated in Figure 5.9. A rough estimate

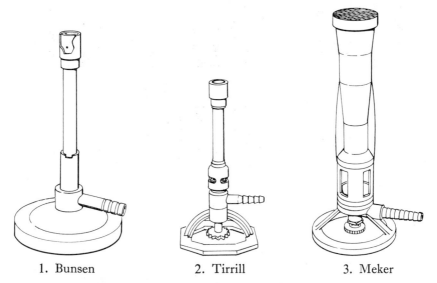

1. Bunsen   2. Tirrill   3. Meker

Fig. 5.9 Common Laboratory Burners.

of the temperature of an object can be gained from observation of the color of light it is emitting; Table 5-7 will serve as a guide.

Muffle furnaces. Muffle furnaces are heavy duty electric furnaces capable of maintaining temperatures of 1100° C or higher with a control superior to that obtainable with a burner. When introducing objects into these furnaces, special long-handled tongs and asbestos gloves are often required to protect against the intense heat.

Table 5-7

ESTIMATION OF TEMPERATURE BY COLOR[23]

Temperature, °C	Approximate Color of Object at this Temperature
700	dull red
900	cherry red
1100	orange
1300	white

[23] Adapted from T. B. Smith, *Analytical Processes*, 2d ed., London: Edward Arnold (Publishers) Ltd., 1940.

CHEMICAL REAGENTS

Classifications of Reagents

Technical or commercial grade. Chemicals labeled technical or commercial grade are of indeterminate quality and should only be used where high purity is not of paramount importance. Thus, the potassium dichromate and the sulfuric acid used in the preparation of cleaning solution can be of this grade. In general, however, technical or commercial grade chemicals are not used in analytical work.

Chemically Pure, or C. P. grade. The term, *Chemically Pure*, has little definite meaning. Such reagents are usually more refined than the technical grades, but no specifications define what is meant by the term. Thus, it is prudent to avoid the use of C.P. reagents in analytical work; if this is not possible, testing the reagent for contaminants of importance and also the running of frequent reagent blanks may be necessary.

U.S.P. grade. U.S.P. chemicals have been found to conform to tolerances specified in the United States Pharmacopoeia.[24] The specifications are designed to control the presence of contaminants dangerous to health; thus, chemicals passing U.S.P. tests may still be quite heavily contaminated with impurities that are not physiological hazards.

Reagent grade. For the most part the analytical chemist employs reagent grade chemicals in his work. These have been tested and found to conform to

[24] U.S. Pharmacopoeial Convention, *Pharmacopoeia of the United States of America*, 16th rev. Easton, Pa.: Mack Publishing Co., 1960.

the minimum specifications set down by the Reagent Chemicals Committee of the American Chemical Society.[25] In addition to meeting these requirements the results of the analysis are, in some instances, printed on the label. Thus reagent grade chemicals fall into two categories: namely, those that simply pass the tests and those for which the actual results of the tests are additionally supplied.

Primary standard grade.   A discussion of the full implications of the term *primary standard* is to be found in Chapter 10; we shall merely state here that in addition to possessing desirable chemical properties, these are substances that are obtainable in extraordinarily pure form. Primary standard grade reagents are available commercially; these have been carefully analyzed and the assay value is printed on the label. An excellent source for primary standard chemicals is the National Bureau of Standards. The current edition of *Circular 552* lists these as well as *reference standards*—complex mixtures that have been exhaustively analyzed.[26]

Techniques and Operations

Handling Reagents and Solutions

For successful analytical work the availability of reagents and solutions of established purity is of prime importance. A freshly opened bottle of some reagent grade chemical can be used with confidence in most applications; whether the same confidence is justified when this bottle is half full depends entirely upon the care with which it was handled after being opened. The rules that are given here will be successful in preventing contamination of reagents only if they are conscientiously followed.

1. Select the best available grade of chemical for analytical work. If there is a choice, pick the smallest bottle that will supply the desired quantity of substance.

2. Replace the top of every container immediately after removal of reagent; do not rely on having this done by someone else.

3. Stoppers should be held between the fingers and should never be set on the desk top.

4. Unless specifically directed to the contrary, never return any excess reagent or solution to a bottle; the minor saving represented by the return of an excess is indeed a false economy compared to the risk of contaminating the entire bottle.

[25] American Chemical Society, Committee on Analytical Reagents, *Reagent Chemicals, American Chemical Society Specifications*. Washington D.C.: American Chemical Society, 1955.

[26] United States Department of Commerce, *Standard Materials Issued by the National Bureau of Standards, Circular 552*, 3d ed. Washington D. C.: Government Printing Office, 1959.

5. Again, unless specifically instructed otherwise, do not insert spoons, spatulas, or knives into a bottle containing a reagent chemical. Instead, shake the bottle vigorously with the cap in place to dislodge the contents; then pour out the desired quantity.

6. Keep the reagent shelf and the laboratory balances clean. Immediately clean up any spilled chemicals, even though others may be making the same transfer of reagent in the same area.

Preliminary Operations

Cleaning of equipment. Care must be taken that all glass and porcelain ware are thoroughly clean. The use of detergents is recommended for this purpose. After thorough cleaning, an extensive rinse with tap water followed by a rinse with several small portions of distilled water should yield an object that shows uniform and unbroken wetting on its inner surfaces. Glassware seldom needs drying before use; this practice should, in fact, be discouraged on the grounds of being wasteful of time and potentially the source of contamination.

If a grease film should remain after thorough cleaning as above, employment of a cleaning solution may be indicated. The most common of these consists of sodium or potassium dichromate in very strong sulfuric acid. For best results it is used warm, 70° C being a satisfactory temperature. Several points must be borne in mind with regard to cleaning solution. First, while often very effective in removing oxidizable and acid soluble stains from glass surfaces, cleaning solution is otherwise a rather poor cleaning agent. Second, very thorough rinsing is required to remove the last traces of dichromate that adhere strongly to glass and porcelain surfaces. Finally, cleaning solution, especially when hot, is a potentially dangerous preparation that will attack plant and animal matter with great speed. It must be handled with due respect. Any spillages should be promptly diluted with large volumes of water.

> **Preparation of cleaning solution.** In a 500-ml heat-resistant conical flask mix 10 to 15 grams of potassium dichromate with about 15 ml of water. Add concentrated sulfuric acid slowly and with thorough stirring between increments. The contents of the flask will become a semisolid red mass; add just enough sulfuric acid to bring the mass into solution. Allow to cool somewhat before attempting to transfer to a soft glass bottle. The solution may be reused until it acquires the green color of chromium (III) ion, at which time it must be discarded.

Marking of equipment. Most chemical analyses are run in duplicate; it is therefore necessary to mark all beakers and crucibles to indicate the location of each sample. The identification of beakers is simple—the etched area on each beaker can be semipermanently marked with a pencil. Special marking inks are available for porcelain surfaces; after the marking is made, it is baked permanently into the glaze. A saturated solution of ferric chloride can be used in the same fashion although it is not as satisfactory as the commercial preparations.

Preparation of a desiccator for service.   The filling of a desiccator is readily accomplished by employing a crude paper funnel to direct the flow of desiccant to the bottom section. Any dust adhering to the upper walls should be scrupulously wiped off before putting the desiccator into service. The ground glass surfaces should be sufficiently greased to allow a tight seal; this is indicated when the contacting surfaces appear transparent when viewed from above. An excess of grease is a nuisance and should be avoided. It is good practice to label the desiccator, indicating the date of filling and the drying agent employed.

Manipulations Associated with the Weighing Operation

Use of the desiccator.   Whether it is being replaced or removed, the lid of the desiccator is properly moved by a sliding, rather than a lifting, motion. An air-tight seal is achieved by slight rotation and direct downward pressure upon the positioned lid.

When a heated object is placed in a desiccator, the increased pressure of the enclosed air is often sufficient to break the seal between lid and base; if heating has caused the grease on the ground-glass surfaces to soften, there is the further danger that the lid may slide off and break. Upon cooling, the opposite effect is likely to occur, the interior of the desiccator being under a partial vacuum.

Fig. 5.10 (*Right*) Arrangement for the Drying of Samples.

Fig. 5.11 (*Below*) The Weighing Operation. A convenient method for transfer of a solid for weighing by difference.

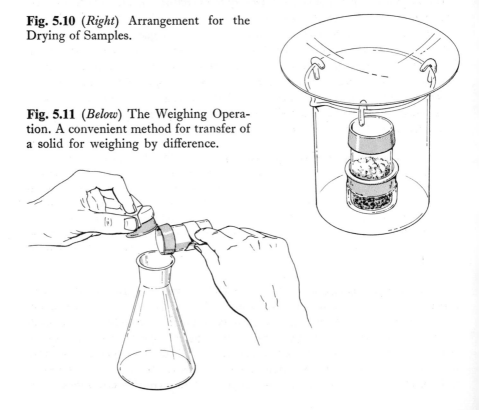

Both of these conditions are undesirable because of the danger of disrupting or contaminating the sample being stored. While it in part defeats the purpose of the desiccator, it is wise to allow some cooling to occur before finally sealing the lid, and also to break the seal several times during cooling in order to relieve any vacuum that may have developed. It is also prudent to lock the lid in place with one's thumbs while moving the desiccator to ensure against accidental breakage.

Very hygroscopic materials should be stored in containers equipped with snugly fitting covers, the covers remaining in place during storage in the desiccator. Other substances should be stored with container covers removed.

Manipulation of weighing bottles. Heating of many solid materials at 105° to 110° C for about an hour is sufficient to remove moisture bound to the surface of the particles. Figure 5.10 depicts the arrangement recommended for the drying of a sample in a weighing bottle. Note that the bottle is contained in a labeled beaker, which in turn is covered by a watch glass supported on glass hooks. In this manner the sample is protected from accidental contamination while maintaining a free access of air.

This arrangement also satisfactorily accommodates crucibles containing precipitates that can be freed of moisture by simple drying.

Weighing data will often be significantly affected by the moisture picked up as a consequence of handling a dried weighing bottle with one's fingers. In such cases the bottle should be manipulated with tongs or with strips of clean paper. The latter technique is illustrated in Figure 5.11, which shows the sequence followed in weighing a sample by difference. The sample is transferred from the bottle to the container, the utmost care being taken to avoid losses during transfer; gentle tapping of the weighing bottle with its top provides adequate control over the process.

Tared watch glasses. When weighing out a precisely known quantity of solid, use may be made of a pair of tared watch glasses. These have been carefully matched according to weight; a rest point that corresponds closely to the zero point will be observed with one of these glasses on each pan of the balance. In practice, this rest point is taken, the desired weight is introduced to the right pan watch glass, and the required amount of solid is then added to the left pan watch glass by means of a spatula, care being taken to avoid spillage. When balance has been achieved, the solid is transferred from the watch glass to the desired container, a camel's hair brush or a stream of water being employed to assure quantitative transfer. This technique is suitable only for the weighing of substantially nonhygroscopic materials.

The weighing of hygroscopic substances. Many substances are hygroscopic to a relatively limited extent. These often equilibrate very rapidly, taking up moisture almost to capacity in a few moments' time. In cases where this effect is pronounced, individual weighing bottles should be used, the approximate weight of a sample being introduced to each. After drying and cooling, the exact weight is determined by difference, care being taken to remove the sample and replace the top of the weighing bottle as rapidly as possible.

Some gravimetric precipitates are quite hygroscopic following ignition and are difficult to weigh. It is recommended that the approximate weight be determined, after which the crucible is reignited. The weights used are returned to the pan of the balance while the crucible is cooling. Since these will require but a slight adjustment, the second weighing will require less time for completion. The process should be repeated until satisfactory agreement between consecutive weighings has been achieved.

The weighing of liquids. The weight of a liquid is always obtained by difference. Samples that are noncorrosive and relatively nonvolatile can be weighed into a tared container fitted with a snugly fitting cover, such as a weighing bottle. If the sample is extremely volatile or corrosive, it should be sealed in a tared glass ampoule before weighing. To fill, the bulb of the ampoule is first heated; the neck is immersed in the sample and as cooling occurs the liquid is drawn into the bulb. The neck is then sealed off with a small flame. After cooling, the bulb and contents are weighed, any glass removed during the sealing being also included in this weighing. The ampoule is then broken in the vessel where the sample is desired; for very precise work, the preparation of dilute solutions in volumetric flasks by this method requires a small correction for the volume of the glass in the ampoule.

Fig. 5.12 Arrangement for the Evaporation of Liquids.

Manipulations Associated with Solutions

Evaporation. An arrangement such as that illustrated in Figure 5.12 is satisfactory for most evaporations. The process is sometimes difficult to control owing to the tendency of some solutions to superheat locally. The bumping that results, if sufficiently violent, can lead to the physical loss of part of the sample. This danger is minimized by careful and gentle heating; the introduction of glass beads, where permissible, is also helpful.

Elimination of common interferences by evaporation. The evaporation process is often employed to rid a solution of unwanted constituents. For example, chloride and nitrate ions can be removed by evaporating a solution containing sulfuric acid until the copious white fumes of sulfur trioxide are observed. Nitrate ion and nitrogen oxides can also be removed by adding urea to an acidic solution and heating. Ammonium salts are often removed by evaporation to dryness and gentle ignition of the residue. When large quantities of ammonium chloride are to be removed, it is better to add concentrated nitric acid to the solution that has been first reduced to a small volume. Rapid oxidation of the ammonium ion occurs upon heating; the solution is then evaporated to dryness.

Unwanted organic substances are frequently removed by evaporation of the solution with sulfuric acid until fumes of sulfur trioxide are observed. Nitric acid may be added at this point to hasten the oxidation and destruction of the organic matter.

Manipulations Associated with the Filtration and Ignition Processes

Preparation of crucibles. Irrespective of type, a crucible employed in converting a precipitate into a form suitable for weighing must maintain a substantially constant weight throughout the drying or ignition process; it is thus necessary to demonstrate that this condition applies at the outset.

As a first step, each crucible should be inspected for defects; this is of particular importance where the crucible has previously been subjected to high temperatures. A porcelain crucible should be placed upright on a hard surface and gently tapped with a pencil. A clear, ringing tone indicates an intact crucible while a dull sound is characteristic of one that is cracked and should be discarded.

The crucible is then cleaned thoroughly. Filtering crucibles are conveniently cleaned by backwashing with suction.

Finally, the crucible should be brought to constant weight using the same heating cycle as will be required for the precipitate. For most purposes, agreement within 0.2 mg between consecutive weighings can be considered as constant weight.

Decantation. The actual filtration process may be considered to consist of the three operations of decantation, washing, and transfer. Decantation is the process of gently pouring off the liquid phase while leaving the precipitated solid essentially undisturbed. It is a valuable timesaver: the pores of any filtering medium become clogged with precipitate and the longer the transfer of precipitate can be delayed, the more rapid will be the over-all process. To this end, the liquid phase is decanted through the filter, a stirring rod being employed to direct the flow (see Figure 5.13). Wash liquid is then added to the beaker and is thoroughly mixed with the precipitate; after allowing the solid to settle, this too is decanted through the filter. It can be seen that the principal washing of the precipitate is carried out before the solid is transferred, a procedure to be highly recommended since it results in a more thoroughly washed precipitate and a more rapid filtration.

Many precipitates have the exasperating property of spreading over wetted surfaces against the force of gravity; this is known as *creeping*. It is because of this phenomenon that the cone of a paper filter is never filled more than three quarters full at any time. Care must also be taken to avoid excessive amounts of liquid in filter crucibles.

Washing.   Several washings with small volumes of liquid are more effective in removing soluble contaminants than the same total volume used in one washing. Consider the case of a precipitate in contact with a volume which is c formal in soluble foreign matter. After decantation, let v ml of this liquid remain associated with the precipitate. Now introduce a volume, V, of wash liquid; the total volume thus becomes $V + v$ and the concentration of the contaminant is reduced to

$$c\left(\frac{v}{V + v}\right)$$

After decantation, v ml of this solution will remain in contact with the precipitate; washing with another volume V will again yield a total volume $V + v$ and the contaminant concentration will be further reduced to

$$c\left(\frac{v}{V + v}\right)\left(\frac{v}{V + v}\right) = c\left(\frac{v}{V + v}\right)^2$$

In general, then, after the nth washing with V ml, the concentration of contaminant remaining will be given by

$$c\left(\frac{v}{V + v}\right)^n \cong c\left(\frac{v}{V}\right)^n \qquad \text{when } V \gg v \qquad (5\text{-}7)$$

To illustrate, Table 5-8 gives the factor by which c is reduced for various schedules of washing with a given volume of liquid. These data were computed with the approximate form of the foregoing equation.

Table 5-8

Effect of Multiple Washings in Removing Soluble Contaminants
(Residual volume, v, is taken as 0.5 ml, total volume, V, is 200 ml)

Number of Washings	Volume of Each Wash, ml	Factor by which Original Concentration is Reduced
1	200	2.5×10^{-3}
2	100	2.5×10^{-5}
4	50	1.0×10^{-8}
5	40	3.0×10^{-10}
8	25	2.6×10^{-14}
10	20	9.5×10^{-17}

(1)

(2)

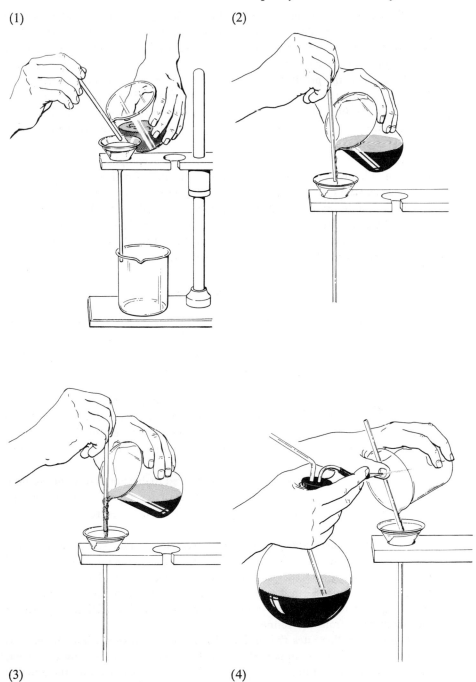

(3)

(4)

Fig. 5.13 The Filtering Operation. Techniques for decantation and transfer of precipitates are illustrated.

We will see in Chapters 7 and 8 that other factors may act to render washing considerably less efficient than this equation predicts. The fact remains, however, that multiple washings with small volumes provide the most efficient removal of soluble contaminants.

Transfer of the precipitate. Figure 5.13 shows the method whereby the decantation and transfer operations are performed. With respect to the latter, the bulk of the precipitate is moved from beaker to filter by means of suitably directed streams of wash liquid; as always, a stirring rod is used to provide direction to the flow of liquid to the filtering medium.

Removal of the last traces of precipitate is accomplished with a *rubber policeman*. This is a small section of tubing, one end of which is crimped shut; it is fitted on the end of a stirring rod and is used to scrub those surfaces to which precipitate may cling. Any solid collected is added to the main portion on the filter. An alternate procedure which is preferable where appropriate is to use small pieces of ashless paper to collect the last traces of solid; this is most conveniently accomplished with the aid of clean forceps.

Specific Directions for the Use of Ashless Filter Paper

Preparation of a filter paper. Figure 5.14 illustrates the sequence followed in folding a filter paper and seating it in a funnel. The paper is first folded exactly in half. The second fold is made so that the corners fail to coincide by about $\frac{1}{8}$ inch in each dimension. A small triangular section is torn from the short corner; this permits a better seating of the filter in the funnel. The paper is next opened out so that a cone is formed and is then gently seated in the funnel with the aid of water from a wash bottle. When properly seated, there will be no leakage of air between paper and funnel; as a result, the stem of the funnel will be filled with an unbroken column of liquid, a condition that markedly increases the rate of filtration.

A gelatinous precipitate should not be allowed to dry out before the washing cycle is complete because the mass shrinks and develops cracks on drying; any liquid subsequently added merely passes through these cracks and accomplishes little or no washing.

Transfer of paper and precipitate to crucible. Upon completion of the filtration and washing steps, the filter paper and its contents must be transferred from the funnel to a tared crucible. Because ashless paper has very low wet strength, considerable care must be exercised in performing this operation. The danger of tearing can be reduced considerably if partial drying occurs prior to transfer from the funnel.

Figure 5.15 illustrates the preferred method of transfer. First, the cone is gently flattened along its upper edge and the corners are folded inward. Next, the top is folded over. Finally, the paper and contents are eased into the crucible so that the bulk of the precipitate is near the bottom.

Ashing of a filter paper. If a heat lamp is available, the crucible is placed on a clean, nonreactive surface; an asbestos pad covered with a layer of aluminum

(1)   (2)

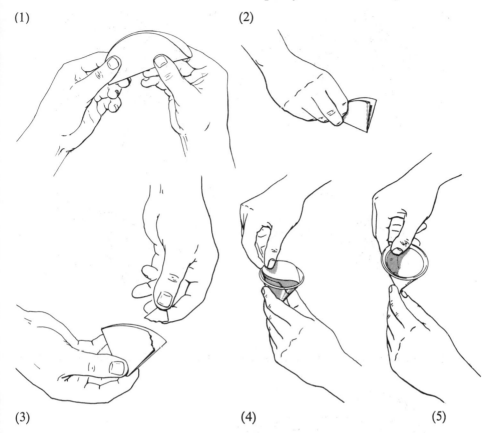

(3)   (4)   (5)

Fig. 5.14 Technique for Folding and Seating of a Filter Paper.

foil is very satisfactory. The lamp is then positioned about $\frac{1}{2}$ inch from the top of the crucible and is turned on. Charring of the paper will take place without further intervention; the process is considerably accelerated if the paper can be moistened with no more than one drop of strong ammonium nitrate solution. Removal of the remaining carbon is accomplished with a burner as described in the following paragraphs.

Considerably more attention must be paid the process when a burner is employed to ash a filter paper. Since the burner can produce much higher temperatures, there exists the danger of expelling moisture so rapidly in the initial stages of heating that mechanical loss of the precipitate occurs. A similar possibility arises if the paper is allowed to flame. Finally, as long as there is carbon present there is also the possibility of chemical reduction of the precipitate; this is a serious problem where reoxidation following ashing of the paper is not convenient.

In order to minimize these difficulties the crucible is placed as illustrated in Figure 5.16. The tilted position of the crucible allows for the ready access

(1)

(2)

(3)

(4)

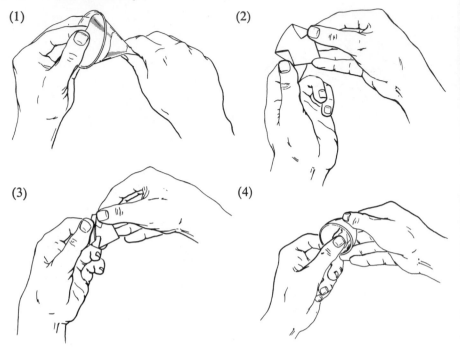

Fig. 5.15 Method for Transfer of a Filter Paper and Precipitate to a Crucible.

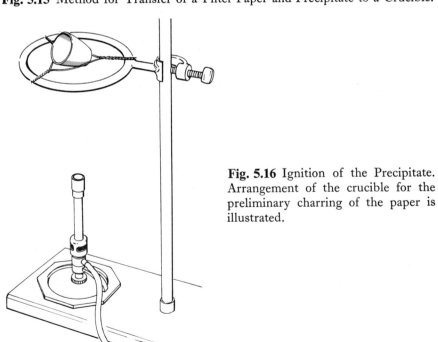

Fig. 5.16 Ignition of the Precipitate. Arrangement of the crucible for the preliminary charring of the paper is illustrated.

of air. A clean crucible cover should be located nearby, ready for use if necessary. Heating is then commenced with a small burner flame. This is gradually increased as moisture is evolved and the paper begins to char. The smoke that is given off serves as a guide with respect to the intensity of heating that can be safely tolerated. Normally this will appear to come off in thin wisps. If the volume of smoke emitted rapidly increases, the burner should be temporarily removed; this condition indicates that the paper is about to flash. If, despite precautions, a flame does appear, it should be immediately snuffed out with the crucible cover. (The cover may become discolored owing to the condensation of carbonaceous products; these must ultimately be removed by ignition so that the absence of entrained particles of precipitate can be confirmed.) Finally, when no further smoking can be detected, the residual carbon is removed by gradually lowering the crucible into the full flame of the burner. Strong heating, as necessary, can then be undertaken. Care must be exercised to avoid heating the crucible in the reducing portion of the flame.

The foregoing procedure must precede final ignition of the sample in a muffle furnace, a reducing atmosphere being equally undesirable here.

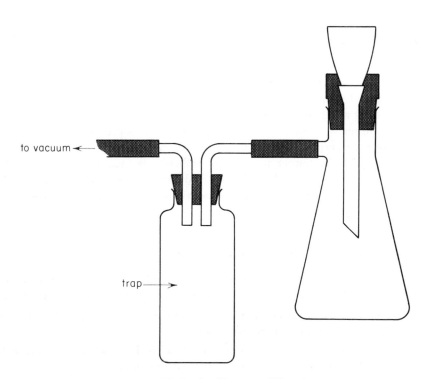

to vacuum

trap

Fig. 5.17 Train for Vacuum Filtration.

Specific Directions for the Use of Filtering Crucibles

Preparation of a Gooch crucible. A white, silky grade of long fibered asbestos is required for use in Gooch crucibles. There should be few, if any, inclusions of iron in the material. In preparing it for use, asbestos should be cut into 0.5 cm lengths; any large clumps of fibers should be manually broken apart. It should then be digested overnight in 4 F hydrochloric acid. Following this, the fibers should be collected, washed free of acid, and stored as an aqueous slurry.

An asbestos mat is prepared by arranging the Gooch crucible in a vacuum filtration train similar to that illustrated in Figure 5.17. A few milliliters of the well-mixed and diluted suspension are poured into the crucible and allowed to stand for about a minute. In this interval the heavier filaments tend to settle and form the basis for supporting the finer fibers; most of the liquid also drains off. Next suction is applied to set the mat. The mat is then washed until no further loss of asbestos can be detected in the washings; several hundred milliliters may be required. Finally, the crucible is dried and ignited to constant weight.

A proper mat is one through which the pattern of holes in the crucible can barely be discerned when viewed against a strong light. Once set, a vacuum must always be applied when introducing material to the crucible. Care must be taken to avoid destroying the mat through the direct pouring of liquids. The flow is always directed with a stirring rod down the wall of the crucible; the mat can be further protected through introduction of a small perforated disk called a *Witt plate*.

Ignitions with a filtering crucible. In general, the use of a filtering crucible tends to shorten the time necessary for the ignition of a precipitate. The porous bottom of such a crucible, however, greatly increases the danger of reducing the precipitate with the burner flame. This difficulty is readily circumvented by placing the filtering crucible inside an ordinary crucible and igniting both.

The reader is again cautioned against subjecting crucibles to unnecessarily abrupt changes in temperature.

Rules for the Manipulation of Heated Objects

1. A crucible that has been subjected to the full flame of a burner should be allowed to cool in place momentarily before being moved to the desiccator.

2. Hot objects should not be placed directly on the desk top but should be set on an asbestos pad or a clean wire gauze.

3. Manipulations should be practiced first to demonstrate that adequate control can be maintained with the implements employed.

4. The tongs and forceps employed in handling heated objects should be kept scrupulously clean. The tips, particularly, should not be allowed to come in contact with the desk top.

Rules for Keeping the Laboratory Notebook

1. The notebook should be permanently bound with consecutively number-ed pages.

2. Entries should be legible and well spaced from one another. Most notebooks have more than ample room; crowding of data is unnecessary.

3. The first few pages of the notebook should be reserved for a table of contents, which should be conscientiously kept up to date.

4. All data should be entered directly into the notebook, and in ink.

(a) Entries should be liberally identified with labels. If a series of weights refers to a set of empty crucibles, they should be labeled "Empty Crucible Weights," or some such. The significance of an entry is obvious when it is recorded, but rapidly loses this quality with the passage of time.

(b) It is good practice to date each notebook page as it is used.

(c) Erroneous entries should not be erased, nor should they be obliter-ated. If some entry has been erroneously made, it should be crossed out with a single horizontal line; the corrected entry should then be entered as nearby as possible. Numbers should never be written over; in time it may be difficult to recall which was written over which.

(d) Pages should not be removed from the notebook. Rather, a single line drawn diagonally across the page to be disregarded is sufficient. It is proper to make a brief notation of the reason for striking out the page.

Suggested form. A satisfactory format involves the consecutive use of all pages for the recording of data. Upon completion of the analysis, the next pair of facing pages is used to summarize the results. The right-hand page should contain:

(1) The title of the experiment—for example, *The Gravimetric Determination of Chloride.*

(2) A brief statement of the principles upon which the analysis is based.

(3) A summary of the data collected, and the result calculated for each sample in the set.

(4) A report of the best value (mean or median, whichever is adjudged the better) for the set and a statement of the precision attained in the analysis.

A sample summary is shown in Table 5-9.

The left-hand page should show:

(1) Equations for the principal reactions in the analysis.

(2) An equation that shows the calculation employed in computing the results.

(3) The calculations themselves.

Table 5-9

GRAVIMETRIC DETERMINATION OF CHLORIDE

The chloride in a soluble sample was precipitated as AgCl and weighed as such

	1.	2.	3.
Weight of bottle + sample, grams	27.1911	26.7874	26.4065
Weight of bottle	26.7874	26.4065	25.9873
Weight of sample taken, gram	0.4037	0.3809	0.4192
Weight of crucible + AgCl, grams	21.8309	21.4140	21.5743
Weight of crucible	21.3406	20.9527	21.0683
Weight of AgCl found, gram	0.4903	0.4613	0.5060
Percent Cl⁻	30.04	29.96	29.86
Average		29.95	
Deviation	0.09	0.01	0.09
Average deviation		0.06	
(or) Range		0.18	

problems

1. Determine the weight of the following to the nearest 0.1 mg from the accompanying long swing data :

    (a) Swings for empty balance   6.7, 13.8, 6.9, 13.6, 7.1
          Weights on pan and rider   20 grams, 1 gram, 400 mg
              Swings with chain at 67 mg   9.3, 17.6, 9.5, 17.4, 9.6
              Swings with chain at 68 mg   7.5, 12.4, 7.7, 12.2, 8.0

                                         ans. 21.4679 grams

    (b) Swings for empty balance   6.8, 12.6, 7.1, 12.4, 7.3
          Weights on pan and rider   5 grams
              Swings with chain at 21 mg   12.4, 19.3, 12.6, 18.9, 12.9
              Swings with chain at 22 mg   10.0, 15.5, 10.2, 15.3, 10.4
    (c) Swings for empty balance   5.8, 14.6, 6.1, 14.4, 6.3
          Weights on pan and rider   10 grams, 700 mg
              Swings with chain at 2 mg   9.1, 17.3, 9.3, 17.0, 9.5
              Swings with chain at 3 mg   6.5, 13.4, 6.7, 13.2, 6.9

2. Determine the weight of the following to the nearest 0.1 mg from the accompanying short swing data :

    (a) Swings for empty balance   7.8, 12.4
          Weights on pan and rider   5 grams, 2 grams, 300 mg
              Swings with chain at 37 mg   9.4, 16.0
              Swings with chain at 38 mg   7.3, 11.5

                                           ans. 7.3379 grams

(b) Swings for empty balance   7.0, 12.2
   Weights on pan and rider   10 grams, 600 mg
    Swings with chain at 47 mg   7.7, 11.9
    Swings with chain at 48 mg   4.2, 9.0

(c) Swings for empty balance   7.7, 12.5
   Weights on pan and rider   20 grams
    Swings with rider at 7 mg   10.5, 14.3
    Swings with rider at 8 mg   7.6, 11.0

3. Estimate the relative error (in parts per thousand) attributable to the weighing operation when

(a) A 5-gram sample is weighed directly to the nearest 0.1 mg
(b) A 0.1-gram sample is weighed directly to the nearest mg
(c) The weight of a 0.5-gram sample is obtained as the difference between two weighings carried to the nearest 0.1 mg
(d) The weight of a 40-mg sample is obtained as the difference between two weighings carried to the nearest 0.1 mg

4. Brass weights were used to determine the mass of the accompanying samples. Correct each for the effect of buoyancy.

(a) 0.4800 gram of diethyl ether ($d = 0.714$)
(b) 0.2540 gram of hexachloroethane ($d = 2.09$)
(c) 0.8440 gram of vanadium ($d = 5.87$)
(d) 1.110 grams of nickel ($d = 8.90$)
(e) 2.104 grams of mercury ($d = 13.55$)

5. Calculate the volume of a pycnometer from the following data :

  Empty weight (against brass weights)   35.6453 grams
  Weight filled with H_2O at 20.9° C ($d = 0.9970$ gram/ml)   85.0770 grams

6. The left and right terminal knife edges of a balance are respectively located 7.500 cm and 7.499 cm from the central knife edge. What will be the apparent mass of a 2.5000-gram object placed on the left pan of this balance ? Neglect the effects of buoyancy.

7. Determine the corrections for a set of brass weights relative to the 1-gram rider of a chain balance, given the following data :

  Tare on left pan = 0.0021 gram
    R = rider

Nominal weight and designation	Weights used for calibration	Reading of chain, mg
1	R	2.3
2′	$1 + R$	2.0
2″	2′	1.8
5	$R + 2' + 2''$	1.9
10′	$R + 2' + 2'' + 5$	2.5
10″	same	2.6
20	$10' + 10''$	2.1
50	$R + 2' + 2'' + 5$ $+ 10' + 10'' + 20$	1.9

For Problems **8** and **9**, the following weights are available :

Nominal weight, gram	Correction, mg	Nominal weight, gram	Correction, mg
0.01	− 0.1	0.10	− 0.2
0.02′	− 0.1	0.20′	0.0
0.02″	0.0	0.20″	+ 0.1
0.05	+ 0.1	0.50	+ 0.2
		1.00	0.0

8. Determine the corrections for a notched beam and rider from the accompanying data. The sensitivity remains constant at 2.8 divisions per mg throughout the weight range considered.

   With a tare on the left pan and the chain at 1.0 mg, rest point = 10.9
   Tare remains undisturbed throughout the calibration.

	Nominal weights on left pan	Swings		
(a)	0.1	7.9	12.7	
(b)	0.2′	7.6	13.2	ans. + 0.2 mg
(c)	0.1 + 0.2′	7.4	13.8	
(d)	0.2′ + 0.2″	9.0	14.0	
(e)	0.5	8.8	12.8	
(f)	0.1 + 0.5	7.4	14.4	
(g)	0.2′ + 0.5	10.2	13.4	
(h)	0.1 + 0.2′ + 0.5	7.9	13.3	
(i)	0.2′ + 0.2″ + 0.5	9.2	13.6	ans. + 0.1 mg
(j)	1.0	8.9	11.7	

9. Determine the corrections for a chain weight from the accompanying data. The sensitivity remains constant at 2.7 divisions per mg throughout.

	Nominal weights on left pan	Swings		
(a)	0.01	7.8	11.4	
(b)	0.02′	8.3	11.1	ans. + 0.1 mg
(c)	0.01 + 0.02′	7.4	12.6	
(d)	0.02′ + 0.02″	7.6	13.8	
(e)	0.05	7.9	13.6	
(f)	0.01 + 0.05	7.4	13.0	
(g)	0.02′ + 0.05	7.2	12.1	
(h)	0.01 + 0.02′ + 0.05	7.6	12.7	
(i)	0.02′ + 0.02″ + 0.05	8.6	12.2	ans. − 0.1 mg
(j)	0.10	7.0	12.8	

10. Use the accompanying data to compute d, the distance between the fulcrum and the center of mass of an analytical balance.

Weight of beam, pans, and rider	105.1 grams
Distance between terminal knife edges	15.6 cm
Length of pointer	22.5 cm

Spacing between divisions on deflection scale	1.0 mm
Swings for empty balance	6.4, 14.1, 6.6
Swings with a 1-mg load	5.0, 10.1, 5.2

11. Soluble salts are to be removed from a silver chloride precipitate by washing. Decantation reduces the volume of liquid associated with the solid to 0.8 ml. Washing is performed with 25-ml portions of 1 percent HNO_3. If the original decantate was $0.04\ F$ in soluble contaminants, what should be their concentration after the third washing?

12. The concentration of solid contaminants in the 0.5 ml of residual liquid associated with a precipitate is $0.04\ F$. What will be the minimum number of 20-ml portions of wash liquid needed to lower this concentration to $1 \times 10^{-5}\ F$?

chapter 6. *The Solubility of Precipitates*

The ultimate accuracy of a gravimetric method of analysis is limited in part by the loss of precipitate owing to its solubility in the reaction medium and in the wash liquid employed for purification. These losses often assume serious proportions, particularly where extensive washing is necessary to assure a pure precipitate or where an overly soluble precipitate must be used. Although he is powerless to eliminate solubility losses completely, the chemist does have a degree of control over the factors that influence the solubility of precipitates; by the judicious regulation of these and the employment of a sample of suitable size, he is usually able to reduce solubility losses to the point where their effect upon the result of the analysis is a tolerable minimum.

This chapter is devoted to a discussion of the common variables that influence the solubility of precipitates—that is, temperature, pH, reagent concentration, salt concentration, and solvent composition. Some of these are readily treated in quantitative terms by means of the equilibrium law; we shall examine in considerable detail the techniques used for this purpose.

THE APPLICATION OF SOLUBILITY PRODUCT CONSTANTS

The Solubility Product Constant

In Chapter 2 we presented an expression for the solubility product. Thus for the equilibrium

$$A_xB_y \rightleftharpoons xA^{y+} + yB^{x-}$$

the solubility product constant is given by

$$K_{sp} = [A^{y+}]^x [B^{x-}]^y$$

This constant is generally applicable in defining equilibrium conditions for saturated solutions of slightly soluble strong electrolytes.

Solubility product constants have been determined for a large number of inorganic compounds; numerical values for many of these will be found in Table A-2 of the Appendix. The examples that follow demonstrate the use to which such data can be put.

Example. How many milligrams of $Ba(IO_3)_2$ can be dissolved in 150 ml of water at 25° C?

$$Ba(IO_3)_2 \rightleftharpoons Ba^{2+} + 2IO_3^- \qquad K_{sp} = 1.57 \times 10^{-9}$$

Setting x equal to the solubility of $Ba(IO_3)_2$ in formula weights per liter it follows that

$$[Ba^{2+}] = x \text{ mol/liter}$$
$$[IO_3^-] = 2x \text{ mol/liter}^1$$

Substituting these values into the expression for K_{sp}

$$[Ba^{2+}] [IO_3^-]^2 = 1.57 \times 10^{-9}$$
$$x(2x)^2 = 4x^3 = 1.57 \times 10^{-9}$$
$$x = 7.3 \times 10^{-4} \text{ formula weight/liter}$$

Multiplication of this result by the formula weight of $Ba(IO_3)_2$ in milligrams (487×1000) yields the solubility in milligrams per liter.

$$\text{solubility} = 7.3 \times 10^{-4} \times 4.87 \times 10^5 \text{ mg/liter}$$

To get the solubility in mg/150 ml we multiply by 150/1000.

$$\text{solubility} = 7.3 \times 10^{-4} \times 4.87 \times 10^5 \times \frac{150}{1000}$$
$$= 53 \text{ mg/150 ml}$$

---

[1] The student should understand clearly that the $2x$ used in this equation represents the IO_3^- concentration in the solution and *not two times the* IO_3^- *concentration*. This is sometimes a point of confusion; it is readily avoided by recalling the quantity for which x stands, it being the formal solubility of $Ba(IO_3)_2$ in this case.

A useful application of the solubility product constant is in deciding whether or not a precipitate will form under a given set of conditions.

Example. Will a $AgBrO_3$ precipitate form when equal volumes of a 0.001 F $AgNO_3$ solution and 0.02 F $KBrO_3$ solution are mixed?

The precipitation reaction will occur only if the solubility product constant is exceeded by the product of the initial Ag^+ and BrO_3^- concentrations. In this problem, the initial concentration of Ag^+ ion will be 0.0005 F and of BrO_3^- will be 0.01 F since dilution will have occurred in the mixing; the product of these concentrations is given by

$$[Ag^+][BrO_3^-] = (5 \times 10^{-4})(1 \times 10^{-2})$$
$$= 5 \times 10^{-6}$$

The solubility product for $AgBrO_3$ is 6×10^{-5} which is larger than the concentration product obtained above. Therefore we conclude that precipitation will not occur.

Example. What is the minimum silver ion concentration required to initiate the precipitation of $AgBrO_3$ from a 0.01 F solution of $KBrO_3$?

Precipitation will occur when the product of the initial concentrations of Ag^+ and BrO_3^- just exceeds the solubility product. Since we know the BrO_3^- concentration of the solution, we can calculate the equilibrium concentration of Ag^+ that corresponds to this value. This will give the maximum Ag^+ concentration that can exist in solutions with that BrO_3^- concentration.

$$[Ag^+][BrO_3^-] = 6 \times 10^{-5}$$
$$[Ag^+](0.01) = 6 \times 10^{-5}$$
$$[Ag^+] = 6 \times 10^{-3} \text{ mol/liter}$$

The Common Ion Effect

From a knowledge of the LeChatelier principle we would predict that the solubility of an electrolyte should be reduced by the presence in solution of an excess of one of the ions common to the compound. This has been amply verified experimentally. The chemist performing gravimetric analyses takes advantage of this fact by using an excess of precipitating agent to reduce solubility losses. The quantitative aspects of the common ion effect can be treated with the aid of the mass law. Some examples are given below.

Example. What is the solubility of $Ba(IO_3)_2$ in formula weights per liter in a 0.020 F solution of KIO_3? The K_{sp} for $Ba(IO_3)_2$ is 1.57×10^{-9}.

$$Ba(IO_3)_2 \rightleftharpoons Ba^{2+} + 2IO_3^-$$
$$[Ba^{2+}][IO_3^-]^2 = 1.57 \times 10^{-9}$$

We shall represent the solubility of $Ba(IO_3)_2$ in formula weights per liter by x. It follows, then, that

$$[Ba^{2+}] = x \text{ mol/liter}$$

Iodate ions arise from two sources—namely, the KIO_3 and the $Ba(IO_3)_2$. The contribution from the former is $0.02 F$ and the latter $2x$. The total iodate concentration is the sum of these

$$[IO_3^-] = (0.02 + 2x) \text{ mol/liter}$$

Substituting these quantities into the solubility product expression we get

$$x(0.02 + 2x)^2 = 1.57 \times 10^{-9}$$

Since the exact solution for x will involve a cubic equation, it is worthwhile making some approximations in order to simplify the algebra. The solubility of $Ba(IO_3)_2$ is not very great, as indicated by the small numerical value of K_{sp}; therefore, it is reasonable to suppose that the concentration of IO_3^- due to solution of the $Ba(IO_3)_2$ is small relative to the iodate concentration from the KIO_3. If we assume that $2x$ is negligible relative to 0.02, that is,

$$(0.02 + 2x) \cong 0.02$$

the expression then becomes much simpler

$$x(0.02)^2 = 1.57 \times 10^{-9}$$

$$x = \frac{1.57 \times 10^{-9}}{4 \times 10^{-4}} = 3.9 \times 10^{-6} \text{ formula weight/liter}$$

We should now go back and check our assumption.

$$(0.02 + 2 \times 3.9 \times 10^{-6}) \cong 0.02$$

We see that it was quite reasonable and can conclude that the approximate solution is valid.

It is interesting to compare the solubility of barium iodate in $0.02 F$ potassium iodate with the solubility in pure water, calculated previously to be 7.3×10^{-4}. The amount dissolved in water is nearly two hundred times greater, which illustrates the rather large effect the presence of a common ion can have upon solubility.

Example. Calculate the solubility of $Ba(IO_3)_2$ in the solution that results when 100 ml of $0.014 F$ $BaCl_2$ are mixed with 100 ml of $0.030 F$ KIO_3.

As a first step, we evaluate the number of milliformula weights of reactants to determine whether an excess of one will remain after reaction has occurred.

number of milliformula weights $BaCl_2 = 100 \times 0.014 = 1.4$
number of milliformula weights $KIO_3 = 100 \times 0.030 = 3.0$

If the formation of $Ba(IO_3)_2$ is complete

number of milliformula weights KIO_3 remaining $= 3.0 - 2 \times 1.4 = 0.2$

Since the volume of the mixture is 200 ml, the formal concentration of KIO_3 is

$$F_{KIO_3} = \frac{0.200}{200} = 1 \times 10^{-3} F$$

The problem now consists of determining the solubility of $Ba(IO_3)_2$ in a $1.00 \times 10^{-3} F$ solution of KIO_3. We will let x be the solubility in formula weights per liter.

Then

$$[Ba^{2+}] = x \text{ mol/liter}$$

and as before

$$[IO_3^-] = (1 \times 10^{-3} + 2x) \text{ mol/liter}$$

$$(x)(1 \times 10^{-3} + 2x)^2 = 1.57 \times 10^{-9}$$

Let us again assume that x is small; then

$$x(1 \times 10^{-3})^2 = 1.57 \times 10^{-9}$$

$$x = 1.57 \times 10^{-3}$$

If we examine the original assumption, however, that

$$(1 \times 10^{-3} + 2 \times 1.57 \times 10^{-3}) \cong 1 \times 10^{-3}$$

we see that it is no longer satisfactory. A more exact solution for the equation must be found. Rearranging the original equation to collect all terms on one side, we find that

$$x(2x + 1 \times 10^{-3})^2 - 1.57 \times 10^{-9} = 0$$

The solution is then obtained by successive approximation. If $x = 0$, the left side of the equation is negative and if $x = 1 \times 10^{-3}$, it is positive; that is

$$9 \times 10^{-9} - 1.57 \times 10^{-9} = 7.4 \times 10^{-9}$$

Therefore the correct solution is somewhat smaller than 1×10^{-3}. Let $x = 0.5 \times 10^{-3}$, which gives

$$2 \times 10^{-9} - 1.57 \times 10^{-9} = 0.4 \times 10^{-9}$$

and while this is closer to a true solution, the correct value of x is still smaller than 0.5×10^{-3}. Assuming $x = 0.4 \times 10^{-3}$, we find

$$1.3 \times 10^{-9} - 1.57 \times 10^{-9} = -0.3 \times 10^{-9}$$

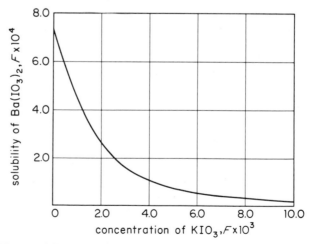

Fig. 6.1 The Theoretical Solubility of Barium Iodate in Solutions of Potassium Iodate.

Since the algebraic sum is now negative, x must lie between 0.4×10^{-3} and 0.5×10^{-3}. By substituting further values for x between these limits, we find that

$$x = 0.44 \times 10^{-3} \text{ formula weight/liter}$$

These calculations suggest that the solubility of a precipitate theoretically decreases rapidly and approaches zero at high concentrations of the common ion. In Figure 6.1 the calculated solubility of barium iodate is plotted against the concentration of iodate ion. From such a plot it might be concluded that use of the largest possible excess of precipitating agent is desirable in order to reduce solubility losses in the analysis to the very minimum. Actually, however, this is not the case because at high concentrations of the common ion other factors start to operate which result in marked departures from the theoretical behavior. As a consequence, actual increases in solubility are often observed. These factors, to be discussed in detail later in this chapter, include complex formation and alteration of the activity coefficients of the ions.

We may summarize by stating that a moderate excess of the precipitating reagent is commonly used to reduce solubility losses in gravimetric analysis, but large excesses generally have a deleterious effect and are avoided.

Effect of *p*H on Solubility

Types of compounds affected by *p*H. The solubilities of most of the precipitates of importance in quantitative analysis are affected by the hydrogen or hydroxyl ion concentration of the solvent. Two types of *p*H dependence can be distinguished. The first of these is a simple common ion effect in which either hydroxyl ion or hydrogen ion is one of the constituents of the precipitate. An example is given below:

$$Mg(OH)_2 \rightleftharpoons Mg^{2+} + 2OH^-$$

In the case of magnesium hydroxide, the solubility decreases with increasing hydroxyl ion concentration, and the solubility at any particular *p*H can be readily calculated by use of the techniques described in the previous section.

The second type of dependence is more complicated and is encountered when the anions or cations of the precipitate are capable of reacting with hydrogen or hydroxyl ions of the aqueous solvent. As an example of this class of compounds we will choose calcium fluoride, a salt of the weak acid, hydrogen fluoride. The solubility of this compound increases with increasing acidity as a result of the following reactions:

$$CaF_2 \rightleftharpoons Ca^{2+} + 2F^-$$
$$+$$
$$2H^+$$
$$\updownarrow$$
$$2HF$$

By the LeChatelier principle, we predict that increases in the hydrogen ion concentration of the solution will result in an increase in the hydrogen fluoride concentration and a decrease in the fluoride ion concentration. The latter change, however, is partially offset by a shift to the right in the solubility equilibrium; the net effect is an increase in the solubility of the precipitate.

Most of the compounds in which we will be interested are salts of weak acids which behave in a manner similar to calcium fluoride. Exceptions are the hydroxides of several of the metallic ions and the salts of strong acids—notably some of the halides and oxyhalides. Quantitative treatment of the solubility of the salts of weak acids requires the consideration of more than one equilibrium and becomes considerably more complex than any of the cases so far encountered.

Quantitative treatment of multiple equilibria. We shall frequently be faced with the problem of calculating the concentration of constituents in a solution in which several competing equilibrium reactions are occurring simultaneously. The dissolution of calcium fluoride in an aqueous solution, cited above, is an example. As a more general case we shall consider the hypothetical precipitate AB which dissolves to give A and B ions.

$$
\begin{array}{ccc}
AB \rightleftharpoons & A & + & B \\
 & + & & + \\
 & C & & D \\
 & \updownarrow & & \updownarrow \\
 & AC & & BD
\end{array}
$$

If A and B are capable of reacting with species C and D to form the soluble substances AC and BD, the introduction of either C or D into the solution will cause a shift in the solubility equilibrium in such a direction as to increase the solubility of AB.

The calculation of the solubility of AB in a case such as this requires information about the formal concentrations of C and D as well as equilibrium constants for all three of the equilibria. Generally several algebraic expressions are needed to describe completely the concentration relationships in such a case; not infrequently the chemist finds the solving of these algebraic equations a more formidable task than setting them up.

One point that should be constantly borne in mind by the student when dealing with multiple equilibria is that the validity and form of a given equilibrium-constant expression is in no way affected by the existence of additional competing equilibria in the solution. For example, in the present case the solubility product expression for AB describes the relationship between the concentrations of A and B in the solution regardless of whether or not C and D are present; that is, the product $[A]$ $[B]$ is a constant under any conditions as long as any solid AB is present. The amount of AB that dissolves is greater in the presence of C or D, but this is not because the product of the concentrations of A and B has changed but rather because some of the precipitate has been converted to AC and BD.

In the following paragraphs we shall endeavor to present a systematic approach by which the student can attack any equilibrium problem involving several equilibria. First we will set down the several steps to be followed and then apply the method to some solubility problems.

Method for the solution of problems involving several equilibria.

1. Write down chemical equations for all of the reactions that appear to have any bearing on the problem.

2. State in terms of equilibrium concentrations what is being sought in the problem.

3. Write equilibrium-constant expressions for all of the equilibria shown in (1).

4. Write mass-balance equations for the system. These are algebraic expressions relating the equilibrium concentrations of the various species in the solution to one another and to the formal concentrations of substances in the solution.

5. Write a charge-balance equation. In any solution the concentrations of the cations and anions must be related to one another in such a way that the solution is electrically neutral. The charge-balance equation expresses this relationship.

6. Count the number of unknown quantities in the equations from steps (3), (4), and (5) and compare with the number of independent equations. If the number of equations is equal to the number of unknown concentrations, the problem can be solved exactly by suitable algebraic manipulations. If there are fewer equations than unknowns, try to see if additional equations can be derived. If this is not possible, it must be concluded that an exact solution to the problem is not possible; it may, however, be possible to arrive at an approximate solution.

7. Make suitable approximations to simplify the algebra or to reduce the number of unknowns so that the problem can be solved.

8. Solve the algebraic equations for those equilibrium concentrations that are necessary to the solution of the problem as given in step (2).

9. With the equilibrium concentrations obtained in (8), check the approximations made in (7) to be sure of their validity.

Step (6) in the above scheme is particularly significant because it indicates whether an exact solution for the problem is theoretically feasible. If the number of equations is as great as the number of unknowns, then the problem becomes the purely mechanical one of obtaining an algebraic solution to several simultaneous equations. On the other hand, if the number of equations is less than the number of unknowns, a search for further equations or an approximation that will reduce the number of unknowns is essential. The student should never undertake the algebraic solution of a complex equilibrium problem without first assuring himself that he has sufficient data; otherwise he may waste a great deal of time to no avail.

Calculation of the solubility of a salt of a weak acid in a solution of known hydrogen ion concentration.

Example. Calculate the solubility of CaC_2O_4 in formula weights per liter in a solution maintained at a hydrogen ion concentration of 1.0×10^{-4}.

Step 1. *Chemical equations.*

$$CaC_2O_4 \rightleftharpoons Ca^{2+} + C_2O_4^{2-} \qquad (6\text{-}1)$$

Since oxalic acid is a weak acid, the oxalate ions will react in part with the hydrogen ions added to maintain the specified hydrogen ion concentration.

$$C_2O_4^{2-} + H^+ \rightleftharpoons HC_2O_4^- \qquad (6\text{-}2)$$
$$HC_2O_4^- + H^+ \rightleftharpoons H_2C_2O_4 \qquad (6\text{-}3)$$

Step 2. *Definition of the unknown.* What is sought? We wish to know the solubility of CaC_2O_4 in formula weights per liter. Since the CaC_2O_4 is ionic, the solubility will be equal to the molar concentration of calcium ion; it is also equal to the sum of the equilibrium concentrations of the oxalate species; that is,

$$\text{solubility} = [Ca^{2+}]$$
$$= [C_2O_4^{2-}] + [HC_2O_4^-] + [H_2C_2O_4]$$

Thus if we can calculate either of these quantities, we will have obtained a solution to the problem.

Step 3. *Equilibrium-constant expressions.*

$$K_{sp} = [Ca^{2+}][C_2O_4^{2-}] = 1.9 \times 10^{-9} \qquad (6\text{-}4)$$

Equation (6-2) above is simply the reverse of the dissociation reaction for $HC_2O_4^-$; we can thus use the value of K_2 for oxalic acid.

$$K_2 = \frac{[H^+][C_2O_4^{2-}]}{[HC_2O_4^-]} = 6.1 \times 10^{-5} \qquad (6\text{-}5)$$

Similarly for (6-3)

$$K_1 = \frac{[H^+][HC_2O_4^-]}{[H_2C_2O_4]} = 6.2 \times 10^{-2} \qquad (6\text{-}6)$$

Step 4. *Mass-balance equations.* Since the only source of Ca^{2+} and the various oxalate species is the dissolved CaC_2O_4, it follows that

$$[Ca^{2+}] = [C_2O_4^{2-}] + [HC_2O_4^-] + [H_2C_2O_4] \qquad (6\text{-}7)$$

Furthermore it is given that at equilibrium

$$[H^+] = 1.0 \times 10^{-4} \qquad (6\text{-}8)$$

Step 5. *Charge-balance equations.* A useful charge-balance equation cannot be written in this case because there has been added to this solution some unknown acid HX to maintain the H^+ concentration at 1.0×10^{-4} mol per liter. In order to write an equation based on the electrical neutrality of the solution, it would be necessary to include the concentration of the anions of the unknown acid $[X^-]$. An equation containing this additional unknown term would be of no help to us.

Step 6. *Comparison of equations and unknowns.* We have four unknowns— namely, $[Ca^{2+}]$, $[C_2O_4^{2-}]$, $[HC_2O_4^-]$, and $[H_2C_2O_4]$. We also have four independent equations : (6-4), (6-5), (6-6), and (6-7) above. Therefore an exact solution is possible and the problem has now become one of algebra.

Step 7. *Approximations.* We are not forced to make any approximations since we have enough data available; we shall attempt an exact solution to the problem.

Step 8. *Solution of the equations.* A convenient way to effect a solution is to make suitable substitutions into equation (6-7) thereby establishing a relationship between $[Ca^{2+}]$ and $[C_2O_4^{2-}]$. We must first derive an expression for $[HC_2O_4^-]$ and $[H_2C_2O_4]$ in terms of $[C_2O_4^{2-}]$. This is done as follows. Substituting the value of 1.0×10^{-4} for $[H^+]$ in equation (6-5), we get

$$\frac{(1.0 \times 10^{-4})\,[C_2O_4^{2-}]}{[HC_2O_4^-]} = 6.1 \times 10^{-5}$$

and

$$[HC_2O_4^-] = \frac{1.0 \times 10^{-4}}{6.1 \times 10^{-5}}\,[C_2O_4^{2-}] = 1.64\,[C_2O_4^{2-}]$$

Substituting the above into equation (6-6)

$$\frac{(1.0 \times 10^{-4})\,(1.64[C_2O_4^{2-}])}{[H_2C_2O_4]} = 6.2 \times 10^{-2}$$

$$[H_2C_2O_4] = \frac{1.0 \times 10^{-4} \times 1.64[C_2O_4^{2-}]}{6.2 \times 10^{-2}} = 0.0026[C_2O_4^{2-}]$$

Substituting these values for $[H_2C_2O_4]$ and $[HC_2O_4^-]$ into equation (6-7)

$$[Ca^{2+}] = [C_2O_4^{2-}] + 1.64\,[C_2O_4^{2-}] + 0.0026\,[C_2O_4^{2-}]$$
$$= 2.64\,[C_2O_4^{2-}]$$

or

$$[C_2O_4^{2-}] = \frac{[Ca^{2+}]}{2.64}$$

Substituting into equation (6-4) gives

$$[Ca^{2+}]\frac{[Ca^{2+}]}{2.64} = 1.9 \times 10^{-9}$$

$$[Ca^{2+}]^2 = 5.02 \times 10^{-9}$$

$$[Ca^{2+}] = 7.1 \times 10^{-5}\ \text{mol/liter}$$

and from step (2) we conclude

solubility of $CaC_2O_4 = 7.1 \times 10^{-5}$ formula weight/liter

Calculation of the solubility of a salt of a weak acid in water. The calculation of the solubility of a slightly soluble salt of a weak acid in pure water is a considerably more difficult problem than the case where the final hydrogen ion concentration of the solution is fixed. For example, a saturated solution of AB will contain the following equilibria:

$$AB \rightleftharpoons A^+ + B^-$$
$$+$$
$$H_2O$$
$$\updownarrow$$
$$HB + OH^-$$

Here B^- is the anion of a weak acid and it will therefore be hydrolyzed to some degree, giving HB and hydroxyl ions. In this case the hydrogen ion concentration of the solution changes as a result of the solubility of the precipitate; thus, a new variable is introduced into the solubility calculation.

In many cases the hydrolytic reaction cannot be neglected in the solubility calculation without causing rather serious errors. The magnitude of the errors will depend both upon the solubility product of the precipitate and the hydrolysis constant for the anion. This is demonstrated in Table 6-1 where the theoretical solubility of a compound AB has been calculated assuming different dissociation constants for HB and therefore different hydrolysis constants for B^-. Two solubility products, 1×10^{-10} and 1×10^{-20}, have been assumed for AB. If hydrolysis is neglected, the solubilities calculated from these constants would be 1×10^{-5} and 1×10^{-10} formula weight of AB per liter; these should be compared with the values in column 4 of Table 6-1 which were obtained by

Table 6-1

CALCULATED SOLUBILITY OF AB FROM VARIOUS ASSUMED VALUES OF K_{sp} AND K_h

Solubility Product Assumed for AB	Dissociation Constant Assumed for HB	Hydrolysis Constant for B^-	Calculated Solubility of AB, formula wt/liter	Calculated Solubility of AB Neglecting Hydrolysis, formula wt/liter
1.0×10^{-10}	1.0×10^{-6}	1.0×10^{-8}	1.03×10^{-5}	1.0×10^{-5}
	1.0×10^{-8}	1.0×10^{-6}	1.2×10^{-5}	1.0×10^{-5}
	1.0×10^{-10}	1.0×10^{-4}	2.4×10^{-5}	1.0×10^{-5}
	1.0×10^{-12}	1.0×10^{-2}	10×10^{-5}	1.0×10^{-5}
1.0×10^{-20}	1.0×10^{-6}	1.0×10^{-8}	1.05×10^{-10}	1.0×10^{-10}
	1.0×10^{-8}	1.0×10^{-6}	3.3×10^{-10}	1.0×10^{-10}
	1.0×10^{-10}	1.0×10^{-4}	32×10^{-10}	1.0×10^{-10}
	1.0×10^{-12}	1.0×10^{-2}	290×10^{-10}	1.0×10^{-10}

taking into account the hydrolysis of the anion. The reader can see that when the hydrolytic reaction is neglected, low values for the solubility of the precipitates are obtained; with the more insoluble precipitate the magnitude of the error becomes larger. A study of Table 6-1 will give the student some idea of the circumstances under which the hydrolysis equilibrium must be taken into account in a solubility calculation.

It is not difficult to write the algebraic equations needed to calculate the solubility of a precipitate of the sort we have been discussing; the exact solution of these, however, is often tedious. An example of such a problem is given below :

Example. Calculate the solubility of $PbCO_3$ in water. The equilibrium reactions that have a bearing on the problem are

$$PbCO_3 \rightleftharpoons Pb^{2+} + CO_3^{2-} \tag{6-9}$$

$$CO_3^{2-} + H_2O \rightleftharpoons HCO_3^- + OH^- \tag{6-10}$$

$$HCO_3^- + H_2O \rightleftharpoons H_2CO_3 + OH^- \tag{6-11}$$

$$H_2O \rightleftharpoons H^+ + OH^- \tag{6-12}$$

The solubility of $PbCO_3$ can be expressed as follows :

$$\text{solubility} = [Pb^{2+}]$$
$$= [CO_3^{2-}] + [HCO_3^-] + [H_2CO_3]$$

The equilibrium expressions are

$$[Pb^{2+}][CO_3^{2-}] = 1.6 \times 10^{-13} \tag{6-13}$$

$$\frac{[HCO_3^-][OH^-]}{[CO_3^{2-}]} = \frac{K_w}{K_2} = \frac{1.0 \times 10^{-14}}{4.4 \times 10^{-11}} = 2.3 \times 10^{-4} \tag{6-14}^2$$

$$\frac{[H_2CO_3][OH^-]}{[HCO_3^-]} = \frac{K_w}{K_1} = \frac{1.0 \times 10^{-14}}{4.6 \times 10^{-7}} = 2.2 \times 10^{-8} \tag{6-15}^2$$

and

$$[H^+][OH^-] = 1.0 \times 10^{-14} \tag{6-16}$$

The mass-balance expression is

$$[Pb^{2+}] = [CO_3^{2-}] + [HCO_3^-] + [H_2CO_3] \tag{6-17}$$

In this case an expression based on the electrical neutrality of the solution can also be written[3]

$$2[Pb^{2+}] + [H^+] = 2[CO_3^{2-}] + [HCO_3^-] + [OH^-] \tag{6-18}$$

[2] The hydrolysis constants have a numerical value equal to K_w/K_d, where K_d is the dissociation constant of the weak acid formed by the hydrolysis (see p. 20). In the first hydrolysis step HCO_3^- is the product; the hydrolysis constant is therefore numerically equal to K_w/K_2, where K_2 is the second dissociation constant for H_2CO_3. The second hydrolysis step gives H_2CO_3 and therefore the constant for this reaction is K_w/K_1.

[3] The concentrations of the doubly charged species are multiplied by two in this expression. If there were triply charged ions, their concentrations would be multiplied by three. The student can best understand this by considering a simple case such as a solution of $BaCl_2$. Here electrical neutrality is preserved because there are two chloride ions for each barium ion in the solution. In other words, to preserve electrical neutrality, the concentration of chloride ions must be twice the concentration of the doubly charged barium ions and we would write $[Cl^-] = 2[Ba^{2+}]$. By analogous reasoning, we conclude that the concentration of both the lead ions and carbonate ions must be multiplied by two in the more complicated case above.

We now have six equations and also six unknowns — namely $[Pb^{2+}]$, $[CO_3^{2-}]$, $[HCO_3^-]$, $[H_2CO_3]$, $[OH^-]$, and $[H^+]$. In theory the problem can be made to yield an exact solution. If we attempt to solve the six equations, however, real difficulty develops insofar as the algebraic manipulations are concerned. Therefore we shall seek an easier way out by attempting some approximations. One fairly obvious assumption is that the concentration of hydrogen ions relative to the other ions is small enough to be neglected. This will be valid if reactions (6-10) and (6-11) proceed to a sufficient extent. A second assumption might be that the second hydrolysis step (the formation of H_2CO_3) is relatively unimportant compared with the first hydrolysis and therefore the concentration of H_2CO_3 is much smaller than the concentration of HCO_3^-. There is some basis for making an assumption such as this since the second hydrolysis constant is but 1/10000 as large as the first.

We shall assume then that $[H^+]$ is smaller than any of the terms in equation (6-18) and can therefore be neglected and also that

$$[HCO_3^-] \gg [H_2CO_3]$$

Equation (6-17) then simplifies to

$$[Pb^{2+}] = [CO_3^{2-}] + [HCO_3^-] \tag{6-19}$$

and equation (6-18) becomes

$$2[Pb^{2+}] = 2[CO_3^{2-}] + [HCO_3^-] + [OH^-] \tag{6-20}$$

Furthermore we will no longer need equations (6-15) and (6-16). Thus we have reduced the number of equations and unknowns to four.

If we multiply equation (6-19) by two and subtract this from (6-20), we obtain

$$0 = [OH^-] - [HCO_3^-]$$

or

$$[OH^-] = [HCO_3^-] \tag{6-21}$$

Substituting this into equation (6-14), we obtain

$$\frac{[HCO_3^-]^2}{[CO_3^{2-}]} = \frac{K_w}{K_2}$$

$$[HCO_3^-] = \sqrt{\frac{K_w}{K_2} [CO_3^{2-}]}$$

With this we can eliminate $[HCO_3^-]$ from equation (6-19)

$$[Pb^{2+}] = [CO_3^{2-}] + \sqrt{\frac{K_w}{K_2} [CO_3^{2-}]} \tag{6-22}$$

From equation (6-13) we have

$$[CO_3^{2-}] = \frac{K_{sp}}{[Pb^{2+}]}$$

Substituting into (6-22) we find

$$[Pb^{2+}] = \frac{K_{sp}}{[Pb^{2+}]} + \sqrt{\frac{K_w K_{sp}}{K_2 [Pb^{2+}]}}$$

Multiplying through by $[Pb^{2+}]$ and rearranging terms

$$[Pb^{2+}]^2 - \sqrt{\frac{K_w}{K_2}} K_{sp} [Pb^{2+}] - K_{sp} = 0$$

Substituting numerical values for the constants we find

$$[Pb^{2+}]^2 - 6.1 \times 10^{-9} [Pb^{2+}]^{\frac{1}{2}} - 1.6 \times 10^{-13} = 0$$

We can readily obtain $[Pb^{2+}]$ by successive approximations, which gives

$$[Pb^{2+}] - 3.3 \times 10^{-6} \text{ mol per liter}$$

and

$$\text{solubility of PbCO}_3 = 3.3 \times 10^{-6} \text{ formula weight/liter}$$

To check the two assumptions that were made, we must calculate the concentrations of most of the other ions in the solution. We can get $[CO_3^{2-}]$ from equation (6-13)

$$[CO_3^{2-}] = \frac{1.6 \times 10^{-13}}{3.3 \times 10^{-6}} = 4.8 \times 10^{-8}$$

From equation (6-19)

$$[HCO_3^-] = 3.3 \times 10^{-6} - 4.8 \times 10^{-8}$$
$$\cong 3.3 \times 10^{-6}$$

From equation (6-21)

$$[OH^-] = [HCO_3^-] = 3.3 \times 10^{-6}$$

From equation (6-15)

$$\frac{[H_2CO_3] [3.3 \times 10^{-6}]}{[3.3 \times 10^{-6}]} = 2.2 \times 10^{-8}$$
$$[H_2CO_3] = 2.2 \times 10^{-8}$$

and finally from equation (6-16)

$$[H^+] = \frac{1.0 \times 10^{-14}}{3.3 \times 10^{-6}} = 3.0 \times 10^{-9}$$

We see that the assumptions should not lead to large errors; the $[H_2CO_3]$ is approximately $1/150$ of the $[HCO_3^-]$ and the $[H^+]$ is about $1/32$ of $2[CO_3^{2-}]$, the next smallest term in equation (6-18).

Finally, if we failed to take into account the hydrolysis reaction, we would have obtained a solubility of 4×10^{-7} which is only about one eighth the value obtained by the above method.

Complex Ion Formation and Solubility

The solubility of a precipitate may be greatly altered by the presence in solution of an ion that will react with the anion or cation of the precipitate to form a soluble complex. For example, the precipitation of aluminum with base is never complete in the presence of fluoride ion although the aluminum hydroxide precipitate formed is an extremely insoluble substance. The explanation for this is found in the fact that the aluminum ion forms sufficiently stable fluoride complexes to prevent quantitative removal of the cation from solution. The equilibria involved can be represented by the following equations:

$$Al(OH)_3 \rightleftharpoons Al^{3+} + 3OH^-$$
$$+$$
$$6F^-$$
$$\Updownarrow$$
$$AlF_6^{3-}$$

Fluoride ions compete with hydroxyl ions for the aluminum cation; as the fluoride concentration becomes larger, more and more of the precipitate is dissolved and converted to fluoroaluminate ions.

Quantitative treatment of the effect of complex formation on the solubility of precipitates. The solubility of a precipitate in the presence of various concentrations of a complexing reagent can be calculated provided the equilibrium constant for the complex formation reaction is known. The techniques used are similar to those discussed in the previous section.

Example. Find the solubility of AgCl in a solution that is $0.100\,F$ in NH_3.

Equilibria

$$AgCl \rightleftharpoons Ag^+ + Cl^-$$
$$Ag^+ + 2NH_3 \rightleftharpoons Ag(NH_3)_2^+$$
$$NH_3 + H_2O \rightleftharpoons NH_4^+ + OH^-$$

Solubility of AgCl $= [Cl^-]$
$$= [Ag^+] + [Ag(NH_3)_2^+]$$

Equilibrium constants

$$[Ag^+][Cl^-] = K_{sp} = 1.82 \times 10^{-10} \tag{6-23}$$

$$\frac{[Ag^+][NH_3]^2}{[Ag(NH_3)_2^+]} = K_{inst.} = 6.3 \times 10^{-8} \tag{6-24}$$

$$\frac{[NH_4^+][OH^-]}{[NH_3]} = K_d = 1.86 \times 10^{-5} \tag{6-25}$$

Mass-balance expressions

$$[Cl^-] = [Ag^+] + [Ag(NH_3)_2^+] \tag{6-26}$$

Since the NH_3 concentration was initially 0.1, we may write

$$0.1 = [NH_3] + 2[Ag(NH_3)_2^+] + [NH_4^+] \tag{6-27}$$

Furthermore

$$[NH_4^+] = [OH^-] \tag{6-28}$$

Charge-balance equation

$$[NH_4^+] + [Ag^+] + [Ag(NH_3)_2^+] = [Cl^-] + [OH^-] \tag{6-29}^4$$

---

[4] We have neglected the $[H^+]$ since its concentration will certainly be negligible in a $0.1\,F$ solution of NH_3.

A close examination of the above seven equations indicates that there are only six independent expressions since (6-29) is the sum of (6-28) and (6-26). There are only six unknowns, however, so a solution is possible.

Approximations

(a) $[NH_4^+]$ is much smaller than the other terms in equation (6-27). This seems reasonable in light of the rather small numerical value of (6-25).

(b) $[Ag(NH_3)_2^+] \gg [Ag^+]$. An examination of the constant for equation (6-24) shows that this is a good assumption except for very dilute solutions of NH_3.

These approximations lead to the simplified equations

$$[Cl^-] = [Ag(NH_3)_2^+] \tag{6-30}$$

$$[NH_3] = 0.1 - 2[Ag(NH_3)_2^+] \tag{6-31}$$

and substituting (6-30) into (6-31)

$$[NH_3] = 0.1 - 2[Cl^-] \tag{6-32}$$

Introducing (6-32) and (6-30) into (6-24)

$$\frac{[Ag^+](0.1 - 2[Cl^-])^2}{[Cl^-]} = 6.3 \times 10^{-8}$$

Now replacing the $[Ag^+]$ in this equation by the equivalent quantity from (6-23) we have

$$\frac{\left(\dfrac{1.82 \times 10^{-10}}{[Cl^-]}\right)(0.1 - 2[Cl^-])^2}{[Cl^-]} = 6.3 \times 10^{-8}$$

$$\frac{(0.1 - 2[Cl^-])^2}{[Cl^-]^2} = 3.5 \times 10^2$$

$$350[Cl^-]^2 = 0.01 - 0.4[Cl^-] + 4[Cl^-]^2$$

This can be rearranged to the following:

$$[Cl^-]^2 + 1.16 \times 10^{-3}[Cl^-] - 2.89 \times 10^{-5} = 0$$

Solving the quadratic equation we obtain

$$[Cl^-] = 4.8 \times 10^{-3} \text{ mol/liter}$$

$$\text{solubility} = 4.8 \times 10^{-3} \text{ formula weight AgCl/liter}$$

A check on the assumptions will indicate that they were valid.

Complex formation involving a common ion of the precipitate. Many precipitates have a tendency to react with one of their constituent ions to form soluble complexes. An example of this is found in the case of silver chloride which is capable of reacting with chloride ions to form complexes believed to be of the composition $AgCl_2^-$, $AgCl_3^{2-}$, etc. The result of this is to counteract the common ion effect and to give rise to increases in solubility at high concentrations of the common ion. This is shown by Figure 6.2 where the experimentally determined solubility of silver chloride is plotted against the log of the potassium chloride concentration of the solution. At chloride concentrations less than $10^{-3} F$, the data found by experiment do not differ greatly from solubilities calculated from the solubility product for silver chloride. At higher concentrations of the salt, however, the calculated solubilities approach zero, while the measured values rise very rapidly. At about 0.3 F potassium chloride, the solubility of silver chloride is the same as in pure water, and in 1 F solution it is approximately eight times this figure. If complete information were available regarding the composition of the complexes and their stability constants, a quantitative description of these effects should be possible.

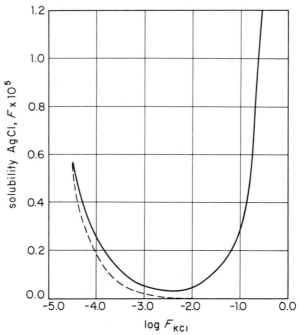

Fig. 6.2 The Solubility of Silver Chloride in Solutions of Potassium Chloride. The dotted line is based on calculation; the solid line represents experimental values of A. Pinkus and A. M. Timmermans, *Bull. Soc. Chim. Belg.*, **46**, 46-73 (1937).

Solubility increases in the presence of large excesses of a common ion are by no means rare. A case of particular interest is that of the amphoteric hydroxides such as those of aluminum and zinc. Upon treatment with base these ions form insoluble precipitates; these redissolve in the presence of excess hydroxyl ions to give the complex aluminate and zincate ions. Equilibria for aluminum are:

$$Al^{3+} + 3OH^- \rightleftharpoons Al(OH)_3$$
$$Al(OH)_3 + OH^- \rightleftharpoons Al(OH)_4^-$$

As with silver chloride, the solubilities of aluminum hydroxide and zinc hydroxide pass through minima and then increase rapidly with increasing concentrations of the common ion. The hydroxyl ion concentration corresponding to the minimum solubility can be calculated readily if the equilibrium constants for the reactions are known.

Example. At what OH concentration is the solubility of $Zn(OH)_2$ a minimum?

$$Zn(OH)_2 \rightleftharpoons Zn^{2+} + 2OH^-$$

$$Zn(OH)_2 + 2OH^- \rightleftharpoons Zn(OH)_4^{2-}$$

Let x be the formal solubility of $Zn(OH)_2$. Then

$$x = [Zn^{2+}] + [Zn(OH)_4^{2-}] \tag{6-33}$$

The equilibrium constants for both reactions have been determined and are

$$K_{sp} = [Zn^{2+}][OH^-]^2 = 4.5 \times 10^{-17} \tag{6-34}$$

$$K = \frac{[Zn(OH)_4^{2-}]}{[OH^-]^2} = 0.13 \tag{6-35}$$

Substituting (6-34) and (6-35) into (6-33) we find

$$x = \frac{K_{sp}}{[OH^-]^2} + K[OH^-]^2 \tag{6-36}$$

Now the OH^- concentration corresponding to the minimum solubility can be obtained by setting the derivative of x with respect to $[OH^-]$ equal to zero. Differentiation of equation (6-36) gives

$$\frac{dx}{d[OH^-]} = -\frac{2K_{sp}}{[OH^-]^3} + 2K[OH^-]$$

When $\dfrac{dx}{d[OH^-]} = 0$

$$\frac{2K_{sp}}{[OH^-]^3} = 2K[OH^-]$$

$$[OH^-] = \left(\frac{K_{sp}}{K}\right)^{1/4}$$

$$= \left(\frac{4.5 \times 10^{-17}}{0.13}\right)^{1/4}$$

$$= 1.4 \times 10^{-4} \text{ mol/liter}$$

In general an excess of a precipitating agent is to be recommended in most gravimetric methods of analysis in order to keep down losses of precipitate by solubility. On the other hand, because of the possibility of complex formation between the precipitate and the common ion, large concentrations are to be avoided.

Separation of Ions by Control of the Concentration of the Precipitating Reagent

When two ions form precipitates of different solubilities with a reagent, the more insoluble of these will form at a lower concentration of the precipitating reagent. If the solubilities are sufficiently different, the quantitative removal of the more insoluble ion from solution may be achieved without the precipitation of the second ion. Such separations require careful control of the concentration of the precipitating reagent at some suitable, predetermined level. A number of important analytical procedures, notably those involving metallic sulfides and hydroxides, are based on this method.

Calculation of the feasibility of separations. An important application of solubility product calculations involves determining the feasibility and the optimum condition for separations based on the control of reagent concentration. The following problem will serve to illustrate such an application.

Example. Is it theoretically possible to separate Fe^{3+} and Mg^{2+} quantitatively from one another in a solution which is 0.1 F in each ion by differential precipitation with OH^-? Assuming that separation is possible, we wish to find the range of OH^- concentration that can be used. Solubility products for the two hydroxides are

$$[Fe^{3+}] [OH^-]^3 = 1.5 \times 10^{-36}$$
$$[Mg^{2+}] [OH^-]^2 = 5.9 \times 10^{-12}$$

The fact that the K_{sp} for $Fe(OH)_3$ is so much smaller than that for $Mg(OH)_2$ indicates that the former will precipitate at a much lower OH^- concentration.

We can answer the questions posed above (1) by calculating the OH^- concentration required to effect the quantitative precipitation of Fe^{3+} from this solution, and (2) by determining the OH^- concentration at which $Mg(OH)_2$ will just begin to precipitate. If (1) is smaller than (2), a separation is feasible and the range of OH^- concentrations to be used will be between the values obtained in (1) and (2).

In order to determine (1), we must first decide what constitutes a quantitative removal of Fe^{3+} from the solution. Under no conditions will we be able to precipitate every ferric ion; we must quite arbitrarily set some limit below which we can, for all practical purposes, neglect the presence of this ion. We shall assume, then, that the precipitation is quantitative when the Fe^{3+} concentration has been reduced to 10^{-6} F. At this level only 1/100,000 of the original quantity of iron will remain in the solution and for most purposes this is certainly a quantitative separation.

We can readily calculate the OH^- concentration in equilibrium with Fe^{3+} having a concentration of 1.0×10^{-6} F by substituting directly into the solubility product expression.

$$(1.0 \times 10^{-6}) [OH^-]^3 = 1.5 \times 10^{-36}$$
$$[OH^-] = 1.1 \times 10^{-10} \text{ mol/liter}$$

Thus, if we maintain the OH^- concentration at 1.1×10^{-10} mol/liter, the

Fe^{3+} concentration will be 1.0×10^{-6} mol/liter. It is of interest to note that quantitative precipitation of $Fe(OH)_3$ can be accomplished in a distinctly acidic solution.

We must now consider question (2) — that is, the determination of the maximum OH^- concentration that can exist in solution without causing formation of $Mg(OH)_2$. Precipitation will occur when the Mg^{2+} concentration multiplied by the square of the OH^- concentration exceeds the solubility product, 5.9×10^{-12}. On the other hand no precipitate formation will take place if the product of these quantities is equal to or less than this number. By substituting 0.1, the molar Mg^{2+} concentration of the solution, into the solubility product expression we can calculate the *maximum* OH^- concentration which can be attained without formation of $Mg(OH)_2$.

$$0.1 \, [OH^-]^2 = 5.9 \times 10^{-12}$$
$$[OH^-] = 7.7 \times 10^{-6} \text{ mol/liter}$$

When the OH^- concentration exceeds this figure, the solution will be super-saturated with respect to $Mg(OH)_2$ and precipitation will begin.

From these calculations we conclude that quantitative separation of $Fe(OH)_3$ can be expected if the OH^- concentration is greater than 1.1×10^{-10} mol/liter, and that $Mg(OH)_2$ will not precipitate until a concentration of 7.7×10^{-6} mol/liter is reached. Therefore it should be possible to separate Fe^{3+} from Mg^{2+} by maintaining the OH^- concentration between these figures.

Sulfide separations. A number of important methods for the separation of metallic ions are based on control of the concentration of the precipitating anion by regulating the hydrogen ion concentration of the solution. Such methods are particularly attractive because of the relative ease with which the hydrogen ion concentration may be controlled and maintained at some prede-termined level.

Perhaps the most well-known methods, wherein the reagent concentration is controlled by acidity, involve the use of hydrogen sulfide as the precipitating reagent. Hydrogen sulfide is a weak acid, dissociating as follows:

$$H_2S \rightleftharpoons H^+ + HS^- \qquad K_1 = \frac{[H^+][HS^-]}{[H_2S]} = 5.7 \times 10^{-8}$$

$$HS^- \rightleftharpoons H^+ + S^{2-} \qquad K_2 = \frac{[H^+][S^{2-}]}{[HS^-]} = 1.2 \times 10^{-15}$$

These may be combined to give an expression for the over-all dissociation of hydrogen sulfide into sulfide ion as follows:

$$H_2S \rightleftharpoons 2H^+ + S^{2-} \qquad K_1K_2 = \frac{[H^+]^2[S^{2-}]}{[H_2S]} = 6.8 \times 10^{-23}$$

It can readily be seen that the constant for this reaction is simply the product of K_1 and K_2.

In sulfide separations, the solutions are ordinarily kept saturated with hydrogen sulfide so that the concentration of the reagent is essentially constant throughout the precipitation. Since hydrogen sulfide is such a weak acid, the actual molar concentration of the compound will correspond rather closely to its solubility in water, which is about 0.1 F. For practical purposes, then, we may assume that throughout any sulfide precipitation

$$[H_2S] \cong 0.1 \text{ mol/liter}$$

Substituting this value into the dissociation constant expression we obtain

$$\frac{[H^+]^2 [S^{2-}]}{0.1} = 6.8 \times 10^{-23}$$

$$[S^{2-}] = \frac{6.8 \times 10^{-24}}{[H^+]^2} \text{ mol/liter}$$

Thus, the molar concentration of the sulfide ion varies inversely as the square of the hydrogen ion concentration of the solution.

This relationship is useful for calculating the optimum conditions for the separation of cations by sulfide precipitation.

Example. We wish to find the conditions under which Pb^{2+} and Tl^+ can be separated quantitatively by H_2S precipitation from a solution 0.1 F in each ion.

The equilibrium constants for the two important reactions are

$$PbS \rightleftharpoons Pb^{2+} + S^{2-} \qquad [Pb^{2+}][S^{2-}] = 7 \times 10^{-28}$$

$$Tl_2S \rightleftharpoons 2Tl^+ + S^{2-} \qquad [Tl^{2+}]^2 [S^{2-}] = 1 \times 10^{-22}$$

PbS will obviously precipitate at lower S^{2-} concentration than the Tl^+. We shall again assume that removal from solution is quantitative when the concentration of Pb^{2+} is equal to or smaller than 10^{-6} F. Substituting this value into the solubility product expression gives the S^{2-} concentration necessary to achieve this level of Pb^{2+} concentration.

$$10^{-6}[S^{2-}] = 7 \times 10^{-28}$$

$$[S^{2-}] = 7 \times 10^{-22}$$

The S^{2-} concentration at which precipitation of Tl_2S begins from a 0.1 F solution can be obtained as follows :

$$(0.1)^2 [S^{2-}] = 1 \times 10^{-22}$$

$$[S^{2-}] = 1 \times 10^{-20}$$

Thus, to achieve a separation, the S^{2-} concentration should be kept between 7×10^{-22} and 1×10^{-20} mol/liter. Now we must compute the H^+ concentrations necessary to hold the S^{2-} concentration within these confines. We will use the relationship derived previously

$$[S^{2-}] = \frac{6.8 \times 10^{-24}}{[H^+]^2}$$

Substituting the two limiting values for S^{2-} concentration we obtain

$$[H^+]^2 = \frac{6.8 \times 10^{-24}}{7 \times 10^{-22}} = 0.97 \times 10^{-2}$$

$$[H^+] = 0.098 \cong 0.1 \text{ mol/liter}$$

and

$$[H^+]^2 = \frac{6.8 \times 10^{-24}}{1 \times 10^{-20}}$$

$$[H^+] = 0.026 \cong 0.03 \text{ mol/liter}$$

By maintaining the H^+ concentration between 0.03 and 0.1 M, it should, in theory, be possible to separate PbS without precipitation of Tl_2S.

ADDITIONAL VARIABLES WHICH AFFECT THE SOLUBILITY OF PRECIPITATES

Temperature

During the process of solution of most solids, heat is absorbed. Therefore, the solubility of precipitates generally increases with rising temperatures; that is, the solubility products for most insoluble compounds become larger numerically at higher temperatures. This is illustrated in Figure 6.3, which

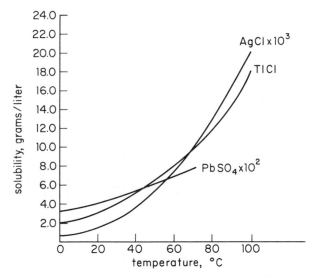

Fig. 6.3 Effect of Temperature on Solubility.

shows the solubilities of three solids as a function of temperature. It should be noted that the slopes of the solubility curves vary considerably among the three compounds. For example, the solubility of silver chloride is 10 times greater at 60° C than at 0° C while that of lead sulfate approximately doubles in the same temperature range.

Generally, it is desirable to form and wash precipitates for gravimetric analysis at elevated temperatures since this tends to enhance their purity. Where the compounds are sufficiently insoluble, the increased losses attending such a procedure do not result in serious errors. A few of the precipitates that we shall discuss are so soluble, however, that treatment in this manner is not possible; with these, the filtering and washing operations must be carried on at room temperature or even ice temperature if serious losses are to be avoided. Examples of such compounds are lead sulfate and magnesium ammonium phosphate.

Electrolyte Concentration

The solubility of most compounds is greater in solutions of an electrolyte than in pure water. This is shown in Figure 6.4 where the solubilities of three substances are plotted as a function of the potassium nitrate concentration of the solution. In the case of barium sulfate the effect is fairly large, the solubility increasing by a factor of about two in going from pure water to $0.02\ F$ potassium nitrate. On the other hand, the solubility of barium iodate increases by a factor of only 1.25 in the same range and that of silver chloride by 1.20.

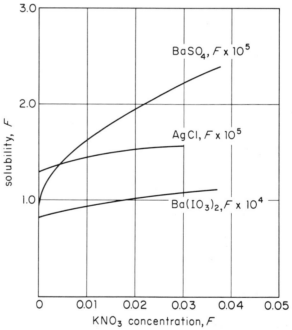

Fig. 6.4 Effect of Electrolyte Concentration on the Solubility of Some Salts.

These effects arise from an increased attraction for the ions of the solute by the dissolved electrolyte in the solvent. As pointed out in Chapter 2 (pp. 22–29), this is not a phenomenon that is unique with solubility equilibria, but can be observed in the behavior of other types of reversible reactions. For example, the degree of dissociation of weak acids is appreciably greater in the presence of electrolytes than in their absence (see Table 2-3).

The application of solubility product constants to the calculation of the solubility of a precipitate in a strong electrolyte solution may result in fairly large errors unless suitable corrections are made for the attractive forces exerted on the solute ions by the electrolyte. Such corrections take the form of activity coefficients which convert the concentrations of the solute ions to activities. We have shown by the example on page 27 how such corrections can be made in a typical case.

In general, the activity coefficients of ions with multiple charges are far more affected by variations in the electrolyte concentrations than singly charged ions. Thus the activity coefficients of silver ion and chloride ion in 0.02 F potassium nitrate are approximately 0.87 compared with the value of about 0.6 for barium and sulfate ions in the same solvent. As a result, the solubility of precipitates made up of ions with multiple charges is much more sensitive to changes in ionic strength than precipitates containing only singly charged species. This is reflected in the slopes of the three curves shown in Figure 6.4.

In all of the solubility calculations made thus far we have assumed that the activity coefficients of the ions were unity. We shall continue to do this for the sake of simplicity. The student should realize however, that this assumption can only lead to approximate results in cases where the electrolyte concentration of the solution is high.

Also, the decreases in solubility of precipitates resulting from the presence of a common ion in the solvent are in part counteracted by the concomittant increase in solubility arising from the larger electrolyte concentration associated with the presence of the salt containing the common ion. This gives added reason for the avoidance of a large excess of precipitating reagent in an analytical precipitation.

Solvent Composition

The solubility of most inorganic compounds is markedly less in aqueous solutions of organic solvents than in pure water. This is illustrated in Table 6-2, where data on the solubility of calcium sulfate as a function of the concentration of ethyl alcohol are given. Occasionally the chemist finds it necessary to take advantage of this effect to reduce losses. Examples include methods based on the precipitation forms of the alkali metals. Indiscriminate use of organic solvents, however, is avoided in gravimetric analysis inasmuch as this may also result in the reduction of the solubility of normally soluble compounds to the point where they will contaminate the precipitate.

Rate of Precipitate Formation

So far no mention has been made of the rate at which reactions approach equilibrium. This is, however, an important consideration in the formation or

Table 6-2

Solubility of Calcium Sulfate in Aqueous Ethyl Alcohol Solution

Concentration of Ethyl Alcohol, weight percent	Solubility of CaSO$_4$, gram Ca SO$_4$/100 grams solvent
0	0.208
6.2	0.100
13.6	0.044
23.8	0.014
33.0	0.0052
41.0	0.0029

T. Yamamoto. *Bull. Inst. Phys. Chem. Res.* (Tokyo), **9**, 352 (1930). W. C. Linke, *Seidell Solubilities of Inorganic Compounds*, 4th ed., Vol. I, 685. Princeton, New Jersey: D. Van Nostrand Company, Inc., 1958.

solution of certain insoluble compounds as well as in some of the other equilibrium reactions to be discussed. No conclusions can be drawn about the rate of a reaction from the magnitude of its equilibrium constant. Numerous examples can be cited of reactions with quite favorable equilibrium constants that approach equilibrium at an imperceptible rate.

Precipitation reactions are often slow, several minutes, or even several hours, being required for completion of the reaction. Occasionally the chemist can take advantage of this to accomplish separations that would not be feasible were equilibrium approached rapidly. For example, calcium can be separated from magnesium by precipitation as the oxalate despite the fact that the latter ion also forms an oxalate of comparable insolubility. The separation is possible because equilibrium for magnesium oxalate formation is approached at a much slower rate than that for calcium oxalate formation; if the calcium oxalate is filtered shortly after precipitation, it is essentially free of contamination by magnesium. If, on the other hand, the precipitate is allowed to remain in contact with the original solvent for very long, it will inevitably be contaminated.

problems

1. The solubility of TlI was found to be 0.52 mg/100 ml. Calculate the solubility product.   ans. 2.5×10^{-10}

2. The compound M$_2$SO$_4$ was found to have a solubility of 1.7×10^{-3} gram/liter. The formula weight of M$_2$SO$_4$ is 320. What is the solubility product of the compound?

3. A saturated solution of Mn(OH)$_2$ was found to have a OH$^-$ concentration of 2.1×10^{-5} mol/liter. What is the solubility product for the compound?

4. Calculate the solubility of the following compounds in mg/100 ml.
   (a) $Pb(OH)_2$ ans. 0.096 mg/100 ml
   (b) $PbCl_2$
   (c) $La(IO_3)_3$

5. Will precipitates be produced in the following solutions? Explain your answer.
   (a) A mixture of 100 ml of $1 \times 10^{-5} F$ $AgNO_3$ and 100 ml of $1 \times 10^{-5} F$ NaCl
   ans. No; in this mixture $[Ag^+] [Cl^-] = 2.5 \times 10^{-11}$. This is smaller than the K_{sp} for AgCl. Therefore, no precipitation will occur.

   (b) A mixture of 100 ml of $0.2 F$ $Pb(NO_3)_2$ and 100 ml of $0.1 F$ NaCl
   (c) A $0.01 F$ solution of Mn^{2+} which was brought to a H^+ concentration of 1×10^{-8} mol/liter (Possible precipitate of $Mn(OH)_2$)

6. Will $Al(OH)_3$ form from a solution which is $0.1 F$ in $AlCl_3$ and which is at all acidic?

7. Calculate the solubility of the following in formula weights per liter in a solution which is $0.0050 F$ in NaCl.
   (a) AgCl
   (b) TlCl   ans. 1.2×10^{-2} formula weight/liter
   (c) $PbCl_2$

8. How much water is required to dissolve 0.10 mg of each of the following compounds?
   (a) $PbCl_2$ ans. 0.012 ml
   (b) $Fe(OH)_3$

9. IO_3^- was precipitated by the addition of a solution of $AgNO_3$. When the precipitation was completed, the supernatant liquid was $0.003 F$ in $AgNO_3$ and had a volume of 200 ml. After filtration the precipitate was washed with 200 ml of water.
   (a) Calculate the milligrams of precipitate lost by solubility in the original solution.   ans. 0.58 mg
   (b) Calculate the milligrams of precipitate lost in the wash liquid. Assume equilibrium is achieved and that the wash water contains no $AgNO_3$.
                                                                ans. 10 mg
   (c) If the precipitate weighed 0.502 gram, what percentage error would result from these solubility losses?   ans. 2.1 percent

10. What weight of $Ba(IO_3)_2$ remains unprecipitated when 100 ml of $0.10 F$ $BaCl_2$ is mixed with 300 ml of $0.10 F$ KIO_3?   ans. 0.49 mg

11. How many milliformula weights of $Mg(OH)_2$ remain unprecipitated in 100 ml of $0.20 F$ NaOH?

12. A Ag_2CrO_4 precipitate was formed in a solution that had a volume of 250 ml. Sufficient $AgNO_3$ was present to make the solution $0.0020 F$ in Ag^+. The precipitate was filtered and washed with 200 ml of water. Neglecting any hydrolysis of CrO_4^{2-}:
   (a) Calculate the solubility loss in the original solvent.
   (b) What was the solubility loss in the wash water, assuming equilibrium and negligible $AgNO_3$ concentration?

13. The solubility product for the compound AB_2 is 2.00×10^{-6}. Calculate the

solubility in formula weights per liter of the compound AB_2 in the following solutions :

(a) pure water;

(b) a 0.00100 F solution of A^{2+};

(c) a 0.00100 F solution of B^-.

14. What is the maximum volume of wash water that may be used for a $BaSO_4$ precipitate if losses during washing are to be kept below 0.2 mg of precipitate ?

15. A solution of Na_2SO_4 is added to a solution which is 0.100 F in Ba^{2+} and 0.100 F in Ca^{2+}.

(a) Which SO_4^{2-} precipitate would form first ?

(b) At what SO_4^{2-} concentration would $CaSO_4$ begin to form ?

(c) What would be the concentration of the less soluble compound when the more soluble precipitate begins to form ?

16. To 1 liter of a solution that was 0.010 F in IO_3^- and 0.010 F in Br^- was added 0.015 formula weight of $AgNO_3$. What is the concentration of Ag^+, Br^-, and IO_3^- in the resulting mixture ?   ans. $[Ag^+] = 6.2 \times 10^{-6}$ mol/liter
$[IO_3^-] = 5.0 \times 10^{-3}$ mol/liter
$[Br^-] = 1.2 \times 10^{-7}$ mol/liter

17. What would be the percentage relative error in a lead determination due to solubility losses if the following conditions prevailed ?

grams Pb^{2+} in solution　0.125

grams Na_2SO_4 added　0.342

final volume of solution after precipitation　150 ml

volume of wash water　75 ml　　　　　　　　ans. 1.7 percent

18. Would it be theoretically possible to separate quantitatively Cu^{2+} and Mn^{2+} by precipitation of the more insoluble hydroxide ? If a separation is feasible what range of OH^- concentration should be employed ? Assume the solution is 0.1 F in each ion.

19. Calculate the solubility of $BaSO_4$ in (a) 0.10 N HCl and (b) a solution of H^+ concentration of 10^{-7} mol/liter.

ans. (a) 3.1×10^{-5} formula weight/liter

(b) 1.0×10^{-5} formula weight/liter

20. Calculate the formal solubility of $Ag_2C_2O_4$ in a solution whose H^+ concentration is (a) 1.0×10^{-5} mol/liter and (b) 1.0×10^{-8} mol/liter.

21. What is the formal solubility of Ag_3AsO_4 in a solution that is maintained at $[H^+] = 1.0 \times 10^{-3}$ mol/liter ?

22. What is the solubility of CaC_2O_4 in a solution which is 0.01 F in $Na_2C_2O_4$ and which has an equilibrium $[H^+] = 1.0 \times 10^{-5}$ mol/liter ?

23. What range of H^+ concentrations can be used for the sulfide separation of cation M^{2+} from N^{2+} in a solution that is 0.1 F in each ion if the solubility products are :

$$K_{MS} = 1.0 \times 10^{-18} \quad \text{and} \quad K_{NS} = 5.0 \times 10^{-29}$$

24. The dissociation constant for the complex, $Ag(S_2O_3)_2^{3-}$ is 6.0×10^{-14}. How many grams of AgI will dissolve in a liter of 1.0 F $Na_2S_2O_3$?　ans. 8.1 grams

25. The dissociation constant for the cuprous complex $Cu(NH_3)^+$ is 6.6×10^{-7}. A solution was prepared that contained 0.01 formula weight of a cuprous com-

pound in 1 liter of 1.0 F NH_3. To a 1-liter portion of this solution was added 0.10 formula weight of NaCl. To a second 1-liter portion was added 0.10 formula weight of NaBr and to a third 0.10 formula weight of NaI. Which, if any, of these solutions would show cuprous halide precipitates? Show how you arrived at your answer.

26. The approximate dissociation constant for $AgCl_2^-$ is $1.0 \times 10^{+5}$ ($AgCl_2^- \rightleftharpoons AgCl + Cl^-$). Calculate the approximate solubility of AgCl in a 1.0 F solution of NaCl.

27. Calculate the solubility of MnS in formula weights per liter of water.

28. How many grams of $Al(OH)_3$ will dissolve in 200 ml of 0.10 F KOH? (K_{inst} for $Al(OH)_4^- = 2.5 \times 10^{-2}$).

29. How many grams of AgBr will dissolve in 1 liter of 6.0 F NH_3?

chapter 7. *The Particle Size of Precipitates*

The size of the particles making up a precipitate profoundly affects the ease and the completeness of the filtering operation. A solid made up of large particles or particle aggregates is quantitatively retained by a coarse filter through which the liquid can be passed at a relatively high speed. On the other hand, a finely divided solid requires the use of a dense medium and a concomitant slow filtration rate.

The results of many experiments have disclosed a tremendous variation in the size of the particles formed by mixing two solutions containing ions that react to form a solid phase. In some instances the individual particles are so small as to be invisible to the naked eye; these tend to remain suspended indefinitely in liquid media and cannot be separated by any of the common filtering devices. A solid-liquid system of this kind is termed a *colloidal suspension.*

By way of contrast other precipitates form as discrete particles that may have dimensions as large as several tenths of a millimeter; these particles rapidly settle out of the solution and are readily separated by the ordinary filtering media. The temporary dispersion of such particles in the liquid medium is called a *crystalline suspension.*

The physical properties of a solid-liquid mixture change continuously as the particle size of the solid phase increases. There are no sharp discontinuities discernible; instead there is a gradual and continuous change in behavior of the mixture as it is transformed from a colloidal suspension to a crystalline one. As a result a range of particle sizes yields suspensions displaying properties intermediate between the defined extremes. Despite this, the classification is a useful one for our purposes. Most of the precipitates encountered in analytical chemistry are readily identified as being in either one or the other category; the way in which the two types of precipitates are treated is entirely different.

MECHANISM OF PRECIPITATE FORMATION

The particle size of a precipitate is determined to some extent by the experimental conditions prevailing at the time of its formation. The temperature, rate of mixing of reagents, concentration of reagents, and the solubility of the precipitate at the time of precipitation are variables affecting particle size over which the chemist possesses some control. All of these can be related to a single property, the *relative supersaturation* of the system. The concept of relative supersaturation and its effect upon particle size was first enunciated by P. P. Von Weimarn shortly after the turn of the century.[1] It was the opinion of this investigator that the particle size of any precipitate was determined by the magnitude of this property only, and was independent of chemical composition.

Supersaturation and Particle Size

Definition of terms. A solution is said to be *supersaturated* when it contains a concentration of solute in excess of that found in a saturated solution. We will define the supersaturation of a solution as

$$\text{supersaturation} = Q - S$$

where Q is the concentration of solute in solution at any instant and S is the equilibrium concentration in a saturated solution of the same solute. The relative supersaturation is given by the equation

$$\text{relative supersaturation} = \frac{Q - S}{S}$$

Since the particle size of any precipitate is intimately related to the supersaturation or the relative supersaturation, we shall explore this property somewhat further.

---

[1] Much of Von Weimarn's original work appeared in the *Journal of the Russian Chemical Society.* A short review in English of his ideas with references to the original work can be found in *Chem. Rev.,* **2,** 217 (1925).

Figure 7.1 is a solubility curve for a typical inorganic compound. The solid line *AB* relates the equilibrium concentration of the solute in a saturated solution to the temperature of the solution. The region on the graph to the right of the line *AB* corresponds to a state of unsaturation, whereas that to the left of the line depicts the region of supersaturation.

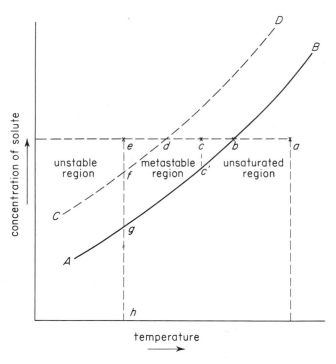

Fig. 7.1 Relationship between the State of a Solution and Temperature.

Stability of supersaturated solutions. A state of supersaturation may be achieved in either of two ways. One method involves lowering the temperature of an unsaturated solution. Consider, for example, the point *a* in Figure 7.1. A solution possessing this concentration of solute is unsaturated at this temperature. Cooling will ultimately result in the solution becoming saturated at the temperature corresponding to point *b*; it will be supersaturated at all lower temperatures. Any supersaturation may be relieved by the formation of a solid phase. Thus, at the temperature corresponding to *c*, a quantity of precipitate may form such that the concentration of the solution is reduced to *c'*; in short, a two-phase system may result with the solid solute in a state of equilibrium with its species in the solution. If the solution is perfectly clean and free of dust or other solid matter, however, precipitation may not occur in any finite length of time. It might be possible, for example, to bring the solution to temperature c or d without the appearance of a precipitate for days, weeks, or even months. Under certain conditions, then, a supersaturated solution behaves as if it were stable. However, if one adds a small amount of the solid solute or, for that matter, certain other solids to the solution at point c, precipitation begins; in a short time the concentration is reduced to the equilibrium concentration c'. The process of causing a precipitate to form from a supersaturated solution by the addition of a solid is called *seeding*.

If no solid is added to the solution, cooling can be continued without immediate precipitation until some point d is reached. At temperatures below this, formation of solid takes place rapidly without seeding. The concentration at which spontaneous precipitate formation will occur is dependent upon the temperature of the solution; the dotted curve CD in Figure 7.1 is a plot of this relationship. The area above and to the left of CD where supersaturation is rapidly relieved by precipitate formation is called the *unstable region*.

The area in Figure 7.1 bounded by the lines AB and CD is called the *metastable region*. Any solution possessing conditions of temperature and concentration represented by this region is clearly supersaturated; at the same time precipitation will not occur rapidly in the absence of seeding.

It should be pointed out that some chemists doubt whether the metastable region is as sharply defined as has been implied here, while others deny its existence altogether. Which of these several viewpoints is correct must await further investigation; here we need only recognize that there is a region representing conditions of slight supersaturation in which the *rate* of spontaneous precipitate formation is low, and further that the virtually instantaneous appearance of a precipitate is characteristic of conditions of high supersaturation.

A second method for preparing a supersaturated solution involves the formation of the solute in solutions at a fixed temperature. This can be accomplished by adding a suitable reagent that will react with a component of the solution to give the desired solute. The path in this case is from h through g to f or e in Figure 7.1.

The production of a precipitate for the purposes of analysis is always preceded by supersaturation of the solvent, and the path followed to reach this state is, of course, the second one mentioned. The physical properties of such a precipitate are largely determined by the degree of supersaturation existing throughout its formation; that is, how far above g in Figure 7.1 the solution is maintained during the precipitation.

Relationship of particle size to supersaturation. Von Weimarn studied, in considerable detail, the effect of relative supersaturation on the particle size of several different compounds. Figure 7.2 shows some of his data for silver sulfate. These are typical and show that a marked decrease in the size of particles occurs with increasing relative supersaturation. This effect can be dramatically demonstrated in the laboratory by precipitating barium sulfate under various conditions of supersaturation. Relative supersaturations of the order of 175,000 can be achieved by rapidly mixing saturated solutions of barium thiocyanate and manganous sulfate (such solutions have concentrations of about 3.5 F);

this will cause a gelatinous barium sulfate precipitate to form in which the colloidal particles are so small they cannot be resolved with a microscope. On the other hand if barium sulfate is formed from very dilute solutions of the same two reagents so that the relative supersaturation is about 25, crystals having a length of about 0.005 mm are obtained. At a relative supersaturation of approximately 3, particles about 0.015 mm in length are observed.

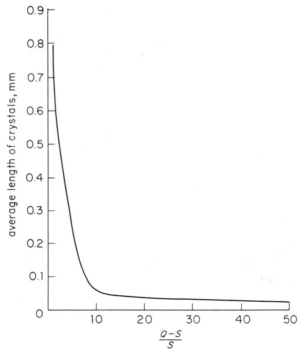

Fig. 7.2 Effect of Relative Supersaturation on the Size of Silver Sulfate Crystals.

Von Weimarn's conclusions relating particle size to supersaturation are useful to the analytical chemist as a guide in finding the precipitation conditions most likely to lead to readily filtered precipitates.

Steps in Precipitate Formation

The formation of a solid phase from a solution involves two processes. One of these we shall call *nucleation* or *nuclei formation* and the second we shall term *particle growth*. The size of the particles of a solid phase is dependent upon the relative rates at which these two competing processes take place.

Nucleation.   For any precipitate there is some minimum number of ions or molecules required to produce a stable second phase in contact with a solution. That is, unless this number of ions or molecules collects together, a solid phase having a finite lifetime will not exist in the solution. We shall call this minimum-sized stable particle a *nucleus*; the first step in the formation of a precipitate involves the generation of many such nuclei. For an insoluble ionic compound AB, nucleus formation involves a reaction that can be written:

$$nA^+ + nB^- \rightleftharpoons (AB)_n$$

where n is the minimum number of A^+ and B^- ions that must combine in order to yield the stable particle $(AB)_n$.

The rate at which nuclei form in a solution is dependent upon the degree of supersaturation. If this is low, the rate may be very low—even zero. This would be the situation for supersaturations in the metastable region shown in Figure 7.1. On the other hand, in a highly supersaturated solution the velocity of nucleation must be very great indeed.

There have been numerous attempts to determine quantitatively how the rate of nucleation varies as a function of supersaturation in typical cases. The experimental difficulties associated with such measurements have proved to be great; no unequivocal answers have as yet been obtained. The bulk of the evidence suggests, however that the rate increases exponentially with the supersaturation. That is,

$$\text{rate of nucleation} = k(Q - S)^x$$

where k and x are constants, x being greater than one.[2]

A plot of a function such as the foregoing is given in Figure 7.3, which shows that there is a range of low supersaturation wherein the nucleation rate is essentially zero. This corresponds to the metastable region in Figure 7.1. Supersaturation corresponding to the rapidly rising portion of the nucleation curve is related to the unstable region in this same graph.

Particle growth. The second process that can occur during precipitation is the growth of particles already present in the solution. This growth can only begin when nuclei or other seed particles are present. In the case of an ionic solid the process involves deposition of cations and anions on appropriate sites.

$$(AB)_n + A^+ + B^- \rightleftharpoons (AB)_{n+1}$$
$$(AB)_{n+1} + A^+ + B^- \rightleftharpoons (AB)_{n+2}, \text{ etc.}$$

Since particle growth can be followed readily in the laboratory, considerably more is known about this process than about nucleation. The rate of growth is found to be directly proportional to the supersaturation and can be expressed by the equation

$$\text{rate of growth} = k'A(Q - S)$$

where A is the surface area of the exposed solid and k' is a constant that is char-

[2] For example, S. H. Bransom, W. J. Dunning, and B. Willard, *Discussions Faraday Society*, **5**, 83 (1949), suggest that $x = 3$ in the case they studied.

acteristic of the particular precipitate. A plot of the rate of growth as a function of supersaturation gives a straight line as is shown in Figure 7.3.

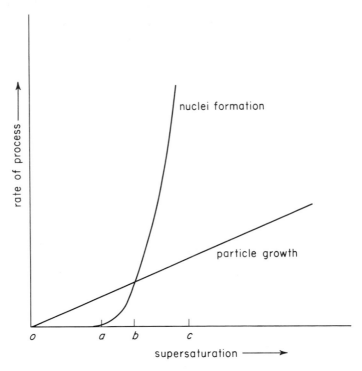

Fig. 7.3 Rate of Nucleation and Particle Growth as a Function of Supersaturation.

The Precipitation Process

We have now seen that precipitation may take place in two ways—either by nucleation or by particle growth. Furthermore, the rates of these two processes are affected differently by the degree of supersaturation of the solution. With this information we can construct a fairly satisfactory picture of the formation of a precipitate and adequately explain Von Weimarn's observations relating particle size to supersaturation.

Initially, reagent is added to the solution causing supersaturation. When this becomes great enough (greater than *a* in Figure 7.3), nucleation occurs at an appreciable rate. If the supersaturation caused by the initial addition of reagent is not large, the number of nuclei formed in a given time will be relatively small. On the other hand, if a high degree of supersaturation is attained (say, above *c* on Figure 7.3), the nucleation rate will be very rapid and a very large number of nuclei will result. Further addition of the reagent will again cause at least momentary supersaturation which will be relieved either by further nuclei formation or by particle growth. Which of these processes predominates will depend upon the degree of supersaturation. If this is low (for example, between o and a on Figure 7.3), the rate of nucleation will be negligible and growth of the nuclei already present will be the more important. We see from Figure 7.3, however, that the rate of nuclei formation increases much more rapidly with supersaturation than does the rate of growth; thus if the supersaturation becomes much greater than b, much of the new precipitate formed will appear as nuclei and only a small part as growth.

This mechanism adequately accounts for the experimental observations relative to the effects of supersaturation on particle size. If the supersaturation is maintained at a low level throughout a precipitation, the relatively few nuclei formed will grow to give a small number of large particles. On the other hand, with high supersaturation many more nuclei are formed initially and nucleation may occur throughout the entire precipitation process. As a result there are many more centers upon which the growth process can take place; none of the particles can become very large. The net effect is a solid phase consisting of a very large number of small particles—in other words, a colloidal suspension. If this interpretation is correct, it should be possible to form a given precipitate with particles of any desired size, ranging from colloidal to crystalline dimensions, provided the supersaturation of the solution can be varied over a sufficient range during precipitation. This has proved to be possible with some precipitates —for example, barium sulfate.

The supersaturation of a solution can be varied by changing either the solubility of the precipitate, S, or the concentration, Q, of solute in solution at any instant. Temperature changes or changes in the composition of the solvent (see Chap. 6) will alter the former. Variations in the concentrations of the reagent solutions and the rate at which these are mixed will control the concentration of solute. Theoretically, then, it should be possible to form all precipitates as easily filtered crystalline solids simply by keeping the supersaturation sufficiently low. In practice, however, this is not always feasible. With very insoluble precipitates, S is so very small relative to Q that the difference $(Q\text{-}S)$ remains large despite all efforts to reduce it. Attempts to decrease Q sufficiently to give low supersaturations result in the use of such extremely dilute solutions and inordinately slow rates of addition of the reagent as to be impractical. Nor can S be increased sufficiently by heat or variation in solvent composition to have much effect on the difference. As a result, all of the very insoluble precipitates, such as the hydrous oxides and sulfides, occur as colloidal suspensions when formed under conditions that are practical for analysis. Only with the more soluble precipitates, such as the oxalates, sulfates, and carbonates, can the supersaturation be kept sufficiently low to yield crystalline precipitates under practical conditions.

Both colloidal and crystalline precipitates are used for gravimetric analysis. However, the treatment of the two types of solids varies considerably; we shall consider the properties and behavior of each in the following sections.

COLLOIDAL SUSPENSIONS

Properties of Colloidal Suspensions

The colloidal suspensions with which we shall be concerned consist of dispersions of finely divided solid particles in a liquid phase. Typically, these will have diameters of 0.001 to 0.1 micron,[3] and will be completely invisible to the naked eye or the ordinary light microscope. As a result, colloidal suspensions or, as they are sometimes called, colloidal solutions, will often appear to be clear and completely homogeneous despite the fact that they may contain several grams of dispersed solid in a few hundred milliliters of the liquid.

The properties of a colloidal solution are quite different from those of a true solution. In the former, the solid is dispersed in the form of aggregates of the ions or molecules making up the precipitate while in the latter, the dissolved solid is homogeneously dispersed as individual ions or molecules. In contrast to the true solution, the solid present in a colloidal suspension has a negligible effect on such properties as freezing point, boiling point, and osmotic pressure. Furthermore, the particles of a colloid are of such dimensions as to scatter visible radiation. Thus when a light beam passes through a colloidal solution, its path can be readily seen. This well-known phenomenon is called the *Tyndall effect*. True solutions do not exhibit this effect because the individual particles present are too small to reflect radiation of the wavelength of light.

Colloidal particles are so small they pass through the ordinary filtering media used by the chemist. They can, however, be retained by certain ultrafilters whose pore sizes are much smaller. Unfortunately, the rate at which liquids pass through these ultrafilters is very slow, and for this reason they cannot ordinarily be used for the filtration of analytical precipitates.

The particles of a colloidal solution are so small they will not settle out as such under the influence of gravity but will remain indefinitely suspended in the liquid in which they were formed. They can, however, be separated from a solution by the use of a very high-speed ultracentrifuge. Under some circumstances the individual particles making up a colloidal suspension can be caused to come together and adhere to one another, the result being a mass of material that will settle out rapidly from solution. The resulting solid is called a *colloidal precipitate* and the process by which it is formed is termed *coagulation* or *agglomeration*. Coagulated colloidal precipitates are quite different in appearance from typical crystalline precipitates. In some cases they are slimy and gel-like and in others rather curdy. They have no regularity in structure and appear as large amorphous masses.

Specific Surface Area of Colloids

Many of the unique properties of colloidal suspensions are due to the fact that the solid in this state of subdivision exhibits a very large surface area. The

---

[3] The micron, μ, is equal to 0.001 mm.

specific surface of a solid is defined as the exposed surface area per unit of weight of the solid, and is often expressed in units of square centimeters per gram. As the particle size of a solid is reduced, its specific surface increases and becomes enormous when dimensions of a typical colloid are reached. For example, in the case of a cube of a solid 1 cm on a side and weighing 3 grams, the surface area of the solid is 6 cm² and the specific surface 2 cm²/gram. This cube could be divided into 1000 cubes, 0.1 cm on a side, or into 1,000,000 cubes, 0.01 cm on a side, and so forth. It is interesting to calculate the specific surface areas in each of these cases in order to see how the surface area increases with subdivision. Table 7-1 gives such data.

Table 7-1

Specific Surface Area of 3 Grams of a Solid in Various States of Subdivision

Number of Particles	Dimensions of Cube, cm	Specific Surface cm²/gram
1	1	2
10^3	0.1	20
10^6	0.01	200
10^{12}	0.0001	20,000
10^{18}	0.000001	2,000,000

The last figures in this table are for particles having the dimensions of a typical colloid; we see that 1 gram of the solid in this state of subdivision has a surface area of 2,000,000 cm², or somewhat more than 2000 ft². A colloidal suspension of only 1 gram of this solid, then, would present a surface that is equivalent to the floor area of a fairly good-sized house. One might expect that any properties of a solid-liquid system that are related to the interface between the solid and liquid would be tremendously amplified in a colloidal suspension where the specific surface area is so great. Actually there are such properties, and we shall discuss these in later sections.

Adsorption of Ions by Colloids

Any solid surface exposed to an aqueous solution of an electrolyte is found to be charged either positively or negatively with respect to the solution. This results from the tendency of some of the ions of the solution to attach themselves to the solid surfaces. With solids of ordinary dimensions, the magnitude of the charge is quite small, and its existence goes undetected unless sensitive means are used to observe it. However, in the case of colloidal suspensions where the surface to weight ratio—that is, the specific surface—is very high, the phenomenon becomes one of primary importance in determining the properties and behavior of the particles.

The process by which ions, or in some cases molecules, are held on the surface of a solid is called *adsorption*. The solid itself is called the *adsorbent* and the substance that is adsorbed is the *adsorbate*.

Adsorption of ions on solid surfaces. We shall consider a case in which a colloidal suspension is in contact with a solution containing an excess of one of the ions making up the lattice structure of the solid. Figure 7.4 is a schematic diagram of one particle in such a medium. Here silver ions are shown as being adsorbed on the surface of a silver chloride particle and, as a result, the particle is positively charged. The adsorbed ions are seen to be held in positions adjacent to some of the chloride ions in the lattice of the solid itself. The forces holding the ions on the surface are the normal chemical bonding forces that bind the ions together in the solid lattice. Undoubtedly a silver chloride particle in such an environment as that shown in the diagram adsorbs other ions from the solution also; these might include hydrogen, sodium, nitrate, etc. The lattice ion, however, would normally predominate.

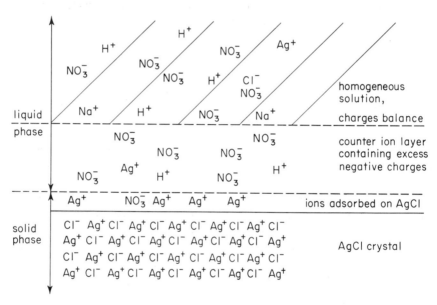

Fig. 7.4 Conditions at the Surface of a Precipitate. Cross-sectional diagram of a silver chloride particle suspended in a solution containing an excess of silver ions.

By no means all of the possible sites are occupied by adsorbed silver ions at any given time; for as the number of cations on the surface increases, the accumulation of positive charges tends to prevent further adsorption. Thus a balance is struck in which the bonding forces are just matched by the electrostatic repulsive forces. It should be understood that the resulting state is a dynamic one in which silver ions are constantly being adsorbed and desorbed at a rapid rate.

The ions adsorbed on a solid are firmly attached and should be considered a part of the solid. This is shown by the fact that the charge resulting from adsorption will cause a colloidal particle to migrate in an electric field; thus in the example in Figure 7.4, the silver chloride particle would move towards a negative electrode and away from a positive one.

Chemical composition of the adsorbed layer. Some ions are much more strongly adsorbed by a given solid than others. Thus, when a given colloid is exposed to a solution containing several ions, the adsorption of one of these will ordinarily predominate. Often it is possible to make a reasonably good guess as to which ion this will be.

As mentioned previously, the adsorbed ions are held on the surface of a solid by normal bonding forces. We might expect, therefore, that there would be a relationship between the tendency for a given ion to be adsorbed by a solid and the tendency of that ion to combine with one of the ions of the solid. This relationship was first suggested by Paneth and later amplified by Fajans,[4] and is often referred to as the *Paneth-Fajans rule*, which states: *Those ions will be strongly adsorbed that form difficultly soluble or weakly dissociated compounds with the oppositely charged ions of the solid lattice.*

An important example of the rule is found where a precipitate is in contact with a solution containing an excess of an ion common to the solid. Since the common ion does form an insoluble compound with the oppositely charged ion of the solid, we would expect the common ion to be strongly adsorbed. This is found to be the case; the common ion will nearly always be adsorbed in preference to all others in the solution. For example, the colloidal particles of a silver chloride precipitate formed in the presence of an excess of silver ions are positively charged owing to the adsorption of silver ions; if sufficient chloride ions are then added to the system so that they are in excess, however, the charge on the particles is found to change from positive to negative as the adsorbed ions change from silver to chloride. Most of the precipitates with which we deal in analytical chemistry are formed in the presence of an excess of a common ion. Thus it is easy to decide what the composition of the adsorbed layer will be.

Counter-ion layer. The presence of a charged particle will lead to an inhomogeneity in the solution immediately surrounding the particle. Thus the positively charged silver chloride particle illustrated in Figure 7.4 will tend to attract anions and repulse cations. As a result there will be a layer of solution in which the concentration of negatively charged ions exceeds that of positively charged ions. We shall call this region the *counter-ion layer*.

A sufficient excess of ions of opposite charge will be within the counter-ion layer of solution to balance that carried by the colloidal particle. The thickness of the layer of solution required to contain this number of ions will be variable and will depend upon the concentration and the nature of the electrolyte. Thus if the solution is dilute, the thickness of the layer having an equivalent number

[4] For a discussion of the Paneth-Fajans rule see H. B. Weiser, *Textbook of Colloid Chemistry*, 2d ed., 112. New York: John Wiley and Sons, Inc., 1949.

of ions of charge opposite to that of the particle will be relatively great. On the other hand, if the solvent has a large concentration of ions, this layer will be thin. In either case, we conceive of the counter-ion layer as being somewhat diffuse and a part of the solution, not the solid.

Effect of concentration on the amount of adsorption.   As we might expect, the number of ions adsorbed on a solid surface bears a direct relationship to the concentration of those ions in the solution. Figure 7.5 is a graph of experimental data showing this relationship for two typical cases. Such curves are called *adsorption isotherms*; they apply only at a single temperature and to a single system. As the concentration of adsorbate is increased, a limiting value is approached for the amount that can be adsorbed. With rising temperatures this quantity decreases markedly.

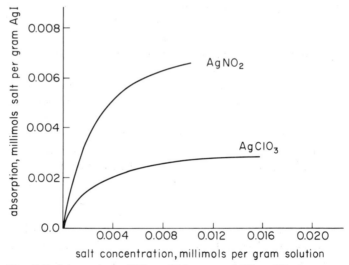

Fig. 7.5 Adsorption of Silver Chlorate and Silver Nitrite upon Silver Iodide.

The data illustrated in Figure 7.5 also demonstrate that the amount of adsorption is a function of the composition of the counter-ion layer. In the two cases shown, the primarily adsorbed ions are silver; however, the counter-ion layers differ. It is clear that the quantity of adsorption is appreciably higher when the counter-ion layer is nitrite than when it is chlorate. This phenomenon has been studied extensively; wide variations in the adsorption of compounds with one ion in common have been observed. On the basis of such studies Hahn[5] proposed that there is an inverse relationship between the amount of adsorption and the solubility of the compound made up of the primarily adsorbed ion and its counter ion. Thus in the case shown in Figure 7.5, the silver nitrite is ap-

[5] O. Hahn, *Applied Radiochemistry*, 90. Ithaca, New York: Cornell University Press, 1936.

preciably less soluble than silver chlorate and therefore the more strongly adsorbed. Subsequent investigation has shown this rule to be far from perfect; however, it has proved to be of some help in the prediction of relative amounts of adsorption.

Coagulation of Colloids

From the viewpoint of the analytical chemist, one of the most important properties of a colloid is its tendency to coagulate into a larger, more readily filtered mass. Generally, this can be brought about by heat or by the addition of an electrolyte. We will examine this process in some detail because it makes possible the use of colloidal suspensions in gravimetric analysis.

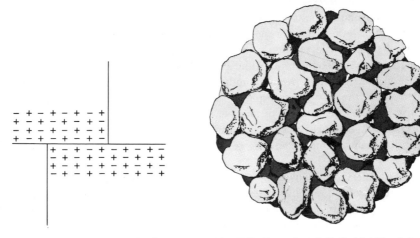

Fig. 7.6 Adhesion of Two Similar Particles to One Another.

Fig. 7.7 Coagulated Colloidal Particles.

Attractive forces between particles. We would expect the particles of a colloid to have a considerable tendency to adhere to one another if brought into close contact, since the same bonding forces that cause adsorption would certainly operate between two like particles, tending to hold these together. Thus, if perfectly plane surfaces of two particles came together, as pictured in Figure 7.6, the positive ions of one could line up opposite the negative ions of the second and vice versa. A strong bond between the two particles would result. Actually the situation pictured in Figure 7.6 is highly idealized in that the particle surfaces are pictured as being perfectly flat and regular. It seems more probable, however, that the colloidal particles formed by chemical reaction are irregular in shape so that only a limited area of contact between any two particles is possible. This is illustrated by Figure 7.7 which shows how several coagulated particles might appear if they could be sufficiently magnified to be seen. Only at areas of contact would there be bonding between the particles.

A solid made of thousands or millions of colloidal particles agglomerated as in Figure 7.7 would be expected to have quite different properties from a crystalline solid. Since the particles are arranged irregularly, the solid phase should be amorphous in appearance; in addition, the apparent density of such a solid should be less than that of a crystalline solid where the ions are tightly packed in a regular arrangement. Finally, this solid should be porous in nature and be capable of containing large volumes of the solvent internally. As a result, a coagulated colloid might be expected to have a considerably larger surface area than would be surmised from the external dimensions of the solid.

Stability of colloidal suspensions. The stability of colloidal suspensions is a direct consequence of the repulsive forces set up among the individual particles by the charged double layers surrounding each particle. These layers prevent the close approach necessary for the adhesive forces between the particles to be effective in bringing about agglomeration. The charged double layer, then, acts as a sort of buffer, causing the particles to sheer away from one another as they travel through the solution.

Mechanism of coagulation. The coagulation process involves a reduction in the forces of repulsion exerted among the colloidal particles by the charged double layers. One way to accomplish this is to reduce the total charge of the layers by reducing the number of ions adsorbed on the particles. As mentioned earlier, this can be done by raising the temperature of the system; experimentally we find that many colloidal solutions are coagulated by heating for short periods of time.

A more effective way of causing a colloidal suspension to agglomerate is to add an electrolyte to the solution. By so doing, the thickness of the counter-ion layer surrounding any one particle is reduced. As a result, the ions of the counter-ion layer approach their counterparts on the solid particle more closely and their effective charge is thus more completely neutralized. This, in turn, allows the particles to approach one another more closely without being deflected by the charged double layer; if the electrolyte concentration is sufficiently great, the layer shrinks to a point where the particles come in contact with one another and adhere.

Experimental data indicate a good deal of variation in the effectiveness of different electrolytes in causing coagulation of a given colloidal suspension. Some generalities can be drawn in this regard. First, the effectiveness is dependent upon the nature of the ion *opposite* in charge to that of the colloidal particles. There is little difference in behavior among the ions having the same charge. Second, the coagulating power of this ion increases greatly as the charge of the ion increases. The foregoing is sometimes called the *Schulze-Hardy rule*; clear experimental substantiation is given by the data in Table 7-2. Here it will be seen that the negatively charged colloid is coagulated by much lower concentrations of salts having a doubly charged cation than by those having a singly charged one. Furthermore, the salts of triply charged cations are even more effective. The nature of the anion appears to make little difference. There are no simple, quantitative explanations for this effect.

Table 7-2

EFFECTIVENESS OF ELECTROLYTES IN THE COAGULATION OF A NEGATIVELY CHARGED
ARSENIC TRISULFIDE COLLOID

Electrolyte	Concentration Required to Cause Coagulation, millimols per liter
LiCl	58.4
NaCl	51.0
KCl	49.5
KNO_3	50.0
K_2SO_4	32.5
$MgCl_2$	0.72
$MgSO_4$	0.81
$BaCl_2$	0.69
$ZnCl_2$	0.68
$AlCl_3$	0.093
$Al(NO_3)_3$	0.095
$Ce(NO_3)_3$	0.080

Peptization of Colloidal Precipitates

The process of coagulation of a colloid is at least partially reversible with many, but not all, colloidal precipitates; that is, a coagulated colloid can often be converted back into its original finely divided state by suitable treatment. This is called *peptization*. A most effective method of bringing about peptization of a precipitate is by washing with water. For example, if a coagulated colloidal silver chloride precipitate is transferred to a filter paper and washed with a few hundred milliliters of water, the washings are first clear but later become opalescent. This is because the silver chloride precipitate has been peptized and converted back into particles of such dimensions that they can pass through the pores of the paper. Of course, if we wish to retain all of the precipitate, as in a quantitative analysis, precautions must be taken to guard against the occurrence of peptization.

In order to understand how peptization occurs, we must acquire a still more complete picture of the constitution of a coagulated colloid, such as the silver chloride precipitate mentioned in the preceding paragraph. A precipitate of this sort consists of a very large number of minute particles adhering together in a random orientation. The solid still has a relatively large surface area exposed to the solution because the particles are irregularly shaped and are not fitted together in any regular fashion. This leaves internal surfaces in the solid exposed to the solvent. Figure 7.7 shows several such inner surfaces interspersed among the particles. On these surfaces are the adsorbed ions that gave the original particles their charge; while the total number of adsorbed ions on the solid has been reduced as a result of agglomeration, there are still large numbers so held.

The counter-ion layer also exists on the coagulated colloid in the film of liquid immediately adjacent to the exposed surfaces. Thus when a colloid is coagulated, it carries down with it appreciable concentrations of compounds consisting of the primarily adsorbed ions and an equivalent number of their oppositely charged counterparts of the counter-ion layer. The solution wetting the surfaces of the coagulated colloid contains, in addition to the counter ions, an appreciable concentration of the electrolyte that was used to coagulate the particles. As a result, a typical coagulated colloidal precipitate is rather heavily contaminated by foreign ions when filtered from solution.

When a precipitate such as silver chloride is washed with water, the wash liquid is not very effective in removing the ions adsorbed on the surface of the solid; the bond between them and the surface is too strong. Nor can the wash liquid remove the counter ions since these are attracted to the primarily adsorbed ions by strong electrostatic forces. However, the wash water will dilute and remove the electrolyte that was used for the coagulation of the colloid. When this occurs, the counter-ion layer tends to become greater in thickness, and as a result the repulsive force between the double layers of the particles again begins to exert itself. If enough of the electrolyte is removed, the force may become sufficient to cause particles to break away from the mass of the precipitate and lead to peptization.

In handling a colloidal precipitate, the chemist is faced with a dilemma. If he chooses to wash it with water to reduce the amount of foreign contaminants, he runs a good chance of losing a part of his precipitate by peptization. On the other hand, while he can prevent this by washing with an electrolyte solution, he will inevitably obtain a solid badly contaminated with that electrolyte. The one solution to this problem is to wash with an electrolyte that can subsequently be removed by volatilization. As an example, a silver chloride precipitate can conveniently be washed with dilute nitric acid which will remove at least part of the nonvolatile contaminants without causing peptization; the nitric acid can later be volatilized by heating the filtered precipitate at about $100°$ C.

CRYSTALLINE PRECIPITATES

A number of the more soluble precipitates used in analysis form as crystalline compounds. With these, filtering characteristics can be improved by increasing the size of the individual crystals or by cementing the single crystals together into larger agglomerates of several such crystals.

Methods for Increasing the Size of Crystals

The reader will recall from earlier discussions that the particle size of a precipitate is increased by reduction of the relative supersaturation of the solution from which it is formed. Experimentally, there are several ways of accomplishing this.

Concentration and rate of mixing of reagents. One method of reducing supersaturation entails the use of dilute solutions so that the solubility product of the compound is only slightly exceeded with the addition of each increment of reagent. Slow addition of the precipitating reagent is also helpful because this allows time for the precipitate to form; the build-up of a large supersaturation is thereby prevented.

There are, of course, practical limitations to the foregoing expedients. As the solutions are made more dilute, solubility losses are larger; in addition the time required for the filtration of the precipitate from the larger volume of liquid is increased. Slow addition of the reagent also increases the time required for the analysis. Thus the conditions actually chosen usually represent a compromise of these several factors and involve dropwise addition of precipitating reagents of concentrations such that no more than a few hundred milliliters of solution have to be filtered.

Variation of solubility of precipitates. The relative supersaturation can often be reduced by increasing the solubility of a precipitate during the period of its formation; this leads to increased particle size. One common way of doing this is to carry out the precipitation at elevated temperatures. Before filtration, the mixture is usually cooled to room temperature, or below, to reduce solubility losses.

With some compounds, large particles are conveniently obtained by precipitating the substance from a solvent medium wherein the solubility is relatively high and the supersaturation correspondingly low. After the reagent has been added, the precipitation is completed by altering the solvent composition so that the solubility is low enough to avoid the concomitant losses of precipitate. Solids whose solubilities are affected by pH are particularly susceptible to this treatment. For example, very large crystals of calcium oxalate can be obtained by forming the bulk of the precipitate in a somewhat acid solution. The precipitation can then be completed by the slow addition of ammonia until the pH is great enough for quantitative removal of the compound; the additional precipitate thus produced forms on the solid already present.

While the foregoing procedures are quite useful for the more soluble compounds, they are of little importance in handling the very insoluble substances that normally separate as colloids. The variations in relative supersaturation that can be attained by these alterations are simply not great enough to convert such solids to crystalline precipitates, nor do they make handling the colloids any easier.

Digestion of Crystalline Precipitates

It has been found experimentally that when a crystalline precipitate is allowed to stand quietly in contact with the solution from which it was precipitated (that is, the so-called mother liquor) it is more easily filtered and requires a less dense filtering medium than the freshly formed precipitate. This improvement in filterability is hastened by heating; common practice is to treat crystalline precipitates in this manner prior to filtration. The process is called *digestion*.

Considerable experimental evidence indicates that the digestion process consists of a cementing together of crystals. The resulting aggregates are appreciably larger than the individual crystals and are therefore more easily retained. We can easily understand how such agglomeration might occur when we recall that the ions making up a precipitate are in dynamic equilibrium with their counterparts in the solution. As a result, solution and reprecipitation of the solid take place constantly. The latter process can occur in such a way as to form bridges between adjacent solid particles thus leading to crystalline aggregates. At elevated temperatures, where the solubility is greater, the process would be expected to take place at an accelerated rate; this is in keeping with the experimental observation that the effectiveness of digestion is greater at higher temperatures.

The improved filterability of precipitates after digestion has also been attributed wholly or in part to growth of large crystals at the expense of small ones. There is some basis for thinking that such a process might occur since, theoretically, small crystals should be more soluble than large ones. Some experimental evidence also exists which appears to demonstrate the greater solubility of small particles.[6] If this is indeed the case, we would expect digestion of a mixture of crystals to result in the disappearance of the more soluble smaller particles and growth of the larger and more insoluble ones.

There is, however, considerable experimental evidence which shows that the crystal-growth process plays but a minor role compared with the aggregation process in improving the filterability of precipitates during digestion. For example, Trimble[7] heated both stirred and unstirred suspensions of freshly precipitated barium sulfate for several hours and found a tremendous improvement in the ease and completeness with which the solid could be filtered from the solutions that were *not* stirred. On the other hand, no noticeable changes in the stirred solutions were observed. We would expect agitation of the solution to interfere with a process involving the cementing of crystals together whereas it should either have no effect or accelerate the transfer of solid from small crystals, to large. Furthermore, microscopic examination of the solid barium sulfate before and after digestion without stirring showed clearly that aggregation of individual crystals had occurred but that little or no change in the size of the individual crystals making up the aggregates was apparent. There was, however, some evidence that the very smallest crystals of the solid may have disappeared during the digestion.

---

[6] For a summary of the theoretical and experimental evidence of the effects of particle size on solubility see H. E. Buckley, *Crystal Growth*, 23. New York: John Wiley and Sons, Inc., 1951. See also M. L. Dundon, *J. Am. Chem. Soc.*, **45**, 2658 (1923), and M. L. Dundon and E. Mack, *J. Am. Chem. Soc.*, **45**, 2479 (1923).

[7] H. M. Trimble, *J. Phys. Chem.*, **31**, 601 (1927).

chapter 8. *Contamination of Gravimetric Precipitates*

A precipitate formed in solution by the combination of suitable reagents is almost invariably contaminated to a greater or lesser extent by the other ions present during the precipitation process. For a gravimetric analysis, the level of such contamination must be kept low in order to avoid any appreciable alteration in the weight of the solid. This is not always possible, and there are a number of precipitates that are not suitable for analytical purposes because of this reason.

Contamination of precipitates can occur by several mechanisms. The most obvious of these is encountered when the solubility product of some compound other than the one desired is also exceeded during addition of the precipitating reagent. Under these circumstances the pollutant is simply codeposited as a second precipitate. This will occur, for example, when silver chloride is formed from a solution containing bromide ion. Ordinarily, simultaneous precipitation of this sort is not of great concern to the analytical chemist because he can readily predict its occurrence provided he has some knowledge of the composition of his solutions. In avoiding such precipitation, he is frequently forced to resort to preliminary separations of one sort or another.

Some precipitates form quite rapidly upon addition of the precipitating reagent; others form rather slowly. As a result it is sometimes possible to separate two ions that form insoluble precipitates with the same reagent by removing the first one before the second has had time to precipitate. For example, calcium oxalate precipitates rapidly while the formation of a magnesium oxalate precipitate may require an hour or more. Here, the difference in rate of precipitation is great enough to permit a quantitative separation of the two ions despite the fact that solubility considerations alone predict that this is impossible. Contamination of the calcium oxalate by magnesium oxalate may occur, however, if the calcium oxalate is not promptly filtered after its formation. Such contamination by subsequent formation of a second insoluble precipitate is called *post precipitation*. We shall consider this further in later pages.

A much more insidious source of error in gravimetric analyses results from the carrying down of normally soluble substances by the insoluble phase. Such contamination occurs to some extent any time a precipitate is formed. The amount of pollutant so introduced is often small and has little effect on the weight of the precipitate. Under some conditions, however, surprisingly large quantities of normally soluble compounds are removed from solution in conjunction with the insoluble solid. In fact, complete removal is occasionally observed.

The phenomenon by which normally soluble substances are carried out of solution by an insoluble precipitate is called coprecipitation. The student should note that this term refers only to the contamination of precipitates by substances that would otherwise remain in solution under the conditions imposed on the solution. The term does not include contamination by simultaneous precipitation or post precipitation of another insoluble compound.

Coprecipitation

Types of Coprecipitation

At least four types of coprecipitation can be recognized. Which of these predominates in a given case depends upon the particle size of the solid as well as the chemical composition of both the solid and the solution. The measures that are effective in reducing analytical errors resulting from coprecipitation are different for the various types. For this reason it is highly desirable that the chemist attempt to recognize which of the types he is most likely to encounter in any given precipitation. They are as follows:

(1) *Surface adsorption*, wherein the contaminant is a compound made up of the ions adsorbed on the surface of the solid and the counter ions of opposite charge in the liquid film immediately adjacent to the particle.

(2) *Isomorphic inclusion* (or mixed crystal formation), in which the coprecipitated compound has dimensions and chemical composition such that it can

fit into the crystal structure of the precipitate without causing appreciable strains or distortions.

(3) *Non-isomorphic inclusion* (or solid solution formation), wherein small quantities of the impurity appear to be dissolved in the precipitate.

(4) *Occlusion*, where the pollutants are found entrapped as imperfections within the crystals of the precipitate.

The foregoing classification is essentially that proposed by Hahn.[1] The reader should be aware, however, that the type of coprecipitation predominant in the contamination of a given precipitate is not always readily apparent.

Surface Adsorption

All precipitates carry down soluble impurities by adsorption on their surfaces. However, only where the specific surface areas are great does coprecipitation of this type have an appreciable effect on the weight of the solid. Therefore we need be concerned with contamination by surface adsorption only when dealing with colloidal precipitates.

Chemical composition of adsorbed contaminants.   As was pointed out in Chapter 7, colloidal particles formed under analytical conditions will always have adsorbed on their surfaces one of the constituent ions of the precipitate. This will, of course, be the one in excess when the precipitation has been completed. Ordinarily, a part of these adsorbed ions remain attached to the solid surface even after coagulation has occurred and are carried down with the solid. Owing to electrostatic forces, an equivalent number of oppositely charged ions from the counter-ion layer are also carried down in the liquid film surrounding the coagulated solid. *Thus a soluble chemical compound consisting of one of the ionic species making up the precipitate and a corresponding number of oppositely charged counter ions is coprecipitated.*

Table 8-1 contains data showing the amounts of various anions adsorbed on colloidal barium sulfate. It is apparent that the quantities are much greater in the presence of excess barium ions than in the presence of excess sulfate. This is consistent with our picture of the adsorption process; with an excess of barium ions, this cation predominates on the surface, and the counter-ion layer will be made up of the anions present in the solution. Thus the coprecipitated compounds are primarily barium salts. When sulfate ions are present in excess, however, coprecipitation will occur primarily in the form of salts of this anion; thus in the second case in Table 8-1 coprecipitation of sodium sulfate, potassium sulfate, or sulfuric acid is undoubtedly large. Coprecipitation of anions other than sulfate is, however, relatively small.

From the foregoing, we see that the nature of one of the adsorbed ions is always fixed in an analytical precipitation by its presence in excess at the end of the precipitation process. The constitution of the counter-ion layer is subject to variation, however, and it is of interest to consider which ion or ions will be

[1] O. Hahn, *Applied Radiochemistry*, 69. Ithaca, New York: Cornell University Press, 1936.

Table 8-1

ADSORPTION OF ANIONS BY BARIUM SULFATE[2]

Anion Present	Precipitate Ion in Excess	Amount of the Anion Adsorbed by 100 Mols BaSO$_4$, mols
Cl$^-$	Ba^{2+}	1.76
MnO$_4^-$	Ba^{2+}	2.85
ClO$_3^-$	Ba^{2+}	5.84
Cl$^-$	SO$_4^{2-}$	0.125
MnO$_4^-$	SO$_4^{2-}$	0.137
ClO$_3^-$	SO$_4^{2-}$	0.227

[2] These data are taken from H. B. Weiser, *A Textbook of Colloid Chemistry*, 2d ed., 110. New York: John Wiley & Sons, Inc., 1949.

contained therein when several different species of proper charge are present in the solution.

Experimental studies on adsorption by a wide variety of precipitates has revealed great differences in the tendency of various ions to be carried down as counter ions. Numerous attempts have been made to relate these differences in adsorbability to various properties of the ions in question; however, no simple, quantitative correlations have been found.

Table 8-2

ADSORPTION OF VARIOUS SILVER SALTS ON SILVER IODIDE[3]
(Equilibrium concentration of silver salt is 0.004 M in each case)

Silver Salt	Quantity Adsorbed, Millimols Salt per gram AgI	Order of Increasing Solubility of Silver Salts
benzoate	0.0070	2
acetate	0.0059	5
nitrite	0.0052	3
bromate	0.0039	1
β-napthalenesulfonate	0.0033	4
benzenesulfonate	0.0027	7
nitrate	0.0023	9
chlorate	0.0020	6
ethyl sulfate	0.0016	8
perchlorate	0.0011	10

[3] Data from J. S. Beekley and H. S. Taylor, *J. Phys. Chem.*, **29**, 942 (1925).

One of the properties that appears to have a bearing on adsorbability is the magnitude of the charge on the ion. Generally ions with multiple charge are more strongly held in the counter-ion layer than are singly charged species. This relationship, however, is far from perfect and many notable exceptions can be found.

Another important effect is related to the solubility of the salt composed of the ion in question and the primarily adsorbed common ion. The Paneth-Fajans rule (see Chap. 7) will provide a rough guide to the extent of adsorption to be expected. The data in Table 8-2 show that predictions based upon this rule are frequently incorrect, however.

In addition to the aforementioned effects such factors as ionic size, deformability, and degree of hydration also appear to influence the relative amounts of various ions that are adsorbed as counter ions. These are difficult to evaluate in quantitative terms.

Amounts of adsorption. The amount of contaminant adsorbed by colloidal precipitates varies tremendously. For example, the hydrous oxides of the heavy metals are nearly always heavily contaminated owing to adsorption of soluble metal hydroxides. Similarly, the sulfides of most of the heavy metals are often found to be seriously contaminated by normally soluble sulfides. By way of contrast, adsorption by the colloidal silver halide precipitates is seldom sufficient to lead to serious errors. Undoubtedly, part of these differences in the adsorption capacity of various precipitates is related to differences in the surface develop-

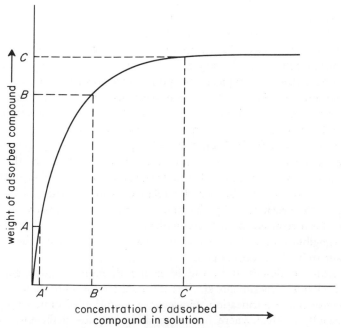

Fig. 8.1 A Typical Adsorption Isotherm.

ment—that is, the specific surface areas. Other specific factors also appear to play a part.

For any given precipitate, the quantity of a compound adsorbed is related to its concentration in solution in a manner shown in Figure 8.1.

Reduction of errors caused by adsorption. Several paths are open to the chemist seeking to reduce the magnitude of error arising from adsorption of contaminants on a colloidal precipitate. Among these are the use of elevated temperatures and dilute solutions, digestion, washing, and reprecipitation.

In Chapter 7 mention was made of the fact that the amount of adsorption by a colloid decreases with increases in temperature. As a consequence, a purer analytical precipitate is produced by the use of elevated temperatures during the formation, filtration, and washing operations. Fortunately, the solubilities of most precipitates forming as colloids are small enough so that such treatment does not result in significant solubility losses.

The quantity of a compound adsorbed by a colloid is directly related to its concentration in the solution. As a result it is advantageous from the standpoint of purity to form a colloidal precipitate from as dilute a solution as possible. There are practical limitations to this technique, however, since filtering and handling large volumes of solutions is tedious and time-consuming.

The purity of some colloidal precipitates is increased by digestion for short periods of time. Silver halide precipitates, for example, benefit from this treatment. Purification in such instances probably results from a decrease in the specific surface area of the coagulated solid and perfection of the particles by the process of solution and reprecipitation. With gelatinous colloids such as the hydrous oxides, digestion is of little or no help.

The adsorbed compounds on a colloidal particle are in a state of dynamic equilibrium with the ions of that compound in solution. As a result, the washing of a coagulated colloid can be effective in reducing the amount of adsorbed contaminants. A study of a typical adsorption isotherm such as shown in Figure 8.1 reveals, however, that the washing process becomes less and less effective as the quantity of adsorbed impurity decreases—that is, as the washing proceeds. For example, if we assume that the first wash of a precipitate results in a concentration, C', of contaminant in the wash liquid, the corresponding quantity, C, is thus left on the precipitate. Further washing would reduce the amount adsorbed to some smaller value, B; the equilibrium concentration in the wash liquid, B', is now considerably smaller relative to the weight of contaminant remaining on the precipitate. By the time the amount of impurity on the precipitate has been reduced to A, the equilibrium concentration, A', in the wash liquid is negligible, and washing is no longer effective in removing the remaining contaminant from the precipitate.

The situation described in the foregoing example is quite general for coagulated colloids. Experiments show that the initial washes of such precipitates result in considerable purification. However, a condition of diminishing return is rather rapidly reached where further washing is quite inefficient. Complete removal of the contaminant by washing is never practical; in some instances the amount of residual impurity is great enough to affect seriously the weight of precipitate. It must constantly be borne in mind that by washing a coagulated colloid with pure water one often runs the risk of peptization of the solid.

There is little the chemist can do to influence the nature of the ion directly adsorbed on the surface of an analytical precipitate; this is fixed through the necessity of having an excess of the precipitating ion in the solution and the strong tendency of the solid to adsorb that ion in preference to all others. The constitution of the counter-ion layer can be varied, however, and in some instances advantage can be taken of this to produce adsorbed compounds that are volatile; for this purpose the choice of electrolyte in the wash liquid is of prime importance. Thus, when silver is determined by precipitation as the chloride, the primary adsorption of chloride ions is inevitable; the counter-ion layer will clearly be cationic in nature. The chemical species in this layer will be determined in part by the relative concentrations of the various cations present in the solution; if the precipitation is made from a nitric acid solution and the solid washed with a dilute solution of this same acid, a fair portion of the counter ions will be hydrogen ions. As a consequence, the coprecipitated compound will be hydrochloric acid; this will have no effect on the final weight of the precipitate since it is volatilized when the solid is dried.

Now, we shall consider the situation when the silver chloride precipitate is employed for the determination of chloride ion. Here the primarily adsorbed ion will necessarily be silver. Since no common volatile silver salts exist, no improvement in the purity of the precipitate can be expected through exchange of the anions of the wash liquid with those present initially in the counter-ion layer. The dilute nitric acid solution commonly employed serves only to prevent peptization. Notice that the analysis of silver by precipitation as silver chloride is inherently more accurate than the determination of chloride using the same precipitate.

Another example, wherein the purity of a precipitate can be improved through the judicious choice of wash liquid, arises when iron is precipitated as the hydrous ferric oxide; here, hydroxyl ions are adsorbed on the surface of the solid. By precipitating from and washing with solutions of ammonium nitrate, the counter-ion layer can often be made to consist largely of ammonium ions. A compound is thus produced that can be volatilized during subsequent ignition steps. Unfortunately, this treatment is not always successful in producing a pure oxide residue. Several of the metal hydroxides exhibit very great tendencies to adsorb on hydrous ferric oxide; these metal ions will predominate even when high concentrations of ammonium ion are maintained in the surrounding solution. Severe coprecipitation errors can result.

One final method for the reduction of adsorption errors involves filtration and washing, followed by solution and reprecipitation of the solid. In most cases only a small fraction of the contaminant present in the solvent is removed by the precipitate. When the solid is redissolved, the new solution will contain a much lower concentration of the contaminant than the original solvent. Consequently, the precipitate formed from this environment will contain much less of the impurity. This procedure is sometimes used to reduce the amount of metal hydroxides coprecipitated with hydrous ferric oxide. In this case the filtered and washed solid can be readily redissolved in acid to give a solution from which the second precipitate can be formed.

Reprecipitation obviously is not possible where good solvents for the precipitate do not exist. Furthermore, the time required for the operations and the possible mechanical losses of precipitate make this method undesirable, particularly where several reprecipitations are necessary.

Isomorphic Inclusion or Mixed Crystal Formation

Substances that crystallize in similar geometric forms are said to be *isomorphous* or *isomorphic*. To be isomorphic, two compounds must (1) have the same type formulas, (2) have components whose sizes relative to one another are about the same, and (3) have their component atoms or ions joined together by the same type of bonding. Such apparently unlike substances as potassium permanganate and barium sulfate are isomorphous because they both have anion to cation ratios of one, have anions and cations whose size ratios are nearly identical, and are both ionic compounds.

When the dimensions of two isomorphic compounds are approximately the same, the two substances are capable of replacing one another in a crystal without causing appreciable distortion in the crystal lattice. Replacement of this sort occurs during the crystallization of a compound in the presence of a substance isomorphic to it; this yields what is known as *mixed crystals* wherein a part of the sites normally occupied by the atoms or ions of the host compound are filled with the atoms or ions of the isomorphic substance.

During the formation of a crystalline precipitate, normally soluble isomorphic substances may be carried out of solution in the form of mixed crystals. The result is a type of coprecipitation that is called *isomorphic inclusion*. For example, if barium sulfate is formed in a solution containing small quantities of lead ion, contamination of the precipitate with lead sulfate occurs despite the fact that the solubility product of the lead compound is not exceeded. Since the two compounds are isomorphic, the lead ions replace some of the barium ions in a random fashion throughout the crystal lattice. This coprecipitation results in no appreciable distortion of the barium sulfate crystals.

When mixed crystal formation occurs under equilibrium conditions, the distribution of the contaminant between the solution and the host crystal is found to obey equilibrium laws over fairly wide concentration ranges. For example, it has been shown[4] that in the aforementioned case the distribution of lead between solvent and solid may be described by the following expression:

$$\frac{[Pb^{2+}]_{solution}}{[Pb^{2+}]_{solid}} = K \frac{[Ba^{2+}]_{solution}}{[Ba^{2+}]_{solid}}$$

[4] I. M. Kolthoff and G. E. Noponen, *J. Am. Chem. Soc.*, **60**, 197, 508 (1938).

Here K is an equilibrium constant and the brackets represent ion concentrations in the solution and the solid.

During a precipitation, formation of the solid is ordinarily so rapid that equilibrium conditions do not prevail; then this expression fails. On the other hand, during digestion and recrystallization, the tendency is for the solid to change its composition so as to approach the condition predicted by the foregoing relationship.

Coprecipitation resulting from mixed crystal formation presents major difficulties to the analytical chemist. There is really little he can do to prevent its occurrence other than to remove the offending ion before precipitation. Digestion is of little help since, at best, the concentration of contaminant will be reduced only to its equilibrium concentration. Washing is of no avail because equilibrium between solid and solvent is too slowly achieved. In theory, at least, reprecipitation ought to reduce the amount of isomorphic inclusion if the distribution constant is not too unfavorable. As pointed out in the previous section, however, this process is usually time consuming and tedious, and should be avoided wherever possible.

Nonisomorphic Inclusions or Solid Solutions

It has been found experimentally that small quantities of *nonisomorphic* substances are sometimes carried out of solution by precipitates in a form which, in many respects, resembles isomorphic inclusions. Such impurities appear to be more or less homogeneously distributed throughout the crystal; they do not cause sufficient distortion of the host crystal to form large imperfections; further, the amount of such coprecipitation appears to be essentially independent of the conditions existing at the time of precipitation and of the manner in which the precipitate is digested and washed. The problems in handling this type of coprecipitation are essentially those described in the section on isomorphic inclusion.

Solid-solution formation has been demonstrated to be the cause of coprecipitation of several compounds with barium sulfate.[5] For example, barium nitrate, ammonium bisulfate, sodium bisulfate, and potassium bisulfate all coprecipitate at least in part in this form.

Occlusion

Occlusion is a type of coprecipitation in which the soluble impurity is enclosed or entrapped within the crystal structure of the solid. As such it acts as a seat of imperfection within the crystal. Occlusion differs from the two previously mentioned types of coprecipitation in several respects. Generally, occluded substances are distributed in a nonhomogeneous fashion throughout the crystal;

[5] G. H. Walden and M. V. Cohen, *J. Am. Chem. Soc.*, **57**, 2591 (1935); G. Walton and G. H. Walden, *J. Am. Chem. Soc.*, **68**, 1742 (1946).

any given site of imperfection may contain a relatively large quantity of the impurity. In addition, the amount of occluded substance is greatly influenced by the conditions and manner of precipitate formation. Finally, digestion is often remarkably effective in reducing the quantity of entrapped contaminant.

Mechanism of occlusion. It is not too difficult to see how the process of occlusion might occur during the formation of crystalline precipitates. We shall consider as an example the precipitation of barium ion as the sulfate. During the early stages, the solution will contain an excess of barium ions, and the solid will have a large number of these ions adsorbed on its surface. With further additions of reagent, the sulfate ions find their way to the surface of the particles, displace the anions in the counter-ion layer, and combine with the adsorbed barium ions. This process undoubtedly takes place very rapidly; the possibility thus exists that some of the foreign counter ions may not be displaced in time to prevent their being completely surrounded by the growing solid. When this occurs, there will be an imperfection within the crystal that will contain internally adsorbed barium ions and counter ions.

If, indeed, adsorption plays an intermediate part in the occlusion process, then we might expect that reversal of the order of mixing would lead to coprecipitation of a different compound. In this case we might very well find coprecipitation of sulfate salts, for under these conditions it would be sulfate ions that were primarily adsorbed; cationic counter ions such as sodium would be entrapped by the rapid growth of the solid. The data shown in Table 8-3 tend to confirm this hypothesis. As expected, the coprecipitation of the anion is considerably greater in every instance where the precipitate was formed with an excess of barium ions in the solution than where an excess of sulfate was present. This situation is reversed for coprecipitation of sodium ions. Appreciable coprecipitation of anions does occur in some cases, however, even where large amounts of sulfate are present in the solution. Thus 217 micromols of nitrate and 107 micromols of chlorate ions are found in the barium sulfate under conditions where little adsorption of these ions is to be expected. This suggests that the mechanism proposed in the foregoing paragraphs is not sufficient to account for all of the coprecipitated impurities. In this connection it is necessary to point out that not all chemists believe that coprecipitation of nitrate ions with barium sulfate is actually the result of occlusion. For example, Walden and Cohen, on the basis of x-ray observations, are of the opinion that this is a solid solution phenomenon.[6] Schneider and Rieman, on the other hand, contend that the bulk of the coprecipitated nitrate shown in Table 8-3 is in the occluded form.[7]

There are some other points worth mentioning in connection with the data in Table 8-3. One of these is the correlation between solubilities of the various barium salts and the amounts coprecipitated. Such an agreement is to be expected from the Paneth-Fajans rule if indeed adsorption plays an important part in the mechanism of occlusion.

[6] G. H. Walden and M. V. Cohen, *J. Am. Chem. Soc.*, **57**, 2591 (1935).
[7] F. Schneider and W. Rieman, III, *Ibid.*, **59**, 356 (1937).

Table 8-3

COPRECIPITATION OF VARIOUS SPECIES IN FRESHLY FORMED BARIUM SULFATE[8]

Coprecipitated Ion	Amount of Coprecipitation; Micromols of Coprecipitated Ion/Four Millimols Barium Sulfate		Solubility of Barium Salt at 30° C, mol/1000 grams Water
	Excess Barium Salt Present	Excess Sodium Sulfate Present	
iodide	1.3	0.20	5.64
chloride	107	18	1.83
chlorate	390	107	1.37
nitrate	783	217	0.455
sodium	163	355	——

[8] Data taken from F. Schneider and W. Rieman, III, *J. Am. Chem. Soc.*, **59**, 354 (1937).

Also of interest is the large quantity of material occluded in some of the cases cited. For instance, if we assume that the nitrate which is coprecipitated in the presence of excess barium ion is brought down as barium nitrate, we can readily calculate that about 0.11 gram of that compound is present in each gram of barium sulfate. Obviously this can lead to serious analytical errors.

Effect of digestion on amount of occlusion. It has been found experimentally that the quantity of a substance occluded in a precipitate is reduced—often greatly—during digestion. This is illustrated by the data in Table 8-4.

The effectiveness of the digestion process undoubtedly arises from the rapid recrystallization and resultant perfection of the lattice that takes place while the solid stands in contact with its mother liquor. During this time, continuous solution and reprecipitation takes place; this opens up the crystal and exposes at least part of the sites containing the trapped impurities. These can then be replaced by the lattice ion present in excess in the mother liquor to give a more perfect crystal structure.

Reduction of errors arising from occlusion. Two somewhat different approaches have been suggested to decrease the amounts of occluded impurities accompanying analytical precipitates. The classical method has been to form the precipitate slowly—that is, under conditions of low supersaturation—and then to digest for a short period at elevated temperatures. If, as we have postulated, occlusion results from entrapping of the ions of the counter-ion layer by the rapidly growing solid, then we might expect a reduction in amount if the rate of growth of the crystal is reduced. The methods for achieving this would be the same as those for growing large crystals—namely, the slow addition of reagent, the use of dilute solutions, and precipitation from hot solutions. That such measures do improve the purity of precipitates has been amply demonstrated

Table 8-4

EFFECT OF DIGESTION ON AMOUNT OF LEAD NITRATE OCCLUDED IN A LEAD SULFATE PRECIPITATE[9]

Condition of Digestion	Nitrate Occluded in Precipitate after Various Times of Digestion, Percent			
	1 minute	1 hour	6 hours	24 hours
solution held at room temperature	0.64	0.42	0.29	0.20
solution held at 90° C		0.40		0.22
solution held at room temperature; made 0.01 N in HNO_3	0.39	0.37	0.29	0.19
solution held at 90° C; made 0.01 N in HNO_3		0.18	0.06	0.01

[9] I. M. Kolthoff and R. A. Halversen, *J. Phys. Chem.*, **43**, 605 (1939).

by many investigations; quite often, however, the digestion process is more effective than these measures in producing a pure solid.

An alternative method, proposed by Kolthoff and Sandell,[10] involves rapid formation of the precipitate from cold concentrated solutions followed by dilution of the resulting mixture and digestion overnight at elevated temperatures. The initial particle size of the solid formed under these conditions is smaller than that by the classical method; the amount of occlusion would be expected to be greater initially. On the other hand, digestion of such precipitates is more effective because more recrystallization occurs as a result of the larger specific surface area of the finely divided solid. Furthermore, with the smaller particles, the impurities can never be trapped as far from the surface; therefore, the likelihood that these will be exposed to the solution during digestion is appreciably greater. With some precipitates, at least, this increased effectiveness of digestion more than outweighs the deleterious effects of rapid formation of the solid. Unfortunately, it is not possible to decide which of these procedures will be the most satisfactory for a given precipitate; this can only be learned by experiment.

The washing of crystalline precipitates is not very effective for removing occluded impurities since these are retained within the crystal structure and therefore do not come in contact with the wash liquid.

Reprecipitation will generally improve the purity of precipitates suffering from contamination by occlusion. During the second precipitation, the amount of contaminant adsorbed will be lower since its concentration in the solution is less. As a result, less will be enclosed within the crystalline structure of the solid when it is again formed. Crystalline calcium oxalate precipitates are often purified by this procedure.

[10] I. M. Kolthoff and E. B. Sandell, *Textbook of Quantitative Inorganic Analysis*, 3d ed., 136. New York: The Macmillan Company, 1952.

Direction of Coprecipitation Errors

Coprecipitated impurities may lead either to high or low values for an analysis. If the contaminant is not a compound of the ion being determined, positive errors will always result. Thus a positive error results when colloidal silver chloride adsorbs silver nitrate during a chloride analysis. On the other hand, where the coprecipitant is a compound of the ion being determined, either positive or negative errors may be observed. In the determination of barium ions by precipitation as barium sulfate, for example, occlusion of barium salts occurs. If the occluded compound is barium nitrate (formula weight 261) a positive error will result since this compound has a greater formula weight than the barium sulfate (formula weight 233) that would have formed had no coprecipitation occurred. If barium chloride (formula weight 208) were the contaminant, however, a negative error would arise since its formula weight is less than that of the sulfate salt.

POST PRECIPITATION

Post precipitation is a rather special type of contamination wherein the impurity is an *insoluble compound* that precipitates after all or a major part of the analytical precipitate has formed. In contrast to coprecipitation, digestion increases the amount of this type of contamination.

Post precipitation arises only when the attempt is made to separate two ions on the basis of the rate at which they precipitate. There are actually several important analytical separations that owe their success to the fact that such rate differences do exist. For example, in the usual scheme for qualitative analysis, mercury and copper are separated from zinc by precipitation of the former pair as sulfides from dilute acid solutions. Under the usual conditions, the solubility product for zinc sulfide is also exceeded; the separation is possible only because precipitation of zinc sulfide from supersaturated solutions is an exceedingly slow process. As a matter of fact, in the absence of solids, supersaturated zinc sulfide solutions are sometimes stable for weeks or even months. However, in the presence of solids, particularly other sulfides, the precipitation is catalyzed. This is illustrated in Table 8-5. It is seen that in the absence of other sulfides almost no zinc sulfide is formed after 4.7 hours. However cadmium, copper, and mercury sulfide all cause the precipitation of part of the zinc sulfide to occur in 1 hour or less. Mercuric sulfide is particularly effective.

Kolthoff and Pearson have demonstrated that it is possible to separate copper sulfide quantitatively from zinc by filtration immediately after precipitation.[11] If the precipitate is allowed to stand in contact with the original solution for any length of time, however, post precipitation of zinc sulfide will cause high results for the recovery of copper.

---

[11] See I. M. Kolthoff and E. A. Pearson, *J. Phys. Chem.*, **36**, 549 (1932).

Table 8-5

POST PRECIPITATION OF ZINC SULFIDE ON OTHER SULFIDES

Time between Precipitation and Filtration, Hours	Percent Zinc Precipitated			
	Solid Sulfides Absent	CdS^{12} Precipitate Present	CuS^{12} Precipitate Present	HgS^{13} Precipitate Present
0.0	0			15
0.5				47
1.0		4.5	18	
1.5	0			
2.0		10		
2.5				
3.0		14	70	
4.7	1			

[12] Data taken from S. Glixelli, *Z. anorg. Chem.*, **55**, 297 (1907). Solutions were 0.5 *F* in $ZnSO_4$.

[13] Data taken from I. M. Kolthoff and R. Moltzau, *J. Phys. Chem.*, **40**, 779 (1936). Solutions were 0.025 *F* in $ZnSO_4$.

THE TECHNIQUE OF HOMOGENEOUS PRECIPITATION[14]

The chemist will frequently wish to form an analytical precipitate from solutions in which the supersaturation is as low as possible; generally this leads to a product that is more easily filtered, and in addition, often gives a precipitate that is less contaminated by coprecipitation. The degree of supersaturation can be reduced by the slow, dropwise addition of the reagent and by the use of dilute solutions of both the ion to be precipitated and the reagent. The obvious practical limits to these expedients are the increased solubility losses in the more dilute solutions and the tedium of the slow addition process. Furthermore, these techniques do not entirely eliminate high supersaturation and rapid precipitation. At the point in the solution where a new drop of reagent first makes contact there is, at least momentarily, a relatively high concentration of reagent. Until stirring disperses the drop, there exists a high degree of local supersaturation and a consequent rapid formation of the solid. Even with the most efficient stirring, this local effect cannot be entirely eliminated.

Homogeneous precipitation is an ingenious method for introducing a precipitating reagent in such a way as to avoid the local excesses that accompany mechanical addition of the reagent. Here precipitation is caused by slowly

[14] For an interesting review article on this subject see L. Gordon, *Anal. Chem.*, **24**, 459 (1952). See also L. Gordon, M. L. Salutsky, and H. H. Willard, *Precipitation from Homogeneous Solution*. New York: John Wiley and Sons, Inc., 1959.

generating the reagent throughout the entire solution by means of a suitable chemical reaction. During the process the solvent remains homogeneous with respect to the concentration of the ion being precipitated and the reagent; if the generating reaction is slow, high supersaturation is avoided and gradual formation of the solid takes place throughout the entire solution.

An important example of homogeneous precipitation involves the use of urea to generate hydroxyl ions for the precipitation of aluminum, ferric iron, thorium, and other heavy metal ions. The precipitating reagent is formed by the following reaction that takes place slowly at temperatures just below the boiling point of water:

$$(NH_2)_2CO + 3H_2O \rightarrow CO_2 + 2NH_4^+ + 2OH^-$$

One or two hours are required to complete the precipitation in a typical case. The physical appearance of the hydrous oxides so formed is dramatically different from that of the same compounds precipitated by slow addition of ammonia. Colloids are obtained by either method, but the densities of the coagulated solids in the first case are much greater than in the second. Generally, the homogeneously formed solids occupy from one tenth to one twentieth the volume of those formed by the classical procedure; as a result they are much easier to filter and wash. In addition the amount of coprecipitation associated with such precipitates is generally appreciably smaller. For example, Gordon[15] reports that a precipitate of the hydrous oxide of aluminum prepared by slow addition of ammonia to a solution containing 0.1 gram of the element and 1 gram of manganese was contaminated by a sufficient quantity of manganese salts to increase the weight of the ignited precipitate by 1.2 mg. When the aluminum was precipitated homogeneously by means of urea, the weight of contaminant was only 0.2 mg. Similar data for other compounds can be found in the literature. It seems probable that the improvement in these cases is a direct result of the decreased surface area associated with the denser precipitates and the consequent reduction in surface adsorption.

Many crystalline precipitates can also be formed homogeneously to give solids of greater particle size. The urea method, for example, can be applied to the precipitation of calcium oxalate. Here the calcium and oxalate ions are brought together in a solution that is sufficiently acid to prevent formation of the solid. Precipitation is then induced through homogeneous neutralization of the acid by heating with urea. Large coarse crystals result.

Barium ions have also been precipitated homogeneously as crystalline barium sulfate by slow hydrolysis of dimethyl sulfate.[16]

$$(CH_3O)_2SO_2 + 2H_2O \rightarrow 2CH_3OH + SO_4^{2-} + 2H^+$$

The particle size is again great since the precipitate is formed under conditions of very low supersaturation.

[15] L. Gordon, *Anal. Chem.*, **24**, 459 (1952).
[16] P. J. Elving and R. E. Van Atta, *Anal. Chem.*, **22**, 1375 (1950).

Table 8-6

METHODS FOR HOMOGENEOUS GENERATION OF ANIONS

Generated Anion	Reagent	Generation Reaction	Elements Precipitated
OH^-	urea	$(NH_2)_2CO + 3H_2O \rightleftharpoons CO_2 + 2NH_4^+ + 2OH^-$	Al, Ga, Th, Fe, Sn, Zr
PO_4^{3-}	triethyl phosphate	$(C_2H_5O)_3PO + 3H_2O \rightleftharpoons 3C_2H_5OH + H_3PO_4$	Zr, Hf
$C_2O_4^{2-}$	ethyl oxalate	$(C_2H_5O)_2C_2O_2 + 2H_2O \rightleftharpoons 2C_2H_5OH + H_2C_2O_4$	Mg, Zn, Ca
SO_4^{2-}	dimethyl sulfate	$(CH_3O)_2SO_2 + 2H_2O \rightleftharpoons 2CH_3OH + SO_4^{2-} + 2H^+$	Ba, Ca, Sr
CO_3^{2-}	trichloroacetic acid	$HC_2Cl_3O_2 + 2OH^- \rightleftharpoons CHCl_3 + CO_3^{2-} + H_2O$	La, Pr

That homogeneous formation of crystalline precipitates will lead to less occlusion than the classical methods seems reasonable when we recall that occlusion is believed to result from envelopment of adsorbed ions and counter ions by the rapidly growing crystalline solid. During homogeneous precipitation, crystal growth occurs less rapidly and there is, therefore, more time for the adsorbed ions to be replaced on the surface by the ions making up the lattice. This, of course, should lead to purer precipitates. Contamination by mixed-crystal or solid-solution formation, on the other hand, should not be greatly changed by the slower rate of precipitate formation.

In Table 8-6 are listed some of the methods that have been used to generate various anions homogeneously.

chapter 9. *The Scope and Applications of Gravimetric Analysis*

In previous chapters we have dealt with the principles underlying gravimetric analysis. A consideration of the more practical aspects of this important topic may now be undertaken.

It was pointed out in Chapter 4 that specificity of behavior is a most desirable property in a precipitating agent. Were a substance to exhibit true specificity, the need for preliminary separations would be entirely eliminated and the analysis would consist simply of isolating the compound formed between the component sought and the reagent. But such specificity is rarely, if ever, encountered. Instead, most precipitating agents are selective in behavior, forming slightly soluble substances with more than one species. Specificity is achieved only after preliminary separation has isolated the desired species from interfering substances. The need for pretreatment of the sample to eliminate potential interferences is more often the rule than the exception.

INORGANIC PRECIPITATING AGENTS

The lack of specificity on the part of most reagents is amply illustrated in Table 9-1, which indicates the scope of applications to which a number of inorganic precipitating agents have been put. Where a given reagent offers a method for the analysis of several ions, each ion represents a potential source of interference with respect to determination of the others.

Table 9-1

SOME APPLICATIONS OF INORGANIC PRECIPITATING AGENTS[1,2]

Precipitating Agent	Element Precipitated (weighing form is indicated in parentheses)
NH_3 (aq)	Be (BeO), Al (Al_2O_3), Sc (Sc_2O_3), Cr (Cr_2O_3)*, Fe (Fe_2O_3), Ga (Ga_2O_3), Zr (ZrO_2), In (In_2O_3) Sn, (SnO_2), U (U_3O_8)
H_2S	Cu(CuO)*, Zn (ZnO, or $ZnSO_4$), Ge (GeO_2), As (As_2O_3, or (As_2O_5), Mo (MoO_3), Sn (SnO_2)*, Sb (Sb_2O_3, or Sb_2O_5), Bi (Bi_2S_3)
$(NH_4)_2S$	Hg (HgS), Co (Co_3O_4)
$(NH_4)_2HPO_4$	Mg ($Mg_2P_2O_7$), Al ($AlPO_4$), Mn ($Mn_2P_2O_7$), Zn ($Zn_2P_2O_7$), Zr ($Zr_2P_2O_7$), Cd ($Cd_2P_2O_7$), Bi ($BiPO_4$)
H_2SO_4	Li, Mn, Sr, Cd, Pb, Ba (all as sulfates)
H_2PtCl_6	K (K_2PtCl_6, or Pt), Rb (Rb_2PtCl_6), Cs (Cs_2PtCl_6)
$H_2C_2O_4$	Ca (CaO), Sr (SrO), Th (ThO_2)
$(NH_4)_2MoO_4$	Cd ($CdMoO_4$)*, Pb ($PbMoO_4$)
HCl	Ag (AgCl), Hg (Hg_2Cl_2), Na (as NaCl from butyl alcohol), Si (SiO_2)
$AgNO_3$	Cl (AgCl), Br (AgBr), I (AgI)
$(NH_4)_2CO_3$	Bi (Bi_2O_3)
NH_4SCN	Cu ($Cu_2(SCN)_2$)
$NaHCO_3$	Ru, Os, Ir (pp'ted as hydrous oxides, reduced with H_2 to metallic state)
HNO_3	Sn (SnO_2)
H_5IO_6	Hg ($Hg_5(IO_6)_2$)
NaCl, $Pb(NO_3)_2$	F (PbClF)
$BaCl_2$	SO_4^{2-} ($BaSO_4$)
$MgCl_2$, NH_4Cl	PO_4^{3-} ($Mg_2P_2O_7$)

[1] Double underline indicates preferred method of analysis; single underline indicates most reliable gravimetric method; asterisk indicates that method is seldom used.
[2] Source: W. F. Hillebrand, G. E. F. Lundell, H. A. Bright, and J. I. Hoffman, *Applied Inorganic Analysis.* New York: John Wiley and Sons, Inc., 1953.

The majority of inorganic precipitating agents function by forming slightly soluble salts or hydrous oxides with the species being determined. In addition,

Table 9-2

GRAVIMETRIC METHODS BASED UPON REDUCTION TO THE ELEMENTAL STATE

Reducing Agent	Analysis
SO_2	Se, Au
$SO_2 + H_2NOH$	Te
H_2NOH	Se
$H_2C_2O_4$	Au
H_2	Re, Ir
HCOOH	Pt
$NaNO_2$	Au
$TiCl_2$	Rh
$SnCl_2$	Hg
Electrolytic reduction	Co, Ni, Cu, Zn, Ag, In, Sn, Sb, Cd, Re, Bi

however, several metal ions can be isolated in elemental form either by electrochemical reduction or by treatment with a suitable chemical reducing agent. Table 9-2 summarizes these applications; organic as well as inorganic reducing agents have been employed for this purpose.

ORGANIC PRECIPITATING AGENTS

Many organic compounds are used as precipitating agents. Of the several ways in which they react, the most important involve either the formation of slightly soluble non-ionic complexes—known also as *coordination compounds*—or the formation of slightly soluble salts.

Reagents which Produce Slightly Soluble Coordination Compounds

In general, a slightly soluble coordination compound is formed from bonding between an organic molecule and a metal ion that results in an uncharged five- or six-membered ring. This requires the presence of two functional groups, properly oriented within the organic molecule, that have the ability to bond with the metal ion. Insofar as the products are nonpolar, they will have low solubility in water and high solubility in organic liquids. Typically, these precipitates are low-density solids; they often possess intense and characteristic colors. Because they are not wetted by water, they are readily freed of moisture at low temperatures. At the same time, the hydrophobic nature of these precipitates endows them with the annoying tendency to creep during the washing and transferring operations.

8-Hydroxyquinoline. This compound, known also as *oxine*, has the structure

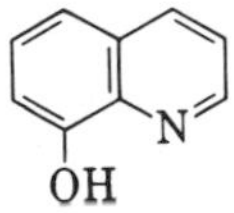

8-hydroxyquinoline

A metal ion is bonded to this molecule through the oxygen and nitrogen atoms. By way of specific example, the product with magnesium can be represented as

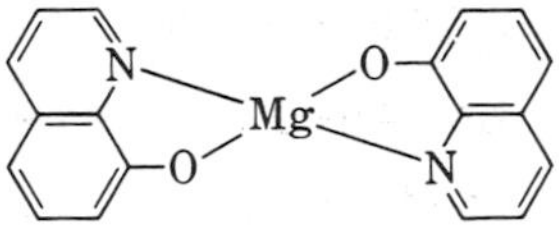

Since hydrogen ions are evolved in the formation of metal oxines, their solubilities will be markedly influenced by the pH of the medium, becoming less soluble as the acidity increases. In addition, some of these precipitates again become soluble in alkaline media. Use may be made of this property to alter somewhat the selectivity of the reagent. Figure 9.1 indicates the pH ranges over which a number of metal oxines are quantitatively precipitated.

Notwithstanding its rather low selectivity, a large number of methods have been developed that make use of 8-hydroxyquinoline in conjunction with other reagents. In most of these the coordination compound itself serves as the weighing form with which the determination is completed. The most widespread applications of the reagent are concerned with the analysis for magnesium and aluminum.[3]

Phenylnitrosohydroxylamine. The ammonium salt of phenylnitroso-hydroxylamine, better known as *cupferron*, is a somewhat more selective organic reagent; it has the structure

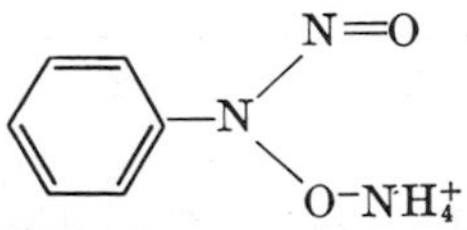

cupferron

Compound formation with a metal ion involves displacement of the ammonium ion in addition to coordination with the nitroso oxygen. Table 9-3 is indicative

[3] For a definitive reference with respect to the applications of 8-hydroxyquinoline, see R. G. W. Hollingshead, *Oxine and its Derivatives*, in 4 volumes. London: Butterworth's Scientific Publications, 1954.

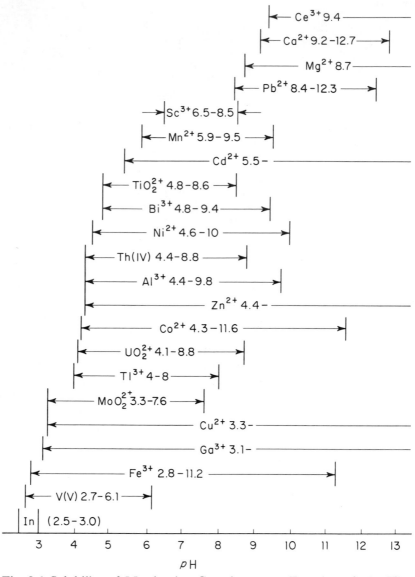

Fig. 9.1 Solubility of Metal-oxine Complexes as a Function of *p*H. The approximate ranges over which quantitative precipitation may be expected are shown.

of the applications of cupferron. The precipitates formed are of uncertain composition; careful ignition to the oxide is a requirement.

Excess cupferron is fairly readily removed from filtrates by heating with concentrated nitric acid. It can thus be advantageously employed for the pre-

Table 9-3

SEPARATIONS EMPLOYING AMMONIUM NITROSOPHENYLHYDROXYLAMINE
(Cupferron)

Medium	Elements Which Form Precipitates	Elements Which Do Not Form Precipitates
Acidic (5-10% HCl or H_2SO_4)	Fe(III), Zr, Ti, Sn(IV) Ta, U(IV), V(V)	Al, Be, P, B, Mn, Ni, U(VI), Cu, Zn, Pb, Ag, Th, Co, alkali, alkaline earths
Neutral to slightly alkaline[4]	Al, Be, Cr(III), Tl(III) In, U(VI), rare earths, Cu^{2+}, Ce^{3+}	

[4] Most useful in acid solution, much of the selectivity of the reagent is lost as solution becomes alkaline.

liminary separation of unwanted cations, and can itself be eliminated before undertaking an analysis with some other reagent.

Dimethylglyoxime. Both oxine and cupferron show a marked lack of specificity. By way of contrast, dimethylglyoxime is virtually without peer with respect to this property. It will precipitate only palladium quantitatively from acid media, and only nickel from weakly basic solutions.

There are three possible geometric isomers of the reagent.

Of these, only the anti form has analytical uses. Owing to its rather low solubility in water, alcoholic solutions of the reagent are ordinarily employed.

Dimethylglyoxime gives a compound with nickel having the structure

This forms as an intense red precipitate which is so bulky that only small quantities of nickel can be conveniently handled. The solid is readily dried at 110° C and is of definite composition.

Experimental directions for the gravimetric determination of nickel with this reagent are given in a later section.

Reagents that Produce Saltlike Precipitates

A number of important organic precipitating agents produce slightly soluble compounds in which the bond with the species precipitated is primarily ionic in character.

Sodium tetraphenylboron. Sodium tetraphenylboron has the formula $(C_6H_5)_4B^-Na^+$, and in cold mineral acid solution is a specific precipitating agent for potassium and ammonium ions. Under these conditions, only mercury (II), rubidium, and cesium interfere and must be removed by prior treatment. The precipitates are stoichiometric, corresponding to the potassium or ammonium salt, as the case may be. Analysis for potassium by this reagent is an improvement over the perchlorate method, where magnesium, calcium, aluminum, iron, cobalt, nickel, manganese, sulfate, and phosphate ions interfere.

Benzidine. Another salt-forming reagent is benzidine

$$H_2N-\hspace{-0.3em}\underset{\text{benzidine}}{\underbrace{\bigcirc\hspace{-0.3em}\bigcirc}}\hspace{-0.3em}-NH_2$$

which precipitates sulfate from a slightly acid medium as $C_{12}H_{12}N_2 \cdot H_2SO_4$. The solubility of the precipitate increases rapidly with temperature and also with the acidity of the environment; both of these variables must be carefully controlled. Instead of weighing as a gravimetric precipitate, benzidine sulfate may be titrated with a standard solution of sodium hydroxide. The method succeeds in the presence of copper, cobalt, nickel, zinc, manganese (II), iron (II), chromium (III), and aluminum ions. Yet another method for completion of the analysis calls for titration of the benzidine with a standard solution of permanganate. Benzidine is well suited to the rapid analysis of sulfate on a routine basis.

Substituted arsonic acids. Substituted arsonic acids have the structure

$$R-As{\overset{\displaystyle OH}{\underset{\displaystyle OH}{=}}}O$$

where R represents an organic radical, such as phenyl or propyl.

A large number of arsonic acids have been found to yield saltlike precipitates with such quadrivalent metal ions as tin, zirconium, titanium, and thorium. The composition of the precipitates generally involves approximately 2 mols of arsonic acid per mol of quadrivalent cation. The nature of the organic portion of the molecule (R) determines, to some extent, the cations that form precipitates as well as the conditions under which they are formed. Whether or not ferric ion will precipitate, for example, depends upon the particular arsonic acid employed.

Because they are difficult to dry without decomposition, metallic arsonates are ignited to their respective oxides.

The reagents discussed in this section represent but a few of literally hundreds that have been proposed as organic precipitating agents.[5] Characteristics of a few others are listed in Table 9-4.

SELECTED METHODS OF ANALYSIS

Determination of Water in Barium Chloride Dihydrate

The water of hydration in a crystalline hydrate such as $BaCl_2 \cdot 2H_2O$ is present in stoichiometric proportions.[6] Thus the determination of the water content of a sample containing such a compound provides a basis for the estimation of that substance. The analysis may be performed either by collecting and weighing the water evolved from a measured quantity of sample or by weighing the residue and determining the weight of water by difference.

Procedure. Throughout this experiment perform all weighings to the nearest 0.1 mg.

Carefully clean two weighing bottles. Dry them for 1 hour at 105 to 110° C, using the beaker and watch-glass arrangement described in Chapter 5 to prevent accidental contamination. After they have cooled to room temperature in a desiccator, determine the weight of each bottle. Repeat this operation until a constant weight has been attained. Next introduce a quantity of unknown into each bottle and determine its weight. Heat the samples for about 2 hours at 105 to 110° C, then cool and weigh as before. Repeat the heating cycle until constant weights for the bottles and their contents have been achieved. Report the percentage of water in the sample.

NOTES:

1. If the unknown consists of the pure dihydrate, $BaCl_2 \cdot 2H_2O$, take samples weighing approximately 1 gram. If the unknown is a mixture of the dihydrate and some anhydrous diluent, obtain the proper sample size from the instructor.

[5] For further information regarding these and other organic precipitating agents see J. F. Flagg, *Organic Reagents Used in Gravimetric and Volumetric Analysis.* New York: Interscience Publishers, Inc., 1948.

[6] See Chapter 31 for a comprehensive discussion of the types of water present in solids.

Table 9-4

SOME ORGANIC PRECIPITATING AGENTS AND THEIR APPLICATIONS

Name	Structure	Analysis	Medium	Interfering Substances	Noninterfering Substances
α-Benzoin oxime		Cu	ammoniacal tartrate	$W(VI)$, $V(V)$	Al, Cd, Co, Fe, Pb, Ni, Zn
o-Aminobenzoic acid (anthranilic acid)		WO_4^{2-}, MoO_4^{2-} Cd, Co, Cu, Zn	strongly acid neutral, to weakly acid	$Cr(VI)$, $V(V)$ heavy metals, in general	Fe alkali, alkaline-earth metals, NH_4^+
α-Nitroso-β-naphthol		Co, also Fe, Zr	acid	Bi, $Cr(III)$, Ag, $Ti(IV)$, $W(VI)$, $U(VI)$, $V(V)$	Al, Be, As, Cd, Cu, Pb, Mn, Ni, Zn, Ga, PO_4^{3-}
4,5-Dihydro-1,4-diphenyl-3,5-phenylimino-1,2,4,-triazole (nitron)		NO_3^-, ClO_4^- ReO_4^-, WO_4^{2-}	slightly acid	ClO_3^-, Br^-, I^-, SCN^-, NO_2^-, CrO_4^{2-}	cations, organic acids

Name	Structure	Analysis	Medium	Interfering Substances	Noninterfering Substances
Salicylaldoxime		Cu, Bi, Pb, Pd, Zn	variable	variable, depending upon pH	variable, depending upon pH
Thioglycolic acid, β-aminonaphthalide (thionalide)		Sb, As, Bi, Cu, Hg, Au, Ag, Pt, Pd, Sn	acid	oxidizing agents	Pd, Cd
		Cd, Cu, Au, Hg, Tl	carbonate + tartrate	same	
		Sb, Bi, Au, Pb, Tl, Sn	carbonate + tartrate + cyanide	same	
		Tl	strongly alkaline	same	

2. Barium chloride can be heated to elevated temperatures without danger of decomposition. If desired, the analysis can be performed in crucibles with a Bunsen flame as the source of heat.

Determination of Chloride in a Soluble Sample

The chloride content of a soluble salt can be determined by precipitation of that ion as its silver salt

$$Ag^+ + Cl^- \rightarrow AgCl$$

The precipitate is collected in a filtering crucible, washed, and brought to constant weight by drying at 105° to 110° C. Precipitation is carried out in acid solution to eliminate potential interference from anions of weak acids (for example, CO_3^{2-}) which form precipitates with silver in neutral media. A moderate excess of silver ion is required to diminish the solubility of the precipitate, but large excesses are undesirable because they will lead to serious coprecipitation.

Silver chloride first precipitates as a colloid; it is coagulated with heat and a relatively high electrolyte concentration. A small quantity of nitric acid is added to the wash liquid to maintain the electrolyte concentration and prevent peptization during washing; the acid is volatilized during the subsequent heat treatment.

In common with other silver halides, silver chloride is susceptible to photo-decomposition, the reaction being

$$AgCl \rightarrow Ag + \tfrac{1}{2} Cl_2$$

The precipitate acquires a violet color due to the accumulation of finely divided silver. If photochemical decomposition occurs in the presence of excess silver ion, the following reaction takes place:

$$3Cl_2 + 3H_2O + 5Ag^+ \rightarrow 5AgCl + ClO_3^- + 6H^+$$

This will cause the analytical results to be high. In the absence of silver ion, the results will be low. Dry silver chloride is virtually unaffected by exposure to light.

Unless elaborate precautions are taken, some photochemical decomposition of silver chloride is unavoidable but, with reasonable care, this effect will not induce an appreciable error in the analysis.

Iodide, bromide, and thiocyanate, if present in the sample, will be precipitated along with silver chloride and cause high results. In addition the chlorides of tin and antimony are likely to hydrolyze and precipitate under the conditions of the analysis.

Procedure.   Clean and dry three fritted glass or porcelain filtering crucibles; bring these to constant weight during periods of waiting in the analysis.

Dry the sample at 105 to 110° C for 1 to 2 hours in a weighing bottle (p. 112). Store in a desiccator while cooling. Weigh (to the nearest 0.1 mg) individual 0.4-gram samples into 400-ml beakers, and to each add about 200 ml of distilled water and 1 to 2 ml of concentrated nitric acid. Slowly, and with good stirring, add 5-percent silver nitrate to the cold solution until the precipitate is observed to coagulate, and then add an additional 3 to 5 ml. Heat the solution almost to boiling, and digest the precipitate at this temperature for about 10 minutes; check for completeness of precipitation by adding a few drops of silver nitrate to the supernatant liquid; should additional silver chloride appear, continue the addition of silver nitrate until precipitation is complete. Store in a dark place for at least 1 to 2 hours, preferably overnight. Then decant the supernatant through a tared filtering crucible. Wash the precipitate several times while it is still in the beaker with a cold solution consisting of 1 to 3 ml of concentrated nitric acid per liter of distilled water; decant these through the filter also. Finally, transfer the bulk of the precipitate to the crucible, using a rubber policeman to dislodge any particles that adhere to the walls of the beaker. Continue washing until these are found to be substantially free of silver ion. Dry the precipitates to constant weight at 105 to 110° C. Report the percentage of chloride in the sample.

NOTES:

1. To determine the approximate amount of silver nitrate needed, calculate the volume required on the assumption that the sample is 100 percent sodium chloride.

2. Washings are readily tested for their silver content by collecting a few milliliters in a test tube and treating with a few drops of hydrochloric acid. Washing is judged complete when little or no turbidity is observed with this test.

Analysis of a Soluble Sulfate

The analysis of a soluble sulfate is based upon precipitation with barium ion

$$Ba^{2+} + SO_4^{2-} \rightarrow BaSO_4$$

The barium sulfate is collected on a suitable filter, washed with water, and strongly ignited.

Superficially, this method appears straightforward. In fact, however, it is subject to a multiplicity of interferences, due chiefly to the tendency of barium sulfate to occlude anions as well as cations. Table 9-5 summarizes many of the more common interferences affecting this analysis. Purification by reprecipitation is not feasible because there is no practical solvent for barium sulfate. It is therefore necessary to eliminate the principal interferences by preliminary treatment of the sample, and then to precipitate the barium sulfate from hot, dilute solutions. Even then the excellent agreement often observed between theoretical and experimental results is due in considerable measure to a cancellation of errors.

Procedure. Dry the unknown for 1 hour at 105 to 110° C. Weigh (to the nearest 0.1 mg) individual 0.5 to 0.7-gram samples into 400-ml beakers and dissolve each in 200 ml of distilled water to which 2 ml of concentrated hydrochloric acid have been added.

For each sample dissolve 1.3 grams of barium chloride dihydrate in 100 ml of distilled water, and filter if necessary. Heat nearly to boiling before

Table 9-5

INTERFERENCES ATTENDING THE GRAVIMETRIC DETERMINATION OF SULFATE AS $BaSO_4$

Effect upon Analysis	Nature of the Interference
Low Results	Excessive amounts of mineral acid present. (Solubility of $BaSO_4$ is appreciably greater in strongly acid media.)
	Coprecipitation of sulfuric acid. (Note that this is not a source of error in a gravimetric determination of barium, since this H_2SO_4 is driven off during ignition.)
	Coprecipitation of alkali metal and calcium ions. (Sulfates of these ions weigh less than the equivalent amount of $BaSO_4$ which should have formed.)
	Coprecipitation of ammonium ion. (Ammonium sulfate is volatilized upon ignition of the precipitate.)
	Coprecipitation of iron as a basic ferric sulfate.
	Partial reduction of $BaSO_4$ to BaS if filter paper is charred too rapidly.
	Presence of trivalent chromium. (May not achieve complete precipitation of $BaSO_4$ due to formation of soluble complex chromic sulfates.)
High Results	Absence of mineral acid. (The slightly soluble carbonate or phosphate of barium can precipitate under these conditions.)
	Coprecipitation of barium chloride.
	Coprecipitation of anions, particularly nitrate, chlorate as barium salts.

quickly adding, with vigorous stirring, to the samples that have also been heated.

Digest the precipitated barium sulfate for 1 to 2 hours. Decant the supernatant through a fine ashless paper (Note 1). Wash the precipitate three times with hot water, decanting the washings through the filter. Finally, transfer the precipitate to the paper. Place papers and contents in porcelain crucibles that have been ignited to constant weight; gently char off the papers. Ignite to constant weight at 900° C. Report the percentage of sulfate in the sample.

NOTE:

Use of Schleicher and Schuell No. 589 White Ribbon, or Whatman

No. 42 paper is recommended. If desired, the precipitate can be collected in a Gooch crucible or a porcelain filtering crucible.

Analysis of Iron in a Soluble Sample

The analysis of iron in a soluble sample is based upon the precipitation of ferric iron as a hydrous oxide, followed by ignition to Fe_2O_3

$$2Fe^{3+} + 6NH_3 + (x + 3)H_2O \rightarrow Fe_2O_3 \cdot xH_2O + 6NH_4^+$$

$$Fe_2O_3 \cdot xH_2O \rightarrow Fe_2O_3 + xH_2O$$

The hydrous oxide forms as a gelatinous mass which rapidly clogs the pores of most filtering media, and thus a very coarse grade of ashless paper is employed. Even here, however, it is best to delay transfer of the precipitate as long as possible and to wash by decantation.

The ultimate employment of Fe_2O_3 as a weighing form requires that all of the iron present be in the ferric state; this is readily achieved by treating the sample with nitric acid before precipitating the hydrous oxide

$$3Fe^{2+} + NO_3^- + 4H^+ \rightarrow 3Fe^{3+} + NO + 2H_2O$$

The complex $FeSO_4 \cdot NO$ sometimes imparts a very dark color to the solution; it decomposes upon further heating.

Owing to its negligible solubility, the precipitate can be safely washed with a hot solution of ammonium nitrate.

> **Procedure—double-precipitation method.** Unless directed otherwise, do not dry the sample. Consult the instructor for the proper sample size; weigh these into 600-ml beakers. Dissolve in 20 to 30 ml of distilled water to which about 5 ml of concentrated hydrochloric acid have been added. Then add 1 to 2 ml of concentrated nitric acid; heat to complete the oxidation and to remove any oxides of nitrogen. Dilute to 350 or 400 ml and slowly add, with good stirring, freshly filtered aqueous ammonia until precipitation is complete (Notes 1 and 2). Digest briefly, and then check for completeness of precipitation with a few additional drops of ammonia. While still warm, decant the clear supernatant through ashless filter paper (Note 3), and wash the precipitate in the beaker with two 30-ml portions of hot 1-percent ammonium nitrate solution.
>
> Return the filter papers to their appropriate beakers. Add about 5 ml of concentrated hydrochloric acid to each and macerate the paper thoroughly with a stirring rod. Dilute to about 300 ml with distilled water and reprecipitate the hydrous ferric oxide as before. Again decant the filtrate through ashless filter paper, and wash the precipitate repeatedly with hot 1-percent ammonium nitrate. When the filtered decantate gives little or no test for chloride ion (Note 4), transfer the precipitate quantitatively to the filter cone. Remove the last traces of precipitate adhering to the walls of the beaker by scrubbing with a small piece of ashless paper. Allow the precipitate to drain overnight, if possible. Then transfer the paper and contents to a porcelain crucible that has previously been ignited to constant weight. Char the paper at low temperature, taking care to allow free access of air (Note 5). Gradually increase the

temperature until all of the carbon has been burned away. Then ignite to constant weight at 900 to 1000° C.

NOTES:

1. Upon prolonged contact, aqueous ammonia solutions attack the glass of their containers and thereby become contaminated with particles of silica. It is a wise precaution to filter the ammonia prior to use.

2. Aqueous ammonia should be added until its odor is unmistakeable over the solution. The precipitate should appear reddish brown. If it is black (or nearly so), the presence of ferrous iron is indicated; the sample is best discarded.

3. A porous grade of ashless paper is required for this gelatinous precipitate. Schleicher and Schuell No. 589 Black Ribbon or Whatman No. 41 are satisfactory papers for this. The time required for filtration is shortened by performing the principal washing of the precipitate while it is still in the beaker. After it has been transferred to the paper, the rate of filtration is markedly slowed.

4. Before making a test for chloride in the washings, the sample of filtrate taken must first be acidified with dilute nitric acid.

5. Heat lamps are convenient for the initial charring of the filter papers.

Procedure—single-precipitation method.   Proceed as directed in the first paragraph of the double-precipitation method, making the following changes:

1. Instead of washing with but two 30-ml portions of 1-percent ammonium nitrate, wash repeatedly until the decantate is essentially free of chloride.

2. Then transfer the precipitate quantitatively to the filter and complete the analysis as before.

Precipitation of Iron by a Homogeneous Precipitation Method

Iron may also be precipitated by a homogeneous precipitation method. This procedure differs from the ordinary technique in that the precipitant, hydroxyl ion, is uniformly and gradually produced throughout the solution as a consequence of the slow decomposition of urea

$$H_2N-\overset{\overset{\displaystyle O}{\|}}{C}-NH_2 + 3H_2O \rightarrow 2NH_4^+ + CO_2 + 2OH^-$$

As hydrogen ions are consumed in this process, the solution gradually becomes alkaline. Ultimately ferric iron is deposited in the form of its hydrous oxide or as a basic salt; in either event, it is filtered, washed, and ignited to Fe_2O_3. The homogeneous generation of hydroxyl ions yields a rather dense precipitate that is less subject to contamination than that formed by conventional means.

The procedure given here represents a modification of the method reported by Willard and Sheldon[7] and involves precipitation of the iron as a basic formate. The sample is first dissolved in hydrochloric acid and treated with nitric acid to oxidize the iron. A requirement for production of a dense precipitate is that the solution be appreciably acidic at the outset. A large excess of hydrogen ions, on the other hand, is undesirable because of the large quantity of urea needed and the extended time required for neutralization. The acidity of the sample is therefore adjusted by the successive addition of ammonia and hydrochloric acid. Then formic acid and ammonium chloride are introduced, the former to yield the desired basic formate precipitate, and the latter to help maintain the solution at the desired pH. Urea is then introduced and heating is commenced. The time required will depend upon the initial acidity and the temperature; ideally precipitation should begin in about 1 hour. Toward the end of the heating period, hydrogen peroxide is added to oxidize any ferrous iron that may have resulted from reaction with formic acid. After filtration the precipitate is ignited to ferric oxide, Fe_2O_3.

Procedure. Carefully weigh replicate samples into 600-ml beakers; the weight taken should be such as will contain approximately 100 mg of iron. Dissolve in about 50 ml of water to which 5 ml of concentrated hydrochloric acid have been added. Then introduce 1 to 2 ml of concentrated nitric acid to each sample and boil gently to remove the oxides of nitrogen.

Dilute to approximately 400 ml with distilled water and add concentrated ammonia slowly until the first permanent appearance of the hydrous oxide is observed. Add dilute hydrochloric acid slowly until the precipitate has definitely been discharged and the solution is again clear. Next introduce about 2 ml of concentrated formic acid, 15 grams of ammonium chloride, and 4 to 5 grams of urea. Heat the samples carefully, maintaining a gentle boil (Note 2). Continue heating until precipitation is complete (Note 3). Toward the end of the heating period introduce about 5 ml of 3-percent hydrogen peroxide.

For filtration use ashless paper suitable for the collection of fine, crystalline precipitates (Note 4). Decant most of the supernatant through the filter, taking care to avoid the undue transfer of precipitate. Wash the precipitate three times with 20 to 30-ml portions of hot 1-percent ammonium nitrate solution and decant the washings through the filter. Then quantitatively transfer the bulk of the precipitate. Shred a half circle of ashless paper into each beaker and pour about 5 ml of concentrated hydrochloric acid slowly down the stirring rod. Scrape the stained surfaces of the beaker with the macerated paper, using the stirring rod to provide direction. Dilute with 75 to 100 ml of distilled water, heat nearly to boiling, and slowly add aqueous ammonia until the solution is definitely alkaline. As before, wash the residue in the beaker by decantation before transferring to the filter paper containing the bulk of the precipitate.

Ignition of the residues is carried out in the manner previously described.

NOTES:

1. These instructions are designed for samples that contain approximately 0.100 gram of iron. With 5 grams of urea, as much as 0.200 gram of iron can be tolerated.

2. Each beaker should be supplied with a special stirring rod to aid in the prevention of bumping. The end of an ordinary stirring rod is heated to the softening point and then firmly pressed into the point of an ordinary

[7] H. H. Willard and J. L. Sheldon, *Anal. Chem.*, **22**, 1162 (1950).

thumbtack. The indentation significantly reduces the danger of loss due to local overheating.

3. Completion of the heating period can be ascertained by measuring the pH of the supernatant. Alternatively, heating may safely be discontinued when it is observed that the supernatant is clear. It is desirable to allow the precipitate to stand overnight prior to filtration.

4. Schleicher and Schuell No. 589 White label, or Whatman No. 42 ashless paper is recommended; 9 or 11 cm circles are convenient.

Determination of Nickel in Steel

The method for determining nickel in steel is based upon the precipitation of nickel from a slightly alkaline solution with an alcoholic solution of dimethyl-glyoxime. Tartaric acid is introduced to prevent interference from iron. The organic-nickel compound serves as a convenient weighing form.

Owing to the bulky character of the precipitate there is a maximum quantity of nickel that can be conveniently handled. The sample weight taken is governed by this consideration. The excess of precipitating agent must be controlled, not only because its solubility in water is low, but also because the nickel compound becomes appreciably more soluble in the presence of alcohol.

Procedure. Weigh individual samples containing between 30 and 35 mg of nickel into 400-ml beakers and dissolve by warming with about 50 ml of 1:1 hydrochloric acid. Carefully introduce about 10 ml of 1:1 nitric acid and boil gently to expel the oxides of nitrogen. Dilute the resulting solution to 200 ml and heat nearly to boiling. Introduce 5 to 6 grams of tartaric acid (Note 1), and neutralize with aqueous ammonia until a faint odor of ammonia can be detected after blowing away the vapors over the solution; add 1 to 2 ml in excess. If the solution is not clear at this stage, proceed as directed in Note 2. Make the solution slightly acid with hydrochloric acid, heat to 60 to 80° C, and add 20 ml of a 1-percent alcoholic solution of di-methylglyoxime. Then, with good stirring, introduce sufficient dilute ammonia until a slight excess is present as indicated by the odor, plus an additional 1 to 2 ml. Digest for 30 to 60 minutes at about 60° C, cool for at least 1 hour, and filter through weighed filtering crucibles. Wash with water until free of chloride. Finally, bring crucibles and contents to constant weight by drying at 110 to 120° C. Report the percentage of nickel in the sample. The precipitate, $NiC_8H_{14}O_4N_4$, contains 20.31 percent nickel.

NOTES:

1. Tartaric acid is conveniently introduced as a strong solution (25 grams diluted to 100 ml). If necessary, this should be filtered prior to use.

2. If a residue is formed upon the addition of base, the solution should be acidified, treated with additional tartaric acid, and again made alkaline. Alternatively, the residue can be filtered through paper. If this is done, thorough washing with a hot, dilute NH_3/NH_4Cl solution is required; the washings should be combined with the rest of the sample.

3. Gooch crucibles, or filtering crucibles of porcelain or fritted glass, may be used for collection of the precipitate.

part 3. *Volumetric Analysis*

chapter 10. *An Introduction to Volumetric Methods of Analysis*

A volumetric method is one in which the analysis is completed by measuring the volume of a solution of established concentration needed to react completely with the substance being determined. Ordinarily, volumetric methods are equivalent in accuracy to gravimetric procedures and are more rapid and convenient; their use is widespread.

DEFINITION OF SOME TERMS

A *titration* is a process wherein the capacity of a substance to combine with a reagent is quantitatively measured. Ordinarily this is accomplished by the controlled addition of a reagent of known concentration to a solution of the substance until reaction between the two is judged to be complete; the volume of reagent is then measured. Occasionally it is convenient or necessary to carry out a volumetric analysis by adding an excess of the reagent and then determining the excess by titration with a second reagent of known concentration. The second titration is called a *back-titration*.

The reagent of exactly known composition used in a titration is called a *standard solution*. The accuracy with which its concentration is known sets a definite limit upon the accuracy of the method; for this reason, much care is taken in the preparation of standard solutions. Commonly the concentration of a standard solution is arrived at in either of two ways: (1) a carefully measured quantity of a pure compound is titrated with the reagent and the concentration calculated from the weight and volume measurements; or (2) the standard solution is prepared by dissolving a carefully weighed quantity of the pure reagent itself in the solvent; this is then diluted to an exactly known volume. In either method, a highly purified chemical compound—called a *primary standard*—is required as the reference material. The process whereby the concentration of a standard solution is determined by titration of a primary standard is called a *standardization*.

The goal of every titration is the addition of standard solution in such amount as to be chemically equivalent to the substance with which it reacts. This condition is achieved at the *equivalence point*. For example, the equivalence point in the titration of sodium chloride with silver nitrate is attained when exactly one formula weight of silver ion has been introduced for each formula weight of chloride ion present in the sample. In the titration of sulfuric acid with sodium hydroxide, the equivalence point occurs when two formula weights of the latter have been introduced for each formula weight of the former.

The equivalence point in a titration is a theoretical concept. In actual fact we can only estimate its position by observing physical changes associated with it in the solution. The point in a titration where such changes manifest themselves is called the *end point*. It is to be hoped that the volume difference between the end point and equivalence point will be small. Differences do arise, however, owing to inadequacies in the physical changes and our ability to observe them. This results in an analytical error called a *titration error*.

One of the common methods of end-point detection employed in volumetric analysis involves the use of supplementary chemical compounds that exhibit changes in color as a result of concentration changes occurring near the equivalence point. Such substances are called *indicators*.

REACTIONS AND REAGENTS USED IN VOLUMETRIC ANALYSIS

Desirable Characteristics of a Volumetric Reaction

In order to be suitable for a volumetric analysis, a chemical reaction should meet certain requirements. (1) The reaction ought to be rapid. Normally, a titration involves addition of the reagent in small increments and observation of the solution for the end point. If the chemical reaction is a slow one, waiting is necessary after each addition and the whole process becomes prohibitively slow and tedious. (2) The reaction should proceed reasonably far towards completion. As we shall show later, this second condition is usually necessary in order for there to be a satisfactory end point for the titration. (3) The reaction must be such that it can be described by a balanced chemical equation; otherwise the weight of the sought-for substance cannot be calculated from the volumetric data. This requirement implies the absence of side reactions between the reagent and the unknown or other constituents of the solution. (4) There must be available a method for detecting the equivalence point in the reaction; that is, a satisfactory end point is required.

Not all volumetric reactions currently in use meet the foregoing requirements perfectly; the most widely accepted procedures, however, are all founded on reactions that closely approach them.

Reaction Types

Volumetric methods may be divided conveniently into four categories based upon reaction type. These include (1) precipitation reactions, (2) complex-formation reactions, (3) neutralization or acid-base reactions, and (4) oxidation-reduction reactions. These categories differ in such things as types of equilibria, kinds of indicators, nature of reagents, kinds of primary standards, and definitions of equivalent weight.

Standard Solutions and Primary Standards

A standard solution is generally used to make several analyses; since the quality of these analyses is directly related to the accuracy with which the concentration of the reagent is known, the chemist ordinarily expends considerable effort to assure himself that the materials and methods used for preparation and standardization will lead to a solution of accurately known concentration.

Requirements of a primary standard. Certain properties of a compound are required if the substance is to serve as a primary standard. (1) It must be of the highest purity, and in order to assure this, established methods should be available for the testing of its purity. (2) A primary standard substance should be stable. It should not be attacked by constituents of the atmosphere. (3) The compound should not be hygroscopic, nor should it be efflorescent; otherwise drying and weighing becomes difficult. (4) A primary standard should be a compound that is readily available and not too expensive. (5) Finally, it should have a reasonably high equivalent weight. We shall see that the number of grams of a compound required to standardize or prepare a solution of a given concentration increases with equivalent weight; since the relative error in weighing decreases with increasing weight, a high equivalent weight will lead to smaller weighing errors in the use of the compound.

There are not many substances that meet or even approach these requirements. As a result the number of good primary standard substances available to the chemist is relatively limited.

Stability of standard solutions. The ideal standard solution would be one whose concentration remains constant for months or years after preparation. A few of the reagents used in volumetric analysis are this stable. Many, however, require frequent restandardization and are used only out of necessity.

END POINTS IN VOLUMETRIC ANALYSES; TITRATION CURVES

The detection of an end point involves observation of some property of the solution that changes in a characteristic way at or near the equivalence point. The properties that have been employed for this purpose are numerous and varied. Some of these include:

(1) Color due to the reagent, the substance being determined, or an indicator substance;

(2) Turbidity resulting from the formation or disappearance of an insoluble phase;

(3) Electric conductivity of the solution;

(4) Electric potential between a pair of electrodes immersed in the solution;

(5) Refractive index of the solution;

(6) Temperature of the solution;

(7) Electric current passing through the solution.

Physical changes such as these arise from alterations in concentration of one of the reactants in the vicinity of the equivalence point. The second and third columns of data in Table 10-1 consist of chloride and silver ion concentrations found in a solution at various stages in the titration of 50 ml of 0.1 F sodium chloride with 0.1 F silver nitrate. These data are plotted in Figure 10.1 where we see that the chloride concentration becomes extremely small as the equivalence point is passed; the silver ion concentration, on the other hand, remains low until the equivalence point, whereupon it begins to rise rapidly.

To illustrate the *relative* changes in the concentration of chloride ion and silver ion as the titration proceeds, the data in the table have been so chosen as to give the volume increments that correspond to each tenfold decrease in the chloride concentration and tenfold increase in the silver concentration. Thus 40.9 ml of reagent are necessary to decrease the chloride ion concentration from the original 0.1 to 0.01 molar; an additional 8.1 ml are required to reduce the concentration from 0.01 to 0.001, and 0.9 ml to give another tenfold decrease. When the data are considered in this light, we see that the equivalence point is marked by very large changes in the relative concentrations of the reactive species for a given volume of the reagent.

The graphical representation of data such as those shown in columns two and three of Table 10-1 is made difficult by the enormous range of concentrations covered by the participants in a titration; this results in a plotting scale

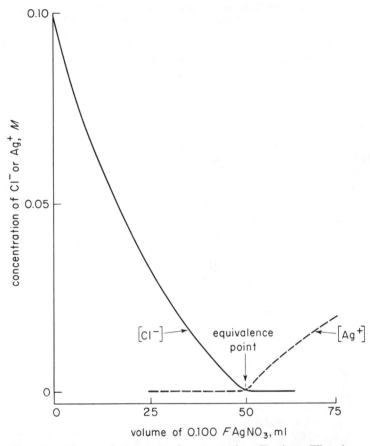

Fig. 10.1 Changes in Reactant Concentrations During a Titration. Plots of $[Ag^+]$ and $[Cl^-]$ as a function of the volume of 0.100 F $AgNO_3$ added to 50.0 ml of 0.100 F NaCl.

so large that changes in the vicinity of the end point are obscured (see Fig. 10.1). This is the region of greatest interest; it is more effectively depicted when the concentrations are expressed in terms of their logarithms. Further, since the majority of concentrations treated in this manner are less than one, the *negative* logarithm is the preferred notation because this allows expression of such concentrations in terms of small positive numbers. These have come to be known as p functions; for our purposes *we will define the p values for a solution as the negative logarithms to the base 10 of the molar concentrations of the ions in that solution.* Values for pAg and pCl are given in Table 10-1.

     Example. Calculate p functions for a solution that is 0.02 F in NaCl and 0.10 F in HCl.

Table 10-1

CONCENTRATION CHANGES DURING TITRATION
(50.0 ml of 0.1 F NaCl titrated with 0.1 F AgNO$_3$)

Volume AgNO$_3$ added, ml	Concentration of Cl$^-$, mol per liter	Concentration of Ag$^+$, mol per liter	$-\log[Cl^-]$ (pCl)	$-\log[Ag^+]$ (pAg)
0	1.0×10^{-1}	0	1.0	——
40.9	1.0×10^{-2}	1.7×10^{-8}	2.0	7.8
49.0	1.0×10^{-3}	1.7×10^{-7}	3.0	6.8
49.9	1.0×10^{-4}	1.7×10^{-6}	4.0	5.8
50.0	1.3×10^{-5}	1.3×10^{-5}	4.89	4.89
50.2	1.0×10^{-6}	1.7×10^{-4}	6.0	3.8
51.7	1.0×10^{-7}	1.7×10^{-3}	7.0	2.8
70.5	1.0×10^{-8}	1.7×10^{-2}	8.0	1.8

1. $[H^+] = 0.10$

$$pH = -\log[H^+] = -\log 0.10$$
$$= 1.0$$

2. $[Na^+] = 0.020 = 2 \times 10^{-2}$

$$pNa = -\log[Na^+] = -\log(2 \times 10^{-2})$$
$$= -\log 2 - \log 10^{-2}$$
$$= -0.3 + 2.0$$
$$= 1.7$$

3. $[Cl^-] = 0.02 + 0.10 = 0.12$

$$= 1.2 \times 10^{-1}$$
$$pCl = -\log 1.2 \times 10^{-1}$$
$$= -\log 1.2 - \log 10^{-1}$$
$$= -0.08 + 1$$
$$= +0.92$$

The p function of an ion will be a negative number if its concentration is greater than 1 molar. For example, calculate the pAg of a 2.0 F solution of AgNO$_3$.

$$[Ag^+] = 2.0$$
$$pAg = -\log[Ag^+] = -\log 2.0$$
$$= -0.3$$

If we plot the values of pAg and pCl given in Table 10-1 against the volume of reagent added, we obtain curves that give a much clearer picture of the changes in the solution near the equivalence point. Figure 10.2 indicates that the pCl shows little change except in the equivalence-point region where it increases rapidly. A sharp decrease in pAg occurs in the same vicinity.

Graphs such as either of those in Figure 10.2 are called *titration curves*. Plots of analogous data for titrations involving neutralization, oxidation-reduction, or complex-formation reactions have the same general appearance. How-

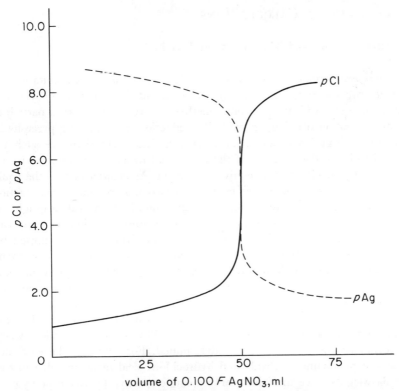

Fig. 10.2 Titration Curve. Plots of pAg and pCl as a function of the volume of $0.100\ F$ AgNO$_3$ added in the titration of 50.0 milliliters of $0.100\ F$ NaCl.

ever, the magnitude of the change in p function at the equivalence point region varies considerably, larger changes being associated with reactions that are more complete and with titrations involving relatively concentrated solutions.

The shape of the titration curve and especially the magnitude of the change in the region of the equivalence point are of great interest to the analytical chemist because these determine how easily and how accurately the equivalence point can be located. The extent of the physical change used to signal an end point is usually directly related to the magnitude of change of p values. In general, then, the most satisfactory end points are to be expected where changes in p functions are large.

It is difficult to overstate the importance of titration curves in providing a rational approach to many of the problems confronting the analytical chemist.

For this reason we shall have frequent occasion to refer to them and to their derivation in future chapters.

VOLUMETRIC CALCULATIONS

Equivalent Weight and Milliequivalent Weight

The most commonly used units of weight in volumetric computations are the *milliequivalent weight* and the *equivalent weight*. The manner in which we shall define these will depend upon whether we are dealing with a participant in a neutralization titration, an oxidation-reduction titration, or a precipitation or complex-formation titration. Furthermore, these definitions are such that we need to know the behavior of the compound in its reaction before we can unequivocally decide upon its equivalent weight. As a matter of fact, the equivalent weight of a given compound may assume two or more values if the reactions of the compound are different. Thus the definition of the equivalent weight of a chemical compound always refers to a specific chemical reaction; without some knowledge of the nature of this reaction, evaluation of this quantity is impossible.

Neutralization titrations.   The *equivalent weight (eq wt)* of a substance participating in a neutralization reaction is that weight which either contributes or reacts with 1 gram formula weight of hydrogen ions in that reaction. The *milliequivalent weight (meq wt)* is 1/1000 of the equivalent weight.

For acids or bases containing but a single reactive hydrogen or hydroxyl ion and for strong acids or bases, the relationship between equivalent weight and formula weight can be readily determined. For example, the equivalent weights of potassium hydroxide and hydrochloric acid must be equal to their formula weights since these have only a single reactive hydrogen or hydroxyl ion. Similarly, we know that only one hydrogen in acetic acid, $HC_2H_3O_2$, is acidic; therefore, the formula weight and equivalent weight for this acid are also the same. Barium hydroxide, $Ba(OH)_2$, is a strong base in which the two hydroxyl ions are identical. Therefore, in any acid-base reaction this compound will necessarily react with two hydrogen ions; by our definition its equivalent weight will be one half its formula weight. In the case of sulfuric acid, the dissociation of the second hydrogen ion is not complete; the bisulfate ion, however, is sufficiently strong so that in all neutralization reactions both hydrogens participate. The equivalent weight of sulfuric acid in a neutralization reaction is, therefore, always one half its formula weight.

The situation is not so simple when two or more hydrogen ions or hydroxyl ions of different strengths are found in a given chemical compound. Phosphoric acid provides a good example of this. If a dilute solution of this acid is titrated with base, certain indicators change color when only one of the protons has been neutralized; that is,

$$H_3PO_4 + OH^- \rightarrow H_2PO_4^- + H_2O$$

On the other hand, with other indicators color changes occur after two hydrogen ions have reacted.

$$H_3PO_4 + 2OH^- \rightarrow HPO_4^{2-} + 2H_2O$$

In the first instance, the equivalent weight of phosphoric acid is equal to its formula weight, and in the second to one half its formula weight. Without knowing which of these reactions is involved, an unambiguous definition of an equivalent weight for phosphoric acid is impossible.

Oxidation-reduction reactions. The *equivalent weight* of a participant in an oxidation-reduction reaction is that weight which, directly or indirectly, consumes or produces 1 mol of electrons. In this case the numerical value of the equivalent weight can be conveniently found by dividing its formula weight by the change in oxidation number of the substance or the compound containing the substance. As an example, consider the oxidation of oxalate ion by permanganate.

$$5C_2O_4^{2-} + 2MnO_4^- + 16H^+ \rightarrow 10CO_2 + 2Mn^{2+} + 8H_2O$$

The change in oxidation number of the manganese, in this instance, is five since the element passes from the $+7$ to the $+2$ state in the course of the reaction. Each carbon atom in the oxalate, on the other hand, exhibits a change of one in going from the $+3$ to the $+4$ state. From a consideration of the number of manganese or carbon atoms contained in any given compound, we can readily calculate the change in oxidation number associated with that component and thus its equivalent weight. This is shown below.

Substance	Equivalent Weight
Mn	$\dfrac{\text{GFW Mn}}{5}$
MnO_4^-	$\dfrac{\text{GFW } MnO_4^-}{5}$
$KMnO_4$	$\dfrac{\text{GFW } KMnO_4}{5}$
$Ca(MnO_4)_2 \cdot 4H_2O$	$\dfrac{\text{GFW } Ca(MnO_4)_2 \cdot 4H_2O}{2 \times 5}$
CO_2	$\dfrac{\text{GFW } CO_2}{1}$
$C_2O_4^{2-}$	$\dfrac{\text{GFW } C_2O_4^{2-}}{2}$

As in neutralization reactions the equivalent weight of a given oxidizing or reducing agent is not an invariant quantity. Potassium permanganate, for example, can react with reducing agents in three different ways depending upon the conditions existing in the solution. The half reactions are

$$MnO_4^- + e \rightarrow MnO_4^{2-}$$
$$MnO_4^- + 3e + 2H_2O \rightarrow MnO_2 + 4OH^-$$
$$MnO_4^- + 5e + 8H^+ \rightarrow Mn^{2+} + 4H_2O$$

The changes in oxidation number of the manganese are one, three, and five; the equivalent weight of potassium permanganate would be the formula weight, one third the formula weight, and one fifth the formula weight, respectively.

Precipitation and complex-formation reactions. In characterizing the equivalent weight of compounds involved in precipitation or complex-formation reactions, it is difficult to devise a definition that is entirely free of ambiguity. As a result, many chemists prefer to avoid the use of equivalent weights when dealing with reactions of this type and in their place use formula weights exclusively. We are in sympathy with this practice; however, we also feel that the student at some time will undoubtedly encounter situations where equivalent or milliequivalent weights of substances involved in precipitation or complex formation will be supplied, and he must therefore have some knowledge of their definition.

The *equivalent weight* of a participant in a precipitation reaction or a complex-formation reaction is that weight of substance which reacts with or provides 1 gram formula weight of the reacting cation if it be univalent, half a gram formula weight if it be divalent, a third of a gram formula weight if trivalent, etc. Our earlier definitions of equivalent weight were based on 1 mol of hydrogen ions or electrons. In the present case the definition is based on 1 mol of a univalent cation or the equivalent thereof. The cation referred to in this definition is always *the cation directly involved in the reaction* and not necessarily the cation contained in the compound whose equivalent weight is being defined.

To illustrate the application of this definition we shall consider the reaction

$$Ag^+ + Cl^- \rightarrow AgCl$$

Here the cation referred to in the definition is the univalent silver ion. The equivalent weights of some compounds that might be associated with this reaction are given below.

Substance	*Equivalent Weight*
Ag^+	GFW Ag^+
$AgNO_3$	GFW $AgNO_3$
Ag_2SO_4	$\dfrac{\text{GFW } Ag_2SO_4}{2}$
NaCl	GFW NaCl
$BaCl_2 \cdot 2H_2O$	$\dfrac{\text{GFW } BaCl_2 \cdot 2H_2O}{2}$
BiOCl	GFW BiOCl

The equivalent weight of the barium chloride dihydrate is one half its formula weight, *not* because such a weight contains one half a formula weight of the divalent barium ions, but rather because this is the weight that reacts with 1 formula weight of silver ions. A consideration of the last example will show why it is important to make this fine distinction. Here we might be tempted to say that the equivalent weight of the bismuth oxychloride is one third of its formula weight since this would be the weight of a compound containing one third of a formula weight of the trivalent bismuth. This would be correct if the bismuth were the cation involved in the reaction. Here, however, the silver ion is the reacting cation and, therefore, the equivalent weight of the bismuth oxychloride must be that weight which reacts with 1 formula weight of silver ions; that is, the formula weight and equivalent weight are identical.

Again, the equivalent weight of a given compound may assume more than one value. For example, silver ion may be titrated with a solution of potassium cyanide and the end point established for either of two reactions

$$CN^- + Ag^+ = AgCN$$

or

$$2CN^- + Ag^+ = Ag(CN)_2^-$$

In the first case the equivalent weight of potassium cyanide would be its formula weight; in the second it would be *twice* the formula weight. This is an example of an equivalent weight that is *greater* than a formula weight.

Equivalent weights of compounds not participating directly in a volumetric reaction. Frequently we need to define the equivalent weight of a compound or element that is only indirectly related to the participants of the titration. In such a case we must know the stoichiometric relationship between that compound and one of the reactants involved in the titration. To illustrate, lead can be determined by an indirect volumetric method in which the lead is first precipitated as the chromate from an acetic acid solution. The precipitate is filtered, washed free of excess precipitant, and then dissolved in dilute hydrochloric acid; this yields a solution of lead and dichromate ions. The latter are then determined by an oxidation-reduction titration. The reactions are

$$Pb^{2+} + CrO_4^{2-} \xrightarrow[\text{HOAc}]{\text{dil}} \underline{PbCrO_4} \text{ (precipitate filtered and washed)}$$

$$2\underline{PbCrO_4} + 2H^+ \xrightarrow[\text{HCl}]{\text{dil}} 2Pb^{2+} + Cr_2O_7^{2-} + H_2O \text{ (precipitate redissolved)}$$

$$Cr_2O_7^{2-} + 6I^- + 14H^+ \rightarrow 2Cr^{3+} + 3I_2 + 7H_2O \text{ (titration)}$$

For the purpose of calculations, we must ascribe an equivalent weight to lead; *since the titration is an oxidation-reduction process, the equivalent weight of lead will have to be based on a change of oxidation number.* Clearly the lead exhibits no such change. It is, however, associated in a 1:1 ratio with chromium, and this element changes from the $+6$ to the $+3$ oxidation state. Therefore, we can say that the change in oxidation state *associated* with each lead is three, and the equivalent weight of that element is, in this case, one third of its atomic weight.

It is often helpful to make an inventory of the chemical relationships existing between the substance whose equivalent weight is sought and one of the participants in the titration. In this instance we see from the equations that

$$2Pb^{2+} \equiv 2CrO_4^{2-} \equiv Cr_2O_7^{2-} \equiv 6e$$

Thus the quantity of each substance associated with an electron change of one is

$$\frac{2 \text{ GFW Pb}^{2+}}{6} \equiv \frac{2 \text{ GFW CrO}_4^{2-}}{6} \equiv \frac{1 \text{ GFW Cr}_2O_7^{2-}}{6} \equiv \frac{6 \text{ mols } e}{6}$$

Let us consider one more example. The nitrogen in the organic compound $C_9H_9N_3$ can be determined by quantitative conversion to ammonia followed by titration with a standard solution of acid. The reactions are

$$C_9H_9N_3 + \text{reagent} \rightarrow 3NH_3 + \text{products}$$

$$NH_3 + H^+ \rightarrow NH_4^+ \text{ (titration)}$$

In this case titration involves a neutralization process; therefore the sought-for equivalent weight must be based on the consumption or production of hydrogen ions. Since the organic compound forms three ammonia molecules, it can be considered responsible for the consumption of three hydrogen ions; the equivalent weight, then, is the formula weight of $C_9H_9N_3$ divided by three. On the other hand, each nitrogen is converted to one ammonia molecule that reacts with one hydrogen ion, and we may also say that the equivalent weight of nitrogen, N, is equal to its formula weight.

Concentration Units Used in Volumetric Calculations

In Chapter 2 we discussed some of the terms used by chemists to express the concentrations of solutions. These included formal concentration, molar concentration, and various types of percentage composition. We now need to define the two additional terms that are the most commonly used for describing solutions used in volumetric analysis. These are the *titer* and the *normality* of a solution.

Titer.   *The titer of a solution is the weight of some substance that is chemically equivalent to 1 ml of that solution.* Thus a silver nitrate solution having a titer of 1.00 mg of chloride would contain just enough silver nitrate in each milliliter to react completely with that weight of chloride ion. The titer might also be expressed in terms of milligrams or grams of potassium chloride, barium chloride, sodium iodide, or other compounds that react with silver nitrate. The use of titer as an expression of concentration is generally confined to volumetric reagents that are to be used for the routine analysis of many samples.

Normality. *The normality, N, of a solution expresses the number of milliequivalents of dissolved solute contained in 1 ml of the solution,* or the number of equivalents contained in 1 liter. Thus a 0.20 N solution of silver nitrate contains 0.20 milliequivalent of this substance in each milliliter of the solution.

Some Important Weight-Volume Relationships

The raw data from a volumetric analysis are ordinarily expressed in units of milliliters, grams, and normality. Volumetric calculations involve conversion of such information into units of milliequivalents followed by reconversion back into the weight in grams of some other chemical species. Two relationships, based on the foregoing definitions, are used for these interconversions.

First we shall examine the process of converting the weight of a chemical compound from units of grams to those of milliequivalents. Such a conversion involves division of the known weight by the weight of 1 meq of the chemical substance; that is,

$$\text{number of meq of substance } A = \frac{\text{weight in grams of } A}{\text{meq weight of } A}$$

Example. A quantity of $BaCl_2 \cdot 2H_2O$ is to be titrated with silver nitrate solution. We wish to calculate the number of milliequivalents contained in 0.367 gram of pure $BaCl_2 \cdot 2H_2O$ (GFW 244).

$$\text{number of meq } BaCl_2 \cdot 2H_2O = \frac{\text{weight in grams of } BaCl_2 \cdot 2H_2O}{\text{meq weight of } BaCl_2 \cdot 2H_2O}$$

$$= \frac{0.367}{244/(2 \times 1000)}$$

$$= \frac{0.367}{0.122} = 3.01 \text{ meq}$$

A second relationship permits calculation of the number of milliequivalents of a reagent contained in a certain number of milliliters of the solution provided we know the normality. In this case, by definition, the normality immediately gives the number of milliequivalents in each milliliter; multiplying this by the volume in milliliters then gives the number of milliequivalents; that is,

$$\text{number of meq of } A = \text{volume of } A \text{ in ml} \times N_A$$

Example. We wish to calculate the number of milliequivalents of $KMnO_4$ used in a titration in which 27.3 ml of 0.200 N $KMnO_4$ were required to reach an end point.

$$\text{number of meq of } KMnO_4 = 27.3 \times 0.200$$
$$= 5.46 \text{ meq}$$

Some examples of the application of such calculations are given below.

Example. How many grams of primary standard $K_2Cr_2O_7$ (GFW 294) are required to prepare exactly 2 liters of a 0.100 N solution of that reagent? (Titrations with dichromate involve $Cr_2O_7^{2-} + 14H^+ + 6e = 2Cr^{3+} + 7H_2O$)

We first calculate the number of milliequivalents of $K_2Cr_2O_7$ required.

$$\text{number of meq } K_2Cr_2O_7 = ml_{K_2Cr_2O_7} \times N_{K_2Cr_2O_7}$$
$$= 2000 \times 0.100 = 200 \text{ meq}$$

To convert this weight in milliequivalents to a weight in grams, we multiply by the milliequivalent weight.

$$\text{weight of } K_2Cr_2O_7 \text{ required} = 200 \times \frac{294}{6 \times 1000}$$
$$= 9.81 \text{ grams}$$

Example. What volume of 0.100 N HCl can be produced by diluting a 150-ml sample of 1.24 N acid?

The number of milliequivalents of HCl must be the same in the two solutions. Therefore we may write

$$\frac{\text{number of meq HCl}}{\text{in diluted solution}} = \frac{\text{number of meq HCl}}{\text{in concentrated solution}}$$

$$\text{ml of dilute solution} \times 0.1 = 150 \times 1.24$$
$$\text{ml of dilute solution} = 1860 \text{ ml}$$

Thus by diluting the 150 ml of 1.24 N acid to exactly 1860 ml, a solution 0.100 N in HCl would be obtained.

A fundamental relationship between quantities of reacting substances. By definition, 1 equivalent weight of an acid contributes 1 formula weight of hydrogen ions to a reaction; also, 1 equivalent weight of a base consumes 1 formula weight of these ions. As a consequence, at the equivalence point in a neutralization titration the number of equivalents of acid and of base will always be numerically equal. Similarly, the equivalent weights of oxidizing and reducing agents are defined in terms of weights that will produce or consume 1 mol of electrons. Thus, at the equivalence point in such a titration the number of equivalents of oxidizing and reducing agents must also be equal. An identical relationship holds for precipitation and complex-formation titrations. To generalize we may state that *at the equivalence point in any titration the number of milliequivalents of standard is exactly equal to the number of milliequivalents of the substance being determined.* Nearly all volumetric calculations based on titration data start with this relationship.

Calculation of Normality of a Solution

The normality of a standard solution is computed either from the data related to its actual preparation or from standardization titrations.

Example. A standard solution of $AgNO_3$ (GFW 169.9) was prepared by weighing exactly 24.15 grams of the carefully prepared solid, dissolving in water, and diluting in a volumetric flask to exactly 2.000 liters. We wish to know the normality of the solution.

Since normality is the number of milliequivalents per milliliter, we may write

$$\text{number of meq } AgNO_3 = \frac{24.15}{169.9/1000}$$

$$N = \frac{\text{no. of meq}}{\text{ml}} = \frac{24.15/0.1699}{2000} - 0.07107$$

Example. A solution of $Ba(OH)_2$ was standardized by titration against 0.1280 N HCl. Exactly 31.76 ml of the base were required to neutralize 46.25 ml of the acid. The normality of the $Ba(OH)_2$ solution is to be calculated.

At the end point in the titration we may say

$$\text{number of meq } Ba(OH)_2 = \text{number of meq HCl}$$

$$ml_{Ba(OH)_2} \times N_{Ba(OH)_2} = ml_{HCl} \times N_{HCl}$$
$$31.76 \times N_{Ba(OH)_2} = 46.25 \times 0.1280$$

$$N_{Ba(OH)_2} = \frac{46.25 \times 0.1280}{31.76} = 0.1864$$

Example. A solution of iodine was standardized against 0.2040 gram of pure As_2O_3 (GFW 197.8). This required 37.34 ml of the reagent. We wish to know the normality of the iodine.

(Reaction: $I_2 + H_2AsO_3^- + H_2O \rightarrow 2I^- + H_2AsO_4^- + 2H^+$)

At the end point

$$\text{number of meq } I_2 = \text{number of meq } As_2O_3$$

The number of milliequivalents of I_2 can be related to the volume and normality. For As_2O_3 the number of milliequivalents can be calculated from the weight of the pure compound. Thus we may write

$$ml_{I_2} \times N_{I_2} = \frac{\text{weight } As_2O_3 \text{ in grams}}{\text{meq weight } As_2O_3}$$

From the chemical equation for the reaction we see that each arsenic atom undergoes a change in oxidation number of 2; therefore a change in oxidation number of 4 is associated with each As_2O_3 molecule making the equivalent weight one fourth of the formula weight. Thus

$$37.34 \, N_{I_2} = \frac{0.2040}{\text{GFW } As_2O_3/4000} = \frac{0.2040}{0.04945}$$

$$N_{I_2} = \frac{0.2040}{37.34 \times 0.04945} = 0.1105$$

Calculation of Results From Titration Data

Example. A sample of an iron ore was analyzed by dissolving 0.804 gram of the material in acid, reducing all of the iron to the ferrous condition, and titrating with a 0.112 N solution of $KMnO_4$. This required 47.2 ml.

We will calculate the percent Fe (GFW 55.9) in the sample, and express the results also as percent Fe_2O_3 (GFW Fe_2O_3 160).

The analytical reaction involves oxidation of Fe^{2+} to Fe^{3+}

$$5Fe^{2+} + MnO_4^- + 8H^+ \rightarrow 5Fe^{3+} + Mn^{2+} + 4H_2O$$

and at the end point

$$\text{number of meq Fe} = \text{number of meq of KMnO}_4$$
$$= 47.2 \times 0.112$$

The weight of Fe in the sample can be calculated by multiplying the number of milliequivalents of Fe by the milliequivalent weight.

$$\text{grams of Fe} = 47.2 \times 0.112 \times \frac{\text{GFW Fe}}{1000}$$

Therefore

$$\text{percent Fe} = \frac{47.2 \times 0.112 \times 55.9/1000}{0.804} \times 100$$
$$= 36.7$$

The percent Fe_2O_3 can be obtained in essentially the same way. We can state that at the equivalence point

$$\text{number of meq of Fe}_2O_3 = \text{number of meq of KMnO}_4$$

and by the same arguments

$$\text{percent Fe}_2O_3 = \frac{47.2 \times 0.112 \times 160/2000}{0.804} \times 100$$
$$= 52.6$$

Example. A 0.475-gram sample containing $(NH_4)_2SO_4$ was dissolved in water and made alkaline with KOH. The liberated NH_3 was distilled into exactly 50.0 ml of 0.100 N HCl. The excess HCl was back-titrated with 11.1 ml of 0.121 N NaOH. We wish to calculate the percent NH_3 (GFW 17.0), and also the percent $(NH_4)_2SO_4$ (GFW 132) in the sample.

At the equivalence point in this titration we may say that the number of milliequivalents of acid and base are equal. In this case, however, there are two bases involved, NaOH and NH_3. Thus

$$\text{number of meq HCl} = \text{number of meq NH}_3 + \text{number of meq NaOH}.$$

After rearranging

$$\text{number of meq NH}_3 = \text{number of meq HCl} - \text{number of meq NaOH}$$
$$= (50.0 \times 0.100 - 11.1 \times 0.121)$$

$$\text{percent NH}_3 = \frac{(50.0 \times 0.100 - 11.1 \times 0.121) \times 17.0/1000}{0.475} \times 100$$
$$= 13.1$$

The number of milliequivalents of $(NH_4)_2SO_4$ is the same as the number of milliequivalents of NH_3, and therefore

$$\text{percent (NH}_4)_2SO_4 = \frac{(50.0 \times 0.100 - 11.1 \times 0.121) \times 132/2000}{0.475} \times 100$$
$$= 50.8$$

Here the milliequivalent weight of $(NH_4)_2SO_4$ is one half the milliformula weight. Thus 1 milliformula weight of the ammonium sulfate contains 2 milliformula weights of ammonia that react in turn with 2 milliformula weights of hydrogen ion.

problems

1. What is the pH and pCl of the following solutions?
   (a) 0.10 F HCl   ans. $pH = 1.0$ $pCl = 1.0$
   (b) 0.10 F BaCl$_2$ ans. $pH = 7.0$ $pCl = 0.7$
   (c) 3.0 F HCl
   (d) A solution that is 0.010 F in HCl and 0.02 F in BaCl$_2$
   (e) A saturated solution of AgCl

2. What is the pAg of the following solutions?
   (a) 0.23 F AgNO$_3$
   (b) 0.0070 F Ag$_2$SO$_4$
   (c) A 0.100 F solution of NaCl saturated with AgCl   ans. 8.74
   (d) A 0.030 F solution of BaCl$_2$ saturated with AgCl
   (e) A saturated solution of AgI

3. In the first column below is a series of reactions used for the volumetric analysis of the compound in the second column. Indicate the type of reaction and the milliequivalent weight of each of these compounds in terms of its formula weight.

Reaction	Compound	
(a) $NH_3 + H^+ \rightarrow NH_4^+$	NH$_4$Cl	ans. $\dfrac{\text{GFW } NH_4Cl}{1000}$
	(NH$_4$)$_2$SO$_4$	ans. $\dfrac{\text{GFW } (NH_4)_2SO_4}{2000}$
	FeSO$_4 \cdot$ (NH$_4$)$_2$SO$_4 \cdot$ 6H$_2$O	
	C$_6$H$_8$N$_2$	ans. $\dfrac{\text{GFW } C_6H_8N_2}{2000}$
	H$_2$SO$_4$	
(b) $Ba^{2+} + CrO_4^{2-} \rightarrow BaCrO_4$	BaCl$_2 \cdot$ 2H$_2$O	
	Ba$_3$N$_2$	
	Cr	ans. $\dfrac{\text{GFW } Cr}{2000}$
	Na$_2$Cr$_2$O$_7$	
(c) $MnO_4^- + 5Fe^{2+} + 8H^+ \rightarrow$ $Mn^{2+} + 5Fe^{3+} + 4H_2O$	Mn	
	MnO$_2$	ans. $\dfrac{\text{GFW } MnO_2}{5000}$
	Mn$_3$O$_4$	
	Fe$_2$O$_3$	ans. $\dfrac{\text{GFW } Fe_2O_3}{2}$
	Fe$_3$O$_4$	

4. Indicate the type of reaction in each of the following and give the equivalent weight of the compounds on the right:

Reaction	Compound	
(a) $Hg^{2+} + 4I^- \rightarrow HgI_4^{2-}$	$HgSO_4$	ans. $\dfrac{GFW\ HgSO_4}{2}$
	Hg	
	Hg_2Cl_2	
	KI	ans. 2 GFW KI
	BaI_2	
	$Ba(IO_3)_2$	
(b) $B_4O_7^{2-} + 2H^+ + 5H_2O \rightarrow 4H_3BO_3$	$Na_2B_4O_7$	
	B	
	B_2O_3	
	HCl	
(c) $H_2S + I_2 \rightarrow 2I^- + S + 2H^+$	H_2S	
	SO_2	
	$Na_2S_2O_3$	
	KI	
	I_2O_5	

5. Give the type of reaction and the equivalent weights of each of the compounds indicated below :

Reaction	Compound
(a) $Al^{3+} + 6F^- \rightarrow AlF_6^{3-}$	Al_2O_3
	SiF_4
	NaF
	C_2F_6
	$AlCl_3$
(b) $Pb^{2+} + SO_4^{2-} \rightarrow PbSO_4$	$KHSO_4$
	$FeOHSO_4$
	$Na_2S_2O_3$
	$Pb(NO_3)_2$
	Pb_3O_4
(c) $2V(OH)_4^+ + H_2C_2O_4 + 2H^+ \rightarrow 2VO^{2+} + 2CO_2 + 6H_2O$	$H_2C_2O_4$
	CO_2
	NH_4VO_3
	V_2O_5

6. Potassium hydrogen iodate, $KH(IO_3)_2$, behaves as a strong acid in aqueous solution and can be used to standardize bases. Solutions of the compound can also be titrated with $AgNO_3$ to give a precipitate of $AgIO_3$. The compound can also participate in an oxidation-reduction reaction in which I_2 is the product. Indicate the equivalent weight of the compound in each of these cases.

7. What weight of silver nitrate should be taken to prepare exactly 2 liters of a solution having a titer of 2.00 mg of KI per ml.   ans. 4.09 grams

8. Describe the preparation of 500 ml of a Na_2CO_3 solution having a titer of 0.500 mg of HCl per ml.

$$\text{(Reaction: } Na_2CO_3 + 2HCl \rightarrow H_2CO_3 + 2NaCl)$$

9. How would you prepare a $K_2Cr_2O_7$ solution having a titer of 3.00 mg of $FeSO_4 \cdot (NH_4)_2SO_4 \cdot 6H_2O$?

$$\text{(Reaction: } 6Fe^{2+} + Cr_2O_7^{2-} + 14H^+ \rightarrow 6Fe^{3+} + 2Cr^{3+} + 7H_2O)$$

10. Calculate the total number of milliequivalents in each of the following:
    (a) 14.2 grams of $AgNO_3$ ans. 83.5
    (b) 30.0 ml of 0.0200 N $Ba(OH)_2$ ans. 0.600
    (c) 13.0 equivalents of KI   ans. 13,000
    (d) 17.0 mg of $FeSO_4 \cdot (NH_4)_2SO_4 \cdot 6H_2O$ ans. 0.0435
           $(Fe^{2+} \rightarrow Fe^{3+} + e)$
    (e) 10.0 liters of 0.200 N $KMnO_4$ ans. 2000
           $(MnO_4^- + 8H^+ + 5e \rightarrow Mn^{2+} + 4H_2O)$
    (f) 375 ml of 0.120 N H_2SO_4 ans. 45.0

11. Calculate the number of milliequivalents in each of the following:
    (a) 100 grams of pure I_2 $(I_2 + 2e \rightarrow 2I^-)$
    (b) 0.500 equivalent of H_2SO_4
    (c) 5 ml of 0.220 N NaOH
    (d) 48.5 mg of Na_2CO_3 $(CO_3^{2-} + 2H^+ \rightarrow H_2CO_3)$
    (e) 3.25 liters of 0.300 N KSCN
    (f) 1.00 liter of 0.0200 F $Ba(OH)_2$

12. Calculate the number of milliequivalents in each of the following:
    (a) 4700 ml of 0.200 N $KMnO_4$
    (b) 0.400 gram of As_2O_3 (arsenic oxidized from $+3$ to $+5$ state in reaction)
    (c) 10 liters of 0.0012 N $Na_2S_2O_3$
    (d) 40 ml of 0.055 F $KMnO_4$ ($KMnO_4$ reduced to Mn^{2+})
    (e) 0.176 gram of pure NaC_2O_4 $(C_2O_4^{2-} \rightarrow 2CO_2 + 2e)$

13. How many grams are contained in each of the following:
    (a) 1.00 meq of $Pb(NO_3)_2$ ans. 0.1656 gram
           $(Pb^{2+} + SO_4^{2-} \rightarrow PbSO_4)$
    (b) 2.00 equivalents of $Ba(OH)_2$ ans. 171.4 grams
    (c) 30.0 ml of 0.100 N $AgNO_3$ ans. 0.510 gram
           $(Ag^+ + Cl^- \rightarrow AgCl)$
    (d) 2.00 liters of 0.330 N $KMnO_4$ ans. 20.8 grams
           $(MnO_4^- + 5e + 8H^+ \rightarrow Mn^{2+} + 4H_2O)$

14. Calculate the number of grams in each of the following:
    (a) 10.0 meq of H_2SO_4
    (b) 130 ml of 0.150 N Br_2 $(Br_2 + 2e \rightarrow 2Br^-)$
    (c) 6.00 liters of 1.00×10^{-3} N $K_2Cr_2O_7$
           $(Cr_2O_7^{2-} + 14H^+ + 6e \rightarrow 2Cr^{3+} + 7H_2O)$
    (d) 0.520 equivalent of KCN $(Ni^{2+} + 4CN^- \rightarrow Ni(CN)_4^{2-})$

15. Calculate the following:
    (a) the number of mols of $Ba(OH)_2$ in 1200 ml of a 0.0100 N solution
    (b) the number of milliequivalents of $Ba(OH)_2$ in 2.0 liters of a 0.0300 N solution
    (c) the number of milliequivalents per milliliter of H_2SO_4 in a 2.0 F solution

    (d) the number of grams of $Ba(OH)_2$ in 150 ml of 0.0400 N $Ba(OH)_2$

    (e) the number of milliequivalents per milliliter of $AgNO_3$ in a solution containing 17.0 grams of $AgNO_3$ per liter

    (f) the normality of a $AgNO_3$ solution containing 17.0 grams of $AgNO_3$ per liter

16. Calculate the following:
    (a) the number of milliequivalents of NaOH in a solution that reacts completely with 17.1 ml of 0.200 N H_2SO_4
    (b) the number of grams of NaOH in the solution in part (a)
    (c) the number of milliequivalents of Na_2CO_3 contained in 2.00 grams of the pure compound $(CO_3^{2-} + 2H^+ \rightarrow H_2CO_3)$
    (d) the number of milliliters of 0.100 N HCl that will react completely with 2.00 grams of Na_2CO_3
    (e) the number of milliequivalents of $KMnO_4$ that will react completely with 14.9 ml of 3.0 N Fe^{2+}
    (f) the weight in grams of $KMnO_4$ in part (e) if the $KMnO_4$ is reduced to Mn^{2+} in the reaction with Fe^{2+}

17. How many grams of solute are required to prepare the following :
    (a) 1500 ml of 0.300 N NaCl
    (b) 4 liters of 0.0500 N $Ba(OH)_2$
    (c) 483 ml of 0.0100 N I_2 $(I_2 + 2e \rightarrow 2I^-)$
    (d) 3900 ml of 0.400 N KCN $(Ag^+ + 2CN^- \rightarrow Ag(CN)_2^-)$
    (e) 6 liters of 0.700 N acetic acid, $HC_2H_3O_2$

18. How would you prepare 500 ml of approximately 0.20 N HCl from
    (a) a solution that was 2.5 N in HCl   ans. dilute 40 ml of 2.5 N to 500 ml
    (b) a constant boiling HCl solution containing 20.2 grams of HCl per 100 grams of solution   ans. dilute 18.1 grams to 500 ml
    (c) concentrated HCl; density 1.18; 37-percent HCl
                         ans. dilute 8.36 ml of conc HCl to 500 ml

19. Describe the preparation of 2 liters of 0.150 N H_2SO_4 from the following:
    (a) a 2.30 N solution of H_2SO_4
    (b) a solution containing 15 grams of H_2SO_4 per 100 grams of solution
    (c) concentrated H_2SO_4; density 1.84; 96-percent H_2SO_4

20. What is the normality of the following:
    (a) a solution containing 1.21 grams of $AgNO_3$ per 750 ml   ans. 0.0095 N
    (b) a concentrated $HClO_4$ solution; density 1.66; 70-percent $HClO_4$
    (c) a solution containing 34.0 grams of Mohr's salt, $FeSO_4 \cdot (NH_4)_2SO_4 \cdot 6H_2O$ in 2400 ml of solution $(Fe^{2+} \rightarrow Fe^{3+} + e)$
    (d) a solution containing 1.0 mg of KOH per ml
    (e) a solution consisting of 30.0 ml of 1.5 N $KMnO_4$ diluted to 1 liter

21. A solution of KSCN was standardized by titration of 25.00 ml of 0.1000 N $AgNO_3$. If 27.14 ml were required what was the normality ?
                              ans. 0.0921 N

22. A solution of $Ba(OH)_2$ was standardized against primary standard potassium acid phthalate $(KHC_8H_4O_4)$, which has a single hydrogen ion available for reaction with the base. A sample of 0.271 gram required 44.1 ml of the base. Calculate the normality.   ans. 0.0301 N

23. A sodium hydroxide solution was standardized against a sample of primary standard potassium acid phthalate ($KHC_8H_4O_4$); the following data were obtained:

    weight $KHC_8H_4O_4$ taken, gram   0.6742   0.7966   0.6736

    volume NaOH required, ml   30.42   36.08   30.46

    Calculate the normality of the NaOH and also the precision of the data.

24. What would be the normality of a solution prepared by dissolving 2.00 grams of oxalic acid, $H_2C_2O_4$, in water and diluting to 250 ml (assume the $H_2C_2O_4$ behaves as a dibasic acid)? How many milliliters of 0.01 N NaOH would be required to react with 37.0 ml of this acid?

25. Exactly 1 ml of an I_2 solution was found to be equivalent to 2.32 ml of a 0.176 N $Na_2S_2O_3$ solution. What was the normality of the iodine?

26. A solution of iodine was standardized against pure As_2O_3. A 0.441-gram sample of the pure compound was dissolved and titrated with 29.7 ml of the iodine. Calculate the normality of the iodine based on the reaction,

$$I_2 + HAsO_3^{2-} + H_2O \rightarrow 2I^- + HAsO_4^{2-} + 2H^+$$

27. A solution of HCl was standardized by treatment of 0.330 gram of pure Na_2CO_3 with exactly 50.0 ml of acid. The solution was boiled to remove all of the CO_2 formed by the reaction, and the excess HCl remaining was back titrated with 2.10 ml of a NaOH solution. Another titration showed that 1.00 ml of the NaOH was equivalent to 1.17 ml of the HCl. Calculate the normality of the acid.   ans. 0.131 N

28. Mercuric oxide is sometimes employed as a primary standard for acids. Upon being treated with an excess of KI, the following reaction proceeds quantitatively: $HgO + 4I^- + H_2O \rightarrow HgI_4^{2-} + 2OH^-$. The hydroxyl ions so formed can then be titrated with the acid. Exactly 0.483 gram of HgO was treated in this manner and 41.2 ml of HCl were required to neutralize the base. What was the normality of the HCl?

29. A 0.612-gram sample of pure $CaCO_3$ was used to standardize a solution of acid. Exactly 40.0 ml of the acid were added to the solid and the solution boiled until all the CO_2 had been evolved. The unreacted HCl was back titrated with 7.41 ml of base, 1.00 ml of which had been found to be equivalent to 0.936 ml of the acid. Calculate the normality of both the acid and the base.

30. What is the normality of a sodium thiosulfate solution standardized by adding an excess of KI to 25.00 ml of 0.1230 N $K_2Cr_2O_7$, the liberated iodine being titrated with 41.40 ml of the thiosulfate solution?

$$Cr_2O_7^{2-} + 6I^- + 14H^+ \rightarrow 3I_2 + 2Cr^{3+} + 7H_2O$$
$$I_2 + 2S_2O_3^{2-} \rightarrow 2I^- + S_4O_6^{2-}$$

       ans. 0.0743 N

31. A 25.0-ml aliquot of a potassium hydrogen oxalate (KHC_2O_4) solution was found to require 31.3 ml of 0.125 N NaOH to reach an end point. Another 25.00-ml aliquot was used to standardize a $KMnO_4$ solution, 48.5 ml of the latter being required. Calculate the normality of the $KMnO_4$ if the chemical reaction was

$$2MnO_4^- + 5C_2O_4^{2-} + 16H^+ \rightarrow 10 CO_2 + 2Mn^{2+} + 8H_2O$$

32. 0.312 gram of an unknown acid required 40.0 ml of 0.150 N NaOH for neutralization.
    (a) What was the equivalent weight of the acid?
    (b) If it were known that the acid contained two titratable hydrogens, what would be the formula weight of the acid?

33. What weight of iron wire (99.8 percent pure) would require a 40.0 ml titration with 0.1175 N $K_2Cr_2O_7$?
$$Cr_2O_7^{2-} + 6Fe^{2+} + 14H^+ \rightarrow 2Cr^{3+} + 6Fe^{3+} + 7H_2O$$

34. A solution of HCl was standardized by precipitating all of the Cl^- from a 25.0-ml aliquot with an excess of silver nitrate. The precipitate was filtered, washed, and dried. It was found to weigh 0.0782 gram. What was the normality of the HCl? ans. 0.0218 N

35. Exactly 75.0 ml of a H_2SO_4 solution produced 0.118 gram of barium sulfate upon precipitation with an excess of $BaCl_2$. What was the normality of the acid?

36. Calculate the milligrams of H_2SO_4 per milliliter in a solution of the acid if 25.0 ml required 37.9 ml of 0.0851 N NaOH for complete neutralization. ans. 6.32 mg per ml

37. What is the percent $BaCl_2 \cdot 2H_2O$ in a sample, 0.412 gram of which consumed 26.4 ml of 0.0500 N $AgNO_3$? ans. 39.1 percent

38. Calculate the percent Fe and the percent Fe_2O_3 if a 0.749-gram sample of the substance consumed 22.2 ml of 0.134 N $KMnO_4$ after suitable treatment.
    (reaction: $MnO_4^- + 5Fe^{2+} + 8H^+ \rightarrow Mn^{2+} + 5Fe^{3+} + 4H_2O$)

39. A 0.641-gram sample of a mixture containing Na_2CO_3 was titrated to a methyl orange end point with 43.0 ml of 0.242 N HCl ($CO_3^{2-} + 2H^+ \rightarrow H_2CO_3$).
    (a) Calculate the percent Na_2CO_3 in the sample.
    (b) How many milligrams of CO_2 would be evolved by boiling the solution during and after the above titration?

40. A sample containing $(NH_4)_2SO_4$ was analyzed by dissolving 1.82 grams in strong base and distilling the liberated NH_3 into 50.0 ml of 0.0804 N HCl. The excess HCl was back-titrated with 9.48 ml of 0.106 N NaOH.
    (a) Calculate the percent $(NH_4)_2SO_4$ in the sample. ans. 10.93 percent
    (b) Calculate the percent N in the sample. ans. 2.32 percent

41. A 0.612-gram sample containing $Ca(ClO_3)_2 \cdot 2H_2O$ was analyzed by reduction of the ClO_3^- to Cl^- which was precipitated by the addition of 25.0 ml of 0.200 N $AgNO_3$. The excess $AgNO_3$ was titrated with 3.10 ml of a 0.186 N KSCN solution ($Ag^+ + SCN^- \rightarrow AgSCN$). Calculate the percent $Ca(ClO_3)_2 \cdot 2H_2O$ in the sample.

42. A 0.500-gram sample of steel was dissolved in acid and the chromium present oxidized to dichromate ($Cr_2O_7^{2-}$) by ammonium persulfate. To the resulting solution was added exactly 1.242 grams of Mohr's salt, $FeSO_4(NH_4)_2SO_4 \cdot 6H_2O$, the ferrous ion of which reduced the dichromate ion to the chromic state (Cr^{3+}). The excess ferrous ion was titrated with 14.1 ml of 0.0463 N $KMnO_4$. Calculate the percent Cr in the steel.

43. A 50.0-ml aliquot of a solution containing uranium in the $+6$ state was passed through a reductor which reduced it to a mixture of the $+3$ and $+4$

states. Bubbling air through the solution converted all of the $+3$ to the $+4$ state which was then oxidized quantitatively back to the $+6$ form with 36.9 ml of 0.0624 N $K_2Cr_2O_7$.

$$3UO^{2+} + Cr_2O_7^{2-} + 8H^+ \rightarrow 3UO_2^{2+} + 2Cr^{3+} + 4H_2O$$

What weight of uranium was contained in a liter of the sample solution?

44. A 2.00-gram sample of chromite $(FeO \cdot Cr_2O_3)$ was fused with sodium peroxide. The resulting mass was dissolved and the excess peroxide destroyed by boiling. After acidification, 50.0 ml of 0.160 N Fe^{2+} was added which reduced the $Cr_2O_7^{2-}$ to Cr^{3+}. A back titration of 3.14 ml of 0.0500 N $K_2Cr_2O_7$ was required to oxidize the excess Fe^{2+}. Calculate (a) the percent Cr in the sample, and (b) the percent chromite in the sample.

45. The routine analysis for H_2SO_4 in an electroplating rinse is to be undertaken. A NaOH solution is to be prepared such that the volume used in titration is numerically 10 times as great as the percent H_2SO_4 in a 20.0-gram sample. What should be the normality of the NaOH solution?

46. The sulfur in an organic compound was determined by combusting a 0.471-gram sample in a stream of oxygen and collecting the resulting SO_2 in a neutral solution of H_2O_2, which converted the SO_2 to H_2SO_4

$$SO_2 + H_2O_2 = H_2SO_4$$

The sulfuric acid was titrated with 28.2 ml of 0.108 N KOH. Calculate the percent sulfur in the sample.

47. An organic mixture was known to contain the compound, $C_6H_4Cl_2$. This compound was analyzed by treating a 1.17-gram sample with metallic sodium which converted the chlorine quantitatively to NaCl. After destruction of the excess sodium metal, the chloride was titrated with 30.1 ml of a 0.0884 N solution of $Hg(NO_3)_2$ $(Hg^{2+} + 2Cl^- \rightarrow HgCl_2)$. Calculate (a) the percent Cl in the sample and (b) the percent $C_6H_4Cl_2$.

48. What should be the normality of a $AgNO_3$ solution in order that its titer be 1.00 mg of KCl?

49. The sulfur in a 5.00-gram sample of steel was evolved as H_2S which was collected in an ammoniacal solution of $CdCl_2$. The CdS formed was treated with 10 ml of 0.0600 N I_2, and the I_2 back-titrated with 4.82 ml of 0.0510 N sodium thiosulfate. The reaction of CdS with I_2 is

$$CdS + I_2 \rightarrow S + Cd^{2+} + 2I^-$$

Calculate the percent S in the steel.

50. How many grams of AgI are formed when 30.0 ml of 0.100 N $AgNO_3$ are mixed with 20.0 ml of 0.180 N KI?

chapter 11. *The Techniques and Tools of Volumetric Analysis*

In Chapter 5 we discussed the common techniques and tools of analytical chemistry, with emphasis on those having particular application to gravimetric methods. We shall now complete the survey by considering three types of apparatus that are indispensible to the performance of a volumetric analysis.

GENERAL CONSIDERATIONS

Units of Volume

The fundamental unit of volume is the *liter*, defined as the volume occupied by 1 kilogram of water at the temperature of maximum density (3.98° C) and at 1 atmosphere of pressure.

The *milliliter* is defined as one one-thousandth of a liter and is widely used in the many instances where the liter represents an inconveniently large volume.

Yet another unit of volume is the *cubic centimeter*. While this and the milliliter can be used interchangeably without effect in most situations, these units are not strictly identical, the milliliter being equal to 1.000028 cubic centimeters. It was originally intended that 1 kilogram of water should occupy exactly 1 cubic decimeter; owing to inadequacies of early experimental measurements, however, this relationship was not realized, and this small difference in units was the result. For all volumetric analyses, the liter or milliliter is used.

Effect of Temperature upon Volume Measurements

The volume of a given mass of liquid varies with temperature. So also does the volume of the various devices employed to measure this quantity. As a consequence, accurate volumetric measurements may require that the effect of temperature be taken into account.

Most volumetric measuring devices are constructed of glass which fortunately has a small temperature coefficient. Thus, for example, an apparatus of soft glass will change in volume by about 0.003 percent per degree; with Pyrex, the change is about one third of this value. Clearly, variations in the volume owing to changes in temperature need be considered only for the most exacting work.

The coefficient of expansion for dilute aqueous solutions is approximately 0.025 percent per degree. The magnitude of this figure is such that a temperature variation of about 5 degrees will measurably affect the precision of ordinary volumetric measurements.

> **Example.** A 25.00-ml sample is taken from a liquid refrigerated at 5° C. We wish to know the volume this sample will occupy at 20° C.
>
> $$V_{20°} = V_{5°} + 0.00025(20 - 5)(25.00) \qquad (11\text{-}1)$$
> $$= 25.00 + 0.094$$
> $$= 25.09 \text{ ml}$$

Volumetric measurements must be referred to some standard temperature, and in order to minimize the need for calculations such as these, 20.0° centigrade, the average room temperature, has been chosen for this reference point. Since the ambient temperature in most laboratories is within a few degrees of this, the need seldom arises for a temperature correction in ordinary analytical work. The coefficient of cubic expansion for many organic liquids, however, is considerably greater than that for water or dilute aqueous solutions. Good precision in the measurement of these liquids may require corrections for temperature variations of a degree or less.

Types of Volumetric Apparatus

The reliable measurement of volume is the common purpose of the *pipet*, the *buret*, and the *volumetric flask*. These can be calibrated either to *deliver* or, alternatively, to *contain* a specified volume. Volumetric equipment is marked by the manufacturer to indicate not only the manner of calibration (usually with a TD, for "to deliver"; or a TC, for "to contain") but also the temperature to which the calibration strictly refers. Ordinarily, pipets and burets are designed and calibrated to deliver specified volumes while volumetric flasks are calibrated on a to-contain basis.

APPARATUS FOR VOLUMETRIC MEASUREMENTS

Pipets

All pipets are designed for the transfer of known volumes of liquid from one container to another. Some deliver a single, fixed volume; these are called *volumetric*, or *transfer pipets*. Others, known as *measuring pipets*, are calibrated in convenient units so that any volume up to the maximum capacity can be delivered.

While all pipets are filled to an initial calibration mark at the outset, the manner in which the transfer is completed is subject to considerable variation. Because of the attraction between most liquids and glass, a drop tends to remain in the tip of a drained pipet. This drop is blown from some pipets, but not others. Table 11-1 and Figure 11.1 summarize the several varieties most likely to be encountered in an analytical laboratory.

Table 11-1

PIPETS

Name	Type of Calibration	Function	Available Capacities, ml	Type of Drainage
Volumetric	TD	delivery of a fixed volume	1—200	free drainage
Mohr	TD	delivery of a variable volume	1—25	drain to lower calibration line
Serological	TD	same	0.1—10	blow out last drop
Serological	TD	same	0.1—10	drain to lower calibration line
Ostwald-Folin	TD	delivery of a fixed volume	0.5—10	blow out last drop
Lambda	TC	to contain a fixed volume	0.001—2	wash out with suitable solvent
Lambda	TD	delivery of a fixed volume	0.001—2	blow out last drop

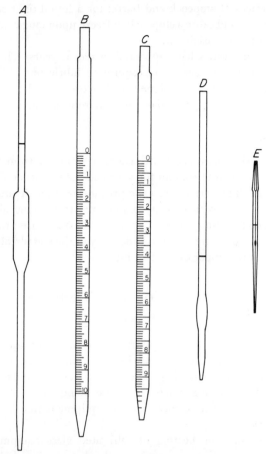

Fig. 11.1 Typical Pipets. (A) Volumetric (B) Mohr (C) Serological (D) Ostwald-Folin (E) Lambda.

Burets

Burets, like measuring pipets, enable the analyst to deliver any volume up to the maximum capacity. The precision attainable with a buret is appreciably better than that with a pipet.

In general, a buret consists of a calibrated tube containing the liquid and a valve arrangement by which flow from a tip can be controlled. Principal differences among burets are to be found in the type of valve employed. The Bunsen valve is the simplest, consisting of a closely fitting glass bead within a short length of rubber tubing. Only when the rubber tubing is deformed can liquid flow past the bead.

Burets equipped with glass stopcocks rely upon a lubricant between the ground-glass surfaces of stopcock and barrel for a liquid tight seal. Some solutions, notably bases, will cause a stopcock to freeze upon long contact; thorough cleaning is indicated after each use.

Valves made of plastics have appeared in recent years. These are inert to attack by most common reagents and require no lubricant.

More elaborate burets are designed so that they may·be filled and leveled automatically; these are of particular value in routine analysis.

Volumetric Flasks

Volumetric flasks are available with capacities ranging from 5 ml to 5 liters, and are usually calibrated to contain the specified volume when filled to the line etched on the neck. They are used in the preparation of standard solutions and in the dilution of samples to known volumes prior to taking aliquot portions with a pipet. Some are also calibrated on a "to-deliver" basis; these are readily distinguishable by two reference lines. If delivery of the stated volume is desired, the flask is filled to the upper of the two lines.

The Measurement of Volume—Techniques and Operations

Only clean glass surfaces will support a uniform film of liquid; the presence of dirt or oil will tend to cause breaks in this film. The appearance of water breaks is a certain indication of an unclean surface. Volumetric glassware is carefully cleansed by the manufacturer before being supplied with markings, and in order for these to have meaning the equipment must be kept equally clean when in use.

As a general rule, the heating of calibrated glass equipment should be avoided. Too rapid cooling can permanently distort the glass and cause a change in volume.

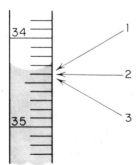

Fig. 11.2 The Liquid in a Buret, Showing the Meniscus. Volume indicated is 34.37 ml.

When a liquid is confined in a narrow tube such as a buret or a pipet, the surface is found to exhibit a marked curvature called a *meniscus*. It is common practice to use the bottom of the meniscus in calibrating and using volumetric ware. This minimum can often be established more exactly if an opaque card or piece of paper is held behind the graduations.

In reading volumetric ware the eye must be level with the liquid; otherwise an error due to *parallax* will arise. Thus if one's eye level is above that of the liquid, it will appear that a smaller volume has been taken than is actually the case. An error in the opposite direction can be expected if the point of observation is too low (see Fig. 11.2).

Directions for the Use of a Pipet

The following instructions pertain specifically to the manipulation of transfer pipets, but with minor modifications they may be used for other types as well. Liquids are usually drawn into pipets through the application of a slight vacuum. The mouth should not be used for suction since, aside from the danger of accidentally ingesting liquids, there is the possibility of contaminating the sample with saliva. Use of a rubber suction bulb or a rubber tube connected to an aspirator pump is strongly recommended (see Fig. 11.3).

Cleaning. Pipets may be cleaned with a warm solution of detergent or with tepid cleaning solution (see page 111). Draw in sufficient liquid to fill the bulb to about one third of its capacity. While holding it nearly horizontal, carefully rotate the pipet so that all interior surfaces are covered. Drain, and then rinse thoroughly with distilled water. Inspect for water breaks, and repeat the cleaning cycle if necessary.

Measurement of an aliquot. As in cleaning, draw in a small quantity of the liquid to be sampled and thoroughly rinse the interior surfaces; repeat this with at least two more portions. Then carefully fill the pipet somewhat past the graduation mark. Quickly place a forefinger over the upper end of the pipet to arrest the outflow of liquid. Make certain that there are no air bubbles in the bulk of the liquid or as a foam at the surface. Tilt the pipet slightly from the vertical and wipe the exterior free of adhering liquid. Touch the tip of the pipet to the wall of a glass vessel (not the actual receiving vessel) and slowly allow the liquid level to drop by partially releasing the forefinger, halting further flow as the bottom of the meniscus coincides exactly with the graduation mark. Place the tip of the pipet well into the receiving vessel and allow the sample to drain. When free flow ceases, rest the tip against an inner wall for a full 10 seconds. Finally, withdraw the pipet with a rotating motion to remove any droplet still adhering to the tip. *The small volume remaining inside the tip is not to be blown or rinsed into the receiving vessel.*

The sequence is illustrated in Figure 11.3.

NOTES:

1. The liquid can best be held at a constant level in the pipet if one's forefinger is slightly moist; too much moisture, however, makes control difficult.
2. It is good practice to avoid handling the pipet by the bulb.
3. Pipets should be thoroughly rinsed with distilled water after use.

Directions for the Use of a Buret

Before being placed in service, a buret must be scrupulously clean. In addition, it must be established that the valve is liquid tight.

Lubrication of a stopcock buret. Carefully remove all of the old grease from the stopcock and barrel. Lightly grease the stopcock, taking care to avoid the area near the hole. Insert the stopcock into the barrel and rotate

(a) Draw liquid past the graduation mark.

(b) Use forefinger to maintain liquid level above the graduation mark.

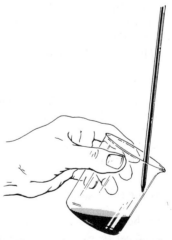

(c) Tilt pipet slightly and wipe away any drops on the outside surface.

(d) Allow pipet to drain freely.

Fig. 11.3 Technique for the Use of a Volumetric Pipet.

it vigorously. When the proper amount of lubricant has been used, the area of contact between stopcock and barrel appears nearly transparent, the seal is liquid tight, and no grease has worked its way into the tip.

Cleaning. Thoroughly clean the tube with detergent and a long brush. If water breaks persist after rinsing, clamp the buret in an inverted position with the end dipped in a beaker of cleaning solution. Connect a hose from the buret tip to a vacuum line. Gently pull the cleaning solution into the buret, stopping well short of the stopcock. Allow to stand for 10 to 15 minutes and then drain. Rinse thoroughly with distilled water and again inspect for water breaks. Repeat the treatment if necessary.

NOTES:

1. Cleaning solution often disperses more stopcock lubricant than it oxidizes and leaves a buret more heavily coated with film than before the treatment. For this reason the cleaning solution should *not* be allowed to come in contact with lubricated stopcock assemblies.

2. Grease films that appear unaffected by cleaning solution may yield to treatment with such organic solvents as acetone or benzene. Thorough washing with detergent should follow such treatment.

3. As long as the flow of liquid is not impeded, fouling of the buret tip with lubricant is not a serious matter. Removal is best accomplished with organic solvents.

4. Before returning a buret to service after reassembly, it is advisable to test for leakage. Simply fill the buret with water and establish that the volume reading does not change with time.

Filling. Make certain that the stopcock is closed. Add 5 to 10 ml of solution and carefully rotate the buret to wet the walls completely; allow the liquid to drain through the tip. Repeat this procedure two more times. Then fill the buret well above the zero mark. Free the tip of air bubbles by rapidly rotating the stopcock and allowing small quantities of solution to pass. Finally, lower the level of the solution to, or somewhat below, the zero marking; after allowing about a minute for drainage, take an initial volume reading.

Titration. Figure 11.4 illustrates the preferred method for manipulation of a stopcock. Any tendency for lateral movement of the stopcock will be in the direction of firmer seatings.

Fig. 11.4 Recommended Technique for Manipulation of a Buret Stopcock.

With the tip well within the titration vessel, introduce solution from the buret in increments of a milliliter or so. Swirl (or stir) the sample constantly to assure efficient mixing. Reduce the volume of the additions as the titration progresses; in the immediate vicinity of the end point, the reagent should be added a drop at a time. When it is judged that only a few more drops are needed, rinse down the walls of the titration vessel before completing the titration. Allow a minute or so to elapse between the last addition of reagent and the reading of the buret.

NOTES:

1. If a particular titration is unfamiliar, many analysts prepare an extra sample. No care is lavished on its titration since its only functions are to reveal the nature of the end point and to provide a rough estimate of titrant requirements. This deliberate sacrifice of one sample often results in an over-all saving of time.

2. Instead of rinsing near the end of the titration, the flask can be carefully tipped and rotated so that the bulk of the liquid picks up any droplets adhering to the walls.

3. Volume increments smaller than a normal drop may be taken by allowing a small volume of liquid to form on the tip of the buret and then touching the tip to the wall of the flask. This droplet is then combined with the bulk of the solution as in Note 2.

Directions for the Use of a Volumetric Flask

Before use, volumetric flasks should be washed with detergent and, if necessary, cleaning solution; only rarely need they be dried. Should drying be required, however, it is best accomplished by clamping the flasks in an inverted position and employing a mild vacuum to circulate air through them.

Weighing directly into a volumetric flask. Direct preparation of a standard solution requires that a known weight of solute be introduced into a volumetric flask. In order to minimize the possibility of loss during transfer, insert a funnel into the neck of the flask. The funnel is subsequently washed free of solid.

Dilution to the mark. After transferring the solute, fill the flask about half full and swirl the contents to achieve solution. Add more solvent, and again mix well. Bring the liquid level almost to the mark, and allow time for drainage. Then use a medicine dropper to make such final additions of solvent as are necessary. Firmly stopper the flask and invert repeatedly to assure uniform mixing. Finally, transfer the solution to a dry storage bottle.

NOTE:

If, as sometimes happens, the liquid level accidentally exceeds the calibration mark, the solution can be saved by correcting for the excess volume. Use a gummed label to mark the actual position of the meniscus. After the flask has been emptied, carefully refill to the mark with water. Then, using a buret, measure the volume needed to duplicate the actual volume of the solution. This volume, of course, should be added to the nominal value for the flask when calculating the concentration of the solution.

CALIBRATION OF VOLUMETRIC WARE

The reliability of a volumetric analysis depends upon agreement between the volumes actually and purportedly contained (or delivered) by the apparatus. Calibration simply verifies this agreement if such already exists, or provides the means for attaining agreement if it is lacking. The latter involves either the assignment of more reliable numerical values to the existing volume markings or the striking of new markings that agree with the existing numerical values.

In general, a calibration consists of determining the mass of a liquid of known density contained (or delivered) by the apparatus. Although this appears to be a straightforward process, a number of important variables must be controlled.

Principal among these is the temperature which influences a calibration in two ways. First and most important, the volume occupied by a given mass of liquid is temperature dependent. Second, the volume of the apparatus itself is variable, owing to the tendency of the glass to expand or contract with changes in temperature.

In Chapter 5 we noted that weighing data must be corrected, under some circumstances, for the volume of air displaced by weights and objects alike, and that the effect of buoyancy upon results becomes most pronounced when the density of the object is significantly lower than that of the weights. As a general rule a buoyancy correction must be applied to weighing data relating to a calibration.

Finally, the liquid employed for calibration requires consideration. Water is the liquid of choice for most work. Mercury is also useful, particularly where small volumes are involved. Because it does not wet glass surfaces, the volume of mercury contained by the apparatus will be identical with that which is delivered. In addition, the convex meniscus of mercury gives rise to a small correction that must be applied to give the corresponding volume for a liquid forming a concave meniscus. The magnitude of this correction is dependent upon the diameter of the apparatus at the graduation mark.

The calculations associated with calibrations, while not difficult, are somewhat involved. The raw weighing data are first corrected for buoyancy. In Chapter 5 we derived an equation that provides a good approximation of this correction

$$W_1 = d_{\text{air}} W_2 \left(\frac{1}{d_1} - \frac{1}{d_2} \right) + W_2 \qquad (11\text{-}2)$$

where W_1 is the corrected mass of the object, d_1 is its density, and W_2 and d_2 are the mass and density of the weights.

The volume of the apparatus at the temperature (t) of calibration is next obtained by dividing the density of the liquid at that temperature into the corrected weight.

Finally, this volume is corrected to the standard temperature of 20° C by means of equation (11-1), page 235.

Example.   A 25-ml volumetric pipet was calibrated against brass weights at 23° C; it was found to deliver an average of 25.08 grams of water. We wish to compute the volume delivered at this temperature and at 20° C.

At 23° C, the density of water is 0.9975; the density of brass is approximately 8.4. For all but the most refined work the density of air can be taken as 0.0012 gram/ml. Introducing these quantities into equation (11-2) we find

$$W_1 = (0.0012)\,(25.08) \left(\frac{1}{0.9975} - \frac{1}{8.4} \right) + 25.08$$

$$= 0.027 + 25.08 = 25.11 \text{ grams}$$

This is the weight of water delivered by the pipet at the temperature of the test. The volume corresponding to this weight is

$$V_{23°} = \frac{25.11}{0.9975} = 25.17 \text{ ml}$$

To determine the volume delivered by this pipet at 20° C, we make use of equation (11-1), taking 0.000025 as the coefficient of cubic expansion for glass.

$$V_{20°} = 25.17 + (0.000025)\,(20 - 23)\,(25.17)$$

$$= 25.17 - 0.0019 \cong 25.17 \text{ ml}$$

Table 11-2 is provided to ease the computational burden of calibration. It tabulates the volume occupied by 1.000 gram of water at various temperatures, as computed by equation (11-2). Provided that brass weights are used in calibration, the corresponding volume is obtained simply by multiplying the weight of water taken by the appropriate factor from this table.

Table 11-2

VOLUME AT 20° C OCCUPIED BY 1.000 GRAM OF WATER IN A GLASS CONTAINER, WEIGHED IN AIR AGAINST BRASS WEIGHTS AT VARIOUS TEMPERATURES

Temperature (°C)	Volume, ml	Temperature (°C)	Volume, ml	Temperature (°C)	Volume, ml
10	1.0016	17	1.0023	24	1.0036
11	1.0017	18	1.0025	25	1.0038
12	1.0018	19	1.0026	26	1.0041
13	1.0019	20	1.0028	27	1.0043
14	1.0020	21	1.0030	28	1.0046
15	1.0021	22	1.0032	29	1.0048
16	1.0022	23	1.0034	30	1.0051

Example. A 10-ml pipet was found to deliver 9.861 grams of water when calibrated against brass weights at 17° C. We need to know the volume this pipet delivers at 20° C.

From Table 11-2 we find that 1 gram of water at 17° C occupies 1.0023 ml at 20° C. Therefore, the volume delivered by the pipet at this temperature will be

$$V_{20°} = (9.861)(1.0023) = 9.88 \text{ ml}$$

General Directions for Calibration Work

All volumetric apparatus should be painstakingly freed of water breaks before being tested. Burets and pipets need not be dried; volumetric flasks should be thoroughly drained.

The water used for calibration should be drawn well in advance of use in order to allow it to reach thermal equilibrium with its surroundings. This condition is best established by noting the temperature of the water at frequent intervals and waiting until no further changes are observed.

An analytical balance is used for calibrations involving 50 ml or less; weighing to the nearest milligram is sufficient for all volumes in excess of 1 ml. Weighing bottles or small, well-stoppered Erlenmeyer flasks are convenient receivers for small volumes.

A two-pan laboratory balance is employed for the calibration of apparatus holding volumes larger than can be accommodated by an analytical balance. Since the arm lengths of such a balance can no longer be assumed to be equal, weighing by substitution is the preferred method.

Calibration of a Volumetric Pipet

Determine the empty weight of the receiver. Introduce an aliquot from the pipet, weigh receiver and contents to the nearest milligram, and calculate the weight of water delivered from the difference in these weights. Repeat the calibration several times.

Calibration of a Buret

Fill the buret, making certain that no bubbles are entrapped in the tip. Withdraw water until the level is at, or just below, the zero mark. Touch the tip to the wall of a beaker to remove any adhering drop. After allowing time for drainage, take an initial reading of the meniscus, estimating to the nearest 0.01 ml. Allow the buret to stand for 5 minutes and recheck the reading; if the stopcock is tight, there should be no noticeable change. During this interval, weigh (to the nearest mg) a 125-ml Erlenmeyer flask fitted with a rubber stopper.

Once tightness of the stopcock has been established, run approximately 10 ml into the flask at a rate of flow of about 10 ml per minute. Touch the tip to the wall of the flask. Wait 1 minute, record the volume, and refill the buret. Weigh the flask and its contents to the nearest 5 mg; the difference between this and the initial weight gives the mass of water actually delivered. Convert this mass into terms of volume using Table 11-2. Compute the correction in this interval by subtracting the apparent volume from that actually delivered.

Starting again from the zero mark, repeat the calibration using about 20 ml. Test the buret at 10-ml intervals over its entire length. Prepare a plot of the correction to be applied as a function of the volume delivered.

NOTE:

Any correction larger than 0.10 ml should be verified by duplicate determinations before being accepted.

Calibration of a Volumetric Flask

Weigh the clean, dry flask, placing it on the right-hand pan of a laboratory balance. Set a beaker on the left pan and add lead shot until balance is achieved. Remove the flask, and in its place substitute known weights until the same point of balance is reached. Carefully fill the flask with water of known temperature until the meniscus concides with the graduation mark. Return the flask to the right-hand pan and the beaker to the left. Repeat the process of counterweighing with lead shot followed by substituting weights for the flask. The difference between the two weighings gives the mass of water contained by the apparatus; calculate the corresponding volume with the aid of Table 11-2.

NOTE:

A glass tube with one end drawn to a tip is useful in making final adjustments of the liquid level.

Calibration of a Volumetric Flask Relative to a Pipet

The calibration of a flask relative to a pipet makes possible an excellent method for partitioning a sample into aliquots. The following directions pertain specifically to a 50-ml pipet and a 500-ml flask; other combinations are equally convenient.

With a 50-ml pipet, carefully transfer 10 aliquots to a 500-ml volumetric flask. Mark the location of the meniscus with a gummed label; coat the label with paraffin to assure permanence. When a sample is diluted to this mark, an aliquot taken with the same 50-ml pipet will constitute exactly one tenth of the total sample.

chapter 12. *Precipitation Titrations*

Volumetric methods based on the formation of a slightly soluble product are called *precipitation titrations*. Few of the vast number of reactions employed in gravimetric analysis can be satisfactorily adapted to yield corresponding volumetric methods. Although this is perhaps surprising, reasons for it can be seen if we consider precipitation reactions in light of the requirements for a volumetric analysis (page 212). Many stoichiometric precipitation reactions require appreciable periods of time for completion; others fail to yield a single product of definite composition; and for some precipitation reactions that would otherwise be suitable, no indicator is available.

Notwithstanding these limitations, the value of this type of titration has long been recognized; the names of such early chemists as Gay-Lussac, Mohr, and Volhard are associated with specific methods of volumetric-precipitation analysis and attest to the interest they generated in the past. The comparatively recent discovery of indicators that function as a consequence of being adsorbed on the surface of solids has made possible the development of further analytical methods based on these reactions. Finally, the determination of a number of important substances, notably the halides, can be readily and rapidly accomplished in this manner; this fact, as much as any, accounts for the widespread employment of precipitation titrations.

TITRATION CURVES FOR PRECIPITATION REACTIONS

Derivation

The curves presented in Chapter 10 illustrate the changes occurring in reactant concentration during the course of a titration. These are particularly useful in indicating the conditions prevailing at and near the equivalence point; they are thus valuable aids in deciding not only whether a particular substance can be titrated, but also in determining the magnitude of error to be expected from the application of a given indicator system.

To illustrate the derivation of a typical curve, consider the titration of 50.0 ml of 0.1 F sodium chloride with a 0.1 F solution of silver nitrate. A plot of either pAg or pCl with respect to the volume of silver nitrate added will yield a titration curve for the process. The solubility-product constant for silver chloride is 1.82×10^{-10} mol²/liter² at 25° C.

At the outset, the concentration of chloride ion is 0.1 M; pCl will thus be 1.00. Since none has yet been introduced, the silver ion concentration is zero, and pAg is indeterminate. After the addition of, say, 10.0 ml of silver nitrate solution, the chloride ion concentration will be decreased not only because of precipitate formation, but also because of the increase in volume of the solution; it will be given by

$$[Cl^-] = \frac{(50.0 \times 0.1 - 10.0 \times 0.1)}{60} + [Ag^+]$$

The first term in this equation would express the concentration of chloride ion only if the precipitate were absolutely insoluble. In fact, of course, silver chloride is slightly soluble, yielding equal quantities of silver and chloride ions to the solution as a consequence. The second term in the equation accounts for the contribution from this source.

Substituting this equation into the solubility-product expression will provide an exact solution for the silver ion concentration; however, a quadratic equation must be solved. Actually, we can arrive at an entirely satisfactory answer by assuming that the silver ion concentration is numerically small with respect to the chloride ion concentration; we then would find that

$$[Cl^-] \cong \frac{(50.0 \times 0.1 - 10.0 \times 0.1)}{60} = 6.7 \times 10^{-2} \text{ mol/liter}$$

Substituting this value into the solubility-product expression,

$$[Ag^+] = \frac{1.82 \times 10^{-10}}{6.7 \times 10^{-2}} = 2.7 \times 10^{-9} \text{ mol/liter}$$

It is evident that we were justified in making the simplifying assumption. The respective p functions may now be calculated as follows:

$$pCl = -\log 6.7 \times 10^{-2} = 2 - \log 6.7 = 1.17$$
$$pAg = -\log 2.7 \times 10^{-9} = 9 - \log 2.7 = 8.57$$

After 20.0 ml of silver nitrate solution have been added, we find that

$$[Cl^-] = \frac{(50.0 \times 0.1 - 20.0 \times 0.1)}{70} + [Ag^+] \cong 4.3 \times 10^{-2} \text{ mol/liter}$$

Thus

$$[Ag^+] = \frac{1.82 \times 10^{-10}}{4.3 \times 10^{-2}} = 4.1 \times 10^{-9} \text{ mol/liter}$$

Therefore

$$pCl = 2 - \log 4.3 = 1.37$$
$$pAg = 9 - \log 4.1 = 8.39$$

In this manner, then, we can calculate the equilibrium concentrations of chloride ion and silver ion resulting from the addition of any volume of silver nitrate short of the equivalence point. The chloride ion concentration is determined directly, the small contribution from the dissolved silver chloride being negligibly small except in the immediate region of the equivalence point; the silver ion concentration is calculated from the solubility-product expression.

At the equivalence point we have added an amount of silver ion identical to the amount of chloride ion in the sample; the system here consists simply of a saturated solution of silver chloride. The concentrations of the two ions are therefore identical and are readily calculated from the solubility-product expression.

$$[Ag^+][Cl^-] = [Ag^+]^2 = 1.82 \times 10^{-10}$$
$$[Ag^+] = [Cl^-] = 1.35 \times 10^{-5} \text{ mol/liter}$$
$$pAg = pCl = 5 - \log 1.35 = 4.87$$

Beyond the equivalence point, we evaluate the silver ion concentration directly, it now being present in excess; the chloride ion concentration is computed from the solubility-product expression. Thus, for example, after 52.5 ml of silver nitrate solution have been introduced,

$$[Ag^+] = \frac{(52.5 \times 0.1 - 50.0 \times 0.1)}{102.5} + [Cl^-]$$

Here the first term gives the silver ion concentration due to the excess of reagent present, while the second gives that due to the solubility of the precipitate. The assumption that the latter is vanishingly small with respect to the former is justified; after making this approximation, we find that

$$[Ag^+] = \frac{0.25}{102.5} = 2.4 \times 10^{-3} \text{ mol/liter}$$

Thus

$$[Cl^-] = \frac{1.82 \times 10^{-10}}{2.4 \times 10^{-3}} = 7.6 \times 10^{-8} \text{ mol/liter}$$

Therefore

$$pAg = 3 - \log 2.4 = 2.62$$
$$pCl = 8 - \log 7.6 = 7.12$$

Factors Affecting Titration Curves

Reagent concentrations. Table 12-1 is a compilation of data for this titration as well as for two others involving the same reactants at lower concentrations. We see from the table that the sum of pAg and pCl is a constant for each point, being equal to the negative logarithm of the solubility-product constant. Also, the equivalence point for each of these titrations is characterized by invariant silver and chloride ion concentrations; we take advantage of this fact in devising methods for indication of this end point.

Table 12-1

p FUNCTIONS FOR SEVERAL TITRATIONS OF CHLORIDE WITH SILVER ION

Volume AgNO₃ added, ml	50.0 ml of 0.1 F Cl⁻ titrated with 0.1 F AgNO₃		50.0 ml of 0.01 F Cl⁻ titrated with 0.01 F AgNO₃		50.0 ml of 0.001 F Cl⁻ titrated with 0.001 F AgNO₃	
	pAg	pCl	pAg	pCl	pAg	pCl
0	—	1.00	—	2.00	—	3.00
10	8.57	1.17	7.57	2.17	6.57	3.17
20	8.37	1.37	7.37	2.37	6.37	3.37
30	8.14	1.60	7.14	2.60	6.14	3.60
40	7.79	1.95	6.79	2.95	5.79	3.95
45	7.46	2.28	6.46	3.28	5.50*	4.24
47.5	7.15	2.59	6.15	3.50	5.27*	4.47
49.0	6.76	2.98	5.78*	3.96	5.05*	4.69
49.9	5.74	4.00	5.05*	4.69	4.88*	4.86
50.0	4.87	4.87	4.87	4.87	4.87	4.87
50.1	4.00	5.74	4.69	5.05*	4.86	4.88*
51	3.00	6.74	4.20	5.54*	4.64	5.10*
52.5	2.62	7.12	3.62	6.12	4.48	5.26*
55.0	2.32	7.42	3.32	6.42	4.20	5.54*

* Value computed using no approximations.

The pAg data for these titrations are plotted in Figure 12.1 as a function of the volume of silver nitrate added. It is noteworthy that all the curves have the same general shape. The most obvious characteristic is the abrupt change in pAg occurring in the vicinity of the equivalence point; we know from the previous discussion that pCl will show a similar rapid change in the opposite direction. This break in the titration curve becomes more pronounced with increasing reactant concentrations—a factor of considerable importance from the practical standpoint of end-point detection since the most satisfactory indicator behavior is observed in titrations whose curves display sharp breaks in the region of the equivalence point.

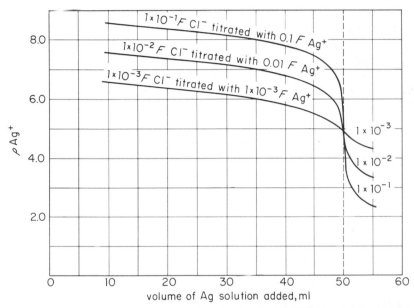

Fig. 12.1 Theoretical Curves for the Titration of Chloride Ion with Silver Ion. Plots of pAg as a function of the volume of AgNO$_3$ added. Note the effect of reagent concentration upon the shape of the curves.

Completeness of reaction. Figure 12.2 illustrates curves obtained when $0.1\ F$ solutions of several anions are titrated with $0.1\ F$ silver nitrate. Inspection of these curves reveals a definite relationship between the sharpness of the end-point break and the solubility of the precipitate formed in the reaction. The greatest change in pAg accompanies the titration of iodide ion which, of all the anions represented, forms the least soluble silver salt. The reaction in this instance is the most complete of all those shown. The poorest break is observed for the reaction that is least complete—that is, in the titration of bromate ion. Silver bromate is the most soluble of the salts depicted. Those reactions that produce silver salts having solubilities intermediate between these extremes yield titration curves that are also of intermediate character.

To summarize, the change in p function in the vicinity of the equivalence point is dependent not only upon reagent concentration, but also upon the completeness of the reaction. The most pronounced breaks are thus observed where the solutions titrated are of relatively high concentration and the chemical reactions proceed to essential completion.

END-POINT DETECTION

End Points Based upon Indicators

A chemical *indicator* is a substance that can react with one of the participants of the volumetric reaction in such a way as to produce an observable change in the appearance of the solution; ordinarily this change involves an alteration in color. The indicator substance, by virtue of its tendency to react with the reagent or the substance titrated, competes with one of the participants of the reaction for the other. In order to minimize the titration error, the reaction involving the indicator must suddenly become either highly favorable or highly unfavorable at the equivalence point as a consequence of the change in p function. Thus, the most satisfactory indicator behavior is observed with the most abrupt changes in solution composition at the equivalence point. For example, consider an indicator that shows a perceptible color change as pAg varies from 4.5 to 5.5. Referring to Figure 12.1 we see that each of these titrations requires a different

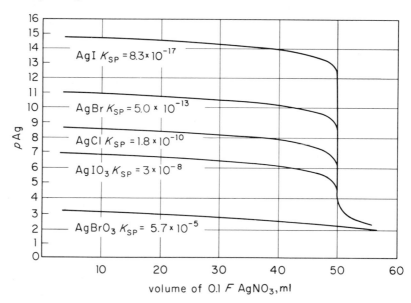

Fig. 12.2 Theoretical Curves for the Titration of Several Anions with Silver Ion. Plots of pAg with respect to the volume of 0.100 F AgNO$_3$ added to 50.0 ml of solution in which the anion concentration is 0.100 F.

volume of titrant to encompass this range; this varies from less than 0.1 ml of 0.1 F silver solution to about 7.5 ml of 0.001 F reagent. Clearly, in the first case a pronounced color change will occur in the region of the equivalence point; in the other, the change will be too gradual to be of use. With 0.01 F silver solutions approximately 0.5 ml will be required and the end point will not be sharp.

We must now consider the effectiveness of this same indicator in detecting the equivalence points for the titrations represented by the curves in Figure 12.2. No change will be observed during the bromate titration, and there will be a very gradual and rather premature change in an iodate titration. The indicator will be satisfactory for the titration of bromide and iodide although the change in color occurs after the point of chemical equivalence; these curves are so steep that the resulting error will be negligible for most purposes.

The formation of a second precipitate; the Mohr method. The formation of a second precipitate of distinctive color is the basis for end-point detection with the *Mohr method*. This procedure has been widely applied to the titration of chloride ion and bromide ion with standard silver nitrate. Chromate ion is the indicator, the end point being signaled by the appearance of brick-red silver chromate, Ag_2CrO_4.

The molar solubility of silver chromate is appreciably greater than that of silver chloride; its solubility-product constant is 1.1×10^{-12} mol³/liter³ at 25° C. The addition of chromate ion in suitable concentration should give rise to a system in which the appearance of the red Ag_2CrO_4 coincides with the equivalence point in a chloride titration. Recalling that the concentration of silver ion at the equivalence point in this titration is fixed at 1.35×10^{-5} M, we can compute the chromate ion concentration required to initiate precipitation of silver chromate in this medium.

$$[CrO_4^{2-}] = \frac{K_{sp}}{[Ag^+]^2} = \frac{1.1 \times 10^{-12}}{(1.35 \times 10^{-5})^2} = 6 \times 10^{-3} \text{ mol/liter}$$

In principle, then, the amount of chromate ion necessary to give this concentration could be added, and the red color of silver chromate would signal the appearance of the first excess of silver ion over its equivalence concentration. There are, however, practical considerations that limit the validity of this estimate. First, the color of a solution containing this concentration of chromate ion effectively masks that of the solid; in practice, an indicator concentration no greater than about 5×10^{-3} M can be employed. Second, with a lowering of the chromate ion concentration, the formation of silver chromate requires a silver ion concentration in excess of that in a saturated silver chloride solution. In addition, a finite amount of silver is needed to produce a detectable quantity of the solid. All these effects tend to cause overconsumption of the reagent. Referring again to Figure 12.1, we see that these difficulties are most serious where dilute solutions are involved. With 0.1 N solutions, however, they do not give rise to serious error. A correction may be made for the second effect by determining an indicator blank—that is, by determining the silver ion consumption for a suspension of chloride-free calcium carbonate in about the same volume and with the same quantity of indicator as the sample. An alternative that largely eliminates both errors is to use the Mohr method to standardize the silver nitrate solution against pure sodium chloride; the "working normality" obtained for the solution will compensate not only for these effects but also for the acuity of the analyst in detecting the color change.

Attention must be paid to the acidity of the medium. The equilibrium

$$2CrO_4^{2-} + 2H^+ \rightleftharpoons Cr_2O_7^{2-} + H_2O$$

is displaced to the right as the hydrogen ion concentration is increased; since silver dichromate is considerably more soluble than the chromate, the indicator reaction in acid solution requires far higher silver ion concentrations, if indeed it occurs at all. If the medium is made strongly alkaline, there is danger that silver will precipitate as its oxide.

$$2Ag^+ + 2OH^- \rightleftharpoons 2AgOH \rightleftharpoons Ag_2O + H_2O$$

Thus, the determination of chloride by the Mohr method must be carried out in a medium that is neutral or nearly so (pH 7 to 10). The addition of either sodium bicarbonate or of borax to the solution tends to maintain the hydrogen ion concentration within suitable limits.

Formation of a colored complex; the Volhard method.   A standard solution of thiocyanate may be used to titrate silver ion.

$$Ag^+ + SCN^- \rightleftharpoons AgSCN$$

Ferric ammonium sulfate serves as the indicator; this compound imparts a red coloration to the solution with the first slight excess of thiocyanate as a consequence of the reaction

$$Fe^{3+} + SCN^- \rightleftharpoons Fe(SCN)^{2+}$$
$$\text{red}$$

The titration must be carried out in acid solution to prevent hydrolysis of the ferric iron. These reactions provide the basis for the *Volhard method*. The titration error in the Volhard method is small because the indicator is highly sensitive to thiocyanate ions. Thus, 1 or 2 ml of a saturated ferric ammonium sulfate solution (about 40 percent) per 100 ml will impart a faint orange color to a solution upon addition of about 0.1 ml of 0.01 N thiocyanate. In order to avoid a premature end point in the titration, however, the solution must be shaken vigorously and the titration continued until the indicator color is permanent. This precaution is necessitated by the strong tendency of silver thiocyanate to adsorb silver ions from the solution, thus inhibiting the rate at which they combine with the thiocyanate.

The most important application of the Volhard method is for the indirect determination of chloride, as well as the other halide ions. A known excess of standard silver nitrate solution is added to the sample, and the excess is determined by back titration with a standard thiocyanate solution. The requirement of a strongly acid environment represents a distinct advantage for the Volhard technique over other methods for halide analysis.

An interesting problem in connection with the Volhard determination of chloride ion stems from the greater solubility of silver chloride as compared with silver thiocyanate. As a consequence, we can predict that the reaction

$$AgCl + SCN^- \rightleftharpoons AgSCN + Cl^-$$

will occur when a silver chloride precipitate is exposed to a solution containing an excess of thiocyanate ion. In a Volhard determination of chloride, this is the situation at the end point; the reaction, of course, is undesirable since it will cause the thiocyanate back titration to be too high and the resulting percentage chloride value to be too low.

The equilibrium constant for this reaction can be obtained by dividing the solubility-product expression for silver chloride by that for silver thiocyanate; that is,

$$K = \frac{[Cl^-]}{[SCN^-]} = \frac{[Ag^+][Cl^-]}{[Ag^+][SCN^-]} = 1.65 \times 10^2$$

As mentioned earlier, about 0.1 ml of 0.01 N thiocyanate is required to produce a color in 100 ml of solution when the usual amount of indicator is employed. Thus the thiocyanate concentration must be 10^{-5} F. The equilibrium chloride ion concentration of the solution at this point is then

$$[Cl^-] = K[SCN^-] = 1.65 \times 10^2 \times 10^{-5}$$
$$= 1.6 \times 10^{-3} \text{ mol/liter}$$

Inasmuch as this chloride concentration results primarily from the reaction of the thiocyanate reagent with the silver chloride, we can readily calculate the volume of the overtitration that results. Thus, with a 0.1 N thiocyanate solution and a total volume of 100 ml at the end point

$$\frac{1.6 \times 10^{-3} \times 100}{0.1} = 1.6 \text{ ml}$$

In actual practice the overconsumption of reagent is often greater.

A number of schemes have been developed to circumvent this source of error. Filtration, followed by titration of an aliquot of the filtrate, yields excellent results provided the precipitated silver chloride is first briefly digested; the time required for filtration is, of course, a disadvantage. Probably the most widely employed modification is that of Caldwell and Moyer[1] which consists of coating the silver chloride precipitate with nitrobenzene, thereby substantially removing it from contact with the solution. The coating is accomplished by shaking the titration mixture with a few milliliters of the organic liquid prior to back titration.

Swift and co-workers[2] have shown that these expedients are quite unnec-

[1] J. R. Caldwell and H. V. Moyer, *Ind. Eng. Chem., Anal. Ed.*, **7**, 38 (1935).

[2] E. H. Swift, G. M. Arcand, R. Lutwack, and D. J. Meier, *Anal. Chem.*, **22**, 306 (1950).

essary provided a sufficiently high concentration of ferric ion is employed. Their arguments are worth considering.

At the *equivalence point* in the Volhard method, the silver ion concentration should equal the sum of the concentrations of all the soluble thiocyanate- and chloride-containing species; that is,

$$[Ag^+] = [Cl^-] + [SCN^-] + [Fe(SCN)^{2+}]$$

For the end point and equivalence point to be identical, the $Fe(SCN)^{2+}$ concentration will be the minimum detectable to the eye. *Experimentally*, this is found to be 6.4×10^{-6} M. Thus if the titration error is to be zero, the following relationship must exist in the solution:

$$[Ag^+] = [Cl^-] + [SCN^-] + 6.4 \times 10^{-6} \qquad (12\text{-}1)$$

We may also write equilibrium-constant expressions for the various reactions taking place in the solution.

$$[Ag^+][Cl^-] = 1.8 \times 10^{-10} \qquad (12\text{-}2)$$

$$[Ag^+][SCN^-] = 1.1 \times 10^{-12} \qquad (12\text{-}3)$$

$$\frac{[Fe^{3+}][SCN^-]}{[FeSCN^{2+}]} = 7.2 \times 10^{-3}$$

Substituting 6.4×10^{-6} in the denominator for the dissociation constant of $FeSCN^{2+}$ and rearranging, we find that

$$[Fe^{3+}][SCN^-] = 7.2 \times 10^{-3} \times 6.4 \times 10^{-6}$$
$$= 4.6 \times 10^{-8} \qquad (12\text{-}4)$$

The four numbered equations contain four unknowns, and we should, therefore, be able to calculate equilibrium concentrations of all the species at the equivalence point. We shall compute the ferric ion concentration since this will give the indicator concentration needed if the end point and equivalence point are to coincide.

Substituting (12-2) into (12-1)

$$[Ag^+] = \frac{1.8 \times 10^{-10}}{[Ag^+]} + [SCN^-] + 6.4 \times 10^{-6}$$

Replacing $[Ag^+]$ by $[SCN^-]$ from (12-3)

$$\frac{1.1 \times 10^{-12}}{[SCN^-]} = \frac{1.8 \times 10^{-10}[SCN^-]}{1.1 \times 10^{-12}} + [SCN^-] + 6.4 \times 10^{-6}$$

Substituting equation (12-4) into the above

$$\frac{1.1 \times 10^{-12}[Fe^{3+}]}{4.6 \times 10^{-8}} = \frac{1.8 \times 10^{-10} \times 4.6 \times 10^{-8}}{1.1 \times 10^{-12}[Fe^{3+}]} + \frac{4.6 \times 10^{-8}}{[Fe^{3+}]} + 6.4 \times 10^{-6}$$

This may be rearranged to the quadratic form

$$[Fe^{3+}]^2 - 0.27\,[Fe^{3+}] - 0.30 = 0$$

Thus

$$[Fe^{3+}] = 0.4 \ \ \text{mol/liter}$$

These calculations indicate that removal of silver chloride is unnecessary provided the concentration of ferric ion is sufficiently high. The calculated figure of 0.4 F is about 20 times greater than the iron concentration normally recommended for the Volhard method. Swift has shown that satisfactory titrations and stable end points are obtained if indicator concentrations of this order are used.

The Volhard method may be applied to the analysis of any anions that form slightly soluble salts with silver nitrate. Steps must be taken to prevent interference from those precipitates more soluble than silver thiocyanate.

Adsorption indicators. Substances that impart a distinctive color to the surface of a precipitate are known as *adsorption indicators*. Under properly chosen circumstances, the adsorption (or the reverse desorption process) can be made to occur at or near the equivalence point in the titration; thus the appearance or disappearance of a color on the precipitate signals the end point.

An example of an adsorption indicator is the organic dye, *fluorescein*, which is employed as an indicator for the titration of chloride ion with silver nitrate. In aqueous solution this compound partially dissociates into hydrogen ions and negatively charged fluoresceinate ions which impart a yellowish-green color to the medium. The fluoresceinate ion forms a highly colored silver salt of limited solubility; in its application as an indicator, however, the concentration of the dye is *never large enough to exceed the solubility product of the silver fluoresceinate*.

At the outset of the titration of chloride with silver ions, the dye anion is not appreciably adsorbed by the silver chloride precipitate; it is, in fact, repelled from the surface by the negative charge resulting from adsorbed chloride ions. When the equivalence point is passed, however, the precipitate particles become positively charged by virtue of the strong adsorption of excess silver ions; under these conditions, retention of the fluoresceinate ions in the counter-ion layer is observed. This adsorption of silver fluoresceinate is marked by the appearance of the red silver-fluoresceinate color *on the surface of the precipitate*. This is an *adsorption*, not a precipitation, process inasmuch as the solubility product of the silver fluoresceinate is not exceeded. The process is reversible, the dye being desorbed upon back titration with chloride ion.

Not all dyes act as adsorption indicators, nor will a single indicator serve all precipitation titrations. We may distinguish four requirements of dye and precipitate upon which successful indicator action depends.

(1) Since this is a surface phenomenon, the precipitate should be produced in a highly dispersed state; this is one of the few instances where the analytical chemist is interested in producing and preserving a colloid.

(2) The precipitate must strongly adsorb its own ions. We have seen (Chapter 7) that this is usually the case.

(3) The dye must be strongly held by the primarily adsorbed ion. In general, strong adsorption correlates with low solubility of the salt formed between dye and lattice ion. At the same time, the solubility of the dye salt must be sufficiently great to prevent its precipitation.

(4) Since most adsorption indicators are anions of weak acids, the dye concentration will be dependent upon the pH of the solution in which it is contained. As a consequence, these indicators can only be used in solutions that are not too acid. A few cationic adsorption indicators are known, however, that can be employed in strongly acid solutions. With these, adsorption of the dye and coloration of the precipitate occurs in the presence of an excess of the anion of the precipitate (that is, when the precipitate particles are negatively charged).

Titrations employing adsorption indicators are rapid and reliable. The indicator error is often vanishingly small. At the same time, however, there are distinct limitations to their use. In general, there is a lack of titrations for which an adsorption indicator is available. Furthermore, successful application requires rather careful control of pH. The method may fail in the presence of high electrolyte concentrations, owing to the tendency of the precipitate to coagulate. Finally, in the case of titrations involving silver and halide ions, the presence of some adsorption indicators sensitizes the precipitate toward photodecomposition.

Procedures employing adsorption indicators are often called *Fajans procedures*, in recognition of this investigator's contributions to the technique.

Miscellaneous indicators.   If the precipitation reaction is accompanied by the consumption or release of hydrogen ions, or if the precipitating agent is an acid or a base, an acid-base indicator may be useful in detecting the end point. Similarly, oxidation-reduction indicators may be employed where the precipitating agent possesses oxidizing or reducing properties.

Other Methods of End-point Detection

The cessation of precipitation is sometimes used as an end point in a precipitation titration. The *Gay-Lussac method* for the titration of chloride requires sufficient time between additions of silver ion to allow the precipitate to settle; the titration is continued until further additions of reagent fail to produce a turbidity. This is called the *clear point*. An objection to the use of the clear point in a chloride titration is that a perceptible turbidity attends the addition of either silver ion or chloride ion to a saturated solution of silver chloride; thus, the clear point actually occurs slightly beyond the equivalence point. The *Mulder* modification of the Gay-Lussac method takes advantage of the fact that equal turbidity results from treatment of a saturated solution of silver chloride with either silver ion or chloride ion, provided only that the concentrations and volumes of these solutions are the same. Here, small quantities of supernatant liquid are withdrawn from the titrated solution and divided, each half being treated with identical volumes of a solution of one of the ions; the equivalence point is taken when equal turbidity is observed. This is the most reliable of all volumetric methods for the determination of chloride; because it is time consuming, however, it is employed only where the greatest accuracy is required.

With more insoluble precipitates, the clear point often provides a practical indication of the equivalence point.

In later chapters we shall describe several electroanalytical methods whereby the end points for some precipitation reactions may be detected.

APPLICATIONS OF PRECIPITATION-TITRATION ANALYSIS

Many volumetric precipitation methods make use of a standard solution of silver nitrate. The term *argentometry* refers to the employment of this reagent in volumetric analysis. Table 12-2 provides an indication of the variety of argentometric methods available to the analyst.

Table 12-2

ARGENTOMETRIC PRECIPITATION METHODS

Substance Determined	End Point	Remarks
AsO_4^{3-}, Br^- I^-, CNO^-, SCN^-	Volhard	
CO_3^{2-}, CrO_4^{2-}, CN^-, Cl^- $C_2O_4^{2-}$, PO_4^{3-}, S^{2-}	Volhard	removal of silver salt required before back titration of excess Ag^+
BH_4^-	modified Volhard	basis: titration of excess Ag^+ following $BH_4^- + 8Ag^+ + 8OH^- \rightleftharpoons 8Ag + H_2BO_3^- + 5H_2O$
Epoxide	Volhard	basis: titration of excess Cl^- following hydrohalogenation
K^+	modified Volhard	basis: precipitation of K^+ with known excess of $B(C_6H_5)_4^-$, addition of excess Ag^+ which precipitates $AgB(C_6H_5)_4$ and back titration of this excess
Br^-, Cl^-	Mohr	
I^-	clear point	
Br^-, Cl^-, I^-, SeO_3^{2-}	Fajans	
$V(OH)_4^+$, fatty acids, mercaptans	electroanalytical	direct titration with Ag^+

Table 12-3 lists applications of other precipitating agents to volumetric analysis.

Table 12-3

VOLUMETRIC PRECIPITATION METHODS

Precipitating Agent	Element Determined	Product	End–point Detection	Remarks
$K_4Fe(CN)_6$	Zn	$K_2Zn_3[Fe(CN)_6]_2$	diphenylamine	reverse titration also feasible
	In, Ga	$M_4[Fe(CN)_6]_3$	electroanalytical	
	Ag, Co, Ni, Mn		starch I_2	
	Hg(II)	$Hg_4[Fe(CN)_6]$	electroanalytical	
$Hg_2(NO_3)_2$	Cl^-, Br^-	Hg_2X_2	diphenylcarbazone	
	$Fe(CN)_6^{3-}$	$KHg_2Fe(CN)_6$	electrometric	titrate in 20-percent ethanol
$HgCl_2$	I^-	HgI_2	starch I_2	
$Pb(NO_3)_2$	SO_4^{2-}		Fajans	erythrosin B indicator
$Pb(OAc)_2$	PO_4^{3-}, $C_2O_4^{2-}$			dibromofluorescein for PO_4^{3-}, fluorescein for $C_2O_4^{2-}$
$Th(NO_3)_4$	F^-		Fajans	sodium alizarin sulfonate indicator
	$C_2O_4^{2-}$			alizarin red indicator

PRECIPITATION ANALYSIS WITH SILVER ION

Following are specific directions for the argentometric determination of chloride ion by the Mohr, the Volhard, and the Fajans methods. These may be used with little or no modification for the titration of other anions; Table 12-2 will serve as a guide in this respect.

Silver Nitrate and Its Solutions

Silver nitrate may be obtained in primary-standard purity. It has a high equivalent weight (169.89) and is readily soluble in water. Both the solid and aqueous solutions must be scrupulously protected from dust and other organic materials, and from sunlight. Metallic silver is produced by chemical reduction in the former instance, and by photodecomposition in the latter. The reagent is relatively expensive.

Silver nitrate crystals may be freed of surface moisture by drying at 110° C for about an hour. Some discoloration of the solid may result, but the amount of decomposition occurring in this time is ordinarily negligible.

Procedure for preparation of standard 0.1 *N* silver nitrate. Use a laboratory balance to weigh approximately 17 grams of silver nitrate into a clean, dry weighing bottle. Heat bottle and contents for 1 hour at 110° C; store in a desiccator. When cooled, determine the weight to the nearest mg with an analytical balance. Carefully transfer the bulk of the solid to a 1-liter volumetric flask and dilute to the mark; reweigh the bottle and residual solid. From the difference in weights compute the normality of the solution.

NOTES:

1. Weighing silver nitrate to only the nearest milligram will incur a maximum error of 1 part in 17,000 upon the value of the normality; since uncertainties in the subsequent analyses commonly exceed this figure, more accurate weighing is of little value.

2. If desired, silver nitrate solutions prepared to approximately the desired strength can be standardized against sodium chloride.

3. Once prepared, and when not actually in use, the solution should be stored in a dark place.

Determination of Chloride by the Mohr Method

Procedure. Carefully weigh 0.25 to 0.35-gram samples to the nearest 0.1 mg; dissolve each in about 100 ml of water. Add a pinch of sodium bicarbonate, making further additions, if necessary, until no further effervescence is noted. Then introduce 1 to 2 ml of 5-percent potassium chromate and titrate with standard silver nitrate solution to the first permanent appearance of the buff, silver chromate color.

NOTE:

The solubility of silver chromate increases with rising temperatures; its sensitivity as an indicator in this titration undergoes a corresponding decrease. Satisfactory results using the Mohr method require titration at room temperature.

Determination of Chloride by the Volhard Method

As applied to the analysis of chloride ion, the Volhard method is an indirect method involving the addition of a measured excess of standard silver nitrate solution to the sample and back titration of the excess with a standard potassium thiocyanate solution.

Procedure for preparation and standardization of 0.1 N potassium thiocyanate. Dissolve approximately 9.8 grams of KSCN in about 1 liter of water. Mix well.

Accurately measure 25 to 30-ml samples of standard $AgNO_3$ solution into Erlenmeyer flasks and dilute to approximately 100 ml. Add about 2 ml of concentrated HNO_3, followed by 2 ml of a saturated solution of ferric ammonium sulfate (Note 3). Titrate with the KSCN solution, with vigorous swirling of the flask, until the red-brown color of $FeSCN^{2+}$ is permanent for 1 minute. Calculate the normality of the KSCN solution. Results from duplicate standardizations should show agreement within 2 to 3 parts per thousand; if this precision has not been attained, perform further standardization titrations.

NOTES:

1. Potassium thiocyanate is usually somewhat moist; the direct preparation of a standard solution is not ordinarily attempted. However, Kolthoff and Lingane[3] report that gentle fusion followed by storage over calcium chloride yields a product that can be used for the direct preparation of standard solutions.

2. A potassium thiocyanate solution retains its titer over extended periods of time.

3. The indicator is readily prepared by placing some lumps of ferric alum, $NH_4Fe(SO_4)_2 \cdot 12H_2O$, in a bottle and shaking with dilute nitric acid. The acid prevents hydrolysis of the ferric ion.

Procedure for the analysis of chloride. Dry the sample at 100° to 110° C for 1 hour. Weigh several 0.25 to 0.35-gram samples to the nearest 0.1 mg into numbered 250-ml Erlenmeyer flasks. Dissolve each sample in 100 ml of distilled water. Introduce an excess of standard silver nitrate, being sure to note the volume taken. Acidify with 2 ml of concentrated nitric acid; add 2 ml of ferric alum indicator and 5 ml of chloride-free nitrobenzene. Shake vigorously. Titrate the excess silver with standard thiocyanate until the color of $FeSCN^{2+}$ is permanent for 1 minute.

[3] I. M. Kolthoff and J. J. Lingane, *J. Am. Chem. Soc.*, **57**, 2126 (1935).

NOTES:

1. With the concurrence of the instructor, a larger quantity of unknown can be weighed into a volumetric flask and diluted to known volume. The determination can then be made upon aliquot portions of this solution.

2. To obtain an approximation of the volume of standard $AgNO_3$ constituting an excess, calculate the amount that would be required for one of the samples assuming that it is 100-percent NaCl. When actually adding silver solution to the sample, swirl the flask vigorously, and add 3 or 4 ml in excess of the volume required to produce the clear point.

3. Nitric acid is introduced in order to improve observation of the clear point. Since the lower oxides of nitrogen tend to attack thiocyanate, the acid should be freshly boiled.

4. Nitrobenzene is a hazardous chemical and must be handled with respect. Poisoning can result not only from prolonged breathing of its vapors, but also from absorption of the liquid through the skin. In the event of spillage on one's person, the affected areas should be promptly and thoroughly washed with soap and warm water. Clothing soaked with the liquid should be removed and laundered.

5. At the outset of the back titration an appreciable quantity of silver ion is adsorbed on the surface of the precipitate. As a result there is a tendency for a premature appearance of the end-point color. Since success of the method depends upon an accounting for all of the excess silver ion, thorough and vigorous agitation is essential to bring about desorption of this ion from the precipitate.

Other applications and limitations of the Volhard method.   As indicated in Table 12-2, the Volhard method may be employed for the determination of a fairly extensive variety of substances. Where the solubility of the silver salt formed is increased by the presence of strong acid, a filtration is mandatory prior to back titration. The use of nitrobenzene or of filtration can be omitted in the few instances where the salts formed are less soluble than silver thiocyanate in acid media; silver bromide and silver iodide are the only common examples.

The method cannot be employed in the presence of oxidizing agents because of the susceptibility of thiocyanate to attack. It is often possible to eliminate this source of interference by prior treatment of the sample with a reducing agent. The method also fails in the presence of cations that form slightly soluble thiocyanates, notably palladium and mercury. Again, preliminary treatment may eliminate these sources of interference.

Determination of Chloride by the Fajans Method

The Fajans method employs a direct titration with organic substance, dichlorofluorescein, as indicator. Only a standard silver nitrate solution is required.

Procedure.   Dry the unknown for 1 hour at 100° to 110° C. Weigh individual 0.20 to 0.29-gram samples (to the nearest 0.1 mg) into 500-ml flasks. Dissolve in 175 to 200 ml of distilled water. Introduce 10 drops of dichlorofluorescein solution and about 0.1 gram of dextrin; immediately titrate with standard silver nitrate to the first permanent appearance of the pink color of the indicator.

NOTES:

1. Kolthoff[4] states that the chloride concentration should be within the range of 0.025 to 0.005 M; the directions given yield solutions approaching this maximum concentration only when pure sodium chloride at the upper weight limit is taken. If the approximate percentage of chloride in the sample is known, a corresponding decrease in volume or increase in sample size can be tolerated.

2. Silver chloride is particularly sensitive to photodecomposition in the presence of the indicator; the titration will fail if attempted in direct sunlight. Where this problem exists, the approximate equivalence point should first be ascertained in a trial titration, this value being used to calculate the volume of silver nitrate required for the other samples; the addition of indicator and dextrin should be delayed until the bulk of the silver nitrate has been added to the samples, after which the titration should be completed without further delay.

3. The use of polyethylene glycol is reportedly superior to dextrin for stabilizing the colloid.[5] The *indicator* may be prepared in a 50-percent aqueous solution of this preparation. In contrast to dextrin, polyethylene glycol is not susceptible to attack by microorganisms.

4. The indicator solution may be prepared as a 0.1-percent solution of dichlorofluorescein in 70-percent alcohol, or as a 0.1-percent aqueous solution of the sodium salt.

problems

1. A standard $AgNO_3$ solution is prepared by diluting 4.675 grams of the pure salt to exactly 1 liter. This solution is to be used for the titration of soluble Cl^- samples; express its concentration in terms of its
   (a) normality
   (b) titer, as mg Cl^-/ml
   (c) titer, as mg NaCl/ml

2. Calculate the normality of an $AgNO_3$ solution if 39.80 ml were required to titrate 0.1497 gram of pure NaCl.

3. What is the normality of an $AgNO_3$ solution for which the titer is 1.0 mg of Cl^-/ml? What weight of pure $AgNO_3$ would be required to produce 2 liters of this solution?

[4] I. M. Kolthoff, W. M. Lauer, and C. J. Sunde, *J. Am. Chem. Soc.*, **51**, 3273 (1929).
[5] R. B. Dean, W. C. Wiser, G. E. Martin, and D. W. Barnum, *Anal. Chem.*, **24**, 1638 (1952).

4. Samples of impure NaCl are to be determined by the Volhard method. Calculate the minimum volume of 0.0800 N AgNO$_3$ that must be added to 0.200-gram samples in order to assure an excess of the reagent.

5. A 50.00 ml aliquot of 0.0492 N AgNO$_3$ was introduced to a 0.410-gram sample of impure KBr. Titration of the excess Ag$^+$ required 7.50 ml of 0.0600 N KSCN. Calculate the percentage of KBr in the sample.

6. The effluent from a manufacturing process is to be analyzed for its Cl$^-$ content by a Fajans titration. This will ordinarily range from 0.40 to 0.75 mg/ml, expressed as NaCl. Titration of 100-ml aliquots should not exceed 40 ml of standard AgNO$_3$.

(a) What should be the normality of the AgNO$_3$ solution?
(b) What will be its titer, expressed as mg NaCl/ml?
(c) What is the minimum volume of this solution that will ordinarily be required?

7. Silver ion readily replaces a hydrogen attached to an acetylenic carbon atom according to the reaction

$$HC{\equiv}CR + 2Ag^+ + NO_3^- \rightleftharpoons AgC{\equiv}CR \cdot AgNO_3 + H^+$$

The sample is treated with an excess of AgNO$_3$ in an ammoniacal medium; the product is essentially insoluble in water (and violently explosive when dry). The analysis is completed by titration of the excess Ag$^+$ in the supernatant liquid. Calculate the percentage of 3-butyne-1-ol (GFW 70) in a crude sample on the basis of the following information:

weight of sample taken	0.1860 gram
volume of 0.1120 N AgNO$_3$ used	50.00 ml
volume of 0.1080 N KSCN required for back titration	7.40 ml

8. A potassium ferrocyanide solution was standardized against a 0.1830-gram sample of ZnO, 38.80 ml being required for the titration. The same solution was then used to determine the percentage of zinc carbonate in a 0.2380-gram sample of the mineral smithsonite; this titration required 30.60 ml. Both processes may be represented by the reaction

$$2Fe(CN)_6^{4-} + 3Zn^{2+} + 2K^+ \rightleftharpoons K_2Zn_3[Fe(CN)_6]_2$$

Calculate the percent ZnCO$_3$.

chapter 13. *Volumetric Methods Based on Complex-formation Reactions*

Metal ions enter into two types of bonding. The first is electrostatic in character, being due simply to the attraction between ions of opposite charge. Metal ions can also bond groups by covalent forces. The extent of this type of bonding is determined by several factors, principal among which is the electronic configuration of the metal ion itself. It is this tendency to enter into covalent bonding that gives rise to the complex species we shall discuss in this chapter.

In covalent bonding, the metal ion invariably acts as an electron-pair acceptor; the donor species, or *ligand*, must therefore have at least one unshared electron pair available for donation. The water molecule, the ammonia molecule, the chloride ion, and the cyanide ion are common examples of simple ligands. The product of such a union is known as a *coordination compound* if it is electrically neutral, or a *complex ion* if it carries an electrostatic charge.

While exceptions are known, a given species of cation usually accommodates

266

coordination bonds to a maximum of two, four, or six, its *coordination number* being simply a statement of this maximum. In common with all covalent bonding, coordinate bonds have directional character. Thus a linear configuration is characteristic of species with a coordination number of two. With a coordination number of four, the bonds are directed toward the corners of a square in some instances, and toward those of a regular tetrahedron in others. The configuration of a regular octahedron is commonly associated with a coordination number of six.

The species produced as a consequence of coordination bonding can be electrically positive, neutral, or negative. For example, cupric ion, with a coordination number of four, forms a cationic complex with ammonia, $Cu(NH_3)_4^{2+}$; a neutral compound with glycine, $Cu(H_2NCH_2COO)_2$; and is incorporated in an anionic complex with chloride ion, $CuCl_4^{2-}$.

Complex-formation reactions have been used to advantage in quantitative analysis for at least a century. The truly remarkable growth in analytical applications is of recent origin, however, and is due to a particular class of coordination compounds known as *chelates*. These compounds result from the reaction between a metal ion and a ligand that contains two or more donor groups; their properties frequently differ markedly from the parent cation.

A chelating agent containing two groups that coordinate with the metal ion is classified as *bidentate*, while one with three such groups is called *terdentate*. *Quadri-*, *quinque-*, and *sexadentate* chelating agents are also encountered.

TITRATION CURVES FOR COMPLEX-FORMATION REACTIONS

Any point in a complex-formation titration is characterized by a condition of equilibrium among the species participating in the reaction. If constants for the equilibria involved are known, we can derive titration curves that relate the concentration of one of the reactants to the volume of titrant introduced. For complex ions, numerical values for equilibrium constants will be found tabulated in the literature either as *instability constants* or as *formation (or stability) constants*. These constants were defined in Chapter 2.

Consider the equilibrium between a metal ion M, possessing a coordination number of four, and the quadridentate ligand D;

$$MD \rightleftharpoons M + D$$

We have intentionally avoided supplying either species with electrostatic charge since, for present purposes, these are not of importance.

In a similar manner, the equilibrium between M and the bidentate ligand B can be represented by

$$MB_2 \rightleftharpoons M + 2B$$

Here, however, the equation and the equilibrium constant derived from it refer to the *over-all* process, whereas the actual dissociation is stepwise involving the intermediate formation of the substance MB. Thus,

$$MB_2 \rightleftharpoons MB + B \tag{13-1}$$

$$MB \rightleftharpoons M + B \tag{13-2}$$

Instability constants can be expressed for these individual processes.

$$K_1 = \frac{[MB]\,[B]}{[MB_2]} \qquad\qquad K_2 = \frac{[M]\,[B]}{[MB]}$$

Addition of equations (13-1) and (13-2) yields the expression for the over-all process. The equilibrium constant for the over-all reaction is given by the product of the individual steps.

$$K_{\text{over-all}} = \frac{[M]\,[B]^2}{[MB_2]} = K_1 K_2$$

The complex produced between M and the simple ligand A results in the over-all equilibrium

$$MA_4 \rightleftharpoons M + 4A$$

Again, this process occurs in a stepwise fashion; the equilibrium constant for the over-all reaction is therefore numerically equal to the product of the constants for the four constituent processes.

Unless otherwise noted, instability- (or formation-) constant data tabulated in the literature refer to the over-all reaction. This is of paramount importance when considering a system as the basis of a titration.

Derivation of Titration Curves

For the derivation of a titration curve we shall use the least involved case of complex formation—namely, a system in which the reactants combine in a 1:1 ratio. Specifically, consider the titration of 60.0 ml of a 0.02 F solution of M ion with a 0.02 F solution of the quadridentate ligand D. The equilibrium-constant expression for the system can be expressed as

$$K_{\text{inst}} = \frac{[M]\,[D]}{[MD]}$$

If the numerical value for this instability constant is 1×10^{-20} mol/liter, we can compute the equilibrium concentrations of all species at any point in the titration from the volume of titrant added. A plot of pM or pD against the volume of D will yield the desired titration curve. We shall use pM.

Before the addition of any reagent, the concentration of M is 2×10^{-2} molar. The initial pM for the system is thus equal to 1.70.

After 10.0 ml of D have been added, the volume of the solution will have increased to 70 ml. The solution will contain relatively large quantities of M and

MD, but the concentration of D will be small since its only source will be the dissociation of the complex. We may express the concentration of M as follows:

$$[M] = \frac{60 \times 0.02 - 10 \times 0.02}{70} + [D]$$

Assuming that D is small with respect to the first quantity, we find that

$$[M] \cong \frac{1.2 - 0.2}{70} = 1.43 \times 10^{-2} \text{ mol/liter}$$

Thus the value for pM at this point is

$$p\text{M} = -\log 1.43 \times 10^{-2} = 2 - \log 1.43$$
$$= 1.84$$

In order to check the assumption employed in the above calculation, we must estimate a value for D. We first calculate the concentration of the complex

$$[MD] = \frac{10 \times 0.02}{70} - [D]$$

$$\cong \frac{0.20}{70} = 2.86 \times 10^{-3} \text{ mol/liter}$$

The concentration of D may then be found from the instability-constant expression

$$[D] = K_{\text{inst}} \frac{[MD]}{[M]} = \frac{1 \times 10^{-20} \times 2.86 \times 10^{-3}}{1.43 \times 10^{-2}}$$

$$= 2 \times 10^{-21} \text{ mol/liter}$$

This is indeed small when compared with the concentration of either M or MD.

The value of pM may be evaluated in this way for any point short of the equivalence point; this provides the means for defining the initial portion of the titration curve.

At the equivalence point the system consists of a 0.01 F solution of MD. Dissociation of the complex is the sole source of M and D ions; consequently, they are present in equal concentration.

$$[M] = [D]$$
$$[MD] = 0.01 - [M] \cong 1 \times 10^{-2} \text{ mol/liter}$$

Substituting into the instability expression

$$\frac{[M]^2}{1 \times 10^{-2}} = 1 \times 10^{-20}$$

$$[M] = 1 \times 10^{-11} \text{ mol/liter}$$

and

$$pM = -\log 1 \times 10^{-11} = 11$$

After passing the equivalence point, values for pM can be calculated from the formal concentration of excess D and the instability constant. For example, after 70 ml of a 0.02 F solution of D have been introduced,

$$[D] = \frac{70 \times 0.02 - 60 \times 0.02}{130} + [M]$$

Assuming that [M] is much smaller than the first term

$$[D] \cong \frac{0.2}{130} = 1.5 \times 10^{-3} \text{ mol/liter}$$

$$[MD] = \frac{60 \times 0.02}{130} - [M]$$

$$\cong \frac{1.2}{130} = 9.2 \times 10^{-3} \text{ mol/liter}$$

Introducing these values into the instability-constant expression gives

$$[M] = \frac{1.0 \times 10^{-20} \times 9.2 \times 10^{-3}}{1.5 \times 10^{-3}}$$

$$= 6.2 \times 10^{-20} \text{ mol/liter}$$

and

$$pM = 19.2$$

The reader should note that the assumptions made in this calculation appear valid in view of the very small value for [M].

The curve for this titration appears in Figure 13.1. As in the case of precipitation titrations, the equivalence-point region is characterized by a marked change in the p function. Shown also are curves representing the titration of M with the bidentate ligand B to produce a 2:1 complex

$$K_1 = 1 \times 10^{-8}, \quad K_2 = 1 \times 10^{-12}$$

and with the unidentate A to yield a 4:1 complex

$$K_1 = 1 \times 10^{-2}, \quad K_2 = 1 \times 10^{-4}, \quad K_3 = 1 \times 10^{-6}, \quad K_4 = 1 \times 10^{-8}$$

Although the units vary, the numerical value for the over-all equilibrium constant is 1×10^{-20} in each of the three cases. Thus for any given value of the over-all equilibrium constant, the single-step process gives the most definite change in pM at the equivalence point; the presence of intermediate species of varying stability tends to diminish the magnitude of this break. It is in this respect that chelating agents show great superiority over other complexing agents, since fewer intermediate complexes are involved.

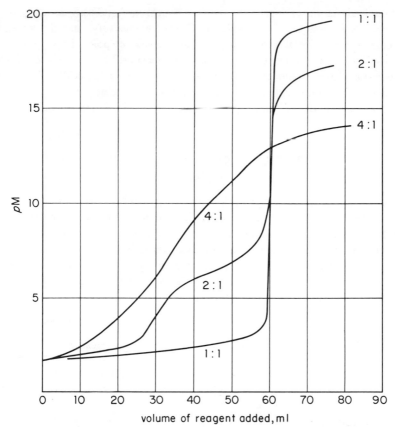

Fig. 13.1 Theoretical Titration Curves for Typical Complex Formation Reactions. Plots of pM as a function of the volume of 0.100 F solutions of simple, bidentate, and quadridentate ligands added to 60.0 ml aliquots of 0.100 F M. For each process the over-all equilibrium constant is 1×10^{-20}.

METHODS OF END-POINT DETECTION

We have seen that end-point detection is based upon the establishment of auxiliary equilibria that yield a tangible indication of the marked change in p function, or upon the physicochemical measurement of some property of the system. We shall consider examples from the former only and defer discussion of the latter to later chapters.

Formation or Disappearance of a Solid Phase

The *Liebig* determination of cyanide provides an example of an end point based upon the appearance of a solid phase. Addition of standard silver nitrate solution leads to formation of the soluble complex, $Ag(CN)_2^-$. As long as any free cyanide is present, the complex is formed exclusively, and the silver ion concentration remains at a low level. At the equivalence point, however, a rapid increase in silver ion concentration occurs, and reaction between this species and the complex then produces the slightly soluble silver cyanide, AgCN (or $Ag[Ag(CN)_2]$). The end point, then, is signaled by the appearance of the first faint turbidity in the solution. The equations for the titration are as follows:

$$Ag^+ + 2CN^- \rightleftharpoons Ag(CN)_2^- \qquad \text{(analytical reaction)}$$

$$Ag^+ + Ag(CN)_2^- \rightleftharpoons Ag[Ag(CN)_2] \qquad \text{(indicator reaction)}$$
$$\text{white precipitate}$$

The disappearance of turbidity can also be employed as an end point in a complex-formation titration. An example is in a routine method for the titration of nickel with a standard solution of potassium cyanide. The analysis is performed in an ammoniacal medium containing just enough suspended silver iodide to impart a slight turbidity to the solution; the end point is signaled by the disappearance of this turbidity, resulting from the rapid increase in the cyanide concentration. The reactions may be formulated as follows :

$$Ni(NH_3)_4^{2+} + 4CN^- \rightleftharpoons Ni(CN)_4^{2-} + 4NH_3 \qquad \text{(analytical reaction)}$$

$$AgI + 2CN^- \rightleftharpoons Ag(CN)_2^- + I^- \qquad \text{(indicator reaction)}$$

Since the end point and equivalence point do not coincide exactly, common practice is to standardize the cyanide reagent against a known quantity of nickel.

Acid-base Indicators

The electron-donor groups of most common ligands tend to combine not only with metallic ions but, to a lesser extent, with protons as well. As a consequence, the equivalence point in a complex-formation titration is often accompanied by a rather marked change in pH which can be detected with an acid-base indicator. For example, the sodium salt of ethylenediaminetetraacetic acid, a widely used complexing reagent having the general formula Na_4Y, can be employed to titrate several cations, the analytical reaction being

$$Y^{4-} + M^{2+} \rightleftharpoons MY^{2-}$$

where MY^{2-} is the soluble complex ion. When the equivalence point has been passed, the solution becomes basic as a result of the hydrolysis of the excess Y^{4-} ions

$$Y^{4-} + H_2O \rightleftharpoons HY^{3-} + OH^-$$
$$HY^{3-} + H_2O \rightleftharpoons H_2Y^{2-} + OH^-$$
etc.

The consequent rise in hydroxyl ion concentration can be detected with an acid-base indicator.

Formation or Disappearance of a Soluble Complex

In Chapter 12 we noted that the Volhard method utilizes the formation of a colored complex to signal the end point. Analogous indicator reactions are often employed for complex-formation titrations. In recent years, a large number of reagents that form colored complexes with certain metal ions have been developed expressly for use as indicators in these titrations.

Metal-indicator systems function in two ways. If the cation being titrated produces a color with the indicator, the end point will be characterized by the disappearance of this color. Where this cation does not give a colored complex, a second cation that does is introduced; the first excess of titrant then decolorizes this complex. Clearly, the complex formed in the analytical reaction must be sufficiently stable to prevent decomposition of the indicator-metal complex until the equivalence point has been passed.

The selection of a metal indicator for a particular titration is often a matter of trial and error—the principal reason being the rather formidable number of equilibria that exert an appreciable influence upon the course of the titration and the behavior of the indicator. The active species in both titrant and indicator are often pH dependent; the hydrogen ion concentration of the medium is thus an important experimental variable. Provided numerical values for all constants are known, it is possible to predict in advance whether a given metal indicator will provide a satisfactory color change. Reilley and Schmid[1] describe in detail a method for ascertaining the optimum reaction conditions for the use of metal indicators in chelate titrations.

APPLICATION OF ORGANIC CHELATING REAGENTS TO VOLUMETRIC ANALYSIS

A number of tertiary amines containing carboxylic acid groups form remarkably stable complexes with a variety of metal ions. These compounds are marketed under such trade names as Complexones or Versenes (Dow Chemical Co.). Their application to analysis was first suggested by Schwarzenbach in 1945, and since that time the amount of investigative effort devoted to their application has been truly phenomenal. We shall therefore consider titrations employing these reagents in some detail.

Reagents

Ethylenediaminetetraacetic acid—often abbreviated EDTA—has the structure

$$\begin{array}{ccc} HOOC-CH_2 & & CH_2-COOH \\ & \diagdown N-CH_2-CH_2-N \diagup & \\ HOOC-CH_2 & & CH_2-COOH \end{array}$$

[1] C. N. Reilley and R. W. Schmid, *Anal. Chem.*, **31**, 887 (1959).

EDTA is the most widely used of this class of compounds. It is a weak acid for which $pK_1 = 2.0$, $pK_2 = 2.67$, $pK_3 = 6.16$, and $pK_4 = 10.26$.[2] These values indicate that two of the protons are lost much more readily than the remaining two. In addition to the four acidic hydrogens, each nitrogen atom has an unshared pair of electrons; the molecule thus has six potential sites for bonding with a metal ion.

The abbreviations H_4Y, H_3Y^-, H_2Y^{2-}, etc., are often employed in referring to EDTA and its ions.

Because of its limited solubility, the free acid, H_4Y, is not often employed for the preparation of standard solutions. Likewise, the tetrasodium salt, Na_4Y is not very satisfactory owing to its extensive hydrolysis in solution and the resulting high alkalinity. The disodium salt, Na_2H_2Y, is most useful for analytical purposes, being obtainable in high purity as the dihydrate. Recrystallization of a commercial preparation of this salt is advisable before considering it to be of primary standard quality.[3] Solutions of this substance are stable for months when stored in plastic or borosilicate glass containers.

Another common reagent is nitrilotriacetic acid—abbreviated NTA— which has the structure

$$\text{HOOC—CH}_2 \diagdown \atop \text{N} \diagup \begin{matrix} \text{CH}_2\text{—COOH} \\ \\ \text{CH}_2\text{—COOH} \end{matrix}$$

Aqueous solutions of this quadridentate ligand are prepared from the free acid which is commercially available.

Many related substances have also been investigated; these have not been widely employed for volumetric analysis, however. We shall confine our discussion to the applications of EDTA.

Reactions of EDTA with Metal Ions

In general, 1:1 complexes are formed with metallic ions; the reactions of EDTA, for example, can be summarized as follows:

$$M^{2+} + H_2Y^{2-} \rightleftharpoons MY^{2-} + 2H^+$$
$$M^{3+} + H_2Y^{2-} \rightleftharpoons MY^- + 2H^+$$
$$M^{4+} + H_2Y^{2-} \rightleftharpoons MY + 2H^+$$

Obviously the extent to which these complexes form is markedly affected by pH. In addition, there is a rough correlation between charge on the cation and stability of the resultant complex. Welcher states this relationship as follows:[4]

[2] The negative logarithm of an equilibrium constant is called its pK. Thus, in this instance, $K_1 = 1.00 \times 10^{-2}$, and therefore $pK_1 = 2.0$.

[3] W. J. Blaedel and H. T. Knight, *Anal. Chem.*, **26**, 743 (1954).

[4] F. J. Welcher, *The Analytical Uses of Ethylenediaminetetraacetic Acid*. Princeton, N.J.: D. Van Nostrand Company, Inc., 1958.

(1) the complexes of EDTA with divalent cations are very stable in basic or slightly acidic solutions; (2) complexes of the trivalent cations are stable in the pH range of 1 to 2; and (3) complexes of quadrivalent cations are often stable at a pH less than one. A number of exceptions to these generalities can be found.

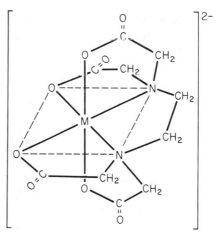

The Versenes are remarkable from the standpoint of the high stability of their metal adducts and the ubiquity of their complexes. All cations react, at least to some extent; with the exception of the alkali metals, the complexes formed are sufficiently stable to form the basis for volumetric analysis. Undoubtedly this great stability arises from the several complexing groups within the molecule giving rise to structures that effectively surround and isolate the cation. This is depicted in Figure 13.2.

Fig. 13.2 Proposed Structure of a Metal-EDTA Chelate.

Titration Methods Using EDTA

Several procedures are employed in the application of EDTA to volumetric analysis. The most common of these are considered in the following paragraphs.

Direct-titration procedure. Direct titration of metal ions by a standard solution of EDTA is feasible when a suitable method for end-point detection is available. For example, magnesium and a number of other divalent ions can be determined in this way by using the metal indicator *Eriochrome black T*. The titration is carried out in somewhat basic solution, the reaction being

$$Mg^{2+} + H_2Y^{2-} \rightleftharpoons MgY^{2-} + 2H^+$$

The indicator is a complex organic chelating agent. In neutral or somewhat basic solutions it is a doubly dissociated ion, HIn^{2-}, that is blue in color. This ion forms red chelate complexes with several metallic ions including magnesium; the indicator equilibrium may be represented as follows:

$$\underset{\text{red}}{MgIn^-} + H^+ \rightleftharpoons Mg^{2+} + \underset{\text{blue}}{HIn^{2-}}$$

When a direct titration of magnesium is performed with Eriochrome black T, the solution is initially red due to the large magnesium ion concentration (low pMg). At the equivalence point there is sharp increase in pMg (see Fig. 13.1), and the above equilibrium is shifted to the right. Clearly, the stability of the metal-indicator complex must be such that EDTA can compete favorably with it for the magnesium ions.

Acid-base indicators are also employed in direct titrations in which Na_4Y is the reagent (see p. 272).

Back titrations. Back-titration procedures are useful for the analysis of metallic ions that form very stable complexes with EDTA and for which a satisfactory indicator is not available. In these instances the excess EDTA is determined by back titration with a standard magnesium solution using Eriochrome black T as the indicator. The metal-EDTA complex must be more stable than the magnesium complex; otherwise the back titrant would tend to decompose it.

This technique is also useful for the analysis of metals in the presence of anions that form insoluble precipitates with the metal under the conditions of the analysis; the presence of EDTA prevents this precipitation.

Displacement titrations. In displacement titrations an excess of a solution containing EDTA in the form of a magnesium or zinc complex is introduced; if the metal ion forms a more stable complex than that of magnesium or zinc, the following reaction occurs:

$$MgY^{2-} + M^{2+} \rightleftharpoons MY^{2-} + Mg^{2+}$$

The liberated magnesium is then titrated with a standard EDTA solution. This technique is useful where no satisfactory indicator is available for the metal ion being determined.

Alkalimetric titrations. In alkalimetric titrations an excess of Na_2H_2Y is added to a solution of the metallic ion in which the pH has first been adjusted to near neutral; the liberated hydrogen ions are then titrated with a standard solution of a base.

Scope of Complexometric Titrations

Complexometric titrations are employed for the analysis of nearly all the metallic ions.[5] The following directions illustrate some of the techniques employed.

Reagents and Solutions

Preparation of 0.01 F EDTA solution—direct method. Dry the purified dihydrate ($Na_2H_2Y \cdot 2H_2O$, GFW = 372.1) at 80° C to remove superficial moisture. After cooling, weigh about 3.8 grams (to the nearest milligram) into a 1-liter volumetric flask and dilute to the mark.

---

[5] For further information regarding specific methods of analysis, the interested reader should consult the works of Welcher, cited previously; S. Chaberek and A. E. Martell, *Organic Sequestering Agents*. New York: John Wiley and Sons, Inc., 1959; and G. Schwarzenbach, *Complexometric Titrations*. London: Methuen and Co. Ltd., 1957.

NOTES:

1. Specific instructions for the purification of commercial preparations of the disodium salt are to be found in the reference by Blaedel and Knight (see p. 274).

2. Direct preparation of standard solutions requires the total exclusion of polyvalent cations. If any doubt exists regarding the quality of the distilled water, a pretreatment by passage through a cation-exchange resin is recommended.

3. If desired, the anhydrous salt (GFW = 336.1) may be employed instead of the dihydrate. The weight taken should be adjusted accordingly.

4. Alternatively, a solution of approximately the desired concentration may be prepared and standardized against a Mg^{2+} solution of known strength.

Preparation of the magnesium complex of EDTA, 0.1 F solution. To 37.2 grams of $Na_2H_2Y \cdot 2H_2O$ in 500 ml of distilled water add an equivalent quantity (24.65 grams) of $MgSO_4 \cdot 7H_2O$. Introduce a few drops of phenolphthalein followed by sufficient sodium hydroxide to turn the solution faintly pink. Dilute the solution to 1 liter. When properly prepared, portions of this solution should assume a dull violet color when treated with pH-10 buffer and a few drops of Eriochrome black T (Erio T) indicator. Furthermore, a single drop of 0.01 F Na_2H_2Y should cause a color change to blue, while an equal quantity of 0.01 F Mg^{2+} should cause a change to red. The composition of the solution should be adjusted by addition of Mg^{2+} or Na_2H_2Y until these criteria are met.

Preparation of Eriochrome black T solution. Dissolve 200 mg of the solid in a solution consisting of 15 ml of triethanolamine and 5 ml of absolute ethanol.

Preparation of buffer, pH 10. Dilute 570 ml of aqueous NH_3 (sp gr 0.90) and 70 grams of NH_4Cl to approximately 1 liter.

Applications

Determination of magnesium by direct titration. The sample will be issued as an aqueous solution; transfer it to a clean 500-ml volumetric flask. Dilute to the mark. Take 50.00-ml aliquots, treating each with 1 to 2 ml of pH-10 buffer and 2 to 4 drops of Erio T indicator solution. Titrate with 0.01 F Na_2H_2Y to a color change from red to blue. Express the results of the analysis in terms of milligrams of Mg^{2+}/liter of solution.

NOTES:

1. The color change of the indicator is slow in the vicinity of the end point. Care must be taken to avoid overtitration.

2. Other alkaline earths, if present, will also be titrated with magnesium and should be removed prior to the analysis. $(NH_4)_2CO_3$ is a suitable reagent for this. Most polyvalent cations also interfere, and should be precipitated as hydroxides.

3. Since this reaction may be expressed as

$$Mg^{2+} + H_2Y^{2-} \rightleftharpoons MgY^{2-} + 2H^+$$

the normality of the chelating solution is twice its formal concentration. For a 40-ml titration, each aliquot should contain about 10 mg of magnesium ion.

Determination of calcium by substitution titration. Weigh the sample into a 500-ml volumetric flask and dissolve in a minimum quantity of dilute HCl. Neutralize the solution with sodium hydroxide (Note 2) and dilute to the mark. Take 50.00-ml aliquots for titration, treating each as follows: add approximately 2 ml of pH-10 buffer, 1 ml of the magnesium-chelate solution, and 2 to 4 drops of Erio T indicator. Titrate with standard 0.01 F Na_2H_2Y to a color change from red to blue. Report the percentage of calcium oxide in the sample.

NOTES:

1. The weight of sample taken should be such as will contain about 150 to 160 mg of Ca^{2+}.

2. To neutralize the solution, introduce a few drops of methyl red and add base until the red color is discharged.

3. The amount of MgY^{2-} solution added is not critical. Its presence is required because the indicator does not exhibit a satisfactory color change when Ca^{2+} is present alone. Since the calcium chelate is more stable, the reaction

$$Ca^{2+} + MgY^{2-} \rightleftharpoons CaY^{2-} + Mg^{2+}$$

takes place. Introduction of titrant leads to the preferential formation of the calcium chelate for the same reason. At the equivalence point, the reaction is actually between H_2Y^{2-} and Mg^{2+}, for which Erio T is an excellent indicator.

4. Interferences with this method are substantially the same as with the direct determination of magnesium, and are eliminated in the same way.

Determination of calcium and magnesium in hard water. Acidify 100-ml aliquots with a few drops of HCl and boil gently for a few minutes to remove CO_2. Cool, add a few drops of methyl red, and neutralize the solution with NaOH. Introduce 2 ml of pH-10 buffer, 2 to 4 drops of Erio T indicator, and titrate with standard 0.01 F Na_2H_2Y to a color change from red to blue. Report results of the analysis in terms of milligrams of $CaCO_3$/liter of water.

NOTE:

If the color change of the indicator is sluggish, the absence of magnesium is indicated. In this event 1 to 2 ml. of standard 0.1 F MgY^{2-} solution should be added.

Determination of calcium by back titration. Prepare the sample as directed for the substitution analysis of calcium. To each 50.0-ml aliquot taken add about 2 ml of pH-10 buffer and 2 to 4 drops of Erio T indicator. Run in an excess of 0.01 F Na_2H_2Y solution from a buret and record the volume taken. Titrate the excess chelating agent to a color change from blue to red with standard 0.01 F $MgSO_4$ solution.

NOTE:

The magnesium sulfate solution is conveniently prepared by dissolving about 2.5 grams of the heptahydrate in 1 liter of distilled water. The equivalence of this solution with respect to the Na_2H_2Y solution should be determined by a direct titration.

problems

1. A solution is prepared by dissolving 5.200 grams of KCN in water and diluting to 1 liter. Express the concentration of this solution in terms of its
   (a) formality
   (b) normality with respect to the reaction $\qquad$ $Ag^+ + 2CN^- \rightleftharpoons Ag(CN)_2^-$
   (c) normality with respect to the reaction $\qquad$ $Ni^{2+} + 4CN^- \rightleftharpoons Ni(CN)_4^{2-}$
   (d) titer, in terms of mg Ag^+/ml
   (e) titer, in terms of mg Ni^{2+}/ml

2. A solution contains 18.60 grams of $Na_2H_2Y \cdot 2H_2O$ (GFW = 372) per liter. Express the concentration of this solution in terms of its
   (a) formality
   (b) normality with respect to the reaction $\qquad$ $Ca^{2+} + H_2Y^{2-} \rightleftharpoons CaY^{2-} + 2H^+$
   (c) titer, in terms of mg Ca^{2+}/ml

3. The excess cyanide in a silver-plating bath was determined by titration with standard $AgNO_3$ solution. A 50.00-ml aliquot required 23.4 ml of 0.0920 N Ag^+. Calculate the weight of free potassium cyanide present per liter of this solution.

4. Another 50-ml sample of the plating bath in Problem 3 was heated with HNO_3 to expel all of the cyanide as HCN. The residual $AgNO_3$ was rendered ammoniacal, after which 23.6 ml of 0.105 F KCN were introduced; this was sufficient to produce a clear solution. Titration of the excess KCN required 1.62 ml of 0.0920 N Ag^+. On the basis of the data in this and the preceding problem calculate the following:
   (a) the weight of Ag^+ present per liter of the plating solution
   (b) the total KCN content of this solution (in grams per liter)

5. The Ni in a 1.020-gram alloy sample was titrated with 28.40 ml of KCN solution. A 25.00-ml aliquot of solution containing 3.94 mg Ni^{2+}/ml required 36.30 ml of this cyanide solution. Calculate the following:
   (a) the formal concentration of the KCN solution
   (b) the titer of the KCN solution in mg Ni^{2+}/ml
   (c) the normality of the standard Ni^{2+} solution
   (d) the formal concentration of the standard Ni^{2+} solution
   (e) the normality of the cyanide solution with respect to this reaction
   (f) the percentage of Ni in the alloy

6. A standard Ca^{2+} solution was prepared by dissolving 0.4644 gram of $CaCO_3$ in HCl and diluting to 1 liter. Calculate the concentration of this solution in terms of parts Ca^{2+}/million.

7. A 50.00-ml aliquot of the solution in Problem 6 was titrated with 31.40 ml of an EDTA solution. Express the titer of this solution in terms of mg Ca^{2+}/ml EDTA.

8. A 50.00-ml aliquot of hard water required 19.80 ml of the EDTA solution in Problem 7. Express the hardness of the water in terms of parts Ca^{2+}/million.

chapter 14. *Theory of Neutralization Titrations of Simple Systems*

All end points employed in neutralization titrations are based upon the sharp changes in pH that occur near the equivalence points; the range over which they occur varies considerably. In order to make intelligent use of the available end points, the chemist must be able to derive a titration curve for any particular reaction. We shall now show how this is done.

ACID-BASE EQUILIBRIA — pH CALCULATIONS

Aqueous solutions always contain hydrogen ions[1] as well as hydroxyl ions as a consequence of the dissociation of water.

$$H_2O \rightleftharpoons H^+ + OH^-$$

Certain solutes, however, cause tremendous changes in the concentrations of the two species, often with profound effects as far as the chemical behavior of

---

[1] More exactly, hydronium ions, resulting from the reaction $2H_2O \rightleftharpoons H_3O^+ + OH^-$.

the solution is concerned. The chemist must, therefore, have a clear idea of those factors that influence the hydrogen or hydroxyl ion content of a solution.

Ion Product Constant for Water

As shown on page 21, application of the mass law to the dissociation of water leads to the expression

$$K_w = [H^+] [OH^-]$$

where K_w is a constant called the *ion product constant* for water. At 25° C the numerical value of this constant is 1.0×10^{-14} mol²/liter². The dissociation of water is an endothermic process; the extent to which it occurs, therefore, increases with temperature. Thus at 60° C the value of K_w is approximately 1×10^{-13} mol²/liter²; at 100° C it has increased to about 5×10^{-13}.

In pure water or in the presence of a solute that does not react to give hydrogen or hydroxyl ions, the concentrations of these two species are identical, and therefore their concentrations must equal the square root of the ion product constant—that is, 1.0×10^{-7} mol/liter at 25° C. A solution in which the concentration of hydrogen and hydroxyl ions is the same is said to be *neutral*. The hydrogen and hydroxyl ion concentrations of a neutral solution increase with temperature.

A solution is said to be *acidic* when the hydrogen ion concentration exceeds that of the hydroxyl ions; it is *basic* when the reverse is the case. Such changes take place as a result of the presence of a solute that reacts to produce or consume one of the two ions.

A useful relationship is obtained by taking the negative logarithm of both sides of the ion product constant expression. Thus

$$- \log K_w = - \log [H^+] [OH^-] = - \log [H^+] - \log [OH^-]$$

and

$$- \log K_w = pH + pOH$$

at 25° C, $- \log K_w$, or pK_w, is equal to fourteen and we may write

$$pH + pOH = 14$$

Example. We shall calculate the pH and pOH of a solution having a hydrogen ion concentration of 2.0×10^{-3}.

$$[H^+] = 2.0 \times 10^{-3} \text{ mol/liter}$$
$$pH = - \log 2.0 \times 10^{-3} = - \log 2 - \log 10^{-3}$$
$$= - 0.3 - (-3) = 2.7$$
$$pOH = 14 - 2.7 = 11.3$$

Solutions of Strong Acids and Strong Bases

A strong acid is defined as one that dissociates completely in a given solvent; a strong base is defined in an analogous way. The calculation of the pH or pOH

of solutions of such reagents is a straightforward matter since the hydrogen or hydroxyl ion concentration is directly related to the formal concentration of the substance. Such calculations are shown in the following examples.

Example. We shall calculate the pH and pOH of a 0.050 F solution of HCl.

Since this acid is completely dissociated, the hydrogen ion concentration is numerically equal to the formal concentration of HCl in the solution.

$$[H^+] = 0.05 = 5 \times 10^{-2} \text{ mol/liter}$$

Here the hydrogen ions arising from the dissociation of water are negligible as compared to the large number from the HCl. Then

$$pH = - \log 5 \times 10^{-2} = 1.3$$
$$pOH = 14 - 1.3 = 12.7$$

Example. Calculation of the pH, pOH, and the hydrogen and hydroxyl ion concentrations of a $3.2 \times 10^{-4} F$ solution of $Ba(OH)_2$ will serve as another example.

$Ba(OH)_2$ is a strong base containing 2 mols of hydroxyl ions for each formula weight of base. Thus

$$[OH^-] = 2 \times 3.2 \times 10^{-4} = 6.4 \times 10^{-4} \text{ mol/liter}$$
$$pOH = - \log 6.4 \times 10^{-4} = 3.19$$
$$pH = 14 - 3.19 = 10.81$$
$$[H^+] = \frac{K_w}{[OH^-]} = \frac{1.0 \times 10^{-14}}{6.4 \times 10^{-4}}$$
$$= 1.56 \times 10^{-11} \text{ mol/liter}$$

Solutions of Weak Acids and Bases

Weak acids and bases are only partially dissociated in solution. As a consequence the hydrogen or hydroxyl ion concentration in such solutions will be less than the formal concentration of the reagent. Calculation of the pH or pOH requires a knowledge of the magnitude of the dissociation constant for the substance.

Calculation of the pH of solutions of weak, monoprotic acids. To consider the relationships existing in a solution of the weak acid HA, we may write

$$HA \rightleftharpoons H^+ + A^-$$

and

$$K_a = \frac{[H^+][A^-]}{[HA]} \tag{14-1}$$

where K_a is the so-called *dissociation constant* or *ionization constant* of the acid. The only other equilibrium of importance in such a solution involves the dissociation of water.

$$H_2O \rightleftharpoons H^+ + OH^-$$

Since the number of negatively charged particles in the solution must equal the number that are positively charged, we may write

$$[H^+] = [OH^-] + [A^-]$$

Under most circumstances, however, since K_w is small, $[OH^-]$ will be very small compared to $[A^-]$, and the presence of hydrogen ions from HA will tend to repress the dissociation of water. As a result, in dealing with solutions of weak acids we can generally neglect the $[OH^-]$ contribution from the dissociation of water and write

$$[H^+] = [A^-] \tag{14-2}$$

Furthermore, if we know the formal concentration of HA in the solution, we can write

$$F_{HA} = [A^-] + [HA] \tag{14-3}$$

since the dissolved weak acid is present either as HA or A^-. Thus we have three equations and three unknowns ($[H^+]$, $[A^-]$, and $[HA]$), assuming K_a and the formal concentration of HA (F_{HA}) are known. For convenience we shall say that

$$[H^+] = [A^-] = x$$

According to (14-3) we may write

$$[HA] = (F_{HA} - x) \tag{14-4}$$

Substituting into (14-1) we find

$$\frac{x^2}{(F_{HA} - x)} = K_a \tag{14-5}$$

which can be rearranged to give

$$x^2 + K_a x - K_a F_{HA} = 0$$

The solution to this quadratic equation is then

$$x = [H^+] = \frac{-K_a + \sqrt{K_a^2 + 4K_a F_{HA}}}{2} \tag{14-6}$$

The calculation of the hydrogen ion concentration of a weak acid solution can often be greatly simplified without introducing a serious error. By assuming that the concentration of undissociated acid is not greatly different from its formal concentration—that is, that x is small compared to F_{HA}—equation (14-4) becomes

$$[HA] \cong F_{HA} \quad \text{when} \quad x \ll F_{HA}$$

Equation (14-5) then simplifies to

$$\frac{x^2}{F_{HA}} = K_a$$

and

$$x = [H^+] = \sqrt{K_a F_{HA}} \tag{14-7}$$

The magnitude of the error introduced by this assumption will become greater as the concentration of acid becomes smaller and as the dissociation constant of the acid becomes larger. This is illustrated by the data in Table 14-1.

In general, it is good practice to make this assumption, thereby obtaining a trial value for x that may be compared with F_{HA} in equation (14-5). If this trial value alters [HA] by an amount smaller than the allowable error in the calculation, the solution may be assumed to be satisfactory. Otherwise the quadratic equation must be solved to give a more exact value for x.

Table 14-1

ERRORS INTRODUCED BY ASSUMING x SMALL RELATIVE TO F_{HA} IN EQUATION (14-5)

Value of K_a	Value of F_{HA}	Value for $x = [H^+]$ using assumption	Value for $x = [H^+]$ by more exact equation
1×10^{-2}	1×10^{-5}	3.16×10^{-4}	0.1×10^{-4}
	1×10^{-3}	3.16×10^{-3}	0.92×10^{-3}
	1×10^{-1}	3.16×10^{-2}	2.70×10^{-2}
1×10^{-4}	1×10^{-5}	3.16×10^{-5}	0.92×10^{-5}
	1×10^{-3}	3.16×10^{-4}	2.70×10^{-4}
	1×10^{-1}	3.16×10^{-3}	3.11×10^{-3}
1×10^{-6}	1×10^{-5}	3.16×10^{-6}	2.70×10^{-6}
	1×10^{-3}	3.16×10^{-5}	3.11×10^{-5}
	1×10^{-1}	3.16×10^{-4}	3.16×10^{-4}

Example. We shall calculate the concentration of hydrogen ions in a 4.00×10^{-2} F solution of formic acid, and also the pH of the solution. The equilibrium for the formic acid solution is

$$HCOOH \rightleftharpoons HCOO^- + H^+$$

From Table A-3 of the Appendix

$$K_a = \frac{[H^+][HCOO^-]}{[HCOOH]} = 1.7 \times 10^{-4}$$

Now, let $x = [H^+]$. From a consideration of charge balance in the solution

$$x = [H^+] = [HCOO^-] + [OH^-] \cong [HCOO^-]$$

This assumption is probably valid since the $[OH^-]$ of this acidic solution will be very small. Then, material balance requires that

$$[HCOOH] = (F_{HCOOH} - x) = (4.00 \times 10^{-2} - x)$$

Substituting into the dissociation-constant expression, we find

$$\frac{x^2}{(4.00 \times 10^{-2} - x)} = 1.74 \times 10^{-4}$$

We shall first assume x is small relative to 4.00×10^{-2} and obtain a trial value of x; that is,

$$x^2 = 4.00 \times 10^{-2} \times 1.74 \times 10^{-4}$$
$$x = 2.64 \times 10^{-3}$$

We may now test the assumption by comparing 2.64×10^{-3} with 4.00×10^{-2}; we see that the latter value is about 7 percent of the former which gives a maximum value for the error in the trial value of x. If we desire a more accurate figure, we must solve the quadratic equation; we then find

$$x = [H^+] = 2.55 \times 10^{-3} \text{ mol/liter}$$
$$pH = -\log 2.55 \times 10^{-3} = 2.6$$

Calculation of the pH of solutions of weak bases.   The techniques discussed in the previous sections are readily adaptable to the calculation of the hydroxyl ion concentration in solutions of weak bases; from this, the pH can be obtained. In contrast to the large variety of common weak acids, the number of weak bases is quite small, consisting of ammonia, several organic amines, and a few miscellaneous compounds. Of these only ammonia is generally encountered by the chemist.

Aqueous solutions of ammonia are basic by virtue of the reaction

$$NH_3 + H_2O \rightleftharpoons NH_4^+ + OH^-$$

The predominant species in these solutions has been clearly demonstrated to be NH_3. Despite this, such solutions are frequently called ammonium hydroxide, this terminology being vestigial from the time when the substance NH_4OH rather than NH_3 was believed to be the undissociated form of the base. Application of the mass law to this equilibrium yields the expression

$$K_b = \frac{[NH_4^+][OH^-]}{[NH_3]}$$

By analogy to the case of the weak acid, the constant K_b is commonly called a *basic dissociation constant* despite the fact that this terminology does not describe the reaction very well. The magnitude of this constant is independent of the formula used in the denominator, be it the more correct NH_3 or the historical NH_4OH.

Equilibrium constants for other weak bases are formulated in a similar fashion. For example, the equilibrium for dimethylamine can be written

$$(CH_3)_2NH + H_2O \rightleftharpoons (CH_3)_2NH_2^+ + OH^-$$

for which

$$K_b = \frac{[(CH_3)_2NH_2^+][OH^-]}{[(CH_3)_2NH]}$$

The hydroxyl ion concentration in a solution of a weak base is readily calculated from its dissociation constant by the same techniques and assumptions as used for weak acids.

> **Example.** We wish to know the pH of a 0.075 F solution of NH_3. From the table of basic dissociation constants (Appendix)

$$K_b = 1.86 \times 10^{-5} = \frac{[NH_4^+][OH^-]}{[NH_3]}$$

In this instance, we shall let $x = [OH^-]$. The condition of electrical neutrality demands that

$$[NH_4^+] + [H^+] = [OH^-] = x$$

In this case since $[H^+] \ll [NH_4^+]$, we may write

$$[NH_4^+] = [OH^-] = x$$

From the standpoint of material balance

$$[NH_4^+] + [NH_3] = 7.5 \times 10^{-2}$$
$$[NH_3] = (7.5 \times 10^{-2} - x)$$

Substituting these quantities into the dissociation-constant expression

$$\frac{x^2}{(7.5 \times 10^{-2} - x)} = 1.86 \times 10^{-5}$$

Finally, we shall assume that x is small relative to 7.5×10^{-2}. Then

$$x^2 = 7.5 \times 10^{-2} \times 1.86 \times 10^{-5} = 13.95 \times 10^{-7}$$
$$x = [OH^-] = 1.18 \times 10^{-3} \text{ mol/liter}$$

Comparing this result with 7.5×10^{-2}, we see that the maximum error in x will be less than 2 percent. Were a better value for x desired, it could be obtained by solution of the quadratic equation. To obtain pH, we first compute pOH.

$$pOH = -\log 1.18 \times 10^{-3} = 2.93$$

Then

$$pH = 14 - 2.93 = 11.07$$

Calculation of the pH of dilute solutions of very weak acids. In determining the pH of a weak acid solution by these techniques the assumption was made that the concentration of hydrogen ion resulting from the dissociation of the acid so far exceeded the concentration arising from the dissociation of water that the latter contribution could be neglected. A similar assumption was made regarding calculation of the hydroxyl ion concentration of weak base solutions. As shown by the following example, this assumption is not valid when dealing with dilute solutions of very weak acids or bases.

Example. We wish to calculate the pH of a $5 \times 10^{-5} F$ solution of phenol ($K_a = 1.05 \times 10^{-10}$).

First we shall employ the techniques already developed for the calculation of the pH of such a solution.

Using ROH as the symbol for phenol we write

$$\text{ROH} \rightleftharpoons \text{RO}^- + \text{H}^+$$

$$K_a = 1.05 \times 10^{-10} = \frac{[\text{H}^+][\text{RO}^-]}{[\text{ROH}]}$$

and assuming that the only source of hydrogen ions is the phenol we can say

$$x = [\text{H}^+] = [\text{RO}^-]$$
$$(5 \times 10^{-5} - x) = [\text{ROH}]$$

Then substituting these values into the dissociation-constant expression and solving in the usual way, we find

$$x = [\text{H}^+] = 7.25 \times 10^{-8} \text{ mol/liter}$$

The above answer is patently impossible, however, since a solution of even the weakest acid must have a hydrogen ion concentration greater than pure water. The difficulty here lies in the assumption that $[\text{H}^+]$ and $[\text{RO}^-]$ are identical; this would only be true if there were no other significant source of hydrogen ions. We can arrive at a more exact description of the system by writing an equation based on the electrical neutrality of the solution—namely,

$$[\text{H}^+] = [\text{RO}^-] + [\text{OH}^-]$$

Remembering that $[\text{H}^+][\text{OH}^-] = K_w$, we may write

$$[\text{RO}^-] = ([\text{H}^+] - [\text{OH}^-]) = \left([\text{H}^+] - \frac{K_w}{[\text{H}^+]}\right)$$

Since the phenol is present either as ROH or RO$^-$

$$[\text{ROH}] + [\text{RO}^-] = 5 \times 10^{-5}$$

We shall again assume that $[\text{RO}^-]$ is small and that

$$[\text{ROH}] \cong 5 \times 10^{-5}$$

Substituting into the equilibrium expression, we obtain

$$\frac{[\text{H}^+]\left([\text{H}^+] - \frac{K_w}{[\text{H}^+]}\right)}{5 \times 10^{-5}} = 1.05 \times 10^{-10}$$

$$[\text{H}^+]^2 - K_w = 5.25 \times 10^{-15}$$
$$[\text{H}^+] = 1.23 \times 10^{-7} \text{ mol/liter}$$
$$p\text{H} = 6.91$$

This same technique can be applied to the calculation of the pH of a solution of a very weak base.

When confronted with problems involving very dilute solutions—particularly of weak acids or bases—the chemist should always check the validity of the assumption that the reagent is the only significant source of hydrogen ions or hydroxyl ions. This may be done by computing a trial value using ordinary assumptions and comparing this value with the hydrogen or hydroxyl ion content of pure water. If the two figures are of the same order of magnitude, a recalculation, as shown above, should be made.

Solutions of Salts of Weak Acids and Bases

The solution of a salt consisting of the anion of a weak acid and the cation of a strong base is alkaline because of the tendency of the anion to react with water giving hydroxyl ions. This phenomenon is known as *hydrolysis*. As an example, a solution of sodium acetate is basic by virtue of the hydrolytic reaction

$$OAc^- + H_2O \rightleftharpoons HOAc + OH^-$$

Solutions of salts of weak bases and strong acids are acidic, on the other hand, because of the tendency of the cation to react to produce hydrogen ions. Thus when ammonium chloride is dissolved in water, the following reaction occurs:

$$NH_4^+ \rightleftharpoons NH_3 + H^+$$

The situation is somewhat more complex in the case of a salt made up of the anion of a weak acid and the cation of weak base. Ammonium acetate is such a salt; upon solution of this compound both reactions occur. Whether the solution is acidic or basic depends upon which reaction takes place to the greater extent; this, in turn, is related to the relative magnitudes of the acidic and basic dissociation constants for acetic acid and ammonia.

The ability to calculate the pH of solutions containing various types of salts is of fundamental importance to the analytical chemist.

Salts of weak acids and strong bases. Consider a solution prepared by dissolving the salt NaA in water. If HA is a weak acid, the following reaction takes place:

$$A^- + H_2O \rightleftharpoons HA + OH^-$$

The equilibrium constant for this reaction, often called a *hydrolysis constant*, takes the form

$$K_h = \frac{[HA][OH^-]}{[A^-]} \tag{14-8}$$

We saw on page 20 that the numerical value of this constant is simply related to the dissociation constant, K_a, for the weak acid HA and the ion-product constant for water, K_w; that is,

$$K_h = \frac{K_w}{K_a} = \frac{[HA][OH^-]}{[A^-]} \tag{14-9}$$

Now we shall relate the equilibrium concentration of the various constituents in a solution of NaA to one another and to the formal concentration of NaA. Since the anion of the dissolved salt is present either as A^- or HA, material balance requires that

$$F_{NaA} = [A^-] + [HA] \qquad (14\text{-}10)$$

We may also write an equation to account for the electrical neutrality of the solution; namely,

$$[Na^+] + [H^+] = [OH^-] + [A^-]$$

Now, the concentration of sodium ions in the solution is equal to the formal concentration of the salt, inasmuch as these ions in no way react with water. Thus we may write

$$F_{NaA} + [H^+] = [OH^-] + [A^-] \qquad (14\text{-}11)$$

Subtracting equation (14-11) from (14-10) and rearranging, we find

$$[HA] = [OH^-] - [H^+] \qquad (14\text{-}12)$$

If the hydrolytic reaction proceeds to an appreciable extent, the hydroxyl ion concentration becomes considerably greater than the hydrogen ion concentration.

This allows a very convenient simplification of (14-12); that is, if $[H^+] \ll [OH^-]$,

$$[HA] = [OH^-] \qquad (14\text{-}13)$$

The foregoing assumption is valid for all but the least hydrolyzed salts. In cases of doubt, the assumption should be made and then checked when the hydroxyl ion concentration has been calculated.

If we now set x equal to the equilibrium hydroxyl ion concentration from (14-13) we may say

$$x = [OH^-] = [HA]$$

Equation (14-10) then rearranges to

$$[A^-] = (F_{NaA} - x)$$

and by substituting these values into (14-9) we obtain

$$K_h = \frac{K_w}{K_a} = \frac{x^2}{(F_{NaA} - x)} \qquad (14\text{-}14)$$

This equation is similar in form to equation (14-5) on page 284 and, as in that case, an exact or an approximate solution can be found. The exact solution will take the form

$$x = [OH^-] = \frac{-K_h \pm \sqrt{K_h^2 + 4K_h F_{NaA}}}{2} \qquad (14\text{-}15)$$

If, as is frequently the case, $\dfrac{K_w}{K_a}$ is small or if F_{NaA} is large, the assumption may be made that x is small with respect to F_{NaA}, whereupon (14-14) becomes

$$\frac{K_w}{K_a} = \frac{x^2}{F_{NaA}}$$

Solving for x we find

$$x = [OH^-] = \sqrt{\frac{K_w}{K_a} F_{NaA}} \qquad (14\text{-}16)$$

Here, again, a trial value of x via equation (14-16) is advisable before undertaking a more rigorous solution to the problem. We can then readily determine whether or not the assumption made was a valid one.

Equation (14-15) or (14-16) indicates that salts of weaker acids will yield solutions that are more basic than salts of stronger acids.

Example. We wish to calculate the pH of a 0.100 F solution of NaCN.

The equilibrium of interest is

$$CN^- + H_2O \rightleftharpoons HCN + OH^-$$

for which

$$K_h = \frac{K_w}{K_a} = \frac{[HCN]\,[OH^-]}{[CN^-]}$$

The dissociation constant for HCN is 2.1×10^{-9}. Setting $x = [OH^-]$, we may write

$$x = [OH^-] = [HCN]$$

and

$$(0.1 - x) = [CN^-]$$

Therefore,

$$\frac{x^2}{(0.1 - x)} = \frac{1.0 \times 10^{-14}}{2.1 \times 10^{-9}} = 4.8 \times 10^{-6}$$

If we now assume $x \ll 0.1$, then

$$\frac{x^2}{0.1} = 4.8 \times 10^{-6}$$

$$x = [OH^-] = 6.9 \times 10^{-4} \text{ mol/liter}$$

We see that x is indeed small relative to 0.1; for most purposes, then, this approximate solution is entirely adequate.

$$p\text{OH} = -\log 6.9 \times 10^{-4} = 3.16$$
$$p\text{H} = 14 - 3.16 = 10.84$$

Salts of weak bases and strong acids. The calculation of pH for a solution of a salt of a weak base is analogous to the calculation just completed.

Example.   We shall calculate the pH of a 0.200 F solution of NH_4Cl. The equilibrium in this instance can be written

$$NH_4^+ \rightleftharpoons NH_3 + H^+$$

for which

$$K = \frac{[NH_3]\,[H^+]}{[NH_4^+]}$$

It is readily shown that this constant is numerically equal to K_w/K_b, where K_b is the basic dissociation constant for NH_3; that is,

$$\frac{K_w}{K_b} = \frac{[H^+]\,[OH^-]}{[NH_4^+]\,[OH^-]/[NH_3]} = \frac{[NH_3]\,[H^+]}{[NH_4^+]} = K$$

If we combine a material-balance equation and a charge-balance equation as in the previous section, we find that

$$[NH_3] = [H^+] - [OH^-]$$

and since $[OH^-] \ll [H^+]$

$$[NH_3] = [H^+] = x$$

Also

$$[NH_4^+] = (0.200 - x)$$

Substituting these quantities into the equilibrium expression

$$\frac{x^2}{(0.2 - x)} = \frac{1.0 \times 10^{-14}}{1.86 \times 10^{-5}}$$

Now assuming x to be small relative to 0.2

$$\frac{x^2}{0.2} = \frac{1.0 \times 10^{-14}}{1.86 \times 10^{-5}}$$

and

$$x = [H^+] = 1.04 \times 10^{-5}\,\text{mol/liter}$$
$$pH = 4.98$$

Salts of a weak acid and a weak base.   When the salt of a weak acid and a weak base is dissolved in water, two reactions occur that affect the pH. For example, in a solution of ammonium acetate, both the following equilibria exist:

$$OAc^- + H_2O \rightleftharpoons HOAc + OH^-$$
$$NH_4^+ \rightleftharpoons NH_3 + H^+$$

Whether the resulting solution is basic or acidic depends upon which of these equilibria proceeds further to the right. Calculation of the pH of such a solution must take both processes into account; how this is accomplished is considered in the next chapter.

Solutions of a Weak Acid and Its Salt or a Weak Base and Its Salt

The addition of a salt of a weak acid to a solution of that acid has the effect of markedly reducing the hydrogen ion concentration of the solution. This effect is predicted by the principle of Le Chatelier. Thus, if a salt NaA is added to a solution of HA, the equilibrium

$$HA \rightleftharpoons H^+ + A^-$$

will be displaced to the left. As a result, a decrease in the hydrogen ion concentration occurs. Similarly, additions of ammonium chloride to a solution of ammonia will shift the equilibrium

$$NH_3 + H_2O \rightleftharpoons NH_4^+ + OH^-$$

in such a direction as to remove hydroxyl ions from the solution. The magnitude of these effects is readily calculated.

Calculation of the pH of a solution of a weak acid and its salt. We shall consider a solution having a formal acid concentration F_{HA} and a formal salt concentration F_{NaA}. The important equilibrium in this solution involves dissociation of the acid, the constant for which is

$$K_a = \frac{[H^+][A^-]}{[HA]}$$

We shall first set $x = [H^+]$, and then express the concentrations of A^- and HA in terms of their respective formal concentrations and x. Since the substance A is present either as the ion A^- or as HA, we may write a material-balance equation

$$F_{NaA} + F_{HA} = [A^-] + [HA] \tag{14-17}$$

From electrical neutrality considerations we may write

$$[Na^+] + [H^+] = [A^-] + [OH^-]$$

But

$$[Na^+] = F_{NaA}$$

Therefore,

$$F_{NaA} + [H^+] = [A^-] + [OH^-] \tag{14-18}$$

Subtracting equation (14-18) from (14-17), we find

$$F_{HA} - [H^+] = [HA] - [OH^-] \tag{14-19}$$

On the *assumption* that $[OH^-]$ is small relative either to $[A^-]$ or to $[HA]$, equations (14-18) and (14-19) can be rearranged and written

$$[A^-] = F_{NaA} + x \tag{14-20}$$
$$[HA] = F_{HA} - x \tag{14-21}$$

It is important to appreciate the significance of these statements. The first simply states that the anion concentration, $[A^-]$, is equal to the formal concentration of the salt plus the concentration of A^- resulting from dissociation of the acid (which is equal numerically to $[H^+]$). The undissociated acid concentration, $[HA]$, on the other hand, is equal to the formal concentration of the acid minus that lost by dissociation (which is also equal to $[H^+]$).

Substituting equations (14-20) and (14-21) into the dissociation-constant expression, we obtain

$$K_a = \frac{x(F_{NaA} + x)}{(F_{HA} - x)} \qquad (14\text{-}22)$$

Rearrangment yields the quadratic equation

$$x^2 + (F_{NaA} + K_a)x - F_{HA}K_a = 0$$

which can be solved to give

$$x = \frac{-(F_{NaA} + K_a) \pm \sqrt{(F_{NaA} + K_a)^2 + 4 F_{HA}K_a}}{2}$$

In nearly every case the solution of (14-22) can be greatly simplified, however, by assuming x is small relative to either F_{NaA} or F_{HA}; then (14-22) becomes

$$K_a = x\frac{F_{NaA}}{F_{HA}}$$

when $x \ll F_{NaA}$ and F_{HA} or

$$x = [H^+] = K_a \frac{F_{HA}}{F_{NaA}} \qquad (14\text{-}23)$$

As in the earlier calculations in this chapter, this assumption should always be made and a trial value of x obtained, which is then used to test the validity of the assumption.

A few comments are in order regarding equation (14-23) and the solutions it describes. Within the limits of the assumptions required to convert (14-22) to (14-23), the hydrogen ion concentration is dependent only upon the ratio of the formal concentrations of the two components of the solution. Furthermore, this ratio remains *independent of the dilution* of the solution since the concentration of each component changes in a proportionate manner upon dilution. Thus *the hydrogen ion concentration of a solution of a weak acid and its salt is independent of dilution and depends only upon the ratio of the number of formula weights of each component present.* In this respect solutions made up of a mixture of a weak acid and its salt differ from solutions containing either of the components alone.

The following example will illustrate the technique for calculating the *p*H of a solution of this type.

Example. We wish to know the pH of a solution that is $4.00 \times 10^{-2} \ F$ in formic acid and $0.100 \ F$ in sodium formate.

The equilibrium governing the hydrogen ion concentration in this solution is

$$HCOOH \rightleftharpoons H^+ + HCOO^-$$

for which

$$K_a = \frac{[H^+][HCOO^-]}{[HCOOH]} = 1.74 \times 10^{-4}$$

We shall let x equal the equilibrium hydrogen ion concentration $[H^+]$. Then

$$[HCOO^-] = (0.100 + x)$$
$$[HCOOH] = (0.0400 - x)$$

and

$$\frac{x \, (0.100 + x)}{(0.0400 - x)} = 1.74 \times 10^{-4}$$

Making the assumption that x is small compared with 0.1 and 0.04, the equation simplifies to

$$\frac{x \, (0.1)}{(0.04)} = 1.74 \times 10^{-4}$$

and

$$x = [H^+] = 6.96 \times 10^{-5} \ mol/liter$$

Comparing this value with 0.04 and 0.1, we see that the assumption is quite reasonable for most purposes. The pH of this solution is given by

$$pH = - \log 6.96 \times 10^{-5} = 4.16$$

If we compare the pH of this solution with that of a solution which is $0.04 \ F$ in formic acid only (see page 285), we see that the addition of the formate has raised the pH from 2.6 to 4.16 as a result of the common ion effect.

Calculation of the pH of a solution of a weak base and its salts. The following example will show that the calculation of pH in this instance is analogous to that discussed in the previous section.

Example. We wish to know the pH of a solution that is $0.28 \ F$ in NH_4Cl and $0.070 \ F$ in NH_3. The equilibrium of interest is

$$NH_3 + H_2O \rightleftharpoons NH_4^+ + OH^-$$

for which $K_b = 1.86 \times 10^{-5}$.

The concentration of NH_4^+ will be equal to the formal concentration of the salt plus any NH_4^+ formed in the equilibrium reaction. Thus if we let $x = [OH^-]$, we may write

$$[NH_4^+] = (0.28 + x)$$

The NH_3 concentration will be equal to its formal concentration minus the loss due to the above reaction, or

$$[NH_3] = (0.07 - x)$$

Substituting these values into the equilibrium-constant expression,

$$\frac{x\,(0.28 + x)}{(0.07 - x)} = 1.86 \times 10^{-5}$$

If we now assume $x \ll 0.28$ and 0.07, then

$$\frac{0.28\,x}{0.07} = 1.86 \times 10^{-5}$$
$$x = [OH^-] = 4.65 \times 10^{-6} \text{ mol/liter}$$
$$pOH = -\log 4.65 \times 10^{-6} = 5.33$$
$$pH = 14 - 5.33 = 8.67$$

Here again, the assumption regarding the magnitude of x was justified.

BUFFER SOLUTIONS

Buffer solutions or *buffers* tend to resist changes in pH even upon addition of strong acids or bases or upon dilution. Buffers are of great importance to the chemist.

Types of Buffers

One type of buffer solution is readily prepared by dissolving a weak acid and a salt of the same acid in water. We have just considered the calculation of the pH of such a solution. Additions of hydrogen ions to this buffer cause a shift in the principal equilibrium governing the pH of the solution.

$$HA \rightleftharpoons H^+ + A^-$$

Thus a portion of the added hydrogen ions is consumed by the reservoir of anions, and the shift in pH is considerably less than would be found with an unbuffered solution. Hydroxyl ions, if added, are consumed by hydrogen ions from the weak acid. This also causes a shift in the principal equilibrium in a direction tending to counteract the effects of the added base. Consequently a smaller change in pH occurs than would be observed were the buffer not present.

Another type of buffer is prepared from a weak base and its salt. The mechanism of its buffering action is analogous; added hydroxl ions are partially consumed by reaction with the cation to form the undissociated base, while hydrogen ions are taken up by the hydroxyl ions of the solution.

A solution can also be buffered by saturating it with a slightly soluble acid or base and providing an excess of the substance as a second phase. A soluble salt of the acid or base is also added to the solution. In a zinc chloride solution saturated with zinc hydroxide, for example, the principal equilibrium would be

$$Zn(OH)_2 \rightleftharpoons Zn^{2+} + 2OH^-$$

Hydroxyl ions added to this mixture would be largely consumed by the formation of more precipitate. Added hydrogen ions, on the other hand, would consume the hydroxyl ions; the latter, however, would be partially replaced as some of the excess solid dissolved.

In this section we shall discuss only the first two kinds of buffers since these are the types most generally encountered.

Properties of Buffer Mixtures

Effect of addition of acids or bases. The resistance of a buffer mixture to changes in pH is readily demonstrated.

Example. We wish to compare the pH before and after the addition of 10 milliformula weights of NaOH to 1 liter of a buffer solution that is 0.100 F in HOAc and 0.200 F in NaOAc.

$$\frac{[H^+][OAc^-]}{[HOAc]} = 1.75 \times 10^{-5}$$

The pH prior to addition of the strong base is readily found by assuming that $[OAc^-]$ and $[HOAc]$ are equal to the formal concentrations. Then

$$\frac{0.2\,x}{0.1} = 1.75 \times 10^{-5}$$
$$x = [H^+] = 8.75 \times 10^{-6} \text{ mol/liter}$$
$$pH = 5.06$$

Upon addition of base, some of the HOAc will be neutralized; as a result, the concentration of weak acid will be diminished while that of the salt will be increased. We can obtain the pH of the solution by calculating the new formal concentrations. The number of milliformula weights of HOAc will be equal to the original number less the number of milliformula weights of base added; that is,

$$\text{number of milliformula weights HOAc} = 1000 \times 0.1 - 10 = 90$$

$$F_{\text{HOAc}} = \frac{90}{1000} = 0.090$$

The number of milliformula weights of NaOAc will have increased by this number. Thus

$$\text{number of milliformula weights NaOAc} = 1000 \times 0.2 + 10 = 210$$

$$F_{\text{NaOAc}} = \frac{210}{1000} = 0.210$$

We may now calculate the pH of this solution using the same technique as before; that is,

$$\frac{0.210\,x}{0.090} = 1.75 \times 10^{-5}$$
$$x = [H^+] = 7.5 \times 10^{-6} \text{ mol/liter}$$
$$pH = 5.12$$

In this example the addition of 10 milliformula weights of sodium hydroxide causes an increase in pH of about 0.06. In contrast to the behavior of this buffered solution, we can easily show that the addition of 10 milliformula weights of the base to an unbuffered solution would raise the pH to a value of about 12.

We shall now see what happens to the same buffer solution when a strong acid is added.

Example. We shall calculate the pH of 1 liter of the foregoing buffer solution after addition of 10 milliequivalents of HCl.

We may predict an increase in the concentration of undissociated acetic acid as well as a decrease in the acetate concentration owing to the reaction between the latter and the hydrogen ions from the added HCl. As before, we can readily calculate the pH of the solution after first calculating the new formal concentrations of acetic acid and acetate ion. Thus

$$F_{HOAc} = \frac{0.1000 \times 1000 + 10}{1000} = 0.110$$

$$F_{NaOAc} = \frac{1000 \times 0.2 - 10}{1000} = 0.190$$

We can then calculate the $[H^+]$ by employing the same technique as above:

$$[H^+] = 1.01 \times 10^{-5} \text{ mol/liter}$$
$$pH = 4.99$$

We see that the acetate buffer solution also resists changes in pH when a strong acid is added. In contrast, the addition of the same quantity of hydrochloric acid to an unbuffered solution of this pH would lower this value to about 2.

Effect of dilution. On page 294 we stated that for all practical purposes, the hydrogen ion concentration of a solution of a weak acid and its salt is directly dependent upon the ratio of the formal concentrations of the two components. Since this *ratio* does not change with dilution, the hydrogen ion concentration will be constant despite the addition of solvent. This independence of pH on dilution is typical of all buffer solutions.

This statement, however, is true as a first approximation only; with very concentrated and very dilute buffers, small pH changes do occur when dilution takes place. In very dilute solutions, such changes become noticeable when the molar concentrations of the components can no longer be assumed to be the same as the formal concentrations; that is, when the assumptions made in proceeding from equation (14-22) to (14-23) on page 294 are no longer valid. Small pH changes are also observed in more concentrated solutions. These variations arise from the change in ionic strength of the solution accompanying the dilution process; as indicated earlier (p. 22), changes in the electrolyte concentration have an effect on the degree of dissociation of an acid and hence the pH of a solution of that acid.

Buffer Capacity

In the previous example we considered 1 liter of solution containing 0.1 mol of acetic acid and 0.2 mol of sodium acetate. Since the pH of a buffer is governed by the ratio between acid and salt, we would expect a solution in which the concentrations were ten times that of the example to have substantially the same pH; the same should be true for a solution having concentrations one tenth as great. The number of equivalents of acid or base that each of these could tolerate without material alteration of the pH will vary appreciably, however. For example, we have seen that 1 liter of the original buffer changes pH from 5.06 to 5.12 upon addition of 10 milliequivalents of sodium hydroxide; 100 milliequivalents of the strong base could be added before the same pH change would occur in 1 liter of the more concentrated buffer. On the other hand, the most dilute of the three buffers would require only 1 milliequivalent of base to cause an identical pH change. Thus, while the three solutions have the same pH, they are quite different in what is called their *buffer capacity*—that is, in the quantity of acid or base they are capable of consuming.

In more quantitative terms, buffer capacity is often defined as *the number of mols of strong base required to cause a unit increase in pH in 1 liter of a buffer solution*. This quantity can be calculated if the composition of the buffer is known.

Ordinarily a high buffer capacity is desirable, and one obvious way of achieving this is by using high reagent concentrations. Less apparent, perhaps, but also effective is the proper choice of the buffering system. Generally, the maximum buffer capacity is achieved when the dissociation constant of the weak acid is numerically identical to the desired hydrogen ion concentration. Under these circumstances the ratio of the concentration of the salt to the concentration of the acid approaches unity. It can easily be shown that the maximum in buffer capacity is associated with a 1:1 concentration.[2]

[2] We shall assume that a buffer is prepared by adding NaOH to the weak acid HA. If m formula weights of the base are added to n formula weights of the acid, the formal concentration of the salt and the acid will be given, approximately, by

$$F_{NaA} = \frac{m}{v}$$

$$F_{HA} = \frac{n - m}{v}$$

where v is the volume of the solution in liters. If the dissociation of the acid is small, we may write

$$K_a = \frac{[H^+]F_{NaA}}{F_{HA}} = \frac{[H^+](m/v)}{(n-m)/v}$$

$$[H^+] = \frac{K_a(n-m)}{m}$$

$$pH = -\log K_a - \log(n-m) + \log m$$

Differentiating this expression with respect to m gives the rate of change of pH with added base.

$$2.303 \frac{d(pH)}{dm} = \frac{1}{n-m} + \frac{1}{m} = \frac{n}{m(n-m)}$$

To find when this function is a minimum, we set $\dfrac{d^2(pH)}{dm^2}$ equal to zero; that is,

Preparation of Buffer Solutions

When faced with the need for preparing a buffer solution, the chemist ordinarily finds a host of reagents from which to pick. In making a choice among these he will try to avoid substances that will react with other components of the system he wishes to buffer. Insofar as possible, he will attempt to attain the maximum buffer capacity for a given concentration of reagent by employing a system in which the formal ratio of salt to acid or base is near unity.

Recipes for preparation are readily available in chemical handbooks and reference works.[3] Because of their widespread employment, two types of buffer systems deserve specific mention. McIlvaine buffers cover a pH range from about 2 to 8 and are prepared by mixing solutions of citric acid with disodium hydrogen phosphate. Clark and Lubs buffers, which include a pH range from 2 to 10, make use of three systems—namely, phthalic acid, potassium hydrogen phthalate; potassium dihydrogen phosphate, dipotassium hydrogen phosphate; and boric acid, sodium borate.

ACID-BASE INDICATORS

There is a wide variety of substances whose color in solution is dependent on the pH of the medium. Many occur naturally in various plants; their properties have been recognized and used for thousands of years for the determination of the alkalinity or acidity of water. These compounds, called *acid-base indicators*, are widely employed by the modern chemist to estimate the pH of solutions and to signal the end point in acid-base titrations. We shall consider briefly the theory of their behavior as well as the chemical constitution of some of the most common of these substances.

Theory of Indicator Behavior

Acid-base indicators are generally fairly high molecular weight, complex organic compounds. In water or other solvents they behave as weak acids or

NOTE[2] (*cont.*)

$$\frac{d^2(pH)}{dm^2} = \frac{n(2m-n)}{m^2(n-m)^2} = 0$$

Then

$$m = \frac{n}{2}$$

Thus a minimum in the rate of change of pH with added base is found when the number of formula weights of added base is equal to exactly one half the number of formula weights of acid originally present—that is, when the ratio of formula weights of salt to acid is one. We see from equation (14-23) that the hydrogen ion concentration will be equal to the dissociation constant of the acid at this point.

[3] See, for example, I. M. Kolthoff and C. Rosenblum, *Acid-Base Indicators*, 239. New York: The Macmillan Company, 1937.

bases and thus participate in equilibrium reactions involving the hydrogen ion. Accompanying the dissociation or association reactions of these compounds are complex, internal structural rearrangements to which they owe their change in color. These rearrangements are responsible for the indicator properties of these compounds.

Without discussing the nature of the structural rearrangements we can symbolize the association or dissociation reaction of an acid-base indicator as follows :

$$HIn \rightleftharpoons H^+ + In^-$$

(acid color)   (base color)

or

$$In + H_2O \rightleftharpoons InH^+ + OH^-$$

(base color)   (acid color)

In the first instance the indicator (which may be ionic or molecular in constitution) behaves as a weak acid that yields an anion, In^-, and a hydrogen ion upon dissociation. In the second case, the indicator acts as a weak base capable of combining with hydrogen ions to form the conjugate acid of the base. In either case, the two species involved in these equilibria differ from one another in color. Which species, and hence which color, predominates depends upon the pH. Thus, in the first instance, HIn will be the major constituent in strongly acid solutions and will be responsible for the "acid color" of the indicator, whereas In^- will represent its "basic color"; in the second case the species In will predominate in basic solutions and thus be responsible for the "basic color" of this indicator, while InH^+ will constitute the "acid color."

Equilibrium expressions for these processes take the following forms:

$$K_a = \frac{[H^+][In^-]}{[HIn]}$$

and

$$K_b = \frac{[InH^+][OH^-]}{[In]}$$

Rearrangement of these expressions will show the dependence of the ratio of indicator species not only upon the value for the equilibrium constant but also upon the hydrogen ion concentration; in this respect, indicators behave as typical weak acids and bases.

The color of an indicator will vary according to the pH of its environment. Experiments show that the change will occur gradually over a range that typically embraces 2 pH units. The ability of the human eye to discriminate between colors is not overly acute, and therefore something on the order of a tenfold excess of one of the forms of the indicator is needed before the observer can state that the color due to this form predominates. In other words, the subjective "color change" of the indicator typically involves a major alteration in the position of the indicator equilibrium from a tenfold excess of one of the species to a similar excess of the other. For example, in the first instance cited we may write that the indicator exhibits its acid color to the average observer when

$$\frac{[\text{In}^-]}{[\text{HIn}]} \leq \frac{1}{10}$$

and its basic color when

$$\frac{[\text{In}^-]}{[\text{HIn}]} \geq \frac{10}{1}$$

At ratios between these two values the color appears to be intermediate between the two. These numerical estimates, of course, represent average behavior only; some indicators require smaller ratio changes and others larger. Furthermore, a good deal of variation among observers can be found, the extreme being the case of the color-blind person.

If we now substitute the two concentration ratios into the dissociation-constant expression for the indicator, we can find the variation of hydrogen ion concentration necessary to effect the indicator color change. Thus for the acid color

$$\frac{[\text{H}^+][\text{In}^-]}{[\text{HIn}]} = \frac{[\text{H}^+]}{10}\frac{1}{} = K_a$$

$$[\text{H}^+] = 10\,K_a$$

and, similarly, for the basic color

$$\frac{[\text{H}^+]\,10}{1} = K_a$$

$$[\text{H}^+] = \frac{1}{10}K_a$$

Converting these to pH, we get

$$pH = -1 - \log K_a \text{ for the acid color}$$
$$pH = +1 - \log K_a \text{ for the basic color}$$

or

$$pH \text{ range} = pK_a \pm 1$$

Thus an indicator having an acid dissociation constant of 1×10^{-5} would show a color change when the pH of the solution in which it was dissolved changed from 4 to 6. A similar relationship is easily derived for an indicator of the basic type.

Types of Acid-base Indicators

A list of compounds having acid-base indicator properties is very large and comprises a variety of organic structures. An indicator covering almost any desired pH range can ordinarily be found. A few of the most common indicators are given in Table 14-2.

The large majority of acid-base indicators can be classified into perhaps half a dozen categories based upon structural similarities.[4] Three of these classes are described below.

Table 14-2

SOME IMPORTANT ACID-BASE INDICATORS[5]

Common Name	Transition Range, pH	Color Change		Indicator Type[6]
		Acid	Base	
Methyl violet	0.5 — 1.5	yellow	blue	
Thymol blue	1.2 — 2.8	red	yellow	2
	8.0 — 9.6	yellow	blue	
Methyl yellow	2.9 — 4.0	red	yellow	3
Methyl orange	3.1 — 4.4	red	yellow	3
Bromcresol green	3.8 — 5.4	yellow	blue	2
Methyl red	4.2 — 6.3	red	yellow	3
Chlorophenol red	4.8 — 6.4	yellow	red	2
Bromthymol blue	6.0 — 7.6	yellow	blue	2
Phenol red	6.4 — 8.0	yellow	red	2
Neutral red	6.8 — 8.0	red	yellow-orange	
Cresol purple	7.4 — 9.0	yellow	purple	2
	1.2 — 2.8	red	yellow	
Phenolphthalein	8.0 — 9.6	colorless	red	1
Thymolphthalein	9.3 — 10.5	colorless	blue	1
Alizarin yellow	10.1 — 12.0	colorless	violet	3

[5] Taken from I. M. Kolthoff and H. A. Laitinen, *pH and Electro Titrations*, 29. New York: John Wiley and Sons, Inc., 1941.

[6] See text: 1. phthalein; 2. sulfonphthalein; 3. azo.

Phthalein indicators. Most phthalein indicators are colorless in moderately acidic solutions and exhibit a variety of colors in alkaline media. In strongly alkaline solutions their colors tend to fade slowly which is, at times, an inconvenience. As a group the phthaleins are insoluble in water but quite soluble in alcohol; the latter is the preferred solvent in preparing solutions of these indicators.

The best-known example of a phthalein indicator is *phenolphthalein*, whose structures are given below:

[4] See I. M. Kolthoff and C. Rosenblum, *Acid-Base Indicators*, Chap. 5. New York: The Macmillan Company, 1937.

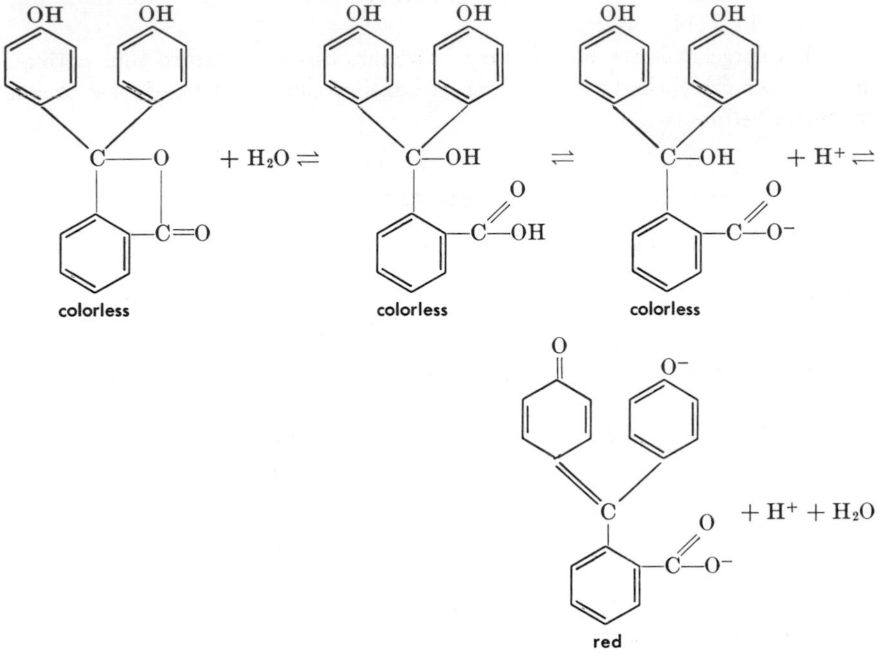

The last equilibrium in this series predomin ates in the pH range 8.0 to 9.8 A major structural change involving the formation of a quinoid structure is associated with this reaction; it is this quinoid structure that is responsible for the red color imparted to the solution.

The other phthalein indicators differ in that the phenolic rings contain various additional functional groups. In the case of *thymolphthalein*, for example, there are two alkyl groups on each ring. The basic structural changes associated with the color change of this indicator are no different from phenolphthalein.

Sulfonphthalein indicators. Many of the sulfonphthaleins exhibit two useful color-change ranges. One color change occurs in rather acidic solutions and the other in neutral or moderately basic media. In contrast to the phthaleins, the basic color shows good stability toward strong alkali.

Indicator solutions of these compounds can be prepared in approximately 20-percent alcohol; however, such reagents are often sufficiently acidic to alter slightly the pH of the media into which they are introduced. To avoid this problem, the sodium salt of the indicator is frequently prepared by dissolving a weighed quantity of the solid indicator in a suitable volume of dilute aqueous sodium hydroxide.

The simplest example of this class of indicators in *phenolsulphonthalein* or *phenol red*. The principal equilibria in solutions of this compound are shown below:

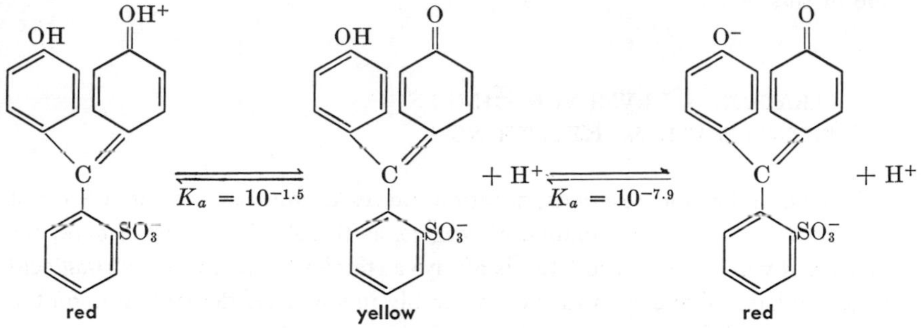

Only the second of the two color changes, occurring in the pH range 6.4 to 8.0, is commonly employed.

Substitution of halogens or alkyl groups for the hydrogens in the phenolic rings of the parent compound leads to a large variety of indicators that differ in color and pH range.

Azo indicators. Most of the azo indicators exhibit a color change from red to yellow with increasing basicity. Their transition range is generally on the acid side. The best-known compounds in this class are *methyl orange* and *methyl red*; the behavior of the former is shown below for illustrative purposes:

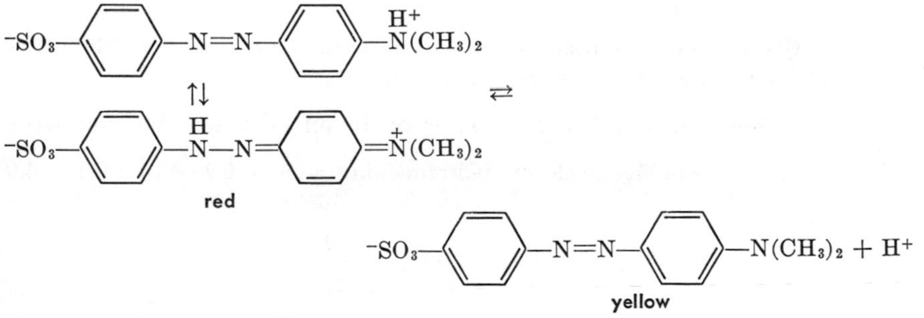

Methyl red is similar to methyl orange with the exception that the sulfonic acid group is replaced by a carboxylic acid group. Variations in the substituents on the amino nitrogen and in the rings lead to a series of indicators with slightly different properties.

Variables that Influence the Behavior of Indicators

Several factors play a part in determining the pH interval over which a given indicator exhibits a color change. Among these are temperature, electrolyte concentration, the presence of organic solvents, and the presence of colloidal particles. Some of these effects, particularly the last two, can cause a shift in the color range of one or more pH units. A discussion of these effects is beyond the scope of this book.[7]

TITRATION CURVES FOR SIMPLE NEUTRALIZATION REACTIONS

For neutralization reactions, titration curves customarily consist of a plot of pH as the ordinate and volume of reagent as the abscissa. The reagent, for reasons that will be discussed later, is always a standard solution of a strong acid or a strong base. The curves vary considerably in character depending upon the concentration of the reactants and the completeness of the reaction.

Titration of Solutions of Strong Acids and Bases

The simplest neutralization titrations are encountered in the analysis of strong acids and bases. In this instance the reaction involves

$$H^+ + OH^- \rightleftharpoons H_2O$$

and only this equilibrium need be considered in the derivation of titration curves.

Titration curves of strong acids. We shall first see how a curve may be derived for the titration of 50.0 ml of a 0.100 N solution of hydrochloric acid with a standard 0.100 N solution of sodium hydroxide.

Calculation of initial pH. Before addition of base the solution is 0.100 N in HCl. Therefore the pH is 1.00.

Calculation of pH after addition of 10 ml of base. We may write

number of milliequivalents HCl remaining $= 50 \times 0.1 - 10 \times 0.1 = 4.0$

thus

$$N_{HCl} = [H^+] = \frac{4.0}{60}$$

$$pH = -\log\frac{4}{60} = 1.18$$

In these calculations, the dissociation of water has been neglected since the contribution of hydrogen ions from that source is negligible compared with the high concentration from the strong acid.

Calculations analogous to this provided the data in Table 14-3 for all additions short of the equivalence point.

Calculation of pH after addition of 50.0 ml of base. When 50.0 ml of base have been added, the titration is at the equivalence point and neither acid nor base is present in excess. Therefore the pH is 7.00.

[7] For a further discussion see H. F. Walton, *Chemical Analysis*, 258-266. New York: Prentice-Hall, Inc., 1952.

Calculation of pH after addition of 50.01 ml of base. The addition of 50.01 ml of base results in a slight excess of base in the solution; we can obtain the pH from its concentration.

$$N_{\text{NaOH}} = [\text{OH}^-] = \frac{50.01 \times 0.1 - 50.0 \times 0.1}{100.01}$$

$$= 1.00 \times 10^{-5}$$

$$pOH = 5.00$$

$$pH = 14.00 - 5.00 = 9.00$$

Additional calculations such as this yielded the remaining data shown in Table 14-3. Clearly the equivalence point is signaled by a very marked pH change. Thus a 0.02 ml addition of reagent in this region causes the pH to rise from 5 to 9 which corresponds to a ten-thousandfold change in hydrogen ion concentration. These data are plotted as Curve A of Figure 14.1 which further emphasizes the characteristic equivalence point behavior of the system.

Table 14-3

CHANGES IN pH DURING THE TITRATION OF 50 ML OF 0.100 N HCl
WITH 0.100 N NaOH

Volume NaOH Added, ml	pH
0.0	1.00
10.0	1.18
25.0	1.48
40.0	1.95
49.0	3.00
49.9	4.00
49.99	5.00
Equivalence → 50.00	7.00
point 50.01	9.00
50.1	10.00
51.0	11.00
60.0	11.96
75.0	12.30
100.0	12.52

Effect of concentration. Curve B of Figure 14.1 is the theoretical titration curve for 50 ml of 0.001 N hydrochloric acid with 0.001 N base. It is apparent that a reduction in concentration of reactants results in a decrease in the pH change associated with the equivalence point.

Indicator choice. The transition intervals of three common acid-base indicators are also shown in Figure 14.1. Any of the three should produce a sharp end point with a negligible titration error in the case of the 0.1 N reactants; this is no longer true, however, with the titration involving 0.001 N

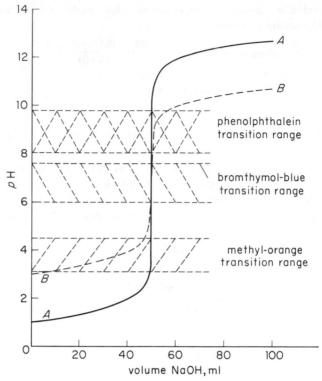

Fig. 14.1 Curves for the Titration of 50.0 ml Aliquots of Hydrochloric Acid with Sodium Hydroxide. Plots of pH as a function of the volume of NaOH added. *A*. Reactant concentrations are 0.100 *N*; *B*. Reactant concentrations are 0.00100 *N*.

solutions. Here, only *bromthymol blue* could be employed, and even with this indicator the end point would be troublesome. That a pH change of 6.0 to 8.0 would occur in the interval between 49.9 and 50.1 ml is easily calculated. Thus, nearly 0.2 ml of reagent would be required to produce a change in color. Clearly, titrations of solutions more dilute than this could not be performed with accuracy.

Titration curves of strong bases. The uppermost curve of Figure 14.4 is typical for a solution of a strong base with a strong acid. The derivation of this curve is analogous to that for a strong acid, involving only the calculation of the concentration of excess acid or base at any point in the titration.

Titration of Solutions of Weak Acids or Bases

The derivation of a titration curve for a solution of a weak acid or base is somewhat more complicated because the dissociation equilibrium of the acid or base must be taken into account. In arriving at a complete titration curve, four types of calculations must be employed, corresponding to four distinct parts of the curve. At the outset, the solution contains only a weak acid or a weak base and the calculation is based on the formal concentration of that substance. Addition of various increments of the reagent (in amounts up to but not including an equivalent amount) produces a series of buffers; the pH of each can be calculated from the formal concentrations of the salt and the residual weak acid or base. At the equivalence point, the solution contains simply the salt of the weak acid or base being titrated, and the pH can be calculated from the formal concentration of that substance. Finally, beyond the equivalence point an excess of the strong acid or base is present which represses the hydrolysis of the salt; thus, the pH is governed largely by the concentration of the excess reagent.

Titration curves of weak acids. To illustrate, we shall consider the derivation of a titration curve for the neutralization of 50.0 ml of 0.100 N acetic acid $(K_a = 1.75 \times 10^{-5})$ with 0.100 N sodium hydroxide.

Initial pH. Here we must calculate the pH of a 0.1 N solution of HOAc; using the method shown on page 284, we obtain a value of 2.88.

pH after addition of 10 ml of reagent. A buffer solution of NaOAc and HOAc has now been produced; to calculate the pH we must know the formal concentrations of the two constituents. These can be ascertained as follows :

$$F_{HOAc} = \frac{50 \times 0.1 - 10 \times 0.1}{60} = \frac{4.0}{60}$$

$$F_{NaOAc} = \frac{10 \times 0.1}{60} = \frac{1.0}{60}$$

Then letting $x = [H^+]$, and substituting appropriate numbers for the various terms in the dissociation-constant expression for acetic acid, we obtain

$$\frac{x \cdot 1/60}{4/60} = K_a = 1.75 \times 10^{-5}$$
$$x = [H^+] = 7.0 \times 10^{-5} \text{ mol/liter}$$
$$pH = 4.15$$

Calculations similar to this will delineate the curve in the entire buffer region. Data from such calculations are given in Table 14-4. Note that when the acid has been 50 percent neutralized—in this instance, after an addition of exactly 25.0 ml of base—the formal concentrations of acid and salt are identical; within the limits of the approximations ordinarily used, so also are their molar concentrations. Thus these terms cancel one another out in the equilibrium-constant expression, and the hydrogen ion concentration is equal to the dissociation constant; that is, the pH is equal to the pK_a. One can use this relationship to ascertain quickly the pH at the midpoint in the titration of a weak acid. In the case of a weak base, the hydroxyl ion concentration at the midpoint in its titration is equal numerically to the dissociation constant of the base.

Table 14-4

<small>CHANGES IN *p*H DURING THE TITRATION OF 50 ML OF 0.100 *N* ACETIC ACID WITH</small>
0.100 *N* NaOH

Volume NaOH Added, ml	*p*H
0.0	2.88
10.0	4.15
25.0	4.76
40.0	5.36
49.0	6.45
49.9	7.45
50.0	8.73
50.1	10.0
51.0	11.00
60.0	11.96
75.0	12.30
100.0	12.52

Equivalence point *p*H. At the equivalence point in the titration the acetic acid has been converted to sodium acetate; the solution is therefore identical to one formed by dissolving that salt in water. After calculation of the formal concentration of the salt, the *p*H calculation is similar to that described on page 290; that is,

$$F_{NaOAc} = \frac{50 \times 0.100}{100.0} = 0.05$$

The equilibrium that determines the *p*H at this point is the hydrolysis reaction,

$$OAc^- + H_2O \rightleftharpoons HOAc + OH^-$$

and letting $x = [OH^-]$ we find

$$\frac{x^2}{(0.05 - x)} \cong \frac{x^2}{0.05} = \frac{1 \times 10^{-14}}{1.75 \times 10^{-5}}$$
$$x = [OH^-] = 5.36 \times 10^{-6} \text{ mol/liter}$$
$$pOH = 5.27$$
$$pH = 8.73$$

_p_H after addition of 50.10 ml of base. When 50.10 ml of base have been added, we can write

$$F_{NaOH} = \frac{50.10 \times 0.1 - 50.00 \times 0.1}{100.1} = 1.0 \times 10^{-4}$$

In this case hydroxyl ions arise both from the excess sodium hydroxide and the hydrolysis of the acetate ion. The contribution of the latter is small enough to be neglected, however, since the excess base will tend to repress this reaction. This becomes evident when we consider that the hydroxyl ion concentration arising from hydrolysis was only 5.36×10^{-6} mol/liter in the previous case, when no excess base was present. Once an excess of strong base has been added, the contribution from the hydrolytie reaction will be even smaller, so that we may say

$$[OH^-] \cong F_{NaOH} = 1.0 \times 10^{-4}$$
$$pOII = 4.0$$
$$pH = 10.0$$

Thus the titration curves for a weak acid and a strong acid become identical in the region slightly beyond the equivalence point. This is confirmed by the identical data found in Tables 14-3 and 14-4 for the final points in the titrations.

Before discussing the curve just derived, we should consider the calculation of the pH beyond the equivalence point under circumstances where the hydrolysis of the salt cannot be neglected. This would be the situation if the excess of base were very small, or if the hydrolysis constant of the salt were relatively large.

Example. We shall calculate the pH of a mixture of 50.00 ml of 0.100 N HA ($K_a = 1.0 \times 10^{-7}$) and 50.05 ml of 0.100 N NaOH.

Upon calculating the formal concentrations of the constituents, we find

$$F_{NaA} = \frac{50.0 \times 0.100}{100.05} = 0.0500$$

$$F_{NaOH} = \frac{50.05 \times 0.100 - 50.00 \times 0.100}{100.05} = 5.0 \times 10^{-5}$$

Inasmuch as HA is a rather weak acid, we might suspect that a significant fraction of the OH^- ions would arise from hydrolysis of its salt. If we let y be the concentration of HA formed by the hydrolysis reaction, we can express the equilibrium concentrations of the various species as follows :

$$[HA] = y$$
$$[A^-] = 0.0500 - y$$
$$[OH^-] = (5 \times 10^{-5} + y)$$

The last equation simply states that hydroxyl ions originate both from the presence of the strong base (5×10^{-5} N) and the hydrolysis (y) of the salt. Substituting these into the hydrolysis expression,

$$\frac{y (5 \times 10^{-5} + y)}{(0.0500 - y)} = \frac{1 \times 10^{-14}}{1 \times 10^{-7}}$$

gives, on rearrangement,

$$y^2 + 5.01 \times 10^{-5}y - 5.0 \times 10^{-9} = 0$$

Solving this quadratic equation for y,

$$y = [HA] = 5.0 \times 10^{-5} \text{ mol/liter}$$
$$[OH^-] = (5 \times 10^{-5} + y) = 1.0 \times 10^{-4} \text{ mol/liter}$$
$$pOH = 4.0$$
$$pH = 10.0$$

The data from Table 14-4 are plotted as Curve A in Figure 14.2. Contrasting this with the curve for a strong acid, we see that one obvious difference lies in the pH at the equivalence point, which is about 8.7 for the weak acid and 7.0 for the strong. Furthermore, the pH of the solution at all points prior to the equivalence point is higher. The net effect is to reduce the magnitude of the pH change occurring in the equivalence-point region.

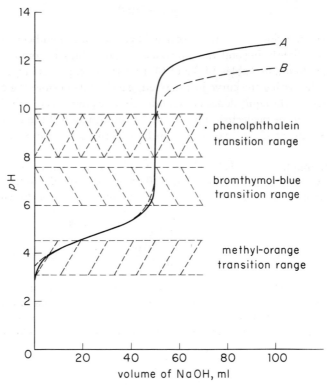

Fig. 14.2 Curves for the Titration of 50.0-ml Aliquots of Acetic Acid with Sodium Hydroxide. *A.* Reactant concentrations are 0.100 N; *B.* Reactant concentrations are 0.0100 N.

Effect of concentration. Curve B in the Figure 14.2 is the theoretical curve for a 0.01 N solution of acetic acid titrated with 0.01 N base. In this instance we see that the initial pH is higher, but in the buffer region the two curves are identical, a consequence of the fact that the pH of buffer solutions is nearly independent of dilution. The equivalence point pH for the more dilute solutions can be shown to be about half a unit lower, or 8.2, while beyond the equivalence point the two curves differ by about one pH unit. Thus the change in pH associated with the equivalence point becomes less with dilution.

Indicator choice; feasibility of titration. These curves clearly indicate that the choice of indicators for the titration of a weak acid is more limited than for a strong acid. *Methyl orange* is totally unsuited for this titration; nor would *bromthymol blue* be satisfactory, since its color change requires the addition of 1 or 2 ml of reagent. Only an indicator having a transition range in the basic region would be appropriate; *phenolphthalein* is an excellent example.

Figure 14.3 illustrates that the end-point problem becomes more critical, for weaker acids; in this figure are shown titration curves for a series of weak acids of differing strengths. Because of the limited sensitivity of the eye to color changes, we may conclude that 0.1 N solutions of acids with dissociation constants much less than 1×10^{-6} could not be titrated accurately; actually, the limiting figure for errors of $\pm$ 0.2 percent is about 5×10^{-7}. Somewhat weaker acids can be titrated by preparing a color standard to compare against the solution being titrated. The standard should have about the same volume and the same quantity of indicator as the solution being analyzed; obviously the pH of the standard must be that of the equivalence point. With these added precautions, 0.1 N solutions of acids with dissociation constants as low as 1×10^{-8} can be determined with an error no greater than about 0.2 percent. By working with more concentrated solutions titration of even weaker acids is possible.

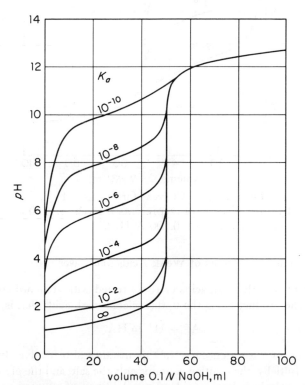

Fig. 14.3 Titration Curves for Acids of Various Strengths. Each curve represents titration of 50.0 ml of 0.100 N acid with 0.100 N NaOH.

Titration curves of weak bases. Figure 14.4 shows the theoretical titration curves for a series of bases of differing strengths. These curves were derived by techniques similar to those described for the case of acetic acid. With weak bases, an indicator having an acid transition range is obviously required.

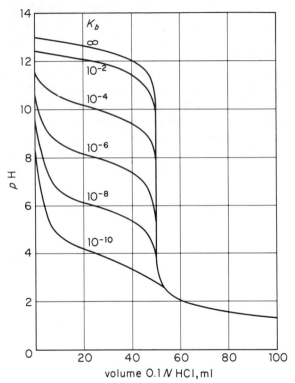

Fig. 14.4 Titration Curves for Bases of Various Strengths. Each curve represents titration of 50.0 ml of 0.100 N base with 0.100 N HCl.

Titration of Solutions of Salts of Weak Acids and Bases

Salts of sufficiently weak acids can be titrated with standard solutions of a strong acid, the product being the undissociated weak acid; that is,

$$A^- + H^+ \rightleftharpoons HA$$

The derivation of a titration curve in this instance also involves four types of calculations. Initially, the solution contains only the salt, and the pH is calculated by taking its hydrolysis into consideration. Addition of a strong acid produces a buffer; the pH can be obtained from the formal concentrations of the residual salt and the weak acid formed. At the equivalence point, the solution is made up of a weak acid plus the salt of the strong acid. Since the latter has no appreciable effect, the pH can be calculated from the formal concentration of the weak acid. Finally, beyond the equivalence point the solution is made up of a mixture of a weak and a strong acid. When the concentration of the latter becomes sufficiently great, the dissociation of the weak acid is repressed to the point where the pH can be calculated directly from the formal concentration of the strong acid.

Titration curves of salts of weak acids. By way of illustration, consider the titration of 50 ml of 0.1 F sodium cyanide with 0.1 N hydrochloric acid. Hydrogen cyanide is a weak acid with a dissociation constant of 2.1 $\times$ 10^{-9}.

Initial pH. By the technique shown on page 290 the calculated pH will be 10.84.

Buffer region pH. Additions of HCl to the solution result in the formation of HCN, thus giving a buffer mixture. For example, after the addition of 10 ml of HCl, we may write

$$F_{HCN} = \frac{10 \times 0.1}{60} = \frac{1}{60}$$

$$F_{NaCN} = \frac{50 \times 0.100 - 10 \times 0.100}{60} = \frac{4.0}{60}$$

The pH of this buffer mixture is obtained in the usual way and has a value of 9.28.

Equivalence point pH. At the equivalence point the salt is converted to the weak acid; thus we may say

$$F_{HCN} = \frac{50 \times 0.100}{100.0} = 0.0500$$

Calculation of the pH of a 0.0500 F solution of HCN will give 4.99.

Postequivalence point pH. When 50.01 ml of reagent have been added

$$F_{HCl} = \frac{50.01 \times 0.1000 - 50.00 \times 0.1000}{100.01} = 1.0 \times 10^{-5}$$

$$F_{HCN} = \frac{50.0 \times 0.1000}{100.01} = 5.0 \times 10^{-2}$$

Hydrogen ions are present from the excess HCl and from the dissociation of the HCN. If we let x be the concentration of cyanide arising from dissociation, we may write

$$[CN^-] = x$$
$$[H^+] = 1.0 \times 10^{-5} + x$$
$$[HCN] = 0.05 - x$$
$$\frac{(1.0 \times 10^{-5} + x)(x)}{(0.05 - x)} = 2.1 \times 10^{-9}$$

If we assume x is small relative to 0.05 and 1.0×10^{-5}, we obtain a trial value of 1×10^{-5} which makes it apparent that we must solve the quadratic equation

$$x^2 + 1.0 \times 10^{-5}x - 1.05 \times 10^{-10} = 0$$

giving

$$x = [CN^-] = 0.64 \times 10^{-5} \text{ mol/liter}$$
$$[H^+] = 1.0 \times 10^{-5} + 0.64 \times 10^{-5} = 1.64 \times 10^{-5} \text{ mol/liter}$$
$$pH = 4.78$$

As the excess HCl increases, the contribution of the HCN to the acidity rapidly becomes smaller; soon after the equivalence point it can be neglected and the pH calculated from the formality of the HCl. Thus after the addition of 50.5 ml of acid, the formality of HCl is 4.78×10^{-5}. By assuming this to be the hydrogen ion concentration, a pH of 3.32 is obtained.

Figure 14.5 illustrates the titration curve derived from these calculations. Clearly none of the indicators shown is well suited for the titrations; a compound such as *methyl red* (transition range 4.1 to 6.2) would, however, be satisfactory.

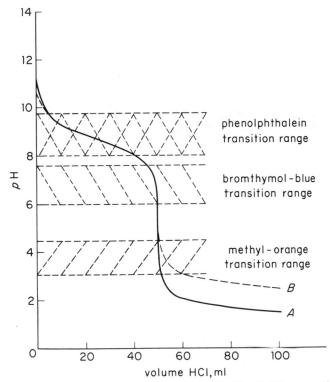

Fig. 14.5 Curves for the Titration of 50.0 ml Aliquots of Sodium Cyanide Solutions with Hydrochloric Acid. *A*. Reactant concentrations are 0.100 *N*; *B*. Reactant concentrations are 0.0100 *N*.

Curve B illustrates the effect of dilution. Obviously a solution of this salt more dilute than 0.01 F cannot be titrated with a very high degree of accuracy.

We have already demonstrated that end points become less sharp for solutions of weaker acids or bases. A similar situation applies in the case of salts. Here the measure of the base or acid strength of the compound is the hydrolysis constant; as this becomes smaller (that is, as the dissociation constant of the acid or base becomes larger) the titration curves are the less well defined. In other words, the salts of very weak acids or bases can be titrated successfully, while those of stronger reagents cannot. Titration curves for a series of salts of acids of differing hydrolysis constants will have an appearance similar to the curves in Figure 14.4. Only those salts having hydrolysis constants larger than about 10^{-7} are readily titrated with a high degree of accuracy. These, of course, would be salts of acids that have dissociation constants smaller than about 10^{-7}.

These observations also apply to the titration of salts of the weaker bases with a standard solution of a strong base. In this case the end points occur in basic solutions, the titration curves being similar in appearance to those for the weak acids shown in Figure 14.3. Good end points are found only in those cases where the salt is derived from a fairly weak base—that is, a base with a dissociation constant of 10^{-7} or smaller.

problems

1. The value for K_w at $0°$, $50°$ and $100°$ C is 1.15×10^{-15}, 9.61×10^{-14}, and 5.13×10^{-13}, respectively. Calculate the pH of a neutral solution at each of these temperatures.   ans. $pH = 7.47$ at $0°$ C

2. Calculate the $[H^+]$, $[OH^-]$, pH, and pOH of (a) 0.200 N $HClO_4$ and (b) 0.0035 F KOH.
        ans.   (a) $[H^+] = 0.200$; $[OH^-] = 5 \times 10^{-14}$; $pH = 0.7$; $pOH = 13.3$

3. Calculate the $[H^+]$, $[OH^-]$, pH, and pOH of (a) 0.0084 F $Ba(OH)_2$ and (b) 3.7×10^{-4} N $Ba(OH)_2$.

4. What is the pH of a 0.500 N solution of KOH at $0°$ C (K_w at $0°$ C $= 1.15 \times 10^{-15}$)?

5. Calculate the pH of a 2.0 F solution of HCl.

6. What is the $[H^+]$ of a solution having (a) a pH of 3.42, (b) a pH of 9.67, (c) a pOH of 13.16?   ans. (a) $[H^+] = 3.8 \times 10^{-4}$ mol/liter

7. What is the pH of a solution containing 0.0731 gram of NaOH in 1 liter?

8. What is the pH of a solution prepared by mixing 17.4 ml of 0.200 N $HClO_4$ with 36.0 ml of 0.109 N NaOH?   ans. $pH = 11.92$

9. What is the pH of a mixture of 30.0 ml of 0.0300 N HCl and 6.4 ml of 0.080 N KOH?

10. Calculate the pH of a 1.0×10^{-7} N solution of HCl.   ans. $pH = 6.8$

11. Calculate the pH of a 2.0×10^{-7} F solution of NaOH.

12. Calculate the pH of a 0.050 F solution of the weak acid HA having a dissociation constant of 3.5×10^{-6}.   ans. $pH = 3.38$

13. What is the pH of a 0.100 F solution of hydroxylamine?   ans. $pH = 9.51$

14. Calculate the pH of a 0.0020 F solution of formic acid.   ans. $pH = 3.30$

15. What is the pH of a 1.50×10^{-4} F solution of methylamine?

16. What is the change in pH of the following solutions upon twofold and upon one-hundredfold dilution: (a) a 0.010 F solution of HCl and (b) a 0.010 F solution of HOAc?

17. What is the pH of a solution that is 0.0200 F in sodium benzoate and 0.0100 F in benzoic acid?   ans. $pH = 4.5$

18. What is the pH of a solution that is 0.710 F in NH_4Cl?   ans. $pH = 4.71$

19. Calculate the pH of a buffer prepared by mixing 50 ml of 1.07 F NH_3 with 100 ml of 0.25 F NH_4Cl.

20. Describe the preparation of 1 liter of a buffer of pH 4.0 from 1.0 F solutions of HOAc and NaOAc.
     ans. Mix 149 ml of the NaOAc with 851 ml of the HOAc

21. How many grams of NH_4Cl should be added to 2 liters of 0.30 F NH_3 to give a buffer having a pH of 9.0?

22. What is the pH of a 0.0010 F solution of aniline hydrochloride?
     ans. $pH = 3.79$

23. To 400 ml of 1.00 F formic acid was added 21 grams of sodium formate. What was the pH of the resultant buffer?

24. How many milliliters of 1.0 N NaOH should be added to 1.0 liter of 0.60 F HOAc to give a pH 5.0 buffer?

25. What is the equivalence point pH in the titration of 0.200 F NH_3 with 0.200 F HCl?

26. What is the pH in the titration in Problem 25 when half the NH_3 has been neutralized?

27. What is the pH of a solution obtained by mixing 19.0 ml of 0.20 F sodium benzoate with 20 ml of 0.050 F HCl and diluting to 100 ml?

28. What is the pH of a mixture of 100 ml of 0.100 F formic acid to which has been added 0.2 ml of 1.00 F HCl?   ans. $pH = 2.29$

29. A mixture containing HOAc and NaOAc had a pH of 4.12. Calculate the ratio of the formal concentrations of the two species.

$$\text{ans. } \frac{F_{NaOAc}}{F_{HOAc}} = 0.231$$

30. A 1.00-liter mixture of NH_4Cl and NH_3 had a pH of 9.64. The total number of formula weights of the two substances was 0.80. Calculate the formal concentration of each in the solution.

31. A buffer is 0.500 F in NaOAc and 0.75 F in HOAc. To 1 liter of this is added 1.0 ml of 1.0 N NaOH. Calculate the change in pH of the solution.

32. To 1 liter of the buffer in Problem 31 is added 1.0 ml of 1.5 N HCl. Calculate the change in pH.

33. Calculate the pH of the following mixtures:
    (a) 10.0 ml of 0.0100 N $Ba(OH)_2$ and 5.0 ml of 0.0200 N HCl
    (b) 10.0 ml of 0.0100 F $Ba(OH)_2$ and 5.0 ml of 0.0200 N HCl
    (c) 10.0 ml of 0.0100 N $Ba(OH)_2$ and 10.0 ml of 0.0200 N HCl
    (d) 10.0 ml of 0.150 N formic acid
    (e) 10.0 ml of 0.150 N formic acid and 10.0 ml of 0.100 N NaOH
    (f) 10.0 ml of 0.150 N formic acid and 15.0 ml of 0.100 N NaOH
    (g) 10.0 ml of 0.150 N formic acid and 20.0 ml of 0.100 N NaOH

34. Calculate the pH of the solutions obtained by the addition of the following volumes of 0.100 N HCl to 25.0 ml of 0.200 F methylamine:
    (a) 0.0 ml
    (b) 10.0 ml
    (c) 25.0 ml
    (d) 45.0 ml
    (e) 49.9 ml
    (f) 50.0 ml
    (g) 50.1 ml
    (h) 51.0 ml
    (i) 60.0 ml

35. Sketch a titration curve from the data obtained in Problem 34 and choose an indicator from those in Table 14-2 that would be suitable for this titration.

36. For 25.0 ml of 0.100 N HA ($K_a = 1.0 \times 10^{-7}$) with 0.100 N KOH derive a theoretical titration curve. Suggest an indicator for this titration.

37. Calculate the pH of the solutions obtained by mixing the following volumes of 0.150 N HCl with 30.0 ml of 0.100 F NaOCl.
    (a) 0.0 ml
    (b) 5.0 ml
    (c) 10.0 ml
    (d) 19.0 ml
    (e) 19.9 ml
    (f) 20.0 ml
    (g) 21.0 ml
    (h) 25.0 ml

38. Sketch a titration curve from the data in Problem 37 and suggest an indicator for this titration.

39. Derive a theoretical titration curve for the titration of 20.0 ml of 0.100 F pyridinum chloride with 0.0500 N NaOH. Suggest an indicator for the titration.

40. Calculate the pH of the solution 0.1 ml before and 0.1 ml after the equivalence point in the following titrations:
    (a) 25.0 ml of 0.200 N HNO_2 with 0.100 N KOH
    (b) 25.0 ml of 0.00200 N NaOH with 0.00180 N $HClO_4$
    (c) 30.0 ml of 0.0300 F NaCN with 0.0500 N HCl
    (d) 25.0 ml of 0.0100 N NH_3 with 0.0100 N HCl

41. What is the pH of a solution which is 0.100 F in $ZnCl_2$ and saturated with $Zn(OH)_2$?

chapter 15. *Theory of Neutralization Titrations in Complex Systems*

Thus far we have considered titration curves for only those rather simple systems in which a single equilibrium predominates. We must now discuss more complex systems involving several equilibria. Although we will draw specifically upon weak acid and acid-salt mixtures for examples, the methods of calculation demonstrated are equally adaptable to weak base and base-salt systems.

CALCULATION OF pH IN COMPLEX SYSTEMS

In approaching the problems of pH calculation where several equilibria are involved, two types of assumptions are made to facilitate computation. The first requires chemical knowledge since it consists of identifying the equilibria that will determine the composition of the solution and disregarding those having little or no effect. The second type is mathematical in nature and consists of the approximations made with respect to the relative magnitude of quantities appearing in equations.

Mixtures of a Weak and Strong Acid

A mixture of a strong and a weak acid contains two potential sources for hydrogen ions. In many instances, however, the presence of the strong acid is sufficient to repress dissociation of the weak acid to the point where the contribution from the latter may be neglected. Under these circumstances the normal concentration of the strong acid will provide an adequate estimate of the hydrogen ion concentration.

This assumption is not justified and will lead to serious errors if the dissociation constant for the weak acid is relatively large or if the formal concentration of the strong acid is small compared to the weak acid. We considered an example of the latter situation on page 315; we shall now consider the former case.

Example. We wish to know the hydrogen ion concentration of a $0.0100\ F$ solution of H_2SO_4.

This solution may be treated as if it consisted of two acids of different strengths. One of these, H_2SO_4, is a strong acid, completely dissociated into H^+ and HSO_4^- ions. The second is the weak acid HSO_4^-. The concentration of hydrogen ions from the first will be 0.0100 mol/liter; that from the second can be obtained from the K_a for HSO_4^- which has a value of 1.2×10^{-2}. The concentration of hydrogen ions from the dissociation of HSO_4^- will equal numerically the concentration of SO_4^{2-} in the solution; thus, we may write

$$[SO_4^{2-}] = x$$

$$[H^+] = (0.0100 + x)$$

$$[HSO_4^-] = (0.0100 - x)$$

Substituting into the dissociation-constant expression for HSO_4^-,

$$\frac{(0.01 + x)x}{0.01 - x} = 1.2 \times 10^{-2}$$

Assuming x to be small relative to 0.01, we obtain a trial value of 0.012 for this quantity; obviously the assumption is not justified, and therefore we must solve the quadratic equation

$$x^2 + 0.022x - 1.2 \times 10^{-4} = 0$$

from which

$$x = [SO_4^{2-}] = 0.0045\ \text{mol/liter}$$

$$[H^+] = 0.01 + 0.0045 = 0.0145\ \text{mol/liter}$$

In this instance the weak acid makes an important contribution to the hydrogen ion concentration of the solution.

Example. We wish to obtain the hydrogen ion concentration of a solution that is 0.0100 F in HCl and 0.0200 F in HOAc. As in the foregoing example we may say

$$[OAc^-] = x$$
$$[H^+] = (0.01 + x)$$
$$[HOAc] = (0.02 - x)$$

Substituting into the dissociation-constant expression for HOAc,

$$\frac{(0.01 + x)x}{(0.02 - x)} = 1.75 \times 10^{-5}$$

If we assume x is small relative to 0.01 and 0.02, we find that x is approximately 3.50×10^{-5} mol/liter, a value that makes the assumption reasonable. Thus

$$[H^+] = 0.01 + 3.5 \times 10^{-5} \cong 0.01 \text{ mol/liter}$$

In this case the weak acid contributes insignificantly to the pH of the solution.

Polybasic Weak Acids

Many of the weak acids encountered in analytical chemistry contain more than one ionizable proton, and we may write two or more equilibrium reactions for a solution of such an acid.

$$H_2A \rightleftharpoons H^+ + HA^-$$
$$HA^- \rightleftharpoons H^+ + A^{2-}$$

Application of the mass law to these reactions yields the expressions

$$K_1 = \frac{[H^+][HA^-]}{[H_2A]} \tag{15-1}$$

$$K_2 = \frac{[H^+][A^{2-}]}{[HA^-]} \tag{15-2}$$

Furthermore, from material-balance considerations we may say

$$F_{H_2A} = [H_2A] + [HA^-] + [A^{2-}] \tag{15-3}$$

Since there must also be a charge balance in the solution

$$[H^+] = [HA^-] + 2[A^{2-}] \tag{15-4}$$

In equation (15-4) we assumed that the concentration of hydroxyl ions in the acidic solution is negligibly small relative to the other terms. The reason for doubling the concentration of A^{2-} is discussed in footnote 3, page 139.

The four equations shown above are sufficient to provide a rigorous determination of the hydrogen ion concentration, since they contain only four unknowns. The algebra is somewhat awkward, however, involving the solution of a cubic equation; in almost every case a simplifying assumption can be made without great loss of accuracy.

We shall assume that the concentration of hydrogen ions from the first dissociation step is great enough to repress the second dissociation to negligible proportions. Then the concentration of A^{2-} will be small with respect to the concentration of H^+ or HA^- and equation (15-4) becomes

$$[H^+] \cong [HA^-] \tag{15-5}$$

Furthermore, equation (15-3) simplifies to

$$F_{H_2A} \cong [H_2A] + [HA^-] \tag{15-6}$$

Substituting (15-5) and (15-6) into (15-1) gives

$$\frac{[H^+]^2}{F_{H_2A} - [H^+]} = K_1 \tag{15-7}$$

This, of course, is simply the equation for calculating the hydrogen ion concentration of a simple weak acid of dissociation constant K_1; it may be solved by the quadratic formula where necessary or by assuming $[H^+]$ is small relative to F_{H_2A} as demonstrated on page 285. Thus, in making assumption (15-5) we are simply ignoring the second dissociation reaction.

Fortunately it is easy to judge the validity of this assumption; if we substitute (15-5) into (15-2) we find

$$[A^{2-}] \cong K_2$$

This gives an approximate value for $[A^{2-}]$ that can readily be compared with the trial value of $[H^+]$ (or $[HA^-]$). The validity of the assumption made in going from (15-4) to (15-5) can then be judged. In general, only in the few instances where K_1 and K_2 approach one another or where the solution is very dilute will neglect of the second dissociation give rise to serious errors. A rigorous treatment will then be required.

Example. We shall calculate the pH of a $0.100 \ F$ solution of oxalic acid

$$K_1 = 6.5 \times 10^{-2} = \frac{[H^+][HOx^-]}{[H_2Ox]}$$

$$K_2 = 6.1 \times 10^{-5} = \frac{[H^+][Ox^{2-}]}{[HOx^-]}$$

Disregarding the second dissociation step and setting $x = [H^+]$ we find

$$x = [H^+] = [HOx^-]$$
$$(0.1 - x) = [H_2Ox]$$

and substituting these into the expression for K_1 gives

$$\frac{x^2}{(0.1 - x)} = 6.5 \times 10^{-2}$$

which must be solved by the quadratic formula. This gives

$$x = [H^+] = 0.055 \text{ mol/liter}$$

We must now check the assumption implied in this calculation, namely, that $[Ox^{2-}]$ is small compared with $[HOx^-]$ with the result that $[HOx^-] = [H^+]$. From the expression for K_2 we get

$$\frac{[H^+][Ox^{2-}]}{[HOx^-]} = [Ox^{2-}] = 6.1 \times 10^{-5} \text{ mol/liter}$$

Thus $[Ox^{2-}]$ is indeed small compared with the calculated value of 0.055 for $[H^+]$ and $[HOx^-]$; the assumption was therefore a reasonable one.

Salts of Polybasic Acids

A number of weak-acid salts of the type Na_2A are of importance in analytical chemistry. For solutions of these, the following equilibria may be written:

$$A^{2-} + H_2O \rightleftharpoons HA^- + OH^-$$
$$HA^- + H_2O \rightleftharpoons H_2A + OH^-$$

Hydrolysis constants for each take the form

$$K_{h'} = \frac{K_w}{K_2} = \frac{[HA^-][OH^-]}{[A^{2-}]} \tag{15-8}$$

$$K_{h''} = \frac{K_w}{K_1} = \frac{[H_2A][OH^-]}{[HA^-]} \tag{15-9}$$

where K_1 and K_2 are the first and second dissociation constants of H_2A. We may also write two additional equations, one based on material balance and the other on electrical neutrality.

$$F_{Na_2A} = [A^{2-}] + [HA^-] + [H_2A] \tag{15-10}$$
$$[Na^+] + [H^+] = 2[A^{2-}] + [HA^-] + [OH^-]$$

The sodium ion concentration will equal twice the formal concentration of the salt; in addition, the hydrogen ion concentration in the basic solution is probably very small. The last equation thus simplifies to

$$2F_{NaA} = 2[A^{2-}] + [HA^-] + [OH^-] \tag{15-11}$$

Here again is a problem with four equations containing four unknowns; the hydroxyl ion concentration of the solution can therefore be computed. By analogy to the case of a weak dibasic acid, however, it is likely that simplification of the algebra involved is possible without undue sacrifice of accuracy by making the further assumption that the second hydrolysis step occurs to a negligible extent only. Where this is justified, the concentration of H_2A may be disregarded, and equation (15-10) simplifies to

$$F_{Na_2A} = [A^{2-}] + [HA^-] \tag{15-12}$$

Equating this with (15-11) gives

$$[HA^-] = [OH^-] \qquad (15\text{-}13)$$

Substituting (15-13) and (15-12) into (15-8) gives an expression identical to that for a simple salt as shown on page 290.

$$\frac{[OH^-]^2}{(F_{Na_2A} - [OH^-])} = \frac{K_w}{K_2}$$

Verification of the assumption used in deriving this equation is readily made by substituting equation (15-13) into (15-9) which then gives an approximate concentration for H_2A.

$$\frac{K_w}{K_1} = \frac{[H_2A][\cancel{OH^-}]}{[\cancel{HA^-}]} \cong [H_2A]$$

Comparison of this concentration with the computed trial value of $[HA^-]$ and $[OH^-]$ will indicate if it is indeed small compared with $[HA^-]$ as we assumed.

Example. We shall calculate the hydroxyl ion concentration of a 0.100 F solution of Na_2CO_3, given that $K_1 = 4.6 \times 10^{-7}$ and $K_2 = 4.4 \times 10^{-11}$. We may write

$$CO_3^{2-} + H_2O \rightleftharpoons HCO_3^- + OH^-$$

$$HCO_3^- + H_2O \rightleftharpoons H_2CO_3 + OH^-$$

$$\frac{K_w}{K_2} = \frac{1.0 \times 10^{-14}}{4.4 \times 10^{-11}} = \frac{[OH^-][HCO_3^-]}{[CO_3^{2-}]}$$

$$\frac{K_w}{K_1} = \frac{1.0 \times 10^{-14}}{4.6 \times 10^{-7}} = \frac{[OH^-][H_2CO_3]}{[HCO_3^-]}$$

Neglecting the second hydrolysis,

$$x = [OH^-] \cong [HCO_3^-]$$

$$(0.100 - x) = [CO_3^{2-}]$$

$$\frac{x^2}{(0.100 - x)} = \frac{1.0 \times 10^{-14}}{4.4 \times 10^{-11}} = 2.27 \times 10^{-4}$$

Assuming further that x is small with respect to 0.100, we find

$$x = [OH^-] = 4.75 \times 10^{-3} \text{ mol/liter}$$

An estimate of the concentration of H_2CO_3 in this solution will indicate whether we were justified in neglecting the second hydrolysis. If we assume that $[OH^-] = [HCO_3^-]$ we obtain, upon substitution into the second hydrolysis equation,

$$[H_2CO_3] = \frac{1.0 \times 10^{-14}}{4.6 \times 10^{-7}}$$

$$[H_2CO_3] = 2.2 \times 10^{-8} \text{ mol/liter}$$

This figure is indeed small with respect to 4.75×10^{-3} mol/liter, the value calculated for the concentration of OH^- and HCO_3^-.

Solutions of Acid Salts

Solutions of weak-acid salts of the type NaHA present a unique problem in equilibrium calculations in that the anion may react in two ways:

$$HA^- + H_2O \rightleftharpoons H_2A + OH^-$$

and

$$HA^- \rightleftharpoons H^+ + A^{2-}$$

One of these reactions produces hydrogen ions while the other yields hydroxyl ions. Whether the solution is acidic or basic will be determined by the relative magnitude of the equilibrium constants for these processes.

$$K_h = \frac{K_w}{K_1} = \frac{[H_2A]\,[OH^-]}{[HA^-]} \qquad (15\text{-}14)$$

$$K_2 = \frac{[H^+]\,[A^{2-}]}{[HA^-]} \qquad (15\text{-}15)$$

If K_h is greater than K_2, the solution will be basic; otherwise it will be acidic.

Equations describing the solution in terms of material balance and charge balance can be written. Thus we may state

$$F_{NaHA} = [HA^-] + [H_2A] + [A^{2-}] \qquad (15\text{-}16)$$

and

$$[Na^+] + [H^+] = [HA^-] + 2[A^{2-}] + [OH^-]$$

Since the sodium ion concentration is equal to the formal concentration of the salt, we may rewrite the last equation as

$$F_{NaHA} = [HA^-] + 2[A^{2-}] + [OH^-] - [H^+] \qquad (15\text{-}17)$$

Often we cannot neglect either the hydroxyl or the hydrogen ion concentration in solutions of this type, and we need, therefore, a fifth equation to take care of the five unknowns. The ion-product constant for water will serve this purpose.

$$K_w = [H^+]\,[OH^-]$$

The derivation of a rigorous expression for the hydrogen ion concentration from these five equations is difficult; however, a reasonably good approximation, applicable in most cases, can be obtained fairly easily. Equating (15-17) with (15-16) yields

$$[H_2A] = [A^{2-}] + [OH^-] - [H^+]$$

With the aid of (15-14) and [15-15] we can express $[H_2A]$ and $[A^{2-}]$ in terms of $[HA^-]$; that is,

$$\frac{K_w[HA^-]}{K_1[OH^-]} = \frac{K_2[HA^-]}{[H^+]} + [OH^-] - [H^+]$$

Replacing $[OH^-]$ by the equivalent expression $K_w/[H^+]$ gives

$$\frac{[H^+][HA^-]}{K_1} = \frac{K_2[HA^-]}{[H^+]} + \frac{K_w}{[H^+]} - [H^+]$$

Multiplying through by $[H^+]$ and rearranging yields

$$[H^+]^2 \left(\frac{[HA^-]}{K_1} + 1\right) = K_2[HA^-] + K_w$$

This is easily converted to

$$[H^+] = \sqrt{\frac{K_1K_2[HA^-] + K_1K_w}{[HA^-] + K_1}} \tag{15-18}$$

Equation (15-18) is useful only where the following is valid:

$$[HA^-] \cong F_{NaHA} \tag{15-19}$$

This is a reasonable assumption provided that neither the hydrolytic nor the dissociation reaction is extensive and provided that the formality of the salt is not too low.

Substituting (15-19) into (15-18) yields

$$[H^+] = \sqrt{\frac{K_1K_2F_{NaHA} + K_1K_w}{F_{NaHA} + K_1}} \tag{15-20}$$

K_1 will often be so much smaller than F_{NaHA} that it can be neglected in the denominator, thus permitting further simplification of equation (15-20). Furthermore, K_1W_w will frequently be much smaller than $K_1K_2F_{NaHA}$. When both of these conditions exist, equation (15-20) simplifies to

$$[H^+] = \sqrt{K_1K_2} \text{ when } K_1 \lll F_{NaHA} \text{ and } K_1K_w \lll K_1K_2F_{NaHA} \tag{15-21}$$

Thus, in many instances, the pH of a solution of a salt of the type NaHA will be independent of the concentration of the salt over a considerable concentration range.

Example. We wish to obtain the hydrogen ion concentration of a 0.100 F solution of $NaHCO_3$.

We shall first examine the assumptions necessary to use equation (15-21). From the table of dissociation constants we find that K_1 for H_2CO_3

is 4.6×10^{-7} and K_2 is 4.4×10^{-11}. Obviously K_1 is much smaller than F_{NaHA}; $K_1 K_w$ is equal to 4.6×10^{-21}, while $K_1 K_2 F_{NaHA}$ has the value of about 2×10^{-18}; we can therefore use the simplified equation. Thus

$$[H^+] = \sqrt{4.6 \times 10^{-7} \times 4.4 \times 10^{-11}}$$
$$[H^+] = 4.5 \times 10^{-9} \text{ mol/liter}$$

Example. We shall calculate the hydrogen ion concentration of a $1.0 \times 10^{-3} F$ solution of Na_2HPO_4.

Here we are concerned with the acid salt HPO_4^- for which the pertinent dissociation constants (those containing $[HPO_4^-]$) are K_2 and K_3 for H_3PO_4. These have the values of 6.2×10^{-8} and 4.8×10^{-13}, respectively. Looking again at the assumptions implicit in equation (15-21), we find that 6.2×10^{-8} is indeed much smaller than the formality of the salt. On the other hand, $K_2 K_w$ is by no means much smaller than $K_2 K_3 F_{Na_2HPO_4}$, the former having a value of 6.2×10^{-22} and the latter of about 3×10^{-23}. We should, therefore, make use of equation (15-20).

$$[H^+] = \sqrt{\frac{6.2 \times 10^{-8} \times 4.8 \times 10^{-13} \times 1.0 \times 10^{-3} + 6.2 \times 10^{-22}}{1.0 \times 10^{-3} + 6.2 \times 10^{-8}}}$$
$$= 8.1 \times 10^{-10} \text{ mol/liter}$$

Use of equation (15-21) would have yielded a value of 1.7×10^{-10} mol/liter.

Example. We wish to find the hydrogen ion concentration of a $0.0100 F$ solution of NaH_2PO_4.

In this case the two dissociation constants of importance are K_1 and K_2 for H_3PO_4; these have values of 7.5×10^{-3} and 6.2×10^{-8}. Because K_1 is not much smaller than $F_{NaH_2PO_4}$, we again apply equation (15-20).

$$[H^+] = \sqrt{\frac{7.5 \times 10^{-3} \times 6.2 \times 10^{-8} \times 1.0 \times 10^{-2} + 7.5 \times 10^{-17}}{0.0075 + 0.010}}$$
$$= 1.6 \times 10^{-5} \text{ mol/liter}$$

Salts of a Weak Acid and a Weak Base

Solutions of salts of a weak acid and a weak base are analogous to the solutions just considered inasmuch as hydrogen ions and hydroxyl ions are produced by competing equilibria. For example, in the case of a solution of ammonium acetate, we write

$$NH_4^+ \rightleftharpoons NH_3 + H^+$$
$$H_2O + OAc^- \rightleftharpoons HOAc + OH^-$$

The equilibrium constants for these reactions are

$$\frac{[NH_3][H^+]}{[NH_4^+]} = \frac{K_w}{K_b} \tag{15-22}$$

$$\frac{[HOAc][OH^-]}{[OAc^-]} = \frac{K_w}{K_a} \tag{15-23}$$

From a consideration of charge balance we write

$$[NH_4^+] + [H^+] = [OAc^-] + [OH^-] \tag{15-24}$$

Furthermore, mass balance requires that

$$F_{NH_4OAc} = [NH_4^+] + [NH_3] \tag{15-25}$$

$$F_{NH_4OAc} = [OAc^-] + [HOAc] \tag{15-26}$$

Equating (15-25) and (15-26) and subtracting (15-24) from the resultant gives the relationship

$$[NH_3] - [H^+] = [HOAc] - [OH^-]$$

Substituting (15-22) and (15-23) to eliminate $[NH_3]$ and $[HOAc]$, and also $\dfrac{K_w}{[H^+]}$ for $[OH^-]$ gives

$$\frac{[NH_4^+] K_w}{[H^+] K_b} - [H^+] = \frac{[OAc^-][H^+]}{K_a} - \frac{K_w}{[H^+]}$$

Rearrangement of this will yield

$$[H^+] = \sqrt{\frac{[NH_4^+]K_w K_a + K_b K_w K_a}{[OAc^-]K_b + K_a K_b}}$$

For this equation to be useful, the concentrations of ammonium ion and acetate ion must not differ greatly from the formal concentration of the salt; under these circumstances

$$[H^+] = \sqrt{\frac{K_w K_a(F_{NH_4OAc} + K_b)}{K_b(F_{NH_4OAc} + K_a)}} \tag{15-27}$$

Frequently, K_b and K_a will be much smaller than the formal concentration of the salt; when this is so

$$[H^+] = \sqrt{\frac{K_w K_a}{K_b}} \tag{15-28}$$

As long as this assumption is justified, a solution of this type will have a pH that is independent of the formal concentration of the salt.

Buffer Systems Involving Polybasic Acids

Two buffer systems can be prepared from the weak acid H_2A and its salts NaHA and Na_2A. The first consists of the free acid and the acid salt (H_2A — NaHA); the second makes use of the acid salt and the salt (NaHA — Na_2A). The pH of the latter system will be the higher, since the dissociation constant for the acid salt is always less than that of the weak acid itself.

It is a simple matter to write enough equations to make a rigorous evaluation of the hydrogen ion concentration of either of these systems through use of the two dissociation constants for the acid and the formal concentrations of the components. Solving these simultaneous equations is difficult, however, and it is expedient to introduce the simplifying assumption that only one of the equilibria in the solution is important in determining the hydrogen ion content. Thus, for a buffer consisting of H_2A and NaHA, only the first dissociation of H_2A is considered; the further reaction of HA^- to yield A^{2-} is disregarded on the assumption that the concentration of A^{2-} is small compared with that of HA^- and H_2A. With this simplification the hydrogen ion concentration is easily calculated by the technique described on page 294 for a simple buffer solution. As before, it is an easy matter to check the validity of the assumption by calculating an approximate concentration for A^{2-} and comparing this value with the concentrations of H_2A and HA^-.

For a buffer prepared from NaHA and Na_2A, the second dissociation reaction is assumed to predominate and the hydrolytic reaction

$$HA^- + H_2O \rightleftharpoons H_2A + OH^-$$

is disregarded; the assumption here is that the concentration of H_2A is negligible compared to that of HA^- and A^{2-}. If this is the case, the hydrogen ion concentration can be determined from the second dissociation constant employing the techniques for a simple buffer solution. To test the assumption an estimate of the concentration of H_2A is compared with the concentrations of HA^- and A^{2-}.

Example. We shall calculate the hydrogen ion concentration of a buffer solution that is $0.100\,F$ in phthalic acid (H_2P) and $0.200F$ in potassium acid phthalate (KHP).

From the table of dissociation constants we find

$$\frac{[H^+][HP^-]}{[H_2P]} = K_1 = 1.3 \times 10^{-3}$$

$$\frac{[H^+][P^{2-}]}{[HP^-]} = K_2 = 3.9 \times 10^{-6}$$

We shall assume that the principal equilibrium in this solution is

$$H_2P \rightleftharpoons H^+ + HP^-$$

and that the further dissociation of HP^- is negligible. Implicit in this assumption is that $[P^{2-}] \lll [HP^-]$ and $[H_2P]$. We then proceed as on page 295,

$$x = [H^+]$$
$$[H_2P] = (0.100 - x) \cong 0.100$$
$$[HP^-] = (0.200 + x) \cong 0.200$$

and

$$\frac{0.200x}{0.100} = 1.3 \times 10^{-3}$$

$$x = [H^+] = 6.5 \times 10^{-4} \text{ mol/liter}$$

To find if we were justified in neglecting the second dissociation step we obtain an approximate concentration of P^{2-} from the second dissociation-constant expression. Thus

$$\frac{6.5 \times 10^{-4} [P^{2-}]}{0.200} = 3.9 \times 10^{-6}$$
$$[P^{2-}] \cong 1.2 \times 10^{-3} \text{ mol/liter}$$

This figure is small with respect to the concentrations of H_2P and HP^-; we may therefore conclude that the value obtained for $[H^+]$ is reasonably accurate.

Example. We shall calculate the hydrogen ion concentration of a buffer that is $0.0500 \, F$ in potassium acid phthalate and $0.150 \, F$ in potassium phthalate.

Here we assume that the principal equilibrium is

$$HP^- \rightleftharpoons H^+ + P^{2-}$$

and that the concentration of H_2P in this solution is negligible. Then

$$x = [H^+]$$
$$[HP^-] = (0.050 - x) \cong 0.05$$
$$[P^{2-}] = (0.150 + x) \cong 0.150$$

Substituting into the equation for K_2 we find

$$x = [H^+] = 1.3 \times 10^{-6} \text{ mol/liter}$$

We must now calculate the approximate concentration of H_2P to check the first assumption. Substituting the value calculated for $[H^+]$ into the first dissociation-constant expression, we find

$$\frac{(1.3 \times 10^{-6}) (0.0500)}{[H_2P]} = 1.3 \times 10^{-3}$$
$$[H_2P] = 5 \times 10^{-5} \text{ mol/liter}$$

This result justifies the assumption that $[H_2P]$ is much smaller than $[HP^-]$ and $[P^{2-}]$—that is, that the hydrolytic reaction can be neglected.

In all but a few cases the assumption of a single principal equilibrium, as invoked in these examples, will lead to a satisfactory estimate of the pH of buffer mixtures derived from polybasic acids. Appreciable errors will arise, however, where the concentration of the acid or the salt is very low or where the two dissociation constants are numerically close to one another. A more laborious and rigorous solution of the problem is then necessary. Because this situation is rarely encountered we shall not deal with it.

ACID-BASE TITRATION CURVES FOR COMPLEX SYSTEMS

The titration curves for solutions of polybasic acids or mixtures of acids are ordinarily more complex than those that we have so far considered. The same may be said for mixtures of bases or polyfunctional bases. Titration curves having multiple inflection points are observed when the component acids (or bases) differ appreciably in strength; such systems may have more than one useful equivalence point if the alterations in pH are sufficiently abrupt in the region of these inflections.

The techniques described in the first part of this chapter make possible the derivation, with reasonable accuracy, of theoretical titration curves for mixtures of acids or for polybasic acids provided the ratio between the dissociation constants of the two acids is greater than 10^3. If the ratio is less than this, particularly in the region of the first equivalence point, the error becomes prohibitive, and a more rigorous treatment of the equilibrium relationships is required. We shall consider only those systems in which the acids differ considerably in strength.

Polybasic Acids

Derivation of titration curves. We shall first consider derivation of the theoretical curve for the titration of 25.0 ml of 0.100 F maleic acid with 0.100 F sodium hydroxide. Maleic acid is a typical organic dibasic acid having the empirical formula $H_2C_4H_2O_4$; for the sake of simplicity we will use H_2M to symbolize the free acid. The dissociation equilibria and their constants for maleic acid are as follows:

$$H_2M \rightleftharpoons H^+ + HM^- \qquad\qquad K_1 = 1.5 \times 10^{-2}$$
$$HM^- \rightleftharpoons H^+ + M^{2-} \qquad\qquad K_2 = 2.6 \times 10^{-7}$$

Initial pH. Neglecting the second dissociation reaction and treating the solution as a weak acid with a dissociation constant of 1.5×10^{-2} will make it possible to ascertain the initial pH of the solution. This type of calculation was demonstrated on page 323; in this instance a pH of 1.5 is obtained.

First buffer region. The initial addition of base forms a buffer, consisting of the acid H_2M and its salt HM^-; their formal concentrations are required to determine the pH. Thus when 5.00 ml of the reagent are introduced we may say

$$F_{H_2M} = \frac{(25.0 \times 0.1) - (5.00 \times 0.100)}{30.0} = \frac{2.0}{30}$$

$$F_{NaHM} = \frac{5.00 \times 0.100}{30} = \frac{0.5}{30}$$

As shown on page 330, the second dissociation of the acid can be neglected in determining the pH of the solution. Because the acid is relatively strong, however, we must solve for the hydrogen ion concentration via the quadratic expression rather than making the usual simplifying assumption. The pH for this point is 1.72.

Additional points in the first buffer region can be computed by this method. Within a few tenths of a milliliter of the first equivalence point the effect of the second dissociation of the acid begins to assume importance, and the simplified calculation for pH becomes subject to considerable error. Computation of approximate values for M^{2-} will define the extent of this region of uncertainty; as a practical matter, the region is so small the titration curve can ordinarily be defined without the necessity of calculating the points within it.

First equivalence point. When 25.0 ml of base have been added, the solution is $0.0500\ F$ with respect to the acid salt NaHM. The pH is calculated using equation (15-20) (p. 327) and has a value of 4.26.

Second buffer region. Additional base added to the solution of the acid salt forms a new buffer consisting of HM^- and its salt M^{2-}. When enough base has been added so that the hydrolysis of HM^- may be neglected—in this instance a few tenths of a milliliter—the pH of the mixture is readily obtained from K_2. With the introduction of 25.5 ml of NaOH for example, the number of milliformula weights of Na_2M is equal to the number of milliformula weights of base in excess of that required to form NaHM, namely $(25.5 - 25.0)0.100$, and

$$F_{Na_2M} = \frac{(25.5 - 25.0)0.100}{50.5} = \frac{0.050}{50.5}$$

The number of milliformula weights of NaHM remaining will be equal to the amount present at the first equivalence point (25.0×0.100) minus the number of milliformula weights of Na_2M subsequently produced; here

$$F_{NaHM} = \frac{(25.0 \times 0.100) - (25.5 - 25.0)0.100}{50.5}$$

$$= \frac{2.45}{50.5}$$

Expressing $[H^+]$ as x we may write

$$\frac{x\left[(0.050/50.5) + x\right]}{[(2.45/50.5) - x]} = 2.6 \times 10^{-7}$$

A trial solution neglecting x in the parenthetic terms yields a value of 1.29×10^{-5}; this answer is certainly reliable enough for the present purposes.

Similar calculations will yield data to establish the form of the curve up to the second equivalence point.

Second equivalence point. After the addition of 50 ml of sodium hydroxide the solution is $0.0333\ F$ in Na_2M. The hydroxyl ion concentration can be determined as shown on page 324; only the first hydrolysis of the salt need be taken into account. These calculations give a pH of 9.55.

pH beyond the second equivalence point. Further additions of strong base repress the hydrolysis of M^{2-}; the pH can be calculated from the concentration of NaOH added in excess of that required for total neutralization of H_2M (see p. 311).

Titration curves. Figure 15.1 is a plot of the titration curve for a solution of maleic acid from data obtained as shown in the previous section. Two end points are apparent; through the judicious choice of indicator either one or both of the available hydrogen ions of the acid could be titrated. The second end point is clearly the better one to use, however, inasmuch as the pH change is more pronounced.

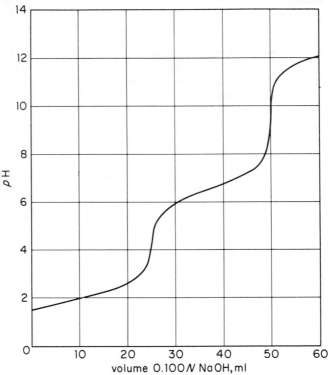

Fig. 15.1 Titration Curve for a Weak Dibasic Acid. Titration of 25.0 ml of 0.100 F maleic acid $(K_1 = 1.5 \times 10^{-2}, K_2 = 2.6 \times 10^{-7})$ with 0.100 N NaOH.

Figure 15.2 shows titration curves for three other dibasic acids; these data illustrate that only where the degree of dissociation of the two acids is considerably different will a well-defined end point corresponding to the first equivalence point be observed. For example, the curve for sulfuric acid (curve C) has no apparent inflection at a point corresponding to neutralization of the first proton, and the graph is not greatly different from that for a simple strong acid.

In the case of oxalic acid (curve B) the ratio of K_1 to K_2 is approximately one thousand. Here the titration curve shows a distinct inflection corresponding to the first equivalence point. The magnitude of the pH change at this point is not sufficiently great, however, to allow an accurate location of this equivalence point; thus, only the second equivalence point can be used for analytical purposes.

Graph *A* illustrates the theoretical titration curve for the tribasic phosphoric acid. The ratio of K_1 to K_2 here is approximately 10^5. This is about 100 times greater than that for oxalic acid, and two well-defined end points are observed, either of which is satisfactory for analytical purposes. With an indicator having an acid transition range, one equivalent of base would be consumed per mol of acid; with an indicator exhibiting a basic color change, two equivalents of

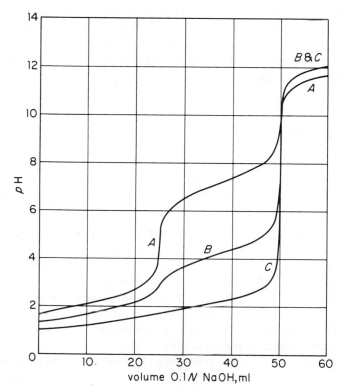

Fig. 15.2 Curves for the Titration of Three Polybasic Acids. A 0.100 *N* NaOH solution is used to titrate: *A*. 25.0 ml of 0.100 *F* H_3PO_4 ($K_1 = 7.5 \times 10^{-3}$, $K_2 = 6.2 \times 10^{-8}$); *B*. 25.0 ml of 0.100 *F* oxalic acid ($K_1 = 6.5 \times 10^{-2}$, $K_2 = 6.1 \times 10^{-5}$); *C*. 25.0 ml of 0.100 *F* H_2SO_4 ($K_2 = 1.2 \times 10^{-2}$).

base would be consumed. The third hydrogen of phosphoric acid is so slightly dissociated ($K_3 = 4.8 \times 10^{-13}$) that it does not yield an end point of any practical value. The effect of the third dissociation is noticeable in curve *A*, however, causing the *p*H to be lower than that for the other two acids in the region beyond the second equivalence point.

In general, the titration of polyfunctional acids or bases yields individual end points that are of practical value only where the ratio of the two dissociation constants is at least 10^5. If the ratio is much less than this, the *p*H change at the first equivalence point will be too small for accurate detection—only the second end point will prove satisfactory for analysis.

A judicious choice of indicators thus makes possible the analysis of the individual components of certain mixtures. Consider, for example a solution containing phosphoric acid and a dihydrogen phosphate salt. If an aliquot is titrated with an indicator having a transition interval in the range of pH 4, the milliequivalents of base consumed are equal to the number of milliformula weights of phosphoric acid alone. Titration of another aliquot with an indicator such as phenolphthalein, on the other hand, requires an amount of base corresponding to the number of milliformula weights of the dihydrogen phosphate salt present plus twice the milliformula weights of the acid. From these data the quantity of each of the constituents in the mixture can be readily computed.

Mixtures of Acids

The techniques involved in the derivation of titration curves for solutions containing more than one acid component do not differ much from those just discussed since a solution of a dibasic acid can be thought of simply as an equiformal mixture of two acids with different dissociation constants.

Titration curves for strong and weak acid mixtures. The derivation of a theoretical titration curve for a mixture of a strong and a weak acid is quite straightforward. The pH of the solution at any point up to the first equivalence point is found from the formal concentrations of the strong acid by using the method described on page 306. Neutralization of the strong acid is complete at the first equivalence point and for purposes of pH calculation, the solution is then treated as a simple weak acid. The remainder of the titration curve is therefore characteristic of the weak acid.

Curves with two inflection points are obtained for a mixture of this type. Where the weak acid has a dissociation constant smaller than approximately 10^{-5}, the pH change at the first equivalence point is well marked; by the proper choice of indicator the concentration of each of the components is readily found. When the weak acid has a dissociation constant much greater than 10^{-5}, however, only the total acid content of the solution can be determined.

Titration curves for mixtures of two weak acids. The derivation of a theoretical titration curve for a mixture of two weak acids is more complex than for a weak dibasic acid inasmuch as the formal concentrations of the two acids are no longer related to each other. In those instances where the ratio of the dissociation constants is greater than about 10^3 or 10^4 and the ratio of the initial concentration of the acids does not differ greatly from one, a curve can be obtained in a fashion analogous to that described for the dibasic weak-acid system. In the region preceding the first equivalence point, the dissociation of the weaker acid is disregarded; beyond this point, hydrolysis of the salt of the stronger acid is not considered. Thus the titration curve approaches a composite of the individual titration curves for each of the acids. It can be shown that the pH at the first equivalence point is approximated by means of the expression[1]

$$[\mathrm{H^+}] = \sqrt{K_l K_m \frac{F_m}{F_l}}$$

where K_l and K_m are the dissociation constants of the stronger and weaker acids, respectively, and F_l and F_m are their initial formal concentrations. When the initial concentrations are identical, this equation reduces to equation (15-21) (p. 327).

Salts of Polybasic Acids

Several salts of polybasic acids are important in volumetric analysis; since the derivation of titration curves for this class of compounds involves no new principles we need not treat the technique in detail.

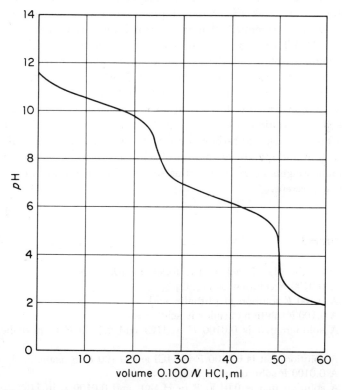

Fig. 15.3 Curve for the Titration of 25.0 ml of 0.100 F Sodium Carbonate with 0.100 N Hydrochloric Acid.

[1] J. E. Ricci, *Hydrogen Ion Concentration*, 180. Princeton, New Jersey: Princeton University Press, 1952.

As an example, consider the titration of a solution of sodium carbonate with standard hydrochloric acid. The important equilibrium constants are

$$H_2CO_3 \rightleftharpoons H^+ + HCO_3^- \qquad\qquad K_1 = 4.6 \times 10^{-7}$$

$$HCO_3^- \rightleftharpoons H^+ + CO_3^{2-} \qquad\qquad K_2 = 4.4 \times 10^{-11}$$

The hydrolysis of the carbonate ion governs the initial pH of the solution; the technique shown on page 324 is used in the pH calculation. With the first additions of acid a carbonate-bicarbonate buffer develops, and the hydrogen ion concentration is found from K_2; the effects of hydrolysis of the bicarbonate to carbonic acid can be neglected. A solution of sodium bicarbonate forms at the first equivalence point; equation (15-21) will give the pH of this solution. With the introduction of more acid, a bicarbonate-carbonic acid buffer governs the pH, and here the hydrogen ion content is obtained from K_1. Upon reaching the second equivalence point, the solution consists of carbonic acid and sodium chloride; the hydrogen ion concentration is calculated in the usual way for a simple weak acid. Finally, as excess hydrochloric acid is introduced, the dissociation of the weak acid is soon repressed to a point where the hydrogen ion concentration is essentially that of the formal concentration of the strong acid.

Figure 15.3 illustrates a titration curve for a solution of sodium carbonate; two end points are observed, the second being a good deal sharper than the first. Clearly, the stoichiometry of the analytical reaction depends upon the transition interval of the indicator chosen for the titration. It is also apparent that mixtures of sodium carbonate and sodium bicarbonate can be analyzed by neutralization methods. Thus, titration to a phenolphthalein end point would yield the number of milliformula weights of carbonate present, while titration to a methyl-orange color change would require an amount of acid equivalent to twice the number of milliformula weights of carbonate plus the number of milliformula weights of bicarbonate present.

problems

1. Calculate the pH of each of the following solutions:
   (a) A $0.0500\ F$ solution of H_3PO_4 ans. 1.80
   (b) A $0.100\ F$ solution of phthalic acid   ans. 1.97
   (c) A $0.100\ F$ solution of salicylic acid
   (d) A solution that is $0.0100\ F$ in HCl and $0.0200\ F$ in phthalic acid
                                                                               ans. 1.92
   (e) A solution that is $0.0100\ F$ in HCl and $0.100\ F$ in oxalic acid
   (f) A $0.0100\ F$ solution of H_2SO_4
   (g) A solution that is $0.0100\ F$ in H_2SO_4 and $0.0100\ F$ in HCl   ans. 1.63
   (h) A solution that is $0.100\ F$ in acetic acid and $0.100\ F$ in boric acid

2. Calculate the pH of the following solutions:
   (a) A buffer solution that is $0.100\ F$ in NaH_2PO_4 and $0.200\ F$ in Na_2HPO_4
                                                                               ans. 7.51

(b) A buffer that is 1.00 F in NaH_2PO_4 and 2.00 F in H_3PO_4 ans. 1.82

(c) A buffer prepared by mixing 50.0 ml of 0.200 F tartaric acid and 50.0 ml of 0.0800 F NaOH   ans. 2.85

(d) A buffer prepared by adding 200 ml of 0.0800 F NaOH to 50.0 ml of 0.200 F tartaric acid   ans. 4.71

(e) A buffer prepared by adding 1.00 ml of 10.0 N HCl to 100 ml of 0.300 F Na_2CO_3

(f) A mixture of 1.00 liter of 0.250 F potassium acid phthalate and 1.00 liter of 0.100 F NaOH

(g) A mixture of 1.00 liter of 0.250 F potassium acid phthalate and 1.00 liter of 0.0500 F HCl

(h) A solution that was 0.100 F in H_2SO_4 and 0.100 F in $NaHSO_4$

3. Calculate the pH of the following solutions:

(a) A 0.200 F solution of potassium acid phthalate   ans. 4.15

(b) A 0.0100 F solution of NaHS   ans. 9.62

(c) A 0.0400 F solution of Na_2HPO_4

(d) A 0.0100 F solution of $NaHSO_3$

(e) A 0.0100 F solution of NaH_2PO_4

4. Calculate the pH of the following solutions:

(a) A 0.100 F solution of Na_2CO_3 ans. 11.68

(b) A solution that is 0.100 F in Na_2CO_3 and 0.0100 F in NaOH   ans. 12.08

(c) A 0.0100 F solution of Na_3PO_4

5. How many ml of 1.00 N HCl must be added to a liter of a 0.100 F solution of Na_2CO_3 to give a buffer having a pH of 10.0?   ans. 69.4 ml

6. How many ml of 1.00 N HCl must be added to 1.00 liter of 0.100 F Na_2CO_3 to give a buffer having a pH of 6.00?

7. How many grams of $NaH_2PO_4 \cdot H_2O$ (GFW = 138) should be added to 1.00 liter of 0.500 F H_3PO_4 to give a buffer of pH 2.00?   ans. 51.8 grams

8. How many grams of potassium acid phthalate must be added to 2.00 liters of 0.400 N NaOH to produce a buffer of pH 7.0?

9. A solution containing 0.500 gram formula weight of H_3PO_4 was brought to a pH of 4.00. The final volume of the solution was 1.00 liter. Calculate the equilibrium concentration of each of the phosphate-containing species.

$$\text{ans. } [H_3PO_4] = 6.6 \times 10^{-3} \text{ mol/liter}$$
$$[H_2PO_4^-] = 0.493 \text{ mol/liter}$$
$$[HPO_4^{2-}] = 3.1 \times 10^{-4} \text{ mol/liter}$$
$$[PO_4^{3-}] = 1.5 \times 10^{-12} \text{ mol/liter}$$

10. Calculate the equilibrium concentrations of each of the phosphate-containing species in a solution of pH 8.00 and in which the total concentration of these species is 0.100 F.

11. Calculate the equilibrium concentrations of each of the phosphate-containing species in a series of solutions in which the total concentration of these species is 0.100 M and that have been brought to the following pH's: 1.0, 2.0, 3.0, 4.0, 5.0, 6.0, 7.0, 8.0, 9.0, 10.0, 11.0, 12.0, and 13.0.

On a single sheet of graph paper plot the concentration of each of the species as a function of pH.

12. Calculate the ratios of the various carbonate species in a sample of human blood having a pH of 7.3.

13. Exactly 25.0 ml of 0.100 F Na_2CO_3 was titrated with 0.100 N HCl. Calculate the pH of the solution after addition of the following volumes of reagent: 0.0, 5.0, 12.5, 20.0, 24.0, 25.0, 26.0, 30.0, 37.5, 45.0, 49.0, 50.0, 51.0, 55.0. Plot the data.

14. Calculate the pH of the following mixtures:
    (a) 100 ml of 0.125 F tartaric acid and 50.0 ml of 0.098 N NaOH
    (b) 100 ml of 0.125 F tartaric acid and 100 ml of 0.125 F NaOH
    (c) 100 ml of 0.125 F tartaric acid and 50.0 ml of 0.300 F NaOH
    (d) 100 ml of 0.125 F tartaric acid and 100.0 ml of 0.300 F NaOH

15. Calculate the pH of the following mixtures:
    (a) 100 ml of 0.0500 F potassium acid phthalate and 20.0 ml of 0.100 N HCl
    (b) 100 ml of 0.0500 F potassium acid phthalate and 100 ml of water
    (c) 100 ml of 0.0500 F potassium acid phthalate and 100 ml of 0.0500 N HCl
    (d) 100 ml of 0.0500 F potassium acid phthalate and 100 ml of 0.0500 N NaOH

16. Calculate the pH of solutions produced by mixing the following with 50 ml of 0.100 F Na_2HPO_4.
    (a) 20 ml of 0.200 N HCl
    (b) 40 ml of 0.200 N HCl
    (c) 200 ml of water
    (d) 20 ml of 0.200 N NaOH
    (e) 25 ml of 0.200 N NaOH

17. A solution that was 0.100 F in acetic acid and 0.100 F in HCl was titrated with a standard 0.100 N solution of KOH. Calculate the pH after the following addition of reagent if the original volume of the acid mixture was 30.0 ml: 0.0, 15.0, 25.0, 29.0, 29.9, 30.0, 30.1, 31, 35, 45.0, 55.0, 59.0, 59.9, 60.0, 60.1, 61.0, 65.0. Construct a titration curve from these data.

18. Calculate the hydrogen ion concentration of a 1.00 and a 0.0100 F solution of H_2SO_4.

chapter 16. *Applications of Neutralization Titrations*

Methods based upon neutralization reactions find extensive employment in volumetric analysis. In general, a standard solution of acid or of base is used to titrate the hydroxyl or hydrogen ions liberated upon solution of the sample in water; in some instances, however, the preliminary treatment of the substance being analyzed results in the evolution of an equivalent amount of one of these ions.

Although space limitations restrict this discussion to neutralizations occurring in aqueous solution, we should recognize that the acidic or basic character of a solute is determined in part by the nature of the liquid in which it is dissolved and that the use of solvents other than water can result in a marked change in these properties. As a result, acids and bases that are too weakly dissociated for titration in water can often be determined successfully and accurately by titration in nonaqueous media.

REAGENTS FOR NEUTRALIZATION TITRATIONS

Strong acids or bases are employed as reagents in neutralization titrations. Standard solutions of the former are commonly prepared from hydrochloric, sulfuric, or perchloric acid. Because of their high solubility, sodium hydroxide and potassium hydroxide are preferred in the preparation of the latter; in addition, barium hydroxide may be used for dilute basic solutions.

Preparation of Indicator Solutions

Many indicators are available for end-point detection in neutralization titrations. Specific directions for the preparation of several of the more common of these follow.

Stock solutions generally contain 0.5 to 1.0 gram of indicator per liter of solution.

Methyl orange and methyl red.   Dissolve the sodium salt directly in distilled water.

Phenolphthalein and thymolphthalein.   Dissolve the solid indicator in a solution that is 80 percent by volume in ethyl alcohol.

Sulfonthaleins.   Dissolve the sulfonthaleins in water by adding sufficient NaOH to react with the sulfonic acid group of the indicator. To prepare stock solutions of these, triturate 100 mg of the solid indicator with the specified volume of 0.1 N NaOH; then dilute to 100 ml with distilled water. The volumes in milliliters of base required are as follows: *bromcresol green,* 1.45; *bromthymol blue,* 1.6; *bromphenol blue,* 1.5; *thymol blue,* 2.15; *cresol red,* 2.65; *phenol red,* 2.85.

Preparation of Standard Solutions of Acid

Hydrochloric acid is the most commonly employed volumetric acid reagent. Dilute solutions of this substance are indefinitely stable, and the reagent can be used in the presence of most cations without interference due to precipitate formation. It is reported that 0.1 N solutions of the reagent can be boiled for as long as 1 hour without loss of acid provided the evaporated water is replaced periodically; 0.5 N solutions are stable for at least 10 minutes at boiling temperature.

Standard solutions of sulfuric or perchloric acid also make satisfactory acid reagents. They may be used to advantage when chloride ion causes precipitation problems. Nitric acid is seldom used because of its oxidizing properties.

There are several ways to prepare acid solutions of known composition, the most common involving suitable dilution of the concentrated reagent followed by standardization against a primary-standard base. In addition, the normality of a concentrated acid can be determined by careful density measurements;

less concentrated solutions of known normality are then prepared by diluting weighed quantities to an exact volume.[1] Finally, a stock solution of known normality of some acids may be prepared by distillation of the commercial reagents under specified conditions; so-called *constant boiling HCl* is prepared in this way. To obtain this reagent, the concentrated acid is distilled at atmospheric pressure, the first three quarters of the distillate being rejected. The final quarter has a constant and definite composition, its acid content being dependent only upon the atmospheric pressure (see Table 16-1). A standard solution of acid can be prepared by dilution.

Table 16-1

COMPOSITION OF CONSTANT BOILING HYDROCHLORIC ACID[2]

Atmospheric Pressure, mm Hg	Concentration of Distillate, grams HCl/100 grams of solution	Weight of Acid to Prepare 1 liter of 1 N HCl, grams
770	20.197	180.407
760	20.221	180.193
750	20.245	179.979
740	20.269	179.766
730	20.293	179.555

[2] Data from C. W. Foulk and M. Hollingsworth, *J. Am. Chem. Soc.*, **45**, 1220 (1923).

Preparation of approximately 0.1 N HCl.   Boil slightly more than 1 liter of distilled water and cool to about room temperature. (This step is necessary only when preparing very dilute solutions unless the distilled water is supersaturated with respect to CO_2). Transfer the water to a 1-liter glass-stoppered bottle and add about 8.1 ml of concentrated HCl. Mix the contents of the bottle thoroughly.

Preparation of constant boiling HCl.   Prepare a hydrochloric acid solution having a density of about 1.1 by diluting approximately 600 ml of the concentrated acid to 1 liter. Place the acid in a distillation flask and distill at a rate of 3 to 4 ml per minute. Reject the first three fourths of the distillate; collect the remainder in a clean dry container until only 50 to 60 ml remain in the distilling flask. Note the barometric pressure. Transfer the distillate to a clean, dry bottle fitted with a tight stopper. The composition of this acid can be determined from the data in Table 16-1.

Preparation of 0.1 N HCl from constant boiling HCl.   Clean and dry a 50-ml glass-stoppered flask and weigh to the nearest 5 mg. Transfer approximately 18 grams of the constant boiling mixture to the flask, being careful to avoid wetting the ground-glass surfaces. Stopper the flask tightly and

[1] See I. M. Kolthoff and V. A. Stenger, *Volumetric Analysis*, **2**, 64. New York: Interscience Publishers, Inc., 1947.

reweigh. Immediately add boiled and cooled distilled water and transfer the contents quantitatively to a calibrated 1-liter volumetric flask. Dilute to volume with boiled and cooled distilled water; mix the contents thoroughly. The normality of the acid can be calculated from the data in Table 16-1.

Standardization of Acids

Dilute acid solutions are standardized in several ways. One method involves determining the anion content of the solution by a gravimetric procedure. Thus, for hydrochloric acid, a silver chloride precipitate is used; for sulfuric acid, standardization by precipitation of barium sulfate is feasible. These methods assume, of course, that the anion concentration is equivalent to the hydrogen ion concentration.

More common procedures involve the titration of a weighed quantity of a primary-standard base. There are several such substances available, the most common being sodium carbonate.

Standardization of acids with sodium carbonate. Sodium carbonate is available commercially in high enough quality for most purposes; it can also be prepared by heating a good grade of sodium bicarbonate for 1 hour at 270° to 300° C.

$$2NaHCO_3 \rightarrow Na_2CO_3 + H_2O + CO_2$$

A titration curve for a solution of sodium carbonate is shown in Figure 15.3 (p. 337). Of the two possible end points, the second, occurring in acid solution, is always employed in standardization procedures since it is a good deal sharper than the first. As a matter of fact, the simple expedient of heating to decompose the reaction product, carbonic acid, sharpens the pH change in the end-point region even further. One of the several ways of doing this involves titration of the solution at room temperature until an acidic indicator (such as bromcresol green, methyl orange, or bromphenol blue) just begins to exhibit its acid color. At this point, a small quantity of unreacted bicarbonate remains in the solution in addition to a large amount of carbonic acid. All of the latter is removed when the solution is boiled for a few minutes.

$$H_2CO_3 \rightarrow CO_2 + H_2O$$

During this process the pH rises to a slightly alkaline level owing to the residual bicarbonate. After cooling, the titration is completed; now, however, the pH change accompanying the final increments of reagent is considerably larger and, as a consequence, the end point is more accurately located.

An alternative method is to run in enough acid initially to provide a small excess. The solution is then boiled to remove the carbon dioxide and the excess acid is back titrated with a dilute solution of base. Any of the indicators suited to a strong acid-base titration can then be employed. Obviously, the combining capacity of the base for the acid must also be determined in order to make a correction for the back titration.

Procedure. Dry a quantity of primary-standard sodium carbonate for 2 hours at 150° C and cool in a desiccator. For standardization of 0.1 N solutions, weigh 0.2- to 0.25-gram portions of the salt (to the nearest 0.1 mg) into 250-ml flasks and dissolve in 25 to 50 ml of water. Add 3 drops of brom-cresol green and titrate until the solution just begins to change from blue to green. Boil the solution for 2 to 3 minutes, cool to room temperature under a water tap, and complete the titration. During the heating process the indicator should change back to blue. If it does not, an excess of acid was added originally; back titrate this excess with standard base. Determine an indicator blank by titrating a similar volume of water containing approximately the same amounts of sodium chloride and indicator as are present at the end point in the solution being titrated.

Other primary standards for acids. Several other primary-standard compounds are recommended for standardization of acids. Among these is sodium tetraborate decahydrate, $Na_2B_4O_7 \cdot 10H_2O$, which can be formed by recrystallization of borax from water followed by washing with alcohol and ether. The air-dried product is then stored in tightly stoppered bottles. The stoichiometry of its reaction with acid is expressed by the equation

$$B_4O_7^{2-} + 2H^+ + 5H_2O \rightarrow 4H_3BO_3$$

Sodium oxalate can also be employed; it is readily obtained in a pure state. Weighed quantities of this compound are ignited in platinum crucibles; the sodium carbonate produced is then titrated by the methods already described. The following equation expresses the ignition reaction:

$$Na_2C_2O_4 \rightarrow Na_2CO_3 + CO$$

Detailed descriptions of the use of these and other standards can be found in the reference work of Kolthoff and Stenger.[3]

Preparation of Standard Solutions of Bases

Sodium hydroxide is the most common basic reagent, although potassium and barium hydroxides are also employed. No basic reagent is obtainable in primary-standard quality; standardization of the solutions after preparation is always necessary. The solutions are all reasonably stable provided care is taken to protect them from atmospheric contamination.

Effect of carbon dioxide on standard-base solutions. The hydroxides of sodium, potassium, and barium, either in the solid or solution form, rapidly absorb carbon dioxide from the atmosphere with the production of the corresponding carbonate.

$$CO_2 + 2OH^- \rightarrow CO_3^{2-} + H_2O$$

[3] I. M. Kolthoff and V. A. Stenger, *Volumetric Analysis*, **2**, 74-93. New York: Interscience Publishers, Inc., 1947.

From the equation we see that this process consumes 2 mols of hydroxyl ions per mol of carbon dioxide; thus we might expect exposure to the atmosphere to alter the combining capacity of a standard-base solution. This is true under some circumstances but not in others.

In the case of barium hydroxide solutions, the absorption of carbon dioxide results in the precipitation of the slightly soluble barium carbonate. Clearly, a loss in normality of the base results. With sodium and potassium hydroxides, however, the carbonate salt is soluble and quite capable of reacting with hydrogen ions. The number of ions consumed, however, is variable and depends upon the type of indicator used in conjunction with the reagent. If the indicator that is used has an acidic transition range (such as methyl orange), the carbonate ion reacts with two hydrogen ions of the acid being titrated; because this is chemically equivalent to the amount of base used up in producing the carbonate ion, the solution will have the same combining capacity for hydrogen ions as before the absorption. If, on the other hand, the analysis calls for an indicator with a basic transition range (such as phenolphthalein), the carbonate ion will have reacted with only one hydrogen ion to form bicarbonate when color change occurs Here the combining capacity of the base will be diminished as a consequence of contamination by the carbon dioxide. In order to avoid this potential decrease in normality care is taken to protect standard solutions of base from contact with the carbon dioxide of the atmosphere.

Another facet of the carbonate problem arises from the presence of that ion in the solids used to prepare standard solutions of base. Often the amount of this ion is high, even in the best quality of solid reagent, due to absorption of carbon dioxide from the atmosphere; as a result a freshly prepared solution may be badly contaminated. The presence of carbonate need not lead to errors in analysis provided care is taken to employ the same indicator used in the standardization; this may often be inconvenient, however, and much of the versatility of the reagent is lost.

Carbonate contamination is generally troublesome because it leads to poorly defined end points. This is particularly true where an indicator such as phenolphthalein must be employed (see Fig. 15.3).

From these remarks we see that carbonate ion in a standard solution of a base can cause errors by altering the normality of the base or by obscuring the end point of the titration. To avoid these *carbonate errors*, common practice is to prepare standard alkali solutions in such a way that none of the contaminant is present initially and to store the reagent so as to protect it from absorption of atmospheric carbon dioxide.

Preparation and storage of standard solutions of base. Several methods are available for the preparation of carbonate-free solutions of base. Barium hydroxide may be employed as the reagent, in which case the carbonate is insoluble, particularly in the presence of a neutral barium salt such as the chloride or nitrate. A barium salt may also be dissolved in a sodium or potassium hydroxide solution in order to remove the carbonate. Often, however, the presence of barium ion is undesirable since it forms precipitates with a number of anions that may be present in the sample to be analyzed. Carbonate-free solutions of the alkali-metal hydroxides may be prepared by direct solution of the clean metals; here precautions must be taken to avoid explosion during the solution process.

One convenient method for the preparation of sodium hydroxide solutions takes advantage of the very low solubility of sodium carbonate in concentrated solutions of the alkali. An approximately 50-percent solution of sodium hydroxide is prepared; after the sodium carbonate has settled out, the supernatant liquid is decanted and diluted to produce a reagent of the desired concentration. An alternative method is to remove the sodium carbonate by filtration through a sintered glass crucible. Details of this procedure follow.

In preparing carbonate-free base the distilled water used must be free of carbon dioxide. Water that is in equilibrium with the atmosphere contains only 1.5×10^{-5} formula weight of carbon dioxide per liter, a negligible concentration in most applications. Often, however, distilled water is supersaturated with respect to carbon dioxide, and removal of the excess is essential. This is readily accomplished by boiling for a few minutes. The water is then cooled to room temperature before introduction of the base since hot alkali solutions absorb carbon dioxide avidly.

An arrangement for the storage of standard-base solutions so that contamination by carbon dioxide of the air will not occur is shown in Figure 21.1 (p. 498). A simpler apparatus is shown in Figure 16.1; the air entering the storage vessel is passed over Ascarite—a solid absorbent consisting of sodium hydroxide deposited on asbestos. With this arrangement some contamination of the base by

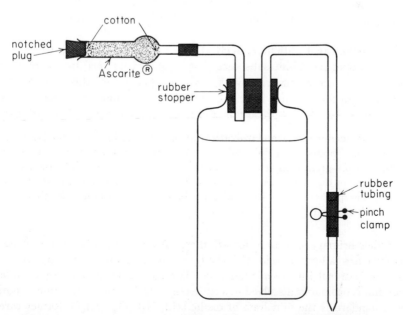

Fig. 16.1 Arrangement for the Storage of Standard Solutions of Base.

atmospheric carbon dioxide will occur while the solution is being transferred to the buret; the extent, however, is generally small. Finally, absorption by the solution in the open buret is also to be expected. Covering the top of the buret with a small beaker or test tube will minimize this.

Solutions of the alkalies should not be stored in glass-stoppered containers, nor should they be left in burets any longer than necessary because the stopper or the stopcock will tend to stick. Rubber stoppers are generally used with containers of alkali. Another problem is the slow attack of glass containers by solutions of base. Where the reagents are to be kept for extended periods, paraffin-lined bottles or polyethylene ware should be employed.

Preparation of carbonate-free 0.1 N NaOH.   Prepare a bottle for protected storage as in Figure 16.1. Boil approximately 1 liter of distilled water and cool to room temperature. Prepare a Gooch crucible and wash thoroughly (p. 122).

Use a platform balance to weigh approximately 10 grams of NaOH pellets into a 125-ml Erlenmeyer flask. Promptly introduce 10 ml of distilled water and swirl to hasten solution of the base (caution!). Allow flask and contents to cool; filter through the Gooch crucible and collect the clear filtrate in a clean test tube placed in the filter flask. Pour about 4 to 5 ml of this into the cooled water and mix thoroughly. Immediately transfer the solution to the storage bottle.

Standardization of Solutions of Base

Several excellent primary standards are available for alkali solutions. Most of these are weak organic acids that require the use of an indicator with a basic transition range.

Potassium acid phthalate, $KHC_8H_4O_4$.   The acid salt of phthalic acid is a nonhygroscopic crystalline substance that is readily obtained in a state of high purity. For the most accurate work, analyzed samples of this compound are available from the National Bureau of Standards. For most purposes, however, the commercial analytical-grade reagent can be used without further purification.

Procedure for standardization of 0.1 N NaOH.   Dry the $KHC_8H_4O_4$ for 1 to 2 hours at 110° C and cool. Accurately weigh (to the nearest 0.1 mg) 0.7- to 0.9-gram samples into 250-ml Erlenmeyer flasks and dissolve in 25 to 50 ml of distilled water (preferably freshly boiled and cooled). Add 2 drops of phenolphthalein and titrate with the base to the first pink color that persists for a half minute.

Other primary standards for alkalies.   Among the other primary standards for bases is benzoic acid, $HC_7H_5O_2$, a substance obtainable in pure form from the National Bureau of Standards. The compound is not very soluble in water and is ordinarily dissolved in a little ethyl alcohol before titration. Another useful standard is the dihydrate of oxalic acid, $H_2C_2O_4 \cdot 2H_2O$. Rather careful preparation and storage of this compound are necessary to ensure its containing exactly 2 mols of water per mol of acid. Potassium bi-iodate, $KH(IO_3)_2$, is an excellent nonhygroscopic primary standard of high equivalent weight. In contrast to the other standards mentioned, it is a strong acid; thus the standardization can be made with any indicator showing a color change in the pH range between four and ten.

SOME APPLICATIONS OF NEUTRALIZATION TITRATIONS

This section contains detailed descriptions of typical applications of neutralization analysis. In all of these experiments an acid-base indicator is used to estimate the equivalence point in the titration. Other means of detecting the equivalence point are available, however, the most important being based upon measurement of the potential of a pH-sensitive electrode immersed in the solution being titrated. Such measurements give the pH directly. From a plot of these data against the volume of added reagent, the equivalence points can be determined. This useful procedure is described in greater detail in Chapter 24; the instructor may choose to substitute this technique for the suggested indicator in one or more of the experiments.

A Preliminary Experiment

A useful preliminary exercise consists of determining the combining ratio for a solution of a strong acid and a strong base. This permits the student to become familiar with acid-base end points; additionally, the data obtained may be used to calculate the concentration of one of the reagents from the normality of the other.

> **Procedure.** Prepare approximately 0.1 N solutions of HCl and carbonate-free NaOH as instructed on page 343 and 348. Fill burets with each of the reagents, and place a small beaker or test tube over the one containing the base. Run a 35- to 40-ml portion of the acid into a 250-ml Erlenmeyer flask, touch the inside of the flask with the buret tip and rinse down with a little water. Add 2 drops of phenolphthalein and run in the NaOH until the solution turns distinctly pink. Now add the HCl dropwise until the solution is decolorized, rinse down the sides of the flask, and carefully add base until the faintest pink color that persists for 30 seconds is observed. In making the final adjustment, add fractional drops of the base by forming a part of a drop on the buret tip, touching this to the side of the flask, and rinsing down with distilled water. The end point will slowly fade as carbon dioxide is absorbed from the atmosphere.
>
> Repeat the experiment and calculate the ratio of the volume of acid to the volume of base. Duplicate runs should agree to within 2 to 3 parts per thousand.
>
> Repeat this experiment with an indicator that shows an acidic transition range, such as bromcresol green. In this particular case, adjust the solution until the faintest tinge of green is imparted to the solution.

Analysis of Weak Acids

The experiments in this section make use of a standard solution of a strong base. The preparation and standardization of this reagent were described on page 348. Although a solution of hydrochloric acid is not absolutely necessary for these titrations, it is very helpful. With this acid the end point may be located with greater certainty; in addition, an overtitrated sample can be saved. The combining capacity of the acid for the base must be determined as described in the previous section. In each analysis a weak acid is being titrated; thus, an indicator with a basic transition range must be employed.

Determination of potassium acid phthalate.   The sample is a mixture of potassium acid phthalate and a neutral salt; the percentage of the former is to be determined.

> **Procedure.**   Dry the sample for 2 hours at 110° C and weigh out suitable-sized samples into 250-ml Erlenmeyer flasks. Add 50 ml of boiled water and 2 drops of phenolphthalein to each. Titrate with standard NaOH to the first pale pink color that persists for 30 seconds. If a standard HCl solution is available, run in a slight excess of base and find the end point exactly by addition of small increments of acid and base.
>
> Calculate the percent potassium acid phthalate, $KHC_8H_4O_4$, in the sample. The duplicate analyses should agree to within 3 to 4 parts per thousand.

Determination of the equivalent weight of an acid.   The equivalent weight of a weak acid is useful for identification purposes; it is readily determined by titration of a weighed quantity of the pure compound.

> **Procedure.**   Weigh 0.3-gram samples of the purified acid into 250-ml flasks and dissolve in 25 to 50 ml of boiled water. In some cases heating may be required to dissolve the sample. If necessary, use ethyl alcohol or alcohol-water mixtures as a solvent.
>
> Add 2 drops of phenolphthalein and titrate with standard 0.100 N base to the first persistent pink color. Calculate the equivalent weight of the acid.

Determination of the acid content of vinegar.   The principal acid in vinegar is acetic acid which is present in concentrations of about 4 percent. The total acid content of the substance is readily obtained by titration with standard base. Ordinarily the results are reported as percent acetic acid despite the fact that other acids are present in small quantities.

> **Procedure.**   Pipet 25 ml of vinegar into a 250-ml volumetric flask and dilute to the mark with boiled and cooled water. Mix thoroughly and pipet 50-ml aliquots into 250-ml flasks. Add 50 ml of water, 2 drops of phenolphthalein, and titrate to the first permanent pink color with standard 0.1 N NaOH.
>
> Calculate the grams of acetic acid per 100 ml of sample.

Analysis of Carbonate Mixtures

An interesting application of neutralization titrations is found in the qualitative and quantitative determination of the constituents of a solution containing sodium carbonate, sodium bicarbonate, and sodium hydroxide alone or admixed. No more than two of these components can exist in appreciable concentrations in any solution, inasmuch as reaction will occur to eliminate the third. For example, the addition of sodium hydroxide to a solution of sodium bicarbonate results in sodium carbonate forming until either the sodium hydroxide or the sodium bicarbonate is consumed. In the former case, a mixture of bicarbonate and carbonate will result; in the latter, the solution will contain only carbonate and hydroxide ions in analytical concentrations.

In order to analyze solutions of this sort, two titrations with a standard acid are performed on identical samples. One titration employs an indicator such as phenolphthalein that has a transition interval at a pH of about 9. The other makes use of an indicator that has an acidic range— methyl orange, for example. As shown in Table 16-2, the constitution of the solution can then be judged from the relative volumes of acid required for the two titrations. After deduction of the qualitative constitution of the solution from the data, the concentrations of the component or components can be derived from the normality of the acid and the volumes of reagent.

Table 16-2

VOLUME RELATIONSHIP IN THE ANALYSIS OF
CARBONATE—BICARBONATE—HYDROXIDE MIXTURES

Constituents Present	Relationship between Volume of Acid to Reach a Phenolphthalein End Point, V_{ph}, and a Methyl Orange End Point, V_{mo}
NaOH	$V_{ph} = V_{mo}$
Na_2CO_3	$V_{ph} = \frac{1}{2} V_{mo}$
$NaHCO_3$	$V_{ph} = 0$
NaOH, Na_2CO_3	$V_{ph} > \frac{1}{2} V_{mo}$
Na_2CO_3, $NaHCO_3$	$V_{ph} < \frac{1}{2} V_{mo}$

Sample calculation. A solution contained one or more of the following constituents. Na_2CO_3, $NaHCO_3$, and NaOH. A 50.0-ml portion consumed 22.1 ml of 0.100 N HCl when titrated to a phenolphthalein end point. Methyl orange was then added to the solution and the titration continued, the solution being boiled near the second equivalence point in order to remove CO_2. An additional 26.3 ml of the HCl was used. We wish to know the formal composition of the solution.

Had the solution contained only NaOH, no appreciable volume of acid would have been required to go from the phenolphthalein to the methyl-orange end point (that is, $V_{ph} \cong V_{mo}$). On the other hand, if only $NaHCO_3$ had been present, the solution would have been acidic to the phenolphthalein at the outset, and acid would have been used only in the second titration. Finally, if Na_2CO_3 were the sole constituent, the volume of acid to reach the phenolphthalein end point would have been exactly one half the total volume required to achieve a methyl-orange end point (that is, $V_{ph} = \frac{1}{2} V_{mo}$). In fact, however, a total of 48.4 ml was required. Since less than half of this amount was involved in the first titration, the solution must have contained some $NaHCO_3$ in addition to Na_2CO_3. We can now calculate the concentration of the two constituents. When the phenolphthalein end point was reached, the CO_3^{2-} originally present was converted to HCO_3^-; thus we may say that

$$\text{number of milliformula weights } Na_2CO_3 = 22.1 \times 0.100 = 2.21$$

The titration from the phenolphthalein to the methyl-orange end point involved both the bicarbonate ion originally present and that formed by titration of the carbonate; thus

$$\begin{array}{l} \text{number of milli-} \quad \text{number of milli-} \\ \text{formula weights} + \text{formula weights} = 26.3 \times 0.1 \\ \qquad NaHCO_3 \qquad\qquad Na_2CO_3 \end{array}$$

Hence

$$\frac{\text{number of milliformula}}{\text{weights } NaHCO_3} = 2.63 - 2.21 = 0.42$$

From these data we calculate the formal concentrations:

$$F_{Na_2CO_3} = \frac{2.21}{50.0} = 0.0442 \text{ GFW/liter}$$

$$F_{NaHCO_3} = \frac{0.42}{50} = 0.0084 \text{ GFW/liter}$$

The analysis given in the above example is not entirely satisfactory because the first end point, involving the formation of the bicarbonate ion, cannot be fixed with certainty; no indicator will give a sharp end point over the small pH change associated with this equivalence point (see Fig. 15.3, p. 337). In order to minimize the titration error, the color of the solution being titrated can be matched against a comparison solution that contains an identical concentration of indicator and an approximately equivalent concentration of sodium bicarbonate. Even with this precaution, errors of 1 percent or more are to be expected.

The slight solubility of barium carbonate can be used to improve the titration of carbonate-hydroxide or carbonate-bicarbonate mixtures. The *Winkler* method for the determination of carbonate-hydroxide mixtures involves titration of both components in an aliquot using an indicator with an acidic transition range; this titration will closely resemble the standardization of an acid against sodium carbonate (p. 345). An excess of neutral barium chloride is then added to a second aliquot to remove all of the carbonate ion, following which the hydroxide ion is titrated to a phenolphthalein end point. If the concentration of the excess barium ion is approximately 0.1 M, the solubility of the barium carbonate will be too low to offer interference in this titration.

To achieve an accurate analysis of a carbonate-bicarbonate mixture, the total equivalence of the two components is measured by titration with a standard acid following the procedure for an acid standardization against carbonate (p. 345). The bicarbonate concentration is then determined in a second aliquot by the addition of a large excess of barium chloride and a measured excess of standard base. Precipitation of the carbonate ion from the original salt as well as that formed by reaction of the bicarbonate with the base leaves only the unreacted sodium hydroxide. This sodium hydroxide can be determined by a back titration with standard acid. Filtration of the barium carbonate is unnecessary if an indicator such as phenolphthalein is employed; the end point is easily discerned in the presence of the white solid.

Analysis of a sample of impure sodium carbonate. The method given below is for a sample containing sodium carbonate and various neutral salts.

Procedure. Dry the sample for 2 hours at 150° C and then cool in a desiccator. Weigh samples of the proper size (see instructor) into 250-ml Erlenmeyer flasks. Dissolve in 25 to 50 ml of boiled water and add 2 drops of bromcresol-green indicator. Titrate with standard 0.1 N HCl until the indicator just begins to change to its green color. Boil the solution for 2 to 3 minutes, cool, and complete the titration. If additional acid is not required after boiling, the end point may have been passed. In this case discard the sample; alternatively, determine the excess acid by a back titration with a standard alkali solution.

Calculate the percentage of sodium carbonate in the sample.

This analysis is conveniently carried out at the same time as the standardization of the acid (see p. 345).

Analysis of a sodium carbonate—sodium bicarbonate mixture. The following instructions are for a sample containing both sodium carbonate and sodium bicarbonate as well as inert neutral salts.

Procedure. If the sample is a solid, weigh dried portions and dissolve in 25 to 50 ml of water. Transfer the solutions quantitatively to 250-ml volumetric flasks and dilute to the mark with boiled and cooled water. If the sample is a solution, pipet suitable portions into 250-ml volumetric flasks and dilute to volume. Mix thoroughly.

To determine the total number of milliequivalents of the two components, transfer a 25-ml aliquot of each of the solutions to 250-ml Erlenmeyer flasks and titrate with standard 0.1 N HCl following the directions given in the previous section for a sodium carbonate sample.

To determine the bicarbonate content of the sample, pipet a second 25-ml portion of each solution into an Erlenmeyer flask and add a carefully measured volume of standard 0.1 N NaOH to the solution (conveniently 50.00 ml). Immediately add 10 ml of 10-percent $BaCl_2$ and 2 drops of phenolphthalein indicator. Titrate the excess NaOH at once with a standard 0.1 N HCl solution to the disappearance of the pink color. Titrate a blank consisting of 25 ml of water, 10 ml of the $BaCl_2$ solution, and *exactly* the same volume of NaOH as used with the samples. The difference in volume of HCl for the blank and sample corresponds to the $NaHCO_3$ present.

Calculate the percentage of $NaHCO_3$ and of Na_2CO_3 in the sample.

Analysis of a sodium carbonate—sodium hydroxide mixture.   The sample is an aqueous solution containing sodium carbonate and sodium hydroxide.

Procedure. Transfer the sample to a 250-ml volumetric flask immediately upon receipt and dilute to the mark with boiled and cooled distilled water. Keep the flask tightly stoppered to avoid absorption of CO_2.

Pipet 25-ml aliquots of the sample solution into 250-ml Erlenmeyer flasks, add 2 drops of bromcresol green, and titrate with 0.1 N HCl following the procedure given on page 353 for the analysis of carbonate. This will give the total number of milliequivalents of NaOH and Na_2CO_3.

To determine the quantity of NaOH present, transfer a 25-ml aliquot of the sample solution into a 250-ml Erlenmeyer flask. Slowly add 10 ml of a neutral 10-percent $BaCl_2$ solution and 2 drops of phenolphthalein. Titrate immediately with standard 0.1 N HCl. To avoid contamination of the samples by atmospheric CO_2, complete this part of the analysis as rapidly as possible.

Calculate the weight of NaOH and Na_2CO_3 in the original sample.

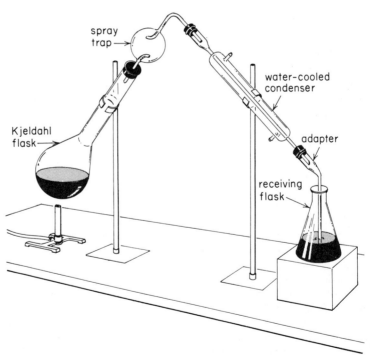

Fig. 16.2 Apparatus for the Distillation and Collection of Ammonia.

Analysis of Ammonium Salts

Neutralization methods provide a rather simple means for determining the concentration of ammonium salts. Three steps are involved in the process: first the compound is decomposed with an excess of strong base; the liberated ammonia is then distilled from the mixture and collected quantitatively; finally the amount of ammonia is determined by neutralization titration. Since few substances other than ammonium salts give a volatile basic compound under the conditions employed, the method is relatively specific.

Distillation of ammonia. Figure 16.2 illustrates a distillation apparatus for the analysis of ammonia. The long-necked, round-bottomed flask, called a Kjeldahl flask, is connected to a spray trap that serves to prevent small droplets of the strongly alkaline solution from being carried over in the vapor stream. A water-cooled condenser is provided; during distillation the end of the adapter tube extends below the surface of an acidic solution in the receiving flask.

Titration of ammonia. One method of completing the analysis is to place a measured quantity of standardized strong acid in the receiver flask of the apparatus. After the distillation is complete, the excess acid is back titrated with a standard solution of base, using an indicator with an acidic transition interval. An indicator with a neutral or basic range is not suitable because of the presence of the ammonium salt.

A convenient modification of the procedure—it requires only a single standard reagent—employs an unmeasured quantity of very weak acid solution to retain the ammonia; a 4-percent solution of boric acid ($K_a = 5.8 \times 10^{-10}$) serves this purpose very well.

$$NH_3 + HBO_2 \rightleftharpoons NH_4^+ + BO_2^-$$

Because it is the salt of a very weak acid, the borate formed may be titrated with a standard solution of hydrochloric acid.

$$BO_2^- + H^+ \rightleftharpoons HBO_2$$

At the equivalence point, the solution contains boric acid and ammonium chloride; an indicator with an acidic transition interval is therefore required.

> **Procedure.** The sample should contain 2 to 4 milliequivalents of ammonium salt. Introduce the sample into a 500-ml Kjeldahl flask and add enough water to give a total volume of about 200 ml.
>
> Set up a distillation apparatus similar to that shown in Figure 16.2. Use a buret or pipet to measure precisely 50 ml of 0.1 N HCl into the receiver flask; clamp the flask so that the tip of the adapter reaches just below the surface of the standard acid. Start the cooling water through the jacket of the condenser.
>
> For each sample prepare a solution of 45 grams of NaOH in about 75 ml of water. Cool this solution to room temperature before use. With the

Kjeldahl flask tilted, slowly pour the caustic down the side of the container so that little mixing occurs with the solution in the flask (Note 1). Add several pieces of granulated zinc (Note 2) and a small piece of litmus paper. *Immediately* connect the flask to the spray trap; very *cautiously* mix the solution by gentle swirling. After mixing is complete, the litmus paper should indicate that the solution is basic.

Immediately bring the solution to a boil and distill at a steady rate until one half to one third of the original solution remains. Watch the rate of heating during this period to prevent the receiver acid from being sucked back into the distillation flask. After the distillation is judged complete, lower the receiver flask until the tip of the adapter is well out of the standard acid; then remove the flame, disconnect the apparatus, and rinse the inside of the condenser with a small amount of water. Disconnect the adapter and rinse it thoroughly. Add 2 drops of bromcresol green or methyl red and titrate to the color change of the indicator.

This procedure can be modified to use about 50 ml of 4-percent boric acid solution in place of the standard HCl in the receiver flask. The distillation is then carried out in an identical fashion; when complete, the ammonium borate present is titrated with a standard 0.1 N HCl solution using 2 to 3 drops of bromcresol-green indicator.

Calculate the percent nitrogen in the sample.

NOTES:

1. The more dense caustic solution should form a second layer on the bottom of the flask. Mixing is avoided at this point in order to prevent loss of the volatile ammonia; manipulations should be carried out as rapidly as possible.

2. The granulated zinc is added to reduce bumping during the distillation. It reacts slowly with the alkali to give small bubbles of hydrogen gas that minimize superheating of the liquid.

Analysis of nitrates.   An important modification of the method just discussed makes possible the analysis of inorganic nitrates. Here the nitrate is reduced to an ammonium salt which is then determined by the distillation and titration procedure. The most common reducing agent for this purpose is Devarda's alloy which consists of 50-percent copper, 45-percent aluminum, and 5-percent zinc. The finely powdered metal is introduced into a strongly alkaline solution of the sample; the ammonia is distilled after allowing time for the evolution of hydrogen to cease. Arnd's alloy, which is 60-percent copper and 40-percent magnesium can also be used for the reduction.

Determination of Nitrogen in Organic Compounds; the Kjeldahl Method

Nitrogen occurs in a variety of important organic substances including proteins, peptides, synthetic drugs, and fertilizers. As a consequence, procedures for the analysis of this element find widespread use in industry and research.

Methods for nitrogen analysis.   Basically two methods are employed for the determination of nitrogen in organic compounds—the *Dumas* and the

Kjeldahl methods. The former, which is suitable for the analysis of nearly all types of organic nitrogen compounds, consists of mixing the sample with powdered cupric oxide and igniting in a stream of carbon dioxide gas in a combustion tube. At elevated temperatures the organic substance is oxidized to carbon dioxide and water by the cupric oxide. Any nitrogen in the compound is converted primarily to the elemental state although nitrogen oxides may also result. These oxides are reduced to elemental nitrogen by passing the gases over a bed of hot copper. The products of the ignition are then swept into a gas buret filled with highly concentrated potassium hydroxide; this completely absorbs the carbon dioxide, water, and other products of the combustion, such as sulfur dioxide and hydrochloric acid. The elemental nitrogen remains undissolved in the caustic, however, and its volume is directly measured.

The second method, for which there are a number of modifications, bears the name of its inventor, Kjeldahl. The procedure, first conceived in 1883, has undoubtedly been one of the most widely used of all analyses. In its original form, oxidation of the sample was accomplished with hot concentrated sulfuric acid; it was presumed that the organic nitrogen was completely converted to ammonium sulfate. The latter compound was then decomposed with strong base, the liberated ammonia being distilled into a standard acid solution where it was determined by back titration with standard base.

The Kjeldahl procedure is in much wider use than the Dumas method because it requires no special equipment and because it is more easily adapted to the routine analysis of large numbers of samples. It has become the standard method for determining the protein nitrogen of grains, meats, and other biological materials.

Sample oxidation in the Kjeldahl method. Undoubtedly the most critical step in the Kjeldahl procedure is the sulfuric acid oxidation of the organic compound. During this operation the carbon in the sample is converted to carbon dioxide and the hydrogen to water; the fate of the nitrogen is, however, highly dependent upon the form in which it occurs in the original compound. Where the element is bound as an amide or an amine, as in proteinaceous materials, quantitative conversion to ammonium ion nearly always results, and it is retained as such in the sulfuric acid. On the other hand, where the nitrogen occurs in a more highly oxidized form, such as in nitro, azo, or azoxy groups, losses of the element result because the oxidation product is elemental nitrogen or nitrogen oxides; certain heterocyclic nitrogen compounds behave similarly. In these cases, the Kjeldahl procedure leads to erroneously low results unless the precaution of introducing a reducing agent prior to the digestion step is taken; this assures conversion of the element to an oxidation state that gives ammonium ion upon treatment with the sulfuric acid. One method for prereduction calls for the addition of sodium thiosulfate and salicylic acid to the concentrated sulfuric acid solution containing the sample; the digestion is then carried out in the usual way.

Rate of the Kjeldahl oxidation. The most time-consuming step in the Kjeldahl analysis is the oxidation process, which may require several hours to complete; numerous modifications have been proposed to improve the kinetics of the process. One such modification, involving the introduction of a neutral salt such as potassium sulfate, is now almost universally employed. This very simple expedient, proposed by Gunning in 1899, is effective because it raises the boiling point of the sulfuric acid and thus the temperature at which the oxidation can be carried out. Care is necessary, however, for oxidation of the ammonium ion occurs if the salt concentration is too great—a serious problem if evaporation of the sulfuric acid is excessive during digestion.

Catalysts of many sorts have been suggested as a means of hastening the digestion process, the most common of these including elemental mercury, copper, selenium, or compounds of these elements. Mercury or mercury compounds appear to be the most effective. The ions of mercury or copper, if present, should be precipitated as sulfides prior to the distillation step; otherwise some ammonia remains as an ammine complex of the metallic ion.

Many attempts to hasten the Kjeldahl oxidation by the addition of stronger oxidizing agents such as perchloric acid, potassium permanganate, and hydrogen peroxide have failed because oxidation of the ammonium ion to volatile nitrogen oxides occurs.

The procedure given below is for the analysis of protein nitrogen in samples such as dried blood or wheat flour. Prereduction is unnecessary. A simple modification will allow the analysis of samples that contain oxidized forms of nitrogen[4].

Procedure.   Weigh out three samples of 0.25 to 2.5 grams, depending upon the nitrogen content, and wrap each in a 9-cm filter paper. Drop each into a 500-ml Kjeldahl flask (the paper wrapping will prevent the sample from clinging to the neck of the flask). Add 25 ml of concentrated sulfuric acid and 10 grams of powdered K_2SO_4. Add catalyst (Note 1) and clamp the flask in an inclined position in a hood. Heat the mixture carefully until the H_2SO_4 is boiling. Continue the heating until the solution becomes colorless or light yellow; this may take as much as 2 to 3 hours. If there is much reduction in the volume of acid, replace that lost by evaporation.

Remove the flame and allow the flask to cool; swirl the flask if the contents begin to solidify. Cautiously dilute with 200 ml of water and cool under a water tap to room temperature. If mercury or copper was used as the catalyst, introduce 25 ml of a 4-percent sodium sulfide solution. Proceed as in paragraph 2 of the procedure on page 355 for the analysis of an inorganic ammonium salt.

Calculate the percent nitrogen in the samples.

NOTES:

1. As a catalyst, one may use a drop of mercury, 0.5 gram of HgO, a crystal of $CuSO_4$, 0.1 gram of Se, etc; or the catalyst may be omitted.

2. For best results, a blank should also be carried through the digestion and distillation steps.

[4] See *Official and Tentative Methods of Analysis*, 5th ed., 27. Washington, D.C.: Association of Official Agricultural Chemists, 1940.

ACID-BASE TITRATIONS IN NONAQUEOUS SOLVENTS[5]

Since accurate titrations are feasible only in those instances where the analytical reaction is relatively complete, volumetric methods are limited to those weak acids and bases having dissociation constants greater than about 10^{-8} in aqueous solution. Many acids or bases that are too weak for determination in an aqueous medium, however, become susceptible to titration in appropriate nonaqueous solvents. As a consequence, there is now an imposing list of neutralization methods that call for solvents other than water.[6] Unfortunately current empirical knowledge of these applications has far outstripped the fund of basic information regarding the properties of acids and bases in these media. As a result, quantitative treatment of nonaqueous titrations analogous to that presented for aqueous sytems in previous chapters is seldom possible. On the other hand, a good qualitative understanding of acid-base processes in nonaqueous solvents can be gained with the aid of the Brønsted theory (see pp. 11 to 14).

Solvent Classification

We shall distinguish among three types of solvent behavior. *Amphiprotic* solvents are those which possess both acidic and basic properties as a result of self-dissociation, or *autoprotolysis*. Although water is the most common amphiprotic solvent, many other substances exhibit analogous behavior; note, for example, the resemblance among the following equilibria:

$$2H_2O \rightleftharpoons H_3O^+ + OH^-$$

$$2C_2H_5OH \rightleftharpoons C_2H_5OH_2^+ + C_2H_5O^-$$

$$2HOAc \rightleftharpoons H_2OAc^+ + OAc^-$$

or, in general,

$$2SH \rightleftharpoons SH_2^+ + S^-$$

where SH represents the amphiprotic solvent molecule and SH_2^+ the solvated proton.

A second class, known as *aprotic* or inert solvents, has no appreciable acidic or basic properties and does not undergo autoprotolysis to any detectable extent. Benzene, carbon tetrachloride, and pentane fall into this category.

Finally, there are a number of solvents with basic properties but essentially no acidic tendencies. These include ketones, ethers, esters, and pyridine. Such solvents do not undergo autoprotolysis.

[5] For a good discussion of the application of nonaqueous titrations see J. S. Fritz and G. S. Hammond, *Quantitative Organic Analysis*, Chap. 3. New York: John Wiley and Sons, Inc., 1957.

[6] See J. A. Riddick, *Anal. Chem.*, **24**, 41 (1952); **26**, 77 (1954); **28**, 679 (1956); **30**, 793 (1958); **32**, 172R (1960).

Titration of Weak Bases

In an amphiprotic solvent the analytical reaction occurring in the titration of a weak base may be expressed as follows :

$$B \; + \; SH_2^+ \; \rightleftharpoons \; BH^+ \; + \; SH$$

$$\underset{\substack{\text{base}_1 \\ \text{(substance} \\ \text{titrated)}}}{} \quad \underset{\substack{\text{acid}_1 \\ \text{(titrant)}}}{} \qquad \underset{\text{acid}_2}{} \qquad \underset{\substack{\text{base}_2 \\ \text{(solvent)}}}{}$$

Where the solvent is water, the equation takes the familiar form

$$B + H_3O^+ \rightleftharpoons BH^+ + H_2O$$

and where it is glacial acetic acid, this becomes

$$B + H_2OAc^+ \rightleftharpoons BH^+ + HOAc$$

The position of equilibrium in these reactions depends on the relative acidic or basic strengths of the reactants and products. If base$_1$ and acid$_1$ are strong, and if base$_2$ and acid$_2$ are weak, the equilibrium will lie far to the right; if the reverse is true, the equilibrium will lie far to the left. The reader will recall that the ease and accuracy of equivalence-point detection increase as analytical reactions become more complete. Thus the best reagent for the titration of a weak base should be a strong acid which, when dissolved, is completely converted to the solvated proton. Another expedient is to employ a less basic solvent; this would have the effect of rendering base$_2$ in the preceding equation weaker, which in turn would result in a more complete reaction. This is an effective approach. For example, a weak base must have a dissociation constant greater than about 10^{-8} for successful titration in aqueous solution. However, accurate determination of compounds with basic dissociation constants (in water) of 10^{-12} and smaller is possible with a less basic solvent such as glacial acetic acid. Other acidic solvents would operate in the same manner. Inert solvents such as dioxane or benzene, being less basic than water, also serve in the analysis of very weak bases. Solvents more basic than water should, of course, be avoided.

Titrations in glacial acetic acid. Much work has been done with glacial acetic acid as a solvent for the determination of weak bases. Commonly, standard perchloric acid solutions are used as the titrant. These are prepared by dissolving the 72-percent aqueous reagent in glacial acetic acid; enough acetic anhydride is then introduced to react with all of the water, giving more acetic acid.

End points can be determined with suitable acid-base indicators chosen largely on the basis of empirical knowledge of their behavior in the solvent. More commonly, a pH-sensitive electrode system is employed, the end point being determined from potential measurements (see Chap. 24). In most cases, the ordinary glass electrode is a satisfactory indicator electrode.

Nonaqueous titrations in glacial acetic acid or in mixtures of acetic acid and various inert solvents are used for the determination of a variety of primary, secondary, and tertiary amines. Many aromatic amines as well as heterocyclic nitrogen compounds (such as pyridine, quinoline, and the alkaloids) are also titrated successfully. Amino acids are determined in the solvent, the carboxylic acid group being inert in the presence of the large excess of acetic acid.

A number of weak-acid salts that react incompletely in aqueous solutions (salts with hydrolysis constants smaller than 10^{-7}) are titrated with standard perchloric acid in acetic acid. The analytical reaction can be written as follows:

$$X^- + H_2OAc^+ \rightleftharpoons HX + HOAc$$
$$\text{base}_1 \quad \text{acid}_1 \quad \text{acid}_2 \quad \text{base}_2$$

Reactions of this sort are generally more complete than the corresponding aqueous reactions because acetic acid is a weaker base than water. Thus titration of the alkali salts of many carboxylic acids (such as sodium acetate, citrate, benzoate, tartrate, etc.) is successful in this medium whereas in water it is not.

Titrations in other solvents. Other solvents employed for the titration of very weak bases include trifluoroacetic acid, acetic anhydride, acetonitrile, and various mixtures of glycols with a hydrocarbon solvent. The last have proven particularly useful for the analysis of alkali-metal salts of many organic acids. Ordinarily the solvent consists of a 1:1 mixture of ethylene or propylene glycol and an aliphatic alcohol, a hydrocarbon, or a chlorinated hydrocarbon. Most salts are readily soluble in such mixtures and can be titrated with a standard solution of perchloric acid in the solvent.

Titration of Weak Acids

We have seen that titration of very weak bases is successful in solvents less basic than water. As might be expected, very weak acids, which also do not give satisfactory end points in aqueous solution, can be analyzed in a solvent that is less acidic than water. The equation for the neutralization reaction in water is as follows:

$$HA + OH^- \rightleftharpoons H_2O + A^-$$
$$\text{acid}_1 \quad \text{base}_1 \quad \text{acid}_2 \quad \text{base}_2$$

Although water is a relatively weak acid, its concentration in aqueous solution is so great that it tends to inhibit completion of the reaction. Thus, if the water is replaced wholly or in part by some less acidic solvent, a shift in the above equilibrium to the right is to be expected.

Solvents and reagents for titrations of acids. Perhaps the best solvents for titrations of weak acids are pyridine, dimethylformamide, ethylenediamine, and acetonitrile. Aliphatic alcohols, as well as butylamine and acetone, have also proven useful.

Several strong bases are used as standard reagents. Potassium hydroxide, for example, can be dissolved in ethanol or in one of the other aliphatic alcohols. Sodium or potassium methoxide in methanol or methanol-benzene mixtures is also employed to give standard solutions as concentrated as 0.1 N. Such a reagent is a strong base, its reaction with weak acids being

$$HA + OCH_3^- \rightleftharpoons HOCH_3 + A^-$$

Sodium methoxide is also used in conjunction with such solvents as ethylene-diamine, butylamine, and dimethylformamide. It does need protection from water vapor and from atmospheric carbon dioxide.

Solutions of the strong base tetrabutylammonium hydroxide, $(C_4H_9)_4NOH$, may be prepared in benzene-methanol, isopropyl alcohol, or ethyl alcohol to give a standard base suitable for nonaqueous titrations.

Applications. Empirically chosen acid-base indicators may be employed for titrations of weak acids in nonaqueous media. The potentiometric end point (Chap. 24) with a glass or antimony indicator electrode is also useful.

Most carboxylic acids are easily determined in nonaqueous solvents. Many phenols are also susceptible to analysis by the procedure. A number of salts of weak bases, including ammonium salts as well as those of aliphatic and aromatic amines, give satisfactory end points when titrated in ethylenediamine or dimethylformamide. Certain enols and imides can also be titrated in these more basic solvents.

problems

1. Describe the preparation of the following solutions:
   (a) 5 liters of approximately 0.01 N $Ba(OH)_2$
   (b) 500 ml of approximately 0.3 N HCl from the concentrated reagent (sp gr = 1.18; percent HCl = 37)
   (c) 2 liters of about 0.2 N H_2SO_4 from the concentrated reagent (sp gr = 1.83; percent H_2SO_4 = 96)
   (d) 10 liters of approximately 0.05 N KOH from the solid reagent
   (e) 5 liters of 0.2 N carbonate-free NaOH from a 50-percent solution of the reagent (sp gr = 1.53)
   (f) 1 liter of about 0.1 N NaOH from metallic Na
   (g) 1 liter of exactly 0.0500 N $HClO_4$ from a 0.275 N solution of the acid

   ans. (a) Dissolve 4.3 grams $Ba(OH)_2$ in 5 liters of water
   (b) Dilute 12.5 ml of the reagent to 500 ml

2. A constant-boiling solution of HCl was prepared by distillation of the concentrated reagent at a pressure of 770 mm of Hg. How many grams of this mixture should be taken to prepare exactly 2.000 liters of 0.1500 N HCl? (see Table 16-1)   ans. 54.16 grams

3. Describe the preparation of exactly 1.000 liter of 0.2500 N HCl from constant-boiling HCl distilled at 735 mm pressure. (See Table 16-1)

4. A solution of $HClO_4$ was standardized against 0.2127 gram of pure Na_2CO_3. Exactly 35.00 ml of the acid was added, the solution boiled, and the excess acid back titrated with 1.84 ml of NaOH. In a separate titration, 27.10 ml of the NaOH was found to be equivalent to 25.00 ml of the $HClO_4$. Calculate the normality of the acid and the base.   ans. 0.1205 N $HClO_4$; 0.1112 N NaOH

5. A solution of HCl was standardized against 0.1750 gram of pure Na_2CO_3. Exactly 47.00 ml of the acid was added to the carbonate and the solution boiled to remove the CO_2. The excess acid was back titrated with 2.12 ml of

0.01740 N NaOH to a bromcresol-green end point. What was the normality of the acid? ans. 0.06947 N HCl

6. Exactly 2.000 grams of Na_2CO_3 were weighed out, dissolved in water, and diluted to 250.0 ml. A 50.05-ml aliquot of this solution was titrated with a solution of HCl; CO_2 was removed by boiling near the end point. Calculate the normality of the acid if 30.15 ml were required to produce a methyl-orange end point.

7. A second 50.05-ml aliquot of the Na_2CO_3 solution in Problem 6 was treated with 35.00 ml of a solution of $HClO_4$. After boiling, the excess acid was titrated with 1.45 ml of a solution of NaOH (1.000 ml NaOH = 1.097 ml of the $HClO_4$). Calculate the normality of the $HClO_4$.

8. Exactly 34.10 ml of a solution of HCl was required to titrate 0.5000 gram of borax, $Na_2B_4O_7 \cdot 10\ H_2O$. What was the normality of the acid?

9. A 0.3320-gram sample of $Na_2C_2O_4$ was decomposed to Na_2CO_3 and titrated with 24.76 ml of HCl to a methyl-orange end point. What is the normality of the acid?

10. A 0.2031-gram sample of potassium acid phthalate required 47.63 ml of a $Ba(OH)_2$ solution to reach a phenolphthalein end point. What was the normality of the base?

11. A 1.00-gram sample of a carbonate was analyzed by titration with 0.202 N HCl. The solution was boiled near the end point to remove CO_2. (a) Calculate the percent CO_2 in the sample if 24.2 ml of acid were required. (b) What would be the error in parts per thousand if the end point were exceeded by 0.15 ml?

ans. (a) 10.8 percent
(b) 6.2 parts/1000

12. Suggest a range of sample weights for the standardization of 0.2 N NaOH against potassium acid phthalate.

13. 50.0 ml of a solution of HCl yielded 0.964 gram of a pure precipitate of AgCl.
 (a) What was the normality of the acid?
 (b) How many ml of this acid would be required to prepare 1 liter of 0.0500 N acid?

14. A 50.0-ml sample of vinegar (density = 1.06) was diluted to 250 ml and a 25.0-ml aliquot titrated with 0.136 N NaOH to a phenolphthalein end point. If 28.1 ml of base were required, what was the percent acetic acid in the sample?

15. A 0.0770-gram sample of a pure amine was dissolved in glacial acetic acid and titrated with a 0.0820 N solution of $HClO_4$ in acetic acid. If 10.1 ml of acid were required, what was the equivalent weight of the organic compound?

ans. 93 grams/equivalent

16. A 1.500-g sample of substance containing $CaCO_3$ was boiled with 50.0 ml of a solution of HCl. The excess acid was back titrated with 13.1 ml of 0.292 N NaOH. From a separate titration it was learned that 1.000 ml of the HCl was equivalent to 1.407 ml of the NaOH.
 (a) Calculate the percent $CaCO_3$ in the sample.
 (b) Calculate the percent CaO in the sample.
 (c) What volume of CO_2 in ml (STP) was liberated from the sample?

17. A 0.167-gram sample of pure organic acid was titrated with 15.2 ml of 0.202 N NaOH. What was the equivalent weight of the acid?

18. What weight of sample should be taken for the analysis of Na_2CO_3 so that the volume of 0.2760 N $HClO_4$ used will equal the percent Na_2CO_3 in the sample ?
    ans. 1.46 grams

19. Calculate the percent NH_3 in a crude ammonium salt that was analyzed by treating a 0.500-gram sample with concentrated NaOH and distilling the NH_3 into 40.0 ml of 0.116 N HCl. The excess HCl was back titrated with 0.97 ml of 0.143 N NaOH.

20. Compute the percent $(NH_4)_2SO_4$ in the sample described in Problem 19.

21. A 0.746-gram sample of crude urea, $CO(NH_2)_2$ was analyzed by the Kjeldahl procedure. The liberated NH_3 was collected in 50.0 ml of 0.200 N H_2SO_4. The excess acid required 4.24 ml of 0.100 N base. Calculate the percent urea in the sample.

22. A 1.00-gram sample of an organic mixture containing the compound $C_6H_{12}N_4$ was analyzed by the Kjeldahl procedure. The liberated NH_3 was collected in boric acid and titrated with 27.1 ml of 0.320 N HCl. Calculate the percent $C_6H_{12}N_4$ (GFW = 140) in the sample.

23. A 0.500-gram organic sample was burned in a stream of O_2 in a combustion tube. The SO_2 formed was collected in a solution of H_2O_2 which resulted in quantitative formation of H_2SO_4. The latter was titrated with 37.7 ml of 0.262 N NaOH. Calculate the percent S in the sample.

24. A fertilizer sample is to be analyzed for nitrogen by the Kjeldahl method using boric acid to absorb the ammonia. Assuming the sample contains 5-percent N what range of sample weights could be taken in order that the volume of 0.150 N HCl used would be between 25 and 40 ml ?

25. A 100-ml sample of a natural water when titrated to a methyl-orange end point consumed 31.2 ml of 0.0138 N acid. A similar sample consumed 1.80 ml of the acid to reach a phenolphthalein end point. Calculate the mg per liter of CO_3^{2-} and HCO_3^- in the water.
    ans. 14.9 mg/liter CO_3^{2-} ; 232 mg/liter HCO_3^-

26. A series of solutions known to contain one or more of the following was analyzed: Na_2CO_3, $NaHCO_3$, and NaOH. 100-ml aliquots of the solution were titrated with 0.100 N HCl. Given below are the volumes of acid required to reach the phenolphthalein and the methyl-orange end points, respectively, in each case. Calculate the grams per liter of each constituent.
    (a) 27.3 and 30.7   ans. 0.36 gram/liter Na_2CO_3; 0.96 gram/liter NaOH
    (b) 0.0 and 18.1   ans. 1.52 grams/liter $NaHCO_3$
    (c) 4.60 and 12.2
    (d) 27.1 and 27.1
    (e) 15.1 and 30.2
    (f) 15.6 and 48.1
    (g) 20.2 and 49.6

27. A 5.000-g sample of a mixture containing Na_2CO_3, $NaHCO_3$, and inert substances was dissolved in water and diluted to exactly 250 ml. A 25.0-ml aliquot was treated with 50.0 ml of 0.120 N HCl and boiled to remove CO_2. The excess acid was back titrated with 7.60 ml of 0.100 N NaOH to a bromcresol-green end point. A 50.0-ml aliquot was then treated with 50.0 ml of

the base and an excess of neutral $BaCl_2$ added which precipitated all of the carbonate as $BaCO_3$. The excess base was back titrated to a phenolphthalein end point with 26.2 ml of the 0.120 N acid. Calculate the percent $NaHCO_3$ and Na_2CO_3 in the sample.

ans. percent $Na_2CO_3 = 45.6$

percent $NaHCO_3 = 15.6$

28. A 3.50-gram sample containing both $NaOH$ and Na_2CO_3 was dissolved and diluted to 250 ml. A 50.0-ml aliquot required 41.7 ml of 0.0860 N HCl to reach a methyl-orange end point. A second 50.0-ml aliquot was treated with an excess of $BaCl_2$ to precipitate all of the carbonate. The solution then required 7.60 ml of the acid to reach a phenolphthalein end point. Calculate the percentage of each of the constituents.

29. A solution containing both H_3PO_4 and HCl was analyzed by titration of two 25.0-ml aliquots with 0.0500 N base. To reach an end point at a pH of 4 to 5 required 28.6 ml, while 46.1 ml were required to reach an end point at pH 9 to 10. What was the formal concentration of each of the acids in the sample?

30. A solution containing both H_3PO_4 and NaH_2PO_4 was titrated in the manner described in Problem 29. Two 25-ml aliquots required 12.2 and 39.1 ml of 0.0500 N base to reach the end points at pH 4 to 5 and 9 to 10 respectively. What were the formal concentrations of the two constituents?

31. A series of solutions known to contain one or more of the following constituents was titrated with 0.100 N base: HCl, H_3PO_4, and NaH_2PO_4. The data shown below give the volumes of reagent required to reach an end point at pH 4 to 5 (product NaH_2PO_4) and pH 9 to 10 (product Na_2HPO_4) when 25.0-ml aliquots were titrated. Calculate the composition of each solution in milligrams per 100 ml.
    (a) 10.2 and 20.4
    (b) 10.2 and 10.2
    (c) 10.2 and 30.7
    (d) 0.0 and 19.1
    (e) 20.2 and 27.6

32. A sample containing both HPO_4^{2-} and $H_2PO_4^-$ was analyzed by titration of two 25.0-ml portions. The first was titrated with 0.197 N HCl using methyl-orange indicator until the color matched that of a solution containing the same amount of indicator and KH_2PO_4. 17.4 ml of acid were required. The second aliquot was titrated with 0.174 N NaOH using thymolphthalein as an indicator. The end point was taken when the color matched that of a solution of Na_2HPO_4 containing the indicator. 30.7 ml of base were required. Calculate the formal concentrations of HPO_4^{2-} and $H_2PO_4^-$.

chapter 17. *Equilibrium in Oxidation-reduction Systems*

Many chemical processes are characterized by the transfer of electrons between the reacting species; these are known as *oxidation-reduction*, or *redox*, reactions. More volumetric analytical methods are based upon reactions of this type than any other; for this reason we shall examine the application of the equilibrium law to these processes in some detail.

Oxidation involves the loss of electrons by a substance, and *reduction* is the process wherein electrons are gained. In any oxidation-reduction reaction the ratio of mols of substance oxidized to mols of substance reduced is such that the number of electrons lost by the one species is equal to the number gained by the other. This fact must always be taken into account when balancing equations for oxidation-reduction reactions.

Separating an oxidation-reduction reaction into its two component parts—that is, into *half reactions*—is a convenient way to indicate clearly which species gains electrons and which loses them. The electrons symbolized in these equations are, of course, cancelled out when the two half reactions are combined. For example, consider the oxidation-reduction reaction that occurs when a piece of metallic zinc is immersed in a solution of cupric sulfate. The over-all reaction is expressed by the equation

$$Zn + Cu^{2+} \rightleftharpoons Zn^{2+} + Cu$$

which can be resolved into two half reactions, the one depicting the oxidation of zinc metal

$$Zn \rightleftharpoons Zn^{2+} + 2e$$

and the other indicating the reduction of cupric ions

$$Cu^{2+} + 2e \rightleftharpoons Cu$$

The same rules apply to balancing equations for half reactions as for ordinary reactions; that is, the number of atoms of each element as well as the number of charges on either side of the equation must be equal.

The tendency of various substances to gain or lose electrons differs immensely. Those substances possessing a strong affinity for electrons cause other substances to be oxidized by abstracting electrons from them. Reagents that function in this manner are called *oxidizing agents*. In acting as oxidizing agents the substances are themselves reduced. In the above reaction, for example, cupric ion acts as the oxidizing agent and is consequently reduced to metallic copper. Similarly potassium permanganate, a strong oxidizing agent, functions in acid solution by being reduced to manganous ion

$$MnO_4^- + 8H^+ + 5e \rightleftharpoons Mn^{2+} + 4H_2O$$

Reducing agents are substances that readily give up electrons thereby causing some other species to be reduced. In this process the reducing agent is itself oxidized. Metallic zinc acts as a reducing agent in the presence of cupric ion; ferrous iron often behaves similarly

$$Fe^{2+} \rightleftharpoons Fe^{3+} + e$$

Most of the oxidation-reduction reactions that we shall consider are reversible and thus subject to the law of mass action. The position of equilibrium in any case is determined by the relative tendencies of the reactants to acquire or lose electrons. Thus the mixture of a strong oxidizing agent with a strong reducing agent will lead to an equilibrium in which the products are overwhelmingly favored. A less complete reaction results when weaker reagents react. We can measure the relative tendencies of substances to gain or lose electrons. Since these data constitute a quantitative measure of the driving force for individual half reactions, they are invaluable for the calculation of equilibrium constants for oxidation-reduction processes.

Oxidation-reduction reactions may result from the direct transfer of electrons from the donor to the acceptor. Thus, upon immersing a piece of zinc in a cupric sulfate solution, cupric ions migrate to and are reduced at the surface of the metal. One of the unique aspects of these reactions, however, is that the transfer of electrons—and hence the same over-all reaction—can be accomplished when the donor and acceptor are quite remote from one another. With an arrangement such as illustrated in Figure 17.1 (p. 371), direct contact between the metallic zinc and cupric ions is prevented by a porous barrier. Despite this separation, electrons are transferred by means of the external metallic conductor. The half reactions here are the same as before, electrons being transferred from the metallic zinc to the cupric ions. This transfer can be expected to continue until the cupric and zinc ion concentrations achieve levels corresponding to equilibrium for the reaction

$$Zn + Cu^{2+} \rightleftharpoons Zn^{2+} + Cu$$

When this point is reached no further net flow of electrons will be observed. *The reaction and its position of equilibrium is the same regardless of the manner in which the process is carried out.*

The device illustrated in Figure 17.1 is, of course, an electrochemical cell; it is capable of producing electric energy because of the tendency of the reacting species to transfer electrons and thus achieve the condition of equilibrium. The electrical potential existing between the zinc and the copper electrodes is a measure of this driving force and is easily evaluated by a voltmeter, V, placed in the circuit as shown in Figure 17.1. We shall see that the voltage produced by an electrochemical cell is directly related to the equilibrium constant for the particular oxidation-reduction process involved and that, in fact, measurement of these potentials constitutes an important source of numerical values for these constants. We shall, therefore, examine more closely the construction and behavior of electrochemical cells as well as the measurement of the potentials they develop.

FUNDAMENTALS OF ELECTROCHEMISTRY

Basic Laws

Ohm's law. In 1827, G. S. Ohm formulated the now well-known law that the strength of a current flowing in a conductor is directly proportional to the difference in potential between the ends of the conductor and inversely proportional to its resistance. The law may be written

$$i = \frac{E}{R}$$

where i is the strength of the current in amperes, E is the potential difference in volts, and R is the resistance in ohms.

Faraday's laws. Faraday's laws, first enunciated in 1833, relate the quantity of electricity passing through an electrolyte solution to the quantity of material reacting with this current. The quantity of electricity flowing is given by the product of its strength and the time of its flow; it represents the passage of a definite number of electrons. Faraday's laws may be stated as follows:

1. *The amount of chemical change produced by an electric current is proportional to the quantity of electricity passed.* Thus, if a certain quantity of current deposits x grams of copper, twice this amount will deposit $2x$ grams under the same conditions.

2. *The amounts of different substances deposited or dissolved by the same quantity of electricity are proportional to the equivalent weights of these substances.* The equivalent weight is that weight of substance which, directly or indirectly, consumes or produces 1 mol of electrons (see Chap. 10). Thus, for the half reactions

$$Ag^+ + e \rightleftharpoons Ag$$

$$Au^{3+} + 3e \rightleftharpoons Au$$

the same quantity of electricity will deposit amounts of these metals in proportion to their equivalent weights, or

$$\frac{\text{weight Ag}}{\text{deposited}} : \frac{\text{weight Au}}{\text{deposited}} :: \frac{\text{GFW Ag}}{1} : \frac{\text{GFW Au}}{3}$$

Electrical Units

The ampere. The ampere is the unit of strength of an electric current; it is a measure of the number of electrons flowing through a conductor per second. Until recently, the international ampere was employed as the standard in the United States; it is defined as the unvarying direct current that will deposit 0.00111800 gram of silver per second from a silver nitrate solution under certain specified conditions. By an act of Congress in 1948 the absolute ampere was adopted as the unit of current strength; it is defined in terms of a force produced by the current. Fortunately the difference between the international and the absolute ampere is vanishingly small for most purposes, being less than 1 part in 10,000.

The ohm. The international ohm is a unit of electric resistance; it is the resistance, at 0° C, of a uniform column of mercury 106.300 cm in length, weighing 14.4521 grams. Such a column of mercury has a cross-sectional area of about 1 sq mm. The absolute ohm, now the standard in the United States, differs but slightly from the international ohm; it is defined in terms of inductance and frequency.

The volt. The international volt is defined by means of Ohm's law; it is the potential required to produce a current flow of one international ampere through a resistance of one international ohm. The absolute volt is defined in terms of work and varies but slightly from the above.

The coulomb, q. The coulomb is a unit of quantity of electricity, being that amount which corresponds to a constant current of 1 ampere flowing for 1 second. The coulomb then represents a certain number of electrons (6.24×10^{18}).

The faraday, *F*.   The faraday is also a unit of quantity of electricity, being that amount which will cause one equivalent of chemical change at an electrode. The faraday, then, is 6.02×10^{23} electrons, or 96,493 coulombs.

A typical application of these definitions and laws is given in the following example.

Example.   A constant current of 0.500 ampere was passed through a solution of $CuSO_4$ for exactly 30.0 minutes. Copper was deposited at one electrode and oxygen at the other; the weight of each formed in the electrolysis is to be calculated.

To solve this problem we must first know the quantity of electricity in faradays that was passed through the solution.

quantity of current in coulombs = current (amperes) $\times$ time (seconds)

$$= 0.500 \times 30 \times 60$$

$$= 900$$

$$\text{quantity of current} = \frac{900}{96,493} = 9.32 \times 10^{-3} \text{ faraday}$$

According to the foregoing definitions, the number of faradays of current is equal to the number of equivalents of copper deposited as well as the number of equivalents of oxygen evolved. The respective half reactions are

$$Cu^{2+} + 2e \rightleftharpoons Cu$$

$$2H_2O \rightleftharpoons O_2 + 4H^+ + 4e$$

Therefore

$$\text{weight of Cu deposited} = \text{number of equivalents of Cu} \times \frac{\text{GFW Cu}}{2}$$

$$= 9.32 \times 10^{-3} \times 31.8$$

$$= 0.296 \text{ gram}$$

$$\text{weight } O_2 \text{ formed} = 9.32 \times 10^{-3} \times \frac{\text{GFW } O_2}{4}$$

$$= 9.32 \times 10^{-3} \times 8.00$$

$$= 0.0746 \text{ gram}$$

Cells

A *cell* consists of a pair of conductors or electrodes, usually metallic, immersed in an electrolyte. When the electrodes are connected by an external conductor and a passage of current ensues, a chemical oxidation occurs at the surface of one electrode and a reduction at the surface of the other.

Galvanic and electrolytic cells.   When a cell is operated in such a way as to produce electric energy, it is called a *galvanic*, or sometimes, a *voltaic* cell. A cell that consumes electric energy is termed an *electrolytic* cell. Thus the cell shown in Figure 17.1 is capable of behaving as a galvanic cell. When the two electrodes are connected by a wire, electrical energy is produced, and a spontaneous flow of electrons from the zinc electrode to the copper occurs.

The same cell could also be operated as an electrolytic cell. To do this it would be necessary to introduce a battery in the external circuit that would force electrons to flow in the opposite direction through the cell. Under these circumstances, zinc would deposit and copper would dissolve; these processes would consume energy from the battery.

Reversible and irreversible cells. Often, as in the foregoing case, altering the direction of flow of current through a cell simply causes a change in direction of the chemical reactions at the electrodes. In such a case the cell is said to be *reversible*. In some instances, however, reversing the current flow results in completely different reactions at one or both electrodes. The cell is then called *irreversible*.

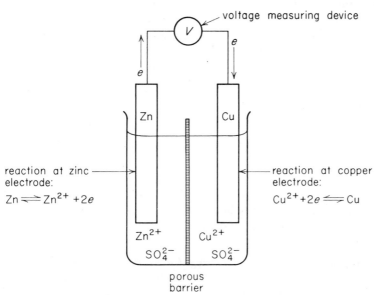

Fig. 17.1 Schematic Diagram of a Simple Electrochemical Cell.

Passage of current through a cell. We shall again consider the galvanic cell shown in Figure 17.1. The electrode reactions that occur when the strips of copper and zinc are connected by a wire are

$$Zn \rightleftharpoons Zn^{2+} + 2e$$

$$Cu^{2+} + 2e \rightleftharpoons Cu$$

As a result, the solution in the cell compartment containing the zinc electrode shows an increase in concentration of zinc ions, while the solution surrounding the copper electrode is depleted in cupric ions. If no internal contact existed between the two parts of the cell, a charge imbalance in the neighborhood of the electrodes would result, an excess of positive ions being found near the zinc and an excess of negative ions near the copper. Such a charge inhomogeneity within the solutions would prevent the further flow of current in the external circuit by virtue of the electrostatic forces set up between these charges and the electrons within the metal.

In an actual cell, redistribution of ions can occur as a result of their mobility within the electrolyte solution. As a result, regions of unbalanced charge do not develop. In the above example, for instance, zinc ions as well as other cations can migrate from the solution surrounding the zinc electrode toward the copper electrode; anions can also migrate but will do so in the opposite direction. The result of these ionic movements is to restore the balance of charge in the two compartments and thus allow a continued flow of current.

The passage of current through a cell, then, involves a migration of ions within the solution, and the current may be considered as conveyed by these ions. Not only the electrode-reactive ions but the total ions in the solution participate in carrying the current, the contribution of each being determined by its relative concentration and inherent mobility in the medium. Only at the electrode surface—the site of transfer of electrons from solution to metallic conductor—is the current passage limited to the reacting species.

Electrode Processes

We have thus far considered a cell as being composed of two *half cells*, each of which is associated with the process occurring at one of the electrodes. This treatment forms an extremely useful approach to many of the problems encountered in electrochemistry. In using it, however, we run the risk of thinking of individual half cells as real, separate entities, capable of independent existence. This, of course, is not so; it is impossible to operate a half cell in the absence of a second half cell, nor is it feasible to measure the potential of a half cell without reference to another.

Anode and cathode. *In either a galvanic or electrolytic cell, the electrode at which oxidation occurs is called the anode, whereas the electrode at which reduction takes place is the cathode.*

Some typical cathodic reactions. There are several common types of cathodic half reactions. These include:

1. Deposition of metals on an electrode surface. Examples would include the formation of copper or silver metal according to the half reactions

$$Cu^{2+} + 2e \rightleftharpoons Cu$$

$$Ag^+ + e \rightleftharpoons Ag$$

Many metal ions can be removed from solution in this way.

2. Formation of hydrogen gas. The evolution of hydrogen gas is often observed at inert electrodes in cells that do not contain other more easily reduced species. The half reaction for this process is

$$2H^+ + 2e \rightleftharpoons H_2$$

3. Change in oxidation state of an ion in solution. A common reaction that can occur at nonreactive metal electrodes is the reduction of an ion to a lower oxidation state. Thus, for example, the following reaction can occur at a platinum electrode:

$$Fe^{3+} + e \rightleftharpoons Fe^{2+}$$

Some common anodic reactions. Several types of reactions are listed below.

1. Oxidation of a metal electrode. Where a readily oxidized metal is used as an anode, consumption of the electrode frequently occurs. For example

$$Cd \rightleftharpoons Cd^{2+} + 2e$$

$$Zn \rightleftharpoons Zn^{2+} + 2e$$

2. Oxidation of halide ions. With an inert metal as an anode, oxidation of halide ions, if present, is common. As an example

$$2Cl^- \rightleftharpoons Cl_2 + 2e$$

3. Change in oxidation state of an ion in solution. Often the anode reaction in a cell involves oxidation of an ion to a higher oxidation state. This will take place at a nonreactive metal electrode. The oxidation of ferrous ion is typical of this process.

$$Fe^{2+} \rightleftharpoons Fe^{3+} + e$$

4. Evolution of oxygen. In the absence of easily oxidizable species, oxidation of water may occur at the anode giving oxygen as a product.

$$2H_2O \rightleftharpoons O_2 + 4H^+ + 4e$$

Electrode signs. A positive or a negative sign is customarily assigned to an electrode to indicate the direction of flow of electrons as the electrode operates as part of a cell. Unfortunately some ambiguities are associated with this practice, particularly when attempts are made to relate these signs to anodic or cathodic processes. In the case of a galvanic cell, the negative electrode is the one from which electrons are obtained. For the cell shown in Figure 17.1, the electrons would be produced at the zinc *anode*. In the case of an electrolytic cell, however, the negative electrode is the one into which electrons are forced. Thus in a cell used for removal of cadmium by electrolysis, the negative electrode is the one upon which metallic cadmium plated out; that is, the *cathode*. Clearly, the sign assumed by, say, the cathode depends upon whether we are dealing with an electrolytic or a galvanic cell. Irrespective of sign, the anode is the electrode at which oxidation occurs and the cathode is the site for reduction; use of these terms, rather than positive and negative, in describing electrodes is unambiguous and much preferred.

Electrode Potential

Turning again to the galvanic cell shown in Figure 17.1 we see that the driving force for the chemical reaction

$$Zn + Cu^{2+} \rightleftharpoons Cu + Zn^{2+}$$

manifests itself in the form of measurable electric force or voltage between the two electrodes. We may consider this force as being the sum of two forces or potentials, called *half-cell potentials* or single-electrode potentials; one of these is associated with the half reaction occurring at the anode and the other with the half reaction taking place at the cathode.

It is not difficult to see how we might obtain information regarding the *relative* magnitudes of half-cell potentials. For example, by replacing the left-hand side of the cell in Figure 17.1 with a cadmium electrode immersed in a solution of cadmium ions, the voltmeter reading would be approximately 0.4 volt less than that of the original cell. Since the copper cathode remains unchanged, we logically ascribe this voltage decrease to the alteration in the anode; thus we conclude that the half-cell potential for the oxidation of cadmium is 0.4 volt less than for the oxidation of zinc. Other substitutions in the left-hand side of the cell would make possible the comparison of the chemical driving forces of a variety of half reactions by means of simple electrical measurements.

Such a technique does not lead to absolute values for half-cell potentials and, as a matter of fact, there is no way to determine such quantities since all voltage-measuring devices must make contact both with the electrode in question and also with the solution. This latter contact, however, inevitably involves a solid-solution interface and hence acts as a second half cell at which a chemical reaction must take place. Thus we do not obtain a measurement of the desired half-cell potential but rather a sum consisting of the potential of interest and the half-cell potential for the contact between the voltage measuring device and the solution.

Relative half-cell potentials. The lack of absolute potentials for half-cell processes is not a serious handicap, however, because relative half-cell potentials are just as useful for our purposes. When combined, these relative potentials give cell potentials; in addition they are useful for the calculation of equilibrium constants for oxidation-reduction processes.

To obtain relative half-cell potentials a common reference electrode is needed against which all other electrodes are compared. The electrode chosen for this purpose is the standard hydrogen electrode.

The standard hydrogen electrode. The electrode selected as reference for the evaluation of relative half-cell potentials should be relatively easy to construct, should exhibit reversible behavior, and should give constant and reproducible potentials for a given set of experimental conditions. The standard hydrogen electrode meets these requirements, and is used almost universally as such a reference. Figure 17.2 illustrates a typical hydrogen electrode. It consists, in essence, of a piece of platinum immersed in a solution with a hydrogen ion activity of one. Hydrogen gas is bubbled across the surface of the platinum in an uninterrupted stream to assure that the electrode is continuously in contact with both the solution and the gas; in order to achieve the largest possible surface area, the electrode is coated with a finely divided layer of platinum called *platinum black*. The partial pressure of hydrogen is maintained at one atmosphere above the liquid phase.

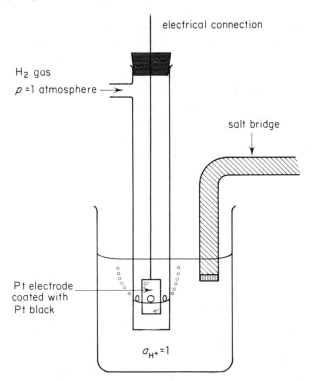

Fig. 17.2 The Standard Hydrogen Electrode.

An electrode of this type is called a *gas electrode*. The platinum takes no part in the electrochemical reaction, serving only as an aid in the transfer of electrons. The half-cell reaction responsible for the transmission of current across the interface is

$$H_2 \text{ (gas)} \rightleftharpoons 2H^+ + 2e$$

Depending upon the type of half cell with which the hydrogen electrode is coupled, it may behave either as an anode or a cathode. In the one case, oxidation of the hydrogen gas to hydrogen ions occurs; in the other, the reverse reaction takes place. Under proper conditions, then, the hydrogen electrode is reversible in its behavior.

The potential of the hydrogen electrode is dependent upon temperature, the concentration of hydrogen ions in the solution, and the pressure of the hydrogen gas at the surface of the electrode. These must be carefully specified if the half cell is for reference purposes. The conditions chosen are unity in atmospheres for the partial pressure of the gas and unit molarity for the activity of the hydrogen ions in the solution. The temperature used is 25° C. When these specifications are met, the electrode is said to be a *standard hydrogen electrode* or a *normal hydrogen electrode*. *Quite arbitrarily* the potential of this half cell is assigned the value of exactly 0 volts.

Measurement of electrode potentials.   Electrode potentials relative to the standard hydrogen electrode can be measured by means of a cell such as that shown in Figure 17.3. Here, one half of the cell consists of the standard hydrogen electrode and the other half is the electrode whose potential is to be determined. Connecting the two cells is a salt bridge, a device that consists of a tube containing a concentrated solution of an electrolyte—most often a saturated potassium chloride solution. This bridge provides electrical contact between the two halves of the cell while preventing mixture of the contents of the half cells; the passage of current takes place by ionic migration, as previously described. In general, the effect of a salt bridge upon the cell voltage is vanishingly small.

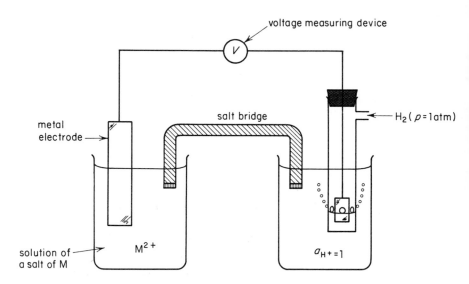

Fig. 17.3 Schematic Diagram of an Apparatus for Measurement of Electrode Potentials Against the Standard Hydrogen Electrode.

The half cell in Figure 17.3 consists of a pure metal in contact with a solution of its ions. The electrode reaction then is

$$M \rightleftharpoons M^{2+} + 2e$$

If we assume for the moment that the metal is cadmium and that the solution is approximately 1 molar in cadmium ions, the voltage indicated by the measuring device V will be about 0.4 volt. Further the metal electrode behaves as an anode so that electrons tend to flow from the metal electrode to the hydrogen electrode via the external circuit. Thus the half-cell reactions for the galvanic cell can be written as

$$Cd \rightleftharpoons Cd^{2+} + 2e \qquad \text{anode}$$

$$2H^+ + 2e \rightleftharpoons H_2 \qquad \text{cathode}$$

The over-all reaction is the sum of these, or

$$Cd + 2H^+ \rightleftharpoons Cd^{2+} + H_2$$

If the cadmium electrode were replaced by a zinc electrode immersed in a solution of zinc ions, an emf of slightly less than 0.8 volt would be observed on the voltmeter. Again the metal electrode would behave as the anode. The larger voltage here reflects the greater tendency of zinc to be oxidized and the difference gives a quantitative measure of the relative strengths of these two metals as reducing agents.

As mentioned earlier, a value of zero is assigned as the potential for the hydrogen electrode. The electrode potentials for the two half cells then become 0.4 and 0.8 volt with respect to this reference. Since this is an arbitrary assignment, however, these so-called half-cell or electrode potentials are indeed potentials of cells involving the hydrogen electrode as a common reference.

If the half cell in Figure 17.3 were a copper electrode in a 1-molar solution of cupric ions, we would find the potential of the cell to be about 0.3 volt. However, we would note a marked difference in behavior of this electrode as compared with the two earlier examples. Here, the copper would be plated out from solution and the external electron flow would be from the hydrogen electrode to the copper electrode. Obviously the cell reaction is

$$Cu^{2+} + H_2 \rightleftharpoons Cu + 2H^+$$

the reverse of the two earlier cases. Thus metallic copper is a much less potent reducing agent than either the zinc or the cadmium. Again the observed potential is a quantitative measure of this strength; in comparing this potential with those for the half reactions of zinc and cadmium, however, we need to indicate the differences in behavior of the electrode systems relative to the reference electrode. This can be done conveniently by assigning a positive or negative sign to the potentials thus making the sign of the potential for the copper half reaction opposite to that of the other two.

Sign conventions for electrode potentials. There are two possible sign conventions that can be applied to electrode potentials. For example we could say above that the potential for the oxidation of cadmium is $+0.4$ volt and that for copper is -0.3 volt, or the reverse could be used. Unfortunately chemists have not been able to agree on a single convention and this often leads to some confusion.

We shall call the potential of an electrode reaction positive if the reaction proceeds spontaneously relative to the standard hydrogen electrode. A nonspontaneous reaction will have a negative potential. Thus for the three half reactions already considered we would write

$$Zn^{2+} + 2e \rightleftharpoons Zn \qquad E = -0.8 \text{ volt}$$

$$Cd^{2+} + 2e \rightleftharpoons Cd \qquad E = -0.4 \text{ volt}$$

$$Cu^{2+} + 2e \rightleftharpoons Cu \qquad E = +0.3 \text{ volt}$$

The reaction involving reduction of cupric ions to copper is spontaneous relative to the standard hydrogen electrode and is, therefore, given a positive sign. Within this same sign convention we may also write

$$Zn \rightleftharpoons Zn^{2+} + 2e \qquad E = +0.8 \text{ volt}$$

$$Cd \rightleftharpoons Cd^{2+} + 2e \qquad E = +0.4 \text{ volt}$$

$$Cu \rightleftharpoons Cu^{2+} + 2e \qquad E = -0.3 \text{ volt}$$

In this case the negative sign for copper indicates the nonspontaneity of the oxidation of copper relative to the hydrogen half cell, whereas the positive sign for the zinc and cadmium potentials indicates that these reactions proceed spontaneously as written; that is, that the reaction

$$M + 2H^+ \rightleftharpoons M^{2+} + H_2$$

is the one which takes place in a galvanic cell under the prescribed conditions. The first set of potential data given above can be called *reduction potentials*; the second set, consisting of the same half-cell reactions written in the opposite sense are *oxidation potentials*. The two differ only in signs.

Clearly, an electrode potential datum has significance only when the half reaction to which it refers is explicitly indicated. This information is usually conveyed by writing the half reaction, as was done above, beside the potential of the half cell.[1] Also, the sign of the electrode potential is employed here to indicate the direction of the spontaneous chemical reaction. *This sign should not be confused with the sign used to indicate the polarity of the actual physical electrode in a galvanic or electrolytic cell.*

Effect of concentration on electrode potentials. The electrode potential for a half reaction is a measure of the chemical force tending to drive that reaction toward equilibrium. Consequently the electrode potential is zero when a system is at equilibrium, but becomes larger as it departs further and further from this state. Thus the potential of a piece of metallic zinc immersed in pure water is large relative to a piece of the same metal immersed in a 1 formal solution of zinc sulfate. In general, then, the concentration of the reactants and

---

[1] The International Union of Pure and Applied Chemistry has proposed the convention that the term "electrode potential" be reserved for half reactions written as reductions. Thus the electrode potential for a zinc electrode is the potential for the reaction

$$Zn^{2+} + H_2 \rightleftharpoons Zn + 2H^+$$

This reaction is not spontaneous and thus the sign of the electrode potential for zinc is negative.

products of a half reaction will have a marked effect on electrode potentials, and the quantitative aspects of this effect must now be considered.

Consider the generalized, reversible half-cell reaction

$$aA + bB + \cdots + ne \rightleftharpoons cC + dD + \cdots$$

where the capital letters represent formulae of reacting species (whether charged or uncharged), e represents electrons, and the remaining uncapitalized letters indicate the number of mols of each species participating in the reaction. It can be shown theoretically as well as experimentally that the potential, E, for this electrode process is governed by the relation

$$E = E° - \frac{RT}{nF} \ln \frac{[C]^c[D]^d \cdots}{[A]^a[B]^b \cdots} \tag{17-1}$$

where

$E°$ = a constant characteristic of the particular half reaction
R = the gas constant = 8.314 volt coulombs/°K/mol
T = the absolute temperature
n = number of electrons participating in the reaction as defined by the balanced chemical equation for the half-cell reaction
F = the faraday = 96,493 coulombs
$\ln$ = the natural logarithm = 2.303 $\log_{10}$

Substituting numerical values for the various constants into (17-1) and converting to a log to the base 10, equation (17-1) becomes, at 25° C

$$E = E° - \frac{0.059}{n} \log \frac{[C]^c[D]^d \cdots}{[A]^a[B]^b \cdots} \tag{17-2}$$

The symbols in brackets represent the activities of the reacting species. As in the case of equilibrium-constant expressions (Chapter 2) certain approximations for these activities are useful. Thus where the substances are dissolved in a solvent

$$[\quad] \cong \text{concentration in mols/liter}$$

If the reactant is a gas,

$$[\quad] \cong \text{partial pressure of the gas in atmospheres}$$

If the reactant exists in a second phase in the form of a pure solid or liquid, then by definition[2]

$$[\quad] = 1$$

---

[2] The arbitrary assignment of unity to the activity of the second phase was also made in arriving at the solubility-product expression (p. 20). In either case, this assumes that the concentration of the substance in the second phase is constant and the equilibrium is unaffected by the *quantity* of this second phase.

Finally if the reactant is the solvent, water,

$$[H_2O] = 1$$

Here the concentration of water is assumed to be unchanged by the reaction as a consequence of the large number of mols of water present compared with the other reactants. Therefore its constant activity is included in the constant $E°$. We shall apply this equation to a few specific examples.

1. $Zn^{2+} + 2e \rightleftharpoons Zn$ $\qquad E = E° - \dfrac{0.059}{2} \log \dfrac{1}{[Zn^{2+}]}$

Here the activity of the elemental zinc is unity and the electrode potential varies as the log of the molar concentration of zinc ions in the solution.

2. $Fe^{3+} + e \rightleftharpoons Fe^{2+}$ $\qquad E = E° - \dfrac{0.059}{1} \log \dfrac{[Fe^{2+}]}{[Fe^{3+}]}$

The electrode potential, in this instance, could be measured with an inert metal electrode immersed in a solution containing ferrous and ferric ions. The potential is dependent upon the ratio of the molar concentrations of these ions.

3. $2H^+ + 2e \rightleftharpoons H_2$ $\qquad E = E° - \dfrac{0.059}{2} \log \dfrac{p_{H_2}}{[H^+]^2}$

In this example, p_{H_2} represents the partial pressure in atmospheres of hydrogen gas at the surface of the electrode. Ordinarily this will be very close to the atmospheric pressure.

4. $Cr_2O_7^{2-} + 14H^+ + 6e \rightleftharpoons 2Cr^{3+} + 7H_2O$

$$E = E° - \dfrac{0.059}{6} \log \dfrac{[Cr^{3+}]^2}{[Cr_2O_7^{2-}][H^+]^{14}}$$

Here the potential is dependent not only the concentration of the chromic and dichromate ions but also on the pH of the solution.

5. $AgCl + e \rightleftharpoons Ag + Cl^-$ $\qquad E = E° - \dfrac{0.059}{1} \log \dfrac{[Cl^-]}{1}$

In this case the activity of both metallic silver and silver chloride are constant and equal to one; the potential of the silver electrode varies as the chloride ion concentration changes.

Equation (17-1) is often called the *Nernst equation* in honor of a nineteenth-century electrochemist. Before describing its further application, we should consider the significance of the constant, $E°$, that appears in the equation. This is a constant for a particular half-cell reaction and is called the *standard electrode potential*.

The standard electrode potential, $E°$. An examination of equation (17-1) or (17-2) reveals that the constant, $E°$, is equal to the half-cell potential when the logarithmic term is zero. This occurs whenever the activity quotient is equal to unity, one instance being when the activities of all of the reactants and products are unity; thus the *standard electrode potential may be defined as the potential of a half-cell reaction versus the standard hydrogen electrode when the reactants and products are at unit activity.*

The standard electrode potential for a half-cell reaction is a fundamental physical constant that gives a quantitative description of the relative driving force of that half-cell reaction. Several facts should be kept in mind regarding it. First, a standard electrode potential is temperature dependent; if it is to have significance, the temperature at which it is determined must be specified. Second, a standard electrode potential is a relative quantity in the sense that it is really a cell potential where one of the electrodes is a carefully specified reference electrode—that is, the standard hydrogen electrode whose potential is assigned a value of zero volts at 25° C. Third, a standard electrode potential is given a sign dependent upon the direction of the spontaneous reaction with respect to the standard hydrogen electrode under the specified conditions of activity. We may have standard oxidation potentials and standard reduction potentials; for a given half reaction these will differ from one another only in their sign. Thus the numerical value for a standard electrode potential is of no use unless the direction of the reaction is specified in addition to the sign convention being used. Finally, the value of a standard potential is a measure of the intensity of the driving force of a half reaction. As such, it is independent of the number of particles participating in the reaction. Thus the potential for the reaction

$$Ag^+ + e \rightleftharpoons Ag \qquad\qquad E° = + 0.799 \text{ volt}$$

while dependent upon the concentration of silver ions, is the same regardless of whether we deal with 1 mol of reactants or 100 mols

$$100Ag^+ + 100e \rightleftharpoons 100Ag \qquad\qquad E° = + 0.799 \text{ volt}$$

Standard electrode potentials are available for a wide variety of half reactions. Many of these have been determined directly from voltage measurements of various cells in which the standard hydrogen electrode constitutes one of the electrodes. It is possible, however, to calculate values for this important constant from equilibrium studies of oxidation-reduction reactions and from thermochemical data relating to such reactions. Many of the values found in the literature were so obtained. W. M. Latimer[3] published a definitive work that is to be recommended as an authoritative source for standard electrode potential data.

A few standard reduction potentials are given in Table 17-1; a more complete list is found in Table A-1 of the appendix. The relatively large positive potentials for the last two half-cell reactions listed indicate that these exhibit a considerable tendency to proceed in the direction indicated. Thus, the species in the lower left side of the equations are most easily reduced and are therefore

[3] W. M. Latimer, *The Oxidation States of the Elements and Their Potentials in Aqueous Solutions*, 2d ed. Englewood Cliffs, N.J.: Prentice-Hall, Inc., 1952.

the most effective oxidizing agents. Proceeding up the table, each succeeding species is a less effective acceptor of electrons than the one below it. The half-cell reactions at the head of the table have little tendency to take place. On the other hand, they do tend to occur in the opposite sense, as oxidations; the most effective reducing agents, then, are those species that appear in the upper right-hand portion of the table of standard reduction potentials.

Table 17-1

A FEW STANDARD REDUCTION POTENTIALS

(See Appendix for a more extensive list.)

Reaction	$E°$ at 25° C, volts
$Zn^{2+} + 2e \rightleftharpoons Zn$	$- 0.763$
$Cr^{3+} + e \rightleftharpoons Cr^{2+}$	$- 0.41$
$Cd^{2+} + 2e \rightleftharpoons Cd$	$- 0.403$
$2H^+ + 2e \rightleftharpoons H_2$	0.000
$Ag(S_2O_3)_2^{3-} + e \rightleftharpoons Ag + 2S_2O_3^{2-}$	$+ 0.010$
$AgCl + e \rightleftharpoons Ag + Cl^-$	$+ 0.222$
$I_3^- + 2e \rightleftharpoons 3I^-$	$+ 0.536$
$Fe^{3+} + e \rightleftharpoons Fe^{2+}$	$+ 0.771$
$Ag^+ + e \rightleftharpoons Ag$	$+ 0.799$
$Cl_2 + 2e \rightleftharpoons 2Cl^-$	$+ 1.359$
$MnO_4^- + 8H^+ + 5e \rightleftharpoons Mn^{2+} + 4H_2O$	$+ 1.51$

A table of standard potentials gives the chemist a qualitative picture regarding the extent and direction of chemical reactions that involve electron transfer between the tabulated species. On the basis of Table 17-1, for example, we see that zinc is more easily oxidized than cadmium; we conclude, then, that a piece of zinc immersed in a solution of cadmium ions results in the plating out of metallic cadmium. On the other hand cadmium has little tendency to reduce zinc ions. Also from Table 17-1 we see that ferric iron is a better oxidizing agent than the triiodide ion. We may therefore predict that in a solution containing an equilibrium mixture of ferric, iodide, ferrous, and triiodide ions, the latter pair will predominate.

Calculation of half-cell potentials from $E°$ values. We shall now consider some applications of the Nernst equation to the calculation of various half-cell potentials.

Example. We wish to know the potential for the half reaction of a cadmium electrode in a solution that is 0.0100 F in Cd^{2+}?

From Table 17-1 we find

$$Cd^{2+} + 2e \rightleftharpoons Cd \qquad\qquad E° = - 0.403 \text{ volt}$$

and we write

$$E = E^o - \frac{0.059}{2} \log \frac{1}{[Cd^{2+}]}$$

Substituting the Cd^{2+} concentration into the equation we get

$$E = -0.403 - \frac{0.059}{2} \log \frac{1}{(0.01)}$$

$$= -0.403 - \frac{0.059}{2} (+2)$$

$$= -0.462 \text{ volt}$$

The sign for the potential simply indicates the direction of the half reaction when this half cell is coupled with the standard hydrogen electrode. The fact that it is negative shows that the reverse reaction

$$Cd + 2H^+ \rightleftharpoons H_2 + Cd^{2+}$$

would occur spontaneously. The calculated potential is a larger negative number than the standard electrode potential itself. This follows from equilibrium considerations since the half reaction, as written, has less tendency to occur at the lower cadmium ion concentration. Note also that an identical value—except for sign—would be obtained had we used the standard oxidation potential for the calculation; that is,

$$Cd \rightleftharpoons Cd^{2+} + 2e \qquad\qquad E^o = +0.403$$

$$E = +0.403 - \frac{0.059}{2} \log [Cd^{2+}]$$

Substituting 0.01 for $[Cd^{2+}]$, we obtain

$$E = +0.462 \text{ volt}$$

Example. We shall calculate the potential at which Cl_2 is liberated at an inert electrode immersed in a 0.0500 F solution of NaCl, assuming the gas is formed at a pressure of 740 mm of mercury.

Here we shall use the standard *oxidation* potential for Cl^-, that is,

$$2Cl^- \rightleftharpoons Cl_2 + 2e \qquad\qquad E^o = -1.36 \text{ volts}$$

$$E = -1.36 - \frac{0.059}{2} \log \frac{p_{Cl_2}}{[Cl^-]^2}$$

The pressure of chlorine gas in atmospheres is 740/760 or 0.974 atmosphere; therefore,

$$E = -1.36 - \frac{0.059}{2} \log \frac{0.974}{(0.05)^2}$$

$$= -1.44 \text{ volts}$$

Thus a potential of greater than 1.44 volts with respect to the standard hydrogen electrode would have to be applied to the electrode to cause deposition of chlorine gas.

Electrode potentials in the presence of precipitation and complex-forming reagents. We shall examine the effect of reagents that can react with the participants of an electrode process. For example the standard electrode potential for the reaction $Ag^+ + e \rightleftharpoons Ag$ is $+0.799$ volt. This is the potential with respect to the standard hydrogen electrode of a piece of silver metal in a solution of silver ions at unit activity. Addition of chloride ions to such a solution will materially alter the silver ion concentration and hence the electrode potential. Consider the following case.

> **Example.** We shall calculate the potential of a silver electrode in a solution that is saturated with silver chloride and that has a chloride ion activity of exactly 1.00.

$$Ag^+ + e \rightleftharpoons Ag \qquad\qquad E^\circ_{Ag^+ \to Ag} = + 0.799 \text{ volt}$$

$$E = E^\circ_{Ag^+ \to Ag} - 0.059 \log \frac{1}{[Ag^+]}$$

We may calculate $[Ag^+]$ from the solubility-product constant

$$[Ag^+] = \frac{K_{sp}}{[Cl^-]}$$

Substituting this into the Nernst equation

$$E = E^\circ_{Ag^+ \to Ag} - \frac{0.059}{1} \log \frac{[Cl^-]}{K_{sp}}$$

This may be rewritten as

$$E = E^\circ_{Ag^+ \to Ag} + 0.059 \; \log \; K_{sp} - 0.059 \; \log \; [Cl^-] \qquad (17\text{-}3)$$

If we substitute 1.00 for $[Cl^-]$ and use a value of 1.82×10^{-10} for K_{sp}, we obtain

$$E = + 0.222 \text{ volt}$$

This calculation shows that the half-cell potential for the reduction of silver becomes less in the presence of chloride ions. Qualitatively, this is what we expect since removal of silver ions would decrease the tendency of the silver ions to be reduced.

Equation (17-3) relates the potential of a silver electrode to the chloride ion concentration of a solution that is also saturated with silver chloride. When the chloride ion activity is unity, the potential is the sum of two constants; this sum can be called the standard electrode potential for the half reaction

$$AgCl + e \rightleftharpoons Ag + Cl^- \qquad\qquad E^\circ_{AgCl \to Ag} = + 0.222 \text{ volt}$$

where

$$E^\circ_{AgCl \to Ag} = E^\circ_{Ag^+ \to Ag} + 0.059 \log K_{sp}$$

Thus, when in contact with a solution saturated with silver chloride, the potential of a silver electrode can be described either in terms of the silver ion concentration using the standard electrode potential for the simple silver half reaction or in terms of the chloride ion concentration using the standard potential for the silver-silver chloride half reaction.

In an analogous fashion to the foregoing, we can treat the behavior of a silver electrode in a solution of an ion capable of forming a soluble complex with silver ion. For example, in a solution of thiosulfate ion, the half reaction is

$$Ag(S_2O_3)_2^{3-} + e \rightleftharpoons Ag + 2S_2O_3^{2-}$$

The standard electrode potential for this half reaction would be the electrode potential when the activity of the complex and the complexing anion were both unity. Using the same approach as in the previous example, we see that

$$E^{\circ}{}_{Ag(S_2O_3)_2^{3-} \to Ag} = E^{\circ}{}_{Ag^+ \to Ag} + 0.059 \log K_{inst}$$

where K_{inst} is the dissociation constant for the complex ion.

Data for the potential of the silver electrode in the presence of a variety of ions are given in the table of standard electrode potentials (Appendix). Similar information is given for other electrode systems. Such data often simplify the calculation of half-cell potentials.

> **Example.** We shall calculate the potential of a silver electrode in a 1.00 F solution of KCN to which enough silver nitrate is added to make the concentration of the $Ag(CN)_2^-$ complex 0.0500 F.
> From the table of standard potentials
>
> $$Ag(CN)_2^- + e \rightleftharpoons Ag + 2CN^- \qquad E^{\circ} = -0.31 \text{ volt}$$
>
> $$E = -0.31 - \frac{0.059}{1} \log \frac{[CN^-]^2}{[Ag(CN)_2^-]}$$
>
> $$= -0.31 - \frac{0.059}{1} \log \frac{(1.00 - 2 \times 0.05)^2}{(0.050)}$$
>
> $$= -0.38 \text{ volt}$$

An identical result would result by calculating the silver ion concentration of this solution from the instability constant for the complex and using the standard potential for the reaction $Ag^+ + e \rightleftharpoons Ag$. This would require a greater amount of computation, however.

Cells and Cell Potentials

Schematic representation of cells. In order to simplify the description of cells, chemists frequently use a shorthand notation. For example, the cell shown in Figure 17.1 can be represented as follows:

$$Zn \mid ZnSO_4(C_1) \mid CuSO_4(C_2) \mid Cu$$

The single vertical lines represent phase boundaries in the cell; potential differences arise across these and are included in the measured cell potential. The potential difference across the junction of the two solutions—a $ZnSO_4$ solution of concentration C_1 and a $CuSO_4$ solution of concentration C_2—is known as a *liquid junction potential*. Often, the magnitude of this potential can be made to approach zero by interposing between the two solutions a *salt bridge* consisting of a saturated solution of potassium chloride. The presence of a salt bridge is symbolized by vertical double lines. Thus for the cell illustrated in Figure 17.3 we can write

$$M \mid M^{2+}(C_1) \parallel H^+(a = 1) \mid H_2(p = 1 \text{ atm}) \, Pt$$

Here the hydrogen ion activity, a, and the partial pressure of hydrogen gas, p, are given in parentheses. Some chemists prefer to replace the single vertical lines with semicolons or colons.

Calculation of cell potentials. One of the important uses of standard electrode potentials is the calculation of the theoretical potential obtainable from a galvanic cell or required to operate an electrolytic cell. Voltages such as these are for cells through which essentially no current passes; additional considerations are necessary where a current flow occurs.

We shall now consider the cell

$$Zn \mid ZnSO_4(1.00 \ F) \parallel CuSO_4(1.00 \ F) \mid Cu$$

The two half reactions involve oxidation of zinc to zinc ions and reduction of the cupric ions to the metal. Standard electrode potentials for these half reactions are available; however, these are actually potentials for cells in which the chemical reactions are

$$Zn + 2H^+ \rightleftharpoons Zn^{2+} + H_2 \qquad E^\circ = +0.763 \text{ volt}$$
$$Cu^{2+} + H_2 \rightleftharpoons Cu + 2H^+ \qquad E^\circ = +0.337 \text{ volt}$$

Here we reversed the direction of the zinc-zinc ion half reaction from the way it appears in the table of standard potentials; as a consequence, the sign of the standard potential has also been changed. Because the hydrogen electrode acts in the opposite sense in the two cases, the potential of a cell with a zinc anode and copper cathode is obtained by adding these two potentials together. Any effect due to the hydrogen electrodes is cancelled out and we get

$$Zn + Cu^{2+} \rightleftharpoons Cu + Zn^{2+} \qquad E = 1.100 \text{ volts}$$

The calculated potential will be for a cell in which the cupric and zinc ion activities are unity; this should be reasonably close to the potential for the actual cell in question where the concentrations are given as one formal.

A certain amount of care is needed in calculations of this type in order to avoid a confusion of the signs. The procedure given below will avoid this difficulty.

1. Calculate the two half-cell potentials using standard oxidation or reduction potentials and the Nernst equation.

2. Write one of the half reactions as an oxidation and the other as a reduction; also write the potentials for each, being sure that the signs are appropriate for the half reactions as written.

3. Add the potentials and the equations for the half reactions. A positive cell potential will indicate that the cell reaction as written is spontaneous; a negative sign, on the other hand, will mean that a potential greater than this must be applied to the cell in order to cause the reaction to proceed in the manner indicated. From the reaction direction, we may also deduce which electrode behaves as the anode and which as the cathode.

Example. We shall calculate the theoretical potential of the following cell:

$$\text{Pt, } H_2(0.8 \text{ atm}) \mid HCl(0.20 \ F), \quad AgCl(\text{saturated}) \mid Ag$$

The two half-cell reactions and standard reduction potentials are

$$2H^+ + 2e \rightleftharpoons H_2 \qquad\qquad E^\circ = 0.000 \text{ volt}$$
$$AgCl + e \rightleftharpoons Ag + Cl^- \qquad E^\circ = + 0.222 \text{ volt}$$

For the hydrogen electrode

$$E = 0.000 - \frac{0.059}{2} \log \frac{0.8}{(0.2)^2} = - 0.038 \text{ volt}$$

For the silver-silver chloride electrode

$$E = + 0.222 - 0.059 \log 0.20 = + 0.263 \text{ volt}$$

Now we may write

$$2AgCl + 2e \rightleftharpoons 2Ag + 2Cl^- \qquad E = + 0.263 \text{ volt}$$

By reversing the other half reaction and changing the sign for its potential

$$H_2(0.8 \text{ atm}) \rightleftharpoons 2H^+(0.2 \ F) + 2e \qquad E = + 0.038 \text{ volt}$$

we obtain on addition

$$H_2 + 2AgCl \rightleftharpoons 2H^+ + 2Cl^- + 2Ag \qquad E_{cell} = + 0.301 \text{ volt}$$

The positive sign shows that the equation describes the *spontaneous* cell reaction and that the hydrogen electrode is the anode of the galvanic cell.

Example. We shall calculate the potential required to initiate deposition of metallic copper from a solution that is $0.010 \ F$ in $CuSO_4$ and that contains sufficient sulfuric acid to give a hydrogen ion concentration of $1.0 \times 10^{-4} \ M$.

Here the cathode reaction is the deposition of copper. Since there are no easily oxidizable substances present, the anode reaction will involve oxidation of H_2O to give O_2. From the table of standard potentials, we find

$$Cu^{2+} + 2e \rightleftharpoons Cu \qquad\qquad E^\circ = + 0.337 \text{ volt}$$
$$\tfrac{1}{2}O_2 + 2H^+ + 2e \rightleftharpoons H_2O \qquad E^\circ = + 1.23 \text{ volts}$$

Then for the copper electrode

$$E = + 0.337 - \frac{0.059}{2} \log \frac{1}{0.01} = + 0.278 \text{ volt}$$

and for the oxygen electrode

$$E = + 1.23 - \frac{0.059}{2} \log \frac{1}{(1 \times 10^{-4})^2 (1.0)^{1/2}} = + 0.99 \text{ volt}$$

We have assumed that the oxygen is evolved at a pressure of 1.0 atmosphere. Upon rewriting the reactions as they would occur in an electrolytic cell

$$2e + Cu^{2+} \rightleftharpoons Cu \qquad\qquad E = + 0.278 \text{ volt}$$
$$H_2O \rightleftharpoons \tfrac{1}{2}O_2 + 2H^+ + 2e \qquad\qquad E = - 0.99 \text{ volt}$$
$$Cu^{2+} + H_2O \rightleftharpoons \tfrac{1}{2}O_2 + 2H^+ + Cu \qquad E_{cell} = - 0.71 \text{ volt}$$

Thus a potential greater than 0.71 volt would have to be applied to cause the cell reaction to occur.

Chemical Equilibrium in Cells; Equilibrium Constants from Standard Electrode Potentials

An important condition at equilibrium. Consider once again the galvanic cell shown in Figure 17.1 where the electrode reactions are

$$Zn \rightleftharpoons Zn^{2+} + 2e$$
$$Cu^{2+} + 2e \rightleftharpoons Cu$$

We have shown in the previous section that the potential here is equal to the sum of the half-cell potentials as written; that is,

$$E_{cell} = E_{Zn \to Zn^{2+}} + E_{Cu^{2+} \to Cu}$$

When current is drawn from this cell, there is an increase in the zinc ion concentration and a decrease in the cupric ion concentration of the solutions surrounding the electrodes. The effect of this is to lower the potentials of each of the half reactions as they occur in this cell

$$E_{Zn \to Zn^{2+}} = E^\circ_{Zn \to Zn^{2+}} - \frac{0.059}{2} \log [Zn^{2+}]$$

$$E_{Cu^{2+} \to Cu} = E^\circ_{Cu^{2+} \to Cu} - \frac{0.059}{2} \log \frac{1}{[Cu^{2+}]}$$

Thus as current is drawn, the cell potential will become smaller and eventually go to zero. When this occurs, the cell is said to be completely discharged; at this point, *the cell reaction*

$$Zn + Cu^{2+} \rightleftharpoons Zn^{2+} + Cu$$

is at chemical equilibrium; that is, the driving force for the forward reaction has become equal to that of the reverse reaction. We may say, then, that *at chemical equilibrium*

$$E_{cell} = 0 = E_{Zn \to Zn^{2+}} + E_{Cu^{2+} \to Cu}$$

or, writing both half-cell potentials as reduction potentials and rearranging, the equation

$$E_{Zn^{2+} \to Zn} = E_{Cu^{2+} \to Cu}$$

This illustrates an important and quite general relationship—*whenever an oxidation-reduction system is at chemical equilibrium, the reduction potentials (or the oxidation potentials) of the two half reactions of the system will be equal.*

Equilibrium constants for oxidation-reduction reactions. Consider the oxidation-reduction equilibrium

$$aA_{red} + bB_{oxid} \rightleftharpoons aA_{oxid} + bB_{red}$$

where the half reactions may be written

$$aA_{oxid} + ne \rightleftharpoons aA_{red}$$
$$bB_{oxid} + ne \rightleftharpoons bB_{red}$$

When the components of this system are at chemical equilibrium

$$E_A = E_B$$

where E_A and E_B are the reduction potentials for the two half cells. Applying the Nernst equation we may write

$$E_A^\circ - \frac{0.059}{n} \log \frac{[A_{red}]^a}{[A_{oxid}]^a} = E_B^\circ - \frac{0.059}{n} \log \frac{[B_{red}]^b}{[B_{oxid}]^b}$$

Upon rearranging and combining the log terms

$$E_B^\circ - E_A^\circ = \frac{0.059}{n} \log \frac{[A_{oxid}]^a [B_{red}]^b}{[A_{red}]^a [B_{oxid}]^b}$$

The concentrations here are equilibrium concentrations; the quotient containing these is the equilibrium constant for the reaction. Thus, we may write

$$\log K_{equil} = \frac{n(E_B^\circ - E_A^\circ)}{0.059}$$

Example. We shall calculate the equilibrium constant for the reaction

$$MnO_4^- + 5Fe^{2+} + 8H^+ \rightleftharpoons Mn^{2+} + 5Fe^{3+} + 4H_2O$$

From the table of standard reduction potentials we find

$$5Fe^{3+} + 5e \rightleftharpoons 5Fe^{2+} \qquad\qquad E^\circ = +0.771 \text{ volt}$$
$$MnO_4^- + 8H^+ + 5e \rightleftharpoons Mn^{2+} + 4H_2O \qquad E^\circ = +1.51 \text{ volts}$$

At equilibrium

$$E_{Fe^{3+} \to Fe^{2+}} = E_{MnO_4^- \to Mn^{2+}}$$

and

$$E_{Fe^{3+}}^\circ - \frac{0.059}{5} \log \frac{[Fe^{2+}]^5}{[Fe^{3+}]^5} = E_{MnO_4^-}^\circ - \frac{0.059}{5} \log \frac{[Mn^{2+}]}{[MnO_4^-][H^+]^8}$$

Rearranging

$$\frac{0.059}{5} \log \frac{[Mn^{2+}][Fe^{3+}]^5}{[MnO_4^-][Fe^{2+}]^5[H^+]^8} = E_{MnO_4^-}^\circ - E_{Fe^{3+}}^\circ$$

$$\log K_{equil} = \frac{(1.51 - 0.77)\,5}{0.059}$$

$$= 62.7$$

$$K_{equil} = 10^{62.7} = 5 \times 10^{62}$$

Example. A piece of metallic copper is placed in a 0.05 F solution of $AgNO_3$. We wish to find the composition of the solution when equilibrium has been achieved.

The reaction is

$$Cu + 2Ag^+ \rightleftharpoons Cu^{2+} + 2Ag$$

We shall first calculate the equilibrium constant for the reaction and then determine the solution composition from this. From a table of standard potentials, we find

$$2Ag^+ + 2e \rightleftharpoons 2Ag \qquad\qquad E^\circ = +0.799 \text{ volt}$$
$$Cu^{2+} + 2e \rightleftharpoons Cu \qquad\qquad E^\circ = +0.337 \text{ volt}$$

Since, at equilibrium

$$E_{Cu^{2+}} = E_{Ag^+}$$

then

$$E_{Cu^{2+} \to Cu}^\circ - \frac{0.059}{2} \log \frac{1}{[Cu^{2+}]} = E_{Ag^+ \to Ag}^\circ - \frac{0.059}{2} \log \frac{1}{[Ag^+]^2}$$

$$\log \frac{[Cu^{2+}]}{[Ag^+]^2} = \frac{2(E_{Ag^+ \to Ag}^\circ - E_{Cu^{2+} \to Cu}^\circ)}{0.059}$$

and

$$\frac{[Cu^{2+}]}{[Ag^+]^2} = K = 4.6 \times 10^{15}$$

From the size of the equilibrium constant, we conclude that the reaction goes far to the right as written and that nearly all of the Ag^+ is used up. Let the concentration of residual silver ions be x. The Cu^{2+} concentration will be just one half of the concentration of Ag^+ consumed by the reaction; that is,

$$[Cu^{2+}] = \tfrac{1}{2}(0.05 - x)$$

Since the reaction goes nearly to completion we shall assume x is small; then

$$[Cu^{2+}] \cong \tfrac{1}{2}(0.05) = 0.025$$

Substituting

$$\frac{(0.025)}{x^2} = 4.6 \times 10^{15}$$

$$x = [Ag^+] = 2.3 \times 10^{-9} \text{ mol/liter}$$

$$[Cu^{2+}] = \tfrac{1}{2}(0.05 - 2.3 \times 10^{-9}) \cong 0.025 \text{ mol/liter}$$

Determination of Dissociation, Solubility Product, and Instability Constants by Potential Measurements

We have seen that the potential of an electrode is determined by the concentration of those species that participate in the electrode reaction; as a consequence, the measurement of a half-cell potential is often a convenient method for determination of the concentration of solute species. One important aspect of such measurements is that the determination can be made without affecting appreciably any equilibria that may exist in the solution. For example, the potential of a silver electrode in a solution containing the silver cyanide complex depends only upon the silver ion activity; with suitable equipment this potential can be measured with a negligible passage of current. Since the concentration of silver ions in the solution is not sensibly altered under these conditions, the position of the equilibrium

$$Ag(CN)_2^- \rightleftharpoons 2CN^- + Ag^+$$

is not disturbed by the measurement process.

Although space limitations prevent a detailed description of the accurate determination of equilibrium constants by potential measurements, a few examples will demonstrate the process.

Example. To determine the solubility-product constant for the compound CuX_2 by a potentiometric method, we might proceed by saturating a 0.01 F solution of NaX with solid CuX_2. After equilibrium was achieved, this solution would be made part of the following cell:

$$Cu \mid CuX_2 \text{ (sat'd), NaX } (0.01 \ F) \parallel \text{ standard hydrogen electrode}$$

Assume that the potential of the foregoing was found to be $+ 0.010$ volt with the copper electrode acting as the anode. Now the standard potential for the *oxidation* of copper is

$$Cu \rightleftharpoons Cu^{2+} + 2e \qquad\qquad E° = - 0.337 \text{ volt}$$

Substituting the experimental potential into the Nernst expression for this half reaction, we get

$$+ 0.010 = - 0.337 - \frac{0.059}{2} \log [Cu^{2+}]$$

This gives

$$[Cu^{2+}] = 1.7 \times 10^{-12} \text{ mol/liter}$$

Since the X^- concentration is 0.01 mol/liter,

$$K_{sp} = (1.7 \times 10^{-12})(0.01)^2$$
$$= 1.7 \times 10^{-16}$$

Any electrode system that is sensitive to the hydrogen ion concentration of the solution can, in theory at least, be used for the estimation of the dissociation constants of acids and bases. This would include all of the half cells in which hydrogen ion is a participant. However, only a few of these have been applied to the problem.

Example. From the potential of the following cell we shall calculate the dissociation constant of the acid HP.

Pt, H_2 (1 atm) | HP (0.01 M), NaP (0.03 M) || standard hydrogen electrode

The half cell on the left was an anode and the cell potential was 0.295 volt. The anode half reaction is

$$H_2 \rightleftharpoons 2H^+ + 2e \qquad\qquad E° = 0.000 \text{ volt}$$

and we may write

$$+\, 0.295 = 0.000 - \frac{0.059}{2} \log \frac{[H^+]^2}{1}$$

$$= -\frac{0.059}{1} \log [H^+]$$

$$[H^+] = 1.00 \times 10^{-5}$$

$$K_D = \frac{[H^+][P^-]}{[HP]}$$

Since the HP present is largely undissociated

$$K_D = \frac{(1.0 \times 10^{-5})\,(0.03)}{0.01} = 3.0 \times 10^{-5}$$

The instability constants for complex ions can be measured in an analogous fashion. Thus to evaluate the constant for the $Ag(CN)_2^-$ complex we might measure the potential of the cell

$$Ag \mid Ag(CN)_2^-\,(C_1),\ CN^-(C_2) \mid\mid \text{standard hydrogen electrode}$$

Here C_1 and C_2 could be obtained from the method of preparation of the solution and the silver concentration from the potential of the cell. The constant could then be calculated from these data.

Some Limitations to the Use of Standard Electrode Potentials

The preceding sections illustrate many of the uses to which half-cell potentials can be put; the availability of a reliable set of standard electrode potentials is thus of great importance to the analytical chemist. There are, however, practical limits to this tool, and the intelligent application of these data requires a knowledge of them.

Use of concentrations instead of activities. It is generally expedient to use molar concentrations of reactive species in the Nernst equation. This involves the assumption that the activities and the molarities are identical.

In Chapter 2, we saw that the activity and molarity of a solute are identical only in very dilute solutions. With increasing electrolyte concentrations, marked differences appear between these two quantities. Unfortunately most of the systems to which we apply the Nernst equation are those having relatively high ionic strengths. As a consequence, we must expect the results calculated on the basis of molar concentration to be somewhat different from those obtained by experiment.

To illustrate, the standard electrode potential for the half reaction

$$Fe^{3+} + e \rightleftharpoons Fe^{2+}$$

is $+ 0.771$ volt. Neglecting activities we would predict that a platinum electrode immersed in a solution that was *one formal* in ferrous ions, ferric ions, and perchloric acid ought to exhibit a potential numerically equal to this relative to the standard hydrogen electrode. In fact, however, a potential of $+ 0.732$ volt is observed experimentally. The reason for the discrepancy is seen if we write the Nernst equation in the form

$$E = E° - 0.059 \log \frac{C_{Fe^{2+}} \, f_{Fe^{2+}}}{C_{Fe^{3+}} \, f_{Fe^{3+}}}$$

where C is the molar concentration of each species and f is the activity coefficient. The activity coefficients of the two species would be less than one, in this instance, in view of the high ionic strength imparted by the perchloric acid and the iron salts. More important, however, the activity coefficient of the ferric iron would be less than that of the ferrous ion inasmuch as the effects of ionic strength on these coefficients increase with the charge on the ion. As a consequence the ratio of the activity coefficients in the equation would be larger than one and the potential of half cell less than the standard potential.

Data concerning the activity coefficients of ions in solutions of the types commonly encountered in oxidation-reduction titrations and electrochemical work are fairly limited; consequently we are forced to use molar concentrations rather than activities in many calculations. Appreciable errors may result.

Effect of other equilibria. The application of standard electrode-potential data to systems that are of interest to the analytical chemist is further complicated by the occurrence of hydrolysis, dissociation, association, and complex-formation reactions involving the species in which he is interested. Often the equilibrium constants required to correct for these effects are not known. Lingane[4] cites an excellent example of this problem in the case of the ferrocyanide-ferricyanide couple

$$Fe(CN)_6^{3-} + e \rightleftharpoons Fe(CN)_6^{4-} \qquad E° = + 0.356 \text{ volt}$$

The experimentally measured potential for this couple is markedly affected by pH. Thus instead of the expected value of $+ 0.356$ volt, solutions containing

[4] J. J. Lingane, *Electroanalytical Chemistry*, 2d ed., 59. New York: Interscience Publishers, Inc., 1958.

equal formal concentrations of the two species yield potentials of $+0.71$, $+0.56$, and $+0.48$ volt with respect to the standard hydrogen electrode when the measurements are made in media which are respectively 1.0 F, 0.1 F, and 0.01 F in hydrochloric acid. The explanation for this variation is fairly simple. Both the ferrocyanide and ferricyanide ions are known to associate with hydrogen ions, and this occurs to a degree in the presence of hydrochloric acid. The hydroferrocyanic acids, however, are weaker than the hydroferricyanic acids; thus, the concentration of the ferrocyanide ion is lowered more than that of the ferricyanide ion as the acid concentration increases. This in turn tends to shift the above equilibrium to the right and leads to more positive electrode potentials.

A somewhat analogous effect is encountered in the behavior of the potential of the ferric-ferrous couple. As mentioned earlier, an equiformal mixture of these two ions in 1 F perchloric acid exhibits a reduction potential of $+0.73$ volt; substitution of hydrochloric acid of the same concentration, however, alters the observed potential to $+0.70$ volt while in phosphoric acid a value of $+0.6$ volt is observed. These differences arise because ferric ion forms more stable complexes with chloride and phosphate ion than does ferrous ion. As a result, the actual concentration of uncomplexed ferric ions in the above solution is less than the ferrous, and the net effect is a shift in the potential in the direction shown.

Phenomena such as these can be taken into account in potential calculations provided the nature of the equilibria involved are known and constants for the processes are available. Often, however, such information is lacking; the chemist is then forced to neglect such effects and hope that they will not lead to serious errors in his calculations.

In order to compensate partially for activity effects and errors resulting from side reactions, Swift[5] has proposed the use of a quantity called the "formal potential" in place of the standard electrode potential in oxidation-reduction calculations. The *formal potential* of a system is the potential against the standard hydrogen electrode when the reactants and products are at one *formal* concentration and the concentrations of any other constituents of the solution are carefully specified. Thus, for example, the formal potential for the reduction of ferric iron is $+0.731$ volt in 1 F perchloric acid and $+0.700$ volt in 1 F hydrochloric acid; similarly, formal potential for the reduction of ferricyanide ion would be $+0.71$ volt in 1 F hydrochloric acid and $+0.48$ volt in a 0.01 F solution of this acid. If these values are employed in place of the standard electrode potential in the Nernst equation, better values for calculated potentials are obtained provided the electrolyte concentration of the solution approximates that for which the formal potential was measured. Application of formal potentials to systems differing greatly as to kind and concentration of electrolyte can, however, lead to errors greater than those encountered when standard potentials are applied. The table in the appendix contains some formal potentials as well as

[5] E. H. Swift, *A System of Chemical Analysis*, 50. San Francisco: W. H. Freeman and Company, 1939.

standard potentials; in the subsequent chapters we shall use whichever seems the more appropriate.

Reaction rate. Calculations from standard potentials will indicate whether or not a given oxidation-reduction reaction is complete enough for application to an analytical problem. Unfortunately, however, we can predict nothing from such a calculation with respect to the rate at which the equilibrium will be achieved; frequently a reaction that looks extremely favorable from equilibrium considerations may be totally unacceptable when its kinetics are examined. The oxidation of arsenious acid with a solution of quadrivalent cerium in dilute sulfuric acid is an example

$$H_2AsO_3^- + Ce^{4+} + H_2O \rightleftharpoons H_2AsO_4^- + Ce^{3+} + 2H^+$$

The formal potentials for the two systems in acid are

$$2Ce^{4+} + 2e \rightleftharpoons 2Ce^{3+} \qquad\qquad E^f = +1.4 \text{ volts}$$

$$H_2AsO_4^- + 2H^+ + 2e \rightleftharpoons H_2O + H_2AsO_3^- \qquad E^f = +0.56 \text{ volt}$$

and an equilibrium constant of about 10^{28} can be calculated from these data. Despite this very favorable equilibrium constant, solutions of arsenious acid cannot be titrated with ceric ion unless a catalyst is introduced because several hours are required to reach equilibrium. In this instance, a number of good catalysts are available that will increase the rate sufficiently to allow use of the reaction in volumetric analysis.

Reversibility of half reactions. In the table of standard electrode potentials, some data will be found for reactions that are not reversible. For example

$$2CO_2 + 2H^+ + 2e \rightleftharpoons H_2C_2O_4 \qquad E^\circ = -0.49 \text{ volt}$$

This potential has been arrived at indirectly. The reaction is non-reversible, for carbon dioxide shows little tendency to recombine to give oxalic acid. Attempts to apply data of this sort to equilibrium calculations lead to meaningless results.

Calculation of real-cell potentials. Thus far in the treatment of cell potentials we have considered only those potentials associated with the reversible oxidation-reduction reactions. There are, however, other potentials that need to be taken into account in considering real cells, particularly those through which current is passed. These include junction potentials, overvoltage potentials, polarization effects, and iR drop. These will be considered in later chapters.

problems

1. In each of the cases below are given the formulae of reactants and products (exclusive of H^+, OH^- and H_2O) of a chemical reaction. Indicate the oxidizing agent and write a balanced equation for its half reaction. Do the same for the reducing agent.

Reactants	Products
(a) Cl_2, I^-	I_3^-, Cl^-
(b) Cd, Ag^+	Ag, Cd^{2+}
(c) H_2, Fe^{3+}	H^+, Fe^{2+}
(d) $Cr_2O_7^{2-}$, U^{4+}	Cr^{3+}, UO_2^{2+}
(e) $V(OH)_4^+$, V^{3+}	VO^{2+}
(f) HNO_2, MnO_4^-	NO_3^-, Mn^{2+}
(g) O_2, I^-	H_2O, I_3^-
(h) IO_3^-, I^-	I_2
(i) $H_2C_2O_4$, Ce^{4+}	CO_2, Ce^{3+}
(j) Ag, Br^-, Sn^{4+}	$AgBr$, Sn^{2+}

ans. (a) Oxidizing agent $Cl_2 + 2e \rightleftharpoons 2Cl^-$
Reducing agent $3I^- \rightleftharpoons I_3^- + 2e$

2. Give the equivalent weight of each oxidizing agent in Problem 1.

ans. (1-a) eq wt $Cl_2 = \dfrac{GFW\ Cl_2}{2}$

3. Write balanced equations for each of the reactions in Problem 1.

4. A current was passed through several cells connected in series. It caused deposition 0.0159 gram of copper at one of the electrodes. How many grams of the following were deposited at each of the other electrodes: Ag, Pb, Cr (from a $Cr_2O_7^{2-}$ solution), O_2, Tl (from a Tl^{3+} solution)
ans. 0.0539 gram Ag, 0.0518 gram Pb, 0.00433 gram Cr,
0.00400 gram O_2, 0.0341 gram Tl

5. A certain quantity of electricity deposited 13.13 mg of gold from an Au^+ solution. How many grams of the following would be deposited by the same current? Cd, Tl (from a Tl^+ solution), Cr (from a Cr^{3+} solution), H_2, Ni

6. What quantity of current was passed through the cells in Problem 4?
ans. 48.2 coulombs

7. What is the quantity of current in coulombs in Problem 5?

8. What volume of hydrogen gas (at standard conditions) will be liberated at a cathode by passage of a current of 2.00 amperes for 10.0 minutes?
ans. 139 ml

9. How many milliliters of Cl_2 gas (at standard conditions) were evolved at an anode by a constant current of 0.500 ampere operating for exactly 1 hour?

10. A chemical coulometer is a device for measuring quantity of electricity. One type involves the measurement of the volume of hydrogen gas evolved by the current at a cathode. Calculate the number of coulombs of electricity passed through such a coulometer if a volume of 19.2 ml of gas were collected at standard conditions.

11. Calculate the number of coulombs of current passing through a coulometer in which hydrogen is evolved at the cathode and oxygen at the anode. The total volume of the two gases is 49.2 ml at standard conditions.

12. Calculate the reduction potentials of the following half cells against the standard hydrogen electrode.
    (a) Ag | Ag$^+$ (0.0100 M)   ans. $E = + 0.681$ volt
    (b) Ni | Ni^{2+} (0.712 M) $E = - 0.254$ volt
    (c) Pt, H$_2$ (1 atm) | HCl (10^{-5} F) $E = - 0.295$ volt
    (d) Pt | Fe^{3+} (1.00 $\times$ 10^{-4} M), Fe^{2+} (0.100 M) $E - + 0.594$ volt
    (e) Ag | AgBr (sat'd), Br$^-$ (3.00 M) $E = + 0.067$ volt

13. Calculate the reduction potentials of the following half cells against the standard hydrogen electrode.
    (a) Cr | Cr^{3+} (0.00100 M)
    (b) Pb | Pb^{2+} (1.00 $\times$ 10^{-5} M)
    (c) Pt, H$_2$ (1.00 atm) | HCl (4.00 F)
    (d) Pt | V^{3+} (0.100 M), V^{2+} (0.0500 M)
    (e) Hg | Hg$_2$Cl$_2$ (sat'd), Cl$^-$ (0.0300 M)

14. Calculate the reduction potentials for the following systems against the standard hydrogen electrode.
    (a) Pt | V^{3+} (1.00 $\times$ 10^{-3} M), VO^{2+} (1.00 M), HCl (0.100 M)
           ans. $E = + 0.420$ volt
    (b) Pt | MnO$_4^-$ (0.300 M), Mn^{2+} (0.100 M), H$^+$ (0.200 M)
           ans. $E = + 1.45$ volts

15. Calculate the reduction potentials for the following systems against the standard hydrogen electrode.
    (a) Pt | UO$_2^{2+}$ (0.010 M), U^{4+} (0.100 M), H$^+$ (0.100 M)
    (b) Pt | TiO^{2+} (0.500 M), Ti^{3+} (0.100 M), H$^+$ (1.00 $\times$ 10^{-3} M)

16. Indicate whether the following half cells would behave as the anodes or cathodes of the galvanic cell formed when coupled with a standard hydrogen electrode.
    (a) Co | Co^{2+} (1.00 M)   ans. anode
    (b) Ag | AgCl (sat'd), Cl$^-$ (1.00 M)   ans. cathode
    (c) Pb | Pb^{2+} (1.00 $\times$ 10^{-4} M)
    (d) Pt, H$_2$ (1 atm) | H$^+$ (1.00 $\times$ 10^{-5} M)
    (e) Ag | AgI (sat'd), I$^-$ (1.00 M)
    (f) Pt | Sn^{4+} (10^{-5} M), Sn^{2+} (10^{-5} M)
    (g) Pt | CuI (sat'd), I$^-$ (1.00 $\times$ 10^{-10} M)

17. Arrange the following substances in their order of decreasing strengths as oxidizing agents: Fe(CN)$_6^{3-}$, Fe^{3+}, Ag(CN)$_2^-$ in 1 F KCN, F$_2$, Al^{3+}, H$^+$ (in 1 F acid), Ce^{4+}, Cd^{2+}, O$_2$ (in 1 F acid), Cr^{3+}.

18. Arrange the following substances in their order of decreasing strength as reducing agents: Cl$^-$, Fe^{2+}, Ni, Ag (in 1 F KI), V^{2+}, Pb, Ti^{3+}, Na$_2$S$_2$O$_3$, KI, H$_2$.

19. Indicate in which direction the following reactions will go if all substances are initially at unit activity.
    (a) Fe^{3+} + Ag $\rightleftharpoons$ Fe^{2+} + Ag$^+$ ans. left
    (b) Sn^{4+} + 2Ag + 2Cl$^-$ $\rightleftharpoons$ Sn^{2+} + 2AgCl   ans. left
    (c) 2Tl + Cd^{2+} $\rightleftharpoons$ 2Tl$^+$ + Cd

(d) $2Ce^{4+} + 2Br^- \rightleftharpoons 2Ce^{3+} + Br_2$

(e) $5Cl_2 + I_2 + 6H_2O \rightleftharpoons 2IO_3^- + 10Cl^- + 12H^+$

(f) $Ag(CN)_2^- + Cr^{2+} \rightleftharpoons Ag + Cr^{3+} + 2CN^-$

(g) $Ba + 2Na^+ \rightleftharpoons Ba^{2+} + 2Na$

(h) $2Ag + Hg_2Cl_2 \rightleftharpoons 2Hg + 2AgCl$

(i) $4Ce^{4+} + 2H_2O \rightleftharpoons 4Ce^{3+} + O_2 + 4H^+$

20. In 1-F acid solution which is the better oxidizing agent, I_2 or H_3AsO_4? In neutral solution? Explain your answers.

21. Calculate the theoretical potentials of the following cells. Indicate which electrode would act as the anode.

(a) $Pb \mid Pb^{2+}$ (0.100 M) $\parallel Cd^{2+}$ (0.0010 M) $\mid Cd$

(b) $Pt \mid I_3^-$ (0.0100 M), I^- (0.1 M), AgI (sat'd) $\mid Ag$

(c) Pt, H_2 (1 atm) $\mid H^+$ (1.0 $\times$ 10^{-5} M), KCl (0.1 M), $AgCl$ (sat'd) $\mid Ag$

(d) $Pt \mid Tl^{3+}$ (1.0 M), Tl^+ (0.01 M) $\parallel Zn^{2+}$ (0.01 M) $\mid Zn$

ans. (a) 0.335 volt, Cd electrode anode

(b) 0.658 volt, Ag electrode anode

(c) 0.576 volt, Pt electrode anode

(d) 2.13 volts, Zn electrode anode

22. Calculate the theoretical cell potentials for the following. Indicate which electrode would behave as the negative electrode (anode) of a galvanic cell.

(a) $Ag \mid AgBr$ (sat'd), Br^- (0.01 M), H^+ (10 M) $\mid H_2$ (1 atm) Pt

(b) $Pt \mid Fe^{3+}$ (0.2 M), Fe^{2+} (0.01 M) $\parallel Cl^-$ (0.1 M), Hg_2Cl_2 (sat'd) $\mid Hg$

(c) $Cd \mid Cd^{2+}$ (0.305 M) $\parallel H^+$ (0.01 M) $\mid O_2$ (1 atm) Pt

(d) $Pt \mid UO_2^{2+}$ (0.510 M), U^{4+} (0.100 M), H^+ (0.01 M) $\parallel Ni^{2+}$ (0.4 M) $\mid Ni$

23. Calculate the potentials of the following cells. Indicate which electrode is the anode.

(a) $Ag \mid AgCl$ (sat'd), KCl (1.0 M) $\parallel KCl$ (0.001 M), $AgCl$ (sat'd) $\mid Ag$

(b) Pt, H_2 (1 atm) $\mid HCl$ (0.5 M) $\parallel HCl$ (1.0 $\times$ 10^{-4} M) $\mid H_2$ (1 atm) Pt

24. In order to determine the solubility product for the salt AgX, the following cell was prepared:

$Ag \mid AgX$ (sat'd), X^- (0.1 M), H^+ (1.0 M) $\mid H_2$ (1 atm) Pt

The cell had a potential of 0.122 volt and the silver electrode behaved as the anode. Calculate the silver ion concentration of the solution and the solubility product of AgX. ans. $[Ag^+] = 2.5 \times 10^{-16}$ mol/liter

$K_{sp} = 2.5 \times 10^{-17}$

25. From the standard potentials

$$Tl^+ + e \rightleftharpoons Tl \qquad\qquad E° = -0.336 \text{ volt}$$
$$TlCl + e \rightleftharpoons Tl + Cl^- \qquad E° = -0.557 \text{ volt}$$

calculate the solubility-product constant for TlCl. ans. $K_{sp} = 1.8 \times 10^{-4}$

26. Calculate the solubility product for $Mn(OH)_2$ from the following data:

$$Mn^{2+} + 2e \rightleftharpoons Mn \qquad\qquad\quad E° = -1.18 \text{ volts}$$
$$Mn(OH)_2 + 2e \rightleftharpoons Mn + 2OH^- \qquad E° = -1.55 \text{ volts}$$

27. The solubility product for Tl_2S is 1.2×10^{-22}. Calculate $E°$ for the reaction

$$Tl_2S + 2e \rightleftharpoons 2Tl^+ + S^{2-}.$$

28. Calculate the instability constant for $Ni(CN)_4^{2-}$ from the data:

$$Ni^{2+} + 2e \rightleftharpoons Ni \qquad\qquad\quad E° = -0.250 \text{ volt}$$
$$Ni(CN)_4^{2-} + 2e \rightleftharpoons Ni + 4CN^- \qquad E° = -0.82 \text{ volt}$$

29. The standard reduction potential for M^{2+} is
$$M^{2+} + 2e \rightleftharpoons M \qquad\qquad E^\circ = +0.0118 \text{ volt}$$
The following cell:
$$M \mid MX_2 \text{ (sat'd)}, X^- (0.400\ F) \parallel \text{ standard hydrogen electrode}$$
was found to have a potential of $+0.204$ volt with the M electrode as an anode. Calculate the solubility product for MX_2.

30. From the data in Problem 29 calculate E° for the reaction
$$MX_2 + 2e \rightleftharpoons M^{2+} + 2X^-$$

31. The cation M^{2+} forms a stable complex with anion Y^- having the formula MY_4^{2-}. A solution of the complex was prepared by dissolving 0.05 of a formula weight of a soluble salt of M^{2+} in 1 liter of 0.800 formal Y^-. A metallic M electrode in this solution behaved as an anode against a standard hydrogen electrode developing a potential of 0.412 volt. The E° value for $M^{2+} + 2e \rightleftharpoons M$ is $+0.0118$ volt. Calculate the instability constant for the complex MY_4^{2-}.

32. The half cell
$$Pt, H_2 \text{ (1 atm)} \mid HA \text{ (0.2 } F\text{)}, NaA \text{ (0.1 } F\text{)}$$
behaves as an anode when coupled with a standard hydrogen electrode. Its potential is 0.443 volt. Calculate the dissociation constant for the weak acid HA.
ans. $K_D = 1.5 \times 10^{-8}$

33. The half cell
$$Pt, H_2 \text{ (1 atm)} \mid HY \text{ (0.12 } F\text{)}, NaY \text{ (0.36 } F\text{)}$$
develops an anode potential of 0.107 volt against the standard hydrogen electrode. What is the dissociation constant for HY?

34. Calculate the theoretical potential required to begin deposition of nickel from a solution buffered to a pH of 7.0 which is 0.05 M in Ni^{2+}. Assume the anode reaction is evolution of O_2 at a pressure of one atmosphere.

35. Calculate the potential necessary to reduce the nickel concentration to 1.0×10^{-4} M in the cell in Problem 34.

36. Compute the equilibrium constant for the reaction.
$$Br_2 + 2Fe^{2+} \rightleftharpoons 2Br^- + 2Fe^{3+}$$
ans. $\dfrac{[Br^-]^2\, [Fe^{3+}]^2}{[Br_2]\, [Fe^{2+}]^2} = 1.0 \times 10^{10}$

37. Calculate the equilibrium constant for the reaction.
$$O_2 + 4I^- + 4H^+ \rightleftharpoons 2I_2 + 2H_2O$$

38. Calculate the equilibrium constant for the reaction.
$$IO_3^- + 8I^- + 6H^+ \rightleftharpoons 3I_3^- + 3H_2O$$

39. An excess of iron filings are shaken in a 0.02 F solution of $CdSO_4$. What is the equilibrium concentration of Cd^{2+}?

40. Calculate equilibrium constants for the following reactions (unbalanced).
    (a) $MnO_4^- + Tl^+ \rightleftharpoons Mn^{2+} + Tl^{3+}$
    (b) $VO^{2+} + 2I^- \rightleftharpoons V^{3+} + I_2$
    (c) $TiO^{2+} + V^{2+} \rightleftharpoons Ti^{3+} + V^{3+}$

chapter 18. *Theory of Oxidation-reduction Titrations*

The successful application of an oxidation-reduction reaction to volumetric analysis requires, among other things, the means for detecting the point of chemical equivalence in the reaction. We must therefore examine the changes that occur in the course of such a titration, paying particular attention to those that are most pronounced in the region of the equivalence point.

Titration Curves

The titration curves so far considered have taken the form of a plot of the logarithm of the concentration of one of the reacting species against the volume of reagent added. The species chosen for plotting in each case has been the one to which the indicator for the reaction is sensitive. Most of the indicators used for oxidation-reduction titrations are themselves oxidizing or reducing agents that respond not to the changes in concentration of any particular ion in the solution but rather to the change in oxidation potential of the substrate. For this reason, the usual practice is to use a half-cell potential as the ordinate in oxidation-reduction titration curves rather than the logarithm of a concentration. To illustrate, for the volumetric reaction

$$Ce^{4+} + Fe^{2+} \rightleftharpoons Ce^{3+} + Fe^{3+}$$

a typical titration curve will consist of the oxidation or reduction potential for the ceric-cerous system or the ferric-ferrous system plotted as a function of the volume of the reagent. Because the two half reactions are in equilibrium throughout the entire titration, *the reduction potentials for the two systems are numerically identical at every point on the curve.*

Equivalence-point Potential

Of the data that can be computed for an oxidation-reduction system the most useful is the potential at chemical equivalence. To illustrate the calculation we shall take the foregoing reaction. The half reactions are

$$Ce^{4+} + e \rightleftharpoons Ce^{3+}$$

$$Fe^{3+} + e \rightleftharpoons Fe^{2+}$$

and we may write

$$E = E^{\circ}_{Ce^{4+}} - 0.059 \log \frac{[Ce^{3+}]}{[Ce^{4+}]}$$

$$E = E^{\circ}_{Fe^{3+}} - 0.059 \log \frac{[Fe^{2+}]}{[Fe^{3+}]}$$

We will add the two equations, recalling that since the system is in equilibrium, the two values of E are identical. We get

$$2E = E^{\circ}_{Ce^{4+}} + E^{\circ}_{Fe^{3+}} - 0.059 \log \frac{[Ce^{3+}] [Fe^{2+}]}{[Ce^{4+}] [Fe^{3+}]}$$

We know, from the stoichiometric equation, that at the equivalence point

$$[Fe^{3+}] = [Ce^{3+}]$$

$$[Fe^{2+}] = [Ce^{4+}]$$

so that

$$2E = E^{\circ}_{Ce^{4+}} + E^{\circ}_{Fe^{3+}} - 0.059 \log \frac{[\cancel{Ce^{3+}}] [\cancel{Fe^{2+}}]}{[\cancel{Ce^{4+}}] [\cancel{Fe^{3+}}]}$$

and

$$E = \frac{E^{\circ}_{Ce^{4+}} + E^{\circ}_{Fe^{3+}}}{2}$$

The E obtained in this way represents the reduction potential (against the standard hydrogen electrode) of either the ceric-cerous or the ferric-ferrous systems at the concentrations of these species present at the equivalence point.

Now we shall take the somewhat more complicated case

$$5Fe^{2+} + MnO_4^- + 8H^+ \rightleftharpoons 5Fe^{3+} + Mn^{2+} + 4H_2O$$

The respective half reactions may be written

$$Fe^{3+} + e \rightleftharpoons Fe^{2+}$$

$$MnO_4^- + 8H^+ + 5e \rightleftharpoons Mn^{2+} + 4H_2O$$

The equilibrium potential of this system is given either by

$$E = E_{Fe^{3+}}^\circ - \frac{0.059}{1} \log \frac{[Fe^{2+}]}{[Fe^{3+}]}$$

or

$$E = E_{MnO_4^-}^\circ - \frac{0.059}{5} \log \frac{[Mn^{2+}]}{[MnO_4^-][H^+]^8}$$

Multiplying the second equation through by 5 will permit the combination of logarithmic terms.

$$5E = 5E_{MnO_4^-}^\circ - 0.059 \log \frac{[Mn^{2+}]}{[MnO_4^-][H^+]^8}$$

After addition we find that

$$6E = E_{Fe^{3+}}^\circ + 5E_{MnO_4^-}^\circ - 0.059 \log \frac{[Fe^{2+}][Mn^{2+}]}{[Fe^{3+}][MnO_4^-][H^+]^8}$$

Inspection of the stoichiometry shows that at the equivalence point

$$[Fe^{3+}] = 5[Mn^{2+}]$$

$$[Fe^{2+}] = 5[MnO_4^-]$$

Substituting

$$E = \frac{E_{Fe^{3+}}^\circ + 5E_{MnO_4^-}^\circ}{6} - \frac{0.059}{6} \log \frac{\cancel{5}[MnO_4^-]\cancel{[Mn^{2+}]}}{\cancel{5}[Mn^{2+}]\cancel{[MnO_4^-]}[H^+]^8}$$

$$= \frac{E_{Fe^{3+}}^\circ + 5E_{MnO_4^-}^\circ}{6} - \frac{0.059}{6} \log \frac{1}{[H^+]^8}$$

This is an example of a reaction for which the equivalence-point potential is dependent upon the *p*H. There are some instances where the concentration of other ions also affect the equivalence potential—for example, the following reaction:

$$6Fe^{2+} + Cr_2O_7^{2-} + 14H^+ \rightleftharpoons 6Fe^{3+} + 2Cr^{3+} + 7H_2O$$

Proceeding as in the previous example we obtain the expression

$$7E = E_{Fe^{3+}}^\circ + 6E_{Cr_2O_7^{2-}}^\circ - 0.059 \log \frac{[Fe^{2+}][Cr^{3+}]^2}{[Fe^{3+}][Cr_2O_7^{2-}][H^+]^{14}}$$

At the equivalence point

$$[Fe^{2+}] = 6[Cr_2O_7^{2-}]$$

$$[Fe^{3+}] = 3[Cr^{3+}]$$

Upon substituting these, we find that

$$E = \frac{E^{\circ}_{Fe^{3+}} + 6E^{\circ}_{Cr_2O_7^{2-}}}{7} - \frac{0.059}{7} \log \frac{2[Cr^{3+}]}{[H^+]^{14}}$$

In general, the equivalence-point potential will depend upon the concentration of one of the participants in the reaction whenever a molar ratio other than unity exists between the species containing that participant as a reactant and product. Dependence upon the hydrogen ion concentration will also be observed whenever this ion takes part in the reaction.

The equilibrium concentrations of the reacting species can be readily calculated from the equivalence-point potential.

Example. We shall calculate the concentration of the various reactants and products at the equivalence point in the titration of a 0.1 N solution of Fe^{2+} in 1 F H_2SO_4 with 0.1 N Ce^{4+} in 1 F H_2SO_4 at 25° C.

From the foregoing discussion we have seen that the equivalence-point potential in this titration is

$$E = \frac{E^{\circ}_{Ce^{4+}} + E^{\circ}_{Fe^{3+}}}{2}$$

In this case formal potentials for the two half reactions in 1 F H_2SO_4 are available and lead to better results than standard potentials. Substituting these for the E° values we obtain

$$E = \frac{+1.44 + 0.68}{2} = +1.06 \text{ volt}$$

The concentrations of Fe^{2+} and Fe^{3+} can be computed by first evaluating their molar ratio at the equivalence point. This is done with the Nernst equation

$$E = E^{\circ}_{Fe^{3+}} - 0.059 \log \frac{[Fe^{2+}]}{[Fe^{3+}]}$$

Substituting

$$+1.06 = +0.68 - 0.059 \log \frac{[Fe^{2+}]}{[Fe^{3+}]}$$

$$\log \frac{[Fe^{2+}]}{[Fe^{3+}]} = -\frac{0.38}{0.059} = -6.44$$

$$\frac{[Fe^{2+}]}{[Fe^{3+}]} = 3.6 \times 10^{-7}$$

From the size of this ratio we see that most of the Fe^{2+} has been converted to Fe^{3+} at the equivalence point, and as a consequence the Fe^{3+} concentration will be approximately one half the original Fe^{+2} concentration (because of dilution); that is,

$$[Fe^{3+}] = 0.05 - [Fe^{2+}] \cong 0.05 \text{ mol/liter}$$

Then

$$[Fe^{2+}] = 3.6 \times 10^{-7} \times 0.05$$

$$= 1.8 \times 10^{-8} \text{ mol/liter}$$

Finally, consideration of the stoichiometry of the reaction allows the following prediction:

$$[Ce^{4+}] \cong 1.8 \times 10^{-8} \text{ mol/liter}$$

$$[Ce^{3+}] \cong 0.05 \text{ mol/liter}$$

Derivation of Titration Curves

The shape of the titration curve for an oxidation-reduction reaction depends upon the nature of the system under consideration. The derivation of several typical curves will serve to illustrate this.

Titration of ferrous iron with quadrivalent cerium. As an example we shall consider the titration of 100 ml of 0.100 F solution of ferrous iron with 0.100 F solution of quadrivalent cerium. Assuming further that both solutions are 1 F in sulfuric acid, we shall use the appropriate formal potentials in place of standard potentials.

$$Ce^{4+} + e \rightleftharpoons Ce^{3+} \qquad E^f = +1.44 \text{ volts}$$

$$Fe^{3+} + e \rightleftharpoons Fe^{2+} \qquad E^f = +0.68 \text{ volt}$$

1. *Initial potential.* At the outset the solution contains no cerium ions; there will, however, be a small but finite quantity of ferric ion present as a consequence of air oxidation. Because the extent of this is uncertain, we cannot calculate an initial potential which has significance.

2. *Potential after addition of 10 ml of oxidizing agent.* With the introduction of a volume of the oxidizing agent, the solution acquires appreciable concentrations of three of the participating ions; that for the fourth, the ceric ion, will be small.

After 10 ml of quadrivalent cerium have been added, ferrous iron remains in excess and equivalent amounts of ferric and cerous ions have been produced. The equilibrium concentration of ceric ion will be very small. Setting this last equal to x, we have

$$[Ce^{4+}] = x$$

$$[Ce^{3+}] = \frac{10 \times 0.1}{110} - x \cong \frac{1}{110} \text{ mol/liter}$$

$$[Fe^{3+}] = \frac{10 \times 0.1}{110} - x \cong \frac{1}{110} \text{ mol/liter}$$

$$[Fe^{2+}] = \frac{100 \times 0.1 - 10 \times 0.1}{110} + x \cong \frac{9}{110} \text{ mol/liter}$$

Since the equilibrium lies reasonably far to the right, the approximation that the concentration of ceric ion (x) is negligibly small should be satisfactory. Its numerical value could be found from the equilibrium constant for the reaction;

however, we need only know that it is very small since the potential for the system can be calculated with the aid of *either of the two equations*

$$E = E'_{Ce^{4+}} - 0.059 \log \frac{[Ce^{3+}]}{[Ce^{4+}]}$$

$$= E'_{Fe^{3+}} - 0.059 \log \frac{[Fe^{2+}]}{[Fe^{3+}]}$$

The second equation is the more convenient since we know, within acceptable limits, the two ionic concentrations. Therefore, substituting for the ferric and ferrous concentration

$$E = +0.68 - 0.059 \log \frac{9/110}{1/110}$$

$$= +0.68 - 0.059 \log 9$$

$$E = +0.62 \text{ volt}$$

Had we used the formal potential for the ceric-cerous system and the equilibrium concentrations of these ions, an identical potential would have been obtained.

3. *Additional pre-end-point potentials.* Further values for the potentials necessary to define a curve up to the end point can be calculated in a fashion

Table 18-1

REDUCTION POTENTIALS DURING TITRATIONS OF FERROUS SOLUTIONS
(100 ml of 0.1 F Fe^{2+})

Volume of 0.1 N Reagent, ml	Potential, Volts vs the Standard Hydrogen Electrode[1]	
	Titration with Ce^{4+}	Titration with MnO$_4^-$
0.0	—	—
10.0	+0.62	+0.62
30.0	+0.66	+0.66
50.0	+0.68	+0.68
70.0	+0.70	+0.70
90.0	+0.74	+0.74
99.0	+0.80	+0.80
99.9	+0.86	+0.86
100.0	+1.06 ← equivalence point →	+1.37
100.1	+1.26	+1.47
101	+1.32	+1.48
110	+1.38	+1.49
130	+1.40	+1.50

[1] These data have been calculated from formal potentials for the two systems in 1 F H$_2$SO$_4$.

strictly analogous to that in (2). Table 18-1 contains a number of these. The student should check one or two to be sure he understands completely how they were obtained.

4. *Equivalence-point potential.* We have seen that the potential at the equivalence point in this titration is given by

$$E = \frac{E^{\circ}_{Ce^{4+}} + E^{\circ}_{Fe^{3+}}}{2}$$

Substituting formal potentials in this equation yields a numerical value of $+ 1.06$ volts.

5. *Potential after addition of 100.1 ml of reagent.* The solution now contains an excess of quadrivalent cerum in addition to equivalent quantities of ferric and ceric ions. The concentration of ferrous iron will be very small; if we set its value equal to y, then

$$[Fe^{2+}] = y$$

$$[Fe^{3+}] = \frac{100 \times 0.1}{200.1} - y \cong \frac{10}{200.1} \text{ mol/liter}$$

$$[Ce^{3+}] = \frac{100 \times 0.1}{200.1} - y \cong \frac{10}{200.1} \text{ mol/liter}$$

$$[Ce^{4+}] = \frac{100.1 \times 0.1 - 100 \times 0.1}{200.1} + y \cong \frac{0.01}{200.1} \text{ mol/liter}$$

These approximations should be reasonably good in view of the favorable equilibrium constant. As before we could calculate the desired potential from the formal value for the ferric-ferrous system; at this stage in the titration, however, it is more convenient to use the ceric-cerous potential since the concentrations of these species are immediately available. Thus

$$E = + 1.44 - 0.059 \log \frac{[Ce^{3+}]}{[Ce^{4+}]}$$

$$= + 1.44 - 0.059 \log \frac{10/200.1}{0.01/200.1}$$

$$= + 1.26 \text{ volts}$$

The additional postequivalence-point potentials shown in Table 18-1 were calculated in a similar fashion.

A curve for this titration is shown in Figure 18.1. Its shape is quite analogous to the curves encountered in neutralization, precipitation, and complex-formation titrations, the equivalence point being signaled by a large change in the ordinate function. In this case the curve is symmetric about the equivalence point. This will only be the case, however, when the oxidant and reductant react in equimolar amounts. From the calculations we see that the potential values are all independent of dilution; thus a curve involving 0.01 F reactants will be, for all practical purposes, identical with the one that was derived.

Titration of ferrous iron with permanganate.   As a second example we will consider how a curve is derived for the titration of 100 ml of 0.100 F ferrous ion with 0.100 N potassium permanganate. To simplify the calculations we will assume that the hydrogen ion concentration of the solution is maintained at 1 M throughout the titration.

The chemical reaction is

$$5Fe^{2+} + MnO_4^- + 8H^+ \rightleftharpoons 5Fe^{3+} + Mn^{2+} + 4H_2O$$

In all of the computations, the formal concentrations of the manganese species must be employed. Note that these are one fifth of the normal concentrations.

1. *Pre-equivalence-point potentials.*   The pre-equivalence-point potentials are most easily calculated from the concentrations of ferrous and ferric ions in the solution; their values will be substantially identical to those computed for the previous titration with ceric ion.

2. *Equivalence-point potential.* The equivalence-point potential in this case is given by the equation (see p. 402)

$$E = \frac{E^\circ_{Fe^{3+}} + 5E^\circ_{MnO_4^-}}{6} - \frac{0.059}{6} \log \frac{1}{[H^+]^8}$$

Substituting formal potentials for the corresponding standard potentials and introducing a value of 1 for the hydrogen ion concentration gives

$$E = \frac{+0.68 + 5(+1.51)}{6} - \frac{0.059}{6} \log \frac{1}{(1)^8}$$

$$= +1.37 \text{ volts}$$

3. *Postequivalence-point potentials.* We shall consider the case when 100.1 ml of the 0.1 N potassium permanganate have been added. From the stoichiometry we find that

$$[Fe^{2+}] = x$$

$$[Fe^{3+}] = \left(\frac{100 \times 0.1}{200.1} - x\right) \simeq \frac{10}{200.1} \text{ mol/liter}$$

$$[Mn^{2+}] = \frac{1}{5}\left(\frac{10}{200.1} - x\right) \simeq \frac{10}{5 \times 200.1} \text{ mol/liter}$$

$$[MnO_4^-] = \frac{1}{5}\left(\frac{100.1 \times 0.1 - 100.0 \times 0.1}{200.1} + x\right)$$

$$\simeq \frac{0.01}{5 \times 200.1} \text{ mol/liter}$$

We are now in a position to calculate the electrode potential from the formal potential of the manganese system; that is,

$$E = + 1.51 - \frac{0.059}{5} \log \frac{[Mn^{2+}]}{[MnO_4^-] [H^+]^8}$$

$$= + 1.51 - \frac{0.059}{5} \log \frac{(10/200.1) \times 5}{(0.01/200.1) \times 5 (1)^8}$$

$$= + 1.47 \text{ volts}$$

Additional values to establish this part of the titration curve can be arrived at in an analogous fashion. Table 18-1 contains such data.

Figure 18.1 illustrates titration curves for ferrous iron with both permanganate and ceric reagent. Comparing the two, we see first that the plots are alike to within 99.9 percent of the equivalence point; however, the potentials at the equivalence points are quite different. Further, the permanganate curve is

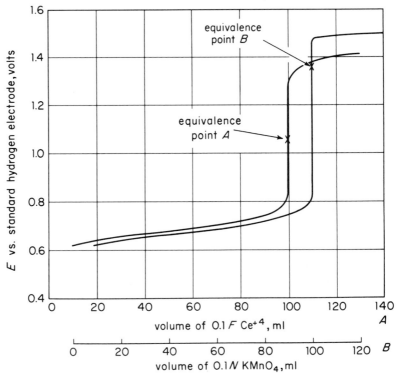

Fig. 18.1 Curves Depicting the Titration of 100 ml of 0.100 F Ferrous Iron Solution with (A) Quadrivalent Cerium and (B) with Permanganate. Note that the horizontal axes have been displaced to permit comparison of the curves.

markedly unsymmetric, the potential increasing only slightly beyond the equivalence point. Finally, the total change in potential associated with equivalence is somewhat greater with the permanganate titration; this is due to the more favorable equilibrium constant for this reaction.

Titration of Mixtures and Substances which Yield Multiple Products

When a mixture of two oxidizing agents or two reducing agents is titrated, a curve having two inflection points results provided the reduction potentials of the two species are sufficiently different. If the difference is greater than about 0.2 volt, end points can ordinarily be obtained that will allow analysis of each of the components. The situation here is quite comparable to the titration of two acids having different dissociation constants or of two ions forming precipitates of different solubilities with the same reagent; curves for these were considered in earlier chapters.

In addition, the titration behavior of a few redox systems is reminiscent of that of polybasic acids—for example, consider the following two half reactions:

$$VO^{2+} + 2H^+ + e \rightleftharpoons V^{3+} + H_2O \qquad E° = +0.361 \text{ volt}$$

$$V(OH)_4^+ + 2H^+ + e \rightleftharpoons VO^{2+} + 3H_2O \qquad E° = +1.00 \text{ volt}$$

The titration curve for V^{3+} with a strong oxidizing agent such as permanganate will have two inflection points; the first will correspond to oxidation of the V^{3+} to VO^{2+} and the second to oxidation to $V(OH)_4^+$. The stepwise oxidation of molybdenum (III) first to the $+5$ oxidation state and subsequently to the $+6$ state is another common example. Here again satisfactory breaks in the curves exist because of a potential difference of about 0.4 volt between the pertinent half reactions.

Derivation of titration curves for either type of system described above is not very difficult if the difference in standard potential is great enough. An example is the titration of a mixture of ferrous and titanous ions by means of potassium permanganate. From the table of standard potentials we find for these systems

$$TiO^{2+} + 2H^+ + e \rightleftharpoons Ti^{3+} + H_2O \qquad E° = +0.1 \text{ volt}$$

$$Fe^{3+} + e \rightleftharpoons Fe^{2+} \qquad E° = +0.77 \text{ volt}$$

The first additions of reagent find the permanganate consumed by the more readily oxidized titanous ion; as long as an appreciable concentration of this species remains in solution, the oxidation potential of the system cannot become high enough to alter greatly the concentration of ferrous ions. Thus, the first part of the titration curve can be defined from the stoichiometric proportions of titanous and titanic ion with the aid of the relationship

$$E = +0.1 - 0.059 \log \frac{[Ti^{3+}]}{[TiO^{2+}][H^+]^2}$$

For all practical purposes, then, the first part of this curve is identical to the titration curve for titanous ion by itself. Beyond the first equivalence point, the solution will contain both ferrous and ferric ions in appreciable concentrations and the points on the curve can be most conveniently obtained from the relationship

$$E = +0.77 - 0.059 \log \frac{[Fe^{2+}]}{[Fe^{3+}]}$$

Throughout this region and beyond the second equivalence point, the titration curve will be essentially identical to that for the titration of ferrous ion by itself (see Fig. 18.1). This, then, leaves only the potential at the first equivalence point. A convenient way of estimating its value is to add the equations for the ferrous and titanous potentials. Since the potentials for all oxidation-reduction systems in solution will be identical at equilibrium, this yields

$$2E = +0.1 + 0.77 - 0.059 \log \frac{[Ti^{3+}][Fe^{2+}]}{[TiO^{2+}][Fe^{3+}][H^+]^2}$$

The ferric ions in the solution at this point arise primarily from the reaction

$$2H^+ + TiO^{2+} + Fe^{2+} \rightleftharpoons Fe^{3+} + Ti^{3+} + H_2O$$

and we may write

$$[Fe^{3+}] \cong [Ti^{3+}]$$

Substituting this into the previous equation for the potential, we obtain

$$E = \frac{+0.87}{2} - \frac{0.059}{2} \log \frac{[Fe^{2+}]}{[TiO^{2+}][H^+]^2}$$

Finally if we assume that $[TiO^{2+}]$ and $[Fe^{2+}]$ are not much different from their formal concentrations, we can compute the equivalence-point potential.

A titration curve for a mixture of ferrous and titanous ions is shown in Figure 18.2.

OXIDATION-REDUCTION INDICATORS

We have seen that the equivalence point in an oxidation-reduction titration is characterized by a marked change in the reduction potential of the system. Several methods are available for detecting such a change, and these can serve to signal the end point in the titration. We shall now consider one of these methods—namely, that based upon the use of oxidation-reduction indicators. Some of the other methods of end-point detection will be encountered in later chapters.

Types of Chemical Indicators

Two types of indicators are encountered in the application of oxidation-reduction titrations.

Specific indicators. Specific indicators are substances that react in a specific manner with one of the participants in the titration to produce a change in color. Perhaps the best known of such indicators is starch, which forms a dark-blue complex with triiodide ion. This complex often serves to detect the end point in titrations in which iodine is either produced or consumed. Another example of such an indicator is potassium thiocyanate, which may be employed in the titration of ferric ion with solutions of titanous sulfate. The end point involves decolorization of the ferric thiocyanate complex due to a marked reduction in ferric ion concentration at the equivalence point.

True oxidation-reduction indicators. There is a class of substances

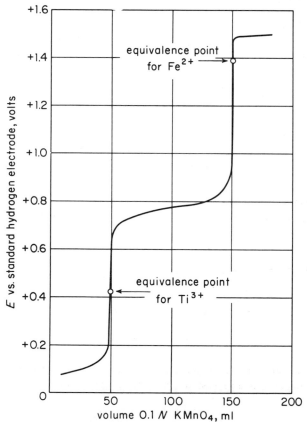

Fig. 18.2 Titration Curve for a Mixture of Titanous and Ferrous Ions with Permanganate. Sample consists of a 50.0-ml aliquot that is 0.100 F with respect to Ti^{3+} and 0.200 F with respect to Fe^{2+}.

whose behavior as indicators depends only upon the change in potential of the system and not specifically upon the change in concentration of one of the reactants. These are the true oxidation-reduction indicators; they have a much wider scope of application than do the specific indicators.

As pointed out by Kolthoff and Stenger[2], a solution of starch containing a little iodine or iodide ion can function in this manner. In the presence of a strong oxidizing agent, the iodine-iodide ratio is high and the blue color of the iodine-starch complex is seen. With a strong reducing agent, on the other hand, iodide ion predominates and the blue color is not seen. Thus, the indicator system changes from colorless to blue in the titration of many strong reducing agents with various oxidizing agents. The change in color in this instance is quite independent of the chemical composition of the reactants, and depends only upon the alteration of the oxidation potential at the equivalence point.

Desirable Properties of Oxidation-reduction Indicators

Intense color.   The indicator substance in one of its forms should impart an intense color to the solution so that the indicator blank is negligible.

Reversibility.   Only if an indicator is reversible in its behavior can a back titration be used to locate an end point exactly. Few available redox indicators are entirely satisfactory in this regard; most are irreversibly oxidized to colorless products, particularly when subjected to high concentrations of strong oxidizing agents.

Sharp color change.   The two colors of the indicator should be sufficiently different to allow detection of a minimum variation in ratio of colored species. As with acid-base indicators, a 50 to 100-fold change in ratio is readily detected with the best indicators.

Solubility and stability.   The indicator should be soluble in water or aqueous solutions of acid or base. The resulting solutions should be stable for reasonable periods.

Transition Range of Indicator

We may write a half reaction for the process leading to a color change for a true oxidation-reduction indicator as follows:

$$In_{oxid} + nH^+ + ne \rightleftharpoons In_{red}$$

With a few indicators no hydrogen ions are consumed by the reduction.

If the foregoing is a reversible process, we may write

$$E = E^\circ - \frac{0.059}{n} \log \frac{[In_{red}]}{[In_{oxid}][H^+]^n}$$

[2] I. M. Kolthoff and V. A. Stenger, *Volumetric Analysis*, 2d ed. 1, 105. New York: Interscience Publishers, 1942.

Typically a color change can be seen when the ratio of reactants shifts by a factor of about 100; that is, when

$$\frac{[In_{\text{red}}]}{[In_{\text{oxid}}]} \geq 10$$

changes to

$$\frac{[In_{\text{red}}]}{[In_{\text{oxid}}]} \leq \frac{1}{10}$$

By substituting these values into the foregoing equation, we find the conditions for color change for a typical indicator.

$$E = E° \pm \frac{0.059}{n} - 0.059 \log \frac{1}{[H^+]} \qquad (18\text{-}1)$$

For simplicity, we have assumed that the indicator reaction does not alter the pH of the solution; this is reasonable since the concentration of the indicator substance is ordinarily quite small.

Thus a typical indicator will exhibit a detectable color change when the volumetric reagent causes a shift in potential of the system of about $0.118/n$ volt. With many indicators, $n = 2$, and a change of 0.059 volt is sufficient.

Table 18-2

A SELECTED LIST OF OXIDATION-REDUCTION INDICATORS[3]

Indicators	Color		Transition Potential, volts	Conditions
	Oxidized	Reduced		
5-nitro-1,10-phenanthroline ferrous complex	pale blue	red violet	+1.25	1 F H$_2$SO$_4$
2,3'-diphenylamine dicarboxylic acid	blue violet	colorless	+1.12	7-10 F H$_2$SO$_4$
1,10-phenanthroline ferrous complex	pale blue	red	+1.11	1 F H$_2$SO$_4$
erioglaucin A	bluish red	yellow green	+0.98	0.5 F H$_2$SO$_4$
diphenylamine sulfonic acid	red violet	colorless	+0.85	dilute acid
diphenylamine	violet	colorless	+0.76	dilute acid
p-ethoxychrysoidine	yellow	red	+0.76	dilute acid
methylene blue	blue	colorless	+0.53	1 F acid
indigo tetrasulfonate	blue	colorless	+0.36	1 F acid
phenosafranine	red	colorless	+0.28	1 F acid

[3] Data taken in part from I. M. Kolthoff and V. A. Stenger, *Volumetric Analysis*, 2d ed. 1, 140. New York: Interscience Publishers, 1942.

The potential at which a color transition will occur depends upon $E°$—the standard potential for the particular indicator system. From Table 18-2 we see that indicators are available which function in any desired potential range up to about $+1.25$ volts.

As shown by equation (18-1), the potential range for many indicators is pH dependent, shifting 0.059 volt for each unit change in pH.

The foregoing treatment of indicator behavior is highly idealized. The oxidized and reduced forms of the indicator are often weak acids or bases and the concentrations of the reactive species are thus pH dependent. Consequently, the transition interval of the indicator varies with pH in a much more complicated manner than that suggested by equation (18-1). Furthermore, in the case of some indicators the color change ensues from a fairly complex set of processes involving several intermediate compounds. Under these conditions, the assumptions used to arrive at equation (18-1) will not be valid. Because of considerations such as these, potential data listed in Table 18-2 are not true standard potentials but rather formal potentials measured when the formal concentrations of In_{oxid} and In_{red} are equal; in order for these data to be significant, acid strength and composition of the solvent must be specified. Formal potentials are sometimes called transition potentials when applied to indicators.

Some Typical Redox Indicators

Ferrous complexes of the orthophenanthrolines.   A class of organic compounds known as 1, 10-phenanthrolines (or orthophenanthrolines) forms stable complexes with ferrous and certain other ions. The parent compound has the structure

1,10-phenanthroline

Each nitrogen atom has a pair of unshared electrons that can coordinate with certain metallic ions to give complexes. In the case of ferrous ion the orthophenanthroline complex is quite stable and is intensely red in color; its structure may be written

The complex is sometimes called ferroin; for convenience we will write its formula as $(Ph)_3Fe^{2+}$.

Ferroin undergoes a reversible oxidation-reduction reaction that may be written

$$(Ph)_3Fe^{3+} + e \rightleftharpoons (Ph)_3Fe^{2+} \qquad E° = + 1.06 \text{ volts}$$
$$\text{pale blue} \qquad\qquad \text{red}$$

The ferric complex is a pale blue; in practice the color change associated with the oxidation is actually from nearly colorless to red. Because of the difference in color intensity the end point is usually taken when only about 10 percent of the indicator is in the ferrous form. This leads to a transition potential of approximately + 1.11 volts in 1 F sulfuric acid.

Despite the apparent simplicity of the indicator reaction, the transition potential for ferroin decreases in strongly acid solutions. Thus in 4 F sulfuric acid a value of + 0.96 volt is observed; this becomes + 0.76 volt in 8 F solution. The explanation for this is not obvious.

Of all of the oxidation-reduction indicators, ferroin approaches most closely the ideal substance. The color change is very sharp, and its solutions are readily prepared and stable. In contrast to many indicators, the oxidized form is remarkably inert towards strong oxidizing agents. The indicator reaction is rapid and reversible.

A number of substituted phenanthrolines have been investigated for their indicator properties, and some of these have proven to be as useful as the parent compound. Among these the 5-nitro and the 5-methyl derivatives are noteworthy, having reduction potentials of + 1.25 volts and + 1.02 volts, respectively.

A compound related to the phenanthrolines is 2,2'-bipyridyl. It also forms

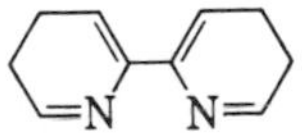

2,2-bipyridyl

a ferrous complex which exhibits indicator properties. The transition potential in this case is + 0.97 volt in 1 F sulfuric acid.

Diphenylamine and its derivatives. One of the first redox indicators to be proposed was diphenylamine.

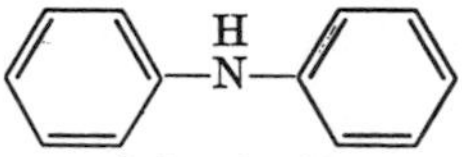

diphenylamine

This was recommended by Knop in 1924 for the titration of ferrous iron with potassium dichromate.

In the presence of a strong oxidizing agent diphenylamine is believed to undergo the following reactions:

diphenylamine   diphenylbenzidine $+ 2H^+ + 2e$

diphenylbenzidine
(colorless)

diphenylbenzidine violet
(violet) $+ 2H^+ + 2e$

The first reaction, involving the formation of the colorless diphenylbenzidine, is nonreversible; the second, however, giving a violet product can be reversed and constitutes the actual indicator reaction.

The reduction potential for the second reaction is about $+ 0.76$ volt, and despite the fact that hydrogen ions appear to be involved, variations in acidity have little effect upon the magnitude of this potential. This may be due to association of hydrogen ions with the colored product.

There are some drawbacks in the application of diphenylamine as an indicator. Because of its low solubility in water, for example, the reagent must be prepared in rather concentrated sulfuric acid solutions. Further, the oxidation product forms an insoluble precipitate with tungstate ion, which renders its use in the presence of this element quite unsatisfactory. Finally, the indicator reaction is slowed by mercuric ions.

These disadvantages are not found with the sulfonic acid derivative of diphenylamine.

diphenylamine   sulfonic acid

Aqueous solutions of the barium or sodium salts of this acid are readily and conveniently prepared. This compound behaves in essentially the same manner as the parent substance. Its color change is somewhat sharper, passing from colorless through green to a deep violet. The transition potential is about $+ 0.8$ volt and again is independent of acid concentration. The sulfonic acid derivative is now widely used in redox titrations.

It might be surmised from the two equations for the indicator reaction of diphenylamine that diphenylbenzidine should behave in an identical fashion and consume less oxidizing agent in its reaction. This is the case; however, the compound has not been widely used because of its very low solubility in water and sulfuric acid. As might be expected, the sulfonic acid derivative of diphenylbenzidine has proved to be a satisfactory indicator.

Selection of Oxidation-reduction Indicators

Theoretically deduced titration curves such as those given in Figures 18.1 and 18.2 suggest the necessary transition range of an indicator for a given titration. For example, in the determination of ferrous iron with ceric ion, an indicator having a transition potential from about $+0.9$ volt to $+1.2$ volts should be satisfactory; the orthophenanthroline-ferrous complex would therefore suffice. In the case of a multicomponent system such as shown in Figure 18.2, it should be possible to differentiate between the two components in the solution by a proper choice of indicators. Thus methylene blue should show an end point upon oxidation of the bulk of the titanous ion while the orthophenanthroline-ferrous complex should not change in color until both the titanous and ferrous ions have been consumed by the reagent.

Summary

Conclusions based on theoretical calculations such as the ones in this chapter are extremely helpful to the analytical chemist as a guide to the choice of proper conditions and indicators for oxidation-reduction titrations; in conclusion, however, it is well to emphasize the theoretical nature of such computations and to point out that these do not take into account all of the factors that determine the applicability and feasibility of a volumetric method. Also to be considered are the rates at which the principal and the indicator reactions occur; the effects of ionic strength, pH, and complexing agents; the presence of colored components other than the indicator in the solution; and the variation among individuals in color perception. The state of the science of chemistry has not advanced to the point where the effects of all of these variables can be determined by computation. Theoretical calculations can only eliminate useless experiments and act as a guide to the ones most likely to be profitable. The final test must always come in the laboratory.

problems

1. Calculate the reduction potential at the equivalence point for each of the following reactions. Where necessary, assume the reactant solutions are initially 0.1 N and that the $[H^+]$ at the equivalence point is unity.
   (a) $2Fe^{3+} + Sn^{2+} \rightleftharpoons 2Fe^{2+} + Sn^{4+}$ ans. $+0.36$ volt
   (b) $2Ce^{4+} + H_3AsO_3 + H_2O \rightleftharpoons 2Ce^{3+} + H_3AsO_4 + 2H^+$
   (c) $2V^{3+} + Zn \rightleftharpoons 2V^{2+} + Zn^{2+}$
   (d) $V(OH)_4^+ + Fe^{2+} + 2H^+ \rightleftharpoons VO^{2+} + Fe^{3+} + 3H_2O$ ans. $+0.88$ volt
   (e) Example (d) when $[H^+] = 0.01$ at the equivalence point
                                     ans. $+0.76$ volt
   (f) $2Ce^{4+} + U^{4+} + 2H_2O \rightleftharpoons 2Ce^{3+} + UO_2^{2+} + 4H^+$
   (g) $Sn^{4+} + Cd \rightleftharpoons Sn^{2+} + Cd^{2+}$
   (h) $5NO_2^- + 2MnO_4^- + 6H^+ \rightleftharpoons 5NO_3^- + 2Mn^{2+} + 3H_2O$

(i) $Br_2 + H_3AsO_3 + H_2O \rightleftharpoons 2Br^- + H_3AsO_4 + 2H^+$

(j) $V(OH)_4^+ + V^{3+} \rightleftharpoons 2VO^{2+} + 2H_2O$

2. Calculate the equivalence-point equilibrium concentration of the first-mentioned species in each of the equations in Problem 1.

3. Calculate the equivalence-point reduction potential in the titration of 50 ml of 0.1 N Cr^{2+} with 0.1 N I_3^-.

$$I_3^- + 2Cr^{2+} \rightleftharpoons 3I^- + 2Cr^{3+} \qquad\qquad \text{ans. } E = +0.24 \text{ volt}$$

4. Calculate the reduction potential at the equivalence point in the titration of 50.0 ml of 0.1 F I^- with 0.1 N $K_2Cr_2O_7$. Assume $[H^+] = 1$.

$$Cr_2O_7^{2-} + 6I^- + 14H^+ \rightleftharpoons 2Cr^{3+} + 3I_2 + 7H_2O$$

5. Choose the indicator from the list in Table 18-2 that would be best suited for each of the titrations in Problem 1.

6. Construct titration curves for each of the following, calculating points at 10, 25, 49, 49.9, 50.0, 50.1, 51, and 60 ml of reagent. Assume $[H^+] = 1$ at all points.

(a) 50 ml of 0.1 N Sn^{2+} with 0.1 N Fe^{3+}

(b) 50 ml of 0.1 N Cr^{2+} with 0.1 N Fe^{3+}

(c) 50 ml of 0.1 N Ti^{3+} with 0.1 N Fe^{3+}

7. Propose an indicator for each of the titrations in Problem 6.

8. Calculate the equilibrium concentrations of the various species (including H^+) in a solution prepared by mixing 50 ml of a solution that is 0.1 N in Fe^{2+} and 0.2 N in acid with 30 ml of a neutral 0.1 N solution of MnO_4^-.

9. Derive a titration curve for 50 ml of 0.1 N Sn^{2+} with 0.1 N I_3^-. Calculate points at 25, 49, 49.9, 50.0, 50.1, 51, and 60 ml of reagent.

10. For the titration shown in Figure 18.2, calculate potential values corresponding to the following additions of 0.1 N $KMnO_4$: 10, 25, 49, 50, 60, 100, 150, and 160 ml.

11. Derive a titration curve for the titration of 50 ml of 0.1 F V^{3+} with 0.1 F Ce^{4+}. Assume that $[H^+] = 1$ at all times. Reactions:

$$Ce^{4+} + V^{3+} + H_2O \rightleftharpoons Ce^{3+} + VO^{2+} + 2H^+$$
$$Ce^{4+} + VO^{2+} + 3H_2O \rightleftharpoons Ce^{3+} + V(OH)_4^+ + 2H^+$$

Calculate potentials for the following additions of the reagent: 10, 25, 49, 50, 51, 60, 75, 99, 100, 101, and 110 ml.

chapter 19. *Oxidizing and Reducing Agents*

Although a complete catalog of substances that have been proposed as reagents for quantitative oxidations or reductions would be imposing, most redox reactions of analytical importance are accomplished with a relatively small number of reagents; we shall consider only those that have found wide application. We shall also consider an important group of substances that are used in the pretreatment of solutions prior to their titration with a standard oxidizing or reducing agent.

STANDARD SOLUTIONS AND PRIMARY STANDARDS

Oxidizing Agents

Most of the common oxidizing agents used in the preparation of standard solutions are listed in Table 19-1. From the table we see that the reduction potentials vary among the reagents from about $+ 0.5$ volt up to $+ 1.7$ volts.

Table 19-1

SOME COMMON OXIDIZING AGENTS USED FOR THE PREPARATION OF STANDARD SOLUTIONS

Reagent	Usual Conditions for Use	Half Reaction	Standard or Formal Reduction Potential, volts	Stability of Solution
Potassium permanganate ($KMnO_4$)	Strongly acid	$MnO_4^- + 8H^+ + 5e \rightleftharpoons Mn^{2+} + 4H_2O$	1.51	Solutions require occasional standardization
	Weakly acid or neutral	$MnO_4^- + 4H^+ + 3e \rightleftharpoons MnO_2 + 2H_2O$	1.69	
	Strongly basic	$MnO_4^- + e \rightleftharpoons MnO_4^{2-}$	0.56	
Quadrivalent cerium (Ce^{4+})	H_2SO_4 solution	$Ce^{4+} + e \rightleftharpoons Ce^{3+}$	1.44	H_2SO_4 solution indefinitely stable
	$HClO_4$ solution	$Ce^{4+} + e \rightleftharpoons Ce^{3+}$	1.70	$HClO_4$ solution requires frequent restandardization
	HNO_3 solution	$Ce^{4+} + e \rightleftharpoons Ce^{3+}$	1.61	HNO_3 solution requires frequent restandardization
Periodic acid (H_5IO_6)	Weakly acid solution	$H_5IO_6 + H^+ + 2e \rightleftharpoons IO_3^- + 3H_2O$	1.6	Occasional standardization necessary
Potassium dichromate ($K_2Cr_2O_7$)	Acid solution	$Cr_2O_7^{2-} + 14H^+ + 6e \rightleftharpoons 2Cr^{3+} + 7H_2O$	1.33	Indefinitely stable
Potassium iodate (KIO_3)	Strong HCl solution	$IO_3^- + 2Cl^- + 6H^+ + 4e \rightleftharpoons ICl_2^- + 3H_2O$	1.23	Stable
Potassium bromate ($KBrO_3 + KBr$)	Dilute acid solution	$BrO_3^- + 5Br^- + 6H^+ \rightleftharpoons 3Br_2 + 3H_2O$ $Br_2 + 2e \rightleftharpoons 2Br^-$	1.05	Indefinitely stable
Ammonium metavanadate (NH_4VO_3)	Strongly acid solution	$V(OH)_4^+ + 2H^+ + e \rightleftharpoons VO^{2+} + 3H_2O$	1.0	Stable
Calcium hypochlorite ($Ca(ClO)_2$)	Neutral or slight alkaline solution	$OCl^- + H_2O + 2e \rightleftharpoons Cl^- + 2OH^-$	0.89	Restandardization necessary
Iodine (I_3^-)	Neutral, dilute acid or base solution	$I_3^- + 3e \rightleftharpoons 3I^-$	0.54	Restandardization necessary

The stronger oxidizing agents have a considerably wider applicability. They tend to be less stable, however, decomposing water to give oxygen; in addition, they are particularly lacking in selectivity. We shall consider some of these compounds in detail in the chapters that follow.

Reducing Agents

In general, standard solutions of reducing agents are used much less frequently than oxidizing agents. This is due, in part, to their lower stability. All are subject to air oxidation to some degree and must be protected from oxygen of the atmosphere or be standardized at frequent intervals. In addition, the more potent reagents reduce hydrogen ions; for this reason also their solutions will change in concentration.

Because of this inherent instability, reducing agents are often used in conjunction with a standard solution of stable oxidizing agent. In these cases an excess of the former is added to the substance being analyzed; after the reaction is complete the excess is determined by titration with the standard oxidant. A blank titration of the reductant then gives its concentration at the time of the analysis.

Table 19-2 gives a list of some of the common reducing agents used for preparing standard solutions.

Primary Standards

In Chapter 10 we noted the desirable properties of a primary standard substance. Table 19-3 lists a large number of oxidizing and reducing agents that meet these requirements.

AUXILIARY REAGENTS

An obvious requirement for a successful redox titration is that the component being determined exist at the outset in a single oxidation state. Since the steps preliminary to the titration frequently result in a mixture of states, a reagent must be introduced to convert the component to the single form. For example, an iron alloy dissolved in an acid reagent almost always results in a mixture of ferrous and ferric ions. Before the iron can be titrated, therefore, a reagent must be added that will convert the element quantitatively either to the ferrous state for titration with a standard oxidizing agent or to the ferric state for titration with a reducing reagent. Such a substance must have certain unique properties. It must be a sufficiently strong oxidizing or reducing agent to convert the substance to be titrated quantitatively to the desired oxidation state. Nevertheless, it should not be strong enough to convert other components of the solution into states in which they will also react with the volumetric reagent. In addition to having the proper reduction potential, the reagent must

Table 19-2

SOME COMMON REDUCING AGENTS USED FOR THE PREPARATION OF STANDARD SOLUTIONS

Reagent	Half Reaction	Standard or Formal Oxidation Potential, volts	Stability
Ferrous ion (Fe^{2+})	$Fe^{2+} \rightleftharpoons Fe^{3+} + e$	-0.77	Not stable unless protected from oxygen
Arsenious oxide solution (H_3AsO_3)	$H_3AsO_3 + H_2O \rightleftharpoons H_3AsO_4 + 2H^+ + 2e$	-0.56	Acid solution stable
Titanous ion (Ti^{3+})	$Ti^{3+} + H_2O \rightleftharpoons TiO^{2+} + 2H^+ + e$	-0.1	Not stable unless protected from oxygen
Sodium thiosulfate ($Na_2S_2O_3$)	$2S_2O_3^{2-} \rightleftharpoons S_4O_6^{2-} + 2e$	-0.08	Restandardization necessary
Chromous ion (Cr^{2+})	$Cr^{2+} \rightleftharpoons Cr^{3+} + e$	$+0.41$	Not stable

Table 19-3

COMMON PRIMARY-STANDARD SUBSTANCES USED IN OXIDATION-REDUCTION REACTIONS

Substance	Formula	Uses
Arsenious oxide	As_2O_3	Preparation of H_3AsO_3 solutions Standardization of MnO_4^-, Ce^{4+}, I_3^-, H_5IO_6, OCl^- solutions
Sodium oxalate	$Na_2C_2O_4$	Standardization of MnO_4^-, Ce^{4+}, $V(OH)_4^+$ solutions
Mohr's salt	$FeSO_4 \cdot (NH_4)_2SO_4 \cdot 6H_2O$	Preparation of Fe^{2+} solutions Standardization of MnO_4^-, Ce^{4+}, $V(OH)_4^+$ solutions
Iron wire	Fe	Standardization of MnO_4^-, Ce^{4+} solutions
Potassium iodide	KI	Standardization of MnO_4^-, Ce^{4+} solutions
Potassium ferrocyanide	$K_4Fe(CN)_6$	Standardization of MnO_4^-, Ce^{4+}, $S_2O_3^{2-}$ solutions
Sodium thiosulfate	$Na_2S_2O_3 \cdot 5H_2O$	Preparation of $S_2O_3^{2-}$ solutions Standardization of I_3^- solutions
Ceric ammonium nitrate	$(NH_4)_2Ce(NO_3)_6$	Preparation of Ce^{4+} solutions
Potassium dichromate	$K_2Cr_2O_7$	Preparation of $Cr_2O_7^{2-}$ solutions Standardization of $S_2O_3^{2-}$, Fe^{2+}, Ti^{3+} solutions
Iodine	I_2	Preparation of I_3^- solutions Standardization of $S_2O_3^{2-}$ solutions
Potassium iodate	KIO_3	Preparation of IO_3^- solutions Standardization of $S_2O_3^{2-}$ solutions
Potassium bromate	$KBrO_3$	Preparation of BrO_3^- solutions Standardization of $S_2O_3^{2-}$ solutions

ordinarily be such that its unreacted portion can readily be removed from the solution, for in nearly every case it will be capable of reacting with the volumetric reagent. Thus, a reagent that would convert iron quantitatively to the ferrous state for titration with a standard solution of permanganate would be a good reducing agent; any excess remaining after the reduction would surely consume permanganate unless removed from the solution.

Some of the reagents that find fairly general application to the pretreatment of samples are described in the following paragraphs.

Oxidizing Reagents

Sodium bismuthate. Sodium bismuthate is an extremely powerful oxidizing agent capable, for example, of converting manganous ion quantitatively to the permanganate state. It exists as an insoluble solid of rather uncertain composition. The formula for the substance is usually written as $NaBiO_3$; upon reaction the bismuth (V) is converted to the more common bismuth (III) state. Ordinarily the solution to be oxidized is boiled in contact with an excess of the solid; the unused reagent is then removed by filtration.

Ammonium persulfate. In acid solutions, ammonium persulfate, $(NH_4)_2S_2O_8$, is a potent oxidizing agent which will convert chromium to dichromate, cerous ion to the ceric state, and manganous ion to permanganate. The oxidations are catalyzed by the presence of a small amount of silver ion. The half reaction is

$$S_2O_8^{2-} + 2e \rightleftharpoons 2SO_4^{2-} \quad E° = 2.01 \text{ volts}$$

The excess reagent is readily removed by boiling the solution for a few minutes

$$2S_2O_8^{2-} + 2H_2O \rightleftharpoons 4SO_4^{2-} + O_2 + 4H^+$$

Sodium and hydrogen peroxide. Peroxide is a convenient oxidizing agent. It can be obtained in the form of the solid sodium salt or as dilute solutions of the acid. Its half reaction in acid solution is written

$$H_2O_2 + 2H^+ + 2e \rightleftharpoons 2H_2O \quad E° = 1.77 \text{ volts}$$

The excess peroxide is readily removed from solution by boiling for a short period.

$$2H_2O_2 \rightleftharpoons 2H_2O + O_2$$

Perchloric acid. Hot concentrated perchloric acid is a powerful oxidizer. It will, for example, quantitatively convert chromium to the hexavalent state and oxidize organic substances with explosive violence. Cool dilute solutions of the acid have no oxidative properties whatsoever and reducing agents such as ferrous ion can be added to these without reaction. Thus, chromium in steels is sometimes determined by solution of the sample in the hot concentrated acid; dichromate and ferric ions are produced. Upon cooling and diluting the dichromate can be titrated with a standard solution of ferrous ion.

Reducing Reagents

Metals. An examination of a table of standard electrode potentials reveals a number of good reducing agents among the pure metals[1]; such elements as zinc, cadmium, aluminum, lead, nickel, copper, mercury, and silver have proved useful for prereduction purposes. The removal of excess reagent is not a difficult problem. Where sticks or coils of the metal are used, the excess

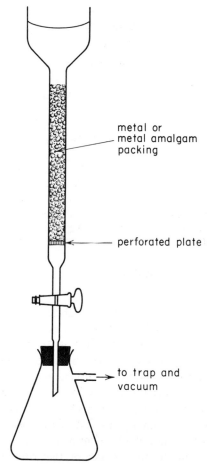

Fig. 19.1 A Metal or Metal Amalgam Reductor.

[1] For a discussion of metal reductants the reader should see I. M. Kolthoff and R. Belcher, *Volumetric Analysis*, 3, 11-23. New York: Interscience Publishers, Inc., 1957; W. I. Stephen, *Ind. Chemist*, **28**, 13, 55, 107 (1952).

reductant is simply lifted from the solution and washed thoroughly. If granular or powdered forms of the metal are employed, filtration will be required. An alternative method involves the use of a *reductor*, an example of which is illustrated in Figure 19.1. Here the solution is passed through a column packed with granules of the metallic reducing agent.

The reducing strengths of the various metals mentioned above vary considerably. Thus, zinc and cadmium are rather powerful reagents compared with silver or mercury. The latter two are usually used in conjunction with hydrochloric acid solutions; since both form insoluble chlorides, they are more effective reducing agents in the presence of that acid. Table 19-4 lists some of the uses of metal reductors.

Table 19-4

A LIST OF SOME OF THE METAL REDUCTORS WHICH HAVE BEEN USED FOR PREREDUCTIONS

Desired Reduction	Metal Reductor
$Fe(III) \rightarrow Fe(II)$	Zn, Cd, Cu, Pb, Hg^2, Ag^2
$V(V) \rightarrow V(II)$	Zn, Cd; not Ag, Hg
$V(V) \rightarrow V(IV)$	Ag, Hg
$Ti(IV) \rightarrow Ti(III)$	Zn, Cd; not Ag
$Cr(III) \rightarrow Cr(II)$	Zn, Cd; not Ag
$Mo(VI) \rightarrow Mo(III)$	Zn, Cd; not Ag
$Mo(VI) \rightarrow Mo(V)$	Ag^2, Hg^2
$U(VI) \rightarrow U(IV)$	Ag^2, Pb, Hg^2
$Cu(II) \rightarrow Cu(I)$	Ag^2

[2] Used in conjunction with HCl solutions

The composition of the solution to be analyzed and the information desired often governs the choice of metal reductant. For example, if a solution containing both ferric iron and titanic ions is brought in contact with metallic zinc or cadmium, reduction to ferrous and titanous ion occurs. Titration with an oxidizing agent then gives the total number of equivalents of iron and titanium present. On the other hand reduction of this same solution with a metallic silver reductor in the presence of hydrochloric acid converts the ferric ions to the ferrous condition but has no effect on the titanic ions. Titration of the reduced solution then gives only the number of equivalents of iron present.

Of the several simple metallic reductors, that containing silver is perhaps the most widely used. This is due to the somewhat selective nature of its reducing properties, as illustrated above; some other applications are given in Table 19-5. The silver reductor is nearly always used with hydrochloric acid solutions.

One disadvantage of the very reactive reductants such as cadmium and zinc arises from their tendency to react with acids to give hydrogen gas. This

Table 19-5

Silver Reductor	Jones Reductor
$e + Fe^{3+} \rightarrow Fe^{2+}$	$e + Fe^{3+} \rightarrow Fe^{2+}$
$e + Cu^{2+} \rightarrow Cu^+$	Cu^{2+} reduced to metallic Cu
$e + H_2MoO_4 + 2H^+ \rightarrow MoO_2^+ + 2H_2O$	$3e + H_2MoO_4 + 6H^+ \rightarrow Mo^{3+} + 4H_2O$
$2e + UO_2^{2+} + 4H^+ \rightarrow U^{4+} + 2H_2O$	$2e + UO_2^{2+} + 4H^+ \rightarrow U^{4+} + 2H_2O$
	$3e + UO_2^{2+} + 4H^+ \rightarrow U^{3+} + 2H_2O^4$
$e + V(OH)_4^+ + 2H^+ \rightarrow VO^{2+} + 3H_2O$	$3e + V(OH)_4^+ + 4H^+ \rightarrow V^{2+} + 4H_2O$
TiO^{2+} not reduced	$e + TiO^{2+} + 2H^+ \rightarrow Ti^{3+} + H_2O$
Cr^{3+} not reduced	$e + Cr^{3+} \rightarrow Cr^{2+}$

[3] Taken from I. M. Kolthoff and R. Belcher, *Volumetric Analysis*, 3, 12. New York: Interscience Publishers, Inc., 1957.

[4] A mixture of oxidation states is obtained; this does not, however, preclude the use of the Jones reductor for the analysis of uranium since any U^{3+} formed can be converted to U^{4+} by shaking the solution with air for a few minutes.

parasitic reaction not only consumes reductant but also introduces large quantities of the metallic ion into the solution; this may be undesirable for the later steps in the analysis. The problem, of course, becomes greater as the pH of the solution decreases and prevents the application of the reactive metal reductants in highly acidic solutions.

Solid amalgams. The problem of hydrogen evolution associated with metallic reductants can be largely overcome by amalgamation of the reducing metal with mercury. The presence of the mercury so inhibits the formation of hydrogen gas that strong reductants such as zinc and cadmium can be used even in quite acid solutions. This effect, known as hydrogen overvoltage, is discussed further in Chapter 22.

An amalgam is ordinarily a somewhat poorer reducing agent than the pure metal itself. For example, the potential for the oxidation of a zinc amalgam

$$Zn(Hg) \rightleftharpoons Zn^{2+} + Hg + 2e$$

is less than that for pure zinc since the activity of the element is less in the amalgam than in the pure metal. In general, the magnitude of the change in reducing power is not very great, amounting to 0.05 to 0.1 volt or less.

The most widely used amalgam reductor—commonly called a Jones reductor—employs zinc as the reducing agent. Figure 19.1 illustrates a typical Jones reductor. The column, which is about 50 cm in length and 2 cm in diameter, is packed with 20- to 30-mesh zinc granules. This is amalgamated by treatment with a solution of mercuric nitrate or chloride; a surface deposit of mercury results

$$Zn + Hg^{2+} \rightleftharpoons Zn^{2+} + Hg$$

A column so prepared can be used for many hundreds of reductions. When not in use, it is filled with water to prevent contact of the packing with air and the resultant formation of basic salts—these tend to clog the column.

When a column is put in use, it is first activated by passage of a dilute acid solution. The solution to be reduced is then pulled through at a rate of about 75 to 100 ml per minute. Finally, the column is washed with more acid. Throughout these operations care is taken to avoid exposure of the packing to the atmosphere.

Table 19-5 lists the principal uses to which the Jones reductor has been put.

Preparation of a Jones reductor. The reductor (see Fig. 19.1) should be about 40 to 50 cm in length and 2 cm in diameter. It should be thoroughly cleaned and a porcelain disk supporting a mat of glass wool or asbestos introduced in the lower end. The mat must be sufficiently thick to prevent passage of zinc granules when the reductor is in use.

The zinc metal should be 20 to 30 mesh and free of impurities such as iron. Cover a suitable quantity of the metal with 1 N HCl and let stand for about 1 minute. Decant the liquid and cover the zinc with a 0.25 F solution of $Hg(NO_3)_2$ or $HgCl_2$. Stir the mixture vigorously for 3 minutes, decant and wash 2 or 3 times with distilled water. Fill the reductor tube with water and slowly add the amalgam until a column of about 30 cm is achieved. Wash with 500 ml of water, being sure that the packing is never uncovered by liquid. Keep the reductor full of water during storage.

Use of the Jones reductor. Solutions containing 3 to 15 percent by volume of HCl or 1 to 10 percent by volume of H_2SO_4 may be used in the reductor. Warm solutions may be passed through the column without harm. Wash the column by drawing through it several 20- to 30-ml portions of an acid solution of about the same composition as the sample. After each addition of wash liquid, drain the column to about 1 cm above the zinc; then add the next portion. Leave the last washing on the column.

Attach a receiver to the column and add the sample to the reservoir at the top. The sample can be pulled through the column at a rate of 75 to 100 ml per minute. When the sample solution has been drawn to within a centimeter of the top of the packing, add 25 ml of a dilute acid as a wash. Follow this with two additional acid washes and then with 100 to 200 ml of water. Leave the column covered with water and place a beaker over its top during storage.

Interferences. Nitrates interfere because they are partially reduced by zinc amalgam to various compounds that subsequently react with the volumetric oxidant. In addition, arsenic, antimony, tin, organic, and polythionic compounds must be absent as they also are reduced to substances that may consume the reagent. Ammoniacal solutions should be avoided because they react with the mercury and very shortly render the packing worthless.

Liquid amalgams. A number of Japanese chemists have systematically investigated the use of liquid amalgams for the prereduction of solutions.[5]

[5] See the following for a brief summary of this work and a list of references to the original work: I. M. Kolthoff and R. Belcher, *Volumetric Analysis*, 3, 18-21. New York: Interscience Publishers, Inc., 1957.

The technique here is different from that discussed in the previous section in that liquid amalgams are employed, and the reduction is performed by shaking the sample with the amalgam in a type of separatory funnel. The recommended apparatus is shown in Figure 19.2. Vessel D is filled with water and stopcock C is closed. The solution to be reduced is introduced through A with 10 to 15 ml of the amalgam. After thorough shaking, stopcock C is opened to allow collection of the amalgam in D. After closing C the titration is performed in the vessel by additions of reagent through A. Where necessary, the solution can be blanketed with an inert gas introduced through B. The amalgam can be used for numerous reductions.

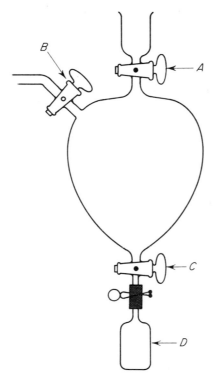

Fig. 19.2 Apparatus for Liquid Amalgam Reductions.

Table 19-6 shows some of the reductions that can be carried out by the method. Amalgams of zinc, cadmium, lead, bismuth, and tin have been used.

Gaseous reductants. Both hydrogen sulfide and sulfur dioxide are fairly good reducing reagents and have been applied to the problem of pre-reduction. With these, the excess reagent can be readily removed by boiling the acidified solution.

The reactions of these reagents are often slow, half an hour or more being required to complete the reduction and rid the solution of excess reagent. In

Table 19-6

REDUCTION PRODUCTS WITH CERTAIN LIQUID AMALGAMS

	Type of Amalgam		
Ion Reduced	Zinc	Bismuth	Tin
Fe(III) →	Fe(II)	Fe(II)	Fe(II)
U(VI) →	U(III), U(IV)	U(IV)	
Mo(VI) →	Mo(III)	Mo(III), Mo(V)	Mo(III)
Sn(IV) →		Sn(II)	Sn(II)
Cu(II) →		Cu(I)	Cu(I)
Ti(IV) →	Ti(III)	Ti(III)	Ti(III)
V(V) →	V(II)	V(IV)	V(II)
W(VI) →	W(III)	W(III)	W(III)

addition to this time disadvantage, the noxious and toxic nature of these gases complicate their use. The employment of other reductants is much preferred.

Sulfur dioxide is used for the reduction of ferric iron to the ferrous condition. Trivalent chromium and titanium are not affected by the reagent and therefore do not interfere. Vanadium(V) is reduced to the quadrivalent state. Pentavalent arsenic and antimony are reduced quantitatively to the trivalent state by sulfur dioxide.

The principal application of hydrogen sulfide has been in the reduction of ferric iron to the ferrous condition.

chapter 20. *The Properties and Applications of Volumetric Oxidizing Agents*

POTASSIUM PERMANGANATE

Potassium permanganate is probably the most widely used of all volumetric oxidizing agents. It is a powerful oxidizer and readily available at modest cost. The intense color of the permanganate ion is sufficient to signal the end point in most titrations; this eliminates the need for an indicator. Several disadvantages attend the use of the reagent. Thus, the tendency of permanganate to oxidize chloride ion represents a serious limitation since hydrochloric acid is often a desirable solvent. The multiplicity of possible reaction products can, at times, result in uncertainty regarding its stoichiometry. Finally, solutions of permanganate have limited stability.

Reactions of Permanganate Ion

Depending upon reaction conditions, permanganate ion is reduced to manganese in the $+2$, $+3$, $+4$, or $+6$ state. Important volumetric methods are based on reactions that yield each of these products.

Reduction to the $+2$ state. In solutions that are 0.1 N or greater in mineral acid, the common reduction product is manganous ion

$$MnO_4^- + 8H^+ + 5e \rightleftharpoons Mn^{2+} + 4H_2O \qquad E^\circ = 1.51 \text{ volts}$$

This is the most widely used of the permanganate reactions. Although the mechanisms involved in the formation of manganous ion are frequently complicated, the oxidation of most substances proceeds rapidly in acid solution. Notable exceptions include the reaction with oxalic acid, for which elevated temperatures are required, and with arsenious oxide, which necessitates the use of a catalyst.

Reduction to the $+3$ state. Solutions of Mn^{3+} are not stable owing to the disproportionation reaction

$$2Mn^{3+} + 2H_2O \rightleftharpoons MnO_2 + Mn^{2+} + 4H^+$$

The manganic ion forms several complexes, however, that are sufficiently stable to permit the existence of $+3$ manganese in aqueous solution. Lingane has made use of this property to titrate manganous ion with permanganate in highly concentrated solutions of pyrophosphate ion[1]; the reaction may be expressed as

$$MnO_4^- + 4Mn^{2+} + 15H_2P_2O_7^{2-} + 8H^+ \rightleftharpoons 5Mn(H_2P_2O_7)_3^{3-} + 4H_2O$$

The titration is carried out in a pH range between 4 and 7.

Reduction to the $+4$ state. In solutions that are weakly acid (above pH 4), neutral, or weakly alkaline, manganese dioxide is the most common reduction product

$$MnO_4^- + 4H^+ + 3e \rightleftharpoons MnO_2 + 2H_2O \qquad E^\circ = 1.70 \text{ volts}$$

Titrations of certain species with permanganate can be carried out to advantage under these conditions. For example, cyanide is oxidized to cyanate; sulfide, sulfite, and thiosulfate are converted to sulfate; manganous ion yields manganese dioxide; and hydrazine is oxidized to nitrogen.

Titrations in which manganese dioxide is the product suffer from the disadvantage that the brown insoluble oxide obscures the end point; time

---

[1] J. J. Lingane and R. Karplus, *Ind. Eng. Chem.*, *Anal. Ed.*, 18, 191(1946).

must be allowed for the solid to settle before the first excess of the permanganate can be detected.

Reduction to the + 6 state. Some important volumetric analyses based on permanganate involve the half reaction

$$MnO_4^- + e \rightleftharpoons MnO_4^{2-} \qquad E^n = 0.56 \text{ volt}$$

This stoichiometry, in which manganate ion is the product, tends to predominate in solutions that are greater than 1 N in sodium hydroxide. Alkaline oxidations with permanganate have proved the most useful in the determination of a number of organic compounds.

Alkaline permanganate oxidations are generally completed in the presence of an excess of barium ions which precipitate the reduction product as barium manganate. This is necessary to minimize further reduction of the manganate ion. It is also desirable to rid the solution of this species because its dark green color renders end-point detection difficult. Even with this precaution, the end point is not easy to see owing to the presence of the finely divided green barium manganate.

End Point

One of the most obvious of the properties of potassium permanganate is its intense purple color; this is commonly used as an indicator for titrations. As little as 0.1 to 0.2 ml of a 0.01 N solution is enough to impart a perceptible color to 100 ml of water. With solutions more concentrated than this, then, the titration error is negligible. For very dilute permanganate solutions, diphenylamine sulfonic acid or orthophenanthroline-ferrous complex will give a sharper end point.

The permanganate end point is not permanent, and gradually fades to give a colorless solution. The decolorization results from the reaction of the excess permanganate ion with the relatively large concentration of manganous ion formed during the titration

$$2MnO_4^- + 3Mn^{2+} + 2H_2O \rightleftharpoons 5MnO_2 + 4H^+$$

The equilibrium constant for this reaction is readily calculated from the standard potentials for the two half reactions and has a numerical value of about 10^{47}. Thus, even in highly acidic solution, the concentration of permanganate in equilibrium with manganous ion is very small. Fortunately, the rate at which this equilibrium is attained is quite slow, with the result that the end point fades gradually.

Measurement of the volume of a permanganate solution in a buret is complicated by the intense color of the reagent. The surface of the liquid, rather than the meniscus, is often taken as the point of reference.

Stability of Permanganate Solutions

Aqueous solutions of permanganate are not completely stable because of the tendency of that ion to oxidize water. In neutral solutions the half reactions may be written

$$4MnO_4^- + 16H^+ + 12e \rightleftharpoons 4MnO_2 + 8H_2O \qquad E^\circ = \quad 1.70 \text{ volts}$$

$$6H_2O \rightleftharpoons 3O_2 + 12H^+ + 12e \qquad\qquad E^\circ = -1.23 \text{ volts}$$

$$4MnO_4^- + 4H^+ \rightleftharpoons 4MnO_2 + 2H_2O + 3O_2$$

From the magnitude of the standard potentials we see that equilibrium position for the over-all reaction should lie to the right even in neutral solutions; it is only by virtue of the very low rate of the reaction that permanganate solutions are stable at all. Experiments show that the decomposition is catalyzed by light, heat, acids, bases, manganous ion, and manganese dioxide. These effects, some of which are demonstrated by the data in Table 20-1, must be taken into account to obtain a stable reagent for analysis.

Table 20-1

STABILITY OF DILUTE $KMnO_4$ SOLUTIONS[2]

Conditions	Decrease in Normality in 7 Months, percent
Solution initially free of MnO_2; stored in dark	0.2
Solution initially free of MnO_2; stored in daylight	0.9
Solution initially free of MnO_2 but contained 0.0002 percent $MnCl_2$; stored in dark	3.2
Solution not freed of MnO_2; stored in dark	4.5
Solution not freed of MnO_2; stored in daylight	6.0
Solution initially free of MnO_2; stored in daylight; solution 2 N in H_2SO_4	95
Solution initially free of MnO_2; stored in daylight; solution 0.04 N in NaOH	10.5

[2] Data from I. M. Kolthoff and I. H. Menzel, *Volumetric Analysis*, 1, 230. New York: John Wiley and Sons, Inc., 1928. With permission.

Solid manganese dioxide greatly accelerates decomposition. Since this is a potential decomposition product we might expect the reaction rate to increase with time as a result of the build-up in amount of the solid; this is indeed observed. It is an example of an *autocatalytic* process, since the reaction product serves to catalyze its own formation.

The photochemical catalysis of the decomposition is often observed when a permanganate solution is allowed to stand in a buret for any extended period; manganese dioxide forms as a brown stain and serves to show that the concentration of the reagent has been altered.

In general, heating of acidic solutions containing an excess of permanganate is to be avoided. This is sometimes recommended for the analysis of substances that are oxidized slowly by the reagent. The practice, however, will inevitably lead to errors that cannot be adequately compensated for by a blank because the blank will contain fewer manganous ions than the sample; clearly this species accelerates the oxidation of water (see Table 20-1). Hot acidic solutions of a reductant can be titrated directly with permanganate without serious error, however, since at no time during the titration is there an appreciable concentration of oxidant in such a solution.

Preparation, Standardization, and Storage of Permanganate Solutions

Preparation. Permanganate solutions are prepared from the potassium salt. This compound is seldom, if ever, of sufficient purity to allow the direct preparation of a standard solution. Furthermore, the water used in preparation of the reagent is frequently contaminated with small amounts of dust, organic compounds, and other oxidizable substances; these alter the concentration by reacting with the permanganate to give manganese dioxide.

Certain precautions must be taken to obtain a permanganate solution of reasonable stability. Perhaps the most important variable affecting stability is the catalytic effect of manganese dioxide. As initially prepared, the solution will always contain some of this compound as a contaminant in the starting material and from the oxidation of organic matter in the water. Removal of manganese dioxide by filtration markedly enhances the life of the standard reagent. Enough time should be allowed for complete oxidation of contaminants in the water before filtration. Boiling the solution may hasten this process. Paper cannot be used for the filtration since it reacts with the permanganate to form the undesirable dioxide.

Standardized solutions should be stored in the dark. If any solid is detected in the solution, filtration and restandardization is necessary. In any case restandardization every few weeks is a good precautionary measure.

Method of preparation. To prepare a 0.1 N solution dissolve 3.2 grams of $KMnO_4$ in about 1 liter of distilled water. Heat to boiling and keep hot for about 1 hour. Cover and let stand overnight. Filter the solution through a fine-porosity sintered glass crucible or through a Gooch crucible with an asbestos mat. Store the solution in a clean, glass-stoppered bottle and keep in the dark when not in use.

Standardization against sodium oxalate. Sodium oxalate of certified purity may be purchased from the National Bureau of Standards. The commercial reagent grades of the salt commonly assay 99.9 percent or better, however, and are entirely adequate standards for all but the most refined work. Regardless of source, sodium oxalate should be oven dried before use.

In acid solution, permanganate oxidizes oxalic acid to carbon dioxide and water

$$2MnO_4^- + 5H_2C_2O_4 + 6H^+ \rightleftharpoons 2Mn^{2+} + 10CO_2 + 8H_2O$$

This is a complex reaction that proceeds only slowly at room temperature. Even at elevated temperatures the reaction is not rapid unless catalyzed by manganous ion. Thus, several seconds are required to decolorize a hot oxalic acid solution at the outset of a permanganate titration. Later, when the concentration of manganous ion has become appreciable, the decolorization becomes very rapid. This is yet another example of autocatalysis.

Even though the mechanism by which oxalic acid reduces permanganate ion has been the subject of extensive studies,[3] there are details about the reaction that remain obscure. It seems clear, however, that an important step in the reaction is the formation of oxalate complexes of $+3$ and $+4$ manganese as a result of oxidation of manganous ion by permanganate. These complex ions then react rapidly with oxalate to give carbon dioxide and manganous ion again. Thus, until a reasonable concentration of manganous ions has been built up in the solution, the formation of these intermediates is inhibited and the reaction remains slow.

The stoichiometry of the reaction has been investigated in great detail by McBride[4] and more recently by Fowler and Bright.[5] The former devised a procedure wherein the oxalic acid is titrated slowly at a temperature of 60° to 90° C until the faint pink color of the permanganate persists. Fowler and Bright have demonstrated that this titration consumes 0.1 to 0.4 percent too little permanganate, due perhaps to air oxidation of a small part of the oxalic acid.

$$H_2C_2O_4 + O_2 \rightleftharpoons H_2O_2 + 2CO_2$$

In the hot solution, the peroxide is postulated to decompose spontaneously to oxygen and water.

Fowler and Bright devised a scheme for standardization in which 90 to 95 percent of the required permanganate is added rapidly to the cool oxalic

---

[3] H. F. Launer, *J. Am. Chem. Soc.*, **54**, 2597 (1932); H. F. Launer and D. M. Yost, *ibid*, **56**, 2571 (1934); and J. M. Malcolm and R. M. Noyes, *ibid*, **74**, 2769 (1952).

[4] R. S. McBride, *J. Am. Chem. Soc.*, **34**, 393 (1912).

[5] R. M. Fowler and H. A. Bright, *J. Research Nat. Bur. Standards*, **15**, 493 (1935).

acid solution. After all of this reagent has reacted, the solution is heated to 55° to 60° C and titrated as before. While this procedure minimizes the air oxidation of oxalic acid and gives data that appear to be in exact accord with the theoretical stoichiometry, it suffers from the disadvantage of requiring a knowledge of the approximate normality of the solution in order to make the proper initial addition of the reagent. In this respect the Fowler-Bright procedure is not as convenient as the McBride method.

For many purposes the method of McBride will give perfectly adequate data (usually 0.2 to 0.3 percent too high). If a more accurate standardization is required, it is convenient to run one titration by this procedure to obtain the approximate normality of the solution; then a pair of titrations employing the Fowler and Bright method can be made. Directions for both procedures follow.

Method of Fowler and Bright. Dry primary-standard grade $Na_2C_2O_4$ for 1 hour at 110° to 200° C. Cool in a desiccator and accurately weigh suitable portions (0.2 to 0.3 gram for 0.1 N $KMnO_4$) into 500-ml beakers. Add 250 ml of 1:19 sulfuric acid that has been boiled for 10 to 15 minutes and cooled to room temperature; stir until dissolved. A thermometer is convenient for this purpose since the temperature must be measured later. Introduce from a buret sufficient permanganate to consume 90 to 95 percent of the oxalate (about 40 ml for a 0.1 N solution and a 0.3-gram sample). A preliminary titration by the McBride method will often provide the approximate volume required. Let stand until the solution is decolorized; then warm to 55° to 60° C and complete the titrations, taking the first pale pink color that persists for 30 seconds as the end point. Determine an end-point correction by titrating 250 ml of 1:19 sulfuric acid at this same temperature. Correct for the blank and calculate the normality.

Method of McBride. Dissolve weighed samples of dried $Na_2C_2O_4$ in a solution prepared by diluting 30 ml of 6 N H_2SO_4 to about 250 ml. Heat to 80° to 90° C and titrate with the $KMnO_4$, stirring vigorously with a thermometer. The first addition of reagent should be made slowly enough so that the pink color is discharged before further additions are made. If the solution drops below 60° C, heat. The end point is the first persistent pink color. Correct the titration for an end-point blank determined by titrating an equal volume of the water and acid.

NOTES:

1. To measure the volume of $KMnO_4$, take the surface of the liquid as a point of reference. Alternatively, use a flashlight or match to provide sufficient illumination for reading of the meniscus in the conventional manner.

2. Solutions of permanganate should not be allowed to stand in burets any longer than necessary, as decomposition to MnO_2 may occur. Freshly formed MnO_2 can be removed from burets and glassware by rinsing with a warm solution prepared by dissolving 1 to 2 grams of $Na_2C_2O_4$ in 6 N H_2SO_4 or with hot, concentrated HCl.

3. Any $KMnO_4$ spattered on the sides of the titration vessel should be washed down immediately with a stream of water.

4. If the addition of $KMnO_4$ is too rapid, some MnO_2 will be produced

in addition to Mn^{2+}. Evidence for this is a faint brown discoloration of the solution. This is not a serious problem as long as sufficient oxalate remains to reduce the MnO_2 to Mn^{2+}; the titration is temporarily discontinued until the solution clears. The solution must be free of MnO_2 at the equivalence point.

Standardization against arsenious oxide. Primary-standard grade arsenious oxide can be obtained from the National Bureau of Standards. A good quality material is sold commercially.

Arsenious oxide is not readily soluble in water or acid; in dilute base, however, solution proceeds at a moderate rate. Alkaline solutions of arsenious oxide are not stable towards atmospheric oxidation. Neutral solutions, on the other hand, can be kept several months without showing a change in reducing capacity. When using arsenious oxide as a primary standard for permanganate, the usual procedure is to dissolve the reagent in sodium hydroxide and then acidify with hydrochloric or sulfuric acid.

The reaction of arsenious oxide solution with permanganate may be written

$$5H_3AsO_3 + 2MnO_4^- + 6H^+ \rightleftharpoons 5H_3AsO_4 + 2Mn^{2+} + 3H_2O$$

The first additions of permanganate to a sulfuric acid solution of tripositive arsenic result in rapid consumption of the reagent. As the titration proceeds, however, the reaction slows and gives rise to yellow or brown products that are probably complex ions made up of tripositive manganese and one of the arsenic species. Even with heating the reaction is not rapid nor is it stoichiometric.

In hot hydrochloric acid solutions the titration of arsenious acid can be carried out successfully. An even better method, however, is to introduce a catalyst such as iodide, iodate, or iodine monochloride to the hydrochloric acid solution; titration at room temperature is then possible. The mechanism of the catalytic activity apparently involves preliminary formation of iodine monochloride which rapidly oxidizes arsenious acid giving iodide. The permanganate then reoxidizes the iodide to the iodine monochloride which then consumes more of the arsenite.

Permanganate end points are particularly transient in hydrochloric acid solutions, owing to the process

$$10Cl^- + 2MnO_4^- + 16H^+ \rightleftharpoons 5Cl_2 + 2Mn^{2+} + 8H_2O$$

Thus the use of an indicator, such as the orthophenanthroline-ferrous complex, is preferable to relying on the color of the excess reagent in this medium.

Method. Dry As_2O_3 for 1 hour at 110° C. Weigh about 0.2-gram portions into 500-ml flasks. Dissolve in 10 ml of 4 N NaOH followed by 20 ml of 6 N HCl. Add about 100 ml of water, 1 drop of 0.0025 F KIO_3, and 1.0 gram of NaCl. Titrate with $KMnO_4$ until the color fades slowly. Then add 2 drops of 0.025 F orthophenanthroline-ferrous complex and complete the titration. The color change is from pink to a faint blue.

NOTES:

1. The orthophenanthroline indicator is prepared by dissolving 1.485 grams of orthophenanthroline in 100 ml of water containing 0.695 gram of $FeSO_4 \cdot 7H_2O$.

2. A 0.0025 F solution ICl may be substituted for the KIO_3 as a catalyst.

Other primary standards. Mohr's salt, $FeSO_4 \cdot (NH_4)_2SO_4 \cdot 6H_2O$, has been suggested for the standardization of permanganate solutions. Because of the possibility for efflorescence, however, the exact composition of this compound is often uncertain; its use is not recommended for standardizations calling for the highest accuracy.

Other compounds that are used for standardization of permanganate solutions include potassium iodide, potassium iodate, potassium ferrocyanide, and metallic iron. Potassium iodate requires preliminary reduction to iodide before titration. This is readily accomplished with sulfur dioxide. Detailed procedures for the use of these standards are found in Kolthoff and Belcher.[6]

Applications of Permanganate Titrations to Acid Solutions

Table 20-2 indicates the multiplicity of analyses that make use of standard permanganate solutions in acidic media. In most cases, the reactions are rapid enough for direct titrations. Manganous ion is the product in each case. Specific directions follow for the determination of iron and for calcium.

Determination of iron in an ore. The main iron ores are hematite (Fe_2O_3), magnetite (Fe_3O_4), and limonite ($3Fe_2O_3 \cdot 3H_2O$). Volumetric methods for iron analysis of samples containing these substances can be broken down into three steps: (1) solution of the sample, (2) reduction of the iron to the ferrous state, and (3) titration with a standard oxidant. In most cases elements tending to interfere with step (3) are either absent or their effects can be avoided without preliminary separations.

Iron ores are often completely decomposed in concentrated hydrochloric acid. The rate of attack by this reagent is increased by the presence of a small amount of stannous chloride, which probably acts by reducing the rather insoluble ferric oxides on the surface of the particles. Hydrochloric acid is a much more efficient solvent than either sulfuric or nitric acids; this is explained, in part, by the tendency of iron to form ferric complexes with chloride ion.

Most iron ores contain silicates that may or may not be decomposed by treatment with hydrochloric acid. Where decomposition is complete, a white residue of hydrated silica remains behind that in no way interferes with the analysis. Incomplete decomposition is indicated by a dark residue remaining after prolonged treatment with the acid. Since this solid residue may contain

[6] I. M. Kolthoff and R. Belcher, *Volumetric Analysis*, 3, 41-59. New York: Interscience Publishers, Inc., 1957.

Table 20-2

APPLICATIONS OF POTASSIUM PERMANGANATE

Direct Titration in Acid Solution

$$MnO_4^- + 8H^+ + 5e \rightleftharpoons Mn^{2+} + 4H_2O$$

Substance Sought	Half Reaction	Condition
I	$I^- + HCN \rightleftharpoons ICN + H^+ + 2e$	In 0.1 F HCN with ferroin indicator
Br	$2Br^- \rightleftharpoons Br_2 + 2e$	Boiling H_2SO_4 solution
As	$H_3AsO_3 + H_2O \rightleftharpoons H_3AsO_4 + 2H^+ + 2e$	KIO_3 or ICl catalyst in HCl solution
Sb	$H_3SbO_3 + H_2O \rightleftharpoons H_3SbO_4 + 2H^+ + 2e$	HCl solution
Sn	$Sn^{2+} \rightleftharpoons Sn^{4+} + 2e$	Prereduction with Zn
H_2O_2	$H_2O_2 \rightleftharpoons O_2 + 2H^+ + 2e$	
Fe	$Fe^{2+} \rightleftharpoons Fe^{3+} + e$	Prereduction with Jones reductor, SO_2, etc.
$Fe(CN)_6^{4-}$	$Fe(CN)_6^{4-} \rightleftharpoons Fe(CN)_6^{3-} + e$	
V	$VO^{2+} + 3H_2O \rightleftharpoons V(OH)_4^+ + 2H^+ + e$	Prereduction with Bi amalgam or SO_2
Mo	$Mo^{3+} + 4H_2O \rightleftharpoons MoO_4^2 + 8H^+ + 3e$	Prereduction with Jones reductor
W	$W^{3+} + 4H_2O \rightleftharpoons WO_4^{2-} + 8H^+ + 3e$	Prereduction with Zn or Cd
U	$U^{4+} + 2H_2O \rightleftharpoons UO_2^{2+} + 4H^+ + 2e$	Prereduction with Jones reductor
Ti	$Ti^{3+} + H_2O \rightleftharpoons TiO^{2+} + 2H^+ + e$	Prereduction with Jones reductor
Nb	$Nb^{3+} + H_2O \rightleftharpoons NbO^{3+} + 2H^+ + 2e$	Prereduction with Zn amalgam
$H_2C_2O_4$	$H_2C_2O_4 \rightleftharpoons 2CO_2 + 2H^+ + 2e$	
Mg, Ca, Zn, Co, La, Th, Ba, Sr, Ce, Ag, Pb	$H_2C_2O_4 \rightleftharpoons 2CO_2 + 2H^+ + 2e$	Insoluble metal oxalates filtered, washed, and dissolved in acid. Oxalic acid titrated.
Se	$SeO_3^{2-} + H_2O \rightleftharpoons SeO_4^{2-} + 2H^+ + 2e$	30 minute reaction time. Excess $KMnO_4$ determined.
HNO_2	$HNO_2 + H_2O \rightleftharpoons NO_3^- + 3H^+ + 2e$	15 minute reaction time. Excess $KMnO_4$ determined.
K	$K_2NaCo(NO_2)_6 + 6H_2O \rightleftharpoons$ $Co^{2+} + 6NO_3^- + 12H^+ + 2K^+ + Na^+ + 11e$	Precipitated as $K_2NaCo(NO_2)_6$. Filtered and dissolved in $KMnO_4$. Excess $KMnO_4$ determined.

Table 20-2 *(continued)*

Substance Sought	Half Reaction	Condition
Na	$U^{4+} + 2H_2O \rightleftharpoons UO_2^{2+} + 4H^+ + 2e$	Precipitated as $NaZn(UO_2)_3(OAc)_9$ $6H_2O$. Filtered, washed, dissolved, and U determined as above

Applications in Essentially Neutral Media
(Reduction product MnO_2 or Mn^{3+})

Substance Sought	Reaction	Condition
Mn	$2MnO_4^- + 3Mn^{2+} + 2H_2O \rightleftharpoons 5MnO_2 + 4H^+$	
Mn	$MnO_4^- + 4Mn^{2+} + 8H^+ \rightleftharpoons 5Mn^{3+} + 4H_2O$	Pyrophosphate solution to complex Mn^{3+}
Tl	$MnO_4^- + 2Tl^+ + 8H^+ \rightleftharpoons$ $Mn^{3+} + 2Tl^{3+} + 4H_2O$	Fluoride solution to complex Mn^{3+}

Applications in Strongly Alkaline Media
$MnO_4^- + e \rightarrow MnO_4^{2-}$

Substance Sought	Half Reaction	Condition
I^-	$I^- + 8OH^- \rightleftharpoons IO_4^- + 4H_2O + 8e$	Ba^{2+} present
IO_3^-	$IO_3^- + 2OH^- \rightleftharpoons IO_4^- + H_2O + 2e$	Ba^{2+} present
CN^-	$CN^- + 2OH^- \rightleftharpoons CNO^- + H_2O + 2e$	Ba^{2+} present
SO_3^{2-}	$SO_3^{2-} + 2OH^- \rightleftharpoons SO_4^{2-} + H_2O + 2e$	Ba^{2+} present
HS^-	$HS^- + 9OH^- \rightleftharpoons SO_4^{2-} + 5H_2O + 8e$	Ba^{2+} present
Certain organic compounds	Products are $CO_2 + H_2O$	See Table 20-3

iron, it must be broken down by more rigorous treatment. The usual procedure entails filtration and ignition of the solid followed by fusion with sodium carbonate. This process converts the metallic components of the residue into carbonates, which can then be dissolved in acid and combined with the solution containing the bulk of the sample.

The dissolution step in an iron analysis will almost inevitably result in part or all of the iron being converted to the ferric state; prereduction of the sample must, therefore, precede final titration with the oxidant. Any of the methods described in Chapter 19 may be used—the Jones reductor, for example. Zinc amalgam, however, will also reduce other elements commonly associated with iron, including titanium, niobium, vanadium, chromium, uranium, tungsten, molybdenum, and arsenic. In their lower oxidation states these, too, react with permanganate; their presence, if undetected, will lead to high results.

A silver reductor has the advantage of being inert with respect to titanium and trivalent chromium, both of which are likely to be present in high concentration. Vanadium continues to interfere.

Perhaps the most satisfactory of all prereductants for iron is stannous chloride. The only other common elements reduced by this reagent are vanadium, copper, molybdenum, tungsten, and arsenic. The excess reducing agent is removed from solution by the addition of mercuric chloride

$$Sn^{2+} + 2HgCl_2 \rightleftharpoons Hg_2Cl_2 + Sn^{4+} + 2Cl^-$$

The insoluble mercurous chloride produced will not consume permanganate nor will the excess mercuric chloride reoxidize ferrous iron. Care must be exerted, however, to prevent occurrence of the alternate reaction

$$Sn^{2+} + HgCl_2 \rightleftharpoons Hg + Sn^{4+} + 2Cl^-$$

Metallic mercury reacts with permanganate to cause a high result. This reaction is favored by appreciable excesses of stannous ion; it is prevented by careful control of this excess and by the rapid addition of a sufficient quantity of mercuric chloride. A proper reduction is indicated by the appearance of a small amount of white precipitate after addition of the mercuric chloride. A gray precipitate indicates the presence of mercury; the sample must then be discarded. The total absence of precipitate indicates that an insufficient amount of stannous chloride was added; again, the sample must be discarded.

The reaction of ferrous iron with permanganate proceeds smoothly and rapidly to completion. In the presence of hydrochloric acid, however, high results are obtained due to oxidation of the chloride ion by the permanganate. This reaction, which normally does not occur rapidly enough to cause serious errors, is induced by the presence of ferrous ion. Its effects are avoided by preliminary removal of chloride by evaporation with sulfuric acid or by use of the Zimmermann-Reinhardt reagent. The latter consists of a solution of manganous ions in fairly concentrated sulfuric and phosphoric acids. The manganous ions inhibit the oxidation of chloride ion while the phosphoric acid complexes the ferric ions produced in the titration and prevents the intense yellow color of the ferric chloride complexes from interfering with the end point.

The detailed procedure that follows provides a choice between two methods. The first consists of a stannous chloride reduction followed by titration in the presence of Zimmermann-Reinhardt reagent. In the second, chloride ion is distilled from a sulfuric acid solution before passing the sample through a Jones reductor.

Analysis of iron in an ore by prereduction with stannous chloride.

1. Special solutions.

(a) Stannous chloride. Dissolve 150 grams of iron-free $SnCl_2 \cdot 2H_2O$ in 1 liter of 1:2 HCl. The solution should be freshly prepared.

(b) Mercuric chloride. Dissolve 5 grams of $HgCl_2$ in 100 ml of water.

(c) Zimmermann-Reinhardt reagent. Dissolve 70 grams of $MnSO_4 \cdot 4H_2O$ in 500 ml of water; cautiously add 125 ml of concentrated H_2SO_4 and 125 ml of 85-percent phosphoric acid. Dilute to about 1 liter.

2. Sample treatment. Dry the ore at 105° to 110°C and weigh individual samples into 250-ml beakers; a sample of optimum size will require 25 to 40 ml of the standard $KMnO_4$. Add 10 ml of concentrated HCl, 3 ml of $SnCl_2$ solution, and heat at just below boiling until the sample is decomposed; this is indicated by the disappearance of all of the dark particles. A pure white residue may remain. A blank consisting of 10 ml of HCl and 3 ml of $SnCl_2$ should be heated for the same length of time. If the solutions become yellow during the heating, add another milliliter or two of $SnCl_2$. After the decomposition is complete, add approximately 0.2 F $KMnO_4$ dropwise until the solution is just yellow. Dilute the solution to about 15 ml. In the case of the blank add $KMnO_4$ until the solution just turns pink; then decolorize with $SnCl_2$ and add 1 drop in excess.

If all dark particles cannot be decomposed, filter the solution through ashless paper, wash with 5 to 10 ml of 6 N HCl and retain the filtrate and washings. Place the paper in a small platinum crucible and ignite. Mix 0.5 to 0.7 gram of finely ground anhydrous Na_2CO_3 with the residue and heat until a liquid melt is obtained. Cool, add 5 ml of water, followed by the cautious addition of an equal volume of 6 N HCl. Warm the crucible and wash the contents into the original filtrate. Evaporate the combined solutions to about 15 ml and proceed as in the following paragraphs.

3. Reduction of iron. Heat the solution of the sample nearly to boiling and add $SnCl_2$ drop by drop until the yellow color disappears. Add 1 *drop* in excess (Note 1). Cool to room temperature and add *rapidly* 10 ml of the $HgCl_2$ solution. A small quantity of white precipitate should appear. If no precipitate forms or if the precipitate is gray, the sample should be discarded. The blank solution should also be treated with 10 ml of the $HgCl_2$ solution.

4. Titration. After 2 to 3 minutes—not much more—transfer the reduced solution quantitatively to a 600-ml beaker containing 25 ml of the Zimmermann-Reinhardt reagent and 300 ml of water. Titrate *immediately* with the $KMnO_4$ to the first faint pink that persists for 15 to 20 seconds. Do not titrate rapidly at any time. Correct the volume of $KMnO_4$ for the blank titration.

Analysis of iron in an ore by prereduction with a Jones reductor.

1. Sample treatment. Dry the sample and weigh out suitable sized portions into 250-ml beakers. Add 10.0 ml of concentrated HCl and heat until the sample is decomposed. Also heat a blank of 10.0 ml of HCl for the same length of time. If solution of the sample is incomplete, filtration and fusion of the dark residue with Na_2CO_3 may be necessary. The method for this is described in the previous procedure.

Add 15.0 ml of 1:1 H_2SO_4 to the solution and heat until fumes of SO_3

are observed. Swirl the solution to dissolve any iron salts that may form on the sides of the beaker. Cool and slowly add 100 ml of water. The first few milliliters should be added dropwise down the sides of the beaker.

2. Reduction of iron. Drain the solution from the Jones reductor to within 1 inch of the top of the packing (Note 2). Pass 200 ml of 2 N H_2SO_4 through the tube, finally draining to about 1 inch from the top of the zinc. Disconnect the receiver and rinse it with distilled water. Again attach the receiver, pass 50 ml of 2 N H_2SO_4 followed by the sample through the reductor at a rate of about 50 to 100 ml/minute. Wash the beaker with five 10-ml portions of 2 N H_2SO_4 passing each through the reductor tube. Finally pass 100 ml of water through the tube. Leave the reductor filled with water. *Throughout the entire process the amalgam should be covered with liquid and never exposed to the atmosphere.*

3. Titration. Disconnect the receiver at once, rinse the tip of the reductor into the flask and add 5 ml of 85-percent H_3PO_4. Titrate. The blank should be treated in the same way as the sample, the volume of $KMnO_4$ consumed being subtracted from that required for the sample.

NOTES:

1. The solution may not become entirely colorless but instead acquire a pale yellow-green hue. This color will not decrease with additional $SnCl_2$. If too much $SnCl_2$ is inadvertently introduced, add 0.2 F $KMnO_4$ until the yellow color is restored and repeat the reduction.

2. If the reductor has not been in constant use, it may be partially clogged with various oxides; the first acid passed through may consume some oxidant. Therefore, test the apparatus before use by passage of about 200 ml of 2 N H_2SO_4 and 100 ml of water. The resulting solution should require no more than 0.06 ml of 0.1 N $KMnO_4$; if it does require more, repeat the operation until a blank of the required size is obtained.

Determination of calcium.   Calcium is conveniently determined by an indirect volumetric method. The element is first precipitated as calcium oxalate. This is filtered, washed, and then redissolved in dilute acid. The liberated oxalic acid is titrated with a standard solution of permanganate or some other oxidizing agent. This method is applicable to samples that contain magnesium and the alkali metals. Most other cations must be absent, however, as they either precipitate or coprecipitate as oxalates and lead to positive errors in the analysis.

In order to obtain satisfactory results by this procedure, the mol ratio of calcium to oxalate must be exactly one in the solution at the time of titration; to assure this, several precautions are necessary. For example, calcium oxalate formed in neutral or ammoniacal solutions is likely to be contaminated with calcium hydroxide or a basic calcium oxalate. The presence of either leads to low results and can be avoided by adding the oxalate to an acid solution of the sample and slowly forming the precipitate by the dropwise addition of ammonia. At a pH of about 4 the solubility of calcium oxalate is sufficiently low to ensure negligible losses provided washing is restricted to freeing the precipitate of excess oxalic acid. A precipitate formed in this way is coarsely crystalline and readily filtered.

Another potential source of positive error in the analysis arises from co-precipitation of sodium oxalate. This occurs when sodium is present in concentrations greater than the calcium. The error from this source can be eliminated by double precipitation; that is, by solution of the precipitate in acid and reprecipitating it as before.

Magnesium, if present in high concentrations, may also contaminate the calcium oxalate precipitate, either by coprecipitation or by post precipitation (see p. 187). This interference is minimized if a sufficient excess of oxalate is provided to form a soluble complex with the magnesium, and if filtration is made shortly after the completion of precipitation. When the magnesium content exceeds that of calcium in the sample, a double precipitation may be required.

The instructions that follow are applicable to samples containing calcium carbonate and moderate amounts of magnesium and the alkali metals. An alternate procedure for the analysis of limestones is also given. This will work well in the presence of moderate amounts of iron and aluminum as well as small amounts of manganese and titanium.

Method for the analysis of calcium in samples of calcium carbonate.
Dry the sample at 110° C. Weigh, into 600-ml beakers, samples large enough to contain about 100 mg of calcium. Cover with a watch glass and add 10 ml of 6 N HCl from a pipet. To avoid losses by spattering the addition should be made slowly and with the watch glass in place. Heat the solution to drive off the CO_2. Wash the watch glass and sides of the beaker with water and dilute to about 150 ml. Heat to 60° to 80° C and add 50 ml of a warm solution containing about 3 grams of $(NH_4)_2C_2O_4 \cdot H_2O$ (if the $(NH_4)_2C_2O_4$ solution is not clear, filter before use). Add 3 to 4 drops of methyl orange; then introduce 1:1 NH_3 dropwise from a pipet until the color changes from red to yellow. Allow the solution to stand for about 30 minutes (but no longer than 1 hour if magnesium is present) without further heating.

Filter the solution through a glass crucible or a Gooch crucible fitted with an asbestos mat. Wash the beaker and precipitate with 10- to 20-ml portions of chilled distilled water until the washings show only a faint cloudiness when tested with an acidified $AgNO_3$ solution (see Note 1). A quantitative transfer of the precipitate is unnecessary.

Rinse the outside of the crucible with water and place it in the beaker in which the precipitate was formed. Add 150 ml of water and 50 ml of 6 N H_2SO_4; heat to 80° to 90° C to dissolve the precipitate. Titrate with 0.1 N $KMnO_4$ with the crucible still in the beaker. The temperature of the solution should not be allowed to drop below 60° C; take as an end point the first pink color that persists for 15 to 20 seconds.

NOTE:

1. If the sample contains very large concentrations of sodium or magnesium ions, more accurate results can be obtained by reprecipitation of the calcium oxalate. To do this, filter the precipitate through paper and wash 4 or 5 times with 0.1-percent $(NH_4)_2C_2O_4$ solution. Pour 50 ml of hot 1:4 HCl through the paper, collecting the washings in the beaker in which the precipitation was made. Wash the paper several times with hot 1:100 HCl and dilute all

of the washings to about 200 ml. Proceed with the precipitation as before, this time collecting the precipitate in a Gooch or glass crucible and washing with cold water. Proceed with the analysis, as before.

Determination of calcium in a limestone. Limestones are composed principally of calcium carbonate. Dolomitic limestones contain large concentrations of magnesium carbonate in addition. Also present in smaller amounts are calcium and magnesium silicates as well as carbonates and silicates of iron, aluminum, manganese, titanium, the alkalies, and other metals.

Hydrochloric acid will often decompose limestones completely; only silica remains undissolved. Some limestones are more readily decomposed if first ignited; a few will yield only to a carbonate fusion.

The following method is remarkably effective for the analysis of calcium in most limestones. Iron and aluminum, in amounts equivalent to the calcium, do not interfere; small amounts of titanium and manganese can be tolerated.[7]

Method for calcium in limestones. Dry the sample for 1 to 2 hours at 110° C. Weigh out 0.25- to 0.3-gram samples. If the material can be readily decomposed with acid, weigh into a 250-ml beaker and cover with a watch glass. Add 5 ml of water and 10 ml of concentrated HCl, taking care to avoid loss by spattering. Proceed with the analysis as given in the next paragraph. If the limestone is not completely decomposed by acid, weigh the sample into a small porcelain crucible and ignite. The temperature should be raised slowly to 800° to 900° C where it is held for 30 minutes. After cooling, place the crucible in a 250-ml beaker, add 5 ml of water, and cover with a watch glass. Carefully add 10 ml of concentrated HCl and heat to boiling. Remove the crucible with a stirring rod, rinsing thoroughly with water.

Add 5 drops of saturated bromine water to oxidize any iron present and boil for 5 minutes to remove the excess bromine. Dilute to 50 ml, heat to boiling and add 100 ml of hot, filtered 5-percent $(NH_4)_2C_2O_4$ solution. Add 3 to 4 drops of methyl orange and precipitate the calcium oxalate by the dropwise addition of 1:1 NH_3. The rate of addition should be 1 drop every 3 or 4 seconds until the solution turns to the intermediate orange-yellow color of the indicator (pH 3.5—4.5). Allow the solution to stand for 30 minutes and filter. A Gooch crucible or a filtering crucible is satisfactory. Wash the precipitate with several 10-ml portions of cold water. Rinse the outside of the crucible and place it in the beaker in which the original precipitation was made. Add 50 ml of water containing 5 to 6 ml of concentrated H_2SO_4.

Heat the solution to 80° to 90° C. If a Gooch crucible was used, stir to break up the asbestos pad. Then titrate with 0.1 N $KMnO_4$. The solution should be kept above 60° C throughout the titration.

Applications of Permanganate Titrations to Alkaline Solutions

The oxidation of some compounds by permanganate takes place more smoothly and rapidly in alkaline solutions than in acid. In general, these are

[7] For further details of the method see J. J. Lingane, *Ind. Eng. Chem., Anal. Ed.*, **17**, 39 (1945).

substances that yield protons to the solution upon being oxidized; for example

$$SO_3^{2-} + H_2O \rightleftharpoons SO_4^{2-} + 2H^+ + 2e$$

Some useful analytical methods are based on permanganate oxidations in alkaline solutions; we shall discuss these briefly.

Nature of the reactions. In basic media, permanganate is reduced to either manganese dioxide, MnO_2, or to manganate ion, MnO_4^{2-}. According to Stamm[8] reactions yielding manganate ion proceed a good deal more rapidly and cleanly than those producing manganese dioxide. For quantitative oxidations, then, there is some advantage to confining the use of permanganate to those reactions in which this ion is formed. This can be accomplished by the use of at least a one- to twofold excess of oxidant and by keeping the solution quite basic (at least 1 to 2 N in sodium hydroxide). Under these conditions, no manganese dioxide is formed.

Generally the oxidations to which alkaline permanganate have been applied are so slow that direct titration is not feasible. This, plus the requirement of a large excess of oxidant makes mandatory an indirect procedure in which an excess of reagent is added and the excess determined by back titration after the oxidation is judged to be complete. Two methods for this back titration have been proposed. In one, a standard solution of sodium formate is employed after the introduction of barium ions into the solution. Here the reaction is

$$HCO_2^- + 2MnO_4^- + 3Ba^{2+} + 3OH^- \rightleftharpoons BaCO_3 + 2BaMnO_4 + 2H_2O$$

The presence of a barium salt is essential to the success of the method since it removes the manganate ions through formation of the slightly soluble green barium manganate. In the absence of barium ions, partial reduction of the manganate to manganese dioxide occurs. The reaction between formate ions and permanganate is reasonably rapid if catalyzed by the presence of a small quantity of nickel; the end point, however, is somewhat difficult to detect because of the presence of the green precipitate.

Another method proposed for back titration involves acidification of the reaction mixture and reduction of the residual permanganate and manganate ions to the manganous state with a standard solution of a reducing agent. Although this method is attractive because of its excellent end point, it does suffer from a serious handicap. The quantity sought represents but a small difference between two large quantities since the reduction product of the analytical reaction (MnO_4^{2-}) consumes four fifths as much standard reducing agent as does the excess permanganate itself. The resulting error in the method is on the order of 1 percent relative.

[8] H. Stamm, in W. Böttger, *Newer Methods of Volumetric Chemical Analysis*, 55-65. New York: D. Van Nostrand Co., 1938 and also H. Stamm, *Angew. Chem.*, **47**, 191 (1934); **48**, 710 (1935).

Applications of alkaline permanganate oxidation. Table 20-3 indicates some applications of alkaline oxidations that have been investigated by Stamm. In general, the required reaction times were 20 minutes or less at 25° C. In a few instances the samples were heated at 40° C for 5 minutes. With the larger organic molecules it was necessary to acidify before back titration in order to complete the oxidation to carbon dioxide; apparently in strongly basic solution some of these reactions tend to stop with the formation of oxalate.

Table 20-3

APPLICATION OF ALKALINE PERMANGANATE OXIDATION[9]

Substance Sought	Equivalents of Oxidant Consumed	Oxidation Product	Method of Back Titration[10]
I^-	8	IO_4^-	B
IO_3^-	2	IO_4^-	B
CN^-	2	CNO^-	B
HPO_3^{2-}	2	PO_4^{3-}	B
Methanol	6	$CO_2 + H_2O$	B
Formaldehyde	4	$CO_2 + H_2O$	B
Sodium formate	2	$CO_2 + H_2O$	B
Ethylene glycol	10	$CO_2 + H_2O$	A
Glycerol	14	$CO_2 + H_2O$	A
Pentoses	20	$CO_2 + H_2O$	A
Hexoses	24	$CO_2 + H_2O$	A
Salicylic acid	28	$CO_2 + H_2O$	A
Phenol	28	$CO_2 + H_2O$	A

[9] Taken from *Newer Methods of Volumetric Analysis*, W. Böttger, Ed., trans. by R. E. Oesper, 64. New York: D. Van Nostrand Co., Inc., 1938. With permission.

[10] B: Back titration with sodium formate in basic solution in presence of barium ions. A: Solution acidified and back titrated.

Many organic compounds are attacked by alkaline permanganate including those containing the following functional groups: $-OH$, $-\overset{\overset{\displaystyle O}{\|}}{C}H$, $-\overset{\overset{\displaystyle O}{\|}}{C}-$, $-NH_2$, $-C{=}C-$ (aliphatic only). Not all compounds of this kind react to give carbon dioxide and water, however; in fact, the number for which a single balanced equation can be written is fairly limited. Thus Kolthoff, in a study of the behavior of 75 compounds, found that more than one third showed a consumption of permanganate that could not be explained by any simple stoichiometry[11].

The usefulness of alkaline permanganate in organic analysis is further limited by the lack of specificity of the reagent. Despite these handicaps, however,

[11] I. M. Kolthoff and R. Belcher, *Volumetric Analysis*, 3, 114-115. New York: Interscience Publishers, Inc., 1957.

a few interesting and important applications can be cited; for example the determination of formic acid or formaldehyde in the presence of the higher aliphatic acids, and the estimation of glycerol or other glycols in aqueous solutions.

Applications of Permanganate Titrations to Neutral Solutions

In solutions that are neutral or nearly neutral, the reduction product of permanganate tends to be manganese dioxide or a complex of manganic ion. A very limited number of volumetric analyses have been based upon such half reactions.

Perhaps the best known of these is the Volhard method for the determination of manganese; the reaction is expressed by

$$3Mn^{2+} + 2MnO_4^- + 2H_2O \rightleftharpoons 5MnO_2 + 4H^+$$

The reaction is carried out at elevated temperatures in a solution buffered with acetate. The titration is slow since the first traces of excess reagent cannot be detected until the brown dioxide settles out.

Manganous ion has also been titrated under circumstances wherein manganic ion is the reaction product. To accomplish this a complexing reagent, such as fluoride or pyrophosphate ion, is added. These stabilize the $+3$ manganese ion and prevent its disproportionation into manganous ion and manganese dioxide.

Titrations in neutral fluoride media have proved useful in the analytical oxidation of thallous ion to the thallic state. Again, the reduction product of the reagent is $+3$ manganese which is stabilized by the excess of fluoride.

QUADRIVALENT CERIUM

A solution of $+4$ cerium in sulfuric acid is very nearly as potent an oxidizing reagent as permanganate; it, therefore, can be substituted for the latter in most of the applications mentioned in the previous section. The reagent is indefinitely stable and does not oxidize chloride ion at a detectable rate; in these respects, it offers considerable advantage over permanganate. Further, only a single reduction product, trivalent cerium, is formed; thus, the stoichiometry of the reaction is less subject to uncertainty. On the other hand, the color of ceric sulfate solutions is not sufficiently intense to serve as an indicator. In addition, the reagent cannot be used in neutral or basic solutions. A final disadvantage is the relatively high cost of ceric compounds. The price differential is quite great because of the large equivalent weight of cerium salts. For example, the cost of the potassium permanganate in 1 liter of 0.1 N solution is about 1 cent; the cost for a similar solution of quadrivalent cerium is about 50 cents. The greater number of man hours required to prepare a permanganate solution may, of course, more than compensate for the difference in cost of the starting materials.

Properties of Solutions of Ceric Ion

Solutions of quadrivalent cerium can be readily prepared from any of several commercially available sulfate or nitrate salts. These have a strong tendency to hydrolyze, even in acid solutions, with production of slightly soluble basic ceric salts. To avoid this the acidity must be 0.1 N or greater. The hydrolytic reaction precludes titration of neutral or basic solutions with the reagent.

Composition of ceric solutions. Acid solutions of quadrivalent cerium are highly complex in their composition. The exact nature of the cerium-containing species present is not yet known. Much of the information available comes from investigation of the effect of the type of acid and its concentration on the ceric-cerous oxidation potential and on the color of these solutions. Table 20-4 provides some typical potential data. Note that stronger oxidizing properties are exhibited by the ceric species present in perchloric acid than in nitric or sulfuric acids. In all three media the formal oxidation potential varies with acid concentration. These data suggest that ceric ions form stable complexes with nitrate and sulfate ions. In addition the existence of such species as $Ce(OH)^{3+}$ and $Ce(OH)_2^{2+}$ seems likely in perchloric acid solution. Finally, the presence of a dimeric ceric ion, particularly in highly concentrated solutions, has been reported. All evidence indicates that the concentration of the simple hydrated ion, $Ce(H_2O)_x^{4+}$, is not great in any of these solutions.

Table 20-4

FORMAL REDUCTION POTENTIALS FOR THE CERIC-CEROUS COUPLE

Acid Concentration, Normality	Formal Potential vs. Standard Hydrogen Electrode, volts		
	$HClO_4$ Solution	HNO_3 Solution	H_2SO_4 Solution
1	1.70	1.61	1.44
2	1.71	1.62	1.44
4	1.75	1.61	1.43
8	1.87	1.56	1.42

We shall express the ceric and cerous species as Ce^{4+} and Ce^{3+} in writing stoichiometric relationships. This notation is one of expedience only, and represents a tremendous oversimplification of the actual state of affairs.

Stability of ceric solutions. Sulfuric acid solutions of quadrivalent cerium are remarkably stable, remaining constant in titer for years. Even solutions heated to 100° C do not change appreciably for considerable periods of time. The same, however, cannot be said for perchloric and nitric acid solutions of the reagent. These, having higher oxidation potentials, decompose water and decrease in normality by 0.3 to 1 percent during a month's storage. The decomposition reaction is light catalyzed.

The oxidation of chloride is so slow that other reducing agents can be titrated without error in the presence of high concentrations of this ion. Hydrochloric acid solutions of ceric ion, however, are not stable enough to be used as standard reagents.

Indicators for ceric titrations. Several of the redox indicators mentioned in Chapter 17 are suitable for use in titrations with ceric solutions. The various phenanthrolines are worthy of particular mention because their transition potentials frequently correspond to the equivalence-point potentials in these reactions.

Preparation and Standardization of Ceric Solutions

Preparation of solutions. Several ceric salts are commercially available.[12] The most common of these are listed below.

Name	Formula	Equivalent Weight
Ceric ammonium nitrate	$(NH_4)_2Ce(NO_3)_6$	548.3
Ceric ammonium sulfate	$(NH_4)_4Ce(SO_4)_4 \cdot 2H_2O$	632.6
Ceric hydroxide	$Ce(OH)_4$	208.2
Ceric bisulfate	$Ce(HSO_4)_4$	528.4

Ceric ammonium nitrate of primary-standard quality can be purchased; from this, standard solutions can be prepared directly by weight. A more common practice, however, is to prepare solutions of approximately the desired normality from one of the less expensive reagent-grade salts and then standardize.

A stable and entirely satisfactory sulfuric acid solution of ceric ions is obtained from ceric ammonium nitrate without removal of the ammonium or nitrate ions. Similarly, perchloric acid solutions are often prepared from this salt by simply dissolving a suitable quantity in that acid. Where a solution free from ammonium and nitrate or sulfate ions is required, ceric hydroxide can readily be obtained from solutions of any of these salts by the addition of ammonia; this can then be filtered through a glass crucible and dissolved in the proper acid solution.

Aqueous solutions approximately $0.5 \ F$ in quadrivalent cerium and $6 \ N$ in perchloric acid are available commercially. These are free of ammonium, sulfate, and nitrate ions but do contain cerous ions in appreciable concentrations.

Directions follow for the preparation of standard solutions of known concentrations from primary-standard ceric ammonium nitrate. Nonstandard solutions can be prepared in approximately the same way.

[12] For further information regarding the preparation, standardization, and use of ceric solutions see the following: G. Frederick Smith, *Cerate Oxidimetry*. Columbus, Ohio: The G. Frederick Smith Chemical Co., 1942; I. M. Kolthoff and R. Belcher, *Volumetric Analysis*, **3**, 121-167. New York: Interscience Publishers, Inc., 1957.

Preparation of an exactly 0.1 N ceric solution in 2 N H$_2$SO$_4$.[13] Dry primary-standard grade ceric ammonium nitrate for 1 to 2 hours at 105° C. Weigh out exactly 54.526 grams into a 1-liter beaker. Add 56 ml of 95-percent H$_2$SO$_4$ and stir for 2 minutes. *Cautiously* and with stirring add 100 ml of water. Stir for 2 minutes before slowly adding another 100 ml of water. Repeat this operation until the volume of solution is 600 to 700 ml and all of the salts have dissolved to give a clear orange solution. Cool and transfer quantitatively to a 1-liter volumetric flask. Dilute to the mark and thoroughly mix after the solution has come to room temperature.

Preparation of approximately 0.1 N ceric solution in H$_2$SO$_4$. Carefully add 30 ml of concentrated H$_2$SO$_4$ to 500 ml of water, and then add 63 grams of Ce(SO$_4$)$_2$ · 2(NH$_4$)$_2$SO$_4$ · 2H$_2$O with continual stirring. Cool, filter if the solution is not clear, and dilute to about 1 liter. Alternatively, reagent (NH$_4$)$_2$Ce(NO$_3$)$_6$ can be employed; in this case use about 55 grams of the reagent and follow the instructions in the previous section.

Standardization against arsenious oxide. Arsenious oxide is perhaps the most satisfactory primary standard for sulfuric, nitric, or perchloric acid solutions of quadrivalent cerium. In the absence of a catalyst the reaction is so slow, particularly in sulfuric acid solution, that ferrous iron can be titrated without interference in the presence of trivalent arsenic. Fortunately good catalysts are available that permit standardization with this useful reagent. The best of these is osmium tetroxide which needs to be present only in very low concentrations (10^{-5} F). Iodine monochloride also catalyzes the reaction.

The orthophenanthroline-ferrous complex is an excellent indicator for sulfuric acid solutions of cerium. It is not as satisfactory for use with nitric or perchloric acid solutions of the reagent because it tends to be momentarily and preferentially oxidized before the equivalence point; a slow titration is needed for unequivocal observation of the end point. Use of the 5-nitro-orthophenanthroline derivative eliminates this difficulty.

Method. Dry the arsenious oxide for 1 hour at 110° C and weigh about 0.2-gram portions into 250-ml flasks. Dissolve in 15 ml of 2 N NaOH, warming to hasten the process. After solution is complete, cool and add 25 ml of 1:5 H$_2$SO$_4$. Dilute to about 100 ml, add 3 drops of 0.01 M osmium tetroxide (Note 1), and 1 drop of ferrous orthophenanthroline (Note 1, p. 439). Titrate to a color change from red to very pale blue or colorless.

NOTE:

1. The catalyst solution may be purchased from the G. Frederick Smith Chemical Co., Columbus, Ohio. It should contain about 0.025 gram of OsO$_4$ in 100 ml of 0.1 N H$_2$SO$_4$.

Standardization against sodium oxalate. Several methods are available for the standardization of sulfuric acid solutions of ceric ion against sodium

---

[13] Directions taken from G. F. Smith and W. H. Fly, *Anal. Chem.*, **21**, 1233 (1949).

oxalate.[14] The directions that follow call for titration at 50° C in hydrochloric acid solution. Iodine monochloride is used as a catalyst and orthophenanthroline as the indicator.

Method. Weigh out 0.3-gram portions of dried sodium oxalate into 250-ml beakers and dissolve in 75 ml of water. Add 20 ml of concentrated HCl and 1.5 ml of 0.017 F ICl (see Note 1). Heat to 50° C and add 2 to 3 drops of ferrous-orthophenanthroline indicator (Note 1, p. 439). Titrate with ceric sulfate until the solution turns pale blue or colorless and the pink does not return within 1 minute. The temperature should be between 45° and 50° C throughout.

NOTE:

1. Iodine monochloride catalyst can be prepared as follows: mix 25 ml of 0.04 F KI, 40 ml of concentrated HCl, and 20 ml of 0.025 F KIO$_3$. Add 5 to 10 ml of CCl$_4$ and shake thoroughly. Titrate with the KI or KIO$_3$ until the CCl$_4$ layer is barely pink after shaking. The KI should be added if the CCl$_4$ is colorless and the KIO$_3$ if it is too pink.

Other primary standards. The primary-standard substances listed on page 439 for permanganate can also be used to standardize solutions of quadrivalent cerium.

Applications of Solutions of Quadrivalent Cerium

A wide variety of applications of ceric solutions are found in the literature. Generally these parallel the uses of permanganate given in Table 20-2.[15] A detailed procedure for one such application follows.

Determination of iron in an ore. Quadrivalent cerium oxidizes ferrous iron smoothly and rapidly at room temperature; orthophenanthroline is an excellent indicator for the titration. In contrast to the analysis based upon oxidation with permanganate, consumption of the reagent by chloride ion is of no concern.

The problems associated with solution of the sample and prereduction of the iron are the same as in the permanganate method (p. 439), the only major difference being that here there is no need for the Zimmermann-Reinhardt solution.

Method for the analysis of iron in an ore.
1. Special solutions. (a) Stannous chloride (see p. 443). (b) Mercuric chloride (see p. 443). (c) Orthophenanthroline indicator (0.025 F) (see Note 1, p. 439). (d) Standard ceric sulfate solution.

[14] I. M. Kolthoff and R. Belcher, *Volumetric Analysis*, 3, 132-134. New York: Interscience Publishers, Inc., 1957.

[15] Information about these methods can be found in I. M. Kolthoff and R. Belcher, *Volumetric Analysis*, 3, 136-158. New York: Interscience Publishers, Inc., 1957.

2. Preparation of the sample. Use the directions on page 443 if a Jones reductor is to be used; see the same page if stannous chloride reduction is to be employed.

3. Titration. If the stannous chloride reduction is used, allow about 2 to 3 minutes after addition of the $HgCl_2$; then add 300 ml of 1 F HCl, a drop of orthophenanthroline, and titrate to the disappearance of the pink color of the indicator. For accurate work a blank should be carried through the entire procedure.

If the Jones reductor is used for prereduction, complete the analysis by titration with the ceric sulfate solution using 1 drop of orthophenanthroline indicator.

Analysis of organic compounds. A number of organic compounds are oxidized quantitatively with sulfuric or perchloric acid solutions of ceric ion. Generally these reactions are slow, requiring a few minutes to several hours for completion; often elevated temperatures are recommended. The usual procedures call for the addition of an excess of the oxidant to the sample; after some prescribed time the excess is determined by titration with a solution of ferrous iron, sodium oxalate, or other reducing reagent.

Sharma and Mehrotra have investigated the oxidation of a number of carboxylic acids by quadrivalent cerium in fairly concentrated sulfuric acid solutions. Their data, some of which are given in Table 20-5, indicate that under the prescribed conditions the acids are converted quantitatively, or nearly so, to carbon dioxide and water.

Table 20-5

OXIDATION OF SOME CARBOXYLIC ACIDS BY STANDARD CERIC SULFATE SOLUTIONS[16]

Acid	Heating Time, minutes	Sulfuric Acid Concentration, vol percent	Equivalent Ce^{4+} consumed/mol acid
Formic	50-90	67	2.00
Tartaric	65-85	50-67	10.00
Malonic	60	66	8.00
Malic	60	66	12.00
Glycolic	65-70	66	6.00
Maleic	60-90	50	11.98
Fumaric	150-180	50	11.93
Benzoic	120-150	66	29.92
Phthalic	180-240	50	29.98
Salicylic	150-180	50	28.01

[16] Data from N. N. Sharma and R. C. Mehrotra, *Anal. Chim. Acta*, **11**, 417, 507 (1954).

Smith and Duke have studied the oxidation of several organic compounds in 4 formal perchloric acid.[17] Under these conditions the ceric species is a very potent oxidizing agent; the method calls for a back titration of the excess reagent with sodium oxalate solution at room temperature. As indicator, 5-nitro-orthophenanthroline is recommended.

Rapid attack of organic compounds was found to occur at points where any two of the following functional groups were adjacent: —COOH, $>$CO, $>$CHOH, and active $>$CH$_2$. Rupture of the carbon to carbon bond between the reactive groups always occurred with the corresponding carboxylic acids being formed; where one of the groups was already an acid, carbon dioxide was the product. Thus, for example,

$$\underset{\text{glycerol}}{\text{H—C—C—C—H}} + 8Ce^{4+} + 3H_2O \rightarrow 3HC\!\!-\!\!OH + 8H^+ + 8Ce^{3+}$$

$$\underset{\text{tartaric acid}}{\text{HO—C—C—C—C—OH}} + 6Ce^{4+} + 2H_2O \rightarrow \ 2CO_2 + 2HC\!\!-\!\!OH + 6Ce^{3+} + 6H^+$$

$$\underset{\text{glyceraldehyde}}{\text{H—C—C—C—H}} + 6Ce^{4+} + 3H_2O \rightarrow 3HC\!\!-\!\!OH + 6H^+ + 6Ce^{3+}$$

$$\underset{\text{malonic acid}}{\text{HO—C—C—C—OH}} + 6Ce^{4+} + 2H_2O \rightarrow 2CO_2 + HC\!\!-\!\!OH + 6H^+ + 6Ce^{3+}$$

$$\underset{\text{biacetyl}}{\text{CH}_3\text{—C—C—CH}_3} + 2Ce^{4+} + 2H_2O \rightarrow 2CH_3C\!\!-\!\!OH + 2H^+ + 2Ce^{3+}$$

The oxidations required 5 to 120 minutes to complete. Most were carried out at room temperature although a few required temperatures between 50° to 60° C. The oxidizing agent did not attack water, formic acid, nor acetic acid under the conditions imposed. Table 20-6 summarizes these findings.

[17] G. F. Smith and F. R. Duke, *Ind. Eng. Chem., Anal. Ed.*, **15**, 120 (1943).

Table 20-6

OXIDATION OF SOME ORGANIC COMPOUNDS BY CERIC ION IN
4 F PERCHLORIC ACID[18]

Compound	Oxidation Time, minutes	Temperature ° C	Equivalents Ce[4+] consumed/mol compound
Ethylene glycol	15	60	6
Glycerol	15	60	8
Glucose	45	26	12
Biacetyl	5	24	2
Acetyl acetone	10	25	6
Tartaric acid	10	26	6
Malonic acid	10	26	6
Citric acid	30	10	14

[18] Data from G. F. Smith and F. R. Duke, *Ind. Eng. Chem., Anal. Ed.*, **15**, 120 (1943).

POTASSIUM DICHROMATE

Potassium dichromate is more limited in application than either potassium permanganate or ceric sulfate owing to its lower oxidation potential and the slowness of some of its reactions. Despite these handicaps, the reagent has proved useful for those analyses where it can be applied. Among the virtues of the reagent are the stability of its solutions and its inertness toward hydrochloric acid. Further, the solid reagent can be obtained in high purity and at modest cost; standard solutions may be prepared directly by weight.

Preparation and Properties of Dichromate Solutions

For most purposes commercially available reagent-grade or primary-standard grade potassium dichromate can be used to prepare standard solutions with no treatment other than drying at 150° to 200° C. If desired, two or three recrystallizations of the solid from water will assure a high quality primary-standard substance.

Solutions of dichromate are orange in hue; the color is not sufficiently intense, however, for end-point determination, and an oxidation-reduction indicator is ordinarily used. Diphenylamine sulfonic acid (p. 415) is most commonly employed. The color change is from the green of the chromic ion to the violet color of the oxidized form of the indicator. An indicator blank is not readily obtained because dichromate oxidizes the indicator only slowly in the absence of other oxidation-reduction systems. Ordinarily, however, this leads to a negligible error. The indicator reaction is reversible and back titration of small excesses of dichromate with ferrous solution is possible. In the presence of a large concentration of the oxidant and at low acidities (above pH 2), diphenylamine is irreversibly oxidized to yellow or red compounds.

In its analytical applications, dichromate ion is reduced to the chromic state

$$Cr_2O_7^{2-} + 14H^+ + 6e \rightleftharpoons 2Cr^{3+} + 7H_2O \qquad E^\circ = 1.33 \text{ volts}$$

The standard potential for this process appears to be appreciably higher than the formal potential for the system during an actual titration; a potential of about 1 volt is a more realistic figure for the latter. The reagent becomes a powerful oxidizing agent in very concentrated acid solutions.

Standard solutions of potassium dichromate are stable indefinitely and may be boiled for long periods without decomposition.

Preparation of a 0.1 N $K_2Cr_2O_7$ solution. Dry primary-standard $K_2Cr_2O_7$ for 2 hours at 150° to 200° C. After cooling, weigh 4.903 grams of the solid, dissolve this in distilled water, transfer quantitatively to a 1-liter volumetric flask, and dilute to the mark.

If the purity of the salt is suspect, recrystallize 3 times from water before drying. Alternatively, prepare a solution of approximate normality by dissolving about 5 grams of the salt in a liter of water, and standardize this against weighed samples of electrolytic iron wire.

Applications of Dichromate

Determination of iron. The principal use of dichromate involves titration of ferrous iron

$$6Fe^{2+} + 2Cr_2O_7^{2-} + 14H^+ \rightleftharpoons 6Fe^{3+} + 2Cr^{3+} + 7H_2O$$

Moderate amounts of hydrochloric acid do not affect the accuracy of the titration. The procedure given below may be readily applied to the analysis of iron after prereduction by stannous chloride or with a Jones reductor (see p. 443).

Method for the analysis of iron. To the prereduced solution add 10 ml of concentrated H_2SO_4 and 15 ml of syrupy H_3PO_4. Add water, if necessary, to bring the volume up to about 250 ml. Cool, add 8 drops of diphenylamine sulfonate indicator (Note 1), and titrate with dichromate to the violet-blue end point.

NOTE:
1. The indicator solution should contain 0.2 gram of sodium diphenylamine sulfonate in 100 ml of water.

Other applications. Uranium can be oxidized from the $+4$ to the $+6$ state by a direct titration with dichromate. Since the color change of diphenylamine is not rapid in the presence of this system, however, a preferable procedure is to titrate the ferrous ion produced when an excess of ferric chloride is added to the quadrivalent uranium

$$UO^{2+} + 2Fe^{3+} + H_2O \rightleftharpoons UO_2^{2+} + 2Fe^{2+} + 2H^+$$

Dichromate solutions are also used for the volumetric determination of sodium ion. The sodium is first precipitated as sodium zinc uranyl acetate,

$$(UO_2)_3NaZn(C_2H_3O_2)_9 \cdot 6H_2O$$

The solid is filtered, washed, and then redissolved in sulfuric acid. The uranium is reduced to the $+ 4$ state in a Jones reductor and titrated as before.

A common method for the determination of oxidizing agents calls for treatment of the sample with a known excess of ferrous ion, followed by titration of the excess with standard dichromate. This technique has been successfully applied to the determination of nitrate, chlorate, permanganate, dichromate, and organic peroxides, among others.

IODINE

A large number of volumetric analyses are based on the half reaction

$$I_3^- + 2e \rightleftharpoons 3I^- \qquad E° = 0.536 \text{ volt}$$

These analyses fall into two categories. The first is made up of procedures that use a standard solution of iodine to titrate easily oxidized substances. These are termed *direct* or *iodimetric methods* and have rather limited applicability since iodine is a relatively weak oxidizing agent. The second class of procedures, called *indirect* or *iodometric methods*, involve the analysis of oxidizing agents. Here the substance to be determined is brought into contact with an excess of iodide ion; a quantity of iodine, chemically equivalent to the amount of the oxidizing agent, is liberated. This is determined by titration with a standard solution of sodium thiosulfate. The quantity of iodide added is not measured; it is only important that enough be present to cause the reaction with the oxidizing agent to proceed to completion.

We shall confine this discussion to direct methods involving a standard iodine solution. In the next chapter we will consider in some detail the indirect iodometric method.

Iodine, as a standard oxidant, has some attractive features. Among these is the selectivity of its behavior which results from its low oxidizing power. Only easily oxidizable substances react with the reagent; in some instances this makes possible an analysis in the presence of components that would react with a more strenuous oxidizing agent. Another factor contributing to the popularity of iodine solutions is the sensitive and reversible indicator available for end-point detection.

One of the limitations of iodine is the lack of stability of its solutions; this makes frequent restandardizations necessary. Also, the low reduction potential of the reagent, while important from the standpoint of selectivity, severely limits the number of analyses to which it can be applied.

Preparation and Properties of Iodine Solutions

Solubility of iodine. Iodine is not very soluble in water, a saturated solution at room temperature being only somewhat greater than 0.001 F. In aqueous solutions of potassium iodide, however, the element dissolves readily as a result of formation of the soluble triiodide complex

$$I_2 + I^- \rightleftharpoons I_3^- \qquad K = 7.1 \times 10^2$$

Advantage is taken of this in preparing solutions for analysis. The concentration of the species I_2 is low in these solutions; from a chemical standpoint, therefore, it would be more proper to refer to them as *triiodide solutions*. As a practical matter, however, they are called *iodine solutions* because of the convenience this affords in writing equations and describing stoichiometric behavior.

Preparation of solutions. Standard iodine solutions can be directly prepared from commercial reagent grades of the element. Where necessary, the solid is readily purified by sublimation.

In weighing out iodine, considerable care must be taken to avoid losses due to the volatile nature of the element. Furthermore, iodine vapor is quite corrosive and can cause serious damage to the balance and other metal equipment. It is, therefore, more prudent to prepare an iodine solution in approximately the desired concentration, and to standardize it against a more tractable primary-standard material.

The rate at which iodine dissolves in potassium iodide solution is slow, particularly where the iodide concentration is low. Because of this, it is common practice to dissolve the solid completely in a few milliliters of a very concentrated iodide solution before diluting to the desired volume. All of the element must be dissolved before dilution; otherwise the normality of the resulting reagent will increase continuously as the remaining iodine slowly passes into solution.

Preparation of an approximately 0.1 N I_2 solution. Weigh about 40 grams of pure KI into a small beaker and dissolve in 10 ml of water. Add 12.7 grams of pure I_2 and stir occasionally until solution is complete. After filtering through an asbestos mat, dilute to about 1 liter. If possible, let the solution stand for 2 to 3 days before standardizing.

Stability. Iodine solutions require restandardization every few days. The instability arises from several sources, one being the volatility of the iodine; even in the presence of a considerable amount of iodide, losses from open containers occur in a relatively short period. Iodine solutions, therefore, should always be stored in bottles with well-seated glass stoppers. While detectable losses will not occur during the short time required for an ordinary titration, iodine solutions should not be left in burets for extended periods.

Iodine will slowly attack rubber or cork stoppers as well as other organic materials. Reasonable precautions must therefore be taken to protect standard solutions of the reagent from contact with these. Contact with organic dust and fumes must also be avoided.

Finally, changes in iodine normality result from air oxidation of the iodide present.

$$4I^- + O_2 + 4H^+ \rightleftharpoons 2I_2 + 2H_2O$$

This reaction is catalyzed by light, heat, and acids. Consequently, it is good practice to store the reagent in a dark, cool place. In contrast to the other effects, air oxidation results in an increase in normality.

Completeness of iodine oxidations. Because iodine is such a weak oxidizing agent, the chemist must often take full advantage of those experimental variables favoring its reduction to iodide by the substance being analyzed. Two effects, pH and the presence of complexing agents, are of particular importance.

In acid solutions, the pH has little influence upon the oxidation potential of the iodine-iodide couple since hydrogen ions do not participate in the half reaction. Many of the substances that react with iodine, however, evolve hydrogen ions in their oxidation; the position of equilibrium may therefore be markedly influenced by pH. An interesting example arises in the case of the arsenite-arsenate system, which has a standard potential differing by only 0.02 volt from that for the iodide-iodine half reaction.

$$H_3AsO_4 + 2H^+ + 2e \rightleftharpoons H_3AsO_3 + H_2O \qquad E^\circ = 0.559 \text{ volt}$$

It can be readily shown that in 1 F acid, the oxidation of arsenious acid by iodine is quite incomplete; as a matter of fact, in very strong acids, the reverse reaction can be made quantitative. On the other hand, in a nearly neutral medium the titration of the arsenious acid with iodine is quite feasible. Although iodine oxidations often become more complete with reduced acidity, care must be taken to avoid hydrolysis of the iodine that tends to occur in alkaline solutions.

$$I_2 + OH^- \rightleftharpoons HOI + I^-$$

The hypoiodite that first forms may disproportionate in part to iodate and iodide as shown by the equation

$$3HOI + 3OH^- \rightleftharpoons IO_3^- + 2I^- + 3H_2O$$

The occurrence of these reactions may lead to serious errors in iodine titrations. In some instances the reaction of the iodate and hypoiodite with the reducing reagent is so slow that overconsumption of iodine is observed. In the oxidation of thiosulfate, these species partially alter the stoichiometry of the reactions (p. 486). In general, then, solutions to be titrated with iodine cannot have pH values much higher than 9; in a few cases a pH greater than 7 is detrimental.

Complexing reagents are also used to force certain iodine oxidations to the point where they are complete enough for analytical purposes. For example, in the oxidation of ferrous iron, a comparison of the standard potential for the reaction

$$Fe^{3+} + e \rightleftharpoons Fe^{2+} \qquad E° = 0.77 \text{ volt}$$

with that of the iodide-iodine couple shows that very little oxidation of ferrous iron by iodine can be expected. However, in the presence of reagents that strongly complex the ferric and not the ferrous ion, the potential for the system is altered in a direction that makes a titration feasible. Pyrophosphate ion and ethylenediaminetetraacetate function in this manner.

The presence of a substance that forms stable complexes with iodide ion tends to keep the concentration of that species low. This, in turn, has the effect of making a more potent oxidizing agent of iodine. For example, it has been found experimentally that the formal potential for the reaction

$$I_2 + 2e \rightleftharpoons 2I^-$$

is increased by better than 0.5 volt in the presence of mercuric sulfate owing to the formation of the very stable mercuric iodide complexes.[19] In the presence of this salt, iodine will oxidize arsenious acid quantitatively even in very acid solution; this is in distinct contrast to its behavior in the absence of mercuric ion.

End points for iodine titrations. Several sensitive methods are available for determining the end point in an iodine titration. For one, the color of the triiodide ion itself is sufficiently intense to serve where colorless solutions are being titrated. Ordinarily, a concentration of about 5×10^{-6} F triiodide can just be detected by the eye; this corresponds to an overtitration of less than one drop of 0.1 N iodine solution in the typical case.

A greater sensitivity can be obtained, at the sacrifice of convenience, by adding a few milliliters of an immiscible organic solvent such as chloroform or carbon tetrachloride to the solution. The bulk of any iodine present is transferred to the organic layer by shaking, and imparts an intense violet color to it. When this end point is used, the titration is carried out in a glass-stoppered flask; after each addition of reagent, the flask is shaken vigorously and then upended so that the organic layer collects in the narrow neck for examination.

The most widely used indicator in iodimetry is an aqueous suspension of starch; this imparts an intense blue color to a solution containing a trace of triiodide ion. The nature of the colored species has been the subject of much speculation and controversy.[20] It is now believed that the iodine is held as an adsorption complex within the helical chain of the macromolecule, β-amylose, one of the components of most starches. Another component, α-amylose, is undesirable because it produces a red coloration with iodine that is not readily reversible in behavior. Interference from α-amylose is seldom serious, however, because the substance tends to settle rapidly from aqueous suspension. Other

---

[19] See N. H. Furman and C. O. Miller, *J. Am. Chem. Soc.*, **59**, 152, 161 (1937).
[20] See R. E. Rundle, J. F. Foster, and R. R. Baldwin, *J. Am. Chem. Soc.*, **66**, 2116 (1944).

starch fractions do not appear to form colored complexes with iodine. Potato, arrowroot, and rice starches contain large proportions of α- and β-amylose and can be employed as indicators. Corn starch is not suitable because of its high content of the former. The so-called *soluble starch* that is commercially available consists principally of β-amylose, the α-fraction having been removed. Indicator solutions are readily prepared from this product.

Aqueous-starch suspensions decompose within a few days, primarily because of bacterial action. The decomposition products may consume iodine as well as interfere with the indicator properties of the preparation. The rate of decomposition can be greatly reduced by preparing and storing the indicator under sterile conditions or by the addition of mercuric iodide which inhibits the growth of bacteria. Alternatively, a fresh indicator solution can be prepared each day an iodine titration is to be carried out.

The indicator properties of a starch suspension depend in part upon the composition of the solution to which it is added. The presence of iodide ion, for example, is essential; if the concentration of this species is less than about $4 \times 10^{-5} F$, a marked decrease in the color intensity is noted. Mercuric ions, which form stable iodide complexes, will often reduce the iodide concentration below this minimum value. Thus starch is not a satisfactory indicator in the presence of appreciable amounts of mercuric salts. Soluble organic alcohols, glycerol, and gelatin decrease the sensitivity of the indicator as does a high hydrogen ion concentration. Starch cannot be used for titrations in which the solution is much above room temperature.

The introduction of starch to a solution containing a high concentration of iodine gives rise to decomposition with the formation of products whose indicator properties are not entirely reversible. Thus, when an excess of iodine is present, addition of the indicator must be postponed until most of the iodine has been titrated, as indicated by the color of the solution.

The sensitivity of starch as an indicator for an iodimetric titration is, perhaps, overrated. Under ideal conditions iodine concentrations as low as $2 \times 10^{-7} F$ can be detected with the aid of starch. In clear colorless solutions, however, nearly this concentration of iodine can be observed from the color of the triiodide complex itself; thus accurate titrations can be made with no indicator at all. In the analysis of many colored solutions, however, the intensity of the blue starch-iodine color offers a real advantage.

> **Preparation of starch indicator.** Make a paste by rubbing about 2 grams of soluble starch and 10 mg of HgI_2 in about 30 ml of water. Pour this into 1 liter of boiling water and heat until a clear solution results. Cool and store in stoppered bottles. For most titrations, about 5 ml of this solution should be used.

Standardization of Iodine Solutions

Iodine solutions are most commonly standardized against arsenious oxide although sodium thiosulfate, tartar emetic, and other materials have also been recommended.

The equilibrium constant for the reaction

$$H_3AsO_3 + I_2 + H_2O \rightleftharpoons H_3AsO_4 + 2I^- + 2H^+$$

has been found experimentally[21] to be 1.6×10^{-1}. From the magnitude of this constant it is clear that a quantitative oxidation of arsenious acid can be expected only if steps are taken to keep the concentration of the products small; from a practical standpoint, the hydrogen ion and the iodide ion concentrations are most readily controlled.

The minimum pH for a quantitative reaction can be estimated by substituting reasonable values for equilibrium concentrations of the other species appearing in the expression

$$K = \frac{[H^+]^2[I^-]^2[H_3AsO_4]}{[I_2][H_3AsO_3]} = 1.6 \times 10^{-1}$$

Assuming a titration of 4 milliequivalents (2 milliformula weights) of H_3AsO_3, 40 ml of 0.1 N (0.05 F) I_2 will be required. At the equivalence point nearly all of the arsenic should be present as H_3AsO_4; if we further assume a volume of 200 ml at this point, we may write that

$$[H_3AsO_4] \cong \frac{2}{200} = 0.01 \, F$$

For the titration to be successful, the arsenious acid concentration must be very small at the equivalence point; if we consider $10^{-6} \, F$ to be a satisfactory value, then

$$[H_3AsO_3] = [I_2] = 1 \times 10^{-6} \, F$$

Iodide ion will arise from reduction of the I_2 and from that present in the iodine reagent. Typically a 0.1 N I_2 solution will be $0.25F$ in KI. Therefore

$$\text{number millimols } I^- \text{ from KI} = 0.25 \times 40 = 10$$
$$\text{number millimols } I^- \text{ from reduction of } I_2 = 0.1 \times 40 = 4$$

then

$$[I^-] = \frac{10 + 4}{200} \cong 0.07 \, M$$

Substituting these values into the equilibrium-constant expression, we obtain

$$\frac{[H^+]^2(0.07)^2 \, (0.01)}{(10^{-6}) \, (10^{-6})} = 1.6 \times 10^{-1}$$
$$[H^+] = 6 \times 10^{-5} \, M$$
$$pH = 4.2$$

On the basis of these assumptions it appears that a pH somewhat greater than 4 is required for a satisfactory oxidation of arsenious acid. Experimentally

---

[21] H. A. Liebhafsky, *J. Phys. Chem.*, **35**, 1648 (1931).

Kolthoff[22] found that a quantitative reaction occurs even at a pH of 3.5; however, the approach to equilibrium becomes prohibitively slow and, in practice, it is advisable to maintain the pH at values in excess of 5. The pH of the medium may be as high as 11 without adverse effects on the results if the arsenite is titrated with iodine. The reverse titration, however, requires a pH less than 9 in order to avoid hydrolysis of the iodine; the iodate and hypoiodite formed react only slowly with the arsenite.

The iodimetric titration of arsenious acid must be carried out in a buffered system so that the hydrogen ions formed in the reaction are consumed; otherwise the pH may decrease below the tolerable limit. Buffering is conveniently accomplished by acidifying the sample slightly and then saturating with sodium bicarbonate. The carbonic acid-bicarbonate buffer so established will hold the pH in a range between 7 and 8.

Procedure for standardization of iodine.

1. Standardization against solid As_2O_3. For a 0.1 N I_2 solution, weigh out about 0.2 gram of primary-standard grade As_2O_3 that has been dried for 1 hour at 110° C. Dissolve in 10 ml of 1 N NaOH. If necessary, warm the solution. Cool, dilute with about 75 ml of water, and add 2 drops of phenolphthalein. Introduce 6 N HCl carefully until the red color disappears; then add about 1 ml of acid in excess. Carefully add 3 to 4 grams of solid $NaHCO_3$, in small portions at first to avoid losses of solution due to effervescence of the CO_2. Add 5 ml of starch indicator and titrate to the first faint purple or blue color that lasts for 30 seconds or more.

2. Standardization against 0.1 N arsenious acid. It is often convenient to prepare a standard arsenious acid solution against which the iodine solution can be standardized periodically. Arsenious oxide solutions, if neutral or slightly acidic, are stable for several months.

To prepare an exactly 0.1 N solution, weigh 4.945 grams of dried As_2O_3 into a beaker and dissolve in 60 ml of 1 N NaOH. Neutralize with 1 N HCl. A small piece of litmus paper can be placed in the solution for this purpose. Transfer quantitatively to a 1-liter volumetric flask and dilute to the mark.

To standardize the iodine solution, withdraw a convenient volume of the arsenious acid, add 1 ml of 6 N HCl, and then proceed as in the foregoing instructions.

Applications of Standard Iodine

Table 20-7 lists the more common analyses that make use of iodine as an oxidizing reagent.

Analysis of sulfides. In the iodimetric determination of hydrogen sulfide the reaction

$$I_2 + H_2S \rightleftharpoons S + 2I^- + 2H^+$$

[22] I. M. Kolthoff and R. Belcher, *Volumetric Analysis*, 3, 217. New York: Interscience Publishers, Inc., 1957.

Table 20-7

$$I_2 + 2e \rightleftharpoons 2I^-$$

Substance Analyzed	Half Reaction
As	$H_3AsO_3 + H_2O \rightleftharpoons H_3AsO_4 + 2H^+ + 2e$
Sb	$H_3SbO_3 + H_2O \rightleftharpoons H_3SbO_4 + 2H^+ + 2e$
Sn	$Sn^{2+} \rightleftharpoons Sn^{4+} + 2e$
H_2S	$H_2S \rightleftharpoons S + 2H^+ + 2e$
SO_2	$SO_3^{2-} + H_2O \rightleftharpoons SO_4^{2-} + 2H^+ + 2e$
$S_2O_3^{2-}$	$2S_2O_3^{2-} \rightleftharpoons S_4O_6^{2-} + 2e$
N_2H_4	$N_2H_4 \rightleftharpoons N_2 + 4H^+ + 4e$
Te	$Te + 3H_2O \rightleftharpoons H_2TeO_3 + 4H^+ + 4e$
Cd^{2+}, Zn^{2+}, Hg^{2+},	$M^{2+} + H_2S \rightleftharpoons 2H^+ + MS$ (filter and wash)
Pb^{2+}, etc.	$MS \rightleftharpoons M^{2+} + S + 2e$

is rapid enough to make direct titration possible. A problem arises, however, having to do with the pH of the solution to be titrated. The use of even slightly alkaline solutions is impractical owing to the partial oxidation of the sulfide to sulfate. At the same time, losses of hydrogen sulfide by volatilization from neutral or acidic solution will occur. To circumvent these problems, an alkaline solution of the sulfide may be added to an excess of acidic standard iodine, followed by back titration with standard thiosulfate. Alternatively, a measured quantity of the iodine solution can be titrated with the solution being analyzed.

The iodimetric determination of hydrogen sulfide is also used for the analysis of metallic ions such as zinc, cadmium, lead, and mercury which form insoluble precipitates with the gas. The sulfides, after precipitation, are filtered, washed free of hydrogen sulfide, and dissolved in an acid solution containing a measured amount of iodine. The excess iodine is then titrated with thiosulfate.

Determination of arsenic, antimony, and tin. The elements, arsenic, antimony, and tin, are conveniently determined by titration with iodine. The reaction of the reagent with arsenious acid was discussed in a previous section.

In acid solutions divalent tin is rapidly oxidized to the quadrivalent state by iodine. The principal difficulty in the use of this reaction arises from the ease with which the stannous ion is oxidized by air. This is avoided by covering the solution with an inert gas during the titration; furthermore, it is necessary to use oxygen-free solutions throughout the analysis. Prereduction of quadrivalent tin is ordinarily accomplished with metallic lead or nickel.

The reaction of trivalent antimony with iodine is quite analogous to that of trivalent arsenic. Here, however, steps are also necessary to prevent precipitation of such basic salts as antimonyl chloride, SbOCl, from the solution as it is neutralized; these react incompletely with iodine and lead to erroneously low results. This difficulty is readily overcome by addition of tartaric acid before dilution; the soluble antimonyl tartrate complex, $SbOC_4H_4O_6^-$, that forms is rapidly and completely oxidized by iodine.

Determination of antimony in stibnite. The analysis of the common antimony ore, stibnite, is a good example of the application of a direct iodimetric method. Stibnite is primarily antimony sulfide containing silica and other contaminants. Provided the material is free of iron and arsenic, determination of its antimony content is a straightforward process. The sample is dissolved in hot, concentrated hydrochloric acid which causes the sulfide to be evolved as hydrogen sulfide. Some care is required in this step to prevent losses of the volatile antimony trichloride; addition of potassium chloride is helpful because it increases the tendency of nonvolatile chloride complexes of antimony to form. These probably have the formulae $SbCl_4^-$ and $SbCl_6^{3-}$.

> **Method.** Dry the sample in an oven. After cooling, weigh into 500-ml Erlenmeyer flasks sufficient quantities of the ore to consume 25 to 35 ml of 0.1 N I_2. Add about 0.3 gram of KCl and 10 ml of concentrated HCl. Heat the mixture to just below boiling and maintain at this temperature until only a white or slightly gray residue of silica remains.
>
> Add 3 grams of solid tartaric acid to the solution and heat for another 10 to 15 minutes. While swirling the solution, slowly add water from a pipet until the volume is about 100 ml. The addition of water should be slow enough to prevent white SbOCl from forming. If reddish Sb_2S_3 forms, stop the addition of water and heat further, adding more acid if necessary.
>
> Add 3 drops of phenolphthalein to the solution, and 6 N NaOH until the first pink color is obtained. Add 6 N HCl dropwise until the solution is decolorized, and then 1 ml in excess. Add 4 to 5 grams of $NaHCO_3$, taking care to avoid losses of solution during the addition. Add 5 ml of starch and titrate to the first blue color that persists for 30 seconds or longer.

POTASSIUM BROMATE

In the presence of acids, bromate ion is nearly as powerful an oxidant as permanganate. Standard solutions of the reagent can be prepared directly from primary standard potassium bromate and are stable for indefinite periods. Such solutions have some unique applications that are of real importance to the analytical chemist.

Reactions of Potassium Bromate

With good reducing agents, bromate ion is reduced to bromide, the half reaction for the process being

$$BrO_3^- + 6H^+ + 6e \rightleftharpoons Br^- + 3H_2O \qquad E^\circ = 1.44 \text{ volts}$$

Bromate ion is, however, capable of oxidizing bromide ion to bromine as is apparent from the standard potential for the half reaction

$$Br_2 + 2e \rightleftharpoons 2Br^- \qquad E^\circ = 1.06 \text{ volts}$$

Thus, after the more potent reducing agent has been consumed in a bromate titration the following reaction will occur if the solution is acidic:

$$BrO_3^- + 5Br^- + 6H^+ \rightleftharpoons 3Br_2 + 3H_2O$$

Ordinarily, then, generation of bromine takes place immediately after the equivalence point is reached in a bromate titration and can act to signal the end point of the process. Several colored organic indicators are available that owe their color changes to reactions with elemental bromine.

With weak reducing agents, conversion of bromate to bromide may be incomplete and mixtures of the latter ion and bromine may be encountered. Sometimes this can be avoided by the introduction of mercuric ions; the tendency for reduction solely to bromide is thus increased owing to formation of the soluble species $HgBr_2$ and $HgBr_4^{2-}$.

By and large, the most important uses of standard bromate solution are not those in which direct reduction to bromide ion occurs, but rather in the employment of the reagent as a stable source of known quantities of elementary bromine.

In order to understand this application we must again consider the reaction

$$BrO_3^- + 5Br^- + 6H^+ \rightleftharpoons 3Br_2 + 3H_2O$$

$$K = \frac{[Br_2]^3}{[BrO_3^-][Br^-]^5[H^+]^6}$$

From the equilibrium-constant expression we see that the concentration of bromine in a solution containing both bromate and bromide is highly dependent upon pH. The magnitude of this constant is such that in neutral solutions little bromine will be found, but at a pH of 1 the reaction will proceed far to the right. This large pH effect can be used to advantage; a neutral solution containing bromate and an excess of bromide is essentially free of bromine and is quite stable. Upon acidification with strong acid, a quantitative release of bromine ensues. Thus, we have a stable solution that can deliver an amount of bromine equivalent to the amount of bromate ion present.

The analytical reactions to which the foregoing have been applied involve combination of the bromine released with organic compounds via addition or substitution reactions. To be sure, a standard bromine solution may be directly employed; such a solution, however, is quite unstable owing to the volatility of bromine. The use of bromate as a source of the element clearly offers a distinct advantage.

Indicators for Bromate Titrations

A number of organic substances have been proposed as indicators for bromate and bromine titrations. Among these are such azo dyes as methyl orange and methyl red which, upon being brominated, are converted from red to yellow compounds. Similarly, bromination of the indigo sulfonic acids causes a change from blue to yellow. Many other dyes are bleached by bromine, and several of these are used in bromate titrations.

Unfortunately such dyestuffs are not entirely satisfactory as indicators because their behavior toward the reagent is totally non-reversible. This precludes any sort of back titration and, more important, makes direct titration more difficult because of the great care required to avoid local excess of the reagent. Thus, successful bromate titrations with these indicators require the slow addition of reagent and thorough mixing after each addition.

Three indicators, α-napthoflavone, *p*-ethoxychrysoidine, and quinoline yellow, are reversible with respect to bromine; these make employment of the reagent more attractive. They are commercially available.

Direct Titrations with Bromate

As shown in Table 20-8, a variety of inorganic and organic substances may be titrated with standard bromate solutions. In most of these applications the reaction medium is 1 N or greater in hydrochloric acid.

Table 20-8

APPLICATION OF BROMATE TITRATIONS

$$BrO_3^- + 6H^+ + 6e \rightleftharpoons Br^- + H_2O$$

Substance Analyzed	Half Reaction	Special Condition
As	$H_3AsO_3 + H_2O \rightleftharpoons H_3AsO_4 + 2H^+ + 2e$	
Sb	$H_3SbO_3 + H_2O \rightleftharpoons H_3SbO_4 + 2H^+ + 2e$	
Sn	$Sn^{2+} \rightleftharpoons Sn^{4+} + 2e$	
Cu	$Cu^+ \rightleftharpoons Cu^{2+} + e$	
Tl	$Tl^+ \rightleftharpoons Tl^{3+} + 2e$	
Se	$Se + 3H_2O \rightleftharpoons H_2SeO_3 + 4H^+ + 4e$	
Fe	$Fe^{2+} \rightleftharpoons Fe^{3+} + e$	$CuCl_2$ catalyst
H_2O_2	$H_2O_2 \rightleftharpoons O_2 + 2H^+ + 2e$	$MnCl_2$ catalyst
N_2H_4	$N_2H_4 \rightleftharpoons N_2 + 4H^+ + 4e$	
Cl, Br, I	$I^- + CN^- \rightleftharpoons ICN + 2e$	
	$Br^- + CN^- \rightleftharpoons BrCN + 2e$	
	$Cl^- + CN^- \rightleftharpoons ClCN + 2e$	
CNS^-	$CNS^- + 4H_2O \rightleftharpoons SO_4^{2-} + CN^- + 8H^+ + 6e$	excess bromate added and back titrated
Dialkyl sulfides	$R_2S + H_2O \rightleftharpoons R_2SO + 2H^+ + 2e$	
Alkyl disulfides	$RSSR + 4H_2O + 2Br^- \rightleftharpoons$ $2RSO_2Br + 8H^+ + 10e$	
$H_2C_2O_4$	$H_2C_2O_4 \rightleftharpoons 2CO_2 + 2H^+ + 2e$	Mn^{2+} catalyst, Hg^{2+} present

Applications of Bromate Reagent to Brominations

Organic compounds react with bromine either by substitution or by addition. The former involves replacement of one or more hydrogen atoms in an aromatic ring by atoms of the halogen. For example, in the bromination of phenol three hydrogen atoms are replaced.

Since this process requires six equivalents of bromine, the equivalent weight of phenol is equal to one sixth of its formula weight. Addition reactions involve the opening of an olefinic double bond. For example, two equivalents of bromine react with ethylene

The equivalent weight of ethylene will clearly be equal to one half of its formula weight in this reaction.

Addition or substitution reactions can proceed in a quantitative or near-quantitative manner with a considerable number of organic compounds; bromination techniques thus offer a good method for analysis. A bromate-bromide mixture serves as the preferred source of bromine for these reactions.

One technique involves a direct bromate titration of an acidified solution (or suspension) of the sample. An excess of bromide is either added to the sample prior to titration or is introduced in the bromate solution itself. As it is liberated, the bromine reacts with the organic compound; the end point is signaled by the appearance of unreacted bromine.

For those compounds that do not react sufficiently rapidly to allow direct titration, an alternative technique must be employed—the addition of a measured excess of a bromate-bromide solution to the sample. The mixture is then acidified; after bromination is judged complete, the excess bromine is determined by back titration. Standard arsenious acid may be used for this. If desired, the analysis may be completed by adding an excess of potassium iodide and titrating the iodine liberated with a standard solution of sodium thiosulfate. Although this iodometric procedure sounds complicated, it is, in practice, quite simple; Chapter 21 should be consulted for further details.

Because this second method involves appreciable quantities of free bromine, the reaction vessels must be tightly stoppered during the bromination, and care must be exercised to minimize volatilization losses during the back titration.

Substitution reactions. Bromination methods have been applied successfully to those aromatic compounds that contain strong *ortho-para* directing substituents in the ring, particularly amines and phenols. Table 20-9 lists some typical examples. There are many organic substances that interfere with the method. These include unsaturated compounds that can add bromine; readily oxidized groups such as mercaptans, sulfides, quinones, etc.; and aliphatic compounds containing carbonyl, carboxylate, and other groups that lead to unwanted substitution reactions.

Table 20-9

SOME ORGANIC COMPOUNDS WHICH CAN BE ANALYZED BY BROMINE SUBSTITUTION[23]

Compound	Reaction Time, minutes	Equivalents Br_2/mol compound	Accuracy, percent theoretical
Phenol	5-30	6	99.89
p-Chlorophenol	30	4	99.87
Salicylic acid	30	6	99.86
Acetylsalicylic acid	30	6	99.84
m-Cresol	1	6	99.75
β-Naphthol	15-20	2	99.89
Aniline	5-10	6	99.92
o-Nitroaniline	30	4	99.86
Sulfanilic acid	30	6	99.98
m-Toluidine	5-10	6	99.87

[23] Data from A. R. Day and W. T. Taggart, *Ind. Eng. Chem.*, **20**, 545 (1928).

An important application of the bromination technique is to the titration of 8-hydroxyquinoline.

The reaction is sufficiently rapid to allow direct titration in hydrochloric acid solution with methyl red as an indicator.

This reaction is of particular interest because 8-hydroxyquinoline is an excellent precipitant for a variety of cations (a list of these is given on p. 196). A convenient way of completing the analysis for these ions thus involves solution of the metal hydroxyquinolate in strong hydrochloric acid and titration with a bromate solution. For example, reactions for the analysis of aluminum are

$$Al^{3+} + \underset{\text{8-hydroxyquinoline}}{3HOC_9H_6N} \xrightarrow{pH\ 4-9} \underset{\text{(filter and wash)}}{Al(OC_9H_6N)_3} + 3H^+$$

$$Al(OC_9H_6N)_3 + 3H^+ \xrightarrow{\text{hot 4 N}\ \text{HCl}} 3HOC_9H_6N + Al^{3+}$$

$$3HOC_9H_6N + 6Br_2 \rightarrow 3HOC_9H_4NBr_2 + 6HBr$$

For each mol of aluminum 6 mols, or 12 equivalents, of bromine are consumed; the equivalent weight of the metal is therefore one twelfth of its formula weight. Also, from the equation for the formation of bromine we see that each mol of bromate corresponds to 3 mols, or 6 equivalents, of bromine; the equivalent weight of bromate, therefore, is one sixth its formula weight in this application.

Addition reactions. Olefinic unsaturation can be determined by the addition of bromine to the double bond. A variety of methods, many involving the use of bromate-bromide mixtures, are found in the literature.[24] Most of these are applied to the estimation of unsaturation in fats, oils, and petroleum products.

POTASSIUM IODATE

Potassium iodate is commercially available in a high state of purity. This product can be used without further treatment, other than drying, for the direct preparation of standard solutions of the reagent. Iodate solutions have a number of interesting and important uses in analytical chemistry.

Reactions of Iodate

One of the important analytical reactions of potassium iodate involves its oxidation of iodide to iodine.

$$IO_3^- + 5I^- + 6H^+ \rightleftharpoons 3I_2 + 3H_2O$$

As in the oxidation of bromide with bromate, the position of equilibrium is quite pH dependent; in neutral solution no appreciable formation of iodine is observed while quantitative reduction of iodate takes place in slightly acid media containing an excess of iodide.

Use is made of this reaction as a convenient source of known amounts of iodine. A measured quantity of iodate is mixed with an excess of iodide in a solution that is 0.1 to 1 N in acid. Exactly 6 equivalents of iodine are liberated for each mol of iodate; the resulting solution can then be used for the standardization of thiosulfate solutions or for other analytical reactions.

The reaction is also used for the standardization of acids against primary-standard potassium iodate. Here a neutral solution containing an excess of iodide and a measured quantity of iodate is titrated with the acid, the iodine liberated being removed by the presence of an excess of sodium thiosulfate. The pH of the solution remains essentially constant until the iodate is used up; then a sharp decrease occurs.

In strongly acid solutions, iodate will oxidize iodide or iodine to the $+ 1$

---

[24] For a review of this subject see A. Polgar and J. L. Jungnickel in *Organic Analysis*, 3. New York: Interscience Publishers, Inc., 1956.

state provided some anion, such as chloride, bromide, or cyanide, is present to stabilize this oxidation state. For example, in solutions that are greater than 3 N in hydrochloric acid, the following reaction proceeds essentially to completion:

$$IO_3^- + 2I_2 + 10Cl^- + 6H^+ \rightleftharpoons 5ICl_2^- + 3H_2O$$

In solutions of hydrogen cyanide, lower acidities (1 to 2 N) are sufficient for the quantitative formation of $+1$ iodine since iodine cyanide, ICN, is more stable than ICl_2^-.

A number of important iodate titrations are performed in strong hydrochloric acid or hydrogen cyanide solutions. In these, the iodate is initially reduced to iodine. As the reducing agent is consumed, however, oxidation of the iodine by the iodate takes place. Generally, the end point for the process is signaled by the complete disappearance of the iodine and the over-all stoichiometry is thus

$$IO_3^- + 6H^+ + 4e \rightleftharpoons I^+ + 3H_2O$$

The equivalent weight of the iodate is therefore equal to one fourth of its formula weight. When used in this manner, iodate is a less powerful oxidant than either permanganate or ceric ion; on the other hand, it is appreciably more potent than iodine itself.

End Points in Iodate Titrations

The first disappearance of iodine from the solution is often employed to indicate the end point in an iodate titration. Iodine is nearly always formed in the initial stages of the reaction; only at the equivalence point is this completely oxidized to the $+1$ state. For titrations carried on in the presence of hydrogen cyanide, starch can serve as the indicator, the equivalence point being signaled by the disappearance of the familiar blue complex. The indicator fails to function, however, in the highly acidic media required for the production of iodine monochloride. Here a small quantity of an immiscible organic solvent is used as the indicator. A few milliliters of carbon tetrachloride, chloroform, or benzene is added at the start of the titration. After each addition of iodate, the mixture is shaken thoroughly; the organic layer is examined after the phases have separated. The bulk of any unreacted iodine remains in the organic layer and imparts a violet-red color to it. The titration is judged complete at the point where this color first disappears from the organic phase. This excellent method for detecting iodine is quite comparable in sensitivity to the starch-iodine color; it suffers, however, from the disadvantage of being time consuming.

Several dyes that are discolored by reactions with iodine have also been proposed as indicators in iodate titrations. These are nonreversible in their behavior and thus less satisfactory than the above indicators.

Applications of Iodate Oxidation[25]

Table 20-10 lists some of the substances that have been determined by iodate titration.

Table 20-10

Analysis by Iodate Titration

$$IO_3^- + 6H^+ + 2Cl^- + 4e \rightleftharpoons ICl_2^- + 3H_2O$$

Substance Analyzed	Half Reaction
As	$H_3AsO_3 + H_2O \rightleftharpoons H_3AsO_4 + 2H^+ + 2e$
Sb	$H_3SbO_3 + H_2O \rightleftharpoons H_3SbO_4 + 2H^+ + 2e$
I-	$I^- + 2Cl^- \rightleftharpoons ICl_2^- + 2e$
I_2	$I_2 + 4Cl^- \rightleftharpoons 2ICl_2^- + 2e$
Sn	$Sn^{2+} \rightleftharpoons Sn^{4+} + 2e$
Tl	$Tl^+ \rightleftharpoons Tl^{3+} + 2e$
Hg_2Cl_2	$Hg_2Cl_2 + 2Cl^- \rightleftharpoons 2HgCl_2 + 2e$
Fe	$Fe^{2+} \rightleftharpoons Fe^{3+} + e$
N_2H_4	$N_2H_4 \rightleftharpoons N_2 + 4H^+ + 4e$
CNS-	$CNS^- + 4H_2O \rightleftharpoons SO_4^{2-} + CN^- + 8H^+ + 6e$
H_2SO_3	$H_2SO_3 + H_2O \rightleftharpoons SO_4^{2-} + 4H^+ + 2e$
$S_2O_3^{2-}$	$S_2O_3^{2-} + 5H_2O \rightleftharpoons 2SO_4^{2-} + 10H^+ + 8e$
$S_4O_6^{2-}$	$S_4O_6^{2-} + 10H_2O \rightleftharpoons 4SO_4^{2-} + 20H^+ + 14e$

PERIODIC ACID

Solutions of periodic acid or its salts are strong oxidizing reagents. The reduction potential for the half reaction is on the order of 1.6 volts.

$$H_5IO_6 + H^+ + 2e \rightleftharpoons IO_3^- + 3H_2O \qquad E^\circ \cong 1.6 \text{ volts}$$

As evidence for their potency, periodate solutions oxidize manganous ion to permanganate and slowly evolve ozone from aqueous solutions.

Periodic acid is of particular importance in the field of organic analysis. The remarkable selectivity of its behavior towards certain functional groups is unusual for so strong an oxidizing agent; this is the property that enhances its value to the analytical chemist. The application of periodates to the selective oxidation of organic compounds was introduced by L. Malaprade[26] in 1928.

---

[25] For details on the application of iodate titrations see I. M. Kolthoff and R. Belcher, *Volumetric Analysis*, **3**, 449-473. New York: Interscience Publishers, Inc., 1957; and R. Lang in W. Böttger, Ed., *Newer Methods of Volumetric Analysis*, 69-98. New York: D. Van Nostrand Inc., 1938.

[26] L. Malaprade, *Compt. rend.*, **186**, 382 (1928).

Periodic Acid Solutions

Composition. Aqueous solutions of $+7$ iodine are complex in nature. In the presence of strong acids the main constituent appears to be paraperiodic acid, H_5IO_6, although other species, such as metaperiodic acid, HIO_4, and such ions as $H_4IO_6^-$, IO_4^-, and $H_3IO_6^{2-}$ are doubtless also present.[27]

Paraperiodic acid behaves as weak acid, having a first dissociation of 2.4×10^{-2} and second of about 5×10^{-9}.

Preparation and properties. Several periodates that can be used for the preparation of standard periodic acid solutions are available commercially. Among these is paraperiodic acid itself, a crystalline, readily soluble, hygroscopic solid. An even more useful compound is sodium metaperiodate, $NaIO_4$, which is soluble to the extent of 12.6 percent by weight at 25° C. Sodium paraperiodate, Na_5IO_6, is not sufficiently soluble for the preparation of standard solutions; however, it is readily converted to the more soluble metaperiodate by recrystallization from hot concentrated nitric acid solutions. Potassium metaperiodate can be used as a primary standard for the preparation of periodate solutions.[28] At room temperature its solubility is only about 5 grams/liter; at elevated temperatures, however, it is readily dissolved and is converted to a more soluble form by the addition of base.[28]

Periodate solutions vary considerably in stability depending on their mode of preparation and storage. A solution prepared by dissolving the sodium metaperiodate in water decomposes at the rate of several percent per week. On the other hand, a solution of potassium metaperiodate in excess alkali was found to change no more than 0.3 to 0.4 percent in 100 days. The most stable periodate solutions appear to be those containing an excess of sulfuric acid; these decrease in normality by less than 0.1 percent in 4 months.[28]

Standardization of periodate solutions. Several iodometric methods are available for the standardization of periodic acid solutions. In the most convenient of these, a solution of arsenious acid of known concentration is employed as the standard of reference. In general, a measured quantity of the periodate is rendered neutral or slightly alkaline with a concentrated borax, bicarbonate, or phosphate buffer. To this is added an excess of iodide ion and iodine is liberated by the following reaction:

$$IO_4^- + 2I^- + H_2O \rightleftharpoons IO_3^- + I_2 + 2OH^-$$

As long as the solution is kept neutral, further reduction of iodate does not occur and the liberated iodine can be titrated directly with the arsenite solution. A standard sodium thiosulfate could also be used for this titration.

An alternative procedure calls for treatment of the buffered periodate solution with an excess of arsenite containing a trace of iodide ion. After reduction of the periodate to iodate is complete, the residual arsenite is titrated with

---

[27] For discussion of the composition of acidic solutions of periodate see C. E. Crouthamel, A. M. Hayes, and D. S. Martin, *J. Am. Chem. Soc.*, **73**, 82 (1951).

[28] H. H. Willard and L. H. Greathouse, *J. Am. Chem. Soc.*, **60**, 2869 (1938).

a standard iodine solution. This procedure suffers from the disadvantage of requiring two standard solutions.

Periodate solutions may also be standardized by adding an excess of iodide to an acidified solution of the periodic acid; under these conditions, reduction to iodine occurs.

$$H_5IO_6 + 7I^- + 7H^+ \rightleftharpoons 4I_2 + 6H_2O$$

The liberated halogen is titrated as before.

Applications of Periodate Solutions

By far the most important applications of periodate are in the field of organic analysis; we will restrict our discussion to these.

Methods used in periodate analyses. The reactions of organic compounds with periodate are never rapid enough to allow direct titration. Therefore, an excess of oxidant is always used, and time is allowed for the reaction to take place. The time required varies considerably; most oxidations are complete, however, in 30 minutes to 1 hour at room temperature. Higher temperatures are ordinarily avoided because this diminishes the selectivity of the reagent and results in partial attack upon a variety of substances.

Oxidations are generally carried out in water although it is possible to obtain quite satisfactory results in aqueous solutions of methanol, ethanol, acetic acid, and dioxane. Where solution in such solvents is not complete, oxidation of a suspension of the substance has been found practical, although this generally calls for a greater reaction time.

The rate of most periodate oxidations is pH sensitive, often attaining a maximum at a value of about four. Iodate ion is the usual reduction product.

Periodate methods for organic compounds fall into two categories. In the first, the outcome is based upon an estimation of the oxidant consumed by the reaction. Here, a known excess of the reagent is introduced; after the reaction is judged complete, the unreacted portion is determined by one of the methods described for the standardization of periodate solutions. In the second category are procedures based upon determination of one or more of the oxidation products. In this instance, the exact amount of periodate used is not significant as long as there is sufficient to complete the oxidation. An obvious requirement here is that the substance being determined yields a compound that is readily analyzed. Carboxylic acids, formaldehyde, acetaldehyde, and ammonia are principal among these.

Compounds attacked by periodate. At room temperature, the only common organic compounds oxidized by periodic acid are those containing the following functional groups *adjacent to one another*: alcohol, $-\overset{\overset{\displaystyle OH}{|}}{C}-$; carbonyl, $-\overset{\overset{\displaystyle O}{\|}}{C}-$ or $-\overset{\overset{\displaystyle O}{\|}}{C}-H$; and primary or secondary amine, $-\overset{\overset{\displaystyle NH_2}{|}}{C}-$ or $-\overset{\overset{\displaystyle NHR}{|}}{C}-$. Organic compounds containing isolated hydroxy, carbonyl, or amine groups are not ordinarily attacked at room temperature; nor are compounds with a carboxylic acid group by itself or adjacent to any of the above. At elevated temperatures this selectivity tends to disappear.

Periodate oxidations of organic compounds follow a regular pattern which makes it rather easy to predict the oxidation products in any given case. The following rules apply:

1. Attack of adjacent functional groups always results in a rupture of the carbon to carbon bond between these groups.

2. In the oxidation process, the carbon atom containing an —OH group is oxidized to an aldehyde or ketone.

3. A carbonyl group is converted to a carboxylic acid group by the oxidation.

4. A carbon atom containing an amine group losses ammonia and is itself converted to an aldehyde.

5. A carbon atom containing a substituted amine group loses the corresponding amine and is converted to an aldehyde.

The following half reactions will serve to illustrate these rules:

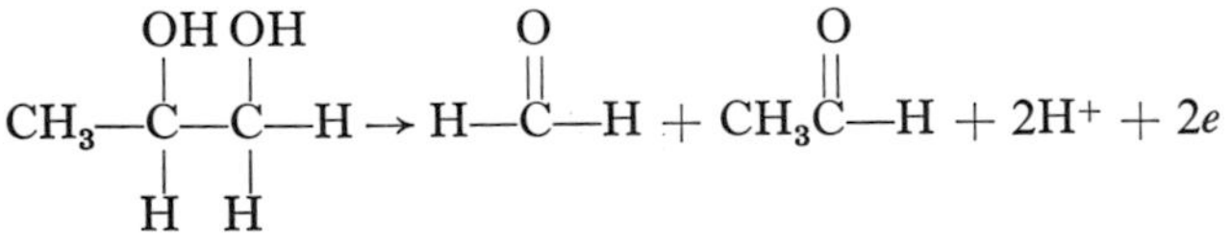

$$CH_3-\underset{\underset{H}{|}}{\overset{\overset{OH}{|}}{C}}-\underset{\underset{H}{|}}{\overset{\overset{OH}{|}}{C}}-H \rightarrow H-\overset{O}{\overset{||}{C}}-H + CH_3\overset{O}{\overset{||}{C}}-H + 2H^+ + 2e$$

propylene glycol

$$H-\underset{\underset{H}{|}}{\overset{\overset{OH}{|}}{C}}-\underset{\underset{H}{|}}{\overset{\overset{OH}{|}}{C}}-\underset{\underset{H}{|}}{\overset{\overset{OH}{|}}{C}}-H + H_2O \rightarrow 2H-\overset{O}{\overset{||}{C}}-H + H-\overset{O}{\overset{||}{C}}-OH + 4H^+ + 4e$$

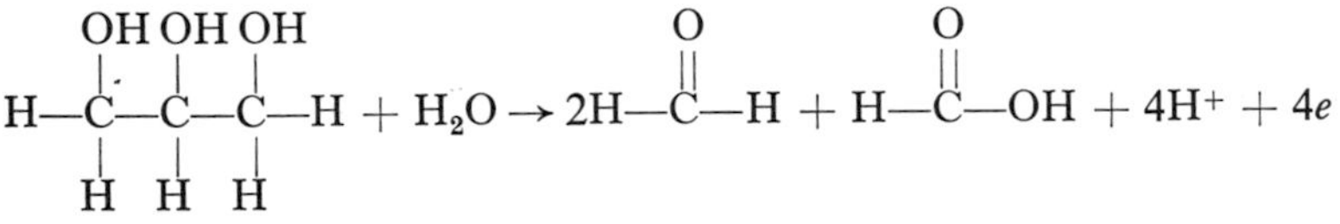

glycerol

For purposes of predicting the reaction products, the first step in the glycerol oxidation can be thought of as producing 1 mol of formaldehyde and 1 mol of glycolic aldehyde, $H-\underset{\underset{H}{|}}{\overset{\overset{OH}{|}}{C}}-\overset{O}{\overset{||}{C}}$. This is then further oxidized and, by the rules given above, produces another mol of formaldehyde plus 1 mol of formic acid.

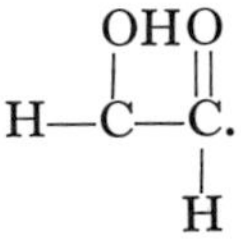

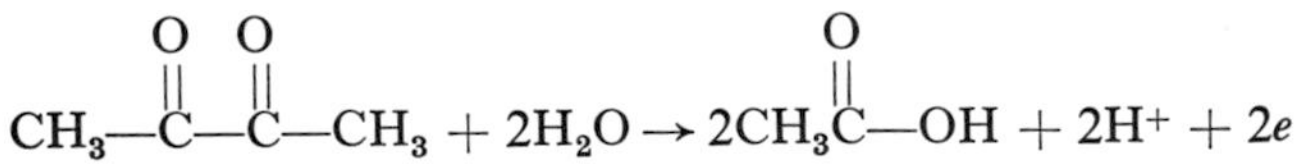

$$CH_3-\overset{O}{\overset{||}{C}}-\overset{O}{\overset{||}{C}}-CH_3 + 2H_2O \rightarrow 2CH_3\overset{O}{\overset{||}{C}}-OH + 2H^+ + 2e$$

biacetyl

$$CH_3-\overset{\overset{O}{\|}}{C}-\overset{\overset{OH}{|}}{C}-CH_3 + H_2O \rightarrow CH_3\overset{\overset{O}{\|}}{C}-OH + CH_3\overset{\overset{O}{\|}}{C}-H + H^+ + e$$

acetoin

$$H-\overset{\overset{OH}{|}}{\underset{\underset{H}{|}}{C}}-\overset{\overset{NH_2}{|}}{\underset{\underset{H}{|}}{C}}-H + H_2O \rightarrow 2H-\overset{\overset{O}{\|}}{C}-H + NH_3 + 2H^+ + 2e$$

ethanolamine

Analysis of glycerol. Glycerol is cleanly oxidized by periodic acid solutions according to the equation

$$CH_2OHCHOHCH_2OH + 2H_5IO_6 \rightarrow$$
$$2HCHO + HCOOH + 2IO_3^- + 5H_2O + 2H^+$$

About one-half hour is required to complete the reaction; the glycerol is readily estimated by iodimetric determination of the periodate consumed.

The following represents a modification of a method reported by Voris, Ellis, and Maynard.[29] It involves reduction of the excess periodate to iodate with a known quantity of arsenite, followed by titration of the excess trivalent arsenic with a standard iodine solution.

Determination of glycerol

Solutions required.
(a) Periodate solution, approximately 0.06 N, based upon a 2-electron reduction. Solution should be approximately 0.1 N with respect to H_2SO_4. Its periodate concentration is conveniently determined concurrently with the analysis.

(b) Standard 0.1 N arsenious acid solution (see p. 464).

(c) Iodine solution, approximately 0.1 N, the volume ratio of which is known with respect to the arsenite solution.

(d) $MgSO_4$ solution, approximately 15 percent by weight.

(e) Starch suspension (see p. 462).

Method. To a 250-ml iodine flask introduce a 10-ml aliquot of dilute glycerol solution and a 25-ml aliquot of periodate solution; allow the reaction to proceed for about 30 minutes at room temperature. Then add 3 drops of $MgSO_4$ solution. Follow this with the dropwise introduction of NaOH until the first faint turbidity of $Mg(OH)_2$ is observed. Add dilute H_2SO_4 until the turbidity is just discharged. Saturate the solution with $NaHCO_3$, and introduce a measured excess of standard arsenite solution. After 5 minutes, add starch and titrate the excess arsenite with iodine solution. Calculate the weight of glycerol in the sample.

[29] L. Voris, G. Ellis, and L. Maynard, *J. Biol. Chem.*, **133**, 491(1940).

NOTE:

1. The $MgSO_4$ solution serves as a rough indicator of the acidity. The concentration of NaOH used to adjust the *p*H may be 3 to 6 *N*, while that of the H_2SO_4 should be somewhat less. Neither concentration is critical, however.

An alternate method for determining the glycerol is based upon the titration of the formic acid produced in the oxidation; this technique warrants discussion since it makes possible the analysis of glycerol in the presence of other oxidizable substances. In essence, the oxidized mixture is titrated with a standard base solution. In order to correct for the acidity of the periodic and iodic acids a blank, identical to the starting mixture in all respects but for the glycerol, is titrated to the same end point. The difference in volume of base required is assumed to be equivalent to the formic acid produced.[30]

Analysis of a mixture of glycerol, ethylene glycol, and propylene glycol. The versatility of periodate oxidation is demonstrated by a procedure developed by Shupe[31] for the analysis of a 3-component mixture. The reagent converts the glycerol to formic acid and formaldehyde, the ethylene glycol to formaldehyde, and the propylene glycol to formaldehyde and acetaldehyde.

$$CH_2OHCHOHCH_2OH + 2H_5IO_6 \rightarrow$$
glycerol
$$2HCHO + HCOOH + 2IO_3^- + 2H^+ + 5H_2O$$

$$CH_2OHCH_2OH + H_5IO_6 \rightarrow 2HCHO + IO_3^- + H^+ + 3H_2O$$
ethylene glycol

$$CH_3CHOHCH_2OH + H_5IO_6 \rightarrow HCHO + CH_3CHO + IO_3^- + H^+ + 3H_2O$$
propylene glycol

The glycerol content of the mixture is obtained by neutralization titration of the formic acid. The amount of propylene glycol in the sample is then assessed by analysis of the acetaldehyde in the oxidation mixture; determination of this compound in the presence of formaldehyde is readily accomplished although we shall not discuss the method here. Finally, ethylene glycol is estimated by

---

[30] Periodic acid ($k_1 = 2.4 \times 10^{-2}$, $k_2 = 5 \times 10^{-9}$), formic acid ($k = 1.7 \times 10^{-4}$), and iodic acid (completely dissociated) will be found in the oxidation mixtures; only the first of these will be present in the blank. The assumption that the sample titration minus the blank represents the volume of base equivalent to the formic acid produced is only valid insofar as the first hydrogen of periodic acid is titrated. This follows from the stoichiometric consideration that 1 mol of periodic acid yields 1 mol of iodic acid.

The indicator choice here requires a compromise. It can be shown that the theoretical equivalence point in the periodic acid titration is about 5 while that for formic acid is about 7.7. Methyl red, the indicator ordinarily used, exhibits its basic color at about a *p*H of 6. This inevitably leads to errors in the analysis although these are ordinarily not large.

[31] I. S. Shupe, *J. Assoc. Offic. Agr. Chemists,* **26,** 249 (1943).

correcting the total periodate consumption, as determined iodimetrically, for the concentration of the other two constituents.

Determination of α-hydroxyamines. As mentioned earlier, periodate oxidation of compounds having hydroxy and amino groups on adjacent carbon atoms results in formation of aldehydes and liberation of ammonia. The latter is readily distilled from the alkaline oxidation mixture and determined by ncutralization titration (see p. 355). This is particularly useful in the analysis of mixtures of the various amino acids found in proteins. Only serine, threonine, β-hydroxyglutamic acid, and hydroxylysine have the requisite structure for liberation of ammonia; the method is thus selective for these compounds.

problems

1. A solution of permanganate was standardized against pure As_2O_3. Exactly 0.200 gram of the standard required 38.1 ml of $KMnO_4$. What was the normality of the solution? The formality?   ans. 0.106 *N*:0.0212 *F*

2. What is the normality of a ceric sulfate solution having a titer of 26.0 mg of Mohr's salt [$FeSO_4 \cdot (NH_4)_2SO_4 \cdot 6H_2O$] per ml?

3. A solution of $KMnO_4$ was standardized against 0.179 gram of pure KI. The titration was carried out in the presence of HCN where the reaction was $2MnO_4 + 5I^- + 5HCN + 11H^+ \rightleftharpoons 2Mn^{2+} + 5ICN + 8H_2O$. Exactly 37.9 ml of reagent was used. What was the normality?

4. A 0.770-gram sample of KIO_3 consumed 41.0 ml of a $KMnO_4$ solution after the IO_3^- was reduced to I^- and titrated as in Problem 3. What was the normality of the $KMnO_4$?

5. A 1.27-gram sample containing Fe_3O_4 was dissolved, the iron reduced in a Jones reductor and then titrated with 26.6 ml of 0.0584 *N* Ce^{4+}. Calculate the percent Fe and the percent Fe_3O_4 in the sample.

6. A 0.814-gram sample of a stibnite ore was decomposed in acid and the + 3 Sb oxidized to the + 5 state with 40.0 ml of 0.119 *N* $KMnO_4$. The excess $KMnO_4$ was back titrated with 3.82 ml of 0.0961 *N* Fe^{2+}. Calculate the percent Sb_2S_3 in the sample.   ans. 45.8 percent

7. A solution of KNO_2 was analyzed by treating exactly 25.0 ml with 50.0 ml of 0.0880 *N* $KMnO_4$. After reaction was complete the solution was heated and the excess $KMnO_4$ back titrated with 1.20 ml of 0.100 *N* Fe^{2+}. What is the formality of the KNO_2 solution? How many grams KNO_2 are in 1 liter of the solution?

8. A 2.02-gram sample of La_2O_3 (GFW = 325.8) was brought into solution and the La precipitated as $La_2(C_2O_4)_3$. The precipitate was filtered, washed, dissolved in acid, and the oxalic acid titrated with 43.2 ml of 0.120 *N* ceric sulfate. Calculate the percent La_2O_3 present.

9. What is the normality of a 0.0500 *F* $KMnO_4$ solution when
   (a) It is used for titrations in strong acid?
   (b) It is used for titrations in neutral solutions where MnO_2 is the product?
   (c) It is used for titrations in strongly basic solutions?

10. A solution of $KMnO_4$ was found to be 0.0761 N when standardized against oxalate in the usual way. The solution was used to determine manganese by the Volhard procedure $(2MnO_4^- + 3Mn^{2+} + 4OH^- \rightleftharpoons 5MnO_2 + 2H_2O)$. A 0.543-gram sample required 29.2 ml of the $KMnO_4$. What percent Mn was contained in the sample?   ans. 6.76 percent

11. A sample containing both iron and titanium was analyzed by solution of a 3.00-gram sample and dilution to exactly 500 ml in a volumetric flask. A 50-ml aliquot of the solution was passed through a silver reductor which reduced the iron to the ferrous state but left the titanium in the titanic state; titration with 0.0750 N Ce^{4+} required 18.2 ml. A 100-ml aliquot was passed through a Jones reductor which reduced both Fe and Ti $(Ti^{4+} \rightarrow Ti^{3+})$. The reduced solution consumed 46.3 ml of the Ce^{4+} solution. Calculate the percent Fe_2O_3 and the percent TiO_2 in the sample.

12. In the presence of a high concentration of fluoride ion, Mn^{2+} can be titrated with MnO_4^-, both reactants being converted to a complex of Mn^{3+}. A 0.312-gram sample containing Mn_3O_4 was dissolved in such a way as to convert all of the manganese to Mn^{2+}. Titration in the presence of fluoride ion consumed 22.7 ml of $KMnO_4$ which was 0.121 N against oxalate. Calculate the percent Mn_3O_4.

13. A sample containing KCN was analyzed by solution of 0.500 gram of the material and treatment with $KMnO_4$ in strongly alkaline solution (unbalanced equation: $MnO_4^- + CN^- + OH^- \rightleftharpoons MnO_4^{2-} + CNO^- + H_2O$). The excess $KMnO_4$ was back titrated with a standard sodium formate solution in the presence of barium ions. Exactly 25.0 ml of 0.107 F MnO_4^- was used and 4.78 ml of 0.0482 F sodium formate were required for the back titration. Calculate the percent KCN.   ans. 14.4 percent

14. A 25.0-ml aliquot of an aqueous methanol solution (CH_3OH) was made strongly alkaline and treated with 40.0 ml of a 0.0200 F solution of $KMnO_4$. After the oxidation of the alcohol to CO_2 and H_2O was complete $(MnO_4^- \rightarrow MnO_4^{2-})$, the mixture was made acidic and 30.0 ml of 0.100 N $Na_2C_2O_4$ added. The mixture was heated to 50° C which reduced both the MnO_4^- and MnO_4^{2-} to Mn^{2+}; the excess oxalate was then titrated with 1.32 ml of the $KMnO_4$ solution. Calculate the milligrams methanol in each milliliter of the sample.   ans. 0.0484 mg/ml

15. How would you prepare 500 ml of exactly 0.200 N Ce^{4+} in 1 F H_2SO_4 from primary-standard grade $(NH_4)_2Ce(NO_3)_6$ (GFW = 548.3)?

16. A 0.412-gram sample containing potassium azide (KN_3) was treated with 50.0 ml of 0.100 N Ce^{4+}. This resulted in oxidation of the azide to N_2 $(2N_3^- \rightarrow 3N_2 + 2e)$. The excess Ce^{4+} consumed 8.11 ml of 0.150 N Fe^{2+}. Calculate the percent KN_3 in the sample.

17. Exactly 5.00 ml of a hydrogen peroxide solution consumed 39.6 ml of 0.250 N Ce^{4+}. What was the weight percent H_2O_2 in the sample? Assume a density of 1.00 g/ml for the solution.

18. A 0.314-gram sample containing K_2SO_4 was treated in such a way as to precipitate the potassium as $K_2NaCo(NO_2)_6 \cdot H_2O$. The precipitate was washed and dissolved in 50 ml of 0.100 N $KMnO_4$ which resulted in oxidation of the

NO_2^- to NO_3^- and reduction of the Co^{3+} to Co^{2+} (see Table 20-2 for the half reaction). The excess $KMnO_4$ was titrated with 22.1 ml of 0.114 N Fe^{2+}.
(a) Write a balanced equation for the reaction between MnO_4^- and $Co(NO_2)_6^{3-}$.
(b) Calculate the percent K_2SO_4 present in the sample.

19. Describe a method for preparation of 2 liters of 0.333 N $K_2Cr_2O_7$.

20. A sample of alkali metal chlorides was analyzed for sodium by dissolving a 0.800-gram sample in water and diluting to exactly 500 ml. A 25.0-ml aliquot of this was treated in such a way as to precipitate the sodium as $NaZn(UO_2)_3$ $(OAc)_9 \cdot 6H_2O$. The precipitate was filtered, dissolved in acid and passed through a lead reductor which converted the uranium to U^{4+}. Oxidation of this to UO_2^{2+} required 19.9 ml of 0.100 N $K_2Cr_2O_7$. Calculate the percent $NaCl$ in the sample.

21. A 0.320-gram sample was analyzed for Cr by solution and oxidation of the Cr^{3+} to $Cr_2O_7^{2-}$ with an excess of ammonium persulfate. After the excess persulfate was destroyed by boiling, exactly 1.00 gram of Mohr's salt (GFW = 392.2) was added and the excess Fe^{2+} titrated with 8.77 ml of 0.0500 N $K_2Cr_2O_7$. Calculate the percent Cr_2O_3 in the sample.

22. A 25.0-ml sample containing ClO_3^- was made strongly acid and the ClO_3^- reduced to Cl^- by the addition of 10.0 ml of 0.250 N ferrous ammonium sulfate. After reaction was complete, the excess Fe^{2+} was titrated with 9.12 ml of 0.100 N $K_2Cr_2O_7$. Calculate the number of milligrams of $NaClO_3$ in the solution.

23. A solution of iodine was standardized against 0.186 gram of pure As_2O_3. The volume of reagent required was 27.7 ml. What was the normality?

24. A 0.402-gram sample of cadmium containing alloy was dissolved and the cadmium precipitated as CdS. After filtration and washing, the precipitate was dissolved in an acidified solution containing 40.0 ml of 0.128 N I_2. The excess iodine remaining after oxidation of the sulfide to sulfur was titrated with 11.6 ml of 0.108 N thiosulfate. What was the percent Cd in the sample?

25. The H_2S concentration of a sample of air was obtained by bubbling 10 liters of the air through an absorption tower containing a solution of Cd^{2+}. The solution was treated with 20.0 ml of 0.0107 N I_2 and acidified, which resulted in the sulfide being oxidized to elemental sulfur. The excess iodine consumed 10.1 ml of 0.0120 N thiosulfate. Calculate the parts per million H_2S present in the air assuming a gas density of 12×10^{-4} gram/ml.

26. What is the titer of a 0.200 N I_2 solution in terms of milligrams of As_2O_3, Sb, SO_2, H_2S, As?

27. How many milliliters of 0.107 N $KBrO_3$ are required to titrate 0.0212 gram of pure KCNS (see Table 20-8 for reaction)?

28. Under suitable conditions, thiourea is oxidized to sulfate by solutions of bromate
$$3CS(NH_2)_2 + 4BrO_3^- + 3H_2O \rightleftharpoons 3CO(NH_2)_2 + 3SO_4^{2-} + 4Br^- + 6H^+$$
A 0.0715-gram sample of the material was found to consume 14.1 ml of 0.0500 N $KBrO_3$. What was the percent purity of the thiourea sample?

29. An aqueous solution of phenol was analyzed by mixture of 25.0 ml of 0.100 N $KBrO_3$, an excess of KBr, and several milliliters of strong acid. Bromination occurred by the following sequence of reactions:

$$BrO_3^- + 5Br^- + 6H^+ \rightarrow 3Br_2 + 3H_2O$$
$$C_6H_5OH + 3Br_2 \rightarrow C_6H_2Br_3OH + 3H^+ + 3Br^-$$

After the bromination was complete 10.0 ml of 0.120 N arsenious acid was added which reduced the excess Br_2

$$Br_2 + H_3AsO_3 + H_2O \rightleftharpoons 2Br^- + H_3AsO_4 + 2H^+$$

Finally the excess H_2AsO_3 was titrated with 1.47 ml of the standard bromate solution. Calculate the number of mg of phenol in the sample. ans. 22.7 mg

30. A sample of Al_2O_3 was analyzed by solution and precipitation with 8-hydroxy-quinoline. After filtration, the precipitate was dissolved and the liberated organic reagent titrated with bromate. The reactions involved are shown on page 470. The data obtained were

weight sample = 0.208 gram
total milliliters of 0.0500 N $KBrO_3$ used = 31.1
milliliters 0.0600 N H_3AsO_3 for back titration = 6.72

Calculate the percent Al_2O_3 in the sample.   ans. 2.36 percent

31. A 0.108-gram sample containing $MgCl_2$ was analyzed by precipitation of $Mg(8\text{-hydroxyquinolate})_2$, filtration, and solution of the precipitate in acid. An excess of bromide and 30.0 ml of 0.100 N $KBrO_3$ were added. The residual bromine was reduced by the addition of 10.0 ml of 0.0500 N arsenious acid and the excess of that reagent titrated with 1.91 ml of the $KBrO_3$. Calculate the percent $MgCl_2$.

32. What is the normality of a 0.0400 F solution of KIO_3 when used for the following purposes?
    (a) As an oxidizing reagent in strong HCN solution.
    (b) As a source of iodine for a redox titration wherein an excess of iodide and acid are present.
    (c) As a standard for the titration of a strong acid (see page 471).

          ans. (a) 0.160 N
          (b) 0.240 N
          (c) 0.240 N

33. What is the titer of a 0.0250 F KIO_3 solution in terms of milligrams of each of the following. Assume in each case the oxidation is carried out in HCl solution where the product is ICl_2^-: As, I_2, KI, N_2H_4, KCNS (see Table 20-10 for the oxidation products).   ans. 3.74 mg As
          8.3 mg KI

34. The mercury in an ore was determined by decomposition and solution of a 0.632-gram sample followed by precipitation as Hg_2Cl_2. The precipitate was filtered and washed; the precipitate and paper were then suspended in a fairly concentrated HCl solution and titrated with 39.2 ml of a 0.0510 N KIO_3 solution (see Table 20-10). A few milliliters of CCl_4 were used to detect the end point. Calculate the percent Hg in the sample.

35. Calculate parts per million of SO_2 in a gas stream (density = 0.0012 gram/ml) that was analyzed by absorption of the compound from a 6.0-liter sample

in sodium hydroxide followed by acidification with HCl and titration of the sulfite with 4.98 ml of 0.0125 N KIO$_3$. The IO$_3^-$ was converted to ICl$_2^-$ in the titration.

36. What is the formality of an iodine solution, 25.0 ml of which consumes 19.3 ml of 0.0200 F KIO$_3$ in the reaction

$$2I_2 + IO_3^- + 6H^+ + 10Cl^- \rightleftharpoons 5ICl_2^- + 3H_2O$$

37. An arsenious acid solution was standardized against 0.127 gram of primary standard KIO$_3$. The procedure involved solution of the IO$_3^-$, acidification, and addition of an excess of KI. The liberated iodine was titrated with 37.4 ml of the arsenious acid after buffering the solution to a pH of 7.5. Calculate the normality of the arsenious acid solution.

38. Strong acids may be standardized against KIO$_3$ (see p. 471). The iodate is dissolved in water containing an excess of KI and sodium thiosulfate. As acid is added, the KIO$_3$ consumes the protons, liberating iodine which reacts with the S$_2$O$_3^{2-}$. Calculate the normality of a H$_2$SO$_4$ solution if 29.10 ml is used in the titration of 0.150 gram of KIO$_3$.

39. A sample of ethylene glycol, CH$_2$OHCH$_2$OH, was analyzed by treatment with 50.0 ml of a solution of periodic acid. After the reaction was complete, the mixture was buffered to pH 7.5 and an excess of iodide added. The liberated iodine reacted with 14.3 ml of a standard 0.100 N arsenite solution. A blank solution containing all but the glycol was found to consume 40.1 ml of the standard arsenite. Calculate the weight of ethylene glycol in the sample if its reaction product is formaldehyde, HCHO.   ans. 0.0801 gram

40. A solution of periodic acid was standardized as follows. Exactly 50.0 ml were neutralized and buffered strongly with a borax buffer. An excess of KI was added and the liberated iodine titrated with 26.4 ml of 0.200 N H$_3$AsO$_3$. Calculate the formality and normality of the solution.
    ans. 0.1056 N; 0.0528 F

41. What is the titer of a 0.100 F solution of periodic acid in terms of milligrams of each of the following compounds: (a) propylene glycol, (b) glycerol, and (c) biacetyl.

42. Write balanced equations for the oxidation of each of the following compounds with periodic acid: (a) ethylene glycol, CH$_2$OHCH$_2$OH, (b) glyoxal, CHOCHO, (c) mannitol, CH$_2$OH(CHOH)$_4$CH$_2$OH, (d) tartaric acid, COOH(CHOH)$_2$-COOH, and (e) glucose, CH$_2$OH(CHOH)$_4$CHO.

43. How many milliliters of a 0.202 N thiosulfate solution will be required to react with the iodine formed from an acidic mixture of 30.0 ml of 0.0200 F periodate and an excess of KI?

44. Write balanced equations for the KMnO$_4$ oxidation of the following substances; assume solutions are strongly acidic (see Table 20-2 for products).
    (a) Br$^-$ (hot H$_2$SO$_4$ solution)
    (b) VO^{2+}
    (c) Mo^{3+}
    (d) H$_2$O$_2$
    (e) I$^-$ (in HCN solution)
    (f) Co(NO$_2$)$_6^{3-}$

45. Write balanced equations for $KMnO_4$ oxidation of the following compounds in strongly basic media.
    (a) I^-
    (b) CN^-
    (c) SO_3^{2-}
    (d) C_2H_5OH (products CO_2 and H_2O)
    (e) $HCOO^-$

46. Calculate the numerical value of the equilibrium constant for the reaction

$$2MnO_4^- + 3Mn^{2+} + 2H_2O \rightleftharpoons 5MnO_2 + 4H^+$$

ans. $K = 2 \times 10^{47}$

47. Assume that the acid concentration at the end point in a $KMnO_4$ titration is 1.0 and the Mn^{2+} concentration is about 0.010. From the value of K in Problem 46 calculate the MnO_4^- concentration in equilibrium with these species.

48. Calculate the equilibrium constant for the chemical reaction involving the decomposition of H_2O by MnO_4^- (see p. 434).

49. Calculate the equilibrium constant for the oxidation of Cl^- to Cl_2 by $KMnO_4$. On the basis of equilibrium considerations alone should it be possible to use $KMnO_4$ for the titration of reducing agents dissolved in HCl?

50. Calculate the equivalence-point potential for the titration of 50.0 ml of 0.1 N Sn^{2+} with (a) 0.1 N $KMnO_4$, (b) 0.1 N $K_2Cr_2O_7$. Assume $[H^+] = 1.0$ where necessary.

51. What is the equilibrium concentration of Fe^{3+} remaining in a solution prepared by passage of a 0.010 F solution of Fe^{3+} in 0.10 F HCl through a silver reductor?

52. What is the equilibrium concentration of Fe^{3+} remaining in a solution prepared by passage of a 0.01 F solution of Fe^{3+} through a silver reductor in the absence of chloride ion?

53. Calculate the equilibrium constant for the reaction

$$BrO_3^- + 5Br^- + 6H^+ \rightleftharpoons 3Br_2 + 3H_2O$$

ans. $K = 4 \times 10^{38}$

chapter 21. *Applications of Reducing Agents*

Aqueous solutions of good reducing agents are not very stable unless precautions are taken to protect them from atmospheric oxidation. Furthermore, the more potent reagents tend to reduce hydrogen ions to the element, thus prohibiting their use in solutions of low pH. As a consequence of these limitations, only a few reducing agents have found widespread use as standard volumetric reagents.

IODOMETRIC METHODS

Iodide ion is a moderately good reducing agent that has been widely employed for the analysis of oxidants. A fair number of substances, in one form or another, have potentials greater than that for iodine.

$$I_2 + 2e \rightleftharpoons 2I^- \qquad E^\circ = 0.54 \text{ volt}$$

All of these are capable of oxidizing iodide to iodine, and hence are potentially susceptible to analysis with this reagent.

The lack of a good method of end-point detection makes direct titration of oxidizing agents by solutions of iodide salts impractical. Consequently, the indirect, iodometric, procedure is always employed. This involves reduction with a moderate and unmeasured excess of potassium iodide. The iodine liberated, equivalent in quantity to the oxidant being determined, is then titrated with a standard solution of a reducing reagent. Sodium thiosulfate is nearly always employed for this purpose, although solutions of arsenious acid or other reducing reagents can be used instead.

The Reaction of Iodine with Thiosulfate Ion

Because of its widespread employment in analysis, the reaction between iodine and thiosulfate ion

$$2S_2O_3^{2-} + I_2 \rightleftharpoons S_4O_6^{2-} + 2I^-$$

warrants discussion in some detail. The production of the tetrathionate ion requires the loss of two electrons from two thiosulfate ions; the equivalent weight of thiosulfate in this reaction must therefore be equal to its gram formula weight.

The quantitative conversion of thiosulfate to tetrathionate ion is somewhat unique with iodine; other oxidizing reagents tend to carry the oxidation, wholly or in part, to sulfate ion. Hypoiodous acid provides an important example of this.

$$4HOI + S_2O_3^{2-} + H_2O \rightleftharpoons 2SO_4^{2-} + 4I^- + 6H^+$$

The pertinence of this reaction becomes apparent when we recall that hypoiodite is a hydrolysis product of iodine itself (see p. 460); its presence in an iodine solution, then, will seriously upset the stoichiometry of the iodine-thiosulfate reaction causing too little thiosulfate or too much iodine to be used in the titration.

The equilibrium constant for the hydrolysis reaction

$$I_2 + H_2O \rightleftharpoons HOI + I^- + H^+$$

is about 3×10^{-13}.[1] From this it can be shown that hypoiodite formation should become significant in titrations where the pH is greater than 7. Kolthoff has demonstrated this to be the case;[2] in the titration of 25 ml of 0.1 N iodine with 0.1 N thiosulfate he found an error of about 4 percent in the presence of 0.5 gram of sodium bicarbonate and of 10 percent in a solution containing about 2 grams of this salt. He recommends that the pH should always be less than 7.6 for titration of 0.1 N solutions, 6.5 or less for 0.01 N solutions, and less than 5 for 0.001 N solutions.

[1] W. C. Bray and E. L. Connolly, *J. Am. Chem. Soc.*, **33**, 1485 (1911).

[2] I. M. Kolthoff and R. Belcher, *Volumetric Analysis*. 3, 214-215. New York: Interscience Publishers, Inc., 1957.

The titration of highly acidic solutions of iodine with thiosulfate yields quantitative results provided care is taken to prevent air oxidation of iodide ion, and provided also that the thiosulfate solution is added slowly to prevent its decomposition.

The end point in the titration is readily established by means of a starch solution. We have discussed the preparation and properties of this indicator in the previous chapter and need only emphasize that starch is partially decomposed in the presence of a large excess of iodine. For this reason the indicator is never added to an iodine solution until the bulk of that substance has been reduced. The change in color of the iodine solution from red to a faint yellow signals the proper time for the addition of the indicator.

Preparation and Properties of Standard Thiosulfate Solutions

Purity of compounds. A pentahydrate, $Na_2S_2O_3 \cdot 5H_2O$, of sufficient purity for the direct preparation of standard solutions can be obtained by equilibrating the recrystallized salt with an atmosphere of appropriate moisture content. Most chemists, however, consider that the effort involved does not justify this convenience and ordinarily use the commercial salt to give a solution of only approximate normality; standardization is readily accomplished.

Anhydrous sodium thiosulfate has also been recommended as a primary standard; the hygroscopic nature of the compound, however, has prevented its wide adoption.

Stability of thiosulfate solutions. Studies on the stability of thiosulfate solutions have been numerous, and there are several instances where the findings from these are in direct conflict. We shall consider only the most pertinent facts since the decomposition process is clearly complex.[3]

Principal among the variables affecting the stability of thiosulfate solutions are the pH, the presence of micro organisms and other impurities, the concentration of the solution, exposure to sunlight, and the presence of atmospheric oxygen. Generally, there is a decrease in iodine titer that may amount to as much as several percent in a few weeks. Occasionally, however, increases in normality are observed. Proper attention to detail makes possible the preparation of standard thiosulfate solutions which need only occasional restandardization.

One of the important factors affecting stability is acidity. In solutions of pH much lower than about 5, the following reaction occurs at an appreciable rate:

$$S_2O_3^{2-} + H^+ \rightleftharpoons HS_2O_3^- \rightarrow HSO_3^- + S$$

The velocity of this reaction increases with the hydrogen ion concentration, and in strongly acid solution, elemental sulfur is formed very rapidly. The bisulfite ion produced is also oxidized by iodine, consuming twice the quantity

[3] For a more complete summary of the subject see I. M. Kolthoff and R. Belcher, *Volumetric Analysis*. **3**, 225-230. New York: Interscience Publishers, Inc., 1957.

of that reagent as the thiosulfate from which it was derived. Clearly, thiosulfate solutions cannot be allowed to stand in contact with acid. On the other hand, iodine solutions that are 3 to 4 N in acid may be titrated without error as long as care is taken to introduce the thiosulfate slowly and with good mixing. Under these conditions the thiosulfate is so rapidly oxidized by the iodine that the slower acid decomposition cannot occur to any measurable extent.

Experiments indicate that the stability of thiosulfate solutions is at a maximum in the pH range between 9 and 10 although, for most purposes, a pH of 7 is adequate. Addition of small amounts of bases such as sodium carbonate, borax, or disodium phosphate is frequently recommended as a means of preserving standard solutions of the reagent. If this is done, the iodine solutions to be titrated must be made sufficiently acidic to neutralize the base; otherwise hydrolysis of the iodine may occur before the equivalence point is attained, resulting in the partial oxidation of the thiosulfate to sulfate.

The most important single cause of instability of thiosulfate can be traced to certain bacteria that are capable of metabolizing the thiosulfate ion, converting it to sulfite, sulfate, and elemental sulfur.[4] Solutions prepared free of bacteria are remarkably stable, and it is common practice to impose reasonably sterile conditions in preparation of standard solutions. Addition of such substances as chloroform, sodium benzoate, or mercuric iodide inhibits the growth of bacteria and is reported to have a salutary effect. Bacterial activity appears to be at a minimum at a pH between 9 and 10 which accounts, at least in part, for the maximum stability of solutions in this range.

Many other variables affect the stability of thiosulfate solutions. Reportedly, decomposition is catalyzed by cupric ions as well as by the decomposition products themselves. As a consequence, solutions that have become turbid should be discarded. Exposure to sunlight increases the rate of decomposition as does atmospheric oxygen. Finally, the decomposition rate is greater in more dilute solutions.

Preparation of approximately 0.1 N sodium thiosulfate. Heat 1 liter of distilled water to boiling in a beaker covered with a watch glass. Boil for at least 5 minutes. Cool, add about 25 grams of $Na_2S_2O_3 \cdot 5H_2O$ and 0.1 gram of Na_2CO_3. Stir until solution is complete, then transfer to a clean glass-stoppered bottle. Store in the dark.

Standardization of Thiosulfate Solutions

Several excellent primary standards are available for the standardization of thiosulfate solutions. In general, these are oxidizing agents that liberate an equivalent amount of iodine when treated with an excess of iodide ion; this can then be titrated with the solution to be standardized.

---

[4] M. Kilpatrick, Jr., and M. L. Kilpatrick, *J. Am. Chem. Soc.*, **45**, 2132 (1923); F. O. Rice, M. Kilpatrick, Jr., and W. Lemkin, *ibid.*, 1361 (1923).

Potassium iodate. As mentioned in the preceding chapter, iodate ion reacts rapidly with iodide in slightly acid solution to give iodine.

$$IO_3^- + 5I^- + 6H^+ \rightleftharpoons 3I_2 + 3H_2O$$

Each mol of potassium iodate furnishes 3 mols of iodine for the standardization process; thus its equivalent weight is one sixth its formula weight since a six-electron change is associated with the reduction of the three iodine molecules, the species actually titrated.

The sole disadvantage of potassium iodate as a primary standard arises from its low equivalent weight (35.67). Only slightly more than 0.1 gram can be taken for standardization of a 0.1 N thiosulfate solution, and the relative error normally incurred in weighing this quantity is somewhat greater than desirable for a standardization. This problem often is circumvented by dissolving a larger quantity of the solid in a known volume and taking aliquots of this solution.

> **Procedure.** Weigh, to the nearest 0.1 mg, about 0.12 gram of pure dry KIO$_3$ into a 250-ml Erlenmeyer flask. Dissolve in 25 ml of water, and add about 2 grams of iodate-free KI. After solution is complete, add 10 ml of 1.0 N HCl and titrate immediately with the thiosulfate solution until the color of the solution becomes pale yellow. Add 5 ml of starch (see p. 462) and titrate to the disappearance of the blue color.
>
> In order to minimize the weighing error, the foregoing can be modified by taking a 0.6-gram sample, dissolving in water, and diluting to exactly 250 ml in a volumetric flask. A 50-ml aliquot of this solution can then be used for each standardization.

Iodine. A reagent grade of iodine is available commercially. The element is readily purified by sublimation should this be necessary. The pure compound is an excellent standard for thiosulfate, and the only problem to be overcome in its use is the prevention of losses due to the volatile nature of the element.

> **Procedure.** Place about 2 to 3 grams of pure, iodate-free KI in a clean weighing bottle and add 0.5 to 1.0 ml of water. Stopper the weighing bottle, dry the outside with a lint-free cloth, and allow to come to room temperature after solution of the KI is complete. Weigh the bottle and solution. Place 0.4 ± 0.05 gram of iodine in a small vial—the weight taken does not need to be known accurately at this point. Open the stoppered weighing bottle containing the KI solution and *quickly* transfer the iodine to the solution. Immediately stopper the weighing bottle and swirl the solution carefully so that all of the iodine is dissolved. Again weigh the bottle.
>
> Add 100 ml of water containing about 1.0 gram of KI to a 500-ml Erlenmeyer flask. Tilt the flask, loosen the stopper on the weighing bottle, and gently slide the bottle and stopper under the surface of the solution. Swirl the solution until homogeneous.

Add 1.0 ml of acetic acid to neutralize any base present in the thiosulfate and titrate until the solution is yellow. Add 5 ml of starch (see p. 462) and complete the titration.

Potassium dichromate. In the preceding chapter we saw that highly pure potassium dichromate is available from commercial sources. Because of its ready availability in pure form, the stability of its solutions, and its reasonably high equivalent weight, it is an ideal primary standard.

In acidic solutions, dichromate oxidizes iodide ion slowly to iodine according to the equation

$$Cr_2O_7^{2-} + 6I^- + 14H^+ \rightleftharpoons 2Cr^{3+} + 3I_2 + 7H_2O$$

For this reaction to be quantitative, control must be maintained over the hydrogen ion concentration, the iodide concentration, and the time of reaction. At the maximum pH for complete oxidation (about 5), the reaction is prohibitively slow; fortunately the rate increases with hydrogen ion content. There is, however, a lower limit to the pH of the solution because the rate at which iodide ion is oxidized by atmospheric oxygen also increases rapidly with increases in hydrogen ion concentration. Experiments indicate that a quantitative oxidation requires 5 minutes when the hydrogen ion concentration is 0.2 molar and the initial iodide concentration is 0.1 molar. Further, for periods of 10 minutes or less, the air oxidation of iodide ion is not appreciable in solutions as acidic as 0.4 molar. Success of the titration, therefore, hinges upon control of the hydrogen ion concentration between these moderately broad limits.

Procedure. Dry the $K_2Cr_2O_7$ for 1 to 2 hours at 100° to 200° C and weigh, to the nearest 0.1 mg, 0.20 to 0.23-gram portions into 500-ml flasks. Dissolve in 50 ml of water. Then add a freshly prepared solution consisting of 3 grams of KI, 5 ml of 6 N HCl, and 50 ml of water. Swirl gently, cover with a watch glass, and let stand in a dark place for 5 minutes. Wash down the sides of the flask, add 200 ml of water, and titrate with the thiosulfate solution. When the yellow color of the iodine can no longer be seen, add 5 ml of starch (see p. 462). Continue the titration until a color change from the blue starch-iodine complex to the green color of the chromic ion is observed.

Other primary standards for sodium thiosulfate include potassium bromate, potassium hydrogen iodate, $[KH(IO_3)_2]$, potassium ferricyanide, and metallic copper. Details for the use of these may be found in various reference works.[5]

Summary of Sources of Errors in Iodometric Methods

Before considering applications of the indirect iodometric method, we shall summarize potential sources of error.

[5] For example, see I. M. Kolthoff and R. Belcher, *Volumetric Analysis.* **3**, 234-243. New York: Interscience Publishers, Inc., 1957.

1. Air oxidation of iodide ion. Considered solely from the standpoint of equilibrium, the reaction

$$4I^- + O_2 + 4H^+ \rightleftharpoons 2I_2 + 2H_2O$$

should make necessary the exclusion of atmospheric oxygen from all iodometric titration mixtures. Fortunately, however, the rate at which this oxidation occurs is sufficiently slow to make such a precaution quite unnecessary in most instances.

Kinetic studies of the reaction have shown that its velocity increases greatly with the hydrogen ion concentration and becomes of real concern in media greater than 0.4 to 0.5 N in acid. Light also catalyzes the process; thus, storage in a dark place is recommended if time is required for completion of the reaction between iodide ion and an oxidizing agent.

Traces of cuprous ion and nitrogen oxides are known to catalyze the oxidation of iodide by air. These latter are potential sources of interference to any iodometric analysis that makes use of nitric acid. The following set of equations gives a simplified picture of the mechanism of this interference.

$$2NO_2 + 4I^- + 4H^+ \rightarrow 2I_2 + 2NO + 2H_2O$$

$$2NO_2^- + 2I^- + 4H^+ \rightarrow I_2 + 2NO + 2H_2O$$

$$2NO + O_2 \text{ (from atmosphere)} \rightarrow 2NO_2$$

Thus, both nitrogen dioxide or nitrite ion react with iodide to produce nitric oxide. The latter is rapidly reoxidized by air to produce more nitrogen dioxide. Over-all, the process consists of the cyclic regeneration of a species that rapidly oxidizes iodide ions.

In the iodometric determination of ferric and vanadate ions, the air oxidation of iodide proceeds at a much greater rate than in the absence of the analytical reactions. The air-oxidation process is thus *induced* by the primary reaction; if allowed to occur, this results in an overconsumption of thiosulfate.

Where air oxidation is believed to be a problem, recourse must be made to an inert atmosphere over the titration mixture. This is conveniently accomplished by periodically adding small portions of sodium bicarbonate (300 mg) to the acid solution as the titration progresses; alternatively, carbon dioxide can be introduced from a tank or in the form of small pieces of dry ice.

2. Volatilization of liberated iodine. Errors from the volatilization of liberated iodine are avoided by using stoppered containers when solutions must stand, by maintaining a goodly excess of iodide ion, and by avoiding elevated temperatures.

3. Decomposition of thiosulfate solutions. The decomposition of thiosulfate solutions has been discussed in a preceding section.

4. Alteration in stoichiometry of the iodine-thiosulfate reaction. As mentioned earlier, alteration in stoichiometry is encountered when basic solutions are titrated.

5. Premature addition of starch indicator. See page 462 for control of the premature addition of starch indicator.

Table 21-1

Some Applications of the Indirect Iodometric Method

$$2I^- \rightleftharpoons I_2 + 2e$$

Substance	Half Reaction	Special Conditions
IO_4^-	$IO_4^- + 8H^+ + 7e \rightleftharpoons \frac{1}{2}I_2 + 4H_2O$	acidic solution
	$IO_4^- + 2H^+ + 2e \rightleftharpoons IO_3^- + H_2O$	neutral solution; titration with arsenite
IO_3^-	$IO_3^- + 6H^+ + 5e \rightleftharpoons \frac{1}{2}I_2 + 3H_2O$	
BrO_3^-	$BrO_3^- + 6H^+ + 6e \rightleftharpoons Br^- + 3H_2O$	
ClO_3^-	$ClO_3^- + 6H^+ + 6e \rightleftharpoons Cl^- + 3H_2O$	strong acid; slow reaction
$HClO$	$HClO + H^+ + 2e \rightleftharpoons Cl^- + H_2O$	
Cl_2	$Cl_2 + 2e \rightleftharpoons 2Cl^-$	
Br_2	$Br_2 + 2e \rightleftharpoons 2Br^-$	
I^-	$I^- + 3Cl_2 + 3H_2O \rightleftharpoons IO_3^- + 6Cl^- + 6H^+$	excess Cl_2 removed by boiling
	$IO_3^- + 6H^+ + 5e \rightleftharpoons \frac{1}{2}I_2 + 3H_2O$	
NO_2^-	$HNO_2 + H^+ + e \rightleftharpoons NO + H_2O$	
H_3AsO_4	$H_3AsO_4 + 2H^+ + 2e \rightleftharpoons H_3AsO_3 + H_2O$	strong HCl
H_3SbO_4	$H_3SbO_4 + 2H^+ + 2e \rightleftharpoons H_3SbO_3 + H_2O$	strong HCl
$Fe(CN)_6^{3-}$	$Fe(CN)_6^{3-} + e \rightleftharpoons Fe(CN)_6^{4-}$	
MnO_4^-	$MnO_4^- + 8H^+ + 5e \rightleftharpoons Mn^{2+} + 4H_2O$	
Ce^{4+}	$Ce^{4+} + e \rightleftharpoons Ce^{3+}$	
$Cr_2O_7^{2-}$	$Cr_2O_7^{2-} + 14H^+ + 6e \rightleftharpoons 2Cr^{3+} + 7H_2O$	
Pb^{2+}, Ba^{2+}, Sr^{2+}	$Pb^{2+} + CrO_4^{2-} \rightleftharpoons PbCrO_4$	filter and wash; dissolve in HClO_4
	$2PbCrO_4 + 2H^+ \rightleftharpoons 2Pb^{2+} + Cr_2O_7^{2-} + H_2O$	
	$Cr_2O_7^{2-} + 14H^+ + 6e \rightleftharpoons 2Cr^{3+} + 7H_2O$	
Fe^{3+}	$Fe^{3+} + e \rightleftharpoons Fe^{2+}$	strong HCl solution; slow reaction
Cu^{2+}	$Cu^{2+} + e \rightleftharpoons Cu^+$	Cu^+ precipitates as CuI
O_2	$O_2 + 4Mn(OH)_2 + 2H_2O \rightleftharpoons 4Mn(OH)_3$	basic solution + iodide; solution finally acidified
	$Mn(OH)_3 + 3H^+ + e \rightleftharpoons Mn^{2+} + 3H_2O$	
O_3	$O_3 + 2H^+ + 2e \rightleftharpoons O_2 + H_2O$	neutral solution
H_2O_2	$H_2O_2 + 2H^+ + 2e \rightleftharpoons 2H_2O$	molybdate catalyst
organic peroxides	$ROOH + 2H^+ + 2e \rightleftharpoons ROH + H_2O$	
MnO_2	$MnO_2 + 4H^+ + 2e \rightleftharpoons Mn^{2+} + 2H_2O$	

Applications of the Indirect Iodometric Method

The number of substances that can be determined by iodometry is large and varied. Some of the more common applications are found in Table 21-1. Detailed instructions follow for the iodometric determination of copper, a typical example of the indirect procedure.

Determination of copper. A cursory examination of electrode potentials suggests that an analysis based upon reduction of cupric ion by iodide would not be feasible; that is,

$$Cu^{2+} + e \rightleftharpoons Cu^+ \qquad\qquad E° = 0.15 \text{ volt}$$
$$I_2 + 2e \rightleftharpoons 2I^- \qquad\qquad E° = 0.54 \text{ volt}$$

In fact, however, the reduction is quantitative in the presence of a reasonable excess of iodide by virtue of the low solubility of cuprous iodide. Thus, when the more appropriate half reaction

$$Cu^{2+} + I^- + e \rightleftharpoons CuI \qquad\qquad E° = 0.86 \text{ volt}$$

is employed, it becomes apparent that the equilibrium

$$2Cu^{2+} + 4I^- \rightleftharpoons 2CuI + I_2$$

lies reasonably far to the right. Here, the iodide ion serves not only as a reducing agent for cupric ion but also as a precipitant for cuprous ion.

Much systematic experimentation has been devoted to establishing ideal conditions for this analysis.[6] These studies established the desirability of an excess of iodide in forcing the reaction to completion; sufficient potassium iodide should be added to give a solution that will be 4 percent or greater with respect to this reagent. Further, a pH of below 4 is expedient; in less acidic solutions, hydrolysis of the cupric ion apparently causes a slower and less complete oxidation of iodide ion. In the presence of cupric ion, hydrogen ion concentrations greater than about 0.3 molar must be avoided to prevent air oxidation of iodide ion.

It has been found experimentally that the titration of iodine by thiosulfate in the presence of solid cuprous iodide tends to yield slightly low results. This has been attributed to the physical adsorption of small but appreciable quantities of the element on the solid. The adsorbed iodine is released only slowly, even in the presence of thiosulfate ion; transient and premature end points result. This difficulty is largely overcome by the addition of thiocyanate ion, which also forms an insoluble precipitate with cuprous ion. Conversion of a part of the cuprous iodide to cuprous thiocyanate occurs at the surface of the solid

$$CuI + SCN^- \rightleftharpoons CuSCN + I^-$$

As a consequence of this reaction, the adsorbed iodine is released to the solution and becomes available for titration. Early addition of thiocyanate must be avoided, however, because there is a tendency for that ion to reduce iodine slowly.

Analysis of copper in an ore. The iodometric method is convenient for the assessment of the copper content of an ore. Ordinarily, the samples dissolve readily in hot, concentrated nitric acid. Care must be taken to volatilize any nitrogen oxides formed in the process since these catalyze the air oxidation

---

[6] See E. W. Hammock and E. H. Swift, *Anal. Chem.*, **21**, 975 (1949).

of iodide. Some samples require the addition of hydrochloric acid during the solution step. If this becomes necessary, the chloride ion must later be removed by evaporation with sulfuric acid because iodide ion will not quantitatively reduce cupric ion from its chloride complexes.

Of the elements ordinarily associated with copper in nature, only iron, arsenic, and antimony interfere with the iodometric procedure. Fortunately, the difficulties caused by the presence of these elements is readily eliminated. Iron is rendered nonreactive by the addition of such complexing agents as fluoride or pyrophosphate; because these form more stable complexes with ferric than with ferrous iron, the potential of this system is altered to the point where oxidation of iodide cannot occur.

We have seen that the arsenite-arsenate couple behaves reversibly with respect to the iodine-iodide system, the position of equilibrium being determined by the pH of the solution. Antimony is similar in deportment. By proper pH control, interference in a copper determination is readily prevented if the two elements are converted to the $+5$ state during solution of the sample. Ordinarily the hot nitric acid will assure this although a small amount of bromine water can be added in case of doubt; any excess bromine is then expelled by boiling. The $+5$ arsenic and antimony will not interfere provided the pH of the solution is kept above 3. As was mentioned earlier, however, incomplete oxidation of iodide by cupric ions is encountered at pH values greater than 4; thus in the presence of arsenic or antimony control of the pH between 3 and 4 is essential. A convenient buffer system for this is obtained by the addition of ammonium bifluoride, NH_4HF_2, to the solution. This salt dissociates as follows:

$$HF_2^- \rightleftharpoons HF + F^- \qquad\qquad K = 0.26$$

$$HF \rightleftharpoons H^+ + F^- \qquad\qquad K = 6.7 \times 10^{-4}$$

The first reaction gives equal quantities of hydrogen fluoride and fluoride ion which then buffer the solution to a pH somewhat greater than 3. In addition to acting as a buffer, the salt also serves as a source of fluoride ions to complex any ferric iron present.

The method that follows is essentially that of Park,[7] and incorporates the details discussed in the preceding paragraphs.

> **Procedure.** Weigh appropriate-sized samples of the finely ground and dried ore (about 1 gram for a sample containing 10- to 30-percent Cu) into 150-ml beakers and add 20 ml of concentrated HNO_3. Heat until all of the Cu is in solution. If the volume becomes less than 5 ml, add more HNO_3. Continue the heating until only a white or slightly gray siliceous residue remains (Note 1). Evaporate to about 5 ml.
>
> Add 25 ml of water and boil to bring all soluble salts into solution. Filter the solution through a small paper, collecting the filtrate in a 250-ml flask. Wash the paper with several small portions of hot 1:100 HNO_3. (If the residue is small and light colored no filtration is necessary.) Evaporate the solution

[7] B. Park, *Ind. Eng. Chem., Anal. Ed.*, **3**, 77 (1931).

to about 25 ml, cool, and slowly add 1:1 NH_3 to the first appearance of the deep blue cupric-ammonia complex. A faint odor of NH_3 should be detectable over the solution. If it is not, add another drop of NH_3 and repeat the test. Avoid an excess (Note 2).

From this point on, treat the samples individually. Add 2.0 ± 0.1 grams NH_4HF_2 and swirl until completely dissolved. Then add 3 grams of KI and titrate immediately with 0.1 N $Na_2S_2O_3$. When the color of the iodine is nearly gone, add 2 grams of KSCN and 3 ml of starch (see p. 462). Continue the titration until the blue starch-iodine color is decolorized and does not return for several minutes.

NOTES:

1. If the ore is not readily decomposed by the HNO_3, add 5 ml of concentrated HCl and heat until the solution is complete and only a small white or grey residue remains. Do not evaporate to dryness. Cool, add 10 ml of concentrated H_2SO_4, and evaporate until copious white fumes of SO_3 are observed. Cool, carefully add 15 ml of water and 10 ml of saturated bromine water. Boil the solution vigorously in a hood until all of the bromine has been removed. Cool and proceed as above, beginning with the filtration step.

2. If too much NH_3 is added, remove the excess by boiling.

Determination of copper in a brass. A brass is an alloy consisting principally of copper, zinc, lead, and tin. In addition, several other elements may be present in minor amounts, including iron and nickel. The iodometric method is convenient for estimating the copper content of such materials.

The method that follows is relatively simple and is applicable to the analysis of brasses containing less than 1 to 2 percent of iron. It involves solution in nitric acid, removal of nitrate by fuming with sulfuric acid, adjustment of the pH by neutralization with ammonia, acidification with a measured quantity of phosphoric acid, and finally the iodometric determination of the copper.

It is instructive to consider the fate of each of the major constituents during the course of this treatment. Tin is oxidized to the $+4$ state by the nitric acid solvent and precipitates slowly as the insoluble hydrous stannic oxide, $SnO_2 \cdot 4H_2O$. This precipitate, which tends to be colloidal in nature, is sometimes called *metastannic acid*. It has a tendency to adsorb cupric and other cations from the solution. Lead, zinc, and copper are oxidized to soluble divalent salts by the nitric acid treatment, and iron is converted to the ferric state. Evaporation and fuming with sulfuric acid redissolves the metastannic acid but may cause precipitation of part of the lead as the sulfate. The copper, zinc, and iron are unaffected. Upon dilution with water, lead is nearly completely precipitated as lead sulfate while the remaining elements are left in solution. None, except copper and iron, is reduced by iodide; interference from the latter is eliminated by complexing with phosphate ion.

Procedure. Weigh about 0.3-gram samples of the clean, dry metal into 250-ml flasks and add 5 ml of 6 N HNO_3. Warm the solution in a hood until decomposition is complete. Then add 10 ml of H_2SO_4 and evaporate to copious white fumes of SO_3. Allow the mixture to cool and then carefully add 20 ml of H_2O. Boil for 1 to 2 minutes and cool.

With good mixing, add concentrated NH_3 dropwise until the first dark-blue color of the cupric-ammonia complex appears. The solution should smell faintly of NH_3. Add 6 N H_2SO_4 dropwise until the dark-blue color just disappears. Then add 2.0 ml of syrupy phosphoric acid. Cool to room temperature.

From this point on treat samples individually. Dissolve 4.0 grams of KI in 10 ml of H_2O and add this to the sample. Titrate immediately with $Na_2S_2O_3$ until the iodine color is no longer distinct. Add 3 ml of starch solution (see p. 462) and titrate until the blue begins to fade. Add 2 grams of KSCN and complete the titration.

FERROUS IRON

Ferrous ion is a poorer reducing agent than iodide ion; nevertheless, it has found a number of interesting and important uses in analytical chemistry.

Preparation and Properties of Ferrous Ion Solutions

Solutions of ferrous ion are most commonly prepared from Mohr's salt, $FeSO_4 \cdot (NH_4)_2SO_4 \cdot 6H_2O$. The use of this substance as a primary standard is open to question since variations from the theoretical composition have frequently been reported.

Standard solutions of Mohr's salt are not very stable owing to the tendency of ferrous ion to react with atmospheric oxygen.

$$O_2 + 4Fe^{2+} + 4H^+ \rightleftharpoons 4Fe^{3+} + 2H_2O$$

The rate of this reaction is a minimum in solutions that are 0.5 to 1.0 N in sulfuric acid; it is accelerated in both neutral solution and more strongly acidic media. Owing to the instability of the reagent, restandardization at the time of use is recommended. A stable standard oxidant is ordinarily employed for this purpose, potassium dichromate or ceric sulfate being particularly useful.

Applications

In most applications, a measured volume of ferrous solution is introduced to the sample, after which the excess is determined with a standard solution of an oxidizing agent such as potassium dichromate, ceric sulfate, or potassium permanganate. This procedure is generally employed even where a direct titration with ferrous ion is feasible. Since a standard oxidizing agent must always be used in conjunction with ferrous solutions, no particular advantage is gained from a direct titration. In addition, the indirect method greatly simplifies the problem of choosing an indicator.

Table 21-2 lists a number of important determinations for which ferrous solutions have proved useful.

Table 21-2

APPLICATIONS OF SOLUTIONS OF FERROUS IRON

$$Fe^{2+} \rightleftharpoons Fe^{3+} + e$$

Substance	Half Reaction	Conditions
ClO_3^-	$ClO_3^- + 6H^+ + 6e \rightleftharpoons Cl^- + 3H_2O$	OsO_4 catalyst
NO_3^-	$NO_3^- + 4H^+ + 3e \rightleftharpoons NO + 2H_2O$	molybdate catalyst
$S_2O_8^{2-}$	$S_2O_8^{2-} + 2e \rightleftharpoons 2SO_4^{2-}$	HF or H_3PO_4 present
H_2O_2	$H_2O_2 + 2H^+ + 2e \rightleftharpoons 2H_2O$	HF or H_3PO_4 present
$V(OH)_4^+$	$V(OH)_4^+ + 2H^+ + e \rightleftharpoons VO^{2+} + 3H_2O$	
Ce^{4+}	$Ce^{4+} + e \rightleftharpoons Ce^{3+}$	
MnO_4^-	$MnO_4^- + 8H^+ + 5e \rightleftharpoons Mn^{2+} + 4H_2O$	
$Cr_2O_7^{2-}$	$Cr_2O_7^{2-} + 14H^+ + 6e \rightleftharpoons 2Cr^{3+} + 7H_2O$	
organic peroxides	$ROOH + 2H^+ + 2e \rightleftharpoons ROH + H_2O$	

TITANOUS AND CHROMOUS SOLUTIONS

Standard solutions of titanous ion and chromous ion are strongly reducing and find use in volumetric analysis as standard reductants. As can be seen from their reduction potentials, these are both more powerful than either iodide ion or ferrous ion.

$$Ti^{3+} + H_2O \rightleftharpoons TiO^{2+} + 2H^+ + e \qquad E° = -0.1 \text{ volt}$$

$$Cr^{2+} \rightleftharpoons Cr^{3+} + e \qquad\qquad\qquad E° = 0.41 \text{ volt}$$

Neither reagent is as convenient to use as any of the volumetric solutions considered thus far. Both react rapidly with atmospheric oxygen, and therefore require special apparatus for storage and titration. Additional precautions are necessary to prevent their reduction of hydrogen ions. Because of these limitations, titanous and chromous solutions are seldom, if ever, applied to a problem where only a single or a few samples are to be analyzed. Only with a large number of analyses does the time and effort expended in preparation become worthwhile.

Figure 21.1 illustrates an apparatus used for storage and delivery of air-sensitive solutions. Hydrogen gas is either supplied from a tank or is generated by the action of sulfuric acid upon zinc in a Kipp generator.

Titanous Solutions

Solutions of titanous ion in hydrochloric acid can be prepared by dilution of the commercial 20 percent titanous chloride. Sulfate solutions have been prepared by dissolving titanous hydride in sulfuric acid or by reduction of a titanic solution electrolytically or with zinc amalgam.

Standardization of titanous solutions is accomplished by titration of ferric ion, which is produced either from a weighed amount of electrolytic iron that has been oxidized by hot, concentrated perchloric acid or from Mohr's salt solution that has been oxidized with a measured amount of primary-standard potassium dichromate. In either case thiocyanate ion is used as an indicator, the end point being signaled by disappearance of the red ferric-thiocyanate complex. These titrations must be done in oxygen-free solvents and in the absence of air. The latter condition is readily achieved by bubbling carbon dioxide or nitrogen through the titration flask.

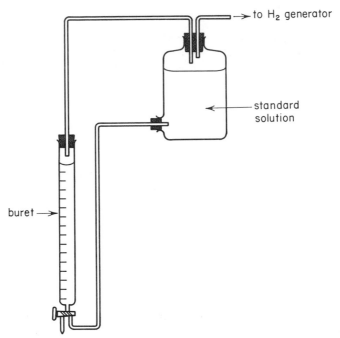

Fig. 21.1 Apparatus for the Storage and Delivery of Air-sensitive Standard Solutions.

Titanous solutions are used for the titration of ferric ion, cupric ion (which gives cuprous chloride), vanadate ion, and chlorate ion (which is reduced to chloride). One interesting use for the reagent is for the determination of aromatic nitro compounds

$$RNO_2 + 6Ti^{3+} + 4H_2O \rightleftharpoons RNH_2 + 6TiO^{2+} + 6H^+$$

In this application, the organic compound is treated with an excess of the titanous solution and, after a few minutes, back titration is undertaken with a standard ferric solution employing thiocyanate ion as an indicator. Many compounds containing nitroso, azo, hydrazo, or sulfoxide groups are also reduced by this procedure and may thus be determined in an analogous fashion.

Chromous Solutions

Chromous ion is the most powerful reductant used in analytical chemistry. A standard solution of the reagent is most easily obtained by reduction of a potassium dichromate solution of known strength first to the chromic state with hydrogen peroxide, and then to the chromous state with zinc amalgam. By storage over zinc amalgam, a solution is obtained that is constant in normality for several days.

Since chromous solutions are extremely labile with respect to atmospheric oxygen, considerable care is necessary to prevent contact with air. Furthermore, the reduction potential of the system is such that chromous ion should evolve hydrogen gas even from solution of fairly high pH. In fact, however, the reaction

$$2Cr^{2+} + 2H^+ \rightleftharpoons 2Cr^{3+} + H_2$$

is slow; provided pure solutions, free of catalysts, are used, this reaction does not occur rapidly enough to be of serious consequence even under acidic conditions.

There has been no systematic search for indicators suitable for titrations with chromous ion, and the potentiometric method (see Chap. 24) has been employed in most studies.

A wide variety of inorganic ions are quantitatively reduced by chromous salts. These include Fe(III), Ti(IV), Cu(II), Mo(VI), U(VI), V(V), Hg(II), Se(IV), Te(IV), Bi(III), Au(III), and Ag(I). Organic compounds containing reducible functional groups are also susceptible to analysis by the reagent. Studies show that aromatic aldehydes, α, β-unsaturated ketones, nitro compounds, maleic and fumaric acids, pyruvic acid, and quinone are reduced by the reagent.

problems

1. What weights of the following compounds, when treated with an excess of KI, are equivalent to 10.0 ml of 0.1 N $Na_2S_2O_3$: (a) $KBrO_3$, (b) $K_2Cr_2O_7$, (c) $CuSO_4$, (d) KIO_3, (e) $(NH_4)_2Ce(NO_3)_6$.
   ans. (a) 0.0278 gram; (b) 0.049 gram

2. A solution of $Na_2S_2O_3$ was standardized by treatment of 0.0600 gram of pure KIO_3 with an excess of KI. After acidification, 32.1 ml of the reagent were required to reach a starch end point. Calculate the normality. ans. 0.0524 N

3. How many grams of $Na_2S_2O_3 \cdot 5H_2O$ are contained in 763 ml of a 0.0500 N solution of that reagent?

4. A sample containing $Ba(ClO_3)_2$ was analyzed by dissolving it in 2 N HCl and treating with an excess of KI. After 5 minutes the liberated iodine was titrated with a 0.0810 N $Na_2S_2O_3$ solution. A 0.418-gram sample required 19.2 ml. What was the percent $Ba(ClO_3)_2$ present? ans. 9.45 percent

5. An excess of KI is added to 50.0 ml of bromine water. The liberated iodine requires 28.6 ml of a 0.0516 N $Na_2S_2O_3$ solution for titration. How many milligrams of Br_2 are contained per milliliter of solution?

6. A 0.661-gram sample containing $BaCl_2 \cdot 2H_2O$ was dissolved and the barium precipitated as $BaCrO_4$. After filtration and washing the precipitate was dissolved in acid and the $Cr_2O_7^{2-}$ formed in this process reduced by the addition of an excess of KI. The iodine liberated required 37.5 ml of 0.121 N $Na_2S_2O_3$. What was the percent $BaCl_2 \cdot 2H_2O$ in the sample? ans. 55.9 percent

7. The lead in a 0.477-gram sample of Pb_3O_4 was precipitated as $PbCrO_4$. After filtration and washing the solid was dissolved in acid and treated with an excess of KI. The liberated iodine consumed 22.2 ml of 0.0916 N $Na_2S_2O_3$. What was the percent Pb_3O_4 in the sample?

8. A solution containing IO_3^- and IO_4^- was analyzed as follows: A 50.0-ml aliquot was buffered with a borax buffer and treated with an excess of KI which resulted in reduction of the IO_4^- to IO_3^-. The liberated iodine consumed 18.4 ml of 0.100 N $Na_2S_2O_3$. A 10.0-ml aliquot was then acidified strongly and an excess of KI added. This required 48.7 ml of the thiosulfate. Calculate the formal concentration of IO_3^- and IO_4^- in the solution.

9. A sample containing a small amount of KI was analyzed by solution of a 1.05-gram portion in water. Chlorine water was added which converted the I^- to IO_3^-. After removal of the excess Cl_2 by boiling, KI was added and the liberated iodine titrated with 11.7 ml of 0.0412 N $Na_2S_2O_3$. What was the percent KI in the sample?

10. A mixture containing $KHSO_4$ and K_2SO_4 was dissolved and treated with an excess of KI and KIO_3. The liberated iodine consumed 26.7 ml of 0.125 N $Na_2S_2O_3$. The sample weight was 0.802 gram. Calculate the percent $KHSO_4$.

11. A fresh solution of $Na_2S_2O_3$ was found to be 0.100 N against iodine. After standing for some time its normality had increased to 0.112. Assuming the only change in the solution was due to the reaction $S_2O_3^{2-} \rightleftharpoons SO_3^{2-} + S$ and that the SO_3^{2-} was oxidized to SO_4^{2-} in the second standardization, calculate the formal concentration of $Na_2S_2O_3$ and Na_2SO_3 in the 0.112 N solution.

12. The nitro groups in nitroglycerin, $C_3H_5O_3(NO_2)_3$, are quantitatively reduced by a titanous solution [the product is $C_3H_5O_3(NH_2)_3$]. A 1-gram sample of this substance was analyzed by dissolving in methanol and diluting to 100 ml. A 10.0-ml aliquot was then treated with 25.0 ml of 0.0509 N $TiCl_3$ solution. The excess titanous ion was back titrated with 10.6 ml of a 0.0421 N solution of Fe^{3+}. Calculate the percent nitroglycerin in the sample.

13. How many grams of pure $K_2Cr_2O_7$ are needed to prepare 800 ml of an exactly 0.100 N solution of Cr^{2+} to be used as a standard reductant?

14. A 0.284-gram sample of sodium peroxide was dissolved in an excess of KI. The liberated iodine consumed 35.4 ml of 0.0896 N thiosulfate. What percent Na_2O_2 was present?

15. A sample containing KI, KBr, and other inert substances was analyzed as follows: A 1.00-gram sample was dissolved in water and diluted to exactly 200 ml. A 25.0-ml aliquot was treated with Br_2 in neutral solution which converted the I^- to IO_3^-. The excess Br_2 was removed by boiling and KI added. After acidification, the liberated iodine was titrated with 40.8 ml of 0.0500 N thiosulfate.

A 50.0-ml aliquot was oxidized with strongly acidic $K_2Cr_2O_7$; the liberated I_2 and Br_2 were distilled and collected in a strong KI solution.

The iodine present after complete reaction required 29.8 ml of the thiosulfate. Calculate the percent KI and KBr in the sample.

16. A 0.941-gram sample containing $Ca_3(AsO_4)_2$ was dissolved in strong HCl. Excess KI was added and the liberated iodine required 27.7 ml of a $Na_2S_2O_3$ solution which had a titer of 2.47 mg of Cu. Calculate the percent $Ca_3(AsO_4)_2$.

17. The hydrolysis constant for the reaction $I_3^- + H_2O \rightleftharpoons HOI + 2I^- + H^+$ is about 4×10^{-16}. What percent of the I_3^- is converted to HOI in a solution which is 0.01 F in I_3^- and 0.02 F in KI and buffered to (a) pH 8.0, (b) pH 9.0, (c) pH 10.0?

18. A 2.31-gram sample of a steel was dissolved in HNO_3; upon addition of $KClO_3$, a quantitative precipitation of MnO_2 occurred. The precipitate was filtered, washed, and dissolved in an acid solution of KI. To titrate the liberated iodine, 41.0 ml of 0.0612 N $Na_2S_2O_3$ were used. What is the percent Mn in the steel?

19. Calculate the solubility-product constant for CuI from the following data:

$$Cu^{2+} + e \rightleftharpoons Cu^+ \qquad\qquad E° = 0.153 \text{ volt}$$
$$Cu^{2+} + I^- + e \rightleftharpoons CuI \qquad\qquad E° = 0.86 \text{ volt}$$

20. Calculate the equilibrium constant for the reaction

$$O_2 + 6I^- + 4H^+ \rightleftharpoons 2H_2O + 2I_3^-$$

part 4. *Electroanalytical Methods*

chapter 22. *An Introduction to Electroanalytical Chemistry*

The field of electroanalytical chemistry encompasses a wide variety of techniques based upon the various phenomena occurring within an electrochemical cell. Many electroanalytical methods can be classified as gravimetric or as volumetric procedures. Thus, the use of an electric current to deposit a substance in a form suitable for weighing can be considered a variety of gravimetric analysis. Similarly, the use of any of several electrical properties will serve to establish the end point in an ordinary volumetric titration. In addition to these, other electroanalytical techniques measure some colligative electrical property that can be related by calibration to the concentration of the species being determined.

This chapter is intended to provide a background for the more detailed consideration of specific electroanalytical methods that follow. A review of Chapter 17 is also suggested.

SOME CELLS AND CELL COMPONENTS USED IN ELECTROANALYTICAL MEASUREMENTS

Reference Electrodes

Many electroanalytical measurements are based upon some phenomenon occurring at only one of the electrodes (*the indicator electrode*) of a cell. The devices available for measuring the electrical consequences of this phenomenon, however, are applicable only to a complete cell consisting of two electrodes. Thus, in addition to the indicator electrode, it is necessary to have present a *reference electrode* whose behavior is reproducible and constant.

Several properties are required of a good reference electrode. It should be relatively easy to assemble from materials and chemicals ordinarily found in the laboratory. In addition, having used reasonable care, the potentials of several such half cells should be identical or nearly so. A constant potential, even after extended storage, is highly desirable. Finally, and most important, the electrode must remain essentially constant in potential during passage of small currents for some length of time. Several electrodes are available that meet these requirements fairly well.

Calomel electrodes. Calomel half cells may be represented as follows:

$$Hg \mid Hg_2Cl_2 \text{ (sat'd), } KCl \text{ } (xF)$$

where x represents the formal concentration of potassium chloride in the solution. The electrode reaction is given by the equation

$$Hg_2Cl_2 + 2e \rightleftharpoons 2Hg + 2Cl^- \qquad E° = 0.2676 \text{ volt}$$

The potential of this cell will clearly vary with the chloride concentration x, and this must be specified in describing the electrode.

Table 22-1 lists the composition and the reduction potential for the three most commonly encountered calomel electrodes. Note that each solution is saturated with mercurous chloride, and that the cells differ only with respect to the potassium chloride concentration. The reduction potential may be calculated by substitution of the desired temperature, in degrees centigrade, for t in the appropriate expression. The reader should note that all of the calomel electrodes behave as cathodes with respect to the standard hydrogen electrode.

The saturated calomel electrode is most commonly used by the analytical chemist because of the ease with which it can be prepared. Compared with the other two, its temperature coefficient is somewhat larger; this is a disadvantage in some applications.

The form and shape of calomel electrodes vary tremendously according to the ingenuity and imagination of their designers. A rather simple, easily constructed saturated electrode is shown in Figure 22.1. The container is a 2-ounce wide-mouth bottle equipped with a stopper that holds two tubes. Sealed into the end of one of these tubes is a small piece of platinum wire;

Table 22-1

<small>SPECIFICATIONS OF CALOMEL ELECTRODES</small>

Name	Concentration of		Reduction Potential (volts) vs Standard Hydrogen Electrode ($Hg_2Cl_2 + 2e \rightleftharpoons 2Hg + 2Cl^-$)
	Hg_2Cl_2	KCl	
saturated	saturated	saturated	$+ 0.242 - 7.6 \times 10^{-4}(t - 25)$
normal	saturated	$1.0\,F$	$+ 0.280 - 2.4 \times 10^{-4}(t - 25)$
decinormal	saturated	$0.1\,F$	$+ 0.334 - 7 \times 10^{-5}(t - 25)$

when the tube is filled with mercury, this offers electric contact between the interior and exterior of the cell. The other tube is simply a bridge for connecting the calomel cell to the indicator half cell; it is ordinarily filled with saturated potassium chloride. A means must be provided to prevent the siphoning of the cell liquid. A fine, fritted glass disk is often sufficient to serve in this capacity.

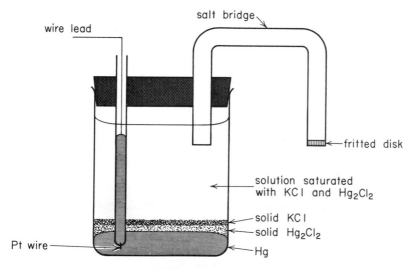

half reaction:
$$Hg_2Cl_2 + 2e \rightleftharpoons 2Hg + 2Cl^-$$

Fig. 22.1 Diagram of a Saturated Calomel Electrode.

The bottom of the cell container is covered with a layer of pure mercury approximately 5 mm in depth. Upon this is placed a layer of a paste prepared by triturating pure mercurous chloride and a little mercury with saturated potassium chloride solution. A layer of solid potassium chloride follows, and then the jar is nearly filled with a saturated solution of that salt. Ordinarily, several days are required for such a cell to come to equilibrium and develop a constant potential.

Several convenient calomel electrodes are available commercially. Typical of these is the one illustrated in Figure 22.2 which consists of a tube 5 to 15 cm in length and 0.5 to 1.0 cm in diameter. The mercury-mercurous chloride paste is contained in an inner tube that is connected to the saturated potassium chloride solution in the outer tube by means of a small opening. Contact with the second half cell is made by means of a porous fiber sealed in the end of the outer tubing. An electrode such as this has a relatively high resistance (2000 to 3000 ohms) and very limited current-carrying capacity.

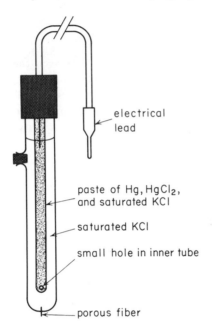

electrical
lead

paste of Hg, $HgCl_2$,
and saturated KCl

saturated KCl

small hole in inner tube

porous fiber

Fig. 22.2 Diagram of a Commercial Fiber-type Saturated Calomel Electrode. (By permission, Beckman Instruments, Inc., Fullerton, California.)

Silver-silver chloride electrodes. A reference electrode system analogous to the foregoing consists of a silver electrode immersed in a solution of potassium chloride that is also saturated with silver chloride.

$$\text{Ag} \mid \text{AgCl(sat'd)}, \quad \text{KCl} \quad (xF)$$

The half reaction is

$$\text{AgCl} + e \rightleftharpoons \text{Ag} + \text{Cl}^- \qquad\qquad E° = 0.222 \text{ volt}$$

Normally this electrode is prepared with a saturated potassium chloride solution, the potential at 25 degrees being $+0.197$ volt with respect to the standard hydrogen electrode.

A simple and easily constructed form of the silver-silver chloride electrode is shown in Figure 22.3. The electrode is contained in a Pyrex tube fitted with a 10-mm fritted disk (tubing containing fritted disks is obtainable from Corning Glass Works). A plug of agar gel saturated with potassium chloride is formed on top of the disk to prevent loss of solution from the half cell. The plug can be prepared by heating 3 or 4 grams of pure agar in 100 ml of water until solution is complete and then adding about 35 grams of potassium chloride. A portion of this, while still warm, is poured into the tube; upon cooling, it solidifies to a gel with low electrical resistance. On top of this is placed a layer of solid potassium chloride and a saturated solution of the salt. A drop or two of 1 F silver nitrate is then added and a heavy gauge (1- to 2-mm diameter) silver wire inserted in the solution.

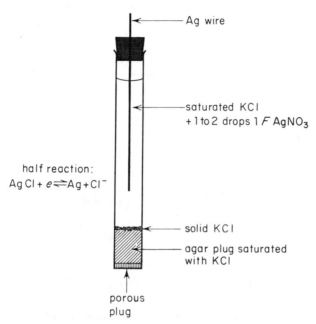

Fig. 22.3 Diagram of a Convenient Silver-silver Chloride Electrode.

Standard Weston Cells

The accurate measurement of potentials for electrochemical work requires the frequent use of a cell whose emf is precisely known. The *Weston cell*, which is used almost universally for this purpose, can be represented as follows:

$$Cd(Hg) \mid CdSO_4 \cdot 8/3\ H_2O(\text{sat'd}), Hg_2SO_4(\text{sat'd}) \mid Hg$$

The half reactions for the cell are

$$Cd(Hg) \rightleftharpoons Cd^{2+} + Hg + 2e$$
$$Hg_2SO_4 + 2e \rightleftharpoons 2Hg + SO_4^{2-}$$

Figure 22.4 shows a typical form of a Weston cell. Its potential, at 25° C, is 1.0183 volts.

The emf of a Weston cell is governed by the activities of the cadmium and mercurous ions in the solution. At any given temperature these quantities are invariant, being fixed by the solubilities of the cadmium sulfate and the mercurous sulfate at that temperature. As a result, this cell will remain constant in voltage for remarkably long periods of time if treated properly.

cell reaction:
$$Cd(Hg) + Hg_2SO_4 \rightleftharpoons Cd^{2+} + 2Hg + SO_4^{2-}$$
$$E = 1.0183 \text{ volts}$$

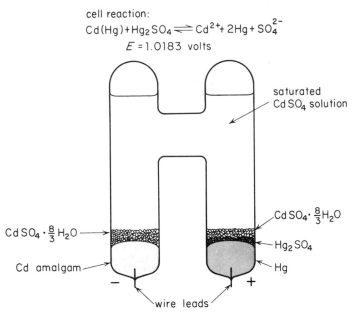

Fig. 22.4 A Weston Standard Cell.

The Weston cell has a temperature coefficient of about $-0.04 \text{ mv}/°$ C due primarily to changes in solubility of the cadmium and mercurous salts. A cell with a coefficient of about one fourth this magnitude is obtained by using a slightly undersaturated solution of cadmium sulfate. This is prepared by using a solvent that has been saturated at 4° C; no solid $CdSO_4 \cdot 8/3H_2O$ is incorporated in the cell itself. Most commercially available cells are of this type and are termed *unsaturated Weston cells*. Their potentials vary from 1.0185 to 1.0195 volts.

Weston cells may be sent to the National Bureau of Standards for calibration and certification.

Some Important Electrical Measurements

In electroanalytical work, the accurate measurement of voltage, resistance, current, and quantity of electricity is frequently important. We must, therefore, consider briefly some of the methods employed in making such measurements.

Voltage Measurements

The electromotive force generated by a galvanic cell cannot be measured very accurately by simply placing a direct-current voltmeter across its electrodes because a small current is required for operation of the meter. When this current is drawn from the cell, a diminution in voltage occurs. This is due, in part, to changes in the concentrations of the reacting species as the cell starts to discharge. In addition, the cell has an internal resistance of its own; when current is passed, there develops an ohmic potential drop (equal to the current times the resistance) which opposes that of the two electrodes. A truly significant value for the output of a cell can thus be attained only if the measurement is made with a negligible passage of current. The most satisfactory method of accomplishing this is with a *potentiometer*. Of the several varieties and modifications of this device, two warrant discussion.

Principles of the potentiometer. First, we will consider the operation of a simple *voltage divider* such as shown in Figure 22.5. This consists of a resistance AB along which there runs a sliding contact C. If a battery is placed across this resistance, a current I will flow; from Ohm's law we may write $E_{AB} = IR_{AB}$ where R_{AB} is the ohmic resistance of the resistor. Of specific interest here is the potential drop between the contact C and one end of the resistor A—that is, E_{AC}. The current passing through the portion of the resistor AC is identical to that through the entire resistor; it follows, then, that $E_{AC} = IR_{AC}$. Dividing this by the foregoing relationship, we get

$$E_{AC} = E_{AB} \frac{R_{AC}}{R_{AB}} \tag{22-1}$$

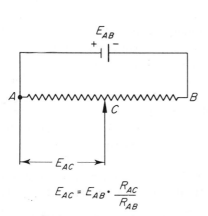

Fig. 22.5 Diagram of a Simple Voltage Divider.

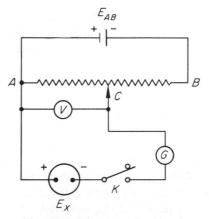

Fig. 22.6 Diagram of a Simple Potentiometer.

Thus, we have a device that will supply a continuously variable voltage from zero (when contact C is at A) to the total output E_{AB} of the power supply (when the contact C has been moved to B). This very simple and useful contrivance is employed in a wide variety of electroanalytical instruments.

A simple potentiometer. A relatively simple instrument that permits the measurement of the emf of galvanic cells with passage of but negligible current is illustrated in Figure 22.6. This consists of a voltage divider powered by a battery whose emf is equal to, or greater than, the emf to be measured, E_x. The galvanic cell is placed across AC in such a way that its output *opposes* that of the working battery. In parallel with the galvanic cell is a tapping key K, which permits momentary closing of the circuit, and a galvanometer G, which serves as a current detector. The direct-current voltmeter V measures the potential drop across AC.

Now, if the potential across the voltage divider E_{AC} is greater than that of the galvanic cell E_x, electrons will be forced from right to left through the latter when K is closed. On the other hand if E_{AC} is smaller than E_x, the electron flow will occur in the opposite direction. Finally, when E_x is equal to E_{AC}, no current will flow *in the circuit containing G, K, and the galvanic cell.*

The process for measuring an unknown voltage E_x with a device such as this consists of adjusting C until the galvanometer G indicates no current flow. This is done by momentarily closing K several times and making suitable adjustments of C as indicated by the galvanometer. The entire process draws only infinitesimal currents from the galvanic cell being measured. When balance is achieved, the voltage drop across AC is read on the meter V and is equal, of course, to E_x. Thus, at balance, the voltmeter is powered by the current from the working battery and not from the unknown galvanic cell.

A potentiometer such as this is limited in accuracy by the precision of the voltmeter V; typically this ranges between about 0.005 and 0.01 volt—entirely adequate for many electroanalytical purposes. Where greater precision is required, recourse must be made to a potentiometer with a linear voltage divider.

Potentiometer employing a linear voltage divider. To achieve the highest precision in potential measurements an instrument such as that shown in Figure 22.7 is utilized. This differs from the previous potentiometer in several important respects. For example, the voltage divider contains an additional variable resistance R to allow adjustment of the voltage across AB. In addition, the resistance AB must now be *linear* in character. That is, the resistance between one end A and any point C is directly proportional to the length AC of that portion of the resistor; then $R_{AC} = kAC$, where AC is expressed in convenient units of length and k is a proportionality constant. Similarly, $R_{AB} = kAB$. A scale, such as that shown in the illustration, can be attached to the resistor to aid in measuring these lengths. Substituting the proportionality relationships into equation (22-1), we obtain

$$E_{AC} = E_{AB} \frac{AC}{AB} \qquad (22\text{-}2)$$

The instrument in Figure 22.7 contains a standard Weston cell E_s; either this or the unknown cell can be placed in the circuit by means of the switch S.

As in the earlier case, E_{AC} will equal the potential of the unknown or the standard cell when no current flow is indicated; with equation (22-2) we may write, then, that

$$E_x = E_{AC_x} = E_{AB} \frac{AC_x}{AB}$$

$$E_s = E_{AC_s} - E_{AB} \frac{AC_s}{AB}$$

where AC_x and AC_s represents the linear distances corresponding to balance when the unknown and standard cells are in the circuit. Dividing these equations, we obtain the relationship

$$E_x = E_s \frac{AC_x}{AC_s} \tag{22-3}$$

Thus, E_x may be obtained from the known emf of the Weston cell and the two measurable quantities AC_x and AC_s.

For convenience, the scale reading of AC_x is ordinarily made to correspond directly to the potential in volts. This is accomplished by first switching the Weston cell into the circuit and positioning the contact so that AC_s corresponds numerically to the output of this standard cell (1.0186 volts). The potential across AB is then adjusted by means of R until no current is indicated by the galvanometer. With the potentiometer adjusted thus, AC_s and E_s will be numerically equal; from equation (22-3), then, AC_x and E_x will also be identical when balance is achieved with the unknown cell in the circuit.

The necessity for a working battery P may be questioned; in principle, there is nothing to prevent a direct measurement

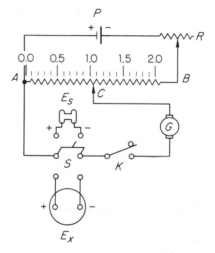

Fig. 22.7 Diagram of a Potentiometer with a Linear Voltage Divider, *AB*.

of E_x by replacing P with the standard cell E_s. It must be remembered, however, that current is being continuously drawn from P; a standard cell could not be expected to maintain a constant potential for long under such usage.

The accuracy of a voltage measurement with an instrument such as the foregoing depends upon several factors. For one thing, it is necessary to assume that the voltage of the working battery P remains constant during the period of time required to balance the instrument against the standard cell and to measure the potential of the unknown cell. Ordinarily this does not lead to appreciable error if P consists of one or two heavy duty dry cells in good condition or a lead storage battery. Calibration of the instrument against the standard cell should, however, be done before each voltage measurement to compensate for possible changes in P.

The linearity of the resistance AB as well as the precision with which distances along its length can be estimated also contribute to the accuracy of the potentiometer. Ordinarily, however, the ultimate precision of a good quality instrument is determined by the sensitivity of the galvanometer relative to the resistance of the circuit. Suppose, for example, that the electrical resistance of the galvanometer plus that of the unknown cell is 1000 ohms—a rather typical figure. Further, if we assume for the moment that the galvanometer is just capable of detecting a current of 1 microampere (10^{-6} ampere), we can readily calculate from Ohm's law that the minimum distinguishable voltage difference will be $10^{-6} \times 1000 = 10^{-3}$ volt or 1 mv. By use of a galvanometer sensitive to 10^{-7} ampere, a difference of 0.1 mv will be detectable. Such a sensitivity is found in an ordinary pointer-type galvanometer; more refined instruments with sensitivities up to 10^{-10} ampere are available.

From this discussion we see that the sensitivity of a potentiometric measurement decreases as the electrical resistance of the cell increases; as a matter of fact, potentials of cells with resistances much greater than a megohm (10^6 ohms) cannot be measured accurately with an instrument employing a galvanometer as the current-sensing device. At first glance this appears to be a very large resistance for a cell. Actually, however, one of the most important electroanalytical applications of potential measurements—that of measuring pH—requires the use of a cell with a resistance in the 1- to 100-megohm range. To detect the very small out-of-balance currents associated with such a high resistance, a direct-current amplifier must be used in place of the galvanometer. This multiplies the current sufficiently to make its detection easy.

Resistance Measurements

Because the measurement of the electrical resistance of a conducting solution is of considerable importance in analytical chemistry, we must examine methods used for this purpose. Generally, a current must be passed through the solution in question; this, of course, requires a pair of electrodes to transmit the current to and from the liquid. Since electrochemical changes at the electrodes result from the passage of current, steps must be taken to prevent these from altering the resistance of the solution, either by changing the electrolyte concentration or through development of a potential that opposes the flow of current. The utilization of an alternating current largely circumvents these problems.

The Wheatstone bridge arrangement, shown in Figure 22.8, is typical of the apparatus used for resistance measurements of electrolytes. The power source S provides an alternating current in the frequency range of 1000 to 10,000 cycles/sec and at a potential of 6 to 10 volts. The resistance of the potential divider AB is linear and is known with precision; thus, the resistances R_{AB} and R_{AC} can be calculated from the position of C. The cell, of resistance R_x, is placed in the upper left arm of the bridge and a precision variable resistance R_s in the right-hand side. A null detector ND is used to indicate the condition of no current flow between D and C.

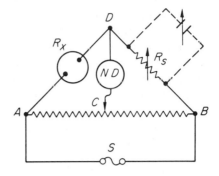

This may consist of a pair of ordinary earphones, because the ear is responsive to frequencies in the 1000-cycle range; alternatively, it may be a "magic eye" tube which, with appropriate circuitry, will indicate the current minima.

When an alternating-current device such as this is used to measure very high resistances, capacitance effects within the latter cause a loss in sensitivity. The origin of this capacitance is beyond the scope of this discussion; in practice, it is compensated for by a variable capacitor (shown in dotted lines) across R_s.

Fig. 22.8 Diagram of a Bridge Circuit for Resistance Measurements.

An unknown resistance R_x is measured by adjusting the contact C until no current passage can be detected through CD. At this point E_C must be equal to E_D. But we may express E_C as follows, employing equation (22-1) for a voltage divider:

$$E_C = E_{AB} \frac{R_{AC}}{R_{AC} + R_{BC}}$$

The circuit ADB may also be treated in this way and

$$E_D = E_{AB} \frac{R_x}{R_x + R_s}$$

Equating these two expressions and rearranging gives

$$R_x = \frac{R_{AC} R_s}{R_{BC}} \tag{22-4}$$

Current Measurements

Current is readily measured with a galvanometer, many types of which are commercially available. These vary tremendously in sensitivity, ruggedness, and cost; the choice among them will depend upon the requirements of the measurement.

A convenient alternate method for the accurate evaluation of small currents makes use of a potentiometer to measure the potential drop resulting from passage of the current through a carefully calibrated resistor. Thus, for example, a 10,000-ohm resistance used in conjunction with a potentiometer sensitive to 10^{-4} volt will detect a current of $10^{-4}/10,000$, or 10^{-8} ampere.

Measurement of Quantity of Electricity

The total quantity of electricity passing through a cell can be determined by means of a device known as a *chemical coulometer* placed in series with that cell. Classically, a coulometer consisted of an electrolytic cell containing a platinum electrode immersed in a silver nitrate solution; from the weight of silver deposited, the quantity of electricity was calculated. The iodine coulometer, based upon the volumetric determination of the quantity of iodine liberated at a platinum anode by the current, is more convenient than the silver coulometer. A direct-reading coulometer can also be constructed in which the total volume of hydrogen and oxygen gas formed at the two electrodes of a cell is measured in a gas buret. This volume is readily related to the quantity of current passing through the circuit.[1]

BEHAVIOR OF CELLS DURING CURRENT PASSAGE

We have, in Chapter 17, considered the computation of theoretical potentials of galvanic and electrolytic cells. In developing methods for these computations, we assumed that current passage through the cell was negligible and that the chemical processes occurring at the electrodes were rapid and reversible. Further, we neglected any potential effects resulting from the use of a salt bridge or from contact between two electrolytes of different composition. Within these limitations the theoretical cell potential is given by the equation

$$E_{cell} = E_{cathode} + E_{anode}$$

where $E_{cathode}$ is the *reduction potential* for the cathode, obtained from the standard potential for the half reaction and the activities of the participating species; E_{anode} is the *oxidation potential* for the anodic half reaction computed in an analogous fashion.

In many analytical uses of electrochemical cells, the conditions imposed by these assumptions are not realized; as a result, discrepancies between calculated and experimentally measured cell potentials are encountered. These are traceable to so-called junction potentials arising at the interface between liquids of different composition; to ohmic potentials required to cause passage of current through the cell; and to polarization effects occurring at one or both electrodes. In the sections that follow we shall consider the cause and, insofar as possible, the magnitude of each of these phenomena.

Liquid Junction Potential

When two electrolyte solutions of different composition are brought in contact with one another, a potential develops at the interface. This *junction*

[1] For a discussion of the hydrogen-oxygen and other coulometers see J. J. Lingane, *Electroanalytical Chemistry*, 452-459. New York: Interscience Publishers, Inc., 1958.

potential, as it is called, arises from an unequal distribution of cations and anions across the boundary due to differences in rates with which these species migrate.

To take the very simplest case, we shall consider the junction formed between a 1 F and a 0.01 F hydrochloric acid solution. We may symbolize this as follows:

$$HCl(1\ F) \mid HCl(0.01\ F)$$

Both hydrogen ions and chloride ions tend to diffuse across this boundary from the more concentrated solution to the more dilute, the driving force for this migration being proportional to the concentration difference. As we shall see in Chapter 25, the rate at which various ions move under the influence of a fixed force varies considerably (that is, their *mobilities* are different); in the present case, hydrogen ions are several times more mobile than chloride ions. As a consequence, there is a tendency for the protons to outstrip the chloride ions as the diffusion takes place; this leads to a charge separation as shown below:

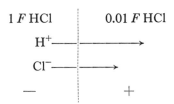

The more dilute side of the boundary becomes positively charged owing to the excess of hydrogen ions; the more concentrated side therefore acquires a negative charge due to the slower moving chloride ions. The charge that develops tends to counteract the differences in mobilities of the two ions; as a consequence, an equilibrium condition soon develops. The resulting charge separation is a potential difference that may amount to several hundredths of a volt or more.

In a simple case such as this, where the behavior of only two ions needs to be considered, the magnitude of the junction potential can be calculated from a knowledge of the mobilities of the ions involved. However, in a more complex situation—often found in many of the cells of electroanalytical importance—the mobilities of several ions may govern the size of the junction potential, and here we do not have sufficient information to estimate its magnitude.

Experiments have shown that the liquid junction potential between two electrolytes can ordinarily be reduced to a small, and often insignificant, quantity by interposing a concentrated electrolyte solution (called a *salt bridge*) between them. The effectiveness of this contrivance increases greatly with increases in concentration of the salt in the bridge. It also becomes more effective as the mobilities of the ions of the salt used approach one another. A saturated potassium chloride solution is good from both of these standpoints, its concentration being somewhat greater than 4 F and the mobility of its ions differing by only 4 percent. With such a bridge the junction potential typically amounts to a few millivolts or less; this is negligibly small for most analytical measurements.

Ohmic Potential; *IR* Drop

The rate of current flow affects the potential of a cell—either electrolytic or galvanic. This is due in part to the necessity of overcoming the resistance of the cell itself to the passage of current. The potential required for this is equal to the product of the current and the electrical resistance of the cell; for obvious reasons this potential is frequently referred to as the *IR drop*.

To illustrate the effect of *IR* drop, we shall consider the behavior of a reversible cell during current passage. We will take as an example

$$\text{Cd} \mid \text{CdSO}_4(1\ F) \parallel \text{CuSO}_4(1\ F) \mid \text{Cu}$$

From the half-cell potentials we can readily calculate that the emf of this cell should be 0.74 volt provided the junction potential is kept small with a suitable salt bridge.

This cell is reversible, which means that it can function either as a galvanic or an electrolytic cell. In the former case, the over-all chemical reaction would be

$$\text{Cd} + \text{Cu}^{2+} \rightleftharpoons \text{Cd}^{2+} + \text{Cu}$$

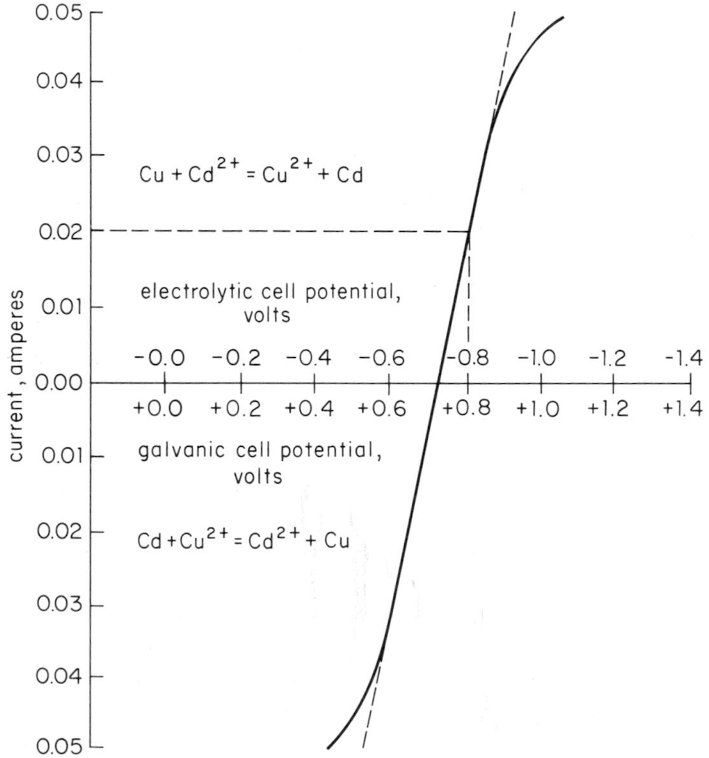

Fig. 22.9 Current-voltage Curve for the Cell Cd | CdSO$_4$(1F) || CuSO$_4$(1F) | Cu. Cell resistance is 4 ohms. Note that the direction of current flow is different above and below the abcissa.

and in the latter, the reverse would occur. Operation as an electrolytic cell would require the application of an emf somewhat greater than the theoretical potential of —0.74 volt. This is shown in the upper half of Figure 22.9 where the current is plotted as a function of the cell potential. The potentials are given a negative sign in this region; according to sign conventions this indicates a nonspontaneous reaction.

The linear relationship between current and the applied potential extends over a considerable current range; within this region the potential required to overcome the resistance of the cell is directly proportional to the current flow. Thus, a potential of —0.08 volt in excess of theoretical would be required to force a current of 0.02 ampere through the cell if its resistance is 4 ohms. Similarly, an excess potential of —0.04 volt would be needed for the passage of 0.01 ampere through this cell; here, a total potential of —0.78 volt would be required.

The portion of the graph below the abcissa describes the current-voltage behavior of the system operating as a galvanic cell. The current now passes in the opposite direction, and the potentials, by convention, are given a positive value. The potential of the cell becomes smaller when current is drawn, the difference from the theoretical again being equal to IR over a considerable range. Thus, the potential of the cell in the preceding paragraph would be +0.66 volt when a current of 0.02 ampere is being drawn from it.

In general, the net effect of IR drop is to increase the potential required to operate an electrolytic cell, and to decrease the output of galvanic cell. By sign convention, therefore, the IR drop is always *subtracted* from the theoretical cell potential; that is,

$$E_{cell} = E_{cathode} + E_{anode} - IR$$

Polarization Effects

The term *polarization* refers to a condition in which cell or electrode potentials exhibit departures, during the passage of current, from values computed on the basis of standard-potential data and the IR drop. Its effects are illustrated in Figure 22.9 where deviations from linearity are observed for large currents. Thus a polarized electrolytic cell requires application of larger potentials than theoretical for a given current flow; similarly, a polarized galvanic cell will produce potentials that are smaller than predicted.

Polarization is an electrode phenomenon; either or both of the electrodes in a cell can be affected. To account for its origin we must examine more closely the nature of electrode processes.

Under some conditions the polarization of a cell may become considerably more extreme than is shown in Figure 22.9. As a matter of fact, the current is frequently observed to become level and essentially independent of the voltage of the cell. Polarization is then said to be complete. Included among the factors affecting the degree of polarization are the size and shape of electrodes, the composition of the electrolyte solution, the agitation of the solution, the temperature, the size of the current, the physical state of the reactants and products of the cell reaction, and the composition of the electrodes.

The polarization of an electrode can arise from any one of several causes. Some of these we understand and are able to describe in quantitative terms. With others, our knowledge is largely empirical. For purposes of discussion polarization phenomena are conveniently classified into the two categories of *concentration polarization* and of *overvoltage*.

Concentration polarization. We shall consider here only those electrode processes that are rapid and reversible. This implies the existence of a thin layer of solution *immediately adjacent to the electrode* that always has the concentration predicted by the Nernst expression. Thus, for a silver electrode in contact with a solution of silver nitrate, we may write

$$E = E^\circ{}_{Ag^+ \to Ag} - 0.059 \log \frac{1}{[Ag^+]}$$

Since the process

$$Ag^+ + e \rightleftharpoons Ag$$

is rapid and reversible, the concentration of silver ion in the film of liquid surrounding the electrode is determined at any instant by the potential of the metal electrode at that instant. If the potential changes, there is an essentially instantaneous alteration of concentration in this film in accordance with the demands of the Nernst equation. This may involve deposition of silver or solution of the electrode.

In contrast to this substantially instantaneous surface process, the rate of attainment of equilibrium between the electrode and the bulk of the solution can be very slow indeed, and depends upon the magnitude of the current passing through the cell.

With these facts in mind consider again the cathodic reaction of the cell whose current-voltage behavior is depicted in the upper half of Figure 22.9; that is,

$$Cd^{2+} + 2e \rightleftharpoons Cd$$

If currents of the indicated magnitudes are to pass, the surface of the cadmium electrode must be replenished with cadmium ions at a rate commensurate with the current demand. As an example we shall calculate this for a current of 0.01 ampere. By definition

$$0.01 \text{ ampere} = 0.01 \text{ coulomb/sec}$$

and

$$0.01 \text{ coulomb/sec} = \frac{0.01}{96,494}$$
$$\simeq 10^{-7} \text{ faraday/sec}$$

Thus, to maintain this current flow it is necessary that 10^{-7} equivalent of cadmium ion be transported to the cathode surface in each second of cell operation. If this rate of mass transfer cannot be met, concentration polarization will set in and lowered currents must result. This type of polarization, then, occurs when the rate of transfer of reactive species from the bulk of the solution to the electrode surface is inadequate.

The transport of material to and from an electrode surface can occur as a result of (a) diffusion forces, (b) electrostatic attractions and repulsions, or (c) mechanical or convection forces. An understanding of the factors that influence the degree of concentration polarization therefore requires that we consider briefly the variables that influence these forces.

Whenever a concentration gradient develops in a solution, the process of diffusion tends to force molecules or ions from the more concentrated to the more dilute regions. The rate at which transfer occurs is proportional to the concentration difference. In the case of an electrolysis, a gradient is set up owing to the removal of ions from the film of solution next to the electrode. Diffusion then occurs, the rate being expressed by the relationship

$$\text{rate of diffusion to the electrode surface} = k\,(C - C_0) \qquad (22\text{-}5)$$

where C is the concentration of the reactant in the bulk of the solution, C_0 is its equilibrium concentration at the electrode surface, and k is a proportionality constant. The value of C_0 is determined by the potential of the electrode and can be calculated from the Nernst equation. As greater potentials are applied to the electrode, C_0 becomes smaller and smaller and the diffusion rate greater and greater.

Electrostatic forces also influence the rate at which an ionic reactant is brought to the electrode surface. Usually the ion and the electrode are of opposite charge, which results in an attractive force between them. This is not always the case, however; for example, iodate, periodate, and dichromate ions can be reduced at a negatively charged cathode. Owing to forces of repulsion, concentration polarization involving these species can be expected to occur sooner than in the reduction of cations. Also the electrostatic attraction (or repulsion) between a particular ionic species and the electrode becomes smaller as the total electrolyte concentration of the solution is increased; it may approach zero when the reactive species makes up but a small fraction of the total concentration of ions of given charge.

Clearly, reactants can be transported to an electrode by mechanical means. Thus stirring or agitation is an aid in preventing concentration polarization; convection currents due to temperature or density differences are also effective.

To summarize, then, concentration polarization occurs when the forces of diffusion, electrostatic attraction, and mechanical mixing are insufficient to transport the reactant to the electrode surface at a rate demanded by the theoretical current flow. This type of polarization is important in several electroanalytical methods. In some instances effort is made to avoid it completely; in others, however, it is essential to the method and every effort is made to promote its occurrence. Experimentally, the degree of concentration polarization can be influenced by (a) the reactant concentration, (b) the total electrolyte concentration, (c) mechanical agitation, and (d) the size of the electrodes; as the area toward which the reactant can be transported becomes greater, polarization effects become smaller.

Overvoltage effects. Sometimes a voltage in excess of theoretical is required to cause the reaction in an electrolytic cell to occur at an appreciable rate even when conditions are such that concentration polarization is not likely. The output voltages of galvanic cells are also occasionally less than theoretical under similar circumstances. This voltage difference is called *overvoltage* and ordinarily results from the slow rate at which the electrochemical oxidation or reduction process occurs at one or both of the electrodes. The phenomenon is observed where additional energy is needed to overcome the energy barrier to the half reaction. In contrast to concentration polarization, the current is controlled to some extent by the rate of the electrode process rather than the rate of mass transfer.

While exceptions can be cited, some empirical generalizations can be made regarding the magnitude of overvoltage. (a) Overvoltage increases with current density (current density is defined as the amperes per square centimeter of electrode surface). (b) It usually decreases with increases in temperature. (c) Overvoltage varies with the chemical composition of the electrode, often being greater for the softer metals such as tin, lead, zinc, and particularly mercury. (d) Overvoltage is often most marked for those electrode processes yielding gaseous products. It frequently is negligible where a metal is being deposited or where an ion is undergoing an alteration of oxidation state. (e) The magnitude of overvoltage in any given instance cannot be specified very exactly because of the number of uncontrollable variables that influence the size of this quantity.

The overvoltage for evolution of hydrogen and oxygen is of particular interest to the chemist. Table 22-2 presents data that will provide a picture of the extent of the phenomenon under some conditions.

The difference in overvoltage of the two gases on smooth and platinized platinum electrodes is of particular interest. This difference is primarily due to the much larger surface area associated with the platinum-black coating on the latter (see p. 375). This results in a much smaller *real* current density than is apparent from the dimensions of the electrode. We see from this why a platinized surface is used in the hydrogen reference electrode; in this application the current density approaches zero as does the overvoltage.

The high overvoltage of hydrogen on several metals is of prime importance to the analytical chemist because it allows him, with proper choice of conditions, to carry out electrolytic reactions without interference from the evolution of hydrogen. For example, it is readily shown from their standard potentials that rapid formation of hydrogen gas should occur well before a potential sufficiently high for deposition of zinc from a neutral solution is reached. In fact, however, a quantitative deposition can be achieved provided a mercury or a copper electrode is used; because of the high overvoltage of hydrogen on these metals, little or no gas is formed during the process.

The magnitude of overvoltage can, at best, be only crudely approximated from empirical information available in the literature. Calculation of cell potentials in which overvoltage plays a part cannot be very accurate. Where it is attempted, the overvoltage is subtracted from the theoretical cell potential.

Table 22-2

OVERVOLTAGE FOR HYDROGEN AND OXYGEN FORMATION AT VARIOUS ELECTRODES AT 25° C [2]

Electrode Com- position	Overvoltage, volts (Current density, 0.001 amp/cm²)		Overvoltage, volts (Current density, 0.01 amp/cm²)		Overvoltage, volts (Current density, 1 amp/cm²)	
	H_2	O_2	H_2	O_2	H_2	O_2
smooth Pt	0.024	0.721	0.068	0.85	0.676	1.49
platinized Pt	0.015	0.348	0.030	0.521	0.048	0.76
Au	0.241	0.673	0.391	0.963	0.798	1.63
Cu	0.479	0.422	0.584	0.580	1.269	0.793
Ni	0.563	0.353	0.747	0.519	1.241	0.853
Hg	0.9*		1.1†		1.1§	
Zn	0.716		0.746		1.229	
Sn	0.856		1.077		1.231	
Pb	0.52		1.090		1.262	
Bi	0.78		1.05		1.23	

* 0.556 volt at 0.000077 amp/cm²; 0.929 volt at 0.00154 amp/cm²
† 1.063 volts at 0.00769 amp/cm²
§ 1.126 volts at 1.153 amp/cm²

[2] National Academy of Sciences, *International Critical Tables of Numerical Data.* **6**, 339-340. New York: McGraw-Hill Book Company, Inc., 1929.

problems

1. From the standard potential for the reaction $Hg_2Cl_2 + 2e \rightleftharpoons 2Hg + 2Cl^-$, calculate the half-cell potential for the reaction in a 0.17 F solution of KCl. Account for the difference between your calculated value and the data given on page 507 for the 0.1 F calomel electrode.

2. An emf of exactly 1.483 volts is applied across a linear resistance that is 100.0 cm in length. What is the potential drop across the shorter section of the resistance when a contact is made at 22.7 cm?   ans. 0.337 volt

3. Three resistances of 25, 50, and 75 ohms, respectively, are placed in series and a potential of 3.0 volts applied across them. What is the potential drop across each?   ans. $E_1 = 0.50$ volt

4. What potential must be applied across a 100.0-cm linear resistance so that each millimeter corresponds to 0.00200 volt?

5. A potentiometer was constructed with a linear resistance that was 60.0 inches in length. When the instrument was balanced with a Weston cell having a potential of 1.018 volt, the contact on the slidewire was found to be at 40.1 inches. With an unknown cell in the circuit, the balance point was at 51.6 inches. What was the potential of the unknown?

6. The galvanometer used for a potentiometer had a sensitivity of 10^{-7} ampere per scale division and a resistance of 100 ohms. Assuming the minimum detectable throw of the galvanometer is 1 scale division, how accurately could one hope to measure the potential of a cell with a resistance of 500 ohms? 5000 ohms? 5,000,000 ohms?

7. The total quantity of current passing through a cell was determined with a silver coulometer. During the current passage 0.1079 gram of silver was deposited. How many coulombs of electricity had passed? ans. 96.5 coulombs

8. An iodine coulometer in which iodide ion is oxidized to I_2 was used to measure a quantity of electricity. The liberated iodine required 14.7 ml of 0.0108 N $Na_2S_2O_3$. Calculate the number of coulombs of electricity.

9. Exactly 34.7 ml (volume corrected for water vapor) of a mixture of H_2 and O_2 were formed by passage of current through a coulometer in which hydrogen was liberated at the cathode and oxygen at the anode. The temperature was 21.0° C and the atmosphere pressure 751.7 mm Hg. How many coulombs of current had passed?

10. A cell with an internal resistance of 10 ohms had a potential of $+$ 1.470 volts when no current was passed. Calculate the potential when 0.010 ampere is drawn. When 0.10 ampere is drawn.

ans. $E = 1.370$ volts; $E = 0.470$ volt

11. Calculate the potential required to pass a 0.01-ampere current through a solution which is 0.100 F in $CuSO_4$ at a pH of 4.0. Assume an internal resistance of 5.0 ohms, an oxygen overvoltage of 0.850 volt, and an oxygen partial pressure of 1 atmosphere.

chapter 23. *Electrogravi-*
metric Methods

Electrolytic precipitation provides a simple method of isolating a number of elements from aqueous solutions. In general, the product consists of the free metal that has been deposited upon a suitable cathode surface. Notable exceptions to this include lead, which is frequently collected upon a platinum anode as lead dioxide, and chloride ion, which may be deposited as silver chloride upon a silver electrode.

The utilization of electrolysis for analytical purposes is not new, having been first suggested as a qualitative tool in the early eighteen hundreds. By the middle of that century quantitative applications had been reported in which the amount of an element was assessed by weighing the electrolytic deposit. Since these early works considerable time and effort has been expended in exploring and improving the art and science of electrolysis. Among the results of these efforts are many useful methods of importance to the modern analytical chemist.

For our purposes most electrogravimetric procedures can be grouped in one of two basic categories. In the first the applied cell potential is continuously adjusted to maintain the potential of the electrode at which deposition occurs at some constant and predetermined value. In the other, no attempt is made to control the potential of the working electrode; instead, the potential applied is simply one that maintains a convenient current flow through the cell. Both techniques will be considered in detail. In addition, we shall discuss briefly a procedure that does not require an external source of power, the metal being deposited at the electrode of a galvanic cell. First, however, we must consider the current-voltage relationships in electrolytic cells under different operating conditions.

CURRENT-VOLTAGE RELATIONSHIP DURING AN ELECTROLYSIS

Decomposition Potential

In a typical analytical electrolysis, a pair of rather large platinum electrodes is immersed in the solution to be analyzed; the potential across these is then increased until a flow of current occurs. The minimum applied potential required for the reaction to begin at an appreciable rate is termed the *decomposition potential*.

We can calculate a theoretical value for the decomposition potential from standard-potential data. For example, consider the electrolysis of a solution that is 0.1 M in cupric ions and 1.0 F in sulfuric acid.

$$Cu^{2+} + 2e \rightleftharpoons Cu \qquad\qquad E° = 0.34 \text{ volt}$$

$$O_2 + 4e + 4H^+ \rightleftharpoons 2H_2O \qquad\qquad E° = 1.23 \text{ volts}$$

Using these we find that the potential for the cell reaction

$$Cu^{2+} + H_2O \rightleftharpoons \tfrac{1}{2} O_2 + 2H^+ + Cu$$

is (+0.31 —1.23) = —0.92 volt.[1] Thus, at applied potentials below —0.92 volt, no current should flow; at higher potentials a current, governed in size by the cell resistance, would be expected.

The dotted line in Figure 23.1 represents the calculated current-voltage curve for this cell. Computation of currents beyond the decomposition potential are based on a cell resistance of 0.5 ohm. The decomposition potential is seen to be the intersection of two straight lines.

In actuality, the current-voltage relationship for this system will more closely resemble the solid curve in Figure 23.1. Here, the overvoltage of oxygen formation on the anode has the effect of displacing the curve to the right. In addition a small current begins to pass as soon as a potential is applied. There are two reasons for this. First, most solutions contain small concentrations

---

[1] In this calculation, we assume that $[Cu^{2+}] = 0.1$, $[O_2] = 1.0$ atmosphere, and $[H^+] = 1.0$. (This is a close approximation since HSO_4^- is a weak acid whose dissociation is strongly repressed by the H^+ present in the solution).

of easily reduced impurities that can diffuse to the electrode surface; dissolved oxygen and ferric iron are two common examples. In addition, part of this initial current flow is due to the reduction of cupric ion itself. This appears to be at odds with what we have said thus far; nothing would indicate that

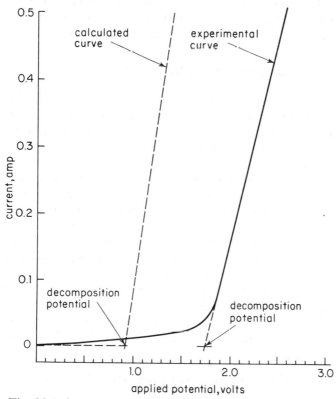

Fig. 23.1 Current-voltage Curve for the Electrolysis of a Cupric Sulfate Solution.

reduction to copper can take place until the theoretical decomposition potential of −0.92 volt has been reached. In calculating this potential, however, the assumption was made that the activity of the metallic copper is constant and equal to one. Thus, for the reduction of copper we wrote

$$E = E° - \frac{0.059}{2} \log \frac{[Cu]}{[Cu^{2+}]} = E° - \frac{0.059}{2} \log \frac{1}{[Cu^{2+}]}$$

This assumption is proper provided the electrode is covered with a metallic copper coating. In the present case, however, the cathode is initially platinum and does not become a copper electrode until deposition has occurred. Experiments have shown that the activity of a metal in a deposit that only partially covers a platinum surface is less than one.[2] Initially, then, the numerator of the logarithmic term in the foregoing equation is infinitely small and approaches one only when a continuous copper film has formed on the platinum surface. Thus, assuming unit activity for copper, a small amount of copper should and does deposit at voltages far more positive than the one calculated. We see from this that the deposition of a metal on a platinum surface does not start suddenly when the decomposition potential is attained; instead, it increases gradually, becoming far more rapid in the region of, and beyond, this potential.

Operation of a Cell at a Fixed Applied Potential

As mentioned previously, one way of performing an analytical electrolysis consists of applying a potential of sufficient magnitude to assure quantitative deposition and to maintain this potential until the reaction is judged complete. The current-voltage relationship during such a process illustrates the limitations of this technique.

To show this, we shall again consider the electrolytic cell described on page 526 in which copper is deposited at the cathode and oxygen is formed at the anode. We have already seen that the theoretical decomposition potential for the cell reaction is —0.92 volt; a considerably larger potential than this is required, however, to cause the reaction to occur at an appreciable rate. Thus if a current of 1.0 ampere is to be employed, an additional 0.5 volt will have to be applied to overcome the ohmic resistance (0.5 ohm) of the cell. Moreover, the high overvoltage of oxygen on platinum must also be taken into account. If the anode area is 100 cm², the current density will be 0.01 ampere/cm². Table 22-2 indicates that an overvoltage of about 0.85 volt would be expected. Thus a reasonable potential to be applied for operation of this cell would be given by the expression

$$E_{cell} = E_{anode} + E_{cathode} - IR - E_{overvoltage} \qquad (23\text{-}1)$$

$$= -0.92 - 0.5 - 0.85$$

$$= -2.3 \text{ volts}$$

We shall now consider the changes that occur as the electrolysis proceeds at this fixed applied potential.

As a consequence of the cell reaction, reduction in the copper concentration and a corresponding increase in the hydrogen ion concentration will take place. This will result in both E_{anode} and $E_{cathode}$ becoming less positive or more negative. For example, by the time the copper concentration is reduced to 10^{-6} molar, we find from the Nernst expression that the theoretical cathode potential has decreased from +0.31 to +0.16 volt; the anode potential will have decreased only slightly to —1.22 volts (the hydrogen ion concentration has changed from 1.0 to 1.2 as a result of the anode reaction). Thus the sum of

---

[2] See L. B. Rogers, et al., *J. Electrochem. Soc.*, **95**, 25, 33, 129 (1949); **98**, 447, 452, 457 (1951).

these is —1.06 volts as compared with the initial value of —0.92 volt. If the cell is operated at a fixed applied potential, the main effect of this change will be to reduce somewhat the *IR* drop and thus the current passing through the cell. Before the electrolysis reaches this point, however, concentration polarization at the cathode will assume significant proportions and will also affect the current passing through the cell.

The causes and effects of concentration polarization were discussed in the previous chapter. In this instance, the phenomenon will appear when the copper concentration is depleted to the point where the supply of cupric ions at the cathode surface is insufficient to meet the demand of a current of 1 ampere. When this occurs, the current must of necessity fall off; this will cause a decrease in both the *IR* drop and the overvoltage. Figure 23.2 illustrates these changes in the cell being considered. The current drops off very rapidly after a few minutes and eventually approaches zero as the electrolysis nears completion.

More important, however, than the changes in current, *IR* drop, and overvoltage are the changes that these induce in the cathode potential. To understand this, we shall again consider equation (23-1) for the over-all cell potential. Here $E_{applied}$ is fixed at —2.3 volts, but at the onset of concentration polarization, $E_{overvoltage}$ and the *IR* drop decrease. Therefore, E_{anode}, $E_{cathode}$, or both, must become more negative. Now, the potential of the anode is stabilized at the equilibrium potential for the oxidation of water because this reactant is present in plentiful supply at the electrode surface. Consequently, it is the cathode potential that changes to more negative values as the *IR* drop and the overvoltage decrease. The effect is shown graphically in Figure 23.3.

We must now consider the possible effects of the sharp change in cathode potential accompanying concentration polarization. First, in an earlier calculation a cathode potential of +0.16 volt versus the standard hydrogen electrode was required for quantitative deposition of copper (that is, for reduction of the copper concentration to 10^{-6} M); from Figure 23.3 it is obvious that a value considerably more negative than this is assured. Given sufficient time, then, complete removal of cupric ion may be expected. Second, there is the possibility of additional electrode reactions taking place at the more negative potentials. Certainly, if other ions which are reduced in the region between 0 to —0.5 volt are also present, codeposition must be expected. Cobaltous ion, with a reduction potential of —0.25 volt, and cadmium ion, which has a standard potential of —0.4 volt, are examples of these. Another possible electrode process at the more negative potential is the formation of hydrogen gas. Under the specified conditions, this process would commence at a cathode potential of about 0 volts were it not for the high overvoltage of hydrogen on the copper-plated cathode. According to Table 22-2, reduction of hydrogen ions may be expected at about —0.5 volt; therefore some gas formation might very well be observed near the end of this electrolysis. This is an undesirable phenomenon.

From this discussion, we should be able to arrive at a clear picture of the limitations of an analytical electrolysis carried out at a fixed applied potential. It is mainly the change in cathode potential toward more negative values that

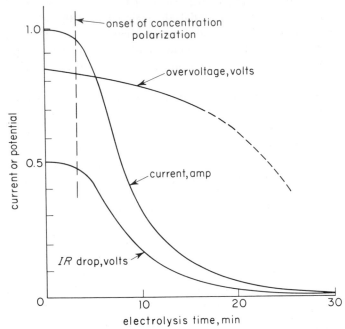

Fig. 23.2 Hypothetical Cell Behavior during Electrolysis. Initial concentrations: $[Cu^{2+}] = 0.100\ M$, $[H^+] = 1.0\ M$. A constant potential of -2.3 volts is applied. The cell has a resistance of 0.5 ohm. Electrode area is $100\ cm^2$.

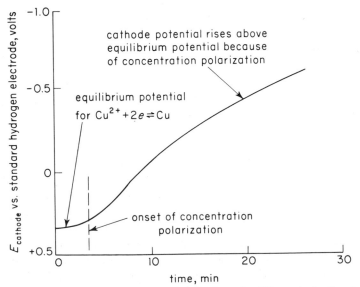

Fig. 23.3 Variation in Cathode Potential during the Electrolytic Precipitation of Copper at a Constant Applied Potential of -2.3 Volts. Data computed from Figure 23.2, assuming an anode potential of -1.2 volts.

prevents the method from being very specific. The magnitude of this change can clearly be minimized by decreasing the initial applied potential; this is done, however, at the expense of a diminution in the initial current and a concomitant increase in the time required for completion of the electrolysis. At best, electrodeposition with a constant applied potential is effective in depositing metals with reduction potentials appreciably less negative than that at which hydrogen evolution occurs. The only feasible separations are from metals that are reduced with difficulty.

Constant-current Electrolysis

An analytical electrodeposition can be carried out by maintaining the current, rather than the applied potential, at a more or less constant level. This requires periodic changes in the applied emf to more negative values as the electrolysis proceeds.

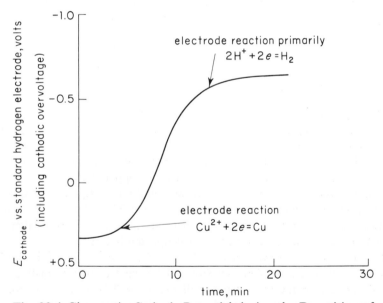

Fig. 23.4 Changes in Cathode Potential during the Deposition of Copper at a Constant Current of 1.0 Ampere.

We have shown in the previous section that the onset of concentration polarization is accompanied by a decrease in current. To offset this effect, the applied potential can be made more negative thus increasing the electrostatic attraction of the cathode for cupric ions. These ions will then be brought to the electrode surface at a greater rate thus maintaining the desired current flow. With continued removal of the ion, however, a point will soon be reached where reduction of cupric ions alone will not hold the current at the desired level.

Further increases in the applied voltage will then result in a rapid change of the cathode potential to a point where codeposition of hydrogen gas occurs. The cathode potential will then be stabilized at a level fixed by the standard potential and the overvoltage for the new electrode reaction; further large increases in the cell potential will no longer be necessary to maintain a constant current. At this point, copper will continue to deposit as cupric ions are brought to the electrode surface; as the bulk of the solution becomes depleted, however, the contribution of this electrode process to the total current will become smaller and smaller. Hydrogen evolution will soon predominate. The changes in cathode potential described above are shown in Figure 23.4.

In electroanalytical work, the vigorous evolution of hydrogen gas results in deposits having poor physical qualities. For this reason, a procedure such as that described above is seldom if ever employed. Instead, the current is permitted to drop in preference to gas formation. Alternatively, the electrolysis is performed in the presence of a substance that is more easily reduced at a lesser potential than hydrogen and whose coreduction does not harm the analytical deposit. In a copper analysis, for example, nitrate ion functions in this fashion, its reduction to ammonium ion taking preference over hydrogen evolution

$$NO_3^- + 10H^+ + 8e \rightleftharpoons NH_4^+ + 3H_2O$$

Constant Cathode-potential Electrolysis

We can easily demonstrate that electrogravimetric methods should be quite selective. For example, the copper concentration of a solution changes from 0.1 formal to 10^{-6} formal as the potential of a platinum cathode immersed in the solution changes from an initial equilibrium value of $+0.31$ volt to $+0.16$ volt. In theory, then, it should be feasible to separate copper from any element that does not precipitate within this 0.15-volt potential range; species that quantitatively precipitate at potentials more positive than $+0.31$ volt could be removed by preliminary deposition while ions depositing at potentials smaller than $+0.16$ volt should not react until copper deposition is complete.

It can similarly be shown that a 0.1-volt difference is required to diminish the concentration of a trivalent ion from 10^{-1} formal to 10^{-6} formal; a singly charged species, on the other hand, will require a 0.3-volt shift in potential to accomplish the same end. Thus, if we are willing to accept a 100-thousandfold lowering of concentration as a quantitative separation, it follows that univalent ions differing in standard potentials by 0.3 volt or greater can, in theory, be separated quantitatively by an electrodeposition method. Correspondingly, 0.2-volt and 0.1-volt differences are required for divalent and trivalent ions.

An approach to these theoretical separation values, within a reasonable electrolysis period, requires a more sophisticated technique than we have so far discussed. This is necessary because concentration polarization will cause a change in the cathode potential which, if unchecked, makes all but the crudest separations quite impossible. The extent of this change is governed by the decrease in *IR* drop; thus, if the applied potential is such that relatively high currents are passed at the outset, the change in cathode potential will be greater than if the initial current is small. Moderately high currents are desirable, however, in order that the electrodeposition be completed in some reasonable length of time. The obvious answer to this dilemma is to vary the applied cell potential, beginning with values sufficiently high to ensure a reasonable current flow; as concentration polarization sets in, the applied potential is then continuously reduced in such a manner as to keep the cathode potential at the level necessary to accomplish the desired separation. Unfortunately, this is not easy to do. It is not feasible to set up a program of applied-potential changes on a theoretical basis because of the uncertainties in overvoltage effects, conductivity changes, etc. Nor does it help to measure the potential across the working electrodes, since this measures only the over-all cell potential. The alternative is to measure the cathode potential against a third electrode whose potential in the solution is known and constant—that is, a reference electrode. The applied potential across the working electrodes can then be adjusted to the level that will give the desired potential at the cathode. This is the principle of a controlled cathode electrolysis.

Experimental details for performing a controlled cathode-potential electrolysis will be presented in a later section. For the present, it is sufficient to say

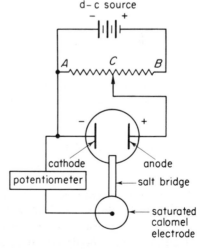

Fig. 23.5 Apparatus for Electrolysis at a Controlled Cathode Potential. Contact *C* is adjusted to maintain the cathode potential at the desired level.

that the potential drop between the reference electrode and the cathode is measured with a potentiometer. A voltage divider is used to control the applied potential across the cathode and the working anode to maintain the cathode potential at a level suitable for the separation. Figure 23.5 shows a diagram of a typical apparatus. Calculation of the approximate cathode potential required for a separation with such an apparatus is illustrated in the following example.

Example. A solution is approximately $0.1 F$ in both zinc and cadmium ions. We shall calculate the cathode potential, with respect to a saturated calomel electrode, that should be used in order to separate these ions by electrodeposition.

From the table of standard potentials we find

$$Zn^{2+} + 2e \rightleftharpoons Zn \qquad\qquad E° = -0.76 \text{ volt}$$
$$Cd^{2+} + 2e \rightleftharpoons Cd \qquad\qquad E° = -0.40 \text{ volt}$$
$$Hg_2Cl_2 + 2e \rightleftharpoons 2Hg + 2Cl^- \text{ (saturated KCl)} \qquad E = +0.24 \text{ volt}$$

The potential necessary for quantitative removal of Cd^{2+} can be obtained if we assume this corresponds to $[Cd^{2+}] = 10^{-6}$. Then

$$E = -0.40 - \frac{0.059}{2} \log \frac{1}{10^{-6}}$$
$$= -0.58 \text{ volt}$$

Precipitation of Zn will begin at potential given by

$$E = -0.76 - \frac{0.059}{2} \log \frac{1}{0.1}$$
$$= -0.79 \text{ volt}$$

Thus, if we maintain the cathode between -0.58 and -0.79 volt with respect to the standard hydrogen electrode, a quantitative separation of cadmium should occur. Suppose we choose a potential of -0.70 volt; against the saturated calomel electrode this will be

$$E_{vs \text{ S.C.E.}} = -0.70 - 0.24 = -0.94 \text{ volt}$$

To maintain this cathode potential, a considerably larger emf will be required across AC in Figure 23.5; its magnitude will depend upon the potential of the anode, the resistance of the solution, and the overvoltage, if any.

An apparatus of the type shown in Figure 23.5 can be operated at relatively high initial applied potentials to give high currents. As the electrolysis progresses, however, the cathode potential will start to change and a lowering of the applied potential across AC will be required. This, in turn, will diminish the current flow. Continued reduction of the applied potential will be required until completion of the electrolysis is indicated by the current approaching zero. These changes in a typical electrolysis are depicted in Figure 23.6. In contrast to the electrolytic methods described earlier, this technique demands constant attention during operation. Unless some provision is made for automatic control, this represents a major disadvantage to the controlled cathode-potential method.

Physical Properties of Electrolytic Precipitates

In the discussion of chemical precipitation we saw that the physical properties of a precipitate in part determine the success or failure of the method. The same holds true for electrogravimetric analysis. Ideally, the deposit should be strongly adherent, dense, and smooth so that the processes of washing, drying, and weighing can be performed without mechanical loss or reaction with the

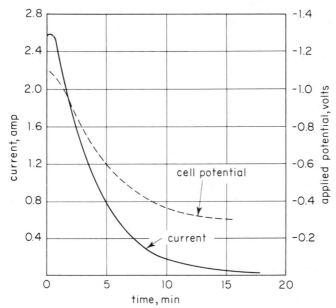

Fig. 23.6 Changes in Applied Potential and Current During the Electrolytic Deposition of Copper at a Cathode Potential of − 0.36 Volt (vs. S. C. E.). (From J. J. Lingane, *Anal. Chim. Acta*, **2**, 589 (1948) with permission.)

atmosphere. Good metallic deposits are fine grained and have a metallic luster; spongy, powdery, or flaky precipitates are likely to be less pure and less adherent.

The principal factors that influence the physical nature of deposits include competing electrode processes, the current density, the temperature, and the presence of complexing agents.

Gas evolution. In general, the formation of a gas during an electrodeposition leads to a spongy and irregular solid. In cathodic reactions, the usual offender is hydrogen, and care must always be taken to keep the cathode potential at a level that will prevent its formation. In some instances the addition of a so-called *depolarizer* is helpful. We have seen that nitrate ion functions as a depolarizer in a copper analysis.

Current density. Electrolytic precipitates resemble chemical precipitates in that their crystal size decreases as the rate of their formation increases—that is, as the current density increases. Here, however, small crystal size is a desirable characteristic; metallic deposits that are smooth, strong, and adherent are constituted from very fine crystals.

While moderately high current densities generally give more satisfactory deposits, extremes should be avoided; these often lead to irregular precipitates of little physical strength that develop as "treelike" structures from a few spots on the electrode. In addition, very high currents also lead to concentration polarization and concomitant gas formation. Ordinarily, a current density between 0.01 and 0.1 ampere/cm² is employed for electroanalytical work.

Stirring. Since good stirring tends to reduce concentration polarization, adequate mixing is desirable in an electrolysis.

Temperature. Although temperature may play an important part in determining the nature of a deposit, predicting its effect is not always possible. On the positive side, temperature increases lead to less concentration polarization by increasing the mobility of the ions and reducing the viscosity of the solvent. At the same time, elevated temperature reduces overvoltage effects; increased gas formation may be observed under these circumstances. Thus, the best temperature for a given electrolysis can only be determined by experiment.

Presence of complexing ions. It is found empirically that many metals form smoother and more adherent films when deposited from solutions in which their ions are present primarily as complexes. Cyanide and ammonia complexes often lead to the best metallic surfaces. The reasons for this are not obvious.

Chemical Factors of Importance in Electrodeposition

The success or failure of an electrolytic determination is often influenced by the chemical environment from which deposition occurs. The *p*H of the medium and the presence of complexing agents deserve particular mention.

Effect of *p*H. Whether or not a given metal can be completely deposited will frequently depend upon the *p*H of the solution. No problem is encountered with such easily reduced species as cupric ion or silver ion; these can be quantitatively removed without interference from quite acidic media. On the other hand, less readily reduced elements cannot be deposited from acid solution owing to the simultaneous evolution of hydrogen; thus, for example, neutral or alkaline media are required for the electrolytic deposition of nickel or cadmium.

Proper *p*H control will sometimes make feasible the quantitative separation of cations. For example, copper is readily separated electrolytically from nickel, cadmium, or zinc in acidic solutions. Even if extreme concentration polarization occurs during the deposition of copper, the resulting change in cathode potential cannot become great enough to cause codeposition of the other metals because hydrogen evolution will occur first. This process will stabilize the cathode potential at a value less negative than that required for the deposition of these metals.

Effect of complexing agents. The deposition of a metal from a solution in which it is present in the form of a complex ion requires a higher applied potential than in the absence of the complexing reagent. The magnitude of this potential shift is readily calculated provided the instability constant for the complex ion is known. The data in Table 23-1 show that these effects are often large and must be taken into account when considering the feasibility of an electrolytic determination or separation. Thus, while copper is readily separated from zinc or cadmium in acidic solution, simultaneous deposition of all three occurs in the presence of appreciable quantities of cyanide ion. The greater potential shifts for silver and copper can be directly attributed to the higher stability of their cyanide complexes.

Table 23-1

EFFECT OF CYANIDE CONCENTRATION ON THE CATHODE POTENTIAL REQUIRED FOR THE DEPOSITION OF CERTAIN METALS FROM 0.1 FORMAL SOLUTIONS

Ion	Calculated Equilibrium Potential		
	No CN⁻ Present	Cyanide Conc. 0.1F	Cyanide Conc. 1 F
Zn^{2+}	—0.79	—1.16	—1.28
Cd^{2+}	—0.43	—0.81	—0.93
Cu^{2+}	+0.31	—0.99	—1.15
Ag^+	+0.74	—0.38	—0.50

Occasionally complexing agents make possible electrolytic separation of ions that would otherwise codeposit. For example, copper in steel can be deposited electrolytically from a solution containing phosphate or fluoride ions. Even though a large amount of iron is present, reduction of ferric ion is prevented by the great stability of its complexes with the anions of the phosphate or fluoride.

Anodic Deposits

Ordinarily, electrogravimetric methods involve deposition of a metallic ion by reduction at the cathode. In a few cases, however, precipitates formed on an anode can also be used for analytical purposes. For example, lead is frequently oxidized to lead dioxide in acid solution

$$Pb^{2+} + 2H_2O \rightleftharpoons PbO_2 + 4H^+ + 2e$$

The physical properties of the deposit make it a suitable weighing form for lead; a gravimetric determination is thus possible. Similarly, cobalt can be formed and weighed as Co_2O_3.

ELECTROLYTIC METHODS WITHOUT CATHODE POTENTIAL CONTROL

Apparatus

A wide variety of equipment is commercially available for conducting an analytical electrolysis. We shall confine our discussion to general aspects only;

the reader should consult manufacturer's instructions for details concerning specific apparatus.

Figure 23.7 shows the components in a typical electrodeposition assembly.

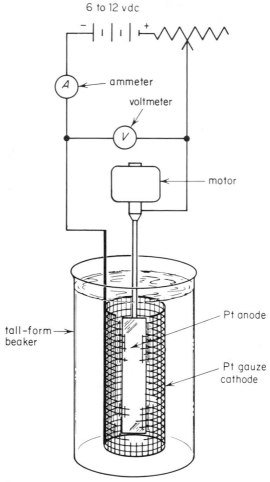

Fig. 23.7 Apparatus for Electrodeposition of Metals.

Electrodes. Electrodes are usually constructed of platinum, although copper, brass, tantalum, and other metals have been used. Platinum electrodes have the advantage of being relatively nonreactive; further than this, they can be ignited to remove any grease, organic matter, or gases which would have a deleterious effect upon the physical properties of the deposit. Certain metals, notably bismuth, zinc, and gallium, cannot be deposited directly onto platinum without causing permanent damage to the electrode. Platinum should always be protected with an electrodeposited coating of copper before undertaking the electrolysis of these metals.

Cathodes are usually formed as gauze cylinders 2 to 3 cm in diameter and perhaps 6 cm in length. This construction minimizes polarization effects by providing a large surface area to which the solution can freely circulate. The anode may also be a gauze cylinder of somewhat smaller diameter so that it can be fitted inside the cathode; alternatively, it may take the form of a heavy wire spiral or a solid paddle.

Electrolytic depositions are ordinarily carried out in tall-form beakers. Efficient stirring is useful in minimizing concentration polarization; this is frequently accomplished by rotating the anode with an electric motor.

Mercury cathode. For certain applications, electrolytic reductions are advantageously performed with a mercury cathode. This is particularly useful in removing certain easily reduced elements as a preliminary step in an analysis. For example, copper, nickel, cobalt, silver, and cadmium are quickly segregated from such ions as aluminum, titanium, phosphates, and the alkali metals, by such a procedure. The precipitated elements dissolve in the mercury; little hydrogen evolution is observed even at high applied potentials because of large

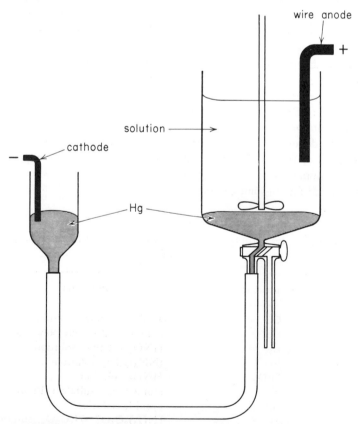

Fig. 23.8 Mercury Cathode for the Electrolytic Removal of Metals from Solution.

overvoltage effects. Ordinarily, no attempt is made to determine the elements deposited in the mercury in this manner, the goal being simply their removal from solution. A cell such as that shown in Figure 23.8 is used.

Electrical apparatus. The apparatus shown in Figure 23.7 is typical of that employed for most electrolytic analyses. The direct-current power supply may consist of a storage battery, a generator, or an alternating-current rectifier. A rheostat is used to control the applied potential; an ammeter and voltmeter are provided to show the approximate current and applied voltage. An entirely adequate electrolysis unit can be assembled from components found in most laboratories; on the other hand, several hundreds of dollars can be spent for more elaborate commercial equipment.

Applications

Without control of the cathode potential, electrolytic methods suffer from a lack of specificity. Despite this limitation, several applications of practical importance make use of this relatively unrefined technique. In general, the species being determined must be the only component in the solution that is more readily reduced than hydrogen ions. Should more than one such element be present, some preliminary treatment is required. The latter frequently involves chemical precipitation of the potential interference, or the addition of some complexing agent that will prevent deposition of the offender without having a serious effect upon the behavior of the element of interest.

This type of electrodeposition is also useful in removing easily reduced ions from solution preliminary to completion of an analysis by some other method. The deposition of interfering heavy metals prior to the quantitative determination of the alkali metals provides an example of this application.

The equipment necessary for these techniques is simple when contrasted to that required for constant cathode-potential methods.

Table 23-2

COMMON ELEMENTS WHICH CAN BE DETERMINED BY ELECTROGRAVIMETRIC METHODS

Ion	Weighed As	Conditions
Cd^{2+}	Cd	alkaline cyanide solution
Co^{2+}	Co	ammoniacal sulfate solution
Cu^{2+}	Cu	HNO_3 -H_2SO_4 solution
Fe^{3+}	Fe	$(NH_4)_2CO_3$ solution
Pb^{2+}	PbO_2	HNO_3 solution
Ni^{2+}	Ni	ammoniacal sulfate solution
Ag^+	Ag	cyanide solution
Sn^{2+}	Sn	$(NH_4)_2C_2O_4^-$ -$H_2C_2O_4$ solution
Zn^{2+}	Zn	ammoniacal or strong NaOH solution

Table 23-2 lists the common elements that can be determined by electro-gravimetric procedures that involve no control over the cathode potential. Specific instructions for two electrolytic analyses are provided in the following paragraphs.

Analysis of copper in a solution. One of the most convenient methods for the determination of copper consists of electrolysis from acidic solutions. Ions that are more readily reduced than copper will interfere and must be removed by preliminary treatment. Metals such as nickel, cobalt, cadmium, and zinc will not deposit provided the acidity is kept sufficiently high.

Ordinarily a mixture of nitric and sulfuric acids is used in the analysis. The presence of the former leads to more satisfactory deposits since it prevents evolution of hydrogen gas (see p. 532). If the nitric acid concentration is too high, however, there is a tendency for the deposition to be incomplete.

Procedure. The solution to be analyzed should be free of chloride ion (Note 1) and contain from 0.2 to 0.3 gram of Cu in 100 ml of water. Transfer the solution to a 150-ml electrolytic beaker, add 2 ml of concentrated H_2SO_4 and 1 ml of freshly boiled and cooled HNO_3.

Prepare the electrodes by immersion in hot $6 N HNO_3$ to which 1 to 2 ml of $3 F KNO_2$ have been added (Note 2). Wash thoroughly with distilled water, rinse several times with small portions of ethyl alcohol or acetone, and dry in an oven at 110° C for 2 to 3 minutes. Cool in a desiccator and weigh the cathode carefully on an analytical balance (Note 3).

Attach the cathode to the negative terminal of the electrolytic apparatus and the anode to the positive one. Elevate the beaker containing the solution so that all but a few millimeters of the cathode is covered. Start the stirring motor and adjust the applied potential so that a current of about 2 amperes passes through the cell (Note 4). When the blue copper color has entirely disappeared from the solution, add sufficient water to raise the level on the cathode by a detectable amount and continue the electrolysis with a current of about 0.5 ampere. If no copper deposit appears on the newly covered portion of the cathode after 15 minutes, the electrolysis is complete. If additional copper deposits, continue testing for completeness of deposition from time to time as directed above.

When no more copper is deposited after a 15-minute period, slowly lower the beaker, while continuously playing a stream of wash water on the electrodes. Rinse the electrodes thoroughly with a fine stream of water. *Do not turn off the applied potential until rinsing is complete* (Note 5). Disconnect the cathode and immerse it in a beaker of distilled water; then rinse with several portions of alcohol or acetone. Dry the cathode in an oven for 2 to 3 minutes at 110° C, cool in a desiccator, and weigh.

NOTES:

1. The presence of much chloride results in attack of the platinum anode; this is not only destructive but will also lead to errors since the dissolved platinum will codeposit with the copper at the cathode.

2. Grease and organic material can be removed from the electrode by heating in a flame to a red heat.

3. Avoid touching the cathode surface with the fingers since grease and oil cause nonadherent deposits.

4. Alternatively, the electrolysis may be carried out without stirring; in this case the current should be kept below 0.5 ampere. Several hours will be required to complete the electrolysis.

5. It is important not to discontinue application of a potential until the electrodes have been removed from the solution and washed free of acid. If this precaution is not observed, solution of the copper may occur.

Separation and determination of copper and nickel in alloys. Copper and nickel are found together in such alloys as Monel and certain coinage metals. Their separation and determination is conveniently accomplished by an electrolytic procedure.

Copper is readily deposited from a nitric-sulfuric acid mixture as described in the preceding section. In such a solution, nickel is not plated out since reduction of nitrate or hydrogen ions will occur preferentially. Upon removal of the copper, nickel can then be deposited by making the solution somewhat basic with ammonia. Before this is done, however, nitrate must be removed by volatilization since it interferes with the quantitative deposition of the nickel.

Metals such as iron and aluminum interfere with the nickel analysis by forming hydrous oxides in the basic medium used for the electrolysis. Unless they are removed by filtration, these precipitates will occlude with the electrolytic deposit and cause high results. Double precipitation of the hydrous oxides is advisable if these form in large amounts; otherwise loss of nickel through adsorption may occur.

Procedure. Weigh into 150- to 200-ml beakers samples that will give 0.1- to 0.3-gram deposits of the two metals; dissolve in a mixture of 10 ml of water, 2 ml of concentrated H_2SO_4, and 2 ml of concentrated HNO_3. Boil the solution, cool, and dilute to about 100 ml.

Deposit the copper as directed in the foregoing procedure; retain the solution and washings for the nickel analysis. After weighing the cathode, keep it in a desiccator without removal of the copper deposit.

Evaporate the solution and washings from the copper electrolysis until fumes of SO_3 are observed. Cool, then carefully add 25 ml of water. Add 1:1 NH_3 until the solution is basic to litmus; if any hydrous oxide precipitates appear, filter through a small paper and collect the filtrate in an electrolysis beaker. Wash with several small portions of water. If much precipitate is formed, redissolve it by pouring a little warm 1:5 H_2SO_4 through the paper and collecting the solution in a fresh beaker. Wash the paper with water and reprecipitate by addition of ammonia to the solution. Filter through the same paper collecting the filtrate in the electrolysis beaker containing the original solution. Wash the precipitate with water. Adjust the volume of the solution to about 100 ml, and add 15 ml of concentrated NH_3. Deposit the nickel on the cathodes used for the copper. Follow the instructions for the electrolysis of the copper solution. Weigh the combined precipitates of copper and nickel.

ELECTROLYTIC METHODS WITH CATHODE-POTENTIAL CONTROL[3]

The full potentialities of electrodeposition as a means of accomplishing separations can be realized only when close control is maintained over the cathode potential. This involves continuous measurement of the cathode potential against a reference electrode and suitable adjustment of the applied cell potential. With close control, elements having reduction potentials differing by only a few tenths of a volt can be quantitatively separated.

Apparatus

The apparatus necessary for a controlled potential electrolysis need not be elaborate. A schematic diagram of the essential equipment is given in Figure 23.5. Measurement of the cathode potential is made against a saturated calomel or silver-silver chloride reference electrode employing a simple potentiometer or vacuum tube voltmeter. An ordinary moving-coil voltmeter is not satisfactory for this purpose because it will draw sufficient current to cause errors in the voltage measurement.

The power supply can be a storage battery or rectifier unit with a well-filtered direct-current output. The voltage divider AB should be of high capacity and have a resistance of no more than 20 or 30 ohms. A tubular-type rheostat, found in most laboratories, is quite suitable. The working electrodes can be similar to those described in the previous section.

Employment of a simple manual device such as this demands the constant attention of the operator. Initially, the applied potential can be quite high (and the currents large). As the electrolysis proceeds, however, continuous reduction of the applied potential is required to maintain a constant cathode potential. During this period the chemist can do little but adjust the equipment. Fortunately, automatic instruments, called *potentiostats*, have been designed to maintain the cathode potential at some constant preset value throughout the electrolysis.[4]

In principle, at least, the conversion of a manual instrument to a potentiostat is not complicated. Consider for example what will occur if the potentiometer in Figure 23.5 is set to the desired emf for the cathode reference-electrode system. Because the two sources of electric energy are in opposition, no current will flow through the potentiometer and reference-electrode circuit as long as the cathode remains at the desired potential. On the other hand, current will

---

[3] This method was first suggested by H. J. S. Sand, *Trans. Chem. Soc.*, **91**, 373 (1907). For many of its applications see H. J. S. Sand, *Electrochemistry and Electrochemical Analysis*. **2**. London: Blackie, 1940. An excellent discussion of applications of modern automatic control to the method can be found in J. J. Lingane, *Electroanalytical Chemistry*, 2d ed., Chaps. 13-16. New York: Interscience Publishers, Inc., 1958.

[4] For details of such instruments see J. J. Lingane, *Electroanalytical Chemistry*, **2d ed.**, 308-339. New York: Interscience Publishers, Inc., 1958.

flow through this circuit whenever the cathode potential differs from that set upon the potentiometer. Since its direction will depend upon which of these potentials is the larger, this current flow may be used as a signal for either increasing or decreasing the applied potential. For example, the current can actuate a switch controlling a reversible motor coupled with the contact of the voltage divider. The contact will then be moved until there is again no current flow in the potentiometer circuit. Alternatively, any of several electronic circuits can be used to achieve this control. Several potentiostats are available commercially.

Applications

The controlled cathode-potential method is a potent tool for the direct analysis of solutions containing a mixture of the metallic elements. For example, Lingane and Jones[5] developed a method for the successive determination of copper, bismuth, lead, and tin. The first three can be plated out from a nearly neutral tartrate solution. Copper is first reduced quantitatively by maintaining the cathode potential at −0.2 volt with respect to a saturated calomel electrode. After weighing, the cathode containing the copper is returned to the solution and the bismuth removed at a potential of −0.4 volt. Following this, lead is plated out quantitatively by increasing the cathode potential to −0.6 volt. Throughout these depositions, the tin is retained in solution by virtue of its very stable tartrate complex. Acidification of the solution after deposition of the lead is sufficient to decompose the complex, converting tartrate ion to the undissociated acid; tin can then be readily deposited at a potential of −0.65 volt. This method can be extended to include zinc and cadmium also. Here, the solution is made ammoniacal after removal of the copper, bismuth, and lead. Cadmium and zinc are then successively plated out and weighed. Finally, the tin is determined after acidification as before.

A procedure such as this is particularly attractive for use with a potentiostat because of the small operator time required for the complete analysis.

Table 23-3 indicates other separations that have been performed by the controlled-cathode method.

Spontaneous Electrogravimetric Analysis (Internal Electrolysis)

An electrogravimetric analysis can occasionally be carried out within a short-circuited galvanic cell. Under these circumstances no external power is required; the deposition takes place as a consequence of the energetics of the cell reaction. For example, the cupric ions in a solution will quantitatively deposit upon a platinum cathode when external contact is made with a zinc anode

[5] J. J. Lingane and S. L. Jones, *Anal. Chem.*, **23**, 1798 (1951).

Table 23-3

SOME APPLICATIONS OF CONTROLLED CATHODE-POTENTIAL ELECTROLYSIS

Element Determined	Other Elements that May Be Present
Ag	Cu and base metals
Cu	Bi, Sb, Pb, Sn, Ni, Cd, Zn
Bi	Cu, Pb, Zn, Sb, Cd, Sn
Sb	Pb, Sn
Sn	Cd, Zn, Mn, Fe
Pb	Cd, Sn, Ni, Zn, Mn, Al, Fe
Cd	Zn
Ni	Zn, Al, Fe

immersed in a solution of zinc ions. The cell reaction may be represented as

$$Zn + Cu^{2+} \rightleftharpoons Zn^{2+} + Cu$$

If allowed to proceed to equilibrium, substantially all of the cupric ions are removed from the solution.

This technique is termed *internal electrolysis* or, perhaps more aptly, *spontaneous electrolysis*. Aside from simplicity insofar as equipment is concerned, it enjoys the advantage of being somewhat more selective than an ordinary electrolysis without cathode-potential control; with a suitable choice of anode system the codeposition of many elements can be positively eliminated. Thus, for example, employment of a lead anode for the deposition of copper will prevent interference from all species with more negative potentials than the lead ion-lead couple.

Apparatus

Figure 23.9 illustrates a typical arrangement for the spontaneous electrolytic determination of copper. The element is deposited on a weighed platinum gauze cathode. An electric stirrer is employed to circulate the solution around the cathode.

The anode is a piece of zinc metal immersed in a zinc sulfate solution. This must be isolated from the solution being analyzed to avoid direct reaction between the zinc and cupric ions; if this is not done, some of the copper will deposit directly on the zinc. The anode may be conveniently isolated with a paper, or porous ceramic cup. A solution of zinc sulfate, or some other electrolyte, is placed in the cup.

Electrolysis is initiated by connecting the platinum and zinc electrodes by means of a wire and is continued until removal of the copper is judged complete.

A constant source of concern in a spontaneous electrolysis is the internal resistance of the cell since this controls the rate at which deposition occurs.

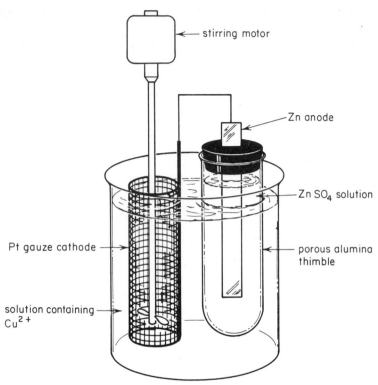

Fig. 23.9 Apparatus for the Spontaneous Electrogravimetric Determination of Copper.

Table 23-4

APPLICATION OF THE INTERNAL-ELECTROLYSIS METHOD

Element Determined	Anode	Noninterfering Elements
Ag	Cu, $CuSO_4$	Cu, Fe, Ni, Zn
Cu	Zn, $ZnCl_2$	Ni, Zn
Bi	Mg, $MgCl_2$	
Pb	Zn, $ZnCl_2$	Zn
Ni	Mg, $MgSO_4$	
Co	Mg, NH_4Cl, HCl	
Cd	Zn, $ZnCl_2$	Zn
Zn	Mg, NH_4Cl, HCl	

If the resistance becomes very high, inordinate lengths of time are required for completion of the reaction. This is in contrast to an ordinary electrolysis where the effects of a high cell resistance can be readily overcome by increasing the applied potential. In this instance large currents can be obtained only if R is kept small. Therefore, the apparatus for this method must always be designed with the view of minimizing the ohmic drop; this involves large electrodes, good stirring, and reasonably high concentrations of electrolyte. Under ideal conditions, depositions can be completed in less than an hour by this method; often, however, several hours are required. This may not be a serious handicap since the cell requires no attention during operation.

Applications

Table 23-4 lists some of the applications of the internal-electrolysis method.

problems

1. How much time is required to deposit 0.375 gram of cobalt as the metal from a solution of Co^{2+} with a constant current of 0.400 ampere? How much time is required to deposit the same amount of cobalt as Co_2O_3 at an anode assuming the same current?

2. In the electrolytic deposition of the copper from a 0.0500 F solution of cupric ion, oxygen was liberated at the anode. If the solution was initially at a pH of 2.0, what was the pH after completion of the deposition?

3. Calculate the applied potential necessary to pass a current of 0.75 ampere through a cell in which the anode reaction involves evolution of oxygen at 0.2 atmosphere and the cathode reaction is the deposition of bismuth metal ($BiO^+ + 2H^+ + 3e \rightleftharpoons Bi + H_2O$). The solution is 0.100 M in BiO^+ and 0.200 M in H^+. The cell resistance is 0.6 ohm and the oxygen overvoltage 0.7 volt.

4. A solution was 0.0500 M in Cd^{2+} and 0.060 M in Co^{2+}. What would be the equilibrium concentration of the more easily reduced ion when the cathode potential had reached a value where the second just began to deposit?
   ans. $[Co^{2+}] = 2.8 \times 10^{-6}$

5. A solution is 0.100 M in Pb^{2+} and Cu^{2+} and the two ions are to be separated by a controlled cathode-potential process. What range of cathode potentials (vs. the saturated calomel electrode) can be used to effect the separation?
   ans. —0.08 to —0.40 volt

6. Calculate the range of cathode potentials (vs. the saturated calomel electrode) that would be used to separate Ni^{2+} from Cd^{2+} in a solution 0.100 M in each species.

7. An internal electrolytic analysis of the silver in a 0.100 F solution of Ag^+ was undertaken with an anode consisting of a zinc electrode immersed in a 0.100 F solution of $ZnSO_4$. What would be the emf of the cell? Assuming the cell

resistance is 10 ohms, what is the maximum current at the beginning of the analysis?   ans. 1.53 volts; 0.153 ampere

8. If the zinc anode in Problem 7 were replaced by an anode made up of metallic copper in a 1.0 F CuSO$_4$ solution, what would be the emf of the cell and the initial current, again assuming a 10 ohms resistance?

9. Calculate the initial potential of an internal electrolysis cell for the analysis of copper in a 0.0100 F solution of CuSO$_4$. A lead electrode immersed in a 1.00 F solution of Pb^{2+} served as the anode. What would be the potential when the Cu^{2+} concentration had been reduced to 1×10^{-5} M if we assume that the concentration of Pb^{2+} did not change appreciably?

chapter 24. *Potentiometric* *Titrations*

We have seen that the potential of a metallic conductor immersed in an electrolyte solution may be sensitive to the concentration of one or more of the components of that solution. An obvious analytical application of this phenomenon is its employment in the measurement of concentration. This is done in either of two ways. The first involves the comparison of the potential developed by a sample with that of standards of known concentration. In a second category the progress of a titration is followed by measurement of the potential of an electrode immersed in the solution being titrated. This electrode must clearly be sensitive to the concentration of one of the participants of the analytical reaction; when this condition exists, an end point can be located by means of the behavior of the electrode.

Procedures of the first sort are called *direct potentiometric methods*; those of the second are termed *potentiometric titrations*. With one important exception, direct potentiometry is not widely employed in quantitative analysis because the technique requires highly precise potential measurements and careful control of variables. In addition, such procedures are often not very selective. Potentiometric titrations, on the other hand, are less demanding insofar as control of experimental conditions and accuracy of potential measurements are concerned; they also tend to be less subject to interference by other components in the solution being analyzed.

The potentiometric end point has been applied to all types of chemical reactions. It can be used with colored or opaque solutions that mask ordinary indicator changes. It is less subjective than indicator methods and inherently more accurate. Finally, the data from a potentiometric titration may reveal the presence of hitherto unsuspected species in the solution that are competing with the unknown for the volumetric reagent. At the same time, however, a potentiometric titration is likely to be more time consuming than the typical volumetric procedure; the necessity for special equipment is another disadvantage.

In this chapter we shall be concerned primarily with potentiometric titrations. However, we must discuss in detail the determination of pH, which represents the one important application of direct potentiometry.

METHODOLOGY

In a potentiometric titration a galvanic cell is formed by immersion of a pair of suitable electrodes in the solution to be analyzed. The emf of this cell is then followed as a function of the volume of reagent added; the rapid changes in potential occurring in the region of the equivalence point serve to indicate completion of the titration.

Apparatus

In addition to the usual equipment needed for a volumetric analysis, a potentiometric titration requires a device for measuring potential, a reference electrode, and an indicator electrode.

Potential measurements. A potentiometric titration requires a device that will measure potentials in the range of 0 to 1.5 volts with an accuracy of at least 0.01 volt and preferably 0.001 to 0.002 volt. The measurement should be accompanied by the passage of essentially no current. The potentiometers shown in Figures 22.6 and 22.7, pages 511 and 513, are entirely adequate for this purpose. A number of instruments embodying the potentiometer principle are available commercially.

A high-resistance vacuum tube voltmeter can also be employed for potentiometric titrations and is convenient because it gives a direct and continuous measure of the cell potential.

Reference electrodes. The most essential property of the reference electrode is that its potential remain constant throughout the course of a single titration. In most applications, this potential need not be known exactly, nor must it necessarily remain constant with storage.

Calomel electrodes and silver-silver chloride electrodes are commonly used as reference electrodes for potentiometric titrations (see pp. 506 and 508). Convenient types are shown in Figures 22.2 and 22.3.

Indicator electrodes. The potential of the indicator electrode must be directly related to the concentration of one or more of the participants or products of the analytical reaction. Then, if the reference electrode output is constant, the changes in the measured emf of the galvanic cell will reflect the concentration changes taking place in the solution as the titration proceeds. Rapid response of the indicator electrode to changes in concentration is a convenience; clearly, its potential behavior should be reproducible within reasonable limits.

Many indicator electrodes are available to the chemist. Where the volumetric process involves the formation of a precipitate or a stable complex, the best electrode is often the elemental form of the cationic participant of the reaction. Thus, for example, a silver electrode can be employed in the titration of silver ion with chloride, bromide, or cyanide since its potential is sensitive to the cationic concentration. Copper, lead, cadmium, and mercury are also satisfactory indicator electrodes for their ions. Some of the harder and more brittle metals, such as iron, nickel, cobalt, tungsten, and chromium, are not; these tend to develop nonreproducible potentials that are due in part to strains or crystal deformations in their structures.

Metal electrodes can also serve as indicator electrodes for those anions that form slightly soluble precipitates with the cation of the metal. Thus the potential of a silver electrode reflects the concentration of chloride ion in a saturated solution of silver chloride; here we may describe the electrode behavior in the following terms:

$$E = E^{\circ}_{AgCl \rightarrow Ag + Cl^-} - 0.059 \log [Cl^-]$$

Such an electrode, then, can serve as the indicator for any volumetric reaction in which chloride ion is a participant. It is an example of an *electrode of the second order* because it measures the concentration of an ion not directly involved in the electron-transfer process. In contrast, the potential of an *electrode of the first order* is a function of the concentration of a direct participant in the electrode process.

For oxidation-reduction reactions, the indicator electrode is most commonly an inert metal such as platinum or gold. We have seen that the potential developed in this instance is a function of the concentration ratio between the oxidized and reduced forms of one or more of the species in the solution.

For a neutralization titration the hydrogen gas electrode may be used; other indicator electrodes for hydrogen ions are also available, and these will be discussed in detail in later sections.

Metal indicator electrodes take a variety of physical forms, often being constructed as coils of wire, as flat metal plates, or as heavy cylindrical billets. Generally, exposing a large surface area to the solution will assure rapid attainment of equilibrium. Thorough cleaning of the metal surface before use is often important; a brief dip in concentrated nitric acid followed by several rinsings with distilled water is satisfactory for many metals.

Figure 24.1 shows a typical apparatus for carrying out a potentiometric titration.

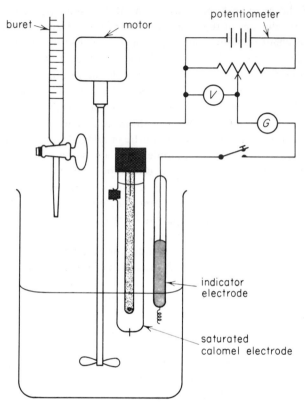

Fig. 24.1 Apparatus for a Potentiometric Titration.

The Titration

A typical potentiometric titration involves measuring and recording the cell potential after each addition of reagent. Initially the standard solution is added in large increments; as the end point is approached (as indicated by larger potential changes per addition) the increments are reduced in size. For some purposes it is convenient to make small and *equal* additions near the equivalence point. Ordinarily the titration is carried well beyond the end point.

Sufficient time must be allowed for the attainment of equilibrium after each addition of reagent; precipitation titrations may require several minutes for this, particularly in the vicinity of the equivalence point. A close approach to equilibrium is indicated when the measured potential ceases to drift by more than a few millivolts. Good stirring is frequently effective in hastening this process.

End-point Determination

The first two columns of Table 24-1 consist of a typical set of potentiometric-titration data. These have been plotted in Figure 24.2. This experimentally obtained graph closely resembles the titration curves derived from theoretical considerations. This is not surprising when we recall that the measured potential is a logarithmic function of the concentration of the ion to which the electrode responds, and that we employed an analogous function in producing the theoretical curves.

Table 24-1

POTENTIOMETRIC TITRATION DATA FOR 2.433 MILLIEQUIVALENTS OF CHLORIDE
WITH 0.1000 N SILVER NITRATE

Vol AgNO$_3$, ml	E vs. S.C.E., volt	$\Delta E/\Delta V$, volt/ml	$\Delta E^2/\Delta V^2$
5.0	0.062		
		0.002	
15.0	0.085		
		0.004	
20.0	0.107		
		0.008	
22.0	0.123		
		0.015	
23.0	0.138		
		0.016	
23.50	0.146		
		0.050	
23.80	0.161		
		0.065	
24.00	0.174		
		0.09	
24.10	0.183		
		0.11	
24.20	0.194		0.28
		0.39	
24.30	0.233		0.44
		0.83	
24.40	0.316		−0.59
		0.24	
24.50	0.340		−0.13
		0.11	
24.60	0.351		−0.04
		0.07	
24.70	0.358		
		0.050	
25.00	0.373		
		0.024	
25.5	0.385		
		0.022	
26.0	0.396		
		0.015	
28.0	0.426		

Graphical methods. Determination of the end point for a potentiometric titration may be accomplished in any of several ways. The most straightforward is a graphical method in which the midpoint in the steeply rising portion of the titration curve is estimated visually. Various mechanical methods have been suggested for evaluation of this point; it seems doubtful, however, that the accuracy is greatly improved by these.

Another graphical approach involves a plot of the change in potential per unit change in volume of reagent ($\Delta E/\Delta$ ml) as a function of the average volume of reagent added. Data for this have been computed from columns 1 and 2 of Table 24-1 and are recorded in column 3; these are plotted in Figure 24.3. The end point is taken as the maximum in the curve and is obtained by extrapolation of the experimental points. Because of the uncertainty in the extrapolation procedure, however, it is questionable whether much is gained in terms of accuracy by such an approach despite the appearance of a spectacular change in the parameter plotted; such a plot is certainly more trouble to obtain

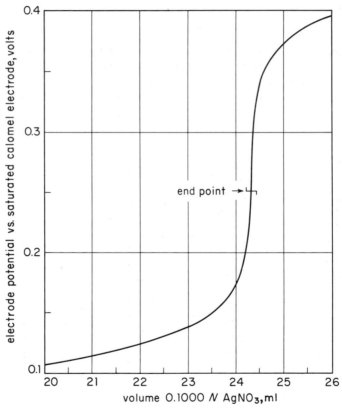

Fig. 24.2 Potentiometric Titration Curve in the Vicinity of the Equivalence Point. Titration of 2.433 milliequivalents of Cl^- with 0.1000 N $AgNO_3$ solution.

With both of these techniques the assumption is made that the titration curve is symmetric about the true equivalence point, and that the inflection in the curve therefore corresponds to that point. This assumption is perfectly valid provided the participants in the chemical process react with one another in a one-to-one molar ratio, and provided also that the electrode process is perfectly reversible. Where these provisions are not met, an appreciable titration error can result from graphical evaluation of the end point.

An obvious example of asymmetry in a titration curve is shown in Figure

18.1. Here the reaction involves 5 ferrous ions for each permanganate ion, and the equivalence point is well above the center of the steep portion of the curve. The maximum in the derivative curve for this titration also occurs before the

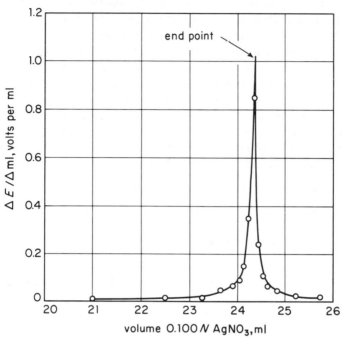

Fig. 24.3 Derivative Curve for the Titration Illustrated in Figure 24.2.

equivalence point is reached. An end point obtained by either graphical method, then, indicates a volume of permanganate that is less than the theoretical. Numerous other reactions with nonsymmetrical titration curves can be cited. For example,

$$Ni^{2+} + 4CN^- \rightleftharpoons Ni(CN)_4^{2-}$$

$$2Ce^{4+} + H_2AsO_3^- + H_2O \rightleftharpoons 2Ce^{3+} + H_2AsO_4^- + 2H^+$$

$$2Ag^+ + C_2O_4^{2-} \rightleftharpoons Ag_2C_2O_4$$

Ordinarily, the titration error arising from this source is quite small and can be neglected. Only when unusual accuracy is desired, or where very small quantities are being determined, must consideration be given to the effect of this source of error upon an analysis. In such cases a correction can be applied. This can be determined empirically by titration of a standard; when the error is due to a nonsymmetrical reaction, it can be also calculated from theoretical considerations.[1]

[1] See I. M. Kolthoff and N. H. Furman, *Potentiometric Titrations*, 2d ed. Chaps. 2, 3. New York: John Wiley and Sons, Inc., 1931.

Analytical determination of the end point. A less time-consuming method of determining the end point is to calculate the values for $\Delta E/\Delta$ ml as the titration proceeds and record these data as shown in column 3 of Table 24-1. The assumption is then made that the function is a maximum at the equivalence point. Thus in the data shown, the end point would clearly be between 24.3 and 24.4 ml; a value of 24.35 would be adequate for many purposes. Lingane[2] has shown that the volume can be fixed more exactly by estimating the point where the second derivative of the voltage with respect to volume (that is, $\Delta E^2/\Delta V^2$) becomes zero. This is easily done if equal increments of solution are added in the vicinity of the equivalence point. In the fourth column of Table 24-1 are the values for $\Delta E^2/\Delta V^2$ that have been calculated by subtracting the corresponding data for $\Delta E/\Delta V$. The function must become zero at some point between the two volumes embracing a change in sign. The volume corresponding to this point can be obtained by interpolation; that is,

at 24.30 $\hspace{4em} \Delta E^2/\Delta$ ml$^2 = +0.44$

at 24.40 $\hspace{4em} \Delta E^2/\Delta$ ml$^2 = -0.59$

$$\therefore \text{End-point volume} = 24.30 + 0.1 \times \frac{0.44}{0.44 + 0.59}$$

$$= 24.34 \text{ ml}$$

It should be apparent that this method is also based upon the same assumption as the graphical methods.

Titration to a fixed potential. Another procedure for carrying out a potentiometric titration consists of adding reagent until the cell potential reaches some predetermined end-point value. This may be the theoretical equivalence point potential calculated from formal potentials or an empirical potential obtained by the titration of standards. Such a method demands that the equivalence-point behavior of the system be entirely reproducible.

PRECIPITATION AND COMPLEX-FORMATION TITRATIONS

Electrode Systems

For a precipitation titration the indicator electrode usually consists of the metal from which the reacting cation was derived. Occasionally, however, an electrode system directly responsive to the anion is employed. Thus, for example, the titration of zinc ion with ferrocyanide may be followed with a platinum electrode, provided a quantity of ferricyanide ion is also present. The potential developed by this electrode is determined by the half reaction

$$\text{Fe(CN)}_6^{3-} + e \rightleftharpoons \text{Fe(CN)}_6^{4-}$$

[2] J. J. Lingane, *Electroanalytical Chemistry*, 2d ed., 93. New York: Interscience Publishers, Inc., 1958.

With the first excess of reagent, the ferrocyanide concentration increases rapidly; this results in a corresponding alteration in the potential of the electrode.

When the anion involved in the precipitation or complex formation reaction is readily hydrolyzed, an indicator electrode sensitive to pH changes can be used (we shall discuss pH electrodes somewhat later in this chapter). An example arises in the titration of calcium with ethylenediaminetetraacetate which was discussed in Chapter 13. The analytical reaction is

$$Ca^{2+} + X^{4-} \rightleftharpoons CaX^{2-}$$

The ethylenediaminetetraacetate ion (X^{4-}), however, is extensively hydrolyzed

$$X^{4-} + H_2O \rightleftharpoons H_2X^{2-} + 2OH^-$$

Consequently, near the equivalence point in the titration there is a sharp rise in hydroxyl ion concentration corresponding to the first excess of X^{4-} ions; this can easily be detected by an acid-sensitive indicator electrode.

Titration Curves

Theoretical curves for a potentiometric titration are often easy to derive. For example, we can readily convert the values shown in Table 12-1 (p. 250) for titration of chloride ion with silver nitrate into potentiometric data for a silver electrode. For this we need the standard potential for the half reaction

$$AgCl + e \rightleftharpoons Ag + Cl^- \qquad E° = 0.222 \text{ volt}$$

and we may describe the potential of the silver electrode during the titration as

$$E_{Ag} = +0.222 - 0.059 \log [Cl^-]$$

From the definition of pCl we may therefore write

$$E = +0.222 + 0.059 \, pCl$$

For any point in the titration, this relationship will allow us to calculate the theoretical potential of the silver electrode *against* the *standard hydrogen electrode* as reference. To obtain the *cell potential* where a saturated calomel electrode is used as a reference, we need to know the potential for that electrode

$$Hg_2Cl_2 + 2e \rightleftharpoons 2Hg + 2Cl^-(\text{sat'd KCl}) \qquad E_{S.C.E.} = +0.242 \text{ volt}$$

Reversing the sign on the potential of the calomel electrode in order to combine this with the silver electrode potential gives

$$E_{cell} = -0.242 + 0.222 + 0.059 \, pCl$$
$$= -0.020 + 0.059 \, pCl$$

With the aid of this equation a theoretical potentiometric titration curve could be computed from pCl data; its shape would be similar to that shown in Figure 24.2.

Effect of adsorption. Experimental curves for precipitation titrations usually correspond closely to the shape predicted from theory. Such discrepancies as are observed can be traced in part to adsorption of ions by the precipitate. This effect can be seen in the titration of iodide ion with silver nitrate. Before the equivalence point, silver iodide strongly adsorbs iodide ions thus reducing their concentration in the solution. This has a noticeable effect on the potential of a silver electrode in the solution particularly near the end point. A similar phenomenon is noted when the equivalence point has been passed; under these conditions adsorption of some of the excess silver ions occurs and results in a lower concentration of these ions in the solution. This in turn causes the experimental potential to differ somewhat from the calculated one. Adsorption phenomena unfortunately tend to produce less sharply defined end points, as may be seen from Figure 24.4. In most instances, however, the extent of adsorption is low enough so that serious interference does not occur.

Titration curves for mixtures. Separate end points are often revealed by a potentiometric titration when more than one reacting species is present. An important application of this arises in the analysis of halide mixtures by titration with silver nitrate. To illustrate the derivation of a theoretical curve for such a system, we shall consider the titration of 25.0 ml of a solution that is 0.100 F in both iodide and chloride ions with a standard 0.100 F silver nitrate solution. From the table of solubility products we find that

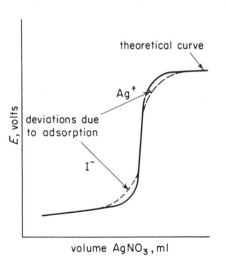

theoretical curve

Ag$^+$

deviations due to adsorption

I$^-$

E, volts

volume AgNO$_3$, ml

Fig. 24.4 Effect of Adsorption upon the Curve for the Potentiometric Titration of Iodide Ion.

$$[Ag^+][I^-] = 8.3 \times 10^{-17}$$

$$[Ag^+][Cl^-] = 1.8 \times 10^{-10}$$

and conclude that the first additions of the reagent will result in precipitation of silver iodide in preference to silver chloride. It is of interest to determine the extent to which this reaction occurs before appreciable chloride precipitation takes place. With the first appearance of silver chloride, both of the foregoing solubility product relations apply; thus, dividing the one by the other we get

$$\frac{[I^-]}{[Cl^-]} = \frac{8.3 \times 10^{-17}}{1.8 \times 10^{-10}} = 4.6 \times 10^{-7}$$

After the first silver chloride forms, this ratio is maintained throughout the remainder of titration. The magnitude of the ratio suggests that nearly all the iodide will have precipitated before the appearance of any of the chloride. Thus, the chloride ion concentration, because of dilution, will be very close to half of its original concentration at the outset of its precipitation. With this assumption we can determine the iodide concentration when the chloride precipitate just appears; that is,

$$\frac{[I^-]}{0.1/2} \cong 4.6 \times 10^{-7}$$

$$[I^-] \cong 2.3 \times 10^{-8} \text{ mol/liter}$$

We can then calculate the percentage iodide unprecipitated at this point in the titration.

$$\text{original number meq } I^- = 25 \times 0.1 = 2.5$$

$$\text{number of meq } I^- \text{ at onset of Cl}^- \text{ precipitate} \cong 50 \times 2.3 \times 10^{-8} = 1.2 \times 10^{-6}$$

$$\text{percent } I^- \text{ unprecipitated} = \frac{1.2 \times 10^{-6}}{2.5} \times 100 = 0.00005$$

These calculations predict that a rather satisfactory separation of iodide from chloride should be possible. Furthermore until within 0.00005 percent of the first end point, the titration curve should be that of a simple iodide solution; the necessary data for plotting the theoretical curve may be calculated on this basis.

Example. We shall calculate the potential of the silver electrode against a saturated calomel electrode after the addition of 5 ml of 0.100 N AgNO$_3$ to the mixture under consideration.

$$[I^-] = \frac{25.00 \times 0.1 - 5.0 \times 0.1}{30.0} = \frac{1}{15}$$

We find in the table of standard potentials

$$AgI + e \rightleftharpoons Ag + I^- \qquad\qquad E^\circ = -0.151 \text{ volt}$$

Thus

$$E = -0.151 - 0.059 \log [I^-]$$
$$= -0.151 - 0.059 \log \frac{1}{15}$$
$$= -0.081 \text{ volt}$$

Combining this with the potential for the saturated calomel electrode as an anode, we obtain

$$E_{\text{cell}} = -0.081 - 0.242$$
$$= -0.323 \text{ volt}$$

The first half of the curve (solid line) shown in Figure 24.5 was obtained from data calculated in this manner. As expected, the curve rises steeply in the vicinity of the iodide equivalence point. This increase in potential, however, is terminated abruptly at a point corresponding to the first formation of silver chloride. This occurs slightly before the iodide equivalence point.

Once precipitation of chloride occurs, the potential can be most conveniently calculated by means of the standard potential for the reaction

$$AgCl + e \rightleftharpoons Ag + Cl^- \qquad\qquad E^\circ = 0.222 \text{ volt}$$

The remainder of the titration curve is thus essentially identical to that for chloride by itself; points defining this portion are calculated as described previously.

Figure 24.5 illustrates that the theoretical titration curve for the mixture

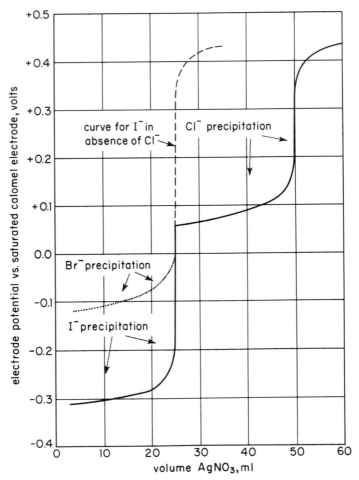

Fig. 24.5 Potentiometric Titration of Halide Mixtures. Solid curve represents titration of 25.0 ml of solution that is 0.100 F with respect to I^- and to Cl^-. Broken curve indicates the course of the titration had sample contained no Cl^-. Dotted curve represents titration of 25.0 ml of a solution that is 0.100 F with respect to Br^- and to Cl^-.

is a synthesis of the two curves for the individual ions, and that marked potential changes are associated with each equivalence point. It is to be expected that the first equivalence point in the titration of a mixture will become less well defined as the solubilities of the two precipitates approach one another. This is shown by the dotted line in Figure 24.5 which represents the theoretical curve for the titration of bromide ion in the presence of chloride. Even here an end point can be distinguished rather readily. As a matter of fact, the solubility of the three silver halides differ sufficiently for a mixture of the three to give a titration curve with three distinct end points.

The obvious conclusion from Figure 24.5 is that the potentiometric method should make possible the analysis of the individual components in halide mixtures. These curves, of course, are theoretical and subject to experimental verification. When this is done, the sharp discontinuities shown in the theoretical curves are not observed; instead, the experimental graphs are curved in these regions. More important, the volume of silver nitrate required to reach the first end point is generally somewhat greater than theoretical while that for the second tends to be low; the total volume, however, approaches the theoretical amount. These observations apply to the titration of chloride-bromide, chloride-iodide, and bromide-iodide mixtures and can be explained by assuming that coprecipitation of the more soluble silver halide occurs during the formation of the less soluble compound. This probably takes the form of a solid solution, and would account for the overconsumption of reagent in the first part of the titration and a corresponding underconsumption during the second.

Despite the coprecipitation error, the potentiometric method is useful for the analysis of halide mixtures. When approximately equal quantities are present, the relative errors can be kept to about 1 to 2 percent.[3]

Neutralization Titrations and the Measurement of pH

The hydrogen ion concentration of aqueous solutions can vary over a tremendous range; yet relatively small changes in this quantity may greatly alter the chemical behavior and properties of a solvent. For example, a tenfold change in hydrogen ion concentration may have a hundredfold or even a thousandfold effect on the solubility of a compound; indeed, there are instances where such a change brings about a millionfold alteration in the concentration of a participant in an oxidation-reduction equilibrium. As a consequence, a knowledge of hydrogen ion concentration is of vital importance to the chemist, and its measurement is one of the most important quantitative analytical processes.

There are actually two measurable quantities of interest relating to the acidity or basicity of a solution: first, the equilibrium hydrogen ion concentration (or hydroxyl ion concentration); and second, the total acidity or basicity—

[3] For further details see I. M. Kolthoff and N. H. Furman, *Potentiometric Titrations*, 2d ed., 154-158. New York: John Wiley and Sons, Inc., 1931.

that is, the concentration of acid or base available for chemical reaction. For a weak acid or base these are quite different quantities, related to each other through the dissociation constant. The first, usually expressed in terms of pH, is most commonly determined by a direct potentiometric measurement. The second, on the other hand, is established by a neutralization titration; this may use a potentiometric determination to establish the end point. A consideration of both of these measurements is clearly important.

Indicator Electrodes

In principle, at least, any electrode process involving the production or consumption of hydrogen ions can be used for potentiometric pH measurements. The number of these, however, is greatly limited by practicality and convenience.

The hydrogen gas electrode. An electrode similar to the standard hydrogen electrode shown on page 375 can clearly be used for the determination of pH. In this application the platinized platinum surface is immersed in the solution to be analyzed and exposed to hydrogen gas of a known partial pressure. The half reaction is

$$\tfrac{1}{2}H_2 \rightleftharpoons H^+ + e$$

and we may write for 25° C

$$E = 0 - 0.059 \log \frac{[H^+]}{(p_{H_2})^{1/2}}$$

where p_{H_2} is the partial pressure of the hydrogen gas expressed in atmospheres. The electrode must be used in conjunction with a reference electrode of known potential such as a calomel half cell; the observed cell potential may be represented as

$$E_{obs} = -0.059 \log \frac{[H^+]}{(p_{H_2})^{1/2}} + E_{Hg_2Cl_2} + E_j$$

where E_j is the junction potential between the reference electrode and the solution to be analyzed. This equation can be rewritten as follows:

$$-0.059 \log [H^+] = E_{obs} - E_{Hg_2Cl_2} - E_j + 0.059 \log \frac{1}{(p_{H_2})^{1/2}}$$

or

$$p H = \frac{E_{obs} - (E_{Hg_2Cl_2} + E_j)}{0.059} + \log \frac{1}{(p_{H_2})^{1/2}}$$

A close examination of the equation in this form reveals the disconcerting fact that it contains two unknown quantities, pH and E_j. Ordinarily E_j cannot be evaluated exactly (see p. 517); it can only be reduced to a fairly small quantity (a few millivolts) through the use of a suitable salt bridge.

In most potentiometric pH measurements, the attempt is made to circumvent the dilemma presented by the existence of a junction potential by first measuring the cell potential with a buffer solution whose pH is reliably known. Then, from the observed potential, the sum $(E_j + E_{Hg_2Cl_2})$ is measured and employed in determining the pH of the unknown. This expedient still contains an assumption namely, that the junction potential associated with the known buffer is identical with that of the unknown solution. This cannot be exactly true and creates a small error which, incidentally, is inherent to all potentiometric pH measurements. This will be considered further in a later section.

Use of the hydrogen electrode is limited to solutions that contain no substances that react directly with hydrogen gas and thus alter the pH of the solvent. Examples of interferences of this kind are permanganate, iodine, ferric iron, and easily reduced organic compounds. Titanous and chromous ions, whose oxidation by hydrogen ions is catalyzed by platinum, must also be absent. Further, a number of other substances such as proteins, hydrogen sulfide, and arsine cause the platinum electrode to behave in a sluggish and erratic manner; such interferences plus the inconvenience and hazard associated with the use of hydrogen severely restrict widespread use of the hydrogen electrode for pH determinations.

The glass electrode.[4] The glass electrode is unquestionably the most important indicator electrode for hydrogen ions and has almost completely displaced all other electrodes for pH measurements. It is convenient to use and subject to few of the interferences affecting other electrodes.

The phenomenon upon which the glass electrode is based was first observed in the early years of this century; its potential application to pH measurement was quickly recognized. Its widespread employment, however, was delayed until suitable electronic voltmeters became available around 1935; without such instrumentation, measurement of the potentials developed was inconvenient.

The experimental observation fundamental to the development of this electrode is that a potential difference develops across a thin, conducting glass membrane interposed between solutions of different pH. The potential can be detected by placing reference electrodes of known and constant potential in the solutions on either side of the membrane, as shown schematically in Figure 24.6. With this arrangement the potential across the reference electrodes varies as a function of the ratio between the hydrogen ion concentrations C_1 and C_2. Thus, at 25° C the following relationship is found to hold:

$$E_{obs} = k - 0.059 \log \frac{C_1}{C_2}$$

where k is a constant.

To measure the pH of an unknown solution, the hydrogen ion concentration of one of the compartments (say C_2) is kept constant. The solution to be measured is placed in the other compartment; if we denote its concentration as C_1, the equation can be rewritten as

$$E_{obs} = K - 0.059 \log C_1$$

$$= K + 0.059 \, pH$$

[4] For a complete discussion of this most useful electrode see M. Dole, *The Glass Electrode*. New York: John Wiley and Sons, Inc., 1941.

where K is a new constant that contains the logarithmic function of C_2. The numerical value of K is obtained by measuring E for a solution whose hydrogen ion concentration, C_1, is known exactly. This equation is identical in form to those for other hydrogen ion indicator electrodes.

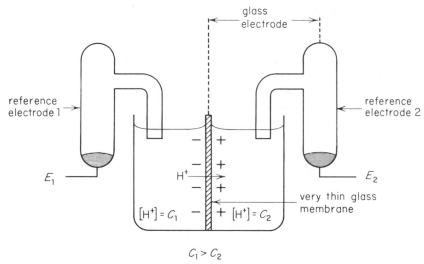

Fig. 24.6 Schematic Diagram of a Cell for pH Measurement. The glass electrode consists of reference electrode 2, the solution of known hydrogen ion concentration C_2, and the glass membrane. The pH of solution C_1 is being determined.

It is important to emphasize that the potentials of the two reference electrodes E_1 and E_2 are quite independent of C_1 and C_2. Thus, any changes in potential between them must arise at some other point in the system—most logically at the glass interface. A reasonable explanation for this phenomenon involves the assumption that the glass membrane of the cell is selectively permeable to hydrogen ions and that these tend to migrate preferentially in the direction of the lesser concentration. The result is that the solution of lower hydrogen ion concentration acquires a positive charge with respect to the solution from which the migration occurred, and a junction potential then develops.

Consistent with this explanation is the experimental fact that passage of current through the glass electrode is accompanied by a transfer of hydrogen ions through the membrane, and that the transfer approaches that predicted by Faraday's law. Also the glass must contain water within its structure if it is to be effective as a pH indicator. When this water is removed by drying in air or by treatment for prolonged periods with a dehydrating agent, the pH sensitivity of the electrode is lost. The obvious implication here is that the water in the glass plays a fundamental part in the proton-transfer process.

Not all glasses function as pH indicators; a soda-lime variety is commonly employed. This responds well except in relatively alkaline (above pH 9 or 10) media where it becomes sensitive not only to hydrogen ions but to sodium ions as well. Other alkali metal ions also interfere, but to a lesser extent. As a consequence, a pH value calculated from theoretical-potential data is likely to be low by several tenths of a unit in solutions containing high concentrations of sodium ion. The alkaline error can be largely avoided by employing a glass membrane constructed of a lithium glass. These are available commercially.[5]

Turning once more to the equation describing the behavior of the glass electrode

$$E = K + 0.059\,p\text{H}$$

it is well to consider what is contained in the constant, K, in addition to the logarithmic function of C_2. As might be expected, the sum of the reference potentials E_1 and E_2 are certainly included. Their sum will, of course, be zero if the two electrodes are identical. The constant K will also take account of the junction potentials between reference electrode 1 and the solution in which it is immersed as well as reference electrode 2 and its solution. The assumption is made that these quantities are constant and independent of the composition of the solution whose pH is being measured. In addition, there is contained in K another term called the *asymmetry potential* which is believed to arise from differences in the strains set up in the two glass surfaces during manufacture. That this exists is evident from the fact that a small potential develops across a glass membrane even when the two solutions are identical. The asymmetry potential changes slowly with time; from a practical standpoint, this means that K must be re-established at regular intervals (every few days) by measuring E_{obs} for a solution of known pH.

Potential measurements with a glass electrode. Even a very thin glass membrane has an electrical resistance amounting to 100 megohms or more, and this complicates the measurement of the potential of a cell containing a glass electrode. The very high resistance precludes the use of a simple potentiometer because the out-of-balance current is reduced to such a small value (page 514). This makes detection of the current impossible even by the most sensitive galvanometer. Two types of instruments overcome this problem. The first, which is entirely adequate for most purposes and convenient to use, is a vacuum-tube voltmeter accurate to about 0.005 volt (or 0.1 pH unit). For more precise measurements, the potentiometric circuit shown on page 513 can be modified by replacing the galvanometer with a direct-current amplifier. This magnifies the very small out-of-balance currents to the point where they can easily be detected with a rugged ammeter. Several instruments of each type, calibrated directly in pH units, are available commercially. These are called pH meters.

Applications of the glass electrode. Glass electrodes, in a variety of shapes and sizes, are available commercially at a relatively low cost. One form is illustrated in Figure 24.7. Here, the electrode consists of a tube of glass with

[5] For a systematic study on glasses for electrode construction see G. A. Perley, *Anal. Chem.*, **21**, 391, 394, 559 (1949).

the pH-sensitive membrane sealed in one end. Within the tube is a dilute hydrochloric acid solution in which is immersed a piece of silver wire coated with silver chloride. This forms a silver-silver chloride reference electrode. As shown in the illustration, the second reference electrode is usually a commercial saturated-calomel electrode.

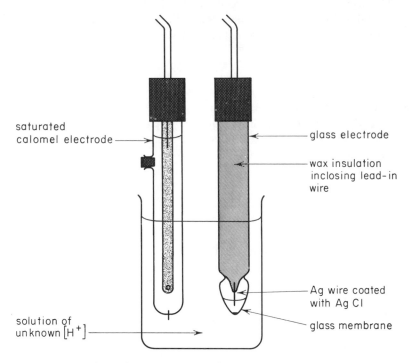

Fig. 24.7 Typical Electrode System for the Potentiometric Measurement of pH.

The glass-calomel electrode system is a remarkably versatile tool for the measurement of pH under a wide variety of conditions. In contrast to other electrode systems, strong oxidants, reductants, proteins, gases, and many other substances do not interfere; the pH of viscous or even semisolid fluids can be determined. Electrodes for a variety of special applications are available. For example, there are micro electrodes for the measurement of a drop or less of solution; systems for insertion in a flowing stream of liquid for the continuous monitoring of pH; and a small glass electrode that can be swallowed to indicate the acidity of the stomach contents (the calomel electrode is kept in the mouth).

We have already mentioned that water in the membrane of a glass electrode is essential to its proper performance as a pH indicator. As a consequence, conditions leading to the dehydration of the glass should be avoided whenever possible. Removal of water will occur upon prolonged exposure of the glass to the atmosphere or to such dehydrating solvents as alcohol or concentrated sulfuric acid. Fortunately, the water is easily restored by soaking the electrode in an aqueous solution for several hours. To avoid dehydration, glass electrodes should be stored in distilled water.

A well-soaked glass electrode does retain enough moisture to permit evaluation of "*p*H numbers" of nonaqueous solutions and, more important, to indicate end points for acid-base titrations carried out in such solvents. While the theoretical interpretation of these numbers is not possible, their empirical application to analysis has proved useful.[6]

Errors in *p*H measurements with the glass electrode. The ubiquity of the *p*H meter and the rather general applicability of the glass electrode tends to lull the chemist into the attitude that any reading obtained with such an instrument is surely correct. It is well to guard against this since there are distinct limitations to the electrode system. These are summarized below.

1. *The alkaline error.* As mentioned before, the ordinary glass electrode becomes somewhat sensitive to the alkali metals at *p*H values greater than 10. Thus, in a solution that is 1.0 *F* in sodium ion, a negative error of nearly 1 *p*H unit is obtained at a *p*H of 12; a 0.1 *F* concentration of this same ion creates an error of about —0.4 unit at this same *p*H. Electrodes designed for high *p*H values are available; with these the alkali error is greatly reduced.

2. *The acid error.* At a *p*H less than zero, values obtained with a glass electrode tend to be somewhat high.

3. *Dehydration.* Dehydration of the electrode may lead to unstable behavior and errors.

4. *Errors in unbuffered neutral solutions.* Equilibrium between the electrode surface layer and the solution is achieved only slowly in poorly buffered, approximately neutral, solutions. Since the measured potential is determined by the surface layer of liquid, errors will arise unless time is allowed for equilibrium to be established; this may take several minutes. In determining the *p*H of such solutions, the glass electrode should be thoroughly rinsed with water before use. Good stirring is also helpful, and several minutes should be allowed to obtain steady readings.

5. *Variation in junction potential.* In the equation relating potential to *p*H we saw that the constant *K* consists of a summation of the reference-electrode potentials, the asymmetry potential, and the junction potentials between the reference electrodes and their respective solutions. Normally *K* is evaluated by measuring the cell potential developed with a buffer solution of known *p*H. The assumption is then made that *K* will have the same value when the unknown solution replaces the buffer. This will not be strictly realized, however, because the junction potential will vary slightly with the composition of the solution being measured, even when a good salt bridge is used. The uncertainty introduced is small, amounting to about ± 1 mv; this corresponds to a *p*H error on the order of 0.02 unit. As shown before, this error is not unique to the glass electrode, but is inherent in all hydrogen ion electrodes since all contain a liquid junction.

[6] For example see J. S. Fritz, *Acid-Base Titrations in Nonaqueous Solvents*. Columbus, Ohio: G. F. Smith Chemical Co., 1952.

The student should appreciate the existence of this fundamental uncertainty in the measurement of pH for which a correction cannot be applied. Values more reliable than 0.01 to 0.02 pH unit are simply unobtainable by potentiometric measurements; it is noteworthy that this corresponds to a relative error in hydrogen ion concentration of about 3 percent.

6. *Error in the pH of the buffer solution.* Since the glass electrode must be regularly calibrated, any inaccuracies in the preparation or changes in composition of the buffer during storage will be reflected as errors in pH measurements.

Potentiometric Acid-base Titrations

In Chapters 14 and 15 we considered the theoretical titration curves for various neutralization titrations in some detail. These curves can be closely approximated experimentally; by measurements of the potential of a pH-sensitive electrode, end points can be determined by inspection, or with the analytical methods described earlier in the chapter. In any case, a study of the theoretical curves will show that the small error inherent in the potentiometric measurement of pH is normally of no consequence insofar as locating the end point is concerned.

Potentiometric acid-base titrations are frequently applied where the sample solutions are colored or turbid. They are particularly useful when mixtures of acids or polyprotic acids (or bases) are to be analyzed since discrimination between the end points can often be made. A numerical value for the dissociation constant of the reacting species can also be estimated from potentiometric titration curves. In theory, this can be obtained from any point along the curve; as a practical matter, it is most easily found from the pH at the point of half neutralization. For example, in the titration of the weak acid HA, we may ordinarily assume that at the midpoint

$$[HA] = [A^-]$$

and therefore

$$K_a = \frac{[H^+][A^-]}{[HA]} = [H^+]$$

or

$$pK_a = pH$$

If desired, the relative concentrations of A^- and HA can be expressed more exactly (see Chap. 14) for a better estimate of K_a.

A value of the dissociation constant and the equivalent weight of a pure sample of an unknown acid can be obtained from a single potentiometric titration; this information is frequently sufficient to identify the acid.

OXIDATION-REDUCTION TITRATIONS

In Chapter 18 we developed techniques for derivation of theoretical titration curves for oxidation-reduction processes. In each example, an electrode potential related to the concentration ratio of the oxidized and reduced forms of either of the reactants was determined as a function of the volume of reagent. These curves can be duplicated experimentally provided an indicator electrode is available that is responsive to one or both of the couples involved in the reaction. Such electrodes are known and available for most, but not all, of the reagents described in Chapters 20 and 21.

Indicator electrodes for oxidation-reduction titration are generally constructed from platinum, gold, mercury, or silver. The metal chosen must be unreactive with respect to the components of the solution—it is merely a medium for electron transfer. Without question, platinum constitutes the most widely employed indicator-electrode material for oxidation-reduction systems. Curves similar to those shown on pages 408 and 411 can be obtained experimentally with a platinum-calomel electrode system. End-point detection can be accomplished from a plot of the experimental data or by the analytical methods already discussed.

Some Special Techniques

This section is devoted to a brief consideration of some special techniques designed to make potentiometric titrations less time consuming or to simplify the equipment needed for these methods.

Differential Titration Methods

We have already seen that a derivative titration curve can be obtained from the data of an ordinary potentiometric titration (see Fig. 24.3), and that such a graph exhibits a marked maximum in the vicinity of the equivalence point. It is possible to acquire titration data directly in the derivative form with rather simple equipment. To do this, the potential of a pair of identical indicator electrodes is measured throughout the titration; one of these is immersed in a portion of the solution that is kept slightly behind the rest of the solution insofar as the progress of the titration is concerned. The principle is illustrated by the original differential method derived by Cox.[7] Here the solution to be titrated is divided into two exactly equal portions that are connected by a salt bridge; identical indicator electrodes are then placed in each. Reagent is slowly added from two burets, the volume from one always being slightly less (ideally 0.1 to 0.2 ml) than that from the other. The difference in potential ΔE for the small volume difference ΔV results directly from measurement of the cell potential. Initially this differential quantity is small; it steadily increases, however, and reaches a maximum at the end point.

Cox's procedure is inconvenient, and were this the only method for performing a differential titration it would probably never be used. Actually, a

[7] D. C. Cox, *J. Am. Chem. Soc.*, **47**, 2138 (1925).

number of ingenious cells have been devised that accomplish the desired purpose in a much simpler manner. In Figure 24.8, which illustrates such a cell, one of the indicator electrodes is partially shielded by a glass tube that is constricted at one end. As a consequence, additions of reagent to the bulk of the solution do not materially alter the composition of the small volume surrounding this electrode. When mixing is desired, the rubber bulb is squeezed. In using this cell, the reagent is added in appropriate increments and the potential difference is measured. The solution is then homogenized by squeezing the rubber bulb several times before the next addition is made. At the equivalence point, the potential measured will be at a maximum as in Cox's procedure. If the volume in the tube enclosing the electrode is kept small (say 1 to 5 ml), the error arising from failure of the final addition of reagent to react with this solution can be shown to be negligibly small.

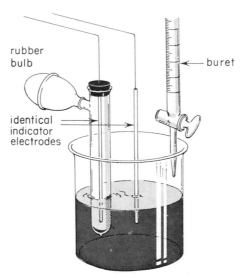

Fig. 24.8 Apparatus for Differential Potentiometric Titrations.

The main advantage of a differential method is the elimination of the reference electrode and salt bridge. The end points are ordinarily very well defined.

Automatic Titrations

In recent years several automatic titrators based on the potentiometric principle have come on the market. These are useful where a large number of routine analyses are to be carried out. Such instruments cannot yield more accurate results than those obtained by manual potentiometric techniques; however, they do decrease the number of man-hours required for the performance of titrations and thus may offer some economic advantages.

Basically, two types of automatic titrators are available. With the first of these, a titration curve is plotted automatically; the end point is then determined by inspection of this curve. In the second type, the titration is stopped automatically when the indicator-reference electrode potential reaches some predetermined value; the volume of reagent consumed can then be read at the operator's convenience.

In general, the addition of reagent in automatic titrators is accomplished either with a buret equipped with an electromagnetic valve or by means of a glass hypodermic syringe. With the latter, the plunger is activated by a motor-driven micrometer screw and the volume delivered is determined with a simple revolution counter. Where the titration is to be stopped at the end point, a potentiometer is used to apply a potential that is equal but opposite in sign to that which will develop across the electrodes at the end point. The small current flow that results can be used to activate a relay controlling the electromagnetic valve or the motor-driven syringe; as a consequence, the reagent flow is stopped at the end point when the current becomes zero.

With either type of titrator, the principle design problem is that of incorporating a device to anticipate the end point and slow the rate of addition of reagent correspondingly. Otherwise, with reasonable flow rates the instrument will overstep the end point; alternatively, the flow must be diminished to the point where prohibitively long times are required to complete the titration. Modern instruments containing this feature perform a titration in much the same way as a human operator; preliminary additions are rapid, and successively decrease in size as the end point is approached.

LABORATORY EXPERIMENTS INVOLVING POTENTIOMETRIC TITRATIONS

This section contains several experiments suitable for demonstrating the potentiometric method. Because these experiments can be performed with any of several types of potentiometer, we have made no attempt to include specific operating instructions.

General Method for Carrying Out a Potentiometric Titration

Although the following general instructions can be applied to most potentiometric titrations, some modification may be required in special cases.

Procedure. 1. A solution of the sample in 50 to 250 ml of water is placed in the titration vessel (usually a 250- to 500-ml beaker). The electrodes are then rinsed well in distilled water and immersed in the solution. A magnetic stirring bar or a glass stirring rod attached to a mechanical stirrer is also introduced. The buret is then filled with reagent and fixed in such a position that the solution can be delivered without splashing. A convenient arrangement is shown on page 552.

2. Start the stirring motor, connect the electrodes to the potentiometer, measure, and record the initial potential.

3. Measure and record the potential after each addition of reagent. Introduce fairly large volumes (1 to 5 ml) at the outset; withhold each succeeding increment until the potential remains constant within 1 to 2 mv (0.03 pH unit) for 30 sec. A stirring motor will occasionally cause erratic potential readings, and it may be advisable to turn off the motor during the actual measuring process. Judge the volume of reagent to be added by calculating an approximate value of $\Delta E/\Delta$ ml for each addition. In the immediate vicinity of the equivalence point, introduce the reagent in exact 0.1-ml increments. Continue the titration 2 to 3 ml past the equivalence point; increase the volumes added as $\Delta E/\Delta$ ml once again decreases.

4. Locate the end point by one of the methods described at the beginning of this chapter.

Potentiometric Analysis of a Chloride-iodide Mixture

Equipment. A polished silver wire or a commercial billet-type silver electrode may be used as an indicator electrode. A commercial fiber-type calomel electrode can be employed as a reference electrode although this will lead to slightly high results as a consequence of diffusion of Cl^- from the salt bridge. An alternative is to place the calomel electrode in a saturated KNO_3 solution and connect this to the solution to be titrated with a KNO_3 bridge. The bridge consists of a bent glass tube filled with a 4-percent agar-saturated KNO_3 solution (see p. 509 for preparation).

Reagent. Prepare a $0.1000\ N$ $AgNO_3$ solution as instructed on page 261.

Procedure. Prepare a solution of the sample that contains a combined total of 2 to 4 milliequivalents of Cl^- and I^- in 100 ± 10 ml of water. Titrate with the silver nitrate, adding the reagent in large increments except near the two end points. Plot the data and determine the end point for each ion. Plot $\Delta E/\Delta V$ vs. V and ascertain the end point. Plot a theoretical titration curve assuming the measured concentrations of the two constituents to be correct. Report the number of milligrams of I^- and Cl^- found in the sample.

Potentiometric Neutralization Titrations

Equipment. Use a glass-calomel electrode system and a commercial pH meter. The electrode system should be calibrated against a buffer solution of known pH.

Reagents. For the titration of weak acids, prepare and standardize a $0.1\ N$ solution of carbonate-free sodium hydroxide according to the directions on page 348. For titration of weak bases, follow the directions on page 343 for the preparation and standardization of a $0.1\ N$ hydrochloric acid solution.

Titration of a weak acid. Prepare a solution containing between 1 and 4 milliequivalents of the acid in about 100 ml of water; a larger volume may be required to dissolve some of the less soluble organic acids. Add 2 drops of phenolphthalein to the solution.

Titrate as directed in the general instructions. Some samples will contain more than one replaceable hydrogen; be alert for more than one break in the titration curve. Note the point at which the indicator changes color.

Plot the titration data and determine the end point or points. Compare these with the phenolphthalein end point. The derivative methods for end-point determinations may also be used (see pp. 554-556).

Calculate the number of milliequivalents of H^+ present in the sample.

Analysis of a carbonate-bicarbonate mixture. Dissolve the sample, which contains a total of 2 to 3 millimols of the two salts, in 200 ml of water. Add 2 drops of phenolphthalein and 2 drops of methyl orange to the solution. Then titrate as directed in the general instructions. Carry the titration 3 to 5 ml beyond the second end point. Note where the two indicators change color.

Plot the data and determine the end points. Compare the potentiometric and indicator end points. Estimate the two dissociation constants for carbonic acid from the curves and compare these with literature values.

Calculate the percent Na_2CO_3 and $NaHCO_3$ in the sample.

Potentiometric Oxidation-reduction Titrations

Titration of ferrous iron with ceric sulfate. The iron in an ore or other iron-containing compound is conveniently determined by a potentiometric titration with quadrivalent cerium. Prereduction of any ferric iron can be accomplished with a stannous chloride solution as described earlier. Removal of the excess stannous ion is not necessary in this case, however, because the potentiometric curve will show two breaks, the first corresponding to the oxidation of the stannous tin and the second to the ferrous oxidation. The difference in volume between these then reflects the amount of iron present.

Equipment. Prepare a platinum-calomel electrode system and connect the leads to a potentiometer.

Procedure. Prepare and standardize a 0.1 N solution of ceric ion in sulfuric acid as directed on pages 452 and 453. Weigh, dissolve, and reduce the iron sample as directed on page 443. Do not, however, remove the excess stannous chloride with mercuric ion. Dilute the solution to about 50 ml and titrate immediately with the ceric sulfate. Two end points will be observed.

Plot the voltage data against the volume of reagent and determine the two end points. Calculate the percent iron in the sample.

Procedure employing a differential end point. In this experiment a pair of platinum electrodes is used; one of these is housed in a glass tube such as shown in Figure 24.8. An ordinary potentiometer may be employed for potential measurements.

Prepare the reagent and sample as directed in the previous section. Add about 1 ml of the reagent and measure the potential. Then homogenize the solution by squeezing the rubber bulb connected to the shielded electrode and add more reagent. Again record the potential and volume of reagent before mixing the solution around the shielded electrode with the bulk of the solution. Continue this process, reducing the volume increments to 0.1 ml in the vicinity of each end point. Carry the titration 2 to 5 ml beyond the second end point.

Plot the data and calculate the percent iron from the difference in reagent volume between the first and second end point.

problems

1. Derive the theoretical potentiometric titration curve for 50 ml of a solution that is 0.05 F each in Br^- and Cl^- ions. The reagent is 0.1 F $AgNO_3$; a silver-saturated calomel-electrode system is used. Calculate the cell potential corresponding to the following volumes of reagent: 10, 20, 24, 24.9, 25.0, 25.1, 26.0, 35, 45, 49, 49.9, 50.0, 50.1, 51.0, and 60.0.

2. Calculate the equivalence-point potential for the titration of 0.2 N I^- with 0.2 N $AgNO_3$. The electrode system consists of a saturated-calomel electrode and a platinum wire. The solution was made 0.02 F in I_2 at the start of the titration (assume the indicator electrode half reaction is $I_2 + 2e \rightleftharpoons 2I^-$).

ans. 0.71 volt

3. What is the potential of a titration cell consisting of a saturated calomel electrode and a hydrogen gas electrode at the point in the titration of acetic acid where half of the acid has been neutralized? Assume the hydrogen gas is at a pressure of 1 atmosphere. Which electrode is the cathode?

4. What is the pH of a solution determined by an emf measurement of a cell consisting of a saturated calomel cathode and a hydrogen gas anode ($p_{H_2} = 725$ mm) if the measured potential is 0.98 volt?

5. What is the emf of the following cell:

saturated calomel electrode $\|$ 0.1 F NH_4Cl, hydrogen electrode

6. What is the dissociation constant of the acid HX if the following cell develops a potential of $+ 0.612$ volt with the calomel electrode as the cathode?

Pt $|$ H_2(1 atm), HX(0.05 F), NaX(0.05 F) $\|$ saturated calomel electrode

chapter 25. *Conductometric*
Titrations[1]

We have seen that the conduction of an electric current through an electrolyte solution involves a migration of positively charged species toward the cathode and negatively charged ones toward the anode. All of the charged particles contribute to the conduction process; the contribution of any given species, however, is governed by its relative concentration and the inherent mobility of its individuals.

When all other factors are held constant, variations in the concentration of a given electrolyte results in a regular change in the ease with which a solution will conduct; it is therefore possible to relate concentration to the property of *conductance*.

The electric conductance of a solution is a nonspecific property. Its employment for the direct quantitative determination of a given ion is relatively limited in scope even though it can be measured with ease and a high degree

[1] For a more detailed discussion of conductometric methods, see the chapter by T. Shedlovsky in A. Weissberger, *Physical Methods of Organic Chemistry*, 3d ed. 1, IV, 3011-3048. New York: Interscience Publishers, Inc., 1960.

of accuracy. The principal applications of direct conductance measurements include the analysis of binary water-electrolyte mixtures and the determination of the total electrolyte concentration of a solution. The latter measurement is particularly important in the examination of distilled waters because it can be used as a criterion of purity.

A more important analytical application of conductance measurements involves their use to signal the end point in a titration. Here, it is the change in conductance of the solution that is important; this imparts a certain degree of selectivity to the analysis that is totally lacking in the direct measurements.

The main advantage to the conductometric end point is its applicability to very dilute solutions and to systems that involve relatively incomplete reactions. For example, while neither a potentiometric nor an indicator method can be used for the neutralization titration of phenol ($K_a = 10^{-10}$) a conductometric end point can be successfully employed.

The method has its limitations. At best, it is subject to many interferences. In particular, it becomes less accurate and less satisfactory as the electrolyte concentration of the solution under examination increases. When the salt concentration becomes high, a conductometric analysis often becomes quite impossible.

ELECTROLYTIC CONDUCTANCE

Some Important Relationships

So far we have related the passage of current at fixed potential to the resistance of an electrolyte solution. It is now more convenient to make use of a closely related property. This is the *conductance L* and is simply the reciprocal of the resistance

$$L = \frac{1}{R}$$

Specific conductance, k. The conductance is directly proportional to the cross-sectional area A and inversely proportional to the length l of a uniform conductor; thus

$$L = k \frac{A}{l} \tag{25-1}$$

where k is a proportionality constant called the *specific conductance*. Clearly it is the conductance when A and l are equal. If these are expressed in terms of centimeters, the specific conductance is the conductance of a cube of liquid, 1 cm on a side. The unit of specific conductance is ohm^{-1}cm^{-1}.

Equivalent conductance, Λ. The equivalent conductance is defined as the conductance of 1 gram equivalent of solute contained between electrodes spaced 1 cm apart. Neither the volume of the solution nor the area of the electrodes is specified; these vary to satisfy the conditions of the definition. For example, a 1.0 N solution (1.0 gram equivalent per liter) would require electrodes with individual surface areas of 1000 sq cm; a 0.1 N solution would need 10,000-sq cm electrodes. The direct measurement of equivalent conductance is thus seldom, if ever, undertaken because of the experimental inconvenience associated with such relatively large electrodes. Instead, this quantity is determined indirectly from specific conductance data. The relationship between these is readily evaluated. In general, the number of cubic centimeters that contain 1 gram equivalent of solute is given by

$$V = \frac{1000}{C}$$

where C is the concentration of the solution in equivalents per liter. The definition of equivalent conductance fixes l at 1 cm so that the electrode area necessary to contact the required volume of solution will be given by

$$A = \frac{1000}{C}$$

Substituting these values into equation (25-1) and remembering that these are the conditions under which $L = \Lambda$, we obtain

$$\Lambda = \frac{1000\,k}{C} \tag{25-2}$$

With the aid of this equation the equivalent conductance of a salt may be obtained from the experimentally measured specific conductance k of a solution of known concentration C.

Equivalent conductance at infinite dilution. From its definition we might conclude that the equivalent conductance for a substance is independent of concentration. Experimentally, however, this quantity increases with increasing dilution. The behavior of sodium chloride may be considered as typical for a strong electrolyte

Concentration of NaCl, equiv/liter	Λ
0.1	106.7
0.01	118.5
0.001	123.7
infinite dilution	126.4

This variation is due, in part, to forces of attraction and repulsion that restrict in some measure the independence of behavior of the individual sodium and chloride ions at finite concentrations.

For a strong electrolyte there is a linear relationship between the equivalent conductance and the square root of the concentration. Extrapolation of this plot to zero concentration yields a value for the equivalent conductance Λ_0 at infinite dilution. A similar plot for a weak electrolyte is nonlinear and difficult to evaluate.

At infinite dilution interionic attractions become nil; the over-all conductance of the solution of a salt simply consists of the sum of the individual equivalent ionic conductances

$$\Lambda_0 = \lambda^0_+ + \lambda^0_-$$

where λ^0_+ and λ^0_- are the equivalent ionic conductances of the anion and cation of the salt at infinite dilution. Values for individual ionic conductances may be determined from electrolysis studies. These allow, among other things, the indirect evaluation of the equivalent conductance at infinite dilution for weak electrolytes. Table 25-1 contains values for the equivalent ionic conductance of a number of frequently encountered species. The reader should note that the symbols $\frac{1}{2} Mg^{2+}$, $\frac{1}{3} Fe^{3+}$, $\frac{1}{2} SO_4^{2-}$, etc., are used to emphasize that the concentration units are in *equivalents* per liter.

Table 25-1 illustrates the appreciable differences that exist in the equivalent ionic conductance of various species. These arise primarily from differences in size of the ions and the degree of their hydration.

Table 25-1

EQUIVALENT IONIC CONDUCTANCES AT 25° C

Cation	λ^0_+	Anion	λ^0_-
H_3O^+	349.8	OH^-	198
Li^+	38.7	Cl^-	76.3
Na^+	50.1	Br^-	78.4
K^+	73.5	I^-	76.8
NH_4^+	73.4	NO_3^-	71.4
Ag^+	61.9	ClO_4^-	68.0
$\frac{1}{2} Mg^{2+}$	53.1	$C_2H_3O_2^-$	40.9
$\frac{1}{2} Ca^{2+}$	59.5	$\frac{1}{2} SO_4^{2-}$	79.8
$\frac{1}{2} Ba^{2+}$	63.6	$\frac{1}{2} CO_3^{2-}$	70
$\frac{1}{2} Pb^{2+}$	73	$\frac{1}{2} C_2O_4^{2-}$	24
$\frac{1}{3} Fe^{3+}$	68	$\frac{1}{4} Fe(CN)_6^{4-}$	110.5
$\frac{1}{3} La^{3+}$	69.6		

The equivalent ionic conductance is a measure of the mobility of an ion under the influence of an electric-force field and is thus a gauge of its current-carrying capacity. For example, the ionic conductances of potassium and chloride ions are nearly identical; therefore, a current passed through a potassium chloride solution will be carried nearly equally by the two species. The situation is quite different with hydrochloric acid; because of the greater mobility of the hydronium ion a greater fraction of the current $[350/(350 + 76)]$ will be carried by that species in an electrolysis.

Ionic conductance data make it possible to predict the relative conductivity of solutions of various substances. Thus, we are justified in saying that a 0.01 F solution of hydrochloric acid will have a greater conductivity than a 0.01 F sodium chloride solution because of the much greater ionic conductance of the hydronium ion. Such conclusions are important in predicting the course of a conductometric titration.

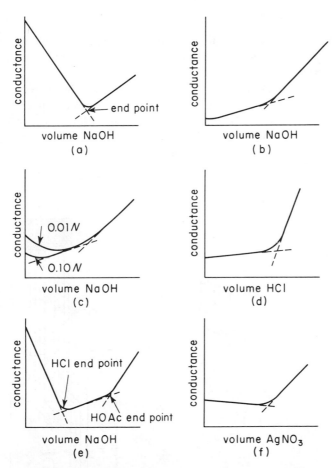

Fig. 25.1 Typical Conductometric Titration Curves. (a) Strong acid. (b) Very weak acid ($K_a \sim 10^{-10}$). (c) Weak acid ($K_a \sim 10^{-5}$). (d) Salt of a weak acid. (e) Mixture of acids (hydrochloric and acetic). (f) Precipitation of chloride ion.

CONDUCTOMETRIC TITRATIONS

A conductometric titration involves measurement of the conductance of the sample after successive additions of reagent. The end point is determined from a plot of either the conductance or the specific conductance as a function of the volume of added titrant. As may be seen from Figure 25.1 these titration curves take a variety of shapes depending upon the chemical system under investigation. In general, however, they are characterized by straight-line portions with dissimilar slopes on either side of the equivalence point.

Methods and Techniques

Conductance measurements. Conductivity measurements are performed with a Wheatstone bridge circuit and an alternating-current source (see p. 515 for further details).

Cells. The electrodes generally consist of a pair of platinum plates that are firmly anchored within the cell to assure constant spacing. They are frequently platinized to increase their effective surface area and minimize capacitance effects. Figure 25.2 illustrates a pair of cell arrangements that are convenient for conductometric titrations.

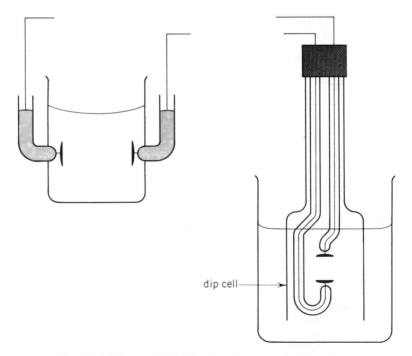

dip cell

Fig. 25.2 Typical Cells for Conductometric Titrations.

Determination of the cell constant. For most conductometric measurements it is the specific conductance k that is desired. According to equation (25-1) this quantity is related to the measured conductance L by a term that is simply the ratio between the distance separating the electrodes and their surface area. This ratio has a fixed and constant value for any given cell, and is known as the *cell constant*. Its value is seldom determined directly; rather, it is evaluated by measuring the conductance L of a solution whose specific conductance is reliably known. Solutions of potassium chloride are commonly chosen for this purpose.[2]

Grams KCl per 1000 grams of solution in vacuum	Specific Conductance at 25° C ohm^{-1} cm^{-1}
71.1352	0.111342
7.41913	0.0128560
0.745263	0.00140877

Having once determined this constant, conductivity data obtained with the cell can easily be converted to terms of specific conductance with the aid of equation 25-1.

Temperature control. The temperature coefficient for conductance measurements is about 2 percent per degree centigrade; as a consequence, some temperature control is ordinarily required during a conductometric titration. Frequently immersion of the cell in a reasonably large bath of water or oil maintained at about room temperature is sufficient. This provides an acceptably constant temperature by removing any heat developed as a consequence of the chemical reaction. Clearly, although a constant temperature is necessary, control at some specific value is not important for a successful titration.

Volume changes. Throughout a titration the volume of the solution is always increasing; unless the conductance is corrected for this effect, nonlinear titration curves result. The correction can be accomplished by multiplying the observed conductance by the factor $(V + v)/V$, where V is the initial volume of solution and v is the total volume of reagent added. The correction presupposes that the conductivity is a linear function of dilution; this is true only to a first approximation. In the interests of keeping v small, the reagent for a conductometric titration is ordinarily several times more concentrated than the solution being titrated. A microburet may then be used for the volumetric measurement.

TITRATION CURVES

To establish a conductometric end point sufficient measurements are necessary to define the titration curve. After correcting for volume change,

[2] G. Jones and B. C. Bradshaw, *J. Am. Chem. Soc.*, **55**, 1780 (1933).

these data are plotted as a function of the volume of reagent added. The two linear portions are then extrapolated, the point of intersection being taken as the equivalence point.

Conductometric titration curves always depart from linearity in the region of the equivalence point. Thus in Figure 25.1, the experimental curves do not show sharp discontinuities; rather, curvilinear behavior is exhibited because of the failure of the chemical reaction to proceed to absolute completion. For example, in a neutralization titration, the hydronium and hydroxide ions remaining at the equivalence point make a significant contribution to the conductance of the medium. The phenomenon becomes more pronounced in the titration of very dilute solutions. A characteristic of conductometric titration curves is that the linear portions can be established from an evaluation of measurements that are far removed from the equivalence point; it is in this respect that the conductometric technique appears to best advantage. In contrast to potentiometric or indicator methods that require experimental observations where the reaction is least complete, the conductometric end point is determined on the basis of data taken in regions where the common-ion effect tends to force the reaction to completion. The method, therefore, is particularly suited for use where the solution is dilute or the equilibrium is relatively unfavorable.

Conductometric end points are employed for all types of volumetric reactions. The number of useful applications to oxidation-reduction titrations is limited, however, because of the high concentrations of other ions—in particular, the hydronium ion—necessary for such processes. These tend to mask conductivity changes associated with the analytical reaction. For this reason we shall discuss only titration curves for other types of reactions.

Acid-base Titrations

Neutralization titrations are particularly adaptable to the conductometric end point because of the very high ionic conductance of the hydronium and the hydroxyl ion and the low conductivity of their product; this results in large changes in the measured quantity and, therefore, well-defined end points.

Titration of strong acids or bases. A typical titration curve for a strong acid with a strong base is shown in Figure 25.1*a*. Initially, the conductance of the solution is high due to the large number of hydronium ions in the solution; as sodium hydroxide is added there is a diminution in their number and a corresponding rise in the number of sodium ions. The latter, however, have a far lower ionic mobility; the net effect is a rather rapid decrease in conductance. After the end point has been passed, a reversal of slope occurs as the hydroxide ion concentration from the excess base increases. With the exception of the immediate equivalence-point region there is an excellent linear variation of conductance with the volume of base added; as a result, only two or three observations on either side of the equivalence point are needed for an analysis.

The percentage change in conductivity during the course of the titration of a strong acid or base is the same regardless of the concentration of the solution. Thus, very dilute solutions can be analyzed with an accuracy comparable to more concentrated ones.

Titration of weak acids or bases. Curve *b* illustrates the application of the conductometric end point to a titration based upon a reaction so incomplete that a potentiometric or an indicator end point would be quite unsatisfactory. A solution of boric acid ($K_a = 6 \times 10^{-10}$) is being titrated with a strong base. In the early stages of the titration a buffer is rapidly established that imparts to the solution a relatively constant and small hydronium ion concentration. The added hydroxide ions are consumed by this buffer and thus do not directly contribute to the conductivity. A gradual increase in this property does result, however, owing to the increase in concentration of sodium and of borate ions. With attainment of the equivalence point, no further borate is produced; further additions of base cause a more rapid increase in conductance due to the increase in concentration of the more mobile hydroxide ion.

Curve *c* illustrates the titration of a moderately weak acid, such as acetic acid ($K_a \simeq 10^{-5}$), with sodium hydroxide. Here the nonlinearity in the early portions of the titration curve causes difficulty in establishing the end point; with concentrated solutions, however, the titration is feasible. As before, we can interpret this curve in light of the ionic changes taking place. Here, the solution initially has a moderate concentration of hydronium ions ($\approx 10^{-3}\ M$). Addition of base results in the establishment of a buffer system, and a consequent reduction in the concentration of this species. This is one of the two factors that determine the conductivity in this region. The other is the increase in the concentration of sodium ion as well as the anion of the acid. These two factors act in opposition to one another. At first the decrease in hydronium ion concentration predominates and a decrease in conductance is observed. As the titration progresses, however, the *p*H becomes stabilized (in the buffer region); the increase in the salt content then becomes the more important factor, and a linear increase in conductance finally results. Beyond the equivalence point the curve steepens because of the greater ionic conductance of hydroxide ions.

In principle, all titration curves for weak acids or bases contain the elements of this behavior. The ionization of very weak species is so slight, however, that little or no curvature occurs with the establishment of the buffer region (see Curve *b*, for example). As the strength of the acid (or base) becomes greater, so also does the extent of the curvature in the early portions of the titration curve. For weak acids or bases with dissociation constants greater than about 10^{-5} this becomes so pronounced that an end point cannot be distinguished.

Titration of salts of weak acids and bases. Curve *d* represents the titration curve for a salt such as sodium acetate with a standard solution of hydrochloric acid. The addition of strong acid results in formation of sodium chloride and undissociated acetic acid. The net effect is a slight rise in conductance owing to the greater mobility of the chloride ion over that of the acetate ion it replaces. After the end point has been passed, a sharp rise in conductance attends the addition of excess hydronium ions.

This method is suitable for the titration of salts that cannot otherwise be determined because of the incompleteness of their reactions.

Curve *e* is typical of the titration of a mixture of two acids that differ in degree of dissociation. The conductometric titration of such mixtures is frequently more accurate than a potentiometric method.

Precipitation and complex-formation titrations. Curve *f* illustrates the titration of sodium chloride with silver nitrate. The initial additions of reagent, in effect, cause an exchange of chloride ions for the somewhat less mobile nitrate ions of the reagent; a slight decrease in conductance results. After the reaction is complete, a rapid rise occurs due to the addition of excess silver nitrate.

This curve is fairly typical for a precipitation titration. The slope of the initial portion of the curve, however, may be either downward or upward depending upon the relative conductance of the ion being determined and the ion of like charge in the reagent. A downward-sloping line will lead to a sharper definition of the end point since this will produce a *V*-shaped titration curve. It is therefore preferable, where possible, to choose a reagent in which the ionic conductance of the nonreactive ion is less than that of the ion being titrated. According to the data in Table 25-1, then, we may predict that lithium chloride would be preferable to potassium chloride as a precipitating agent for silver ion.

Conductometric methods based upon precipitation or complex-formation reactions are not as useful as those involving neutralization processes. Conductance changes during these titrations are seldom as large as those observed with acid-base reactions because no other reagent approaches the great ionic conductance of either the hydronium or the hydroxide ion. Such factors as slowness of reaction and coprecipitation represent further sources of difficulty with precipitation reactions.

SUGGESTED LABORATORY EXPERIMENTS

Since such a variety of measuring equipment and cells is available for conductometric titrations, it is not profitable to give detailed instructions for the operation of any particular combination of these. We shall therefore simply suggest some experiments that are useful for demonstrating the conductometric method.

(1) Titrate a 0.01, 0.001, and 0.0001 *N* solution of HCl with 0.1, 0.01, and 0.001 *N* NaOH, respectively. Plot conductance versus volume of base.

(2) Repeat (1) with solutions of a weak acid.

(3) Titrate various mixtures of acetic acid and hydrochloric acid with 0.1 *N* NaOH and establish the end points for each acid.

(4) Titrate a 0.01 *N* solution of NaCl with 0.1 *N* $AgNO_3$. Repeat with 0.1 *N* silver acetate.

chapter 26. *Coulometric Methods of Analysis*

The direct proportionality between the quantity of current passing through an electrochemical cell and the amount of oxidation and reduction occurring at the electrodes was first recognized by Michael Faraday more than a century ago. This relationship was rapidly put to use in the chemical coulometer, which measures the quantity of electricity through determination of the amount of chemical reaction taking place in a cell. *Coulometric analysis*, however, in which the quantity of electricity is employed to determine the amount of a chemical substance in solution, is of relatively recent origin, dating from about 1940.

The coulometric method of analysis embraces two general techniques. For the first of these the potential of the working electrode is maintained at some fixed value; completion of the analysis is indicated when the current flow approaches zero. The total quantity of electricity required for the electrolysis is evaluated with a chemical coulometer or by integration of the current-time curve for the reaction. The second technique makes use of a constant current that is passed until an indicator signals completion of the reaction. The quantity of electricity required to attain the end point is then calculated from the current and the time of its passage. This method has a wider variety of applications than the other; it is frequently called a *coulometric titration*.

A fundamental requirement of all coulometric methods is that the process or processes involving the species being determined proceed with 100-percent current efficiency. This means that each faraday of current must result in the chemical change of exactly one equivalent in the substance of interest. This does not, however, imply that the species must necessarily participate directly in the electron-transfer process at an electrode. Indeed, more often than not the substance being determined is involved in a reaction that is wholly or in part secondary to the actual electrochemical process. For example, ferrous iron can be oxidized at a platinum anode

$$Fe^{2+} \rightleftharpoons Fe^{3+} + e$$

As the concentration of this species becomes small there is also the possibility of concentration polarization, followed by oxygen formation at this electrode

$$H_2O \rightleftharpoons 2H^+ + \tfrac{1}{2} O_2 + 2e$$

Were this latter process to occur, the current required for the complete oxidation of iron would exceed that called for by theory. It can be prevented by performing the analysis in the presence of a large excess of cerous ion, which is oxidized in preference to water.

$$Ce^{3+} \rightleftharpoons Ce^{4+} + e$$

Any ceric ions produced diffuse rapidly from the anode and in turn oxidize any ferrous ions with which they come in contact.

$$Fe^{2+} + Ce^{4+} \rightleftharpoons Fe^{3+} + Ce^{3+}$$

The net effect is 100-percent current efficiency even though only a fraction of the ferrous iron is directly oxidized at the electrode surface.

The coulometric determination of chloride is based entirely upon a secondary reaction. A silver anode, upon passage of current, supplies silver ions that then precipitate the chloride ion in the sample. A current efficiency of 100 percent is achieved with respect to chloride ion even though no oxidation or reduction of the halide occurs.

COULOMETRIC METHODS AT CONSTANT ELECTRODE POTENTIAL

Coulometric methods employing a controlled potential were first suggested by Hickling[1] in 1942 and have since been exploited by Lingane[2] and others.

---

[1] A. Hickling, *Trans. Faraday Soc.*, **38**, 27 (1942).

[2] For a good discussion of this method see J. J. Lingane, *Electroanalytical Chemistry*, 2d ed., 450-483. New York: Interscience Publishers, Inc., 1958.

These are similar to electrogravimetric methods using potential control (see Chap. 23); they differ only in that a quantity of electricity is measured rather than a weight of deposit.

In contrast to a coulometric titration, a single electrode reaction is required although the species being determined need not react directly at the electrode.

Apparatus and Methods

A controlled potential coulometric analysis requires a potentiostat such as that shown on page 533. In addition, a device is needed to determine the quantity

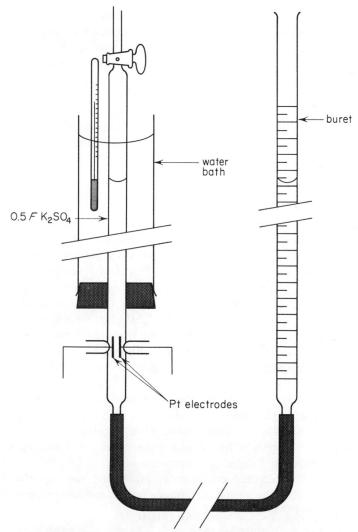

Fig. 26.1 A Hydrogen-oxygen Coulometer.

of electricity required to complete the reaction. A chemical coulometer, placed in series with the working electrode, is commonly employed for this.

Although we have already discussed chemical coulometers briefly in Chapter 22, the hydrogen-oxygen coulometer warrants further consideration because of its particular convenience for the measurement of small quantities of electricity. An example of such a device, designed by Lingane,[3] is illustrated in Figure 26.1. It consists of a tube equipped with a stopcock and a pair of platinum electrodes. This is connected to a buret by a rubber tubing; both are filled with $0.5\ F\ K_2SO_4$. Passage of current through this device liberates hydrogen at the cathode and oxygen at the anode. Both gases are collected, and their total volume is measured by determining the volume of liquid displaced. The water jacket and thermometer provide a means of ascertaining the gas temperature.

The total quantity of current can also be determined graphically. This requires the careful measurement of the current passing through the cell at known time intervals. The area under the curve relating these two variables yields the desired quantity. This procedure is more time consuming and generally less accurate than a chemical coulometer.

A number of mechanical and electronic integrators have been developed that can be adapted to the determination of the current-time integral in a coulometric electrolysis.[4] These are more convenient to use than the chemical coulometer.

Applications of Controlled-potential Coulometric Methods

A coulometric analysis with controlled potential possesses all the advantages of a controlled-cathode potential electrogravimetric method (see p. 532) and in addition is not subject to the limitation imposed by the need for a weighable product. The technique can therefore be applied to systems that yield deposits with poor physical properties and to oxidations or reductions in which no solid product is formed. For example, arsenic may be determined coulometrically by electrolytic oxidation of arsenious acid (H_3AsO_3) to the soluble arsenic acid (H_3AsO_4) at a platinum anode. Similarly, the analytical conversion of ferrous ion to the ferric state can be accomplished with suitable control of the anode potential. Other metallic ions having more than one stable oxidation state can undoubtedly be analyzed in this way also.

The coulometric method has also been advantageously applied to the deposition of metals at a mercury cathode (see p. 539) with which a gravimetric completion of the analysis is inconvenient. Thus, excellent methods have been described for the analysis of lead in the presence of cadmium, copper in the presence of bismuth, and nickel in the presence of cobalt.

The controlled-potential coulometric procedure also appears to offer possibilities for the electrolytic determination of organic compounds. For

[3] J. J. Lingane, *J. Am. Chem. Soc.*, **67**, 1916 (1945).

[4] For a description of several of these see J. J. Lingane, *Electroanalytical Chemistry*, 2d ed., 340-350. New York: Interscience Publishers, Inc., 1958.

example, Meites and Meites[5] have demonstrated that quantitative reductions of trichloroacetic acid and picric acid are possible at a controlled potential mercury cathode.

$$CCl_3COO^- + H^+ + 2e \rightleftharpoons Cl_2HCCOO^- + Cl^-$$

Coulometric measurements make possible the estimation of these compounds with an accuracy of a few tenths of a percent.

COULOMETRIC TITRATIONS

The coulometric-titration procedure was first suggested by Szebelledy and Somogyi in 1938 in a series of classical papers.[6] It was not until 1947, however, that the full potentialities of the method were appreciated by chemists. This came about as a result of the work of E. H. Swift and others. By now, coulometric procedures have been developed for all types of reactions.

In a coulometric titration, the species that reacts with the substance being determined is generated at the electrode. In some instances a single electrode process is involved; the generation of silver ion for the precipitation of a halide represents an example of this. In other cases the direct oxidation or reduction of the substance being analyzed may take place at the generator electrode in addition to formation of the reagent. The coulometric oxidation of ferrous ion— in part by electrogenerated ceric ions and in part by direct electrode oxidation— may be cited as an example. As mentioned before, the net process must approach 100-percent current efficiency with respect to the substance being analyzed.

In contrast to the controlled-potential method, the current during a coulometric titration is carefully maintained at a constant and accurately known level; the product of this current and the time required to reach the equivalence point for the reaction yields the number of coulombs, and through this the number of equivalents involved in the electrolysis. The constant-current aspect of this operation is what ordinarily precludes the quantitative oxidation or reduction of the unknown species entirely by a direct-electrode process; at least part (and often all) of the unknown species must react with an electro-generated species. The coulometric oxidation of iron again may be cited as an example.

[5] T. Meites and L. Meites, *Anal. Chem.*, **27**, 1531 (1955); **28**, 103 (1956).
[6] L. Szebelledy and Z. Somogyi, *Z. anal. Chem.*, **112**, 313, 323, 332, 385, 391, 395, 400 (1938).

A coulometric titration, like a volumetric titration, requires some means of detecting the point of chemical equivalence. Most of the end points applicable to volumetric analysis are equally satisfactory here; colored indicators, potentiometric, amperometric, and conductance measurements have all been successfully applied.

The analogy between a volumetric and a coulometric titration extends well beyond the common requirement of an end point. In both, the amount of the unknown is determined through evaluation of its combining capacity— in the one case with a standard solution and in the other with a quantity of electricity. Similar demands are made of the reactions; that is, they must be rapid, essentially complete, and free of side reactions.

It is of interest to compare volumetric and coulometric analysis from the standpoint of equipment and methods. Figure 26.2 shows a block diagram for a coulometric apparatus; this includes a constant-current source, an electric timer, a switch that simultaneously activates the stop clock and the generator circuit, and a current-measuring device. The direct-current source can be considered analogous to the volumetric reagent, its strength being determined by the magnitude of the current. The electric clock and switch correspond to the buret, the switch performing the same function as the stopcock. During the early phases of a coulometric titration the switch is kept closed for extended periods; as the end point is approached, however, small additions of reagent are achieved by closing the switch for shorter and shorter intervals. The similarity to the operation of a buret is obvious.

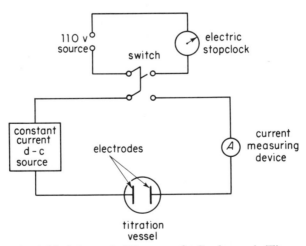

Fig. 26.2 Schematic Diagram of a Coulometric Titration Apparatus.

Some real advantages can be claimed for the coulometric titration when compared with the classical volumetric process. The most obvious of these is that it does away with problems associated with the preparation, standardization, and storage of volumetric solutions. This is particularly important where these consist of such labile substances as chlorine, bromine, or titanous ion. Owing to their instability, their ordinary employment as volumetric reagents poses a problem. Their utilization in coulometric analysis is not beset with these difficulties, since they undergo reaction almost immediately after generation.

Where small quantities of reagent are required, a coulometric titration offers a considerable advantage. By proper choice of current, micro quantities of a substance can be added with considerable ease and accuracy, whereas an equivalent volumetric process would involve the employment of very dilute solutions, a recourse that is always difficult.

A single constant-current source can be employed to generate precipitation, oxidation-reduction, or neutralization reagents. Furthermore, the coulometric method is readily adapted to automatic titrations since control of the passage of a current is easily accomplished.

Coulometric titrations are subject to five potential sources of error: (1) variation in the current during electrolysis; (2) departure of the process from 100-percent current efficiency; (3) error in the measurement of current; (4) error in the measurement of time; and (5) a titration error due to the difference between the equivalence point and the end point. The last of these is common to volumetric methods as well; where the indicator error is the limiting factor, the two methods are likely to be comparable.

With very simple instrumentation, currents constant to 0.2- to 0.5-percent relative are easily achieved; with somewhat more sophisticated apparatus, control to 0.01 percent is obtainable. In general, then, errors due to fluctuations in current need not be serious.

Although generalizations concerning the magnitude of uncertainty associated with the electrode process are difficult, it is certainly true that current efficiency does not appear to be the factor limiting the accuracy of many coulometric titrations.

Errors in measurement of current and of time can be kept small. It is not difficult, even with miniscule currents, to determine the magnitude of the former to 0.01 percent or better. Errors in the latter, however, are likely to be greater and are apt to represent the limiting factor in the accuracy of a coulometric titration. With a good quality, electric stop clock relative time errors of 0.1 percent or smaller can be expected.

To summarize, then, the current-time measurements required for a coulometric titration are inherently as accurate or more accurate than the equivalent volume-normality measurements of classical volumetric analysis, particularly where small quantities of reagent are involved. Often, however, the accuracy of a titration is not limited by these measurements but by the sensitivity of the end point; in this respect the two procedures are equivalent.

Apparatus and Methods

The apparatus for a coulometric titration can be relatively simple in nature compared with that required for the controlled-potential method. The basic components are shown in Figure 26.2 and are discussed in the sections that follow.

Constant-current sources. A variety of constant-current sources for coulometric titrations have been described in the literature. These vary considerably in their complexity and in their performance characteristics. We shall consider only the very simplest type; it is capable of delivering currents to about 20 milliamperes that are constant to approximately 0.5 percent. More complicated devices yield currents of an ampere or greater that vary no more than 0.01 percent over extended periods of time.[7]

A diagram for a constant-current source is shown in Figure 26.3. The power supply consists of two or more high-capacity 45-volt *B* batteries. The current from these passes through a calibrated standard resistance R_1. A potentiometer is connected across this to permit accurate measurement of the potential drop and thus the current via Ohm's law. The resistance of R_1 should be chosen so that IR_1 is about 1 volt; with this arrangement a precise determination of the current I can be obtained even with a relatively simple potentiometer. The variable resistance R_2 has a maximum value of about 20,000 ohms.

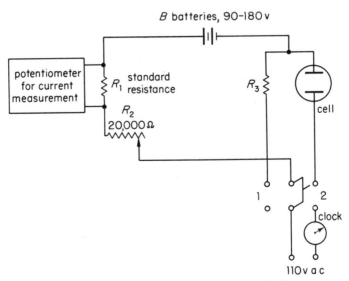

Fig. 26.3 A Simple Apparatus for Coulometric Titrations.

When the circuit is completed by throwing the switch to the number 2 position, the current passing through the cell is

$$I = \frac{E_B - E_{\text{cell}}}{R_1 + R_2 + R_B + R_{\text{cell}}}$$

[7] For a description of constant-current sources see J. J. Lingane, *Electroanalytical Chemistry*, 2d ed., 499-511. New York: Interscience Publishers, Inc., 1958.

where E_B is the potential of the B batteries and E_{cell} consists of the sum of the cathode and anode potentials of the titration cell plus any overvoltage or junction potentials associated with its operation. The resistance of the dry batteries and cell are given by R_B and R_{cell}, respectively.

The potential of the dry cells will remain reasonably constant for short periods of time provided the current drawn is not too large; it is safe to assume, therefore, that E_B as well as R_B will remain constant during any given titration. Variations in I, then, arise only from changes in E_{cell} and R_{cell}. Ordinarily, however, R_{cell} will be on the order of 10 to 20 ohms compared with R_2 which is perhaps 10,000 ohms. Thus, even if R_{cell} were to change by as much as 10 ohms, which is highly unlikely, the effect on the current would be less than one part in a thousand.

Changes in E_{cell} during a titration usually have a greater effect on the current, for the cell potential may be altered by as much as 0.5 volt during the electrolysis. This will result in a variation in I of 0.5 to 0.6 percent if E_B is 90 volts; the same change would cause a variation of only about 0.3 percent if E_B were 180 volts. Experience has shown these to be fairly realistic figures for a simple power source of this kind provided current is drawn more or less continuously from the battery. To accomplish this, a resistance R_3 is employed that is of about the same magnitude as R_{cell}. The switching arrangement shown allows this to be placed in the circuit whenever current is not being passed through the cell.

Measurement of time. The electrolysis time during a titration is best measured with an electric stop clock operated by the same switch as is used to operate the cell. Typically, a titration will involve a time period on the order of 100 to 500 seconds; the clock should therefore be accurate to a few tenths of a second. An ordinary electric stop clock is not very satisfactory because the motor tends to coast when the current is shut off and also lag when started. While the error resulting from one start-stop sequence may be small, a coulometric titration involves many such operations, and the accumulated error from this source can become large. Stop clocks with solenoid-operated brakes eliminate this problem but, unfortunately, are more expensive than the simple laboratory timer.

Another error that may arise in connection with electric timing is caused by variations in the frequency of the 110-volt power supply used to operate the clock. Ordinarily, these variations are less than 0.2 percent and need be considered only when greater accuracies than this are sought.

Cells for coulometric titrations. A typical coulometric titration cell is shown in Figure 26.4. It consists of a generator electrode at which the reagent is formed and a second electrode to complete the circuit. The generator electrode is often a piece or coil of platinum of fairly large surface area; a gauze electrode such as shown on page 538 can also be employed. Frequently the products formed at the second electrode will interfere with the analysis; it is therefore necessary to isolate this from the solution by means of a sintered disk or some other porous medium. Thus, for example, the anodic generation of oxidizing agents is often accompanied by the evolution of hydrogen from the second electrode. Unless this gas is allowed to escape from the solution, reaction between it and the oxidizing agent is likely to occur.

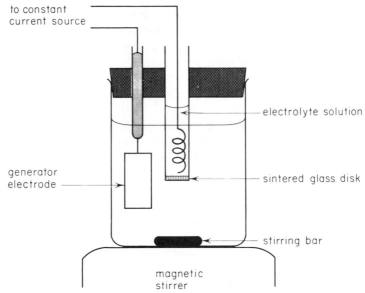

to constant
current source

electrolyte solution

generator
electrode

sintered glass disk

stirring bar

magnetic
stirrer

Fig. 26.4 A Typical Coulometric Titration Cell.

External generation of reagent. Occasionally a coulometric titration cannot be used because of side reactions between the generator electrode and some other constituent in the solution. For example, the coulometric titration of acids involves the formation of base at the cathode

$$2e + 2H_2O \rightleftharpoons H_2 + 2OH^-$$

In the presence of easily reduced substances, a reaction other than this can occur at the generator electrode; departures from the required 100-percent current efficiency thus result. To overcome this problem several ingenious devices for external reagent generation have been devised; Figure 26.5 shows the essential features of one that was designed by DeFord, Johns, and Pitts.[8] During electrolysis an electrolyte solution, such as sodium sulfate, is fed through the tubing at a rate of about 0.2 ml/sec. The hydrogen ions formed at the anode are washed down one arm of the T tube (along with an equivalent number of sulfate ions) while the hydroxyl ions produced at the cathode are transported through the other. The apparatus is so arranged that flow of the electrolyte is discontinued whenever the electrolysis current is shut off, and a flush-out system is provided to rinse the residual reagent from the tube into the titration vessel.

[8] D. D. DeFord, J. N. Pitts, and C. J. Johns, *Anal. Chem.*, **23**, 938 (1951).

Both electrode reactions shown in the illustration proceed with 100-percent current efficiency; thus, the solution emerging from the left arm of the apparatus can be used for the titration of acids, that from the right arm for the titration of bases. A current of about 250 milliamperes is satisfactory for the titration of various acids or bases in the range between 0.2 and 2 milliequivalents; end points are detected with a glass-calomel electrode system. This apparatus has also been used for the generation of iodine by electrolysis of an iodide solution.

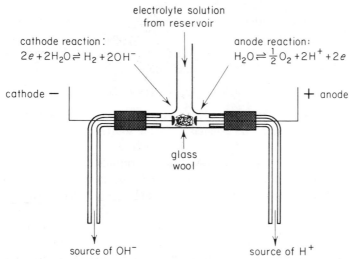

cathode reaction:
$$2e + 2H_2O \rightleftharpoons H_2 + 2OH^-$$

anode reaction:
$$H_2O \rightleftharpoons \tfrac{1}{2}O_2 + 2H^+ + 2e$$

electrolyte solution from reservoir

cathode − + anode

glass wool

source of OH^- source of H^+

Fig. 26.5 Cell for the External Generation of Acid or Base for Coulometric Titrations.

Applications of Coulometric Titrations[9]

Coulometric titrations have been developed for all types of volumetric reactions. Typical applications are described in the following paragraphs.

Neutralization titrations. Both weak and strong acids can be titrated with a high degree of accuracy using electrogenerated hydroxyl ions. The most convenient method, where applicable, involves generation of the reagent at a platinum cathode within the solution. When internal generation can be used, care must be taken to avoid interference from the anode products; with a platinum electrode, these are frequently hydrogen ions. This requires isolation of the anode from the solution being analyzed by some sort of diaphragm arrangement (see Fig. 26.4). A convenient alternative that can often be used involves the addition of chloride or bromide ions to the solution to be analyzed

---

[9] An excellent summary of the applications of the coulometric procedure is found in J. J. Lingane, *Electroanalytical Chemistry*, 2d ed., 536-613. New York: Interscience Publishers, Inc., 1958.

and the use of a silver wire as the anode; the reaction at this electrode then becomes

$$Ag + Br^- \rightleftharpoons AgBr + e$$

Clearly, the product here will not interfere with the neutralization reaction.

Both potentiometric and indicator end points can be employed for these titrations; the problems associated with the estimation of the equivalence point are identical to those encountered in the corresponding volumetric analysis. A real advantage to the coulometric method is that the carbonate problem is far less serious; the only precaution necessary consists of eliminating carbon dioxide from the solution to be analyzed. This is readily accomplished by aeration with a carbon dioxide-free gas before beginning the analysis.

Coulometric titration of strong and weak bases can be performed with hydrogen ions generated at a platinum anode

$$H_2O \rightleftharpoons \tfrac{1}{2}O_2 + 2H^+ + 2e$$

This may be done internally as well as externally; again, if the former method is used, the cathode must be isolated from the solution to prevent interference from the hydroxyl ions produced at that electrode.

Precipitation and complex-formation titrations. A variety of coulometric precipitation titrations involving anodically generated silver ions have been developed; these are summarized in Table 26-1. A cell, such as shown in Figure 26.4, can be employed in which the generator electrode is a piece of heavy silver wire while the cathode is a platinum wire at which hydrogen is evolved. Adsorption indicators or the potentiometric method can be employed for end-point detection. Rather similar applications employing mercurous ion formed at a mercury anode have been described.

Table 26-1

SUMMARY OF APPLICATIONS OF COULOMETRIC TITRATIONS INVOLVING NEUTRALIZATION, PRECIPITATION, AND COMPLEX-FORMATION REACTIONS

Species Determined	Generator Electrode Reaction	Secondary Analytical Reaction
Acids	$2H_2O + 2e \rightleftharpoons 2OH^- + H_2$	$OH^- + H^+ \rightleftharpoons H_2O$
Bases	$H_2O \rightleftharpoons 2H^+ + \tfrac{1}{2}O_2 + 2e$	$H^+ + OH^- \rightleftharpoons H_2O$
Cl^-, Br^-, I^-	$Ag \rightleftharpoons Ag^+ + e$	$Ag^+ + Cl^- \rightleftharpoons \underline{AgCl}$ etc.
Mercaptans	$Ag \rightleftharpoons Ag^+ + e$	$Ag^+ + RSH \rightleftharpoons \underline{AgSR} + H^+$
Cl^-, Br^-, I^-	$2Hg \rightleftharpoons Hg_2^{2+} + 2e$	$Hg_2^{2+} + 2Cl^- \rightleftharpoons \underline{Hg_2Cl_2}$
Zn^{2+}	$Fe(CN)_6^{3-} + e \rightleftharpoons Fe(CN)_6^{4-}$	$3Zn^{2+} + 2K^+ + 2Fe(CN)_6^{4-} \rightleftharpoons$ $\underline{K_2Zn_3[Fe(CN)_6]_2}$
Ca^{2+}, Cu^{2+}, Zn^{2+} and Pb^{2+}	$HgNH_3Y^{2-} + NH_4^+ + 2e$ $\rightleftharpoons Hg + 2NH_3 + HY^{3-}$ (where Y^{4-} is ethylenediaminetetraacetate ion)	$HY^{3-} + Ca^{2+} \rightleftharpoons CaY^{2-} + H^+$, etc.

An interesting coulometric titration makes use of a solution of the mercuric ammonia complex of ethylenediaminetetraacetic acid (H_4Y).[10] The complexing agent is released to the solution as a result of electrolysis at a mercury cathode; the electrode reaction may be written as

$$HgNH_3Y^{2-} + NH_4^+ + 2e \rightleftharpoons Hg + 2NH_3 + HY^{3-}$$

Because the mercury chelate is more stable than the corresponding complexes with calcium, zinc, lead, or copper, complexation of these ions will not occur until the electrode process frees the complexing agent.

Oxidation-reduction titrations. Table 26-2 indicates the large number of reagents that can be generated by the coulometric procedure and the great variety of analyses to which they have been applied. Electrogenerated bromine has proved to be particularly useful among the oxidizing agents, and a host of interesting methods have been developed based upon this substance. Of interest also are some of the unusual reagents not encountered in volumetric titrations because of the instability of their solutions; these include dipositive silver ion, tripositive manganese, and unipositive copper as the chloride complex.

Table 26-2

SUMMARY OF APPLICATIONS OF COULOMETRIC TITRATIONS INVOLVING
OXIDATION-REDUCTION REACTIONS

Reagent	Generator Electrode Reaction	Substance Determined
Br_2	$2Br^- \rightleftharpoons Br_2 + 2e$	As(III), Sb(III), U(IV), Tl(I), I^-, SCN^-, NH_3, N_2H_4, NH_2OH, phenol, aniline, mustard gas, 8-hydroxyquinoline
Cl_2	$2Cl^- \rightleftharpoons Cl_2 + 2e$	As(III), I^-
I_2	$2I^- \rightleftharpoons I_2 + 2e$	As(III), Sb(III), $S_2O_3^{2-}$, H_2S
Ce^{4+}	$Ce^{3+} \rightleftharpoons Ce^{4+} + e$	Fe(II), Ti(III), U(IV), As(III), I^-, $Fe(CN)_6^{-4}$
Mn^{3+}	$Mn^{2+} \rightleftharpoons Mn^{3+} + e$	$H_2C_2O_4$, Fe(II), As(III)
Ag^{2+}	$Ag^+ \rightleftharpoons Ag^{2+} + e$	Ce(III), V(IV), $H_2C_2O_4$, As(III)
Fe^{2+}	$Fe^{3+} + e \rightleftharpoons Fe^{2+}$	Cr(VI), Mn(VII), V(V), Ce(IV)
Ti^{3+}	$TiO^{2+} + 2H^+ + e \rightleftharpoons Ti^{3+} + H_2O$	Fe(III), V(V), Ce(IV), U(VI)
$CuCl_3^{2-}$	$Cu^{2+} + 3Cl^- + e \rightleftharpoons CuCl_3^{2-}$	V(V), Cr(VI), IO_3^-
U^{4+}	$UO_2^{2+} + 4H^+ + 2e \rightleftharpoons U^{4+} + 2H_2O$	Cr(VI), Ce(IV)

[10] C. N. Reilley and W. W. Porterfield, *Anal. Chem.*, **28**, 443 (1956).

problems

1. Lead was separated from cadmium at a mercury cathode by a controlled-potential coulometric method. Calculate the percent lead in a 0.106-gram sample that required 6.13 coulombs to complete the deposition.

    ans. percent Pb $= 6.2$

2. Calculate the theoretical volume of gas at standard conditions that will be formed in a hydrogen-oxygen coulometer with the passage of 1.000 coulomb of current.

3. The quantitative reduction of ferric iron to the ferrous state required 22.8 coulombs of current for a solution having a volume of 25.0 ml. Calculate the formal concentration of iron in the solution.

4. Calculate the grams of acetic acid in a sample that was titrated coulometrically. A current of exactly 0.0631 ampere was used and 112.3 sec were required to reach a phenolphthalein end point.   ans. 0.00442 gram

5. A 0.741-gram sample containing As_2O_3 was dissolved in about 50 ml of a solution that was about 0.1 F in KBr. The $+ 3$ arsenic ions were converted to the $+ 5$ state by titration with electrogenerated Br_2. A constant current of 0.246 ampere was employed and an end point reached after 5.28 min of electrolysis. Calculate the percent As_2O_3 in the sample. ans. 5.39 percent

6. A 0.213-gram sample of an organic chloride was decomposed and the resulting chloride ion was titrated coulometrically with electrogenerated silver ion. With a current of 0.512 ampere, 3.81 minutes were required to reach an end point with fluorescein as an indicator. What was the percent Cl in the sample ?

7. Calculate the equivalent weight of a pure organic acid that was determined by solution of a 0.1516-gram sample in 50 ml of water. With a current of 0.384 ampere, exactly 330 sec were required to reach a phenolphthalein end point with the electrogenerated hydroxyl ions.

8. The Al in a 0.0271-gram sample was precipitated with 8-hydroxyquinoline. The precipitate (AlQ_3) was filtered, washed, and redissolved and the 8-hydroxyquinoline brominated with electrogenerated bromine (see p. 470 for the reaction). With a current of 0.100 ampere, an end point was reached in 403 sec. Calculate the percent Al_2O_3 in the sample.

chapter 27.   *Polarography and Amperometric Titrations*[1]

In this chapter we shall discuss two closely related electroanalytical methods. The first of these, *polarography*, or *voltammetry* was discovered in the early nineteen twenties by Jaroslav Heyrovsky, a Czechoslovakian chemist.[2] Later in the same decade, Heyrovsky and a co-worker described the use of the polarographic technique for the detection of end points in volumetric analyses;[3] such methods have come to be known as *amperometric titrations*.

The majority of the elements, in one form or another, are amenable to analysis by the polarographic procedure. Their behavior is often quite individual, and this imparts a reasonable degree of selectivity to the method. Not only do

[1] There are several excellent monographs dealing with the principles and applications of polarography. Among these are (1) I. M. Kolthoff and J. J. Lingane, *Polarography*, 2d ed. New York: Interscience Publishers, Inc., 1952; (2) L. Meites, *Polarographic Techniques*. New York: Interscience Publishers, Inc., 1955; (3) J. Heyrovsky in W. G. Berl, Ed., *Physical Methods in Chemical Analysis*. **2**. New York: Academic Press, 1951.

[2] J. Heyrovsky, *Chem. listy*, **16**, 256 (1922).

[3] J. Heyrovsky and S. Berezicky, *Collection Czechoslov. Chem. Commun.*, **1**, 19 (1929).

inorganic ions and molecules respond to polarographic analysis, but several organic functional groups exhibit characteristic behavior as well; the method therefore occupies an important role in both inorganic and organic analysis.

While a polarographic analysis is ordinarily performed upon an aqueous solution of the constituent of interest, a variety of organic solvents have also been used. The optimum concentration range for polarographic work lies between 10^{-4} and 10^{-2} F, although the minimum figure can often be reduced by another factor of ten. Analyses can easily be performed on a milliliter or two of solution, and with a little effort quantities as small as a single drop can be handled. Thus, the polarographic method is particularly important in dealing with milligram and microgram amounts although larger quantities can be treated by suitable dilution.

In routine polarographic analyses, relative errors of 1 to 2 percent are to be expected. This is comparable with or better than the accuracy attainable with the classical gravimetric or volumetric methods when these are employed for the determination of very small quantities of a substance.

Fundamental Principles of Polarography

Any discussion of the principles of the polarographic method calls for frequent reference to some of the fundamentals of electrochemistry; thus a review of the material presented in Chapters 22 and 23 may be advisable— particularly the discussion of polarization phenomena found on pages 519 and 528.

A Brief Description of Polarographic Measurements

Polarographic data are obtained by the measurement of the current as a function of the potential applied to a special type of electrolytic cell. A plot of the data gives current-voltage curves, called *polarograms*; these provide both qualitative and quantitative information about the constitution of the solution in which the electrodes are immersed.

Polarographic cells. A polarographic cell consists of a small, easily polarized *microelectrode*, a large nonpolarizable reference electrode, and the solution to be analyzed. The microelectrode, at which the analytical reaction occurs, is an inert metal surface having an area of a few square millimeters. Most commonly it takes the form of the *dropping mercury electrode* shown in Figure 27.1. Here, mercury is forced by gravity through a very fine capillary to provide a continuous stream of identical droplets, each having a maximum diameter of about 0.5 to 1 mm. The lifetime of a drop is typically 2 to 6 sec. We shall see that the dropping mercury electrode has unique properties that make it particularly well suited for polarographic work. Other microelectrodes, however, can be employed; these may take the form of small diameter wires of platinum or other inert metals.

The reference electrode in a polarographic cell should be massive relative to the microelectrode so that its behavior remains essentially constant with the passage of the small currents; that is, it remains unpolarized during the analysis. A saturated calomel electrode and salt bridge, arranged in the manner shown in Figure 27.1 is frequently employed; another common reference electrode consists simply of a large pool of mercury.

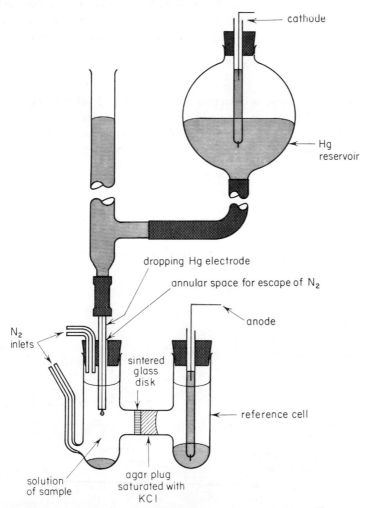

Fig. 27.1 A Dropping Mercury Electrode and Cell. (Reprinted from J. J. Lingane and H. A. Laitinen, *Ind. Eng. Chem., Anal. Ed.*, **11**, 504 (1939) with permission.)

Polarograms. A polarogram is a plot of current as a function of the potential applied to a polarographic cell. In most instances, the microelectrode is attached to the negative terminal of the power supply, and *by convention*, the applied potential is given a negative sign under these circumstances. By convention also, the currents are designated as positive when the flow of electrons is from the power supply into the microelectrode—that is, when that electrode behaves as a cathode.

Figure 27.2 shows two polarograms, the lower being for a solution that is 0.1 F in potassium chloride and the upper for a solution that is 1×10^{-3} F in cadmium chloride in addition. An S-shaped current-voltage curve, called a *polarographic wave*, is produced in the presence of cadmium ion. This results from the cathodic reduction of this species to give cadmium amalgam

$$Cd^{2+} + 2e + Hg \rightleftharpoons Cd(Hg)$$

In both plots a sharp rise in current occurs at about -2 volts; this is associated with reduction of potassium ions to give a potassium amalgam.

For reasons to be considered presently, a polarographic wave suitable for analysis is obtained only in the presence of a large excess of a *supporting electro-*

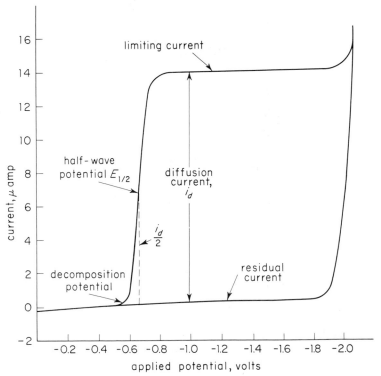

Fig. 27.2 Polarogram for Cadmium Ion. Upper curve is for a solution that is 1×10^{-3} F with respect to Cd^{2+} and 1.0 F with respect to KCl. Lower curve is for a solution containing 1.0 F KCl only.

lyte; potassium chloride serves this function in the present example. Examination of the polarogram for the supporting electrolyte alone reveals that a small current, called the *residual current*, passes through the cell even in the absence of reducible cadmium ions. The voltage at which the polarogram for the electrode-reactive species departs from the residual-current curve is called the *decomposition potential*.

One of the characteristic features of a polarographic wave is the region in which the current levels off after a sharp rise and becomes essentially independent of the applied voltage; this is called a *limiting current*. As we shall show, the limitation is the result of a restriction in the rate at which the participant in the electrode process is brought to the small surface of the electrode; with proper control over experimental conditions, this rate is determined almost exclusively by the velocity at which the reactant diffuses. Under these circumstances the limiting current is given a special name, the *diffusion current*, and assigned the symbol i_d. Ordinarily the diffusion current is directly proportional to the concentration of the reactive constituent and is thus of prime importance from the standpoint of analysis. In Figure 27.1, the limiting current is a diffusion current, and its method of measurement is readily apparent.

We must define one other important term, the *half-wave potential*. This is simply the potential corresponding to a current that is equal to one half the diffusion current. The half-wave potential is usually given the symbol $E_{1/2}$ and is important for qualitative identification of the reactant.

Interpretation of Polarographic Waves

Now we shall undertake a qualitative interpretation of a typical polarographic wave. For this purpose we shall assume that the half reaction at the microelectrode is

$$\text{Ox} + ne \rightleftharpoons \text{Red} \qquad\qquad E° = -0.3 \text{ volt}$$

where Ox represents the substance being reduced and Red is the symbol for the product. The latter may consist of a metal amalgam, a metallic deposit (where a solid microelectrode is employed), or a soluble ion or molecule. We shall further stipulate that the electrode reaction is both rapid and reversible.

We must first consider the chemical changes that occur in the film of liquid immediately surrounding the surface of the microelectrode. The layer under consideration is so thin its constituents may be considered in *instantaneous* equilibrium with the electrode surface; reactant concentrations in this layer are thus determined by the potential of the electrode alone and can be calculated by means of the Nernst equation. Hence

$$E_{\text{cathode}} = E°_{\text{Ox} \rightarrow \text{Red}} - \frac{0.059}{n} \log \frac{[\text{Red}]_0}{[\text{Ox}]_0}$$

where the zero subscripts signify that the indicated activities apply only to the film immediately adjacent to the cathode. If sufficient time were allowed at any given cathode potential, equilibrium would ultimately be established, and these activities would then be applicable for the entire solution. At first, however, equilibrium would only be found in the surface film in which we are interested.

On the basis of the discussion in earlier chapters we may express the applied potential as

$$E_{applied} = E_{cathode} + E_{anode}$$

We are assuming that the junction potential in the cell is negligibly small and that there are no overvoltage effects. Furthermore, we have not taken account of the IR drop inasmuch as the resistance of the cell is ordinarily small and the currents, on the order of 10^{-6} ampere (1 microampere), are minute.

Typically, the anode of a polarographic cell is a saturated calomel electrode; we may therefore write the above equation as

$$E_{applied} = E°_{Ox \to Red} - \frac{0.059}{n} \log \frac{[Red]_0}{[Ox]_0} + E_{S.C.E.}$$

which, in turn, can be rearranged to give

$$\frac{n(E_{applied} - E_{S.C.E.} - E°_{Ox \to Red})}{0.059} = \log \frac{[Ox]_0}{[Red]_0}$$

This allows the calculation of the ratio of the reactant to product activities *at the electrode surface* at any applied potential. It should be noted that this ratio becomes smaller as the applied potential becomes more negative.

We shall now examine the electrode behavior at an applied potential great enough for an appreciable concentration of the product, Red, to be expected at the electrode surface. At an applied potential of —0.6 volt, for example, the ratio of $[Ox]_0$ to $[Red]_0$ calculated from the above equation will be about 0.1 if a value of —0.242 volt is taken for the potential of the saturated calomel electrode and if n is assumed to be 1. At this potential, then, sufficient reduction of Ox will occur essentially instantaneously to establish this concentration ratio in the surface film of solution. The momentary current required for this would diminish rapidly to zero were it not for the fact that more Ox ions or molecules migrate into the surface film from the bulk of the solution. If the electrochemical reaction is essentially instantaneous, the current (that is, the rate of flow of electrons) will be directly dependent upon the rate at which the Ox particles are transported into the surface layer from the bulk of the solution; as a result, we may write

$$i' = k' \times \text{rate of movement of Ox into the surface film}$$

where k' is a proportionality constant and i' is the current at the applied potential of —0.6 volt.

From an earlier discussion (p. 521) we know that movement of ions or molecules in a cell can occur in three ways—by diffusion, by thermal or mechanical convection, and by electrostatic attraction. In polarography every effort is made to eliminate the last two of these; vibration and stirring of the cell solution is avoided and a large excess of a nonreactive supporting electrolyte is employed to reduce the electrostatic attraction between the electrode and the reactant ions. Thus, the rate at which Ox particles are brought into the depleted surface film will depend upon the rate at which this species will diffuse. Since the rate of diffusion of a substance is directly proportional to the concentration difference between the parts of the solution through which this process occurs, we may write

$$\text{rate of diffusion of Ox} = k''([\text{Ox}] - [\text{Ox}]_0)$$

where [Ox] represents the concentration of the reactant in the bulk of the solution and $[\text{Ox}]_0$ is the concentration in the film adjacent to the electrode; k'' is a proportionality constant. Furthermore, if diffusion is the only process responsible for movement of Ox particles to the electrode surface, then

$$i' = k' \times \text{rate of diffusion of Ox}$$

and

$$i' = k([\text{Ox}] - [\text{Ox}]_0')$$

where $[\text{Ox}]_0'$ is the surface concentration of Ox at the applied voltage of -0.6 volt. Therefore, the magnitude of the current is determined by the concentration of the reactant at the electrode surface, which in turn is determined by the electrode potential.

We shall next consider the situation when the applied potential has been increased to -0.7 volt. Here the equilibrium-concentration ratio at the surface of the electrode will be 2×10^{-3}. As a consequence, the concentration of Ox in this region will have decreased to a new value $[\text{Ox}]_0''$, and we may write

$$i'' = k([\text{Ox}] - [\text{Ox}]_0'')$$

The decrease in the surface concentration of Ox results in an increased diffusion rate and thus an increase in the current.

At higher potentials, calculation indicates that the ratio $[\text{Ox}]_0/[\text{Red}]_0$ continues to diminish; it has a value of 1×10^{-6} at -0.9 volt, and 2×10^{-8} at -1.0 volt. With $[\text{Ox}]_0$ becoming smaller and smaller, the difference $([\text{Ox}] - [\text{Ox}]_0)$ assumes an essentially constant value. Thus, as the concentration of Ox at the surface of the electrode approaches zero, the expression for the current in the cell simplifies to

$$i_d = k[\text{Ox}]$$

In this region, then, the diffusion rate is constant and independent of the applied potential. Therefore, the current is also constant. Note that *the magnitude of the diffusion current is directly proportional to the concentration of the reactant in the bulk of the solution.* Quantitative polarography is based upon this fact.

When a limiting current is achieved as a consequence of the limitation of the rate at which a reactant can be brought to the surface of an electrode, a state of complete *concentration polarization* is said to exist. The currents required to reach this condition are very small indeed for a microelectrode—roughly on the order of 3 to 4 microamperes[4] for a solution in which the reactant con-

---

[4] One microampere $= 10^{-6}$ ampere.

centration is 10^{-3} F. This is significant, for it means that the reactant concentration is not altered appreciably by the electrolytic process with an ordinary volume of solution. For example, passage of a 10 microampere current for 10 minutes removes only about 6×10^{-5} milliequivalent of reactant. Thus, the implicit assumption in the foregoing discussion that the concentration of Ox in the bulk of the solution remains constant is reasonable.

Equation for the Polarographic Wave

It is a fairly easy matter to derive an equation for most polarograms; this, in turn, reveals the significance of the half-wave potential. We shall again write the Nernst expression for the process we have been considering

$$E_{\text{applied}} - E_{\text{S.C.E.}} = E° - \frac{0.059}{n} \log \frac{[\text{Red}]_0}{[\text{Ox}]_0} \tag{27-1}$$

We have seen that at any applied potential

$$i = k([\text{Ox}] - [\text{Ox}]_0) \tag{27-2}$$

and that

$$i_d = k[\text{Ox}] \tag{27-3}$$

Subtracting (27-2) from (27-3) gives

$$[\text{Ox}]_0 = \frac{i_d - i}{k}$$

Now, if Red is a soluble substance, its concentration at the surface of the electrode will also be proportional to the current, and we may write

$$i = k_r[\text{Red}]_0$$

Substituting the foregoing into (27-1) gives

$$E_{\text{applied}} - E_{\text{S.C.E.}} = E° - \frac{0.059}{n} \log \frac{i}{(i_d - i)} \cdot \frac{k}{k_r}$$

which can then be written as

$$E_{\text{applied}} = E_{\text{S.C.E.}} + E° - \frac{0.059}{n} \log \frac{k}{k_r} - \frac{0.059}{n} \log \frac{i}{(i_d - i)} \tag{27-4}$$

Now, according to the definition of half-wave potential, when $i = i_d/2$

$$E_{\text{applied}} = E_{1/2}$$

Substituting into (27-4), we obtain

$$E_{1/2} = E_{\text{S.C.E.}} + E° - \frac{0.059}{n} \log \frac{k}{k_r} \tag{27-5}$$

and we see that the half-wave potential is a constant that is related to the standard potential for the half reaction and *independent* of the reactant concentration. Now, we may simplify (27-4) by substituting $E_{1/2}$ for the several equivalent terms;

$$E_{\text{applied}} = E_{1/2} - \frac{0.059}{n} \log \frac{i}{(i_d - i)} \tag{27-6}$$

This defines the relationship between current and applied potential for a reversible reaction involving the formation of a soluble product. It is easy to show that a similar expression results when the half reaction involves reduction of an ion to a metal that is soluble in the mercury electrode.

Half-wave potential. An examination of equation (27-5) reveals the value of the half-wave potential as a reference point on a polarographic wave; it is independent of the concentration of the reactant, but is directly related to the standard potential for the reaction. In practice, the half-wave potential is a useful quantity for the identification of the species responsible for a given polarographic wave.

These remarks and derivations apply only to those electrode reactions that are rapid and reversible. Where this condition is not realized, the half-wave potential may vary considerably with concentration; in addition, equation (27-4) is inadequate for the description of the wave.

Finally, the half-wave potentials of certain types of reversible reactions are concentration dependent.

Effect of complex formation on polarographic waves. We have already seen (p. 384) that the potential for the oxidation or reduction of a metallic ion is greatly affected by the presence of species that form complexes with that ion. It is not surprising, therefore, that similar effects are observed with polarographic half-wave potentials. The data in Table 27-1 indicate that the half-wave potential for the reduction of a metal complex is generally more negative than that for the corresponding simple metal ion.

Table 27-1

EFFECT OF COMPLEXING AGENTS ON POLAROGRAPHIC HALF-WAVE POTENTIALS
AT THE DROPPING MERCURY ELECTRODE

Ion	Noncomplexing Media	1 F KCN	1 F KCl	1 F NH$_3$, 1 F NH$_4$Cl
Cd^{2+}	—0.59	—1.18	—0.64	—0.81
Zn^{2+}	—1.00	NR*	—1.00	—1.35
Pb^{2+}	—0.40	—0.72	—0.44	
Ni^{2+}		—1.36		—1.10
Co^{2+}		—1.45	—1.20	—1.29
Cu^{2+}	+0.02	NR*	+0.04 and —0.22	—0.24 and —0.51

* NR: no reduction before decomposition of the supporting electrolyte.

Lingane[5] has shown that the magnitude of this shift in half-wave potential is related to the stability of the complex as well as the concentration of the complexing reagent; instability constants are often estimated from measurement of such shifts.

Because of the effect of complexing reagents on half-wave potentials, the electrolyte content of the solution should be carefully controlled whenever polarographic data are used for the qualitative identification of the constituents of a solution.

Polarograms for irreversible reactions. Many of the electrode processes used for polarographic analysis are quite irreversible; this is particularly prevalent with organic systems. In these cases the waves are more drawn out and less well defined than those for reversible reactions. The quantitative description of such waves requires an additional term in equation (27-6) to account for the kinetics of the electrode reaction. Figure 27.7 shows a polarogram for a typical irreversible reaction.

Frequently the half-wave potential for an irreversible reaction shows a dependence upon the concentration of the reactant; diffusion currents, however, remain linearly related to concentration and such processes are readily adapted to quantitative analysis.

The Dropping Mercury Electrode

Most of our remarks thus far are applicable to polarograms obtained with all types of microelectrodes. The great majority of polarographic work, however, has been performed with the dropping mercury electrode, and we shall now discuss the unique properties of this device that make it so attractive for polarographic work.

Advantages and limitations of the dropping mercury electrode. The dropping mercury electrode possesses three distinct advantages that have led to its widespread employment in polarography. The first is the large overvoltage for hydrogen gas formation at a mercury surface; this allows polarographic reduction of many ions from acidic solution. Second, the behavior of the electrode is quite independent of its past history, and reproducible current-voltage curves are obtained regardless of how the electrode has been used. This property results from the continuous formation of a new metal surface as the electrode is used. The third unique feature of this device is that reproducible average currents are instantly achieved at any given applied potential; in contrast, a stationary solid microelectrode requires several minutes before steady currents are obtained. This difference in behavior is attributable to the difference in thickness of the diffusion layer in the two cases. In the former, this layer is very thin, since it develops only during the brief lifetime of a drop whereas with the latter a considerably thicker diffusion layer results as the electrolysis proceeds. Current decreases are observed until a stable diffusion layer is formed in the

[5] J. J. Lingane, *Chem. Rev.*, **29**, 1 (1941).

case of the solid electrode. Such decreases are also operating during the formation of a drop with the mercury electrode; these, however, are perfectly reproducible with each drop, and, in addition, tend to be counteracted by an increase in current resulting from the growth of the electrode surface area. Thus the current with a dropping electrode fluctuates during the lifetime of each drop, but these fluctuations are perfectly regular.

The most serious limitation of the dropping mercury electrode arises from the ease with which metallic mercury is oxidized; this severely restricts the use of the electrode as an anode. At applied potentials much above $+0.4$ volt (vs. the saturated calomel electrode) large currents result from the formation of mercurous ions; these mask currents due to other oxidizable species in the solution. Thus, the dropping mercury electrode can be employed for the analysis of only reducible or very easily oxidizable substances.

Current variations with a dropping electrode. We have noted that current fluctuations are observed during the lifetime of a drop. Experimentally, these periodic variations pose no serious measurement problem; by using a well-damped galvanometer, the oscillations are limited to a reasonable magnitude and the average current is readily determined. Figure 27.3 illustrates a polarogram in which the damped current is recorded automatically with a pen and ink recorder.

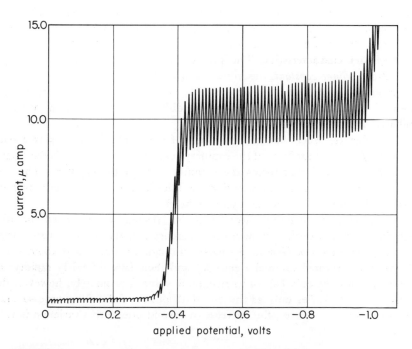

Fig. 27.3 Typical Polarogram Produced by an Instrument that Records Current-voltage Data Continuously and Automatically.

Polarographic Diffusion Currents

As we have seen, the basis of quantitative polarography is the direct proportionality between diffusion current and concentration of the reactive species in the bulk of the solution. We must now consider the variables, other than concentration, that influence the size of this current when a dropping mercury electrode is used.

The Ilkovic equation. In 1934, D. Ilkovic derived a fundamental equation relating the various parameters that determine the size of the diffusion currents obtained with a dropping mercury electrode.[6] This equation, which has been given his name, is shown below

$$i_d = 607 \, n \, D^{1/2} \, m^{2/3} \, t^{1/6} \, C \qquad (27\text{-}7)$$

Here, i_d is the time-average diffusion current in microamperes during the lifetime of a drop. The quantity n is the number of faradays of electricity per mol of reactant; D is the diffusion coefficient for the reactive species expressed in units of square centimeters per second; m is the rate of mercury flow in milligrams per second; t is the drop time in seconds; and C is the concentration of the reactant in millimols per liter. The quantity 607 represents the combination of several constants.

The Ilkovic equation contains certain assumptions that lead to discrepancies of a few percent between calculated and experimental diffusion currents. Corrections to the equation have been derived that yield better correspondence;[7] for most purposes, however, the simple equation gives a satisfactory picture of the factors that influence the current.

Capillary characteristics. The product $m^{2/3} \, t^{1/6}$ in the Ilkovic equation, called the *capillary constant*, describes the influence of the dropping-electrode characteristics upon the diffusion current; since both m and t are readily evaluated experimentally, comparison of diffusion currents from different capillaries is thus possible.

Two factors, other than the geometry of the capillary itself, play a part in determining the magnitude of the capillary constant. One of these is the height of the mercury head that forces the mercury through the capillary. This influences both m and t in such a way that the diffusion current is directly proportional to the square root of the mercury height.

The drop time t of a given electrode is also affected by the applied potential since the interfacial tension between the mercury and the solution varies with the charge on the drop. Generally t passes through a maximum at about —0.4 volt (vs. the saturated calomel electrode) and then falls off fairly rapidly; at —2.0 volts t may be only half of its maximum value. Fortunately, however, the diffusion current varies only as the one sixth power of the drop time so that, over small potential ranges, the decrease in current due to this variation is very small.

---

[6] D. Ilkovic, *Collection Czechoslov. Chem. Commum.*, **6**, 498 (1934).

[7] See J. J. Lingane and B. A. Loveridge, *J. Am. Chem. Soc.*, **72**, 438 (1950).

Diffusion coefficient. The Ilkovic equation indicates that the diffusion currents for various species differ as the square root of their diffusion coefficients D. This quantity gives a measure of the rate at which a given species diffuses through a unit concentration gradient; it is dependent upon such factors as the size of the ion or molecule, its charge, if any, and the vicosity and composition of the solvent. The coefficient for a simple hydrated metal ion is often quite different from that of its complexes; as a result, the diffusion current as well as the half-wave potential may be affected by the presence of a complexing reagent.

Temperature. Temperature has an effect upon several of the variables that govern the diffusion current for a given species, and the over-all effect of this variable is thus complex. The most temperature sensitive of the factors in the Ilkovic equation is the diffusion coefficient which, in most instances, changes by about 2.5 percent per degree. As a consequence, temperature control to a few tenths of a degree is necessary for accurate polarographic analysis.

Kinetic and Catalytic Currents

There are instances where polarographic limiting currents are controlled not only by the diffusion rate of the reactive species but also by the rate of some *chemical reaction* related to the electrode process. These currents are no longer predictable by the Ilkovic equation and are abnormally influenced by temperature, capillary characteristics, and solution composition. They are known as *kinetic currents*. The polarographic behavior of formaldehyde may be cited as an example of this. In aqueous solution, two forms of this compound are in equilibrium

$$CH_2(OH)_2 \rightleftharpoons HCHO + H_2O$$

The hydrated form predominates, but only the unhydrated form is reduced at a dropping electrode. Thus, when a suitable potential is applied, the concentration of the latter is reduced to nearly zero at the electrode surface. This results in a shift in the equilibrium to the right and formation of more unhydrated formaldehyde which then can react. In this instance, however, the rate of the equilibrium shift is slow; thus, the supply of available reactant is controlled by a reaction rate rather than by a diffusion rate. The net effect is a smaller limiting current than would be observed if the process were totally diffusion controlled.

A *catalytic current* represents another type of limiting current that depends upon the rate of a chemical reaction. Here, the reactive species is regenerated by a chemical reaction involving some other component of the solution. The reduction of ferric iron in the presence of hydrogen peroxide provides an example of this phenomenon. The magnitude of the limiting current for iron in the presence of this compound is greatly enhanced even when the applied potential is kept well below the value required to reduce the peroxide. This is readily explained by assuming that the following chemical reaction occurs in the surface layer following the electrolytic formation of ferrous ion:

$$2Fe^{2+} + H_2O_2 \rightarrow 2Fe^{3+} + 2OH^-$$

The current is then controlled in part by the rate of this reaction.

Both kinetic and catalytic currents can be used for analytical purposes. The latter are particularly useful for determining very small concentrations of certain species. Both are quite sensitive to the variables that influence the rates of chemical reactions.

Polarograms for Mixtures of Reactants

Ordinarily, the individual reactants of a mixture will behave independently of one another at a microelectrode so that a polarogram for a mixture is simply the summation of the waves of the individual components. This is illustrated in Figure 27.4 which shows polarograms for a pair of two-component mixtures.

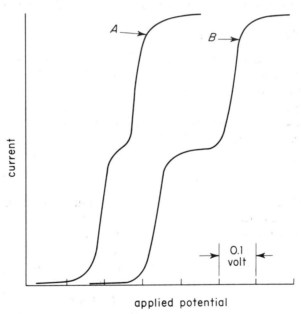

Fig. 27.4 Polarograms for Two-component Mixtures. Half-wave potentials differ by 0.1 volt in curve *A*, by 0.2 volt in curve *B*.

For one, the half-wave potentials of the two reactants differ by about 0.1 volt and for the second by about 0.2 volt. Only in the second case would it be possible to determine accurately the diffusion current of each of the reactants; difficulty would be encountered in making precise current measurements on the basis of the first curve.

Ideally, it is possible to analyze several components of a mixture. In order to achieve this it is necessary that the half-wave potentials of the individual species be sufficiently separated and that their concentrations be approximately the same. For electrode reactions involving a two-electron charge per mol, a difference of 0.1 to 0.2 volt in half-wave potentials will suffice. For one-electron reactions a difference of 0.2 to 0.3 volt is desirable.

Anodic Waves and Mixed Anodic-Cathodic Waves

Anodic waves as well as cathodic waves are encountered in polarography. The former are the less common, however, because of the relatively small range of anodic potentials that can be covered with the dropping mercury electrode before oxidation of the electrode itself commences. An example of an anodic wave is found in Curve 1 of Figure 27.5, where the electrode reaction

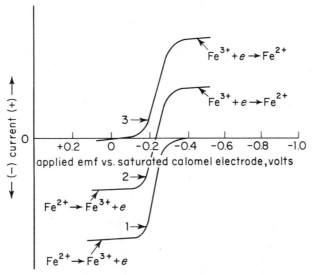

Fig. 27.5 Polarographic Behavior of Ferrous and Ferric Iron in a Citrate Medium. 1. Anodic wave for a solution in which $[Fe^{2+}] = 1 \times 10^{-3}\ F$. 2. Anodic-cathodic wave for a solution in which $[Fe^{2+}] = [Fe^{3+}] = 0.5 \times 10^{-3}\ F$. 3. Cathodic wave for a solution in which $[Fe^{3+}] = 1 \times 10^{-3}\ F$.

consists of the oxidation of $+2$ iron to the $+3$ state in the presence of citrate ion. A diffusion current is obtained at 0 volt (vs. the saturated calomel electrode) which is due to the half reaction

$$Fe^{2+} \rightarrow Fe^{3+} + e$$

As the potential is made more negative, a decrease in the anodic current occurs; at about -0.4 volt the current becomes zero because the oxidation of ferrous ion has ceased.

Curve 3 represents the polarogram for a solution of ferric iron in the same medium. Here a cathodic wave results from the reduction of the $+3$ iron to the

+2 state. The half-wave potential is identical with that of the anodic wave, indicating that the oxidation and reduction of the two iron species is perfectly reversible at the dropping electrode.

Curve 2 is the polarogram of an equiformal mixture of ferrous and ferric iron. The portion of the curve below the zero current line corresponds to the oxidation of the ferrous iron; this reaction ceases at an applied potential equal to the half-wave potential. The upper portion of the curve is for the reduction of the ferric iron present.

Current Maxima

Often the shapes of polarograms are distorted by so-called current maxima (see Fig. 27.6). These are troublesome because they lead to difficulties in the accurate evaluation of diffusion currents and half-wave potentials. The cause or causes of maxima are not fully understood, but fortunately there is a good deal of empirical knowledge of methods for their elimination. These generally involve the addition of traces of any of a variety of high molecular-weight substances such as gelatin, Triton X-100 (a commercial surface-active agent), methyl red and other dyes, and carpenter's glue. The first two of these are particularly useful.

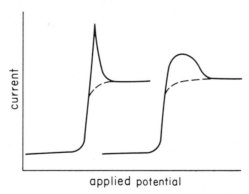

Fig. 27.6 Polarograms that Illustrate Two Common Types of Current Maxima.

Residual Current

As mentioned earlier, the lower concentration limit of the polarographic method is about $10^{-5}\ F$ with respect to the reactive species. Below this figure, diffusion currents have the same magnitude as the residual current, and the uncertainties in measurement of the latter result in major errors in the analyses. Even at higher concentrations, accurate evaluation of the residual current is imperative for good analytical results.

The typical residual current curve, such as that shown in Figure 27.2, has two sources. One of these is the reduction of trace impurities that are almost inevitably present in the blank solution; contributors here would be small amounts of dissolved oxygen, heavy metal ions from the distilled water, as well as impurities present in the salt used as the supporting electrolyte. Because of the high sensitivity of the polarographic method, even the very best grade reagent will often contain sufficient contaminants to affect the residual current. For example, if the solution to be studied is made 1.0 F in potassium nitrate as a supporting electrolyte, as little as a 0.001 percent of reducible impurity in that salt may contribute appreciably to the residual current.

A second component of the residual current is the so-called charging or condenser current arising from a flow of electrons that charge the mercury droplets with respect to the solution; this current may be either negative or positive. At high applied potentials, the droplets are negative with respect to the solution by virtue of the electrons forced to the surface by the high applied voltage. These excess electrons are carried down with the drop as it breaks and since each new drop is charged as it forms, a steady current flow results. The direction of the current in this instance can be given a positive sign. At applied potentials smaller than about —0.4 volt (vs. the saturated calomel electrode) the mercury tends to be positive with respect to the solution; thus, as each drop is formed, electrons are repelled from the surface toward the bulk of mercury. This results in a negative current. At about —0.4 volt, the mercury surface is uncharged and the condenser current is zero.

From the analytical viewpoint, correction of the limiting current for the residual current is necessary if proportionality between current and concentration is to be obtained. For accurate work this involves determination of the polarographic behavior of a blank solution that is as nearly identical as possible to the solution being analyzed except for the component of interest.

Supporting Electrolyte

In order for the electrode process to be diffusion controlled, the presence of a supporting electrolyte is essential. In its absence, limiting currents are still obtained, but their magnitude is sensitive to small variations in the electrolyte content; as a result, these have no analytical value. The data in Table 27-2 illustrates this. The limiting currents for lead ion decrease markedly with the addition of potassium nitrate but finally become independent of the concentration of that salt. Similar behavior is observed when the potassium nitrate is replaced by other salts, and also when other reducible species are studied. At the high salt concentrations, the attractive force between the negative cathode and the positive lead ions is largely masked by the presence of the very high concentration of the electrolyte so that currents dependent upon diffusion alone are observed. At lower salt concentrations, the limiting current is made up of the diffusion current and a current due to electrostatic forces; the latter is sometimes called the *migration current*. For the first entry in Table 27-2, the migration current for lead ion alone in a 9.5×10^{-4} F solution is about 9.2 microamperes (17.6 — 8.45).

Table 27-2

Effect of Supporting Electrolyte Concentration on
Polarographic Currents for Lead Ion[8]

(Solution 9.5×10^{-4} F in $PbCl_2$)

Potassium Nitrate Concentration, F	Limiting Current, microamperes
0	17.6
0.0001	16.2
0.001	12.0
0.005	9.8
0.10	8.45
1.00	8.45

[8] Data from J. J. Lingane and I. M. Kolthoff, *J. Am. Chem. Soc.*, **61**, 1045 (1939). With permission.

It is of interest to note that limiting currents for the reduction of anions (such as iodate or chromate) become larger with added electrolyte since the electrostatic forces are repulsive rather than attractive.

It is general practice to employ a supporting electrolyte concentration that is 50 to 100 times greater than the concentration of the reactive species; in this

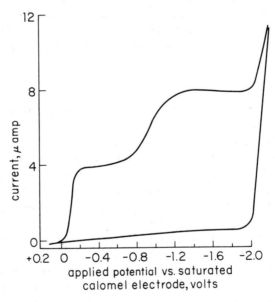

Fig. 27.7 Polarogram for the Reduction of Oxygen in an Air-saturated 0.1 F Potassium Chloride Solution. Curve for 0.1 F KCl alone is also shown.

way, migration currents are avoided and the currents observed are independent of salt concentration.

Oxygen Waves

Dissolved oxygen is readily reduced at the dropping mercury electrode and an aqueous solution saturated with air exhibits two distinct waves attributable to this element (see Fig. 27.7). The first results from the reduction of oxygen to peroxide

$$O_2 + 2H^+ + 2e \rightleftharpoons H_2O_2$$

The second corresponds to the further reduction of the hydrogen peroxide

$$H_2O_2 + 2H^+ + 2e \rightleftharpoons 2H_2O$$

As would be expected from the stoichiometry of the reactions, the two wave heights are alike.

Polarography gives a rather sensitive method for the analysis of oxygen. Of more importance, however, is its interference with the polarographic analysis of other species; removal of oxygen is thus a first step to most analyses. This is readily accomplished by aeration of the solution with an inert gas for several minutes before it is to be studied; a stream of the same gas, usually nitrogen, is passed over the surface during the analysis to prevent reabsorption.

APPLICATIONS OF POLAROGRAPHY

Apparatus

Cells. A typical, general-purpose cell for polarographic analysis is shown in Figure 27.1. Separation of the solution to be analyzed from the calomel electrode is accomplished by means of a sintered disk backed by an agar-agar plug that contains enough potassium chloride to allow passage of current. Such a bridge is readily prepared and will last for extended periods provided a potassium chloride solution is kept in the reaction compartment when the cell is not in use. A capillary side-arm is provided to permit the passage of nitrogen or other gas through the solution. Provision is also made for blanketing the solution with the gas during the analysis.

A simpler cell may be employed in many cases. The nonpolarizable electrode need only be a pool of mercury in the bottom of the sample container. Here, however, the observed half-wave potentials will have little significance since the potential of the mercury pool will be dependent upon the composition of the solution being analyzed.

Dropping electrodes. A dropping electrode, such as that shown in Figure 27.1, can be readily fashioned from "marine barometer tubing".[9] A

[9] Available from Corning Glass Works, Corning, N. Y.

10-cm length of this will ordinarily give a drop time of 3 to 6 seconds under a head of mercury of about 50 cm. In preparing a capillary, the tip should be cut as nearly square as possible. Care should be taken to assure a vertical mounting of the electrode; otherwise erratic and nonreproducible drop times and sizes will be observed.

With reasonable care a capillary can be used for several months or even years. The precautions necessary to ensure such performance involve the use of clean mercury and the unceasing maintenance of a mercury head, no matter how slight. If the head is ever diminished to the point where solution comes in contact with the inner surface of the tip, malfunction of the electrode is to be expected. For this reason, the head of mercury should always be increased to provide a good flow before the tip is immersed in a solution.

Storage of an electrode always presents a problem. One method is to rinse the electrode thoroughly with water, dry, and then carefully reduce the head until the flow of mercury in air just ceases. Care must be taken to avoid lowering the mercury too far. Before use, the head is raised and the tip immersed in 1:1 nitric acid for a minute or so and then washed with distilled water.

Electrical apparatus. To make polarographic measurements it is necessary to have the means for applying a measured voltage that can be varied continuously over the range from 0 to 2.5 volts; the applied potential should usually be known to about 0.01 volt. In addition, it must be possible to measure the cell currents over the range of 0.01 to perhaps 100 microamperes with an accuracy of about 0.01 microampere. A manual apparatus for accomplishing this is easily constructed from equipment available in most laboratories. Rather elaborate devices for recording polarograms automatically are commercially available.

Figure 27.8 shows a circuit diagram of a simple instrument for polarographic work. Two 1.5-volt batteries provide a voltage across the 100-ohm potential divider R, by means of which the voltage applied to the cell can be varied. The magnitude of this voltage can be determined by the means of the potentiometer when the double-pole double-throw switch is placed in the 2 position. The current is measured by determining the potential drop across the precision 10,000-ohm resistance R_2; with the switch in the 1 position, the potentiometer can also be employed for this purpose. For current measurement the null-detecting galvanometer must be damped sufficiently to permit balancing despite the current oscillations associated with the dropping electrode. This may require the introduction of a resistor across the terminals of the galvanometer. An optimum damping resistance can be found empirically that will reduce the galvanometer fluctuations to a point where its average position can be readily determined.

From the standpoint of accuracy of current and voltage measurement, this apparatus is as good or better than the most sophisticated and expensive polarographic equipment available on the market. It is, furthermore, just as convenient and rapid for routine quantitative work, since in such applications current measurements at only two voltages are required (one before the decomposition potential and one in the limiting current region). On the other hand, where the entire polarogram is required, the point-by-point determination of the curve with a manual apparatus is a tedious and time-consuming operation; for this type of work, an automatic recording apparatus is of great value.

Treatment of Data

In measuring currents obtained with the dropping electrode it is common practice to use the *average* value of the galvanometer or recorder oscillations rather than the maximum or minimum values; the measurement thus becomes less dependent upon the damping employed.

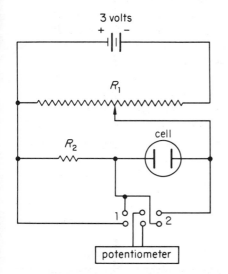 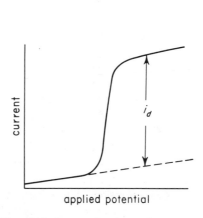

Fig. 27.8 A Simple Circuit for Polarographic Measurements. (From J. J. Lingane, *Anal. Chem.*, **21**, 47 (1949) with permission.)

Fig. 27.9 Determination of the Diffusion Current by Extrapolation of the Residual Current.

Determination of diffusion currents. For analytical work, limiting currents must always be corrected for the residual current. The best way of accomplishing this is to record a residual-current curve as well as the curve of the limiting current. Then the difference between the two can be taken at some potential in the limiting-current region. Because the residual current usually increases nearly linearly with applied voltage, it is often possible to dispense with the residual-current curve and obtain the correction by extrapolation of the residual-current portion of the curve for the sample. This is illustrated in Figure 27.9.

Analysis of mixtures. We have noted that the current-voltage curve for a multicomponent system is a synthesis of the polarographic waves for the individual constituents; if the half-wave potentials for these are sufficiently different, it may be possible to determine each species from a single polarogram. Where a considerable disparity in concentrations exists among the components, however, the accuracy of the analysis may be somewhat diminished. This is encountered in the determination of a minor component that is reduced at a more negative potential than one of the major species. Here, the current sensitivity of the galvanometer or recorder must be low enough to keep the wave for the major constituent within the scale limits of the instrument. The wave for the minor constituent, however, will encompass only a small fraction of the total scale; its diffusion current thus represents a small difference between two large current readings—a situation that tends to impart a large relative error for the difference. This problem does not arise when the minor constituent is the more easily reduced component. Under these circumstances, its diffusion current can be determined at high current sensitivity; then the sensitivity can be reduced to allow measurement of the major species.

There are several ways of treating mixtures containing species in unfavorable concentration ratios. The best method is to cause the wave of the minor component to appear first through alteration in the nature of the supporting electrolyte; with the variety of complexing agents available, this is often feasible. Alternatively, a preliminary chemical separation can be employed. Finally, there is the so-called "compensation technique." Here, the current due to the major constituent is reduced to zero or a very small value by application of a counter emf in the current-measuring circuit. The current sensitivity can then be increased to give a satisfactory signal for the reduction of the minor component. Most modern polarographs are equipped with a compensating device for this purpose.

Concentration determination. The best and most straightforward method for quantitative polarographic analysis involves preliminary calibration with a series of standard solutions; as nearly as possible, these standards should be identical to the samples to be analyzed. They should cover a concentration range within which the unknown samples will likely fall. From such data the linearity of the current-concentration relationship can be assessed, and if it is nonlinear, a curve can be constructed that will allow analysis.

Another useful technique is the standard addition method. The polarogram of an exactly known volume of the sample solution is obtained. Then a carefully measured quantity of a standard solution of the substance of interest is added and the polarogram again obtained. From the increase in wave height and the quantity of standard added, it is an easy matter to calculate the concentration of the original solution; the analyst must assume here that the concentration-current relationship is linear. This procedure is particularly effective when the diffusion current is sensitive to other components of the solution introduced with the sample.

Inorganic Polarographic Analysis

The polarographic method is rather generally applicable to inorganic substances. Most of the metallic cations, for example, are reduced at the dropping electrode to form a metal amalgam or a lower oxidation state. Even the alkali and alkaline-earth metals are reducible provided a supporting electrolyte is used that does not decompose at the high potentials required; the tetraalkyl ammonium halides serve this function well.

The successful polarographic analysis of cations frequently depends upon the employment of a suitable supporting electrolyte. To aid in this selection, "polarographic spectra" have been published that provide half-wave potential data for cations in various supporting media. It is inevitable that some approximations go into the preparation of these "spectra;" therefore they should be used in conjunction with tabular compilations.[10] The judicious choice of anion will often enhance the selectivity of the method. For example, with potassium chloride as a supporting electrolyte, the waves for ferric iron and copper interfere with one another; in fluoride medium, however, the half-wave potential of the former is shifted by about —0.5 volt while the latter changes by only a few hundredths of a volt. The presence of fluoride thus results in the appearance of separate waves for the two ions.

The polarographic method is also applicable to the analysis of such inorganic anions as bromate, iodate, dichromate, vanadate, selenite, and nitrite. In general, polarograms for these substances are affected by the pH of the solution because the hydrogen ion is a participant in the reduction process. As a consequence strong buffering of the solutions at some fixed pH is necessary to obtain reproducible data.

Certain inorganic anions that form complexes or precipitates with the ions of mercury are responsible for anodic waves that form in the region of 0 volt (vs. the saturated calomel electrode). Here, the electrode reaction involves oxidation of the electrode; for example

$$2Hg + 2Cl^- \rightleftharpoons Hg_2Cl_2 + 2e$$

$$Hg + 2S_2O_3^{2-} \rightleftharpoons Hg(S_2O_3)_2^{2-} + 2e$$

Bromide, iodide, thiocyanate, and cyanide also act in this way. The magnitude of the diffusion currents are controlled by the rate of diffusion of the anions to the electrode surface; as a consequence, a linear relationship between current and concentration is observed.

The polarographic method has also been applied to the analysis of a few inorganic substances that exist as uncharged molecules in the solvents used. The determination of oxygen in gases, biological fluids, and water is a

---

[10] A number of polarographic spectra and an annotated summary of half-wave potentials are to be found in L. Meites, *Polarographic Techniques.* New York: Interscience Publishers, Inc., 1955.

most important example of this application. Other neutral substances that react at the dropping electrode include hydrogen peroxide, hydrazine, cyanogen, elemental sulfur, and sulfur dioxide.

For a discussion of the applications of polarography to inorganic analysis the reader is referred to the monograph by Kolthoff and Lingane.[11]

Organic Polarographic Analysis

Almost from its inception, the polarographic method has been used for the study and analysis of organic compounds, and a large number of scientific papers have been devoted to this subject. Several of the common functional groups are oxidized or reduced at the dropping electrode, and a wide variety of compounds containing these groups are thus subject to analysis by the polarographic technique.

In general, the reactions of organic compounds at a microelectrode are slower and more complex than those of inorganic cations. This has made theoretical interpretation of the polarographic data more difficult or even impossible; furthermore, it has required a much stricter adherence to detail where the polarograms are employed for quantitative analysis. Despite these handicaps organic polarography has proved fruitful for the determination of structure, for the qualitative identification of compounds, and for the quantitative analysis of mixtures.

Effect of pH on polarograms. Most organic electrode processes involve hydrogen ions, the most common reaction being represented as

$$R + nH^+ + ne = RH_n$$

where R and RH_n are the oxidized and reduced forms of the organic molecule. Ordinarily, then, half-wave potentials for organic compounds are markedly pH dependent. Furthermore, alteration of pH often results in a change in the reaction products. Thus, for example, when benzaldehyde is reduced in basic solutions, a wave is obtained at about —1.3 to —1.5 volts attributable to the formation of benzyl alcohol

$$C_6H_5CHO + 2H^+ + 2e \rightleftharpoons C_6H_5CH_2OH$$

In solutions of pH less than 2, however, a wave occurs at about —1.0 volt that is just half the size of the foregoing one; here the reaction consists of the production of hydrobenzoin

$$2C_6H_5CHO + 2H^+ + 2e \rightleftharpoons C_6H_5CHOHCHOHC_6H_5$$

At intermediate pH values, two waves are observed indicating the occurrence of both reactions.

It should be emphasized that an electrode process consuming or producing hydrogen ions will tend to alter the pH of the solution *at the electrode surface*;

[11] I. M. Kolthoff and J. J. Lingane, *Polarography*, 2d ed. **2**. New York: Interscience Publishers, Inc., 1952.

unless the solution is well buffered, marked changes in pH can occur in the surface film as the electrolysis proceeds. These will affect the reduction potential of the reaction and lead to drawn out and poorly defined waves. In cases where the electrode process is altered by the pH, nonlinearity in the diffusion current-concentration relationship must also be expected. Thus, in organic polarography, good buffering is a vital necessity if reproducible half-wave potentials and diffusion currents are to be obtained.

Solvents for organic polarography. In organic polarography, solubility considerations frequently demand the use of some solvent other than pure water, and aqueous mixtures containing varying amounts of such miscible solvents as glycols, dioxane, the alcohols, Cellosolve, and glacial acetic acid have been employed. Anhydrous media of acetic acid, formamide, and ethylene glycol have also been used in some instances. Supporting electrolytes are often lithium salts or the tetraalkyl ammonium salts.

Irreversibility of electrode reactions. Few organic electrode reactions are reversible; as a consequence, an equation such as that shown on page 607 does not adequately describe the polarographic waves of organic compounds. In general, the nonreversibility of the electrode process results in a drawn-out wave with a half-wave potential that may be dependent upon the concentration of the reactant. Greater differences in half-wave potentials are required in order to discriminate between waves of more than one substance.

In general, the Ilkovic equation applies to the diffusion currents for electrode reactions that are nonreversible. Thus, the quantitative aspects of organic and inorganic polarography are not much different.

Reactive functional groups. Organic compounds containing any of the following functional groups can be expected to react at the dropping mercury electrode and thus produce one or more polarographic waves.

(1) *The carbonyl group* including aldehydes, ketones, and quinones. In general, aldehydes are reduced at lower potentials than ketones; conjugation of the carbonyl double bond also leads to lower half-wave potentials.

(2) *Certain carboxylic acids* are reduced polarographically although the simple aliphatic and aromatic monocarboxylic acids are not. Dicarboxylic acids such as fumaric, maleic, or phthalic acid, in which the carboxyl groups are conjugated with one another, give characteristic polarograms; this is also true of certain keto and aldehydo acids.

(3) *Most peroxides and epoxides* yield polarograms.

(4) *Nitro, nitroso, amine oxide, azo, and quaternary amine groups* are generally reduced at the dropping electrode.

(5) *Most organic halogen groups* produce a polarographic wave as a result of replacement of the halogen group with an atom of hydrogen.

(6) *The carbon-carbon double bond* is reduced when it is conjugated with another double bond, an aromatic ring, or an unsaturated group.

(7) *Hydroquinones and mercaptans* produce anodic waves.

In addition to these, a number of organic groups lead to catalytic hydrogen waves that can be used for analysis. These include amines, mercaptans, acids, and heterocyclic nitrogen compounds. A large number of applications to biological systems have been reported.[12]

AMPEROMETRIC TITRATIONS

The polarographic method can be employed for the estimation of the equivalence point in a titration provided at least one of the participants or products of the analytical reaction is oxidized or reduced at a microelectrode. Here the current passing through a polarographic cell at some fixed potential is measured as a function of the volume of reagent (or the time in the case of a coulometric titration). Plots of the data on either side of the equivalence point are straight lines of differing slopes so that the end point can be fixed by extrapolation to their intersection.

The amperometric method has a number of advantages. The technique is inherently more accurate than the polarographic method itself and is less dependent upon the characteristics of the capillary and the supporting electrolyte. Furthermore, the temperature need not be fixed accurately during the titration, and must only be kept constant. Finally, the substance being determined does not have to be reactive at the electrode; a reactive reagent or product will suffice.

The amperometric method also has certain advantages when compared with other titration procedures. For one thing, it is less demanding insofar as completeness of the titration reaction is concerned since data can be collected well before and after the equivalence point; in this respect the technique resembles the conductometric procedure and differs from potentiometric and indicator methods. In contrast to the conductometric procedure, however, the presence of a high electrolyte concentration in no way affects the accuracy of the titrations, provided, of course, that the electrolyte does not react at the microelectrode. Very dilute solutions can be titrated by the amperometric method because the indicator electrode is sensitive to very small quantities of the reactants.

Titration Curves

Amperometric titration curves take one of the forms shown in Figure 27.10. Curve *a* represents the case where the substance being analyzed reacts at the electrode while the reagent does not. For example, the titration of lead with sulfate or oxalate ions may be cited. Here, a sufficiently high potential is applied to give a diffusion current for lead; a linear decrease in current results as the lead ions are precipitated from the solution. The curvature near the equivalence point reflects the incompleteness of the analytical reaction at this point. The end point is obtained by extrapolation of the straight lines as shown.

[12] M. Brezina and P. Zuman, *Polarography in Medicine, Biochemistry and Pharmacy.* New York: Interscience Publishers, Inc., 1958.

Curve *b* is typical of a titration in which the reagent reacts at the microelectrode and the substance being analyzed does not. An example of this would be the titration of magnesium with 8-hydroxyquinoline. A diffusion current for the latter is obtained at —1.6 volts (vs. the saturated calomel electrode) whereas the magnesium is inert at this potential.

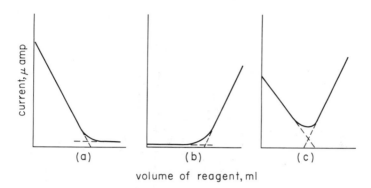

Fig. 27.10 Typical Amperometric Titration Curves. (a) Substance analyzed is reducible, reagent is not. (b) Reagent is reducible, substance analyzed is not. (c) Both substance analyzed and reagent are reducible.

Curve *c* corresponds to the titration of lead ions with a chromate solution at an applied potential greater than —1.0 volt. Both lead and chromate ions give diffusion currents, and a minimum in the curve signals the end point. This system yields a curve of type *b* with zero applied potential since only chromate ions are reduced under these conditions.

Apparatus and Techniques

With relatively simple apparatus, accurate results can be obtained by the amperometric method.

Cells. Figure 27.11 shows a typical cell for an amperometric titration. A reference electrode, such as a calomel half cell, is usually employed as the nonpolarizable electrode; the indicator electrode may be a dropping mercury electrode or a micro wire electrode as shown. The cell should have a capacity of 75 to 100 ml.

Volume measurements. In order to obtain linear plots for fixing the end point of the titration it is necessary to correct for volume changes due to the added titrant. If the original volume is known, the diffusion current can be multiplied by $(V + v)/V$ where V is the original volume and v is the volume of reagent; thus all the measured currents are corrected back to the original volume. An alternative, which is often satisfactory, is to use a reagent that is twenty or more times as concentrated as the solution being titrated. Then v

is always so small with respect to V that the correction is negligible and unnecessary. This does, however, require the use of a microburet so that a total reagent volume of 1 or 2 ml can be measured with a suitable accuracy.

The microburet should be so arranged that its tip can be touched to the surface of the solution after each addition of reagent. This is necessary to remove the fraction of a drop that tends to remain attached.

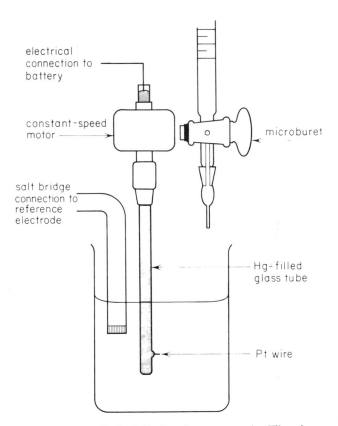

electrical connection to battery

constant-speed motor

salt bridge connection to reference electrode

microburet

Hg-filled glass tube

Pt wire

Fig. 27.11 Typical Cell for Amperometric Titrations Employing a Rotating Platinum Electrode.

Electrical measurements. For amperometric titrations a very simple manual polarograph is entirely adequate. An appropriate voltage is applied to the cell by means of the linear potential divider; the current is measured by a damped galvanometer (which need not even be calibrated) or a low-resistance microammeter. Ordinarily the applied voltage does not have to be known any closer than about ± 0.05 volt. To choose the voltage, a knowledge of the polarographic behavior of the reagent and the substance being titrated is necessary. Then an applied voltage in the diffusion-current region of one or both is employed.

Microelectrodes; the rotating platinum electrode. Many amperometric titrations can be carried out conveniently with a dropping mercury electrode. For those reactions involving oxidizing agents that attack mercury (bromine, silver ion, ferric ion, etc.), a rotating platinum electrode is particularly useful. This microelectrode consists of a short length of platinum wire sealed into the side of a glass tube; mercury inside the tube provides electrical contact between the wire and the lead to the polarograph. The tube is held in a hollow chuck and rotated at a constant speed of above 600 rpm by means of a synchronous motor. Commercial models of the rotating electrode are available. A typical apparatus is shown in Figure 27.11.

Polarographic waves, similar in appearance to those observed with the dropping electrode, are obtained with the rotating platinum electrode. Here, however, the reactive species is brought to the electrode surface not only by diffusion but also by mechanical mixing. As a consequence, the limiting currents are as much as 20 times larger than those obtained with a microelectrode that is supplied by diffusion only. With a rotating electrode, steady currents are instantaneously obtained. This is in distinct contrast to the behavior of a solid microelectrode without stirring (p. 608).

Table 27-3

SOME PRECIPITATION TITRATIONS EMPLOYING THE AMPEROMETRIC END POINT

Dropping Mercury Electrode

Reagent	Substance Determined
K_2CrO_4	Pb^{2+}, Ba^{2+}
$Pb(NO_3)_2$	SO_4^{2-}, MoO_4^{2-}, F^-, Cl^-
8-Hydroxyquinoline	Mg^{2+}, Zn^{2+}, Cu^{2+}, Cd^{2+}, Al^{3+}, Bi^{3+}, Fe^{3+}
Cupferron	Cu^{2+}, Fe^{3+}
Dimethylglyoxime	Ni^{2+}
α-Nitroso-β-napthol	Co^{2+}, Cu^{2+}, Pd^{2+}
$K_4Fe(CN)_6$	Zn^{2+}

Rotating Platinum Electrode

Reagent	Substance Determined
$AgNO_3$	Cl^-, Br^-, I^-, CN^-, RSH

Several limitations restrict the widespread application of the rotating platinum electrode to polarography. The low hydrogen overvoltage prevents its use as a cathode in acidic solutions. In addition, the high currents obtained with the electrode make it particularly sensitive to traces of oxygen in the solution. These two factors have largely confined its employment to anodic reactions.

Limiting currents are not as reproducible as those obtained with the dropping electrode; often these are influenced by the past history of the electrode. For amperometric titrations, however, this need not be a serious handicap.

Application of Amperometric Titrations

The majority of applications of the amperometric end point have been to titrations in which a slightly soluble precipitate is the reaction product. Table 27-3 shows some of these. A variety of organic precipitants appear in this list. In most instances, these reagents are reducible at the dropping electrode.

A few applications of the amperometric method to oxidation-reduction reactions can be found. For example, the technique has been applied to various titrations involving iodine and bromine (in the form of bromate) as reagents.

AMPEROMETRIC TITRATIONS WITH TWO POLARIZED MICROELECTRODES

A convenient modification of the amperometric method involves the use of two stationary microelectrodes (usually platinum) immersed in the well-stirred solution of the sample. A small potential (say 0.01 to 0.1 volt) is applied between these and the current flow is followed as a function of the volume of reagent added. The end point is marked by a sudden current rise from zero, a decrease in the current to zero, or a minimum (at zero) in a V-shaped curve.

Although the use of two polarized electrodes for end-point detection was first proposed just before the turn of the century, almost 30 years were to pass before chemists came to appreciate the potentialities of the method.[13] The name *dead-stop end point* was used to describe the technique, and this term is still used to some extent. It was not until about 1950 that a clear interpretation of dead-stop titration curves was made.

Titration curves.[14] To gain a qualitative understanding of the principles of the method, consider the titration of arsenite with a standard iodine solution. At low applied potentials neither $+3$ nor $+5$ arsenic reacts to any extent at the platinum electrodes; the iodine-iodide half reaction, on the other hand, is perfectly reversible. In the initial stages of the titration, the solution will contain appreciable amounts of arsenite, arsenate, and iodide ion; the iodine concentration will, however, be vanishingly small since the equilibrium constant favors its reduction by the arsenite. If an emf of a few hundredths of a volt is applied to platinum electrodes in such a solution, no current will pass because there is no cathodic reaction that can occur to any extent; the iodine concentration is too small to give an appreciable current, the arsenate is not reduced at this potential, nor for that matter, are hydrogen ions. Current flow is thus

---

[13] C. W. Foulk and A. T. Bawden, *J. Am. Chem. Soc.*, **48**, 2045 (1926).

[14] For an excellent analysis of this type of end point see J. J. Lingane, *Electroanalytical Chemistry*, 2d ed., 280-294. New York: Interscience Publishers, Inc., 1958.

prevented by polarization of the cathode. The anode is not polarized, incidentally, since oxidation of iodide could occur here.

This situation prevails throughout the titration until the end point is passed. Then, with the first excess of iodine depolarization of the cathode occurs. Current can pass as a result of the electrode processes

$$I_2 + 2e \rightarrow 2I^-$$ (cathode)

$$2I^- \rightarrow I_2 + 2e$$ (anode)

The end point is therefore signaled by the sharp rise in current.

The reverse of the foregoing would be observed if a pair of small silver electrodes were used to detect the end point in the titration of silver ion with standard sodium chloride solution. By impressing a small potential, current passage would occur as a result of the reactions

$$Ag \rightarrow Ag^+ + e$$ (anode)

$$Ag^+ + e \rightarrow Ag$$ (cathode)

With the effective removal of silver ion by the analytical reaction, cathodic polarization would occur and the current would approach zero at the end point.

Applications. Amperometric methods with two microelectrodes have not been fully exploited, and there are undoubtedly a number of oxidation-reduction systems to which this end-point technique could be applied with advantage. Most of the published data have been concerned with titrations involving iodine, although a few applications with reagents such as bromine, titanous ion, and ceric ion have been reported. An important use is in the titration of water with the Karl Fischer reagent (see p. 721). The technique has also found a number of applications as an end point in coulometric titrations.

The principal advantage of the twin microelectrode procedure is its simplicity. One can dispense with a reference electrode; the only instrumentation needed is a simple voltage divider powered by a dry cell, and a galvanometer or microammeter for current detection.

Laboratory Exercises Illustrating Polarography and Amperometric Titrations

Because of the wide variety of equipment available for polarographic work, no attempt will be made to provide details concerning the assembly and use of apparatus. For the reader interested in such information, the reference by Meites is recommended.[15]

---

[15] L. Meites, *Polarographic Techniques.* New York: Interscience Publishers, Inc., 1955.

Polarographic Behavior of Simple Metal Ions

Prepare in 50-ml volumetric flasks a series of solutions containing 5 ml of 0.1-percent gelatin, 10 ml of 0.5 F KNO_3, and the following volumes of 5×10^{-3} F solution of Cd^{2+}: 0, 1, 3, 10, and 30 ml. Dilute to the mark. Prepare one solution containing 10 ml of the standard Cd^{2+} solution and 10 ml of the KNO_3, but no gelatin.

Oxygen wave. Rinse the polarographic cell with the solution containing 0 ml of the standard cadmium solution and thermostat in a water bath at about 25° C. Obtain a polarogram from 0 to —2.0 volts. Bubble nitrogen through the solution for 15 min, direct the stream of gas over the surface and again obtain the polarogram. Compare the oxygen wave with Figure 27.7.

Determination of i_d/C and $E_{1/2}$ for Cd^{2+}. Obtain polarograms for each of the prepared solutions after removal of oxygen and thermostatting at 25° C. If a manual polarograph is used, obtain the complete polarogram of the solution containing 10 ml of the cadmium standard only; for the remaining standards, take only the points necessary to determine i_d.

(1) Determine i_d/C in each case making a suitable correction for the residual current.

(2) Calculate $E_{1/2}$ in each case and compare with the literature value.

(3) For the solutions containing 10 ml of the cadmium standard and the gelatin make a plot of E versus $\log i/(i_d - i)$. This should yield a straight line according to equation (27-6). Evaluate n from the slope of this plot.

Polarogram of a mixture of two ions. Prepare 50 ml of a solution that contains 5 ml of the gelatin, 10 ml of the KNO_3, 10 ml of the standard cadmium solution, and 10 ml of 5×10^{-3} F Zn^{2+}. After removal of oxygen, obtain the polarogram.

Determine i_d/C and $E_{1/2}$ for each wave.

Anodic and Cathodic Polarograms

Prepare a saturated solution of oxalic acid. Place exactly 25 ml (or some other carefully measured volume) of this in the polarographic cell, add one drop of methyl-red indicator, and deaerate. Record the polarogram from +0.2 to —0.6 volt versus the saturated calomel electrode. Add 1 ml of a 0.05 F solution of ferric sulfate, deaerate again, and record the polarogram over the same voltage range.

Transfer another 25-ml aliquot of the oxalic acid to the cell, add 1 drop of methyl red, and remove the oxygen. Quickly add 1 ml of a freshly prepared 0.05 F solution of ferrous ammonium sulfate in 0.1 N H_2SO_4 and deaerate for a few minutes. Record the polarogram as before while nitrogen is passed over the surface of the solution.

Finally add 1 ml of the ferric sulfate to the solution in the cell and after a brief aeration, again obtain a polarogram.

Calculate the wave heights and half-wave potentials for the three solutions. Interpret the results.

Amperometric Titration with a Dropping Mercury Electrode

Titration of lead. Prepare a solution which is $0.100 \, F$ in $Pb(NO_3)_2$ and a second solution that is $0.050 \, F$ in $K_2Cr_2O_7$. Also prepare an acetate buffer of pH 4.2 (to prepare, add $0.1 \, F$ NaOAc solution to a solution that is $0.2 \, F$ in HOAc and $0.2 \, F$ in KNO_3 until the proper pH is attained as indicated by a glass electrode).

Transfer 5 ml of the $Pb(NO_3)_2$ to the titration vessel, add 5 ml of 0.1-percent gelatin solution and 40 ml of the acetate buffer. Insert the dropping electrode; the anode should be a saturated calomel electrode. Bubble nitrogen through the solution for 15 min; then measure the current at -1.0 volt after the gas flow has been stopped. Add exactly 1.00 ml of $K_2Cr_2O_7$, bubble with nitrogen, and again determine the current. Continue to add the $K_2Cr_2O_7$ in 1-ml increments until a total of 10 ml has been added. Correct the currents for dilution and plot the titration curve. Determine the end point by extrapolation.

Repeat the titration at 0 applied potential. In this case oxygen removal is unnecessary. Contrast the curves and explain the differences.

Titration of sulfate ion. Transfer 25 ml of a $0.0200 \, F$ solution of K_2SO_4 to the titration vessel and add 25 ml of a 40-percent (by volume) ethyl alcohol solution containing a drop of methyl red. Deaerate the solution and titrate with the $0.1 \, F$ $Pb(NO_3)_2$ following the directions given in the previous section. Determine the end point graphically.

Amperometric Titration with the Rotating Platinum Electrode

Prepare a $0.0300 \, N$ solution of arsenious acid from primary standard As_2O_3 (see p. 464) and a $0.0600 \, N$ solution of $KBrO_3$ ($0.01 \, F$).

The titration apparatus should consist of a rotating platinum electrode, a saturated calomel electrode, and a vessel with a capacity of about 100 ml.

Transfer 10 ml of the arsenious acid solution to the titration vessel and add 40 ml of a solution that is about $2 \, N$ in HCl and $0.1 \, F$ in KBr. Place the bromate solution in a 10-ml buret, and adjust the electrode so that it rotates at about 600 rpm. Apply $+0.2$ volt to the rotating electrode and titrate as described in the foregoing section. Locate the end point graphically.

Dead-stop Method

Assemble a cell consisting of a 100-ml beaker, a good stirrer and a pair of small platinum wire electrodes. Place 5 ml of $0.01 \, N$ iodine in the cell, add 0.1 gram of KI and dilute to about 50 ml. Apply a potential of 0.1 volt to the electrodes, start the stirrer, and titrate the solution with a $0.01 \, N$ solution of $Na_2S_2O_3$. Add the reagent in 0.5-ml increments recording the current after each addition. Plot the data and determine the end point.

part 5. *Optical Methods of Analysis*

chapter 28. *Methods Based on Absorption of Radiation*

The selective absorption of electromagnetic radiation as it passes through a solution causes the emerging beam to differ from the incident one. In the case of visible radiation, this difference is frequently obvious to the naked eye. For example, a white light viewed through a cupric sulfate solution appears blue because the cupric ions interact with and absorb the red components of the beam while transmitting completely the blue portions of the radiation. The application of this type of absorption phenomenon to the qualitative identification of substances is undoubtedly familiar to most readers, for many of the tests in the ordinary qualitative scheme are based upon observation of the color of solutions. Of equal importance is the employment of light absorption for the quantitative measurement of chemical systems.

FUNDAMENTAL CONCEPTS, DEFINITIONS, AND LAWS

Electromagnetic radiation is a form of energy that can be described in terms of its wavelike properties. In contrast to sound waves, electromagnetic waves travel at extreme velocities and do not require the existence of some supporting medium for propagation.

The *wavelength* λ of a beam of electromagnetic radiation is the linear distance traversed by one complete wave cycle. The *frequency* v is the number of cycles occurring per second; it is readily obtained by dividing the wavelength into the velocity of the radiation

$$v = \frac{c}{\lambda}$$

The *velocity c* varies with the medium through which the radiation is passing; it has a value of 3×10^{10} cm/sec when measured in a vacuum.

In some of its interactions with matter, radiation can be shown to behave as if it were composed of discrete packets of energy, called *photons*. The energy of a photon is variable and depends upon the frequency or wavelength of the radiation. The relationship between the energy E of a photon and frequency is given by the relationship

$$E = hv$$

where h is *Planck's constant* and has a numerical value of 6.62×10^{-27} erg sec. The equivalent expression involving wavelength is

$$E = \frac{hc}{\lambda}$$

Clearly short wavelengths are more energetic than long ones.

Electromagnetic radiation exhibits what appears to be a duality of properties. Thus the phenomenon of refraction and diffraction is best explained by assigning it the properties of waves. On the other hand, in its interaction with matter, radiation is best described as having particulate properties, the energy of the particles being equal to hv.

The electromagnetic spectrum covers an immense range of wavelengths. Figure 28.1 depicts qualitatively the major divisions of the spectrum. A logarithmic scale has been employed in this representation; note that the portion to which the human eye is perceptive is small indeed. Such diverse radiations as gamma rays and radio waves are also electromagnetic radiations, differing from visible light only in the matter of wavelength, and hence energy.[1]

---

[1] The units commonly used for describing the wavelength of radiation differ considerably in the various spectral regions. For example, the Ångstrom unit Å, 10^{-7} mm, is convenient for x-ray and ultraviolet radiation; the millimicron mμ, 10^{-6} mm, is employed with visible and ultraviolet radiation; the micron μ, 10^{-3} mm, is commonly employed for infrared radiation.

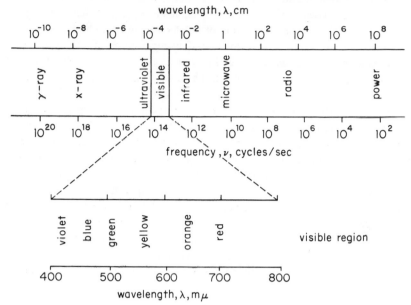

Fig. 28.1 The Electromagnetic Spectrum.

The Absorption Process

We have noted that partial absorption often accompanies the passage of a beam of radiation through a transparent medium, and also that the absorption of electromagnetic radiation is tantamount to the absorption of energy. We shall now examine qualitatively the processes responsible for this absorption, and the ultimate fate of the energy so absorbed.

When an atom, ion, or molecule absorbs a photon, the added energy results in an alteration of state; the species is then said to be *excited*. Excitation may involve any of the following processes:

(1) Transition of an electron to a higher energy level;
(2) A change in the mode of vibration of the molecule;
(3) Alteration of its mode of rotation.

Each of these transitions requires a definite quantity of energy; the probability of occurrence for a particular transition is greatest when the photon absorbed supplies precisely this quantity of energy.

The energy requirements for these transitions vary considerably. In general, promotion of electrons to higher levels requires greater energies than those needed to bring about vibrational changes. Alterations in rotational mode are likely to have the lowest energy requirements of all. Thus, absorptions observed in the microwave and far infrared regions will be due to shifts in rotational level since the energy of the radiation is insufficient to cause other types of transition. Changes in vibrational levels are responsible for absorptions in the near infrared and visible regions. Because a series of rotational states exists for each vibrational level, groups of wavelengths are likely to be absorbed that differ only slightly in energy. Absorption due to promotion of an electron to some higher energy level takes place in the visible, ultraviolet, and x-ray regions of the spectrum. An even greater number of wavelengths is likely to be absorbed in the first two of these regions, since a series of vibrational levels and rotational sublevels exist within each electronic level.[2]

The absorption of the very energetic x-ray radiation involves electronic transitions of the innermost electrons of an atom whereas absorption of ultraviolet and visible radiation involves the outer bond-forming electrons. An interesting consequence of this is that the x-ray absorption spectra of all but the lightest elements are quite unaffected by the form in which the element occurs—that is, whether the element is present in the sample as a compound or in its elemental form. On the other hand, the ultraviolet and visible spectra are profoundly influenced by the manner in which the element is combined. Thus, using manganese as an example, the permanganate ion is red, manganate ion is green, and manganous ion is essentially colorless.

The absorption of radiation by a system can be described by means of a

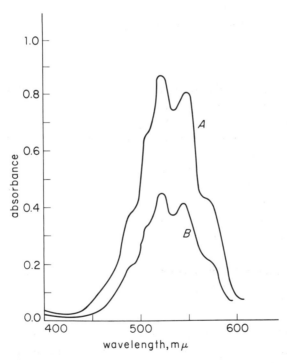

Fig. 28.2 Absorption Spectrum for Potassium Permanganate in the Visible Region. The concentration of MnO_4^- in solution A is twice that of solution B.

[2] Some transitions are more probable than others and some are "forbidden"—that is, highly improbable. The reasons for this are beyond the scope of this book. Recommended for further study is E. J. Bowen, *The Chemical Aspects of Light*, 2d ed. Oxford: The Clarendon Press, 1946.

plot of the absorption as a function of wavelength; such a graph is called an *absorption spectrum* (see Fig. 28.2). Inasmuch as the energies required for the various processes responsible for absorption are unique for a given species, its absorption spectrum is also unique; as a consequence absorption spectra are often helpful for qualitative identification purposes. This is particularly true for the low-energy absorptions that are found in the infrared region.

Irrespective of the amount of energy absorbed, an excited species tends spontaneously to return to its unexcited, or ground, state. To accomplish this, the energy of the absorbed photon must somehow be given up, and this is ordinarily dissipated in the form of heat. In some instances, however, transition to another excited state precedes return to the ground state. Here, the energy of the absorbed photon may be dissipated partially by emission of a photon of lower energy and partially as heat. The phenomenon of fluorescence is accounted for in this fashion.

Beer's Law

We have thus far made no effort to define closely the physical state of the medium in which absorption is occurring. For present purposes, however, absorption by solutions is of paramount interest, and we will henceforth restrict ourselves to this single situation.

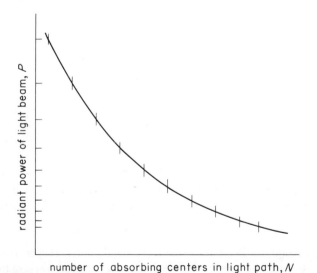

Fig. 28.3 Passage of a Beam of Monochromatic Radiation Through a Solution. Vertical markings indicate attenuation suffered for a unit change in the number of absorbing centers in the light path.

Consideration of the quantitative aspects of radiation absorption requires use of the term *radiant power P*, which is the radiant energy striking a unit area in unit time. Absorption of radiation involves reduction in the power of a beam.

Consider a beam of monochromatic radiation traversing a solution in which there exists a population N of absorbing atoms or molecules. If we were to locate a radiation detector a short distance from the point of entry, absorption by the intervening atoms or molecules would be found to diminish the power of the beam by an amount ΔP. Movement of the detector a greater distance away would result in further diminution of the beam, because of the increased number of absorbing molecules in the path. Measurement of the power at succeeding intervals would indicate further decreases as shown by the plot in Figure 28.3. In this graph the power is initially large, but falls off rapidly as the light path (and, hence, the number of absorbers) increases. Since the radiant power at any point is less than at any preceding point, it is apparent that the amount of absorption ΔP occurring in any interval chosen is dependent not only upon the number of absorbing species ΔN encountered by the beam, but also upon the power of the incident radiation. This may be expressed as

$$\Delta P = -k\,P\,\Delta N \tag{28-1}$$

where k is a proportionality constant. The minus sign is introduced because a diminution in radiant power is occurring. If we make the interval between measurements infinitely small, we may rewrite (28-1) in the following form:

$$\frac{dP}{P} = -k\,dN$$

Integration will give a measure of the total absorption in terms of the number of absorbing bodies. If we designate P_0 as the radiant power of the incident beam, then $P = P_0$ when $N = 0$ and

$$\int_{P_0}^{P} \frac{dP}{P} = -k\int_{0}^{N} dN$$

$$ln\,\frac{P}{P_0} = -kN \tag{28-2}$$

The number of absorbing centers is clearly related to the molar concentration c by means of the volume in which they are contained and the Avogadro number

$$N = c \times 6 \times 10^{23} \times b \times s/1000$$

where b is the length of the path in centimeters and s is its cross section in square centimeters. Substitution of this into equation (28-2) and conversion to Briggsian logarithms gives

$$\log \frac{P_0}{P} = -\frac{k}{2.303} \times 6 \times 10^{23} \times s \times b \times c/1000$$

Collecting the constants yields

$$\log \frac{P_0}{P} = \epsilon\,bc = A \tag{28-3}$$

Equation (28-3) is a fundamental law governing the absorption of all types of electromagnetic radiation. It is known variously as the *Lambert-Beer, Bouguer-Beer*, or most commonly, as *Beer's law*. The logarithmic term on the left side of the equation is called the *absorbance*, and given the symbol A. The constant ϵ is called the *molar absorptivity* when the concentration c is expressed in terms of mols of absorber per liter; it is simply called the *absorptivity* and is given the symbol a when other concentration units are used. For the former the path length of the radiation b is expressed in centimeters. Equation (28-3) shows that the absorbance of a solution is directly proportional to the concentration of absorbing species when the length of light path is fixed and directly proportional to the light path when the concentration is fixed; every quantitative analysis based upon the absorption of radiation makes use of one or the other of these relationships.

Beer's law applies to a solution containing more than one kind of absorbing substance provided there is no interaction among the various species. Thus, for a multicomponent system, we may write

$$A_{\text{total}} = \epsilon_1 b c_1 + \epsilon_2 b c_2 + \epsilon_3 b c_3 \dots.$$

where ϵ represents the molar absorptivity of each of the species and c represents its molar concentration.

Measurement of Absorption

Beer's law, as given by equation (28-3), is not directly applicable to chemical analysis because neither P nor P_0, as defined, can be measured easily in the laboratory. The main reason for this is that the solution to be studied must be held in some sort of container; thus, the beam of radiation must pass through the walls of this container before it can be measured. Interaction between the radiation and the walls is inevitable, producing a loss in power at each interface as a consequence of reflection or possibly absorption. In addition, the beam may suffer a diminution in power during its passage through the solution as a result of scattering by large molecules or inhomogeneities. These phenomena are depicted in Figure 28.4. Reflection losses can be quite appreciable; for example, about 4 percent of visible light is reflected upon vertical passage from air to glass.

The experimental applications of Beer's law requires a correction for these effects. This is most easily done by comparing the power of the beam transmitted through the solution of interest with the power of a beam passing through an identical cell containing the solvent for the sample. An experimental absorbance can then be evaluated which rather closely approximates the true absorbance of the solution; that is,

$$A \simeq \log \frac{P_{\text{solvent}}}{P_{\text{solution}}} \simeq \log \frac{P_0}{P}$$

The term P_0, when used henceforth, will refer to the power of a beam of radiation after it has passed through a cell containing the solvent for the component of interest.

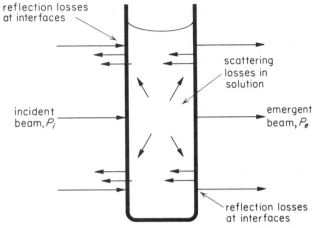

Fig. 28.4 Processes that Attenuate a Beam of Radiation upon Passage through a Solution. The power of the emergent beam P_e is less than that of the incident beam P_i because of (1) reflection at air-wall, wall-solution, solution-wall, and wall-air interfaces; (2) absorption by the two walls; (3) scattering within the solution; (4) absorption by components of the solution.

Terminology Associated with Absorption Measurements

In recent years an attempt has been made to develop a standard nomenclature for the various quantities related to the absorption of radiation. The recommendation of the American Society for Testing Materials is given in Table 28-1 along with some of the alternate names and symbols that are frequently encountered. An important term found in this table is the transmittance T, which is defined as

$$T = \frac{P}{P_0}$$

The transmittance is the fraction of incident radiation transmitted by the solution; it is often expressed as a percent. The transmittance is clearly related to the absorbance as follows:

$$- \log T = A$$

Limitations to the Applicability of Beer's Law

The linear relationship between absorbance and path length at a fixed concentration of absorbing substance is a generalization for which no exceptions

Table 28-1

IMPORTANT TERMS AND SYMBOLS EMPLOYED IN ABSORPTION MEASUREMENT

Term and Symbol[3]	Definition	Alternate Name and Symbol
Radiant power, P, P_0	energy of radiation reaching a given area of a detector per second	radiant intensity, I, I_0
Absorbance, A	$\log \dfrac{P_0}{P}$	optical density, D; extinction, E
Transmittance, T	$\dfrac{P}{P_0}$	transmission, T
Path length of radiation, in cm, b		l, d
Molar absorptivity, ϵ	$\dfrac{A}{bc}$ (c = mol/liter)	molar extinction coefficient
Absorptivity, a	$\dfrac{A}{bc}$	extinction coefficient, k

[3] Taken from H. K. Hughes, et al, *Anal. Chem.*, **24**, 1349 (1952).

have been found. On the other hand, deviations from the direct proportionality between measured absorbance and concentration are quite frequently encountered. Some of these deviations are of such a fundamental nature that they represent a real limitation of the law; others, however, occur as a consequence of the manner in which the absorbance measurements are made or as a result of chemical changes associated with concentration changes; the latter two are sometimes known, respectively, as *instrumental* and *chemical deviations.*

Real limitations to Beer's law. Beer's law is successful in describing the absorption behavior of dilute solutions only; in this sense it is a limiting law. At high concentrations the average distance between solute molecules (or ions) is diminished to the point where each affects the charge distribution of its neighbors. This interaction, in turn, can alter their ability to absorb a given wavelength of radiation. Because the degree of interaction is dependent upon concentration, the occurrence of this phenomenon causes deviations from the linear relationship between absorbance and concentration.

A second fundamental limitation arises from alterations in the refractive index of the solution as a result of high solute concentrations. It has been shown that the quantity $\epsilon \cdot n/(n + 2)^2$ rather than ϵ remains constant as concentration changes (n = refractive index).[4] Where $n/(n + 2)^2$ is altered appreciably by the variation in concentration, deviation from Beer's law will be observed unless a correction is made for this factor. In general, this effect does not become noticeable except in solutions more concentrated than 0.01 F.

[4] G. Kortum and M. Seiler, *Angew. Chem.*, **52**, 687 (1939).

Chemical deviations. Apparent deviations from Beer's law are frequently encountered as a consequence of association, dissociation, or reaction of the absorbing species with the solvent. A classical example of this is observed with unbuffered potassium dichromate solutions, in which the following equilibria exist:

$$Cr_2O_7^{2-} + H_2O \rightleftharpoons 2HCrO_4^- \rightleftharpoons 2H^+ + 2CrO_4^{2-}$$

At most wavelengths, the molar absorptivities of dichromate ion and the two chromate species are quite different. Thus the total absorbance of the solution at any point is dependent upon the ratio of concentrations between the dimeric and monomeric forms. This ratio, however, changes markedly with dilution and causes a pronounced deviation from linearity between the absorbance and the total concentration of chromium. Nevertheless, the absorbance due to the dichromate ion is directly proportional to its molar concentration; the same is true for that of the chromate ion. This is easily demonstrated by making measurements in strongly acidic or strongly basic solution where one or the other of these species will predominate. Thus, deviations in the absorbance of this system from Beer's law are more apparent than real, because they result from shifts in chemical equilibria. These deviations can, in fact, be readily predicted from the equilibrium constants for the reactions and the molar absorptivities of the dichromate and chromate ions.

Instrumental deviation. Strict adherence of an absorbing system to Beer's law is observed only when monochromatic radiation is employed. This is another manifestation of the limiting character of the relationship. Use of a monochromatic beam for absorbance measurements is seldom practical, however, and polychromatic radiation may lead to departures from Beer's law.

Consider a beam made up of radiation of two wavelengths, λ' and λ''. Assuming that Beer's law applies strictly at each of these individually, we may write for radiation λ'

$$A' = \log \frac{P_0'}{P'} = \epsilon' bc$$

or

$$\frac{P_0'}{P'} = 10^{\epsilon' bc}$$

Similarly for λ'' we may write

$$\frac{P_0''}{P''} = 10^{\epsilon'' bc}$$

When an absorbance measurement is made with radiation composed of both wavelengths, the power of the beam passing through the solution is given by $(P' + P'')$ and that of the beam passing through the solvent by $(P_0' + P_0'')$. Therefore the measured absorbance will be

$$A_M = \log \frac{(P_0' + P_0'')}{(P' + P'')}$$

which can be rewritten as

$$A_M = \log \frac{(P_0' + P_0'')}{(P_0' 10^{-\epsilon' bc} + P_0'' 10^{-\epsilon'' bc})}$$

or

$$A_M = \log(P_0' + P_0'') - \log(P_0' 10^{-\epsilon' bc} + P_0'' 10^{-\epsilon'' bc})$$

Now, when $\epsilon' = \epsilon''$, this equation becomes

$$A_M = \epsilon' bc$$

and Beer's law is followed; in all other instances, however, the relationship between A_M and concentration c will certainly not be linear and departures from linearity will become greater as the difference between ϵ' and ϵ'' increases.

Experiments show that deviations from Beer's law due to the use of a polychromatic beam are not appreciable provided the radiation used does not encompass a spectral region in which the absorber exhibits large changes in absorbance as a function of wavelength. This is illustrated in Figure 28.5.

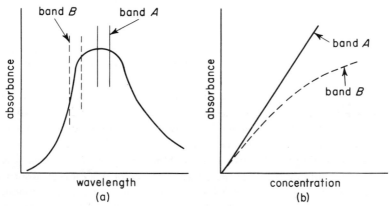

Fig. 28.5 Effect of Polychromatic Radiation upon the Beer's Law Relationship. Band A shows little deviation since ϵ does not change greatly throughout the band. Band B shows marked deviations since ϵ undergoes significant changes in this region.

Analytical Errors in Absorption Measurements

In many (but certainly not all) methods of analysis based upon the absorption of radiation, the major source of indeterminate error lies in the measurement of the absorbance. Determination of this quantity requires evaluation of the power of a beam that has passed through the solution, as well as its power after passage through the solvent only. Because of the logarithmic nature of the relationship between these quantities and concentration, however, the effect on the analytical result of the uncertainties in the power measurements is not immediately obvious. We need, therefore, to examine this question.

In many instances, the measurement of radiant power involves an absolute uncertainty that is independent of the magnitude of the power. As a consequence, the ratio P/P_0 (that is, the transmittance) can be determined with a constant absolute error over a considerable range. We must now see what the effect of this constant error in transmittance has upon the uncertainty of the results of an analysis. To do this we shall write Beer's law in the form

$$-\log T = \epsilon bc \qquad (28\text{-}4)$$

After converting the left-hand side to a natural logarithm, the derivative of this equation is

$$-\frac{0.434}{T} dT = \epsilon b\, dc \qquad (28\text{-}5)$$

Dividing equation (28-5) by (28-4) and rearranging gives

$$\frac{dc}{c} = \frac{0.434}{T \log T} dT \qquad (28\text{-}6)$$

and this may be written as

$$\frac{\Delta c}{c} = \frac{0.434}{T \log T} \Delta T \qquad (28\text{-}7)$$

The quantity $\Delta c/c$ is a measure of the *relative error* in a concentration measurement arising from a given error ΔT in the measurement of T or P/P_0. The reader should note that the relative error in concentration varies as a function of the magnitude of T. This is shown by the data in Table 28-2 in which an absolute error of 0.005 was assumed in a series of transmittance measurements; the relative concentration error goes through a minimum at an absorbance of about 0.4. (By setting the derivative of equation (28-7) equal to zero, it can be shown that this minimum occurs at a transmittance of 0.368 or an absorbance of 0.434). The data in Table 28-2 indicate that wherever possible conditions in an absorption analysis should be arranged so that the absorbance measured lies in the region of 0.1 to 1.0; this will tend to minimize the error in the analysis. Suitable dilution and proper choice of sample size will often accomplish this.

INSTRUMENTS FOR THE MEASUREMENT OF ABSORPTION OF RADIATION

We have seen that the amount of radiation absorbed by a solution is related to the concentration of the absorbing species. We shall now consider some of the ways and means by which the chemist makes use of this relationship for analysis.

The apparatus needed to perform an absorption analysis can be divided into four basic components: (1) a stable source of radiant energy, (2) a device for restricting the band width of radiation employed, (3) transparent containers

Table 28-2

Transmittance, T	Absorbance, A	Percent Error in Concentration $\frac{\Delta c}{c} \times 100$
0.95	0.022	$\pm$ 10.2
0.90	0.046	$\pm$ 4.74
0.80	0.097	$\pm$ 2.80
0.70	0.155	$\pm$ 2.00
0.60	0.222	$\pm$ 1.63
0.50	0.301	$\pm$ 1.44
0.40	0.399	$\pm$ 1.36
0.30	0.523	$\pm$ 1.38
0.20	0.699	$\pm$ 1.55
0.10	1.000	$\pm$ 2.17
0.030	1.523	$\pm$ 4.75
0.020	1.699	$\pm$ 6.38

for the sample and the solvent, and (4) a radiation detector. The nature of these components varies considerably depending upon the region of the spectrum used; their function, however, is similar in each instance. The degree of sophistication and refinement of these components also varies tremendously.

Radiation Sources

In order to be suitable for absorption measurements, the source of radiation must meet certain requirements. First, it must generate a beam with sufficient power to make detection and measurement easy. Second, the radiation should be continuous; that is, its spectrum should contain all wavelengths over the region in which it is to be used. Finally, the source should be stable; to meet this requirement, the power of the radiant beam must remain constant for a period long enough to measure both P and P_0. Only then will the absorbance measurements be reproducible. Some instruments are designed so that P and P_0 are measured simultaneously; with these, fluctuations in the power output are not a problem.

Sources of ultraviolet radiation. The common source of ultraviolet radiation is a hydrogen discharge tube; this consists of a pair of electrodes housed in a glass envelope with a quartz window. The tube contains hydrogen gas at a reduced pressure. Application of a direct-current or an alternating-current voltage to the electrodes causes excitation of the hydrogen molecules and production of continuous radiation in the region between 180 and 350 millimicrons.

Visible radiation. In the simplest types of absorption analysis, ordinary daylight is employed as a source. More commonly, however, a tungsten filament bulb is used. If this is operated by a 110-volt source, a constant-voltage transformer is generally needed to stabilize the radiant-power output. An alternative is to operate the lamp from an ordinary storage battery; this produces a beam that is constant in power for relatively long periods. The tungsten lamp produces continuous radiation in the region between 350 and 2500 millimicrons.

Infrared sources. Continuous infrared radiation is produced by electric heating of an inert solid. A silicon carbide rod, called a *Globar*, is often employed. It is heated to perhaps 1500 degrees by clamping it between a pair of electrodes. Radiant energy in the region of 1 to 40 microns results. A *Nernst glower* produces radiation in the region between 0.4 and 20 microns. This consists of a rod of zirconium and yttrium oxides that is heated to about 1500 degrees by passage of current.

Devices which Isolate a Limited Region of the Spectrum

There are several advantages to be gained from employment of radiation with a limited band width. We have seen that one of these is a greater probability of adherence to Beer's law. In addition, greater specificity of measurement is assured since substances absorbing in other regions of the spectrum will not

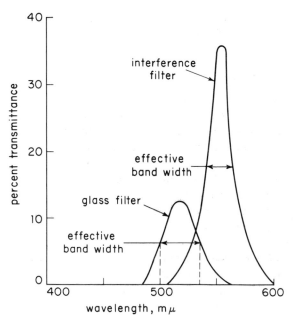

Fig. 28.6 Comparison of the Transmittance Characteristics of Typical Glass and Interference Filters.

interfere. Finally, by restricting the radiation to that portion of the spectrum which is most strongly absorbed by the substance being analyzed, a greater change in absorbance per increment of concentration is observed. This results in increased sensitivity of measurement.

The devices employed for restricting radiation fall into two categories. *Filters* function by absorbing large portions of the spectrum and transmitting relatively limited wavelength regions; these are employed primarily in the visible region of the spectrum. *Monochromators* are more sophisticated devices that isolate a beam of high spectral purity. In contrast to filters, the wavelengths employed can be varied continuously. Monochromators are used for ultraviolet, visible, and infrared radiation.

Filters. It is useful to compare filters in terms of the range of wavelengths over which the transmittance decreases to one half of its maximum value; this is sometimes called the effective band width and is illustrated in Figure 28.6.

From the standpoint of stability, colored-glass filters are preferred over those consisting of dyes suspended in gelatin. The effective band width will vary from filter to filter, but typically will embrace from 20 to 50 millimicrons. Glass filters with transmission maxima throughout the visible spectrum are readily available, and are widely employed in absorption analysis.

An *interference filter* consists of two extremely thin semitransparent metallic films separated by a very thin transparent material. When a perpendicular beam

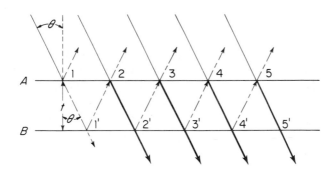

Fig. 28.7 The Interference Filter. At point 1, light strikes the semitransparent film at an angle θ from the perpendicular, is partially reflected, partially passed. The same process occurs at 1′, 2, 2′, etc. For reinforcement to occur at point 2, the distance traveled by the reflected beam must be some multiple of its wavelength, λ. Since the path length between surfaces can be expressed as $t \cos \theta$, the condition for reinforcement is that

$$n\lambda = 2t \cos \theta$$

where n is a small whole number. In practice, θ is made to approach $0°$; under these conditions

$$n\lambda \cong 2t$$

of light strikes this array, a portion passes through the first metallic layer while a portion is reflected; see Figure 28.7. The portion that is passed undergoes a similar partition upon striking the second layer. If the reflected portion from this interaction is of the proper wavelength, it will be partially reflected from the inner side of the first surface in phase with light of the same wavelength that is passing through at this point. The result is that the wavelength is reinforced while all others, being out of phase, suffer destructive interference.

Interference filters provide a degree of spectral purity that is seldom attained in their glass counterparts, effective band widths on the order of 10 millimicrons being readily achieved. Furthermore, a greater amount of light of the desired wave is transmitted. This is shown in Figure 28.6.

Monochromators.   A monochromator serves to resolve the radiation from the source into its component wavelengths and to provide for the isolation of these in very narrow band widths. Light admitted through an entrance slit is collimated with a lens or mirror. It is then dispersed by means of a prism or grating. Any portion of the resulting spectrum can then be focused upon an exit slit, again with a lens or mirror.

The effective band width of radiation emerging from a monochromator will depend upon a number of factors. Among these is the structural nature of the dispersing element. The wavelength under scrutiny is also important with some monochromators. Finally the widths of the entrance and exit slits play an important part, narrower widths leading to smaller effective band widths. Reduction in slit widths, however, results in a decrease in the radiant power emitted from the monochromator and thus the minimum effective band width may be limited by the sensitivity of the device employed to detect the emergent radiation.

Dispersion in a prism arises from the phenomenon of *refraction*. Because its velocity depends upon the medium through which it travels, light is frequently refracted, or bent, as it passes from one medium to another. The extent of this effect depends upon the angle of the oncoming beam with respect to the interface, the wavelength of radiation, as well as the refractive indices of the two media. The refractive index n is the ratio between the velocity of light in a medium v and in a vacuum c

$$n = \frac{v}{c}$$

Extensive refraction is associated with large differences in refractive index.[5]

Dispersion occurs because the refractive index of the prism material is dependent upon the wavelength. Furthermore, this variation in refractive index is most pronounced at short wavelengths. Passage through a prism thus

---

[5] A beam of light striking an interface at an angle θ_1 measured from the perpendicular will be refracted at an angle θ_2 with respect to the same reference. Snell's law states that

$$n_1 \sin \theta_1 = n_2 \sin \theta_2$$

where n_1 and n_2 represent the refractive indices of the respective media.

results in the spreading of a polychromatic beam into its components, with the short wavelengths being dispersed to a greater extent than the longer ones. Figure 28.8 illustrates this schematically.

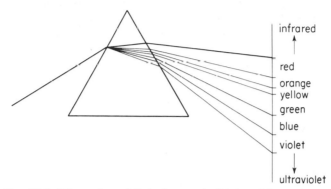

infrared
↑
red
orange
yellow
green
blue
violet
↓
ultraviolet

Fig. 28.8 Dispersion of Polychromatic Light with a Prism.

A Littrow prism consists of a 30° prism that has been silvered on its back face. Light enters and emerges from the same face, having in effect been subjected to the same dispersion as with a full prism. The Littrow prism has the advantage of allowing a more compact instrument design. In addition, any birefringence in the prism material is cancelled out since the light path involves a pass in both directions.

The material used to construct a prism varies considerably depending upon the portion of the spectrum to be studied. For the region between 350 and 2000 millimicrons, glass is often employed; quartz has a greater range of usefulness, extending from about 180 to 4000 millimicrons. In the infrared region where glass and quartz are not transparent, materials such as sodium chloride, lithium fluoride, calcium fluoride, or potassium bromide must be employed. Unfortunately these substances are susceptible to mechanical abrasion and attack by water vapor; consequently special precautions are required for their use.

Polychromatic light can also be dispersed with a *diffraction grating*. The reflection grating, shown schematically in Figure 28.9, is widely employed in spectrophotometry; it consists of a highly polished surface upon which are scribed a large number of equally spaced, parallel grooves. A typical grating will have between 1000 and 2000 of these lines per millimeter, or 25,000 to 50,000 lines to the inch.

When illuminated, a grating partitions the radiation into as many small beams as there are lines; each of these, in effect, behaves as if it were a miniscule light source, sending out radiation in all directions. This accounts for the diffraction, or bending, of light as it strikes a sharp edge or traverses a narrow opening. With many thousands of sources, the possibilities are, enormous for interference and for reinforcement of the diffracted radiation from a grating.

We shall now consider the diffraction of a collimated beam of *monochromatic radiation.* Although it will be diffracted through various angles, reinforcement will occur only where successive light paths differ from each other by integer multiples of the wavelength. For a given line spacing this condition will be satisfied at definite angles (see Fig. 28.9). Radiation diffracted at other angles will suffer destructive interference. A focusing lens placed between the grating and a viewing plate would cause the reinforced beams to appear as a series of bright lines on the latter—each line corresponding to a different value of *n.* The line corresponding to $n = 1$ is said to be *first order,* that for $n = 2$ is second order, etc.

Since reinforcement occurs at different diffraction angles for each wave-

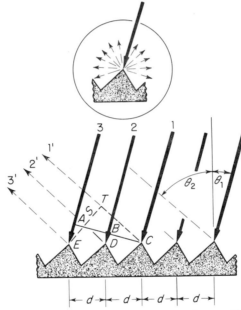

Fig. 28.9 The Diffraction Grating. A wave front ABC strikes the grating at an angle θ_1 and is diffracted (inset). In order for reinforcement with the next wave front, distances CT and $(BD + DS)$ must be integer multiples of the wavelength that is,

$$CT = n\lambda$$

$$(BD + DS) = n'\lambda$$

In terms of the diffraction angle θ_2 and the spacing of the grating, d, the condition for reinforcement is that

$$CT = 2d \sin \theta_2 = n\lambda$$

or, more generally

$$(BD + DS) = d(\sin \theta_1 + \sin \theta_2) = n'\lambda$$

length, illumination of a grating with *polychromatic radiation* will result in the formation of a series of spectra, each corresponding to a different *order*, or value of n (see Fig. 28.9). In contrast to that produced by a prism, these spectra are uniformly dispersed. The fact that several spectra are produced means that overlap of these can occur; this may lead to difficulties. Fortunately, the intensity of the high-order spectra is much less than that of the first order and generally can be removed with suitable filters.

Sample Containers

In common with monochromators, the cells or *cuvettes* that hold the samples must be made from materials that pass radiation in the spectral region of interest. Thus quartz or fused silica is required for work in the ultraviolet region. These materials, as well as glass, can be employed for visible radiation; the infrared region requires windows of such substances as sodium chloride or calcium fluoride. In general, the windows should be perfectly normal to the direction of the beam of radiation in order to minimize losses by reflection. For reasons of economy, however, the simpler instruments employ cells that are cylindrical and thus present a curved surface to the incident beam. With such cells, particular care is needed to reproduce the position of the cell with respect to the beam if reproducible results are to be achieved; it is good practice to mark cylindrical cells so that the same surface is always presented to the radiation front. Variations in the path length will otherwise occur since these cells are seldom perfectly circular in cross section.

Radiation Detectors

To determine the absorbance of a solution it is necessary to have a means for comparing the power of the radiation passing through the sample with that traversing the solvent. Most devices used for this purpose convert the radiant energy into electric energy which can then be measured by conventional equipment.

The device chosen should be responsive over a wide wavelength range. Furthermore it is essential that the electric signal produced by the detector be directly proportional to the power of the beam impinging upon it. When this condition is satisfied, we may write

$$P = kG$$

and

$$P_0 = kG_0$$

Where G and G_0 represent the electrical response of the detector when placed in the path of the radiation passing through the solution and the solvent, respectively. Thus, the absorbance is given by

$$\log \frac{P_0}{P} = \log \frac{kG_0}{kG} = \log \frac{G_0}{G}$$

Detectors for ultraviolet and visible radiation. The *barrier layer* or *photovoltaic* cell is a photoelectric device that is used primarily for the detection and measurement of visible radiation. It consists of a flat copper or iron electrode upon which is deposited a layer of semiconducting material, such as selenium or cuprous oxide. A transparent metallic film of gold, silver, or lead covers this and serves as the second or collector electrode; the entire array is protected by a transparent envelope. The interface between the selenium and the metal film serves as a barrier to the passage of electrons. Irradiation with light, however, provides some electrons with sufficient energy to overcome this barrier, and electrons flow from the semiconductor to the metal film. If the metal film is connected via an external circuit to the plate on the other side of the semiconducting layer, a flow of electrons through this circuit will occur provided its resistance is not too great. Ordinarily this current is large enough to be measured with a galvanometer or microammeter; and under proper conditions its magnitude will be directly proportional to the power of the radiation striking the cell. Currents on the order of 10 to 100 microamperes are typical.

The barrier layer cell constitutes a rugged, low-cost means for measuring radiant power. No external source of electrical energy is required. On the other hand, its output cannot be readily amplified owing to its low internal resistance. Thus, although the barrier cell layer delivers a readily measured response at high levels of illumination, it suffers from lack of sensitivity at low levels when compared with other detectors. Finally, a barrier layer cell exhibits fatigue, its response falling off with time upon prolonged illumination; proper circuit design and choice of experimental conditions largely eliminate this source of difficulty.

A second type of photoelectric device, which may be used in both the ultraviolet and visible regions, is the *phototube*. This consists of a semicylindrical cathode and a wire anode sealed inside an evacuated glass envelope. The concave surface of the cathode supports a layer of photoemissive material, often an alkali-metal or alkaline-earth oxide. The composition of this layer determines the spectral region to which the tube is responsive. A potential is impressed between anode and cathode, the former being positive with respect to the latter.

Electrons emitted from the cathode surface upon illumination are accelerated toward the anode by the applied potential; their number is proportional to the radiant power of the light. The fraction of these electrons collected by the anode becomes greater as the applied potential increases. At so-called saturation, essentially all are collected and further increases in potential have no effect. The current passing through a phototube operated at potentials in excess of that required for saturation will be linear with respect to the radiant power of the light striking it. These relationships are shown graphically in Figure 28.10. Generally, phototubes are operated at an applied potential of about 90 volts.

Instruments employing phototubes are somewhat more complicated than their counterparts equipped with a photovoltaic cell. This results from the need for an external source of electric energy. In addition, the output currents are generally so small that amplification is necessary before they can be measured accurately. Finally, it is necessary to compensate for the small currents, called *dark currents*, that flow even in total darkness owing to the thermal emission of electrons from the cathode. Notwithstanding these drawbacks, however, im-

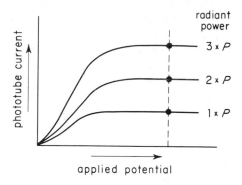

Fig. 28.10 Variation of Phototube Response with Applied Potential at Three Levels of Illumination. Note the linearity of response at the higher applied potentials.

portant advantages accrue from the use of a phototube. Because its output is readily amplified, the phototube is potentially a more sensitive radiation detector than the photocell. Furthermore, phototubes can be constructed that respond to ultraviolet radiation; this property renders absorption measurements feasible in this region.

Figure 28.11 is a schematic diagram of a typical phototube arrangement.

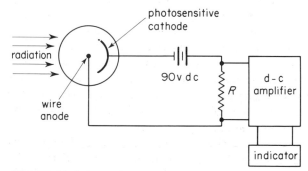

Fig. 28.11 Schematic Diagram of a Phototube and its Accessory Circuit. The current induced by the radiation causes a potential drop across the resistor R; this is amplified and measured by the indicator.

The *photomultiplier tube* is a type of phototube in which the primary signal resulting from photoemission is amplified internally by factors as great as 10^8. In addition to a photosensitive cathode and an anode, this tube contains a number of other electrodes, called dynodes. Each dynode is focused upon the next succeeding one, and each is maintained at a higher positive potential than the preceding one. As a result, a given dynode serves as anode for the one that precedes it and as cathode for the one that follows.

Every electron striking a dynode surface causes the emission of several electrons. These, in turn, are accelerated toward the next dynode where the same process takes place. After this has occurred at six or seven dynodes, each photoelectron emitted by the cathode will have indirectly caused a cascade of 10^6 or more electrons to be collected at the anode. This yields an extremely sensitive photoelectric device.

Detectors for infrared radiation. Generally, infrared radiation is detected by measuring the temperature rise of a blackened material placed in the beam of radiation. The temperature changes resulting from absorption of the radiant energy are minute indeed; thus the ambient temperature must be carefully controlled to avoid large errors in the measurements.

One method of determining the temperature change involves use of a tiny thermocouple or a group of thermocouples called a *thermopile*. With this device the electromotive force developed across a dissimilar metal junction is measured.

A *bolometer* is a second type of temperature detector and consists of a resistance wire or a thermistor whose resistance varies as a function of temperature. Here, it is the change in electrical resistance of the detector that is measured.

Some Typical Instruments and Techniques

Having considered the four basic components required for measuring the absorbance of a solution, we shall now consider how they are combined to give a finished instrument. Inasmuch as each of these components may take a variety of forms, a considerable variation in the design of instruments is to be expected; we shall limit this discussion to a few typical examples.

Absorption methods are commonly classified on the basis of the instruments and techniques employed in the measurements. Thus, a *spectrophotometric method* employs a *spectrophotometer* consisting, in essence, of a light source, a prism or grating monochromator, a photoelectric detector of radiation, and other suitable accessories. Ultraviolet, visible, and infrared spectrophotometers are commonly employed in analytical work.

A *photometer* is a simpler instrument consisting of a photoelectric detector, a light source, and cells. Filters are employed to restrict the radiation from the source. Generally, *photometric* methods are restricted to the visible region of the spectrum; the instruments have also been called *photoelectric colorimeters* or *colorimeters*.

We shall limit use of the term *colorimeter* to those instruments that employ the human eye as the detector for radiation. Colorimetric methods represent the simplest form of absorption analysis.

Colorimetry. In its simplest form, colorimetry consists of visual matching of the color of the solutions of the substance with a set of standards. For such a procedure flat-bottomed tubes called *Nessler tubes* are frequently employed. These are calibrated so that a uniform light path is achieved. Daylight, reflected through the bottoms of the tubes, frequently serves as a radiation source. Ordinarily, no attempt is made to restrict the portion of the spectrum employed.

A somewhat more refined colorimetric procedure involves the comparison of the unknown with a single standard solution. Here, the two solutions are contained in flat-bottomed tubes; the path lengths are varied by means of adjustable transparent plungers that can be moved up and down in the solutions. After balance has been achieved visually, the path lengths are measured and the concentration of the unknown is calculated by assuming that Beer's law is applicable. Thus, at balance

$$A_x = A_s$$

$$\epsilon b_x c_x = \epsilon b_s c_s$$

or

$$c_x = c_s \frac{b_s}{b_x}$$

where x refers to the unknown and s to the standard. A *Duboscq* colorimeter embodies these principles and is equipped with an optical system that permits the ready comparison of the beams passing through an eyepiece with a split field.

Visual colorimetric methods suffer from several disadvantages. A standard or series of standards must always be available. Furthermore, the eye is not capable of matching colors if a second colored substance is present in the solution. Finally, the eye is not as sensitive to small differences in absorbance as a photoelectric device; as a consequence, concentration differences smaller than about 5 percent relative cannot be detected.

Photometric methods and instruments. A considerable increase in sensitivity results from the substitution of a photoelectric detector in place of the eye, and from limiting the range of radiation to those wavelengths that are most strongly absorbed by the sample. Figure 28.12 presents schematic diagrams for two instruments that accomplish this.

The first of these is a simple single-beam photometer, consisting of a tungsten filament bulb, a lens to provide a parallel beam of light, a filter, and photovoltaic cell. The current produced in the last is measured with a micro-ammeter. In most single-beam instruments provision is made for the scale of the meter to read directly in terms of percent transmittance. The face of the meter is scribed with a linear scale from 0 to 100. A cell containing only the solvent is first placed in the light path and the power of the beam adjusted until the meter needle indicates 100. This is accomplished either by varying the voltage applied to the lamp or, as in the diagram, by varying the opening of a diaphragm interposed in the light path. The sample is then placed in the beam; since the signal from the photovoltaic cell is linear, the resultant scale reading will be the percent transmittance (that is, the percent of full scale). Clearly, a logarithmic scale would give the absorbance of the solution directly.

A disadvantage of a single-beam photometer lies in the uncertainty in the measurements resulting from fluctuations in light intensity during the measurement of transmittance. The magnitude of this uncertainty can be reduced by employing a double-beam instrument such as that shown in Figure 28.12.

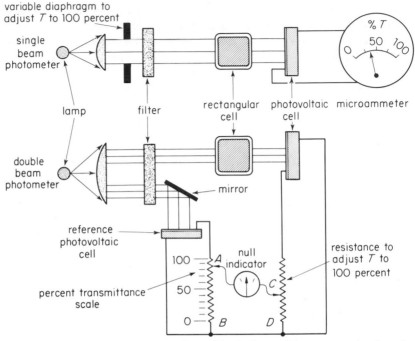

Fig. 28.12 Schematic Diagrams for a Single-beam Photometer (top) and a Double-beam Photometer (bottom).

Here, the light beam is split by some means. A portion passes through the sample or solvent and thence to a detector. The other portion is directed to a second reference detector that continuously monitors the output of the lamp. The output of the working photocell is then compared with that of the reference by a suitable circuit design. In the instrument shown, the currents from the two photovoltaic cells are passed through variable resistances; one of these is calibrated as a transmittance scale in linear units from 0 to 100. A sensitive galvanometer, which serves as a null indicator, is connected across the two resistances. When the potential drop across AB is equal to that across CD, no current will pass through the galvanometer; under all other circumstances a current flow will be indicated. At the outset, the solvent is introduced into the cell and contact A is set at 100; contact C is then adjusted until no current is indicated. Introduction of the sample into the cell results in a reduction of the power and therefore a reduction in the potential drop across CD; this is compensated for by moving A to a lower value. At balance, the percent transmittance is given on the scale.

Filter selection for a photometric analysis. Photometers are generally supplied with an array of filters, each of which transmits a different portion of the spectrum. Selection of the proper one for a given application is important inasmuch as the sensitivity of the measurement is directly dependent upon the filter. Empirically, the choice of the most suitable filter from a group is relatively simple; it should preferably be the color complement of the sample. If several filters possessing the same general hue are available, the one that causes the sample to exhibit the greatest absorbance (or least transmittance) should be used.

Spectrophotometric instruments employing ultraviolet and visible radiation. Several excellent spectrophotometers are available commercially that operate in the visible region of the spectrum; some of these can be used in the ultraviolet region as well, being equipped with quartz optics and phototubes sensitive to this radiation. Less expensive glass optics are employed in those instruments that are for visible radiation only. Among the available instruments, there is a great range not only in wavelength region covered but also in the design, quality of components, performance characteristics, and cost. The highest quality spectrophotometers make possible the use of effective band widths of the order of a few tenths of a millimicron while some of the less refined instruments operate at 10 to 20 millimicron band widths. These latter instruments are considerably less expensive and are entirely adequate for many applications. Recording instruments are also available. These operate on the split-beam principle. The power of the two beams is compared and the ratio recorded as transmittance or absorbance on a recorder chart. The monochromator is motor driven and coupled with the chart drive motor of the recorder; this gives a plot of transmittance or absorbance as a function of wavelength. Instruments of this sort are naturally more expensive than manual ones; they greatly reduce the time required to obtain a complete absorption spectrum of a solution. On the other hand, they save little time in a quantitative analysis where the measurement consists of determining the absorbance at a single predetermined wavelength.

Figures 28.13 and 28.14 illustrate the components of two spectrophotometers that are widely employed by chemists. The first of these is the Beckman DU spectrophotometer which has quartz optics and can be operated in both the ultraviolet and visible regions of the spectrum. The instrument is provided with interchangeable radiation sources, including a hydrogen discharge tube for the lower wavelengths and a battery-operated tungsten filament lamp for the visible region. A pair of mirrors reflect radiation through an adjustable slit into the monochromator compartment. After traversing the length of the instrument, the radiation is reflected into a Littrow prism; by adjusting the position of the prism, light of the desired wavelength can be focused on the slit. The optics are so arranged, however, that the entrance and exit beams are displaced from one another on the vertical axis; thus, the exit beam passes beneath the entrance mirror as it enters the cell compartment. After passing through the sample or solvent, the light passes into the phototube compartment where its power is measured with the more appropriate of a pair of interchangeable phototubes. One of these is sensitive to radiation above 625 millimicrons and the other to shorter wavelengths. The photoelectric current is passed through a fixed resistance, and the potential drop across this resistance measured by means of a potentiometer circuit. Amplification of the out-of-balance currents in the potentiometer circuit is necessary since these are very small.

The Beckman instrument is an example of a high-quality manual spectro-photometer that is capable of operating at narrow band widths. Under optimum conditions, uncertainties in transmittance readings with this instrument can be reduced to a few tenths of a percent.

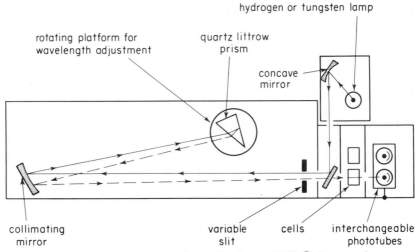

Fig. 28.13 Schematic Diagram of the Beckman DU Ⓡ Spectrophotometer. (By permission, Beckman Instruments, Inc., Fullerton, California.)

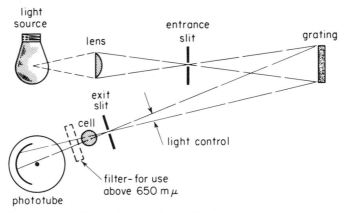

Fig. 28.14 Schematic Diagram of the Bausch and Lomb Spectronic 20 Spectrophotometer. (By permission, Bausch and Lomb Optical Company, Rochester, New York.)

The Bausch and Lomb Spectronic 20, shown schematically in Figure 28.14, may be considered as representative of instruments in which a degree of photometric accuracy is sacrificed in return for simplicity of operation and low cost. Its normal range is 350 to 650 millimicrons although this can be extended to 900 millimicrons by the use of a red-sensitive phototube. The monochromator system consists of a reflection grating, lenses, and a pair of fixed slits.

Because the grating produces a dispersion that is independent of wavelength, a constant band width of 20 millimicrons is obtained throughout the entire operating region. The instrument employs a single phototube. Its output is amplified and used to actuate the pointer of a meter calibrated in terms of both transmittance and absorbance. Instruments of this type find extremely wide application for routine analytical work.

Infrared spectrophotometers. In principal, an infrared spectrophotometer does not differ greatly from instruments for measuring absorption of shorter wavelengths. In detail, however, spectrophotometers for this region are quite dissimilar from those considered in the previous sections. As we have already noted, heated solids are used as sources, and detection of the radiation is accomplished by means of heat-sensitive devices. Prisms are often employed for dispersion of the radiation; these are constructed of such infrared-transparent materials as sodium chloride, lithium fluoride, or calcium fluoride; the first of these is by far the most common. Because most of these materials are attacked by water vapor, much care must be exercised to avoid damage from this source.

Concave mirrors, rather than lenses, are used to focus infrared radiation in order to reduce power losses and also because good achromatic lenses are not available for this type of radiation.

The infrared spectrum of most substances is generally complex when compared with that of the ultraviolet and visible regions; the manual collection of data for such a spectrum is a long and tedious process. Consequently, infrared spectrophotometers are designed as recording instruments that automatically produce a plot of transmittance as a function of wavelength. Thus there are no infrared spectrophotometers commercially available comparable to the simple and inexpensive instrument described in the previous section.

PHOTOMETRIC AND SPECTROPHOTOMETRIC MEASUREMENT

Methods based upon the absorption of radiation represent powerful and useful tools for the analytical chemist. The ultraviolet region is particularly important for the quantitative determination of a variety of organic compounds. Absorption measurements in the visible region provide the means for quantitative determination of trace amounts of most inorganic elements. Spectrophotometric measurement in the infrared is perhaps the most important single tool for the qualitative identification of organic compounds and for the structural determination of such compounds.

Some General Remarks on Quantitative Absorption Spectroscopy

Several important steps precede a quantitative analysis based upon the absorption of radiation; these determine, in no small degree, the ultimate accuracy of the method.

Selection of wavelength to be employed. It is good practice to choose the most suitable wavelength for quantitative measurements from the complete absorption spectrum of the substance to be determined. In some instances, this is available in the literature; often, however, it is preferable to produce the curve experimentally under conditions identical with those to be employed in the analysis.

Ordinarily, a wavelength corresponding to an absorption peak is selected for quantitative measurements. The greatest slope in the curve relating absorbance to concentration will be obtained at an absorption peak; as a result a maximum in sensitivity will be realized. Furthermore, the absorption curve is often relatively flat in the region of a maximum; as a consequence, better conformance to Beer's law is to be expected, particularly when it is necessary to employ a relatively wide effective band width (see p. 644). Finally, if the measurements are made in a region in which absorbance changes little with wavelength, the method will be less sensitive to uncertainties arising from failure to reproduce the wavelength setting from measurement to measurement.

The absorption spectrum, if available, will aid in choosing the most suitable filter for a photometric analysis; if this is lacking, the alternate method for selection given on page 658 may be used.

In order to avoid interference from other absorbing substances, a wavelength other than a peak may be appropriate for analysis. If this is necessary, the region selected should be one in which the change in absorbance with wavelength is not too great.

Variables that influence the absorbance. A number of common variables often influence the absorption spectrum of a substance. The nature of the solvent, the pH of the solution, the temperature, the presence of high electrolyte concentrations, and the presence of certain other substances may be cited as common examples. The effects of these variables must be known and a set of analytical conditions chosen such that the absorbance will not be materially influenced by their uncontrolled variation.

In many instances the absorbing species is produced by a chemical reaction between the substance being determined and a reagent. Here, the influence of reagent concentration and the rate at which the product is formed must also be studied.

Finally the stability of the absorbing substance under the conditions used must be determined.

A description of these effects on a particular system is frequently available in the literature so a detailed investigation is not necessary. In the absence of such information, however, it is not safe to proceed without a preliminary study covering the effects of these variables.

Determination of the relationship between absorbance and concentration. Having decided upon a set of conditions for the analysis, it is necessary to prepare a calibration curve from a series of standard solutions. These standards should approximate the over-all composition of the actual samples and should cover a reasonable range of concentrations with respect to the species being determined. Seldom, if ever, is it safe to assume adherence to Beer's law and use only a single standard to determine the molar absorptivity. It is even more foolhardy to make use of a literature value for the molar absorptivity to calculate the results of an analysis.

Accuracy. The accuracy obtainable by spectrophotometric or photometric procedure varies considerably depending upon the type of instrument employed, the chemistry of the system being studied, and the care taken. In general, it is not difficult to reduce the relative error to 1 or 2 percent. By taking special precautions this figure can often be reduced to 0.2 percent.[6]

Quantitative analysis of mixtures of absorbing substances. We have seen that the total absorbance of a solution at a given wavelength should be equal to the sum of the absorbances of the individual components present. This relationship makes possible the analysis of the individual components of a mixture even when an overlap in their spectra occurs. Consider, for example, the spectra of M and N, shown in Figure 28.15. There is obviously no wavelength at which the absorbance of this mixture is due simply to one of the components; thus an analysis for either M or N is impossible by a single measurement. However, the absorbances of the mixture at the two wavelengths λ_1 and λ_2 may be expressed as follows:

for λ_1

$$A' = \epsilon'_M bc_M + \epsilon'_N bc_N$$

and for wavelength λ_2

$$A'' = \epsilon''_M bc_M + \epsilon''_N bc_N$$

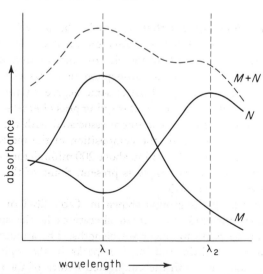

Fig. 28.15 Absorption Spectrum of a Two-component Mixture.

[6] See C. F. Hiskey, *Anal. Chem.*, **21**, 1440 (1949); C. N. Reilley and C. M. Crawford, *ibid*, **27**, 716 (1955).

The four molar absorptivities ϵ'_M, ϵ'_N, ϵ''_M, and ϵ''_N can be evaluated from individual standard solutions of M and of N. The absorbances of the mixture A' and A'' are experimentally determinable as is b, the cell thickness. Thus, from these two equations, the concentration of the individual components in the mixture, c_M and c_N, can be readily calculated. These relationships obviously are valid only if Beer's law is followed. The best accuracy in an analysis of this sort is attained by choosing wavelengths at which the differences in molar absorptivities are large.

Mixtures containing more than two absorbing species can be analyzed, in theory at least, if an additional absorbance measurement is made for each added component. The uncertainties in the resulting data become greater, however, as equations are added. This technique is particularly important when infrared radiation is employed for quantitative analysis; in this region, the spectrum for most compounds is complex and the probability of overlap is great when mixtures are analyzed.

Spectrophotometry in the Ultraviolet Region

The most important application of absorption in the ultraviolet region lies in the quantitative determination of organic compounds containing certain functional groups. In addition, a number of inorganic species, particularly among the rare-earth metals, absorb ultraviolet radiation and are thus susceptible to analysis by this method. Ultraviolet absorption spectra can also be employed at times for qualitative identification; applications for this purpose are limited, however, because of the rather broad and nonspecific nature of the absorption peaks.

Organic functional groups that absorb in the ultraviolet region. Excitation of a molecule by ultraviolet radiation involves electronic transitions and these processes are affected by the electron distribution within the entire molecule. Thus, the position of an absorption band associated with a given functional group is determined, at least to some degree, by the structure of the whole molecule. This is in contrast to absorption peaks in certain portions of the infrared where the position of a maximum associated with a certain functional group is only slightly affected by the composition of the molecule as a whole.

Absorption in the ultraviolet region above 200 millimicrons occurs whenever any of a number of functional groups is present. Some of the more important of these are given in Table 28-3.

Conjugation among the groups shown in Table 28-3 or with olefinic or aromatic double bonds leads to marked alterations in the spectra, generally causing a shift in the peaks to longer wavelengths. The aromatic hydrocarbons in particular show well-defined absorption peaks in the region of 270 millimicrons that are associated with the conjugated nature of their double bonds.[7]

[7] For a more extensive discussion of the relation between structure and absorption spectra, see W. West in A. Weissberger, Ed., *Physical Methods of Organic Chemistry*, 3d ed., **1**, part 3, 1939-1955. New York: Interscience Publishers, Inc., 1960; W. R. Brode,

Table 28-3

SOME FUNCTIONAL GROUPS WHICH ABSORB IN THE ULTRAVIOLET REGION

Functional Group		Approximate Wavelength of Maximum, millimicrons
Carbonyl	$>C-O$	280
Nitro	$-NO_2$	370
Nitrate	$-NO_3$	300
Mercapto	$-SH$	230
Alkyl iodide	$R-I$	250
Alkyl bromide	$R-Br$	200
Acid chlorides	$-\overset{O}{\overset{\|}{C}}-Cl$	240
Thiocarbonyl	$>C=S$	330
Azo	$-N=N-$	370

Quantitative analyses based upon ultraviolet absorption. A principal concern in developing a method employing ultraviolet radiation is the choice of a solvent. Clearly this must be a substance that not only dissolves the sample but also is transparent in the region of interest. Water is excellent from the latter standpoint, transmitting radiation as low as 200 millimicrons but is seldom a satisfactory solvent for organic compounds. Aliphatic hydrocarbons, methyl and ethyl alcohol, and diethyl ether are transparent to ultraviolet radiation and are often employed. Since the solvent may affect the position of a given absorption maximum, the same solvent should be used for calibration and analysis.

Most analyses employing ultraviolet radiation require the use of quartz or fused silica cells.

The reader can obtain an idea of the numerous applications of quantitative ultraviolet spectrophotometry by consulting the series of review articles by Rosenbaum and Hirt.[8]

Application of Absorption Spectroscopy in the Visible Region

The most important applications of photometry and spectrophotometry in the visible region are for the determination of traces of inorganic ions.[9]

Chemical Spectroscopy, 2d ed. New York: John Wiley and Sons, Inc., 1943. Extensive catalogues of ultraviolet spectra can be found in the following: R. A. Friedel and M. Orchin, *Ultraviolet Spectra of Aromatic Compounds*. New York: John Wiley and Sons, Inc., 1951; American Petroleum Institute, *Ultraviolet Spectral Data*, API Research Project 44. Pittsburgh: Carnegie Institute of Technology.

[8] E. J. Rosenbaum, *Anal. Chem.*, **21**, 16 (1949); **22**, 14 (1950); **23**, 12 (1951); **24**, 14 (1952); **26**, 20 (1954); R. C. Hirt, *ibid.*, **28**, 579 (1956); **30**, 589 (1958); **32**, 225R (1960).

[9] The following references describe methods for the majority of the inorganic ions: E. B. Sandell, *Colorimetric Determination of Traces of Metals*, 3rd ed. New York: Interscience Publishers, Inc., 1959; D. F. Boltz, Ed., *Colorimetric Determination of Nonmetals*. New York: Interscience Publishers, Inc., 1958; F. D. Snell and C. T. Snell, *Colorimetric Methods of Analysis*, 3rd ed., 4 vols. New York: D. Van Nostrand Co., Inc., 1948, 1959.

Although a few of these substances are sufficiently colored to make their direct determination possible, the majority do not strongly absorb ultraviolet or visible radiation. Most inorganic ions, however, will combine with various complexing reagents to form intensely colored solutions, thus making their analysis possible.

Colorimetric and spectrophotometric methods are among the most sensitive of all analytical procedures. Thus, determination of metal ions in the concentration range from 10^{-4} to 10^{-5} percent is often feasible by these techniques. Furthermore, many of the chelating agents employed for colorimetric work exhibit a remarkable degree of selectivity; as a result, the analysis for a given ion can frequently be carried out in the presence of overwhelming concentrations of others.

Desirable properties of a colorimetric reagent. A reagent for colorimetric analysis should be stable and reasonably easy to obtain and purify. Solutions of the material should not absorb appreciably in the wavelength region where the measurements are to be made. The solution should react rapidly with the ion being determined to give a stable product that absorbs strongly in the visible region. The reaction itself should be reasonably complete and should involve formation of a single-colored product; otherwise chemical deviations from Beer's law are likely. The absorbing properties of the system should not be critically dependent upon such variables as pH, temperature, or the total electrolyte concentration. Finally, the color-forming reaction should be reasonably selective.

Among the myriad color reagents known,[10] few exhibit all of these desirable properties. There are many, however, that meet most of them. While some of these reagents are inorganic in nature, the majority are organic chelating agents.

Photometric titrations. Photometric or spectrophotometric methods can be employed to locate the end point in a titration where a color change occurs. In this technique, the titration vessel is placed directly in the light path of the instrument; often this will require modification of the cell compartment in order to accommodate larger containers. The absorbance of the solution is then determined after each addition of reagent, and a plot of this quantity as a function of added reagent is prepared. In common with conductometric and amperometric titrations, points well away from the equivalence point are used to construct straight lines; the end point is taken to be the point at which these intersect upon extrapolation. Careful addition of small increments at the end point is therefore unnecessary. Typical titration curves are shown in Figure 28.16.

The straight-line relationships shown in Figure 28.16 are obtained only under certain conditions. Thus, it is necessary that the measured absorbance be corrected for volume change or that the reagent be so concentrated that its addition does not significantly alter the total volume of the solution. In addition, the reaction must be essentially complete in the presence of moderate excesses of either the reagent or the substance titrated (the curvature shown in the equivalence-point region arises from the incompleteness of the reaction).

[10] See, for example, M. G. Mellon, Ed., *Analytical Absorption Spectroscopy*, 55-64. New York: John Wiley and Sons, Inc., 1950.

This technique, however, may be employed for titration of systems with equilibria less favorable than required for an ordinary titration; the data needed are obtained in regions where the common ion effect is operating to force the reaction to completion. A final requirement is that the system follow Beer's law reasonably well, although small deviations can be tolerated.

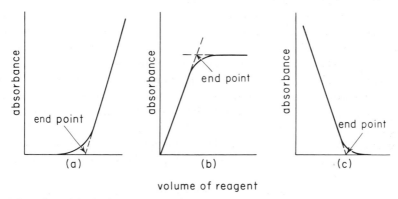

volume of reagent

Fig. 28.16 Typical Photometric Titration Curves. (a) Substance colorless, reagent colored, or substance and reagent colorless with a colorless indicator for reagent present. (b) Reagent and substance colorless, but reaction yields a colored product. (c) Substance colored, but reagent and product colorless.

Photometric titrations are frequently more accurate than an ordinary titration for dilute systems as well as for those involving incomplete reactions. The technique has advantages over direct photometry for the analysis of solutions containing several colored constituents since only the change in absorbance is important. It may also be used where direct photometry is not possible; that is, where only the reagent possesses an appreciable absorbance.

Application of Infrared Spectrophotometry[11]

Infrared spectrophotometry is one of the chemist's most powerful tools for the qualitative identification of organic compounds. It also has some important applications in quantitative analysis but these are overshadowed by its qualitative uses.

We have already pointed out that the instruments used for infrared spectrophotometry are identical in principle to those employed in the ultraviolet and visible regions; but that, in detail, there are marked differences not only in the materials from which the optics are constructed but also in the nature of the

[11] For a more detailed discussion, see D. H. Anderson, N. B. Woodall, and W. West in A. Weissberger, Ed., *Physical Methods of Organic Chemistry*, 3d ed., 1, part 3, 1959-2020. New York: Interscience Publishers, Inc., 1960; L. J. Bellamy, *The Infra-Red Spectra of Complex Molecules*, 2d ed. New York: John Wiley and Sons, Inc., 1958.

sources and detectors. Dissimilarities are also found in the method of handling samples.

Solvents; preparation of samples. No single solvent is transparent throughout the infrared region of the spectrum, nor are there any rugged transparent solids, similar to quartz or glass, that can be employed for cells. As a result, the techniques employed in handling samples differ considerably from those used for ultraviolet or visible work.

Generally, water must be avoided in infrared work; not only does it absorb broad regions of radiation, but it also attacks many of the materials used for constructing cells. From the standpoint of transparency, carbon tetrachloride and carbon disulfide are the two most satisfactory solvents; the former is useful in the region up to about 7.6 microns and the latter from that point to 15 microns. Unfortunately, not all samples will dissolve in these liquids and solvents with more restricted ranges must be employed if a solution technique is to be used. In general, rather concentrated solutions are employed to minimize absorption due to the solvent; this requires narrow cells, thicknesses between 0.01 and 1 millimeter being common.

Because of the difficulty of finding suitable solvents, liquid samples are frequently studied without dilution. This requires that the thickness of the liquid be kept to a minimum. For qualitative work a very thin film of the liquid placed between rock salt plates will often suffice.

Another technique widely employed for solids is to disperse the finely ground sample in a liquid or solid diluent to form what is called a *mull*. Liquid mulls are prepared by grinding the sample in a viscous hydrocarbon having a high refractive index (to reduce scattering of the radiation); the mineral oil Nujol is widely used for this purpose. The slurry is then examined in a cell for liquids. Absorption by the mineral oil does occur in certain spectral regions. A second dispersing medium that is perfectly transparent to infrared radiation is solid potassium bromide. The finely ground sample is mixed with this compound and a solid pellet is formed by compressing the mixture in a hydraulic press at pressures of 10,000 to 20,000 lb/in². The resulting thin disks are held in the beam for study.

Qualitative analysis. The general appearance of an infrared absorption spectrum is quite different from that of the ultraviolet and visible regions. Even for relatively simple compounds there is a bewildering array of sharp peaks and minima. It is this multiplicity of peaks, however, that imparts specificity to the spectrum; no two compounds give identical spectrograms. Thus, identity of spectra between an unknown and a known is widely accepted as proof of identity of composition.

In the shorter wavelength region of the infrared (below about 7.5 microns) are found peaks that are useful for the identification of certain functional groups; the positions of the maxima in this region are only slightly affected by the carbon skeleton to which the groups are attached. Investigation of this portion of the spectrum then gives considerable information regarding the over-all makeup of the molecule under investigation. Table 28-4 gives the positions of characteristic maxima for some common functional groups.[12]

Table 28-4

Functional Group		Wavelength of Absorption Peak, microns
O—H	(alcohol)	2.8 — 3.3
—NH$_2$	(primary amine)	2.9 — 3.0
C—H	(aromatic)	3.2 — 3.3
C—H	(aliphatic)	3.3 — 3.7
C≡N		4.2 — 4.6
C=O	(ester)	5.7 — 5.8
C=O	(acid)	5.8 — 6.0
C=O	(aldehyde and ketone)	5.8 — 6.0
C=C		6.0 — 6.2

In most instances identification of the functional groups in a molecule is not sufficient to permit positive identification of the compound and the entire spectrum must be compared with that of known compounds. Collections of spectra are available for this purpose.[13]

Quantitative analysis. Quantitative measurements in the infrared region are not different in principle from similar measurements in the ultraviolet and visible region. Several practical problems, however, make the attainment of comparable accuracies difficult. Among these is the necessity of using very narrow cell widths which are difficult to reproduce, the complexity of the spectra which leads to a high probability of overlap in absorption among the components in the sample, and the narrowness of the peaks which often leads to deviations from Beer's law.

EXPERIMENTS

Determination of Iron in Water

An excellent and sensitive method for the determination of iron is based upon the formation of the orange-red ferrous-orthophenanthroline complex, the structure of which is shown on page 414. Orthophenanthroline is a weak base, and in acidic solution the principal species present is the phenanthrolium

[12] For more detailed information, see N. B. Colthup, *J. Opt. Soc. Amer.*, **40**, 397 (1950).

[13] American Petroleum Institute, *Infrared Spectral Data*, A.P.I. Research Project 44, Carnegie Institute of Technology, Pittsburgh, Penn.

ion PhH^+. Thus, the complex-formation reaction is best described by the following equation:

$$Fe^{2+} + 3PhH^+ \rightleftharpoons Fe(Ph)_3^{2+} + 3H^+$$

The equilibrium constant for this reaction is 2.5×10^6 at $25°$ C. Clearly, the position of equilibrium is dependent upon the pH; at values lower than 2, incomplete reaction is encountered at the usual reagent concentration. Ordinarily a pH of about 3.5 is recommended for the analysis, but careful control of this variable is not required.

In using this reagent for the analysis of iron, an excess of reducing agent is added to the solution to keep the iron in the reduced state; hydroquinone or hydroxylamine hydrochloride are convenient for this purpose. Once formed, the color of the complex is stable for long periods of time.

Certain ions interfere with the analysis for iron and must therefore be absent. These include colored ions in general; silver and bismuth which form precipitates with the reagent; and cadmium, mercury, and zinc which form colorless soluble complexes with the reagent thus reducing the intensity of the color. Under certain conditions molybdenum, tungsten, copper, cobalt, nickel, and tin may also interfere.[14]

Reagents. *Hydroquinone solution,* 1 percent in water.

Sodium citrate solution, 250 grams/liter of solution.

o-Phenanthroline, 0.5 percent of the monohydrate in water. Warm to effect solution and store in a dark place. Discard the reagent when it becomes colored.

Standard iron solution, 0.1 mg Fe/ml. Dissolve 0.702 gram of analytical reagent grade $FeSO_4 \cdot (NH_4)_2SO_4 \cdot 6H_2O$ in 50 ml of water containing 1 ml of concentrated H_2SO_4. Transfer to a volumetric flask and dilute to exactly 1 liter.

Apparatus. This experiment can be carried out with a spectrophotometer at 508 millimicrons or with a photometer equipped with a green filter. No attempt will be made to describe the operation of these instruments here.

Preparation of calibration curve. Measure a 5-ml aliquot of the iron solution into a beaker, add a drop of bromthymol-blue indicator and add the sodium citrate solution from a pipet until the intermediate color of the indicator is achieved. Note the volume of citrate required and discard the solution. Now measure a second 5-ml aliquot of the iron standard into a 100-ml volumetric flask, and add 1 ml each of the hydroquinone and orthophenanthroline solutions. Introduce the same quantity of citrate solution as was required for the preliminary titration, and allow the mixture to stand 1 hour. Dilute to the mark.

---

[14] See E. B. Sandell, *Colorimetric Determination of Traces of Metals,* 3d ed., 537. New York: Interscience Publishers, Inc., 1959.

Clean the cells for the instrument, rinse one of these with the standard solution and then fill. Rinse and fill the second cell with a blank containing all of the reagents except the iron solution. Carefully wipe the windows of the cells with tissue and place them in the instrument. Measure the absorbance of the standard against the blank.

Prepare at least three other standards so that a range of absorbances of about 0.1 to 1.0 will be covered. Construct a calibration curve for the instrument.

Analysis of sample. Measure 5 ml of the sample into a beaker and adjust the acidity with the sodium citrate solution as before. If the original sample is basic to the indicator, add measured quantities of 0.1 N H_2SO_4 until the color just changes; then proceed as before. Discard the aliquot and transfer 5 ml of the sample to a 100-ml volumetric flask. Add 1 ml of ortho-phenanthroline, 1 ml of hydroquinone, and the amounts of citrate (and H_2SO_4) required to adjust the pH. After 1 hour, dilute to the mark and measure the absorbance. Repeat the analysis using quantities of sample that will give an absorbance in the range of the calibration curve.

Calculate the milligrams of iron per liter of sample solution.

Determination of Manganese in Steel

Small quantities of manganese are readily determined colorimetrically by oxidation of the element to the highly colored permanganate ion. Potassium periodate is effective for this purpose.

$$5IO_4^- + 2Mn^{2+} + 3H_2O \rightarrow 2MnO_4^- + 5IO_3^- + 6H^+$$

Permanganate solutions containing an excess of periodate are relatively stable.

Interferences to this procedure are few. The presence of colored ions can be compensated for by employing as a blank an aliquot of the sample that has not been oxidized by the periodate. This method of correction is not effective in the presence of appreciable quantities of cerous or chromic ions, for both of these are oxidized by the periodate to a greater or lesser extent and their reaction products absorb in the region commonly employed for the permanganate.

The accompanying method is applicable to most steels except those containing large amounts of chromium. The sample is dissolved in nitric acid; any carbon present is removed by oxidation with peroxydisulfate. Phosphoric acid is added to complex the ferric iron and prevent the color of this species from interfering with the analysis. One aliquot of the sample is carried through the entire procedure except that no periodate is added. This serves as a blank to correct for the presence of colored foreign ions.

Apparatus. A spectrophotometer set at 525 millimicrons or a photometer with a green filter may be used for absorbance measurements. Consult the manual of the instrument for operating instructions.

Preparation of calibration curve. Prepare a solution of $KMnO_4$ that contains the equivalent of 0.100 gram Mn/liter by dilution of a standardized solution of $KMnO_4$. Alternatively, exactly 0.100 gram of pure manganese metal can be dissolved in 10 ml of HNO_3. The solution is boiled to remove oxides of nitrogen and diluted to exactly 1 liter. In this case, aliquots of the sample are oxidized with KIO_4 in the same way as the sample; proceed as in the second paragraph for the analysis of steel below.

Transfer 5.00 ml of the permanganate solution to a 50-ml volumetric flask and dilute to the mark with water. Rinse and fill one of the absorption cells of the instrument with this solution; fill a second cell with water and dry the windows of both with a clean tissue. Determine the absorbance of the solution.

Now prepare a series of standards (at least three more) in the same way to cover an absorbance range between 0.1 and 1.0; plot a calibration curve.

Analysis of steel. Weigh out duplicate 0.8-gram samples of the steel and dissolve in 50 ml of 1:3 HNO_3 with boiling. Heating for 5 minutes should suffice. Cautiously add about 1 gram of ammonium peroxydisulfate and boil gently for 10 to 15 minutes. If the solution is pink or contains a brown oxide of manganese, add approximately 0.1 gram of sodium bisulfite or ammonium bisulfite and heat for another 5 minutes. Cool and dilute the solution to exactly 100 ml in a volumetric flask.

Pipet two 25-ml aliquots of the sample into small beakers and add 3 to 5 ml of H_3PO_4. To *one* of the two aliquots add 0.4 gram of KIO_4 and boil the solution for 5 minutes. The second aliquot serves as a blank and is *not treated with periodate*. Cool and dilute both aliquots to exactly 50 ml in volumetric flasks. Determine the absorbance of the periodate-treated sample against the blank that contains no periodate and obtain the milligrams of manganese present from the calibration curve.

Report the percent manganese in the steel.

Spectrophotometric Analysis of a Permanganate—Dichromate Mixture

Both potassium dichromate and potassium permanganate absorb strongly in the visible and ultraviolet regions; their spectra overlap sufficiently so that the presence of one interferes with the quantitative analysis of the other. By employing the techniques described on page 663, however, accurate analysis of a mixture of these compounds is possible.

Reagents. *Standard potassium dichromate solution.* This solution should be about $4 \times 10^{-3} F$; it can be prepared from pure $K_2Cr_2O_7$ by suitable dilution of a weighed quantity of the reagent grade substance. Its concentration should be known to within 1 percent.

Standard potassium permanganate solution. This solution should be about $4 \times 10^{-3} F$. It can be prepared by suitable dilution of a more concentrated solution which has been standardized in the usual way.

Apparatus. A spectrophotometer is needed for this experiment.

Absorption spectra. Transfer 10.0 ml of standard $K_2Cr_2O_7$ solution to a 100-ml volumetric flask and dilute to the mark with approximately $0.5 N$ H_2SO_4. Prepare a solution of $KMnO_4$ in the same way.

Rinse and fill one of the cells of the spectrophotometer with the $K_2Cr_2O_7$ solution, one with the $KMnO_4$, and a third with a 0.5 N solution of H_2SO_4. Measure the absorbance of the $K_2Cr_2O_7$ and the $KMnO_4$ against the H_2SO_4 solution at intervals throughout the wavelength region between 350 and 650 millimicrons. Sufficient data should be collected to define clearly the maxima in the curves.

Plot absorption curves for the two ions.

Calibration curves. Prepare additional standard solutions of the $KMnO_4$ and $K_2Cr_2O_7$ in 0.5 N H_2SO_4 and determine the absorbance of each at two wavelengths chosen as suitable for the analysis of a mixture of the two substances (see the absorption spectrum for each).

Plot absorbance versus concentration for each species at the two wavelengths and calculate the best values of ϵb at these.

Analysis of a mixture. The unknown will consist of an aqueous solution of $K_2Cr_2O_7$ and $KMnO_4$.

Transfer the sample quantitatively to a 100-ml volumetric flask and dilute to the mark with 0.5 N H_2SO_4. Measure the absorbance at the wavelengths at which calibration data were obtained. Calculate the concentration of $KMnO_4$ and $K_2Cr_2O_7$ in the solution.

Report the number of millimols of $KMnO_4$ and the number of millimols of $K_2Cr_2O_7$ in the sample.

problems

1. What would be the color of the following:
   (a) A solution that absorbs strongly all radiation above 500 mμ.
   (b) A solution with a transmission peak at 575 mμ.
   (c) A filter suitable for a photometric analysis based upon the blue copper-ammonia complex.
   (d) A filter for a solution that has an absorption peak at 500 mμ.

2. A photometer consisting of a photovoltaic cell and a microammeter was employed for measurements of light absorption. When a beam of radiation that had passed through the colored solution was caused to fall on the photocell, a current of 16.1 microamperes was observed; the same beam caused a current of 47.1 microamperes when it had passed through the solvent.
   (a) What was the percent transmittance of the solution?   ans. 34.2 percent
   (b) What was the absorbance of the solution?   ans. 0.466
   (c) What would be the microammeter reading if a solution containing twice the concentration of colored substance was placed in the same beam?
                                                   ans. 5.51 microamperes
   (d) What would be the percent transmittance of the solution if a layer of solution half the width of the above were measured?   ans. 58.5 percent

3. A photometer with a linear response to radiation gave a reading of 237 mv when a colored solution was placed in the light path. A reading of 412 mv was obtained when the solvent was interposed in the same beam.
   (a) Calculate the absorbance and percent transmittance of the solution.

(b) What would be the transmittance and absorbance of a solution which contained three fourths the above concentration?

(c) What would be the photometer response in millivolts if the layer of the original solution were increased in thickness by 50 percent?

4. The colored substance M was found to have an absorption peak at 405 mμ. A solution containing 3.00 mg M/liter was found to have an absorbance of 0.842 when examined in a 2.50-cm cell. The formula weight of M is 150.

(a) Calculate the absorptivity of M at 405 mμ.   ans. $1.12 \times 10^2 \text{cm}^{-1}\text{gram}^{-1}$ liter

(b) What is the molar absorptivity of M at 405 mμ?

ans. $1.68 \times 10^4 \text{cm}^{-1}\text{mol}^{-1}$ liter

(c) What weight of M is contained in 100 ml of a solution that has an absorbance of 0.760 at 405 mμ when measured with a 1.00-cm cell?   ans. 0.678 mg

5. A 4.00×10^{-4} M solution of X (GFW = 125) was found to have an absorbance of 0.636 at 500 mμ when examined with a 1.50-cm cell.

(a) What is the molar absorptivity of X at 500 mμ?

(b) Calculate the absorptivity of X at 500 mμ.

(c) What weight of X is contained in 400 ml of a solution having a percent transmittance of 34.8 when a 2.00-cm cell is employed?

(d) A 0.2-gram sample containing approximately 0.05 percent of X was dissolved and diluted to exactly 100 ml. What dilution of this solution should be made so that an absorbance of about 0.5 will be obtained with a 1.00-cm cell?

6. The substance Y has a molar absorptivity of 4,300 at 360 mμ. What is the concentration of Y in a solution having an absorbance of 0.830 at 360 mμ when measured in a 2.50-cm cell?

7. A 0.400-gram sample containing manganese was dissolved, oxidized with periodate, and diluted to 500 ml. The resulting solution was compared with a 2.07×10^{-3} N solution of $KMnO_4$. A visual color match was obtained when the path length for the standard was 5.35 cm and that for the unknown was 4.00 cm. Calculate the percent manganese in the sample.

8. The anion X^{2-} reacts with the metal ion M^{2+} to form the colored complex MX_2^{2-}. In the presence of a hundredfold or greater excess of X^{2-}, the color is found to obey Beer's law. With smaller amounts of reagent, however, deviations are found as a consequence of incompleteness of the reaction.

A 1.00×10^{-3} F solution of M^{2+} was made 0.1 F in X^{2-} and the absorbance was found to be 0.850 with a 1.00-cm cell. A 1.00×10^{-3} F solution which was 10×10^{-3} F in X^{2-} had an absorbance of 0.612. Calculate the instability constant for the complex.   ans. $K = 2.85 \times 10^{-5}$

9. A colored complex ion AB_2^{2-} is formed when A^{2+} is mixed with B^{2-}. Various quantities of Na_2B were added to a 2.00×10^{-3} F solution of A^{2+}; it was found that the absorbance became independent of the quantity of this salt at formal concentrations greater than about 0.2. This was taken to mean that the complex-formation reaction was essentially complete under these conditions. The absorbance in the level region was 0.982.

When the formal concentration of Na_2B was 9.00×10^{-3}, the absorbance was 0.670. Calculate the instability constant for the complex.

chapter 29. *Flame Photometry*

Flame photometry, a relatively simple type of emission spectroscopy, has found a number of useful applications in certain fields of analysis. Before considering the principles and practices of this method we shall consider briefly the more general field of which it is a part.

EMISSION SPECTROSCOPY

Upon being excited by a suitable energy source, the elements contained in a sample will emit visible and ultraviolet radiation. The wavelengths emitted are characteristic of the elements present; the intensity of the radiation is in part dependent upon the concentrations. Thus, both qualitative and quantitative information can be gained through measurement of the wavelengths as well as the intensities of the radiations emitted by an excited sample. The analytical method based on this principle is called *emission spectroscopy*.

Origin of Emission Spectra

The emission spectrum of an excited substance can be divided into three types. First, a *continuous* spectrum is one that is characterized by no sharp discontinuities in radiant energy as a function of wavelength. A continuous spectrum is emitted by all heated bodies. Second, there is a *band spectrum*, which consists of a group of discrete lines of radiation, each of a single wavelength. These lines are adjacent to one another and come closer together at one end of the band. Band spectra arise from the excitation of molecules and for this reason are called *molecular spectra*. The third type consists of the *line spectrum* or *atomic spectrum*, and is characterized by sharply defined individual lines of a single wavelength. This spectrum is characteristic of the elements and is due to the excitation of gaseous atoms or ions. It is the atomic spectrum that is of the greatest importance in analysis.

The atomic spectrum of an element results when sufficient energy is supplied to volatilize the element and to cause the electrons in the atoms or ions to move to higher energy states. The species is then said to be excited. The lifetime of the excited state is short, and the electrons tend to return spontaneously, often in a series of discrete steps, to their lower energy levels. Each of these steps involves the loss of a definite amount of energy in the form of electromagnetic radiation, the wavelength being determined by the magnitude of the energy difference. Inasmuch as there are a large number of possible excited states and a multiplicity of paths by which return to the unexcited state can occur, many radiations of different wavelengths are produced for a given element. However, the collection of radiations for a given species, called its *emission spectrum*, is characteristic of that species alone, and therefore such radiations can be used for identification purposes.

Excitation Methods

Basically, there are two methods for exciting the spectra of the elements for analysis. Electrical excitation is one of these, the needed energy taking the form of a high-potential alternating-current spark or a low-potential direct-current arc. In either case the sample is placed upon or is incorporated in an electrode. An arc or a continuous series of sparks then carries the current between this electrode and a second one placed a short distance away.

Excitation may also be brought about by heating a sample in a hot flame. A familiar example of this is the yellow color produced when a small quantity of a sodium salt is suspended in a gas flame; the yellow color arises from the excited sodium atoms. The excitation energies involved in an ordinary gas-air flame are a good deal lower than those attained by electrical methods, and only the more easily excited elements emit light under these conditions. Hotter flames are obtainable, however, by the use of acetylene or hydrogen with oxygen; these excite a major portion of the elements.

Instrumentation of Spectroscopy

In order to examine the spectrum of an excited substance, a *spectroscope*, a *spectrograph*, or a *spectrophotometer* is required. Each of these contains a slit system and a prism or grating for dispersing the radiation into its component wavelengths. With a spectroscope, the resulting spectrum is observed visually. With a spectrograph the dispersed radiation is displayed on the surface of a plane where it is detected photographically or by photoelectric measurements. The former is the more common, the spectrum appearing on the photographic plate or film as a series of dark lines; the position of each depends upon its wavelength. The intensity of the radiation is determined by measuring the degree of blackening of the plate; from this the concentration of the excited material can be determined. Where photoelectric detection is employed, phototubes are placed in positions corresponding to spectral lines of the elements to be determined. Concentration is estimated from the photoelectric response. With a spectrophotometer, the spectrum is scanned by moving the monochromator to vary the wavelength of radiation striking the exit slit (see Fig. 28.13, p. 660).

The equipment necessary for spectral measurement tends to be complex and expensive. The simplest and least costly is that used in flame photometry.

FLAME PHOTOMETRY

In flame photometry, thermal excitation is accomplished by spraying a solution of the sample into the flame of a burner. The resulting spectra are generally less complex than those obtained by electrical excitation because the energies involved are considerably smaller. When an ordinary gas flame is employed, for example, only those elements, such as sodium or calcium, that have low energy electronic levels are excited. With hotter flames, additional elements give emission lines. Because there are relatively few lines produced in the typical flame spectrum, a rather simple instrument will often give an adequate separation of these for analysis.

Flame Photometer

A flame photometer differs from the ordinary photometer or spectrophotometer described in the previous chapter only in the matter of light source. Conversion of an instrument from one application to the other is a relatively simple matter and involves the replacement of the light source of the photometer with an atomizer and burner.

Atomizer and burner. An atomizer is employed in flame photometry to produce a fine, uniform spray or aerosol of the solution containing the sample. This is carried into the combustion zone of a burner flame by the air or oxygen stream used to support the burning. In order to obtain a reproducible spectrum the atomizer must deliver the solution to the flame at a steady rate and the flame must be stable and constant in temperature. This requires close regulation of the rate of flow of both the fuel and the air or oxygen.

One type of atomizer is shown in Figure 29.1. Here, the solution flows through a capillary tube into a chamber and is broken into fine particles by a stream of air flowing at right angles to the tip of the tube. The larger particles collect on the walls of the chamber while the very fine droplets are carried as an aerosol in the air stream to the burner; the burner itself is an ordinary Meker type which employs commercial gas as a fuel. In an arrangement of this sort, only a small fraction of the sample is atomized, the bulk collecting in the bottom of the chamber as a liquid.

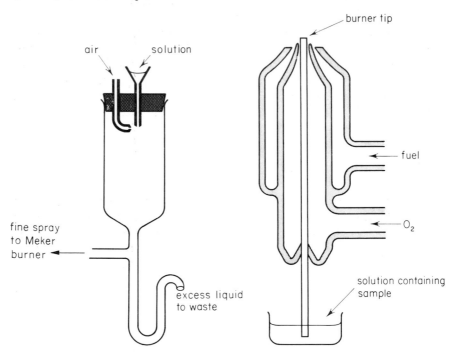

Fig. 29.1 Spray-chamber Type Atomizer.

Fig. 29.2 Diagram of an Atomizer-burner. (Reproduced by permission, Beckman Instruments, Inc., Fullerton, California.)

A second arrangement, shown in Figure 29.2, employs a special burner in which the sample is pulled up a vertical palladium tube by the flow of oxygen gas around the capillary. Upon reaching the upper tip of the capillary, the liquid is broken into droplets. The atomizer is an intergal part of the burner and combustion occurs immediately above the orifice. With this type of burner, all of the sample is drawn into the flame; typically, solution is consumed at the rate of 1 to 2 ml/minute which is in marked contrast to the other type of arrangement which requires 5 to 25 ml of solution/minute.

Many different combinations of fuel and oxidant have been proposed for flame excitation. Illuminating gas, propane, hydrogen, and acetylene are the common fuels; when mixed with air, these give flame temperatures in the range between 1700° and 2000° C. Hotter flames are achieved by the use of oxygen (2700° to 3000° C). The various air-fuel mixtures produce flames that will excite the alkali and alkaline-earth metals but little else. Hydrogen or acetylene, used in conjunction with oxygen, will give sufficiently high temperatures to cause excitation of over half the elements. Quantitative methods based on flame photometry have been proposed for perhaps a third of the elements.

Photometers. When a low-temperature flame is employed as an excitation source, only the most prominent lines of the alkali and alkaline-earth elements appear. As a consequence a filter photometer is frequently adequate to isolate a characteristic line under these conditions. The simplicity and relatively low cost of these devices is attractive. Glass, gelatin, and interference filters have all been employed in flame photometers, the last being the most satisfactory because of their narrow band widths.

Photovoltaic cells and vacuum phototubes are used to measure the power of the radiant energy passing through the filter. With the former, electronic amplification of the electric signal is necessary. Some commercial instruments employ two filters and two detectors to compensate for fluctuation in the output of the source; these are analogous to the two-cell photometers discussed in the previous chapter (p. 658). In this instance an *internal standard* is incorporated into the sample and standard. This consists of an element whose behavior is similar to that of the element being determined. Thus, in the analysis of sodium, a measured quantity of a lithium salt may be introduced. The radiant energy of the excited sample is then split with half the beam passing through a filter that transmits a sodium line and the other half through a filter that removes all but a lithium line. The power of the two beams is then compared by means of a simple bridge circuit such as that shown in Figure 28.12, page 657. The internal standard method is attractive in principle, but difficulty is often encountered in locating an internal standard element that is absent in the original sample and whose excitation behavior is similar to that of the unknown.

There are several commercially available flame photometers.[1] In general, these are satisfactory for the determination of alkali or alkaline-earth metals when these represent major constituents of the sample. Where several elements are present in high concentration, however, the inadequacy of the filter system leads to errors. The principal application of these instruments has been to the determination of sodium and potassium in biological fluids.

Spectrophotometers. The specificity of methods based on flame excitation can be greatly enhanced by employment of a prism or grating mono-

---

[1] For a description of their characteristics see: M. Margoshes and B. L. Vallee in D. Glick, Ed., *Methods of Biochemical Analysis*, **3**, 366-369. New York: Interscience Publishers, Inc., 1956.

chromator to isolate the spectral lines. An instrument incorporating this feature should be called a *flame spectrophotometer*. Several of these instruments are manufactured commercially.[2] A flame photometer is available as an attachment for the Beckman DU spectrophotometer that is described on page 659. A unit consisting of an atomizer and burner replaces the light source housing shown in Figure 28.13.

Flame spectrophotometers are more complex and expensive than the corresponding filter instruments. They do, however, offer wider applications and greater specificity.

Sources of Error in Flame Photometric Measurements

Errors in flame photometry arise from two sources. Instrumental errors are traceable to fluctuations in the behavior of the source or the detector. Errors also arise from differences in composition between the sample and the standards against which they are compared. Errors of the latter type are a good deal more difficult to cope with.

Instrumental errors. A stable flame is necessary to attain reproducible results. This requires that the flow rate of the fuel and oxidant be reproducible to approximately 1 percent. Flow gauges or pressure regulators are employed for this purpose; with properly designed equipment this need not be a serious source of uncertainty.

The atomizer must also perform in a perfectly reproducible manner so that the sample or standard is introduced at a constant rate and in droplets of constant size. Since contamination of the capillary can cause serious errors, every effort must be made to guard against irregularities from this source.

Clearly any drift or fluctuation in the performance of the detector and amplifier will lead to analytical errors.

Errors from radiation of foreign cations. Analytical errors arise when other cationic components of the sample emit radiation of a wavelength that is not completely removed by the monochromator system. The magnitude of the effect is dependent upon the quality of the monochromator, the temperature of the source, and the concentration ratio between the contaminant and the wanted element.

Figure 29.3 shows flame emission spectra of the alkali and alkaline-earth metals. Note that the emission lines for lithium, sodium, and potassium are widely spaced; as a consequence, a simple filter system will prevent the radiation from any two of these elements interfering seriously with the measurement of the third. On the other hand, calcium and strontium emit bands of radiation close to both the sodium and lithium lines; thus, the presence of these elements will lead to serious errors in an alkali analysis unless a very good monochromator system is employed. Where sodium is to be determined and the calcium to

[2] M. Margoshes and B. L. Vallee in D. Glick, Ed., *Methods of Biochemical Analysis*, 3, 366-371. New York: Interscience Publishers, Inc., 1956.

sodium ratio is high, the problem of interference becomes particularly acute. Clearly, sodium and lithium are likely to induce errors in a calcium analysis for the same reason.

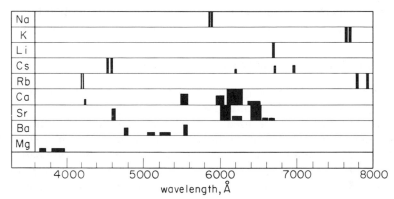

Fig. 29.3 Flame Spectra of the Alkali and Alkaline-earth Elements. (From R. B. Barnes, D. Richardson, J. W. Berry, and R. L. Hood, *Ind. Eng. Chem., Anal. Ed.*, **17**, 605 (1945), by permission.)

As the temperature of the source is increased, the need for a high quality monochromator becomes more acute; this is due not only to the excitation of additional lines for the alkali and alkaline-earth elements, but also to the appearance of radiation from the excitation of other elements. When hydrogen or acetylene is employed in conjunction with oxygen, a prism or grating system is necessary to prevent serious errors from the presence of foreign cations in the sample.

Depression or enhancement of radiation by foreign species. The presence of certain constituents in a solution may affect the intensity of radiation emitted by an element and thus lead to a serious analytical error. For example, it has been found that anions such as oxalate, phosphate, sulfate, and aluminate have a profound effect on the spectra of the alkaline-earth metals; when present in equimolar ratios, decreases of radiant power of 50 percent or greater are observed. The same anions have little or no effect upon the spectra of the alkali metals.

Enhancement of radiation intensity of one metal by the presence of a second has also been reported; sodium and potassium apparently have this mutual effect on one another.

Errors due to variation in background radiation. The line spectra observed by flame excitation are superimposed upon a continuous background radiation arising in part from the flame itself. In addition, the continuous spectra of the various compounds present in the sample will also contribute to this background radiation. The magnitude of the latter effect is dependent upon the kind and amount of salts present; unless a suitable correction is made, it can cause serious analytical errors.

Effect of organic substances. Organic compounds in a sample may influence the emission intensity of a spectral line. This is caused in part by changes in viscosity and surface tension of the liquid which influence the rate at which the sample is brought into the flame. In addition, however, organic compounds affect the temperature of the flame itself through their contribution to the heat of combustion. This usually results in an enhancement of the line intensities.

Analyses by Flame Photometry and Spectrophotometry

From the remarks in the previous section, it is apparent that close control of a number of variables is essential for reliable flame photometric data. Whenever possible, the standards used for calibration should match the over-all composition of the solution of the unknown as closely as possible. Ordinarily it is best to carry out calibration at the same time as the analyses are performed. Even with these precautions, measurements with a filter photometer can be expected to yield good results only where the sample solution is relatively simple in composition and the element being determined is a major constituent.

Several techniques have been suggested for the performance of a flame photometric analysis;[3] two of these are described briefly.

Analyses based upon calibration curves. A series of standards are prepared to contain various known concentrations of the wanted element; the composition of these should approximate that of the unknown. The instrument is then adjusted to give a zero response when pure water is atomized into the burner. Next, the optical or electric system is adjusted to give the maximum reading of the instrument with the most concentrated sample. Without altering this adjustment, meter readings for the remaining standards and the unknown are obtained. It may be necessary to apply a background correction to the calibration data. With a prism instrument this can be done by measuring the intensity of the radiation on either side of the peak at points sufficiently removed so that the line spectrum contributes nothing to the radiation. Otherwise the background is estimated from a blank containing all of the components of the sample except the one being determined. A plot of the corrected instrument response versus concentration of the wanted element is then made; ideally, this will be a straight line, but often it is not. The concentration of the unknown solution is obtained with the aid of this curve.

Standard addition method. In this technique, a calibration curve is prepared as before. Emission readings are then made for an aliquot of the unknown and for an identical aliquot to which a known quantity of the wanted element has been added. The concentration of each of these is then calculated from a calibration curve. The difference in concentration between the two should be equivalent to the quantity added unless there has been an enhancement or depression of the line intensity. If this has occurred, the apparent concentration

[3] See J. A. Dean, *Flame Photometry*, 110-122. New York: McGraw-Hill Book Company, Inc., 1960.

of the unknown is multiplied by the ratio of the true quantity added to the apparent amount added. This correction assumes that the calibration curve is essentially linear in the region being used and that the relative enhancement or depression of the line intensity is the same for the two aliquots.

A somewhat more complicated variant of this procedure is reported to give better results. Equal volumes of the unknown are added to a series of solutions containing varying known quantities of the element of interest. Emission readings for these are then corrected for background and plotted against the concentration of the solution due to the standard present. The resulting line is extrapolated to zero concentration; the concentration of the unknown can then be obtained by dividing the extrapolated reading by the slope of the line.

Application of Flame Photometry

Flame-photometric methods have been applied to the analysis of a wide variety of materials including biological fluids, soils, plant materials, cements, glasses, and natural waters.[4] The most important applications are for the determination of the alkali metals. The detailed instructions that follow describe a procedure for the determination of sodium, potassium, and calcium in natural water.[5] As mentioned earlier, each of these elements affects to some degree the spectral behavior of the others so that some correction procedure is needed to minimize these variations. This is accomplished with a so-called "radiation buffer" which is added to the sample as well as to the standards employed for calibration. These are solutions that have been saturated with all of the elements likely to be present in the sample other than the one being determined. The amounts of the various elements added in this way are ordinarily large relative to the quantities originally present in the sample; thus a uniform level of concentration is established in both standards and samples.

Reagents and Solutions

Standard solutions. Prepare stock solutions from reagent grade chemicals. For 1-liter quantities of 500-ppm standards take 1.2708 grams of NaCl, 0.9535 gram of KCl, or 1.2486 grams of $CaCO_3$. In preparing the last, make a preliminary addition of about 300 ml of water followed by about 10 ml of concentrated HCl. Dilute to volume after solution is complete and the evolution of CO_2 has ceased.

Radiation buffer for sodium determination. Saturate a solution with reagent grade $CaCl_2$, KCl, and $MgCl_2$ in that order.

Radiation buffer for potassium determination. Saturate a solution with reagent grade NaCl, $CaCl_2$, and $MgCl_2$ in that order.

[4] For a summary of applications see J. A. Dean, *Flame Photometry*. New York: McGraw-Hill Book Company, Inc., 1960 .

[5] The method has been adapted from P. W. West, P. Folse, and D. Montgomery, *Anal. Chem.*, **22**, 667 (1950).

Radiation buffer for calcium determination. Saturate a solution with reagent grade NaCl, KCl, and $MgCl_2$ in that order.

Procedure

Preparation of working curves. Introduce 5.00 ml of the appropriate radiation buffer to each of a series of 100-ml volumetric flasks. Then add volumes of standard in the amounts needed to give coverage to the concentration range from 0 to 100 ppm. Dilute to 100 ml with distilled water and mix well. Measure the emission intensity for these samples; take at least three readings for each. Correct the average values for background luminosity and prepare a working curve from these data.

Analysis of a water sample. Prepare the sample as directed above. If necessary, use a standard to calibrate the response of the instrument to the working curve. Then measure the emission intensity for the unknown. After correcting the data for background determine the concentration by comparison with the working curve.

part 6. *The Complete Analysis*

chapter 30. *The Analysis of Real Substances*

Thus far we have concentrated mainly upon the measurement of the concentration of the constituent of interest in a solution uncluttered by interfering components. Frequently this final analytical step is a relatively simple process, because the number of variables to be controlled is small, and the tools available are numerous and refined. Further, we usually have sufficient theoretical knowledge of simple systems to allow definition and solution of any problems arising in the final phase of an analysis. Thus, if every chemical analysis consisted simply of determining the concentration of a single element or compound in a simple and readily soluble homogeneous mixture, analytical chemistry could profitably be entrusted to the hands of a skilled mechanic; certainly a well-trained chemist could find more useful and challenging work for his mind and his hands.

In fact, the materials with which the chemist works are generally not simple, chemically speaking, regardless of whether his talents are employed in academic research or in the laboratories of industry. Rather, most substances that require analysis are complex, consisting of several, or indeed several tens, of elements or compounds. Frequently these materials fall short of the ideal in matters of solubility, volatility, stability, or homogeneity. With such substances, several steps precede the final one of measurement. As a matter of fact the final act is often anticlimactic in a sense, being by far the easiest to perform.

By way of illustration, let us consider the analysis of calcium, an element that occurs widely in nature and is important in many manufacturing processes. Several excellent methods are available for the measurement of the calcium ion concentration of a simple aqueous solution; among these may be cited precipitation as the oxalate, a compound that can be either titrated with a standard solution of permanganate or ignited to the carbonate for a gravimetric measurement. Calcium ion can also be titrated directly with ethylenediamine-tetraacetic acid, or its concentration determined by flame-photometric measurement. All of these methods afford an accurate measure of the calcium content of a simple salt such as the carbonate. The chemist, however, is seldom interested in the calcium content of calcium carbonate. Rather he needs to know the percentage of this element in a sample of animal tissue, in a silicate rock, or in a piece of glass. Here the analysis takes on several new dimensions. First, none of these materials is soluble in water or any of the common aqueous reagents; to obtain the solution needed for the final step requires quite strenuous treatment. Unless precautions are observed, however, such measures may cause losses of the element of interest, or perhaps its introduction from the reagents and vessels employed.

Even after the sample has been decomposed and the calcium put into solution, the excellent procedures mentioned above cannot ordinarily be applied immediately to finish the analysis, for they are all based upon reactions or properties shared by elements other than calcium. Thus a sample of animal tissue, a silicate rock, or a glass would almost surely contain one or more components that would form precipitates with oxalate, react with ethylenediaminetetraacetic acid, or affect the intensity of emitted light in a flame-photometric measurement. As a consequence, steps to free the calcium from potential interferences must usually precede the final measurement; these could well involve several operations.

We have chosen to call substances such as those described in the preceding illustration "real" substances. In contrast, most of the samples encountered in the laboratory course in elementary quantitative analysis definitely are not real, for they are generally homogeneous, usually stable even toward rough handling, readily soluble, and above all chemically simple. In addition, well-established and thoroughly tested recipes exist for their analysis. From the pedagogical viewpoint there is value in introducing the student to analytical techniques via such substances, for they do allow him to concentrate his attention on the mechanical aspects of an analysis. Once these mechanics have been mastered, however, there is little point in the continued analysis of unreal substances; to do so creates the impression that a chemical analysis involves nothing more than the slavish adherence to a well-defined and narrow path, at the end of which is found a number accurate to one or two parts in a thousand. All too many chemists retain this view far into their professional lives.

In truth, the pathway leading to the composition of real substances is often a trying one in which intellectual skills and chemical intuition are more important than mechanical aptitude. Furthermore, the chemist often finds it necessary to draw a compromise between the time he can afford to expend in performing the analysis and the accuracy he thinks he needs. If he is realistic, more often than not he is happy to settle for a part or two in one hundred rather than a part or two in a thousand; with very complex materials even this accuracy will be obtainable only with an expenditure of a great deal of time and effort.

The difficulties encountered in the analysis of real substances arise, of course, from the complexity and variability of their composition. Particularly because of the latter property, the chemist is frequently unable to find in the literature a clearly defined and well-tested route to follow; he is thus forced to modify existing procedures to take the composition of his material into account, or he must blaze a new pathway. In either case, each new component creates several new variables. Using again the analysis for calcium in a sample of calcium carbonate as an example, the number of components is small and the variables likely to affect the results are reasonable in number. Principal among these are the solubility of the sample in acid, the solubility of calcium oxalate as a function of pH, and the effect of the precipitation rate upon the purity and filterability of calcium oxalate. Contrast this with the analysis for calcium in a real sample such as a silicate rock containing a dozen or more other elements. Here the analyst has to consider not only the solubility of the calcium oxalate, but the oxalates of all of the cations present as well; coprecipitation of each of these also becomes a concern. Furthermore, a more drastic treatment is required to dissolve the sample, and additional steps are necessary to separate the interfering ions. With each new step, additional variables arise; and with this increase, a theoretical treatment of the problem becomes difficult or even impossible.

The analysis of a real substance can thus be a challenging problem requiring knowledge, intuition, and experience. Devising a procedure for such materials is not something to be taken lightly even by the experienced chemist.

The remainder of this book will be devoted to a discussion of the analysis of complex substances. In this chapter we shall consider in a general way the manner in which an analytical procedure is chosen or is developed for a real sample and the accuracy that can reasonably be expected in the analysis of such materials. The chapters that follow will give a somewhat more detailed treatment of the steps preceding the final measurement step.

Choosing a Method for the Analysis of Complex Substances

The choice of method for the analysis of a complex substance requires good judgment and a sound knowledge of the strong points and limitations of the various tools of analytical chemistry; a familiarity with the literature on the subject is also essential. We cannot be too explicit concerning this choice because there is no single best way that will apply under all circumstances. We can, however, suggest a somewhat systematic approach to the problem and present some generalities that will aid in making an intelligent decision.

Definition of the Problem

A first step, which should precede any choice of an analytical method, is that of clearly defining the problem at hand. To do this, the chemist should ask certain questions: What is the concentration range of the species to be determined? What sort of accuracy is demanded by the use to which the data are to be put? What other elements or compounds are present in the substance? What are the physical and chemical properties of the sample as a whole? and finally, How many samples are to be analyzed?

The answer to the first question regarding the concentration range of the element or compound of interest is very pertinent, for this may well limit the choice of feasible methods. If, for example, the analyst is interested in an element present in a concentration of a few parts per million, he can generally eliminate gravimetric or volumetric methods and turn his mind to spectrophotometric, spectrographic, and other more sensitive procedures. He knows, furthermore, that in this concentration range he will have to guard against even small losses of an element by coprecipitation and volatility or slight contaminations of samples from reagents. On the other hand, if the substance in which he is interested occurs in large concentrations, these considerations become less important, and he may well want to consider the classical analytical methods.

The answer to the question regarding the accuracy demanded is of vital importance in the choice of an analytical method and the way the procedure is carried out. It is the height of folly to perform physical or chemical measurements with an accuracy much greater than that demanded by the use to which the data are to be put. As we have mentioned earlier, the relationship between time expended and accuracy achieved in an analysis is ordinarily not linear; as a result, a tenfold improvement in the latter often requires a twentyfold, a fiftyfold, or even a one hundredfold increase in the hours required for the measurement. As a consequence, a few minutes spent at the outset of an analysis in careful consideration of what sort of accuracy is really needed represents an investment that a chemist can well afford to make.

The demands of accuracy will frequently dictate the procedure chosen for an analysis. For example, if the allowable error in an aluminum analysis is only a few parts in a thousand, a gravimetric procedure would probably be required. On the other hand if an error of, say, fifty parts per thousand can be tolerated, spectrographic or polarographic procedures should be considered as well. The experimental details of the method are also affected by accuracy requirements. Thus, if precipitation with ammonia were chosen for the analysis of a sample containing 20 percent of aluminum, the presence of 0.2 percent of iron would be of serious concern where an accuracy of a few parts in a thousand was demanded; here, a preliminary separation of the two elements would be necessary. On the other hand, with a limit of error of fifty parts in a thousand, a chemist might well dispense with the separation of iron and thus shorten the analysis considerably. Further, this tolerance would govern his performance in other aspects of the analysis. He would weigh out a 1-gram sample to perhaps the nearest 10 mg and certainly no closer than 1 mg. In addition, he might well be less meticulous in transferring and washing his precipitate and in the other time-consuming operations of the gravimetric procedure. If this is done intelligently, he is not being careless, but rather, realistic in terms of economy of time. The question of accuracy then is one that must be settled in clear terms at the very outset.

The third question to be answered early in the planning stage of an analysis is concerned with the chemical composition of the sample. An answer can frequently be derived from a consideration of the origin of the material; in other cases a partial or complete qualitative analysis must be undertaken. Regardless of its source, however, the information is necessary before an intelligent choice of procedure can be made, because methods for completion of the analysis are based on group reactions or group properties—that is, on reactions or properties shared by several elements or compounds. Thus, a procedure for measuring the concentration of a given element that is simple and straight-forward in the presence of one group of elements or compounds may require many tedious and time-consuming separations before it can be used in the presence of others. A solvent that is suitable for one combination of compounds may be totally unsatisfactory when applied to another. Clearly, a qualitative knowledge of the chemical composition of the sample is a prerequisite for its quantitative analysis.

The chemist must consider the physical and chemical properties of the substance rather closely before attempting to derive a method for its analysis. Obviously he should know whether it is a solid, liquid, or gas under ordinary conditions and whether losses by volatility are likely to be a problem. He should also try to determine whether the material is homogeneous and, if not, what steps can be employed to bring it to this state. Whether or not the sample is hygroscopic or deliquescent is also important information in deciding how to handle the substance. It is essential to know what sort of treatment is sufficient to decompose or dissolve the sample without loss. Preliminary tests of one sort or another may be needed to provide this information.

Finally, the number of samples to be analyzed is an important consideration in the choice of method. If there are many, a good deal of time can be afforded for calibrating instruments, preparing reagents, assembling equipment, and investigating short cuts since the cost of these operations can be spread over the large number of analyses. On the other hand if at most a few samples are to be analyzed, a longer and more tedious procedure involving none of these operations may prove to be the wiser choice from the economic standpoint.

Having answered the questions enumerated above, the chemist is now in a position to consider possible methods for the attack of his problem. At this point he may have a fairly clear idea, based on his past experience, of how he wishes to proceed. He may also find it prudent to speculate a bit on the problems likely to be encountered in the analysis and how they can be solved. He will probably have eliminated some methods from consideration and put others on the doubtful list. Ordinarily, however, he will wish to turn to the analytical literature in order to profit from the experience of others. This, then, is the next logical step in choosing an analytical procedure.

Investigation of the Literature

The literature dealing with chemical analysis is extensive. For the chemist who will take advantage of it, much of value is to be found here. A list of the reference books and journals that deal with various aspects of analytical chemistry appears on page 695. It is not intended to be exhaustive, but rather one that should be adequate for most work. The list is divided into several categories. In many instances the division is arbitrary since some of the works could be logically placed in more than one category.

In the first section are the general reference works into which the most important knowledge in the field has been distilled. Next come the books devoted to the analysis of specific types of substances such as petroleum, ferrous alloys, paints, and agricultural products. This portion of the list is quite incomplete. Following this are those publications devoted primarily to specific methods for completing the analysis; included are monographs on such diverse subjects as polarography, spectroscopy, gas chromatography, chelating agents, and volumetric methods. A portion of many of these is devoted to detailed instructions for the application of the particular technique to commonly encountered substances. Finally, there is included a list of the journals in which most space is devoted to articles concerned with analytical chemistry.

Often the chemist will begin his search of the literature by going to one or more of the general books on analytical chemistry or to those devoted to the analysis of specific types of materials. In addition, he may find it helpful to consult a general reference work on the compound or element in which he is interested. From this study he may get a clear picture of the problem at hand— what steps are likely to be difficult, what separations must be made, what pitfalls must be avoided. In some instances he may find all the answers he needs or even a set of specific instructions for the substance he wishes to analyze. Short of this he may find journal references that will lead directly to this information. In other cases, however, he will come away with only a general notion of how to proceed; he will perhaps have several alternative methods in mind and he may also have some clear ideas on how *not* to proceed. He may then wish to consult the works on specific substances, on specific techniques or go directly to the analytical journals. The monographs on methods for completing the analysis are valuable in deciding among several possible techniques.

One of the main problems in using the analytical journals is that of finding the articles that are pertinent to the problem at hand. The various reference books are useful since most are liberally annotated with references to the original journals. The key to a thorough search of the literature, however, is *Chemical Abstracts*. This journal contains short abstracts of all of the papers found in the various chemical publications. Both yearly and decennial indexes are provided to aid in the search; by looking under the element or compound to be determined and the type of substances to be analyzed, a thorough survey of the methods available can be made. Completion of such a survey involves the expenditure of a good deal of time, however, and is often made unnecessary by the host of good reference books mentioned earlier.

Choosing or Deriving a Procedure

Having defined the problem and investigated the literature for possible solutions, the chemist must next decide upon the route he will follow in the laboratory. In some instances, the choice is simple and obvious, and he can proceed directly to his analysis. In others, however, the decision requires the exercise of a good deal of judgment and ingenuity; here, experience, an understanding of chemical principles and, perhaps, intuition are helpful.

If the substance to be analyzed is one that occurs widely, the literature survey will probably have yielded several alternative recipes for the analysis. In this case economic considerations may well dictate employment of that method which will yield the desired reliability with least expenditure of effort. As mentioned earlier the number of samples to be analyzed will often be a determining factor in this choice.

Investigation of the literature will not invariably produce a method designed specifically for the type of sample in which the chemist is interested. Ordinarily, however, he will have encountered procedures for materials that are at least somewhat analogous in composition to his; he will then need to decide whether the variables introduced by differences in composition are likely to have any effect on the results. This is often difficult and fraught with uncertainty; recourse to the laboratory may be the only way of obtaining an unequivocal answer.

If it is decided that existing procedures are not applicable, the chemist must then consider whether modifications of these will overcome the problems imposed by the variation in composition. Again he may find that his knowledge of the behavior of complex systems is so limited he can propose only tentative alterations; he must go to the laboratory to decide whether these modifications will accomplish their purpose without introducing new difficulties.

After giving due consideration to existing methods and their modifications, the chemist may decide that none of these will fit his problem and he must improvise his own procedure. To do this he will need to marshal all of the facts he has gathered with respect to the chemical and physical properties of the element or compound to be determined and the state in which it occurs. From these he may be able to arrive at several possible ways of performing the desired measurement. Each of the possibilities must then be examined critically, taking into account the behavior of the other components in the sample as well as the reagents that must be used for solution or decomposition. At this point he must try to anticipate various sources of error and possible interferences arising from interactions among the components and reagents; he may very well have to devise methods by which problems of this sort can be circumvented. In the end, it is to be hoped that one or more tentative methods worthy of test will arise. In all probability, the feasibility of one or more of the steps in the procedure cannot be determined on the basis of theoretical considerations alone, and recourse must be had to preliminary laboratory testing of these individual steps. Certainly, critical evaluation of the entire procedure can only come from careful laboratory work.

Testing the Method

Once a procedure for an analysis has been selected, the question usually arises as to whether the method can be employed directly, without testing, to the problem at hand.

The answer to this is not simple and depends upon a number of considerations. If the procedure chosen is one that has not been widely employed, or one to which a single, or at most a few, literature references are to be found, there may be some real point to preliminary laboratory evaluation. With experience the chemist becomes more and more cautious about accepting all of the claims regarding the virtues of a new method. It is not uncommon to find somewhat exaggerated statements regarding the accuracy and general applicability of a new method. As a consequence, a few hours spent in the laboratory testing this type of procedure may be enlightening.

Whenever a major modification of a standard procedure is undertaken or an attempt is made to apply it to a type of sample different from that for which it was designed, a preliminary laboratory test is called for. The effects of such alterations simply cannot be predicted with certainty and the chemist is sanguine indeed who dispenses with such precautions.

Finally, of course, a newly devised procedure must be extensively tested before it is adapted for general use. We must now turn to the methods by which this can be done.

Analysis of standard samples. Unquestionably the best technique for evaluating an analytical method is to employ the procedure for the analysis of one or more standard samples whose composition with respect to the element or compound of interest is known exactly. For this technique to be of value, it is essential that the composition of the standard approximate that of the samples to be analyzed both as to the concentration range of the species of interest and the over-all makeup. In some happy instances standards of this sort can readily be synthesized from weighed quantities of pure compounds. Others may be purchased from sources such as the National Bureau of Standards; these, however, are confined largely to common materials of commerce or widely distributed natural products.

As often as not, the chemist finds himself in the position of being unable to acquire a standard sample that matches closely the substance he wishes to analyze. This is particularly true of complex materials in which the form of the species of interest is unknown or variable and quite impossible to reproduce. In these circumstances the best that can be done is to prepare a solution of known concentration whose composition approximates that of the sample after it has been decomposed and placed in solution. Obviously, such a standard gives no information at all concerning the fate of the substance being determined during the important decomposition and solution steps.

Analysis by other methods. An analytical method can sometimes be evaluated by comparison of one or more of the results of the procedure with some entirely different method. Clearly, a second method must exist and to be useful should differ considerably from the one under examination. If the results from the two are comparable, this serves as presumptive evidence that both are yielding satisfactory results inasmuch as it is unlikely that the same determinate error would affect each. Such a conclusion will not apply to those aspects of the two methods that are similar.

Standard addition to the sample. When the foregoing tests are inapplicable, there is yet another method of evaluation. This consists of carrying out the analysis by the proposed procedure; then a known quantity of the species of interest is mixed with a second portion of the sample, this being, wherever possible, in the form in which it occurs in the original sample. The mixture is then analyzed again and the recovery of the added substance determined from the difference in the two results. Such a test may reveal errors arising from the method of treating the sample or from the presence of the other elements or compounds.

THE LITERATURE OF ANALYTICAL CHEMISTRY

The literature of analytical chemistry can be divided into two broad categories: reference works and papers. The latter may be theoretical or applied in nature and are generally restricted to some specific aspect of analysis. The former are more general in scope and attempt to summarize the current knowledge in the field or in some broad portion of the field. The references and journals in the accompanying list will serve as a point of departure in a search for specific information concerned with analytical chemistry. The classification of reference works employed is certainly imperfect but may be helpful.

Reference Works of a General Nature

American Society for Testing Materials, *ASTM Book of Standards*, 7 vols. Philadelphia: American Society for Testing Materials, 1952.

A. A. Benedetti-Pichler, *Essentials of Quantitative Analysis*. New York: The Ronald Press Co., 1956.

N. H. Furman, Ed., *Scott's Standard Methods of Chemical Analysis*, 5th ed., 2 vols. New York: D. Nostrand Co., Inc., 1939. Applied in nature.

R. C. Griffin, *Technical Methods of Analysis*, 2d ed. New York: McGraw-Hill Book Co., Inc., 1927. Applied in nature.

W. F. Hillebrand, G. E. F. Lundell, H. A. Bright, and J. I. Hoffman, *Applied Inorganic Analysis*, 2d ed. New York: John Wiley and Sons, Inc., 1953.

I. M. Kolthoff and P. J. Elving, *Treatise on Analytical Chemistry*. New York: The Interscience Encyclopedia, Inc., 1959. A multivolume reference work.

I. M. Kolthoff and E. B. Sandell, *Textbook of Quantitative Inorganic Analysis*, 3d ed. New York: The MacMillan Co., 1952.

H. A. Laitinen, *Chemical Analysis*. New York: McGraw-Hill Book Co., Inc., 1960. Devoted primarily to theory.

B. M. Margosches and W. Böttger, Eds., *Die Chemische Analyse*. Stuttgart: Ferdinand Enke, 1907- .

C. N. Reilley, Ed., *Advances in Analytical Chemistry and Instrumentation*. New York: Interscience Publishers, Inc., 1960.

T. B. Smith, *Analytical Processes*, 2d ed. London: Edward Arnold (Publishers) Ltd., 1940.

C. R. N. Strouts, J. H. Gilfillan, and H. N. Wilson, Eds., *Analytical Chemistry*, 2 vols. Oxford: The Clarendon Press, 1955.

F. P. Treadwell and W. T. Hall, *Analytical Chemistry*, 9th ed., 2. New York: John Wiley and Sons, Inc., 1942.

H. F. Walton, *Chemical Analysis*. New York: Prentice-Hall, Inc., 1952. Devoted primarily to theory.

H. H. Willard and H. Diehl, *Advanced Quantitative Analysis*. New York: D. Van Nostrand Company, Inc., 1943.

C. L. Wilson, Ed., *Comprehensive Analytical Chemistry*, several vols. Amsterdam: Elsevier Publishing Co., 1959.

Reference Works Devoted Largely to the Analysis of Specific Types of Substances

Inorganic substances

American Society for Testing Materials, *ASTM Methods for Chemical Analysis of Metals*. Philadelphia: American Society for Testing Materials, 1956.

G. E. F. Lundell and J. I. Hoffman, *Outlines of Methods of Chemical Analysis*. New York: John Wiley and Sons, Inc. 1938.

J. W. Mellor and H. V. Thompson, *A Treatise on Quantitative Inorganic Analysis*. London: C. Griffin and Co., 1938.

E. C. Pigott, *Ferrous Analysis*, 2d ed. New York: John Wiley and Sons, Inc., 1953.

C. J. Rodden, Ed., *Analytical Chemistry of the Manhattan Project*. New York: McGraw-Hill Book Co., Inc., 1950.

W. R. Schoeller and A. R. Powell, *The Analysis of Minerals and Ores of the Rarer Elements*. London: C. Griffin and Co., 1955.

H. S. Washington, *Chemical Analysis of Rocks*. New York: John Wiley and Sons, Inc., 1930.

R. S. Young, *Industrial Inorganic Analysis*. New York: John Wiley and Sons, Inc., 1953.

Organic substances

J. S. Fritz and G. S. Hammond, *Quantitative Organic Analysis*. New York: John Wiley and Sons, Inc., 1957.

J. Grant, Ed., *Quantitative Organic Microanalysis*, 5th ed. London: J. and A. Churchill, 1951.

J. Mitchell, Jr., I. M. Kolthoff, E. S. Proskauer, and A. Weissberger, Eds., *Organic Analysis*, 4 vols. New York: Interscience Publishers, Inc., 1956-60.

S. Siggia, *Quantitative Organic Analysis via Functional Groups*. New York: John Wiley and Sons, Inc., 1949.

S. Siggia and H. J. Stolten, *An Introduction to Modern Organic Analysis*. New York: Interscience Publishers, Inc., 1956.

A. Steyermark, *Quantitative Organic Microanalysis*, 2d ed. New York: Academic Press, Inc., 1961.

K. G. Stone, *Determination of Organic Compounds*. New York: McGraw-Hill Book Company, Inc., 1956.

A. Weissberger, *Techniques of Organic Chemistry*, 3d ed., 1, parts III and IV. New York: Interscience Publishers, Inc., 1960.

Biological substances

D. Glick, Ed., *Methods of Biochemical Analysis*, several vols. New York: Interscience Publishers, Inc., 1954-

Gases

P. W. Mullen, *Modern Gas Analysis*. New York: Interscience Publishers, Inc., 1955.

Agricultural and food products

Association of Official Agricultural Chemists, *Methods of Analysis*, 9th ed. Washington, D. C.: Association of Official Agricultural Chemists, 1960.

M. B. Jacobs, *The Chemical Analysis of Foods and Food Products*. New York: D. Van Nostrand Co., Inc., 1938.

Poisons and pollutants

M. B. Jacobs, *The Analytical Chemistry of Industrial Poisons, Hazards and Solvents*. New York: Interscience Publishers, Inc., 1941.

M. B. Jacobs, *The Chemical Analysis of Air Pollutants*. New York: Interscience Publishers, Inc., 1960.

Water

American Public Health Association, *Standard Methods of Water and Sewage Analysis*. New York: American Public Health Association, 1955.

Reference Works Devoted to Methods for Completion of the Analysis

Volumetric methods

W. Böttger, Ed., *Newer Methods of Volumetric Chemical Analysis*, trans. by R. E. Oesper. New York: D. Van Nostrand Co., Inc., 1938.

I. M. Kolthoff and N. H. Furman, *Potentiometric Titrations*, 2d ed. New York: John Wiley and Sons, Inc., 1931.

I. M. Kolthoff, V. A. Stenger, and R. Belcher, *Volumetric Analysis*, 3 vols. New York: Interscience Publishers, Inc., 1942-57.

J. Mitchell and D. M. Smith, *Aquametry*. New York: Interscience Publishers, Inc., 1948.

G. Schwarzenbach, *Complexometric Titrations*. New York: Interscience Publishers, Inc., 1957.

Organic reagents

J. F. Flagg, *Organic Reagents*. New York: Interscience Publishers, Inc., 1948.

J. H. Yoe and L. A. Sarver, *Organic Analytical Reagents*. New York: John Wiley and Sons, Inc., 1941.

Thermogravimetric methods

C. Duval, *Inorganic Thermogravimetric Analysis*. New York: Elsevier Publishing Co., 1953.

Micro and ultramicro methods

A. A. Benedetti-Pichler, *Microtechnique of Inorganic Analysis*. New York: John Wiley and Sons, Inc., 1942.

P. L. Kirk, *Quantitative Ultramicroanalysis*. New York: John Wiley and Sons, Inc., 1950.

J. B. Niederl and V. Niederl, *Micromethods of Quantitative Organic Analysis*, 2d ed. New York: John Wiley and Sons, Inc., 1942.

Instrumental methods

W. G. Berl, Ed., *Physical Methods in Chemical Analysis*, 2d ed. New York: Academic Press, Inc., 1960.

A. G. Jones, *Analytical Chemistry, Some New Techniques*. London: Butterworths Scientific Publications, 1959.

J. H. Yoe and H. J. Koch, Jr., Eds., *Trace Analysis*. New York: John Wiley and Sons, Inc., 1957.

Electroanalytical methods

P. Delahay, *New Instrumental Methods in Electrochemistry*. New York: Interscience Publishers, Inc., 1954.

J. J. Lingane, *Electroanalytical Chemistry*, 2d ed. New York: Interscience Publishers, Inc., 1958.

H. J. S. Sand, *Electrochemistry and Electrochemical Analysis*, 3 vols. London: Blackie & Son, Ltd., 1946.

Polarography

I. M. Kolthoff and J. J. Lingane, *Polarography*, 2d ed., 2 vols. New York: Interscience Publishers, Inc., 1952.

L. Meites, *Polarographic Techniques*. New York: Interscience Publishers, Inc., 1955.

Colorimetry and absorption spectroscopy

R. B. Barnes, R. C. Gore, U. Liddel, and V. Z. Williams, *Infrared Spectroscopy*. New York: Reinhold Publishing Corp., 1944.

D. E. Boltz, Ed., *Colorimetric Determination of Nonmetals.* New York: Interscience Publishers, Inc., 1958.

M. G. Mellon, Ed., *Analytical Absorption Spectroscopy.* New York: John Wiley and Sons, Inc., 1950.

E. B. Sandell, *Colorimetric Determination of Traces of Metals,* 3d ed. New York: Interscience Publishers, Inc., 1959.

F. D. Snell and C. T. Snell, *Colorimetric Methods of Analysis,* 3d ed., 4 vols. New York: D. Van Nostrand Co., Inc., 1954.

W. West, Ed., *Chemical Applications of Spectroscopy.* New York: Interscience Publishers, Inc., 1956.

Emission spectroscopy and flame photometry

American Society for Testing Materials Committee E-2, *Methods for Emission Spectrochemical Analysis.* Philadelphia: American Society for Testing Materials, 1957.

J. A. Dean, *Flame Photometry.* New York: McGraw-Hill Book Co., Inc., 1960.

N. H. Nachtrieb, *Principles and Practice of Spectrochemical Analysis.* New York: McGraw-Hill Book Co., Inc., 1950.

X-ray spectroscopy

L. S. Birks, *X-Ray Spectrochemical Analysis.* New York: Interscience Publishers, Inc., 1959.

Chromatography

H. G. Cassidy, *Fundamentals of Chromatography.* New York: Interscience Publishers, Inc., 1957.

V. J. Coates, H. J. Noebels, and I. S. Fagerson, Eds., *Gas Chromatography.* New York: Academic Press Inc., 1958.

A. I. M. Keulemans, *Gas Chromatography.* New York: Reinhold Publishing Corp., 1957.

R. L. Pecsok, *Principles and Practice of Gas Chromatography.* New York: John Wiley and Sons, Inc., 1959.

H. H. Strain, *Chromatographic Adsorption Analysis.* New York: Interscience Publishers, Inc., 1945.

Journals

Analytica Chimica Acta
Analytical Abstracts
Analytical Chemistry
Analyst
Chemical Titles
Chimie analitique
Collection of Czechoslovak Chemical Communications
Current Chemical Papers
Journal of the Association of Official Agricultural Chemists
Microchemical Journal
Mikrochimica Acta
Talanta
Zeitschrift für analytische Chemie

ACCURACY OBTAINABLE IN THE ANALYSIS OF COMPLEX MATERIALS

To provide a clear idea of the accuracy that can be expected when the analysis of a complex material is carried out with a reasonable amount of effort and care, data on the determination of four elements in a variety of materials are presented in the tables that follow. The information in these tables was taken from a much larger set of results collected by W. F. Hillebrand and G. E. F. Lundell of the National Bureau of Standards and published in the first edition of their excellent book on inorganic analysis.[1]

The materials analyzed were of naturally occurring substances and items of commerce; these had been especially prepared to give uniform and homogeneous samples and were then distributed among analysts who were, for the most part, actively engaged in the analysis of similar materials. The analysts were allowed to use the methods they considered most accurate and best for the problem at hand. In most instances, special precautions were taken so that the results generally are better than can be expected from the average routine analysis; on the other hand they do not represent the ultimate in analytical perfection.

The value in column 2 of each table is a best value for the measured quantity obtained by the most painstaking analysis. It is considered to be the "true value" for calculations of the absolute and relative errors shown in columns 4 and 5. Column 4 was obtained by discarding results that were extremely divergent, determining the deviation of the remaining individual data from the quantity present (column 2), and averaging these deviations. Column 5, the percent relative error, was obtained by dividing the data in column 4 by the best value found in column 2 and multiplying by 100.

The results for the four elements shown in the tables that follow are typical of the data for 26 elements reported in the original publication. It is to be concluded that analyses as good as a few tenths of a percent relative are the exception when complex mixtures are analyzed by ordinary methods, and that unless he is willing to invest an inordinate amount of time to the analysis, the chemist must accept errors on the order of 1 or 2 percent. If the sample contains less than 1 percent of the element of interest, errors greater than this figure are to be expected.

Finally, it is clear from these data that the accuracy obtainable in the determination of an element is greatly dependent upon the nature and complexity of the substrate. Thus the relative error for the determination of phosphorus in two phosphate rocks was 1.1 percent; in a synthetic mixture it was only 0.27 percent. The error in an iron determination in a refractory was 7.8 percent; in a manganese bronze, having about the same iron content, it was only 1.8

---

[1] W. F. Hillebrand and G. E. F. Lundell, *Applied Inorganic Analysis*, 874-887. New York: John Wiley and Sons, Inc., 1929.

percent. Here, the limiting factor in the accuracy is not the errors in the completion step, but rather those originating in the separation of interferences and solution of the samples.

From these data it is clear that the chemist is well advised to adopt a pessimistic viewpoint regarding the accuracy of an analysis, be it his own or one performed by someone else.

Table 30-1

ANALYSIS OF IRON IN VARIOUS MATERIALS[2]

Material	Iron Present, percent	Number of Analysts	Average Error, absolute	Average Error, relative percent
Soda-lime glass	0.064 (Fe_2O_3)	13	0.01	15.6
Cast bronze	0.12	14	0.02	16.7
Chromel	0.45	6	0.03	6.7
Refractory	0.90 (Fe_2O_3)	7	0.07	7.8
Manganese bronze	1.13	12	0.02	1.8
Refractory	2.38 (Fe_2O_3)	7	0.07	2.9
Bauxite	5.66	5	0.06	1.1
Chromel	22.8	5	0.17	0.74
Iron ore	68.57	19	0.05	0.07

Table 30-2

ANALYSIS OF MANGANESE IN VARIOUS MATERIALS[2]

Material	Manganese Present, percent	Number of Analysts	Average Error, absolute	Average Error, relative percent
Ferro-chromium	0.225	4	0.013	5.8
Cast iron	0.478	8	0.006	1.3
	0.897	10	0.005	0.56
Manganese bronze	1.59	12	0.02	1.3
Ferro-vanadium	3.57	12	0.06	1.7
Spiegeleisen	19.93	11	0.06	0.30
Manganese ore	58.35	3	0.06	0.10
Ferro-manganese	80.67	11	0.11	0.14

Table 30-3

ANALYSIS OF PHOSPHORUS IN VARIOUS MATERIALS[2]

Material	Phosphorus Present, percent	Number of Analysts	Average Error, absolute	Average Error, relative percent
Ferro-tungsten	0.015	9	0.003	20.
Iron ore	0.040	31	0.001	2.5
Refractory	0.069 (P_2O_5)	5	0.011	16.
Ferro-vanadium	0.243	11	0.013	5.4
Refractory	0.45	4	0.10	22.
Cast iron	0.88	7	0.01	1.1
Phosphate rock	43.77 (P_2O_5)	11	0.5	1.1
Synthetic mixtures	52.18 (P_2O_5)	11	0.14	0.27
Phosphate rock	77.56 $(Ca_3(PO_4)_2)$	30	0.85	1.1

Table 30-4

ANALYSIS OF POTASSIUM IN VARIOUS MATERIALS[2]

Material	Potassium Oxide Present, percent	Number of Analysts	Average Error, absolute	Average Error, relative percent
Soda-lime glass	0.04	8	0.02	50.
Limestone	1.15	15	0.11	9.6
Refractory	1.37	6	0.09	6.6
	2.11	6	0.04	1.9
	2.83	6	0.10	3.5
Lead-barium glass	8.38	6	0.16	1.9

[2] The data in these tables were taken from W. F. Hillebrand and G. E. F. Lundell, *Applied Inorganic Analysis*, 874-887. New York: John Wiley and Sons, Inc., 1929. With permission.

chapter 31. *Preparation of a Representative Sample for Analysis*

The use of analytical results requires the tacit assumption that the data derived from a sample is also applicable to the total mass of material from which it was taken. This supposition is justified only insofar as the chemical composition of the sample truly reflects that of the bulk of the material. The term *sampling* is used to describe the operations involved in procuring a reasonable amount of material having this quality.

Sampling is frequently the most difficult step in the entire analytical process. This is certainly true, for example, with many of the raw materials of commerce that are sold in lots weighing hundreds of tons. The value of a given lot is largely based upon the weight of some component rather than the gross weight; the relationship between these quantities must be established by chemical analysis. The end product of the sampling operation will be a quantity of homogeneous material weighing a few grams, or at most, a few hundred grams. Although this may represent less than one fifty millionth of the entire weight of the lot, this sample must approximate as closely as possible the average composition of the total mass. Where, as in the case of an ore, the material consists of non-

703

homogeneous solids, the task of producing a representative sample is indeed formidable. Clearly, the reliability of the analysis cannot exceed that with which the sample was acquired; the most painstaking work upon a poor sample represents a waste of effort.

While a definitive treatment is not possible in the space available, we shall consider the major aspects of the sampling operation. Frequent reference will be made to literature that deals more comprehensively with this important topic.

In general the sampling operation can be divided into three main steps: (1) the collection of a gross sample, (2) the reduction of the gross sample to a size convenient for laboratory work, and (3) the preparation of the laboratory sample for analysis. For any given substance, the details of these steps will vary considerably depending upon the physical character of the material being sampled. In some instances, the gross sample is of a size suitable for laboratory work, and (2) is omitted entirely; in others, considerable treatment is required in order to obtain a quantity of the substance that can be handled in the laboratory.

THE GROSS SAMPLE

Ideally the gross sample is a miniature replica of the bulk of the material to be analyzed. It corresponds to the whole not only in chemical composition but also in particle-size distribution.

In order to obtain a gross sample a certain portion of the whole must be removed in a random fashion; that is, a selection must be carried out in such a way that each portion of the whole has an equal chance of being included in the sample. The technique for obtaining a random sample will vary tremendously depending upon the physical state of the substance, how it is contained, and its total quantity.

Size of the Gross Sample

From the standpoint of convenience and economy it is desirable that the gross sample be no larger than absolutely necessary. Basically, sample size is determined by (1) the certainty required with respect to correspondence between the sample and the whole, (2) the degree of heterogeneity of the substance being sampled, and (3) the level of particle size at which this heterogeneity begins. The last point warrants further amplification. In a well-mixed, homogeneous solution of a gas or a liquid, heterogeneity exists only on a molecular scale, and the size of the molecules themselves will govern the minimum size of the gross sample. At the other extreme, there is the particulate solid such as an ore or a soil where the discrete pieces of solid can be seen to vary in composition. Here the particles referred to may have dimensions of several centimeters or more. Intermediate between these are colloidal materials and solidified metals.

With the former, heterogeneity is first encountered in the particles of the dispersed phase; these typically have diameters in the range of 10^{-5} cm or less. In the latter, heterogeneity is found among the crystal grains of alloys.

In order to obtain a truly representative gross sample a certain number n of the particles referred to in (3) must be taken. The magnitude of this number is dependent upon (1) and (2) and may involve but a few particles, several millions, or even several millions of millions. Where the substance is a homogeneous gas or liquid, such large numbers are of no great concern since the particle size at which heterogeneity begins is of molecular dimensions; thus even a very small weight of sample will contain the requisite number. With a particulate solid, on the other hand, the discrete particles may weigh a gram or more; here the gross sample may comprise several tons of material. Under such circumstances, sampling is a costly, time-consuming procedure at best; determination of the smallest possible quantity of substance required to give the needed information will minimize the expense.

If a random sample is removed from a bulk of material, its composition is governed by the law of chance; thus, by suitable statistical manipulations it should be possible to predict the probability of a given fraction being similar to the whole. A very simple, idealized case will illustrate this. A carload of lead ore is made up of just two kinds of particles of the same size—one of these being galena (lead sulfide) and the other a gangue containing no lead. The car contains, let us say, 100-million particles, and we wish to know what fraction of these are galena. The composition of the car could, of course, be obtained exactly by counting all of the particles since the appearance of the two components is different; this approach, however, would probably involve several lifetimes of work. We must thus settle for the lesser accuracy involved in counting some reasonable fraction of the components. The number contained in this fraction will, of course, depend upon the error we are willing to tolerate in the measurement.

The relationship between the allowable error and the number of particles n to be counted can be stated as follows:[1]

$$n = \frac{(1-p)}{p\sigma^2} \tag{31-1}$$

where p is the fraction of galena particles; $(1-p)$ is the fraction of gangue particles; and σ is the allowable relative standard deviation in the count of the galena particles. Thus, for example, if 80 percent of the particles were present as galena ($p = 0.8$) and the tolerable standard deviation was 1 percent ($\sigma = 0.01$), a random sampling of 2500 particles should be made. A standard deviation of a 0.1 percent would require a sample containing 250,000 particles.

[1] For a discussion of the deviation and significance of this equation and the one that follows see A. A. Benedetti-Pichler in W. G. Berl, Ed., *Physical Methods in Chemical Analysis*, 3, 183-194. New York: Academic Press, Inc., 1956; A. A. Benedetti-Pichler, *Essentials of Quantitative Analysis*, Chap. 19. New York: The Roland Press Company, 1956; or H. A. Laitinen, *Chemical Analysis*, Chap. 27. New York: McGraw-Hill Book Co., Inc., 1960.

We shall now make the problem somewhat more realistic and assume that one of the components in the car contains a high percentage P_1 of lead and the other component a lesser amount P_2. Furthermore the average density of the shipment d differs from the densities d_1 and d_2 of these components. We are now interested in deciding how many particles, and thus what weight, must be taken to assure a sample possessing the average percent lead of the bulk P with a relative standard deviation due to sampling of σ. Equation (31-1) can be extended to include these stipulations.[2]

$$n = p(1 - p) \left(\frac{d_1 d_2}{d^2}\right)^2 \left(\frac{P_1 - P_2}{\sigma P}\right)^2 \tag{31-2}$$

From this equation we see that the demands of accuracy are costly in terms of the sample size required, because of the inverse square relationship between the allowable standard deviation and the number of particles taken. Furthermore a greater number of particles must be taken as the average percentage P of the element of interest becomes smaller. Finally, the degree of heterogeneity as measured by $(P_1 - P_2)$ has a profound effect, the number of particles increasing as the square of the difference in composition of the two components of the mixture.

The problem of deciding upon a gross sample size is ordinarily more difficult than in the case just described because most samples not only contain more than two components but also consist of a range of particle sizes. In most instances the first of these problems can be met by dividing the sample into an imaginary two-component system. Thus, with an actual lead ore, one component might include all of the various lead-bearing minerals of the ore and the other all of the residual components containing little or no lead. After assigning average densities and percentages of lead to each of these, the system would then be treated as if it had but two components.

The problem of variable particle size can be handled by calculation of the number of particles necessary if the sample were made up of particles of a single size. Then the actual sample weight is determined by taking into account the particle-size distribution. One way of doing this is to calculate the necessary weight by assuming all of the particles are the size of the largest. This is not very efficient, however, for it usually means that a larger weight of material than necessary is removed. Benedetti-Pichler gives methods for computing the weight of sample to be chosen under these circumstances.[3]

An interesting conclusion from equation (31-1) is that the number of particles comprising the gross sample is independent of particle size. The weight of the sample, of course, increases directly as the volume (or the cube of the particle diameter) so that reduction in the particle size of a given material has a large effect on the number of pounds required in the gross sample.

[2] *Loc. cit.*

[3] W. G. Berl, Ed., *Physical Methods in Chemical Analysis*, **3**, 192. New York: Academic Press, Inc., 1956.

Clearly a good deal of information must be known about a substance in order to make use of equation (31-2). Fortunately, reasonable estimates of the various parameters in the equation can often be made. These estimates can be based upon a qualitative analysis of the substance, visual inspection of the material, and information from the literature for substances of similar origin. Crude measurements of densities of the various sample components may also be necessary.

Separation of the Gross Sample

The discussion so far has been predicated upon the assumption that random sampling is possible. Thus, having fixed upon a weight for the gross sample, we must now consider how it can be isolated in such a way that each portion of the total has an equal opportunity of being included. The mechanics of this depend greatly upon the physical state of the material, the total amount, and how it is contained. Detailed instructions regarding the best techniques for random sampling of various types of material are to be found in the literature;[4] we can treat this subject only in the most general way.

Homogeneous solutions of liquids and gases. For solutions of liquids or gases the gross sample can be relatively small since nonhomogeneity first occurs at the molecular level, and even small volumes of sample will involve a tremendous number of particles. Whenever possible, the material to be analyzed should be well stirred prior to removal of the sample in order to make sure that homogeneity does indeed exist. With large volumes of solutions this may be impossible; it is then best to sample several portions of the container with a "sample thief," a bottle that can be opened and filled at any desired location in the solution.

In industry, gases or liquids are frequently sampled continuously as they flow through pipes. In such instances, care is taken to insure that the sample collected represents a constant fraction of the total flow and that all portions of the stream are sampled.

Particulate solids. The process of obtaining a random sample from a bulky particulate material is often difficult. It can best be done while the material is being transferred. For example, a certain fraction of the shovelfuls or wheelbarrow loads of the solid may be consigned to a sample pile; or portions of the material may be intermittently removed from a conveyer belt. Alternatively, the material may be forced through a riffle or a series of riffles that provide for the

---

[4] See, for example, N. H. Furman, Ed., *Scott's Standard Methods of Chemical Analysis*, 5th ed., **2**, 1301-1333. New York: D. Van Nostrand Co., Inc., 1939; *Book of Standards*. Philadelphia, Pa.: American Society for Testing Materials (for sampling of paints, fuels, petroleum products, constructional materials, etc.); *Official and Tentative Methods of Analysis*. Washington, D.C.: Association of Official Agricultural Chemists, 9th ed., 1960 (for soils, fertilizers, foods, etc.); *Methods for Chemical Analysis of Metals*, 2d ed., 57-72, Philadelphia, Pa.: American Society for Testing Materials, 1956 (for metals and alloys); F. J. Pettijohn, *Manual of Sedimentary Petrography*. New York: Appleton-Crofts, Inc., 1938 (for soil, outcropping, and rocks).

continuous removal of a small fraction of the stream. Mechanical devices of this sort have been highly developed for handling coals and ores.

Sampling a large pile of a solid is difficult inasmuch as there is seldom certainty that the material has a random distribution throughout. This is particularly true where there is a considerable variation in composition with particle size. Inevitably, small particles tend to collect at the bottom and center of the pile and the larger particles at the outside. Sampling here must involve a division of the material according to some system that will assure the same distribution of particle sizes as in the whole. This involves several stepwise partitions of the total.

A procedure called *coning and quartering* is often employed with amounts of material of the order of 100 pounds or so composed of particles smaller than about 4 mm in diameter. It is also applicable to smaller quantities in the laboratory. In this process, the solid is formed into a cone, each new shovelful being deposited on the apex of the pile. Mixing is accomplished by repeatedly forming cones, the solid from each old cone being removed systematically from around its base. The cone is then flattened by pressing down on its apex with a board or shovel to give a circular layer of material. This is divided into four equal quarters by drawing two perpendicular lines through the center of the circle. Alternate quarters are then discarded; the process is repeated, if necessary, on the residual two quarters.

Another effective way of mixing and dividing a solid weighing up to about 100 pounds involves *rolling and quartering*. In this case a conical pile of the solid, made up of particles that have been reduced to 1 mm or less in diameter, is placed upon a tarpaulin, rubberized sheet, or, for small samples, on a piece of glazed paper. The cone is flattened and the solid mixed by rolling it repeatedly on the sheet. This can be done by pulling first one corner of the sheet over to the opposite corner and then another corner over to its opposite corner. This should be repeated 100 times or more. Finally the material is collected in the center of the sheet by simultaneously raising all four corners. The resultant pile is flattened, quartered, and a pair of opposite quarters rejected. This procedure can also be useful for reducing the size of a laboratory sample.

Metals and alloys. Samples of these materials are obtained by sawing, milling, or drilling. In general it is not safe to assume that chips of the metal removed from the surface will be representative of the entire bulk, and sampling must include solid from the interior of the piece as well as from the surface. With billets or ingots of metal this can be obtained by sawing across the piece at several regularly spaced intervals and collecting the "sawdust" as the sample. Alternatively, the specimen may be drilled, again at various regularly spaced intervals, the drillings being collected as the sample; the drill should pass entirely through the block or halfway through from alternate sides. The drillings can then be broken up and mixed, or melted together in a graphite crucible to give a sample. In the latter case a granular sample can often be formed by pouring the melt into distilled water.

PRODUCTION OF A LABORATORY SAMPLE

We have seen that the gross sample may, in the case of particulate materials, weigh several hundred pounds or more. In such instances a considerable reduction in sample size is desirable before it is brought into the laboratory where, at most, a few pounds are all that can be conveniently handled. The process of reducing the sample volume by a factor of 100 or more is ordinarily multistaged, involving repeated grinding, mixing, and dividing. Diminution in particle size is essential as the weight of sample is decreased because of the theoretical necessity of retaining the same number of particles (see equation 31-2).

Grinding. A variety of mechanical tools are available for reducing the particle size of solids. For coarse materials a jaw crusher is employed; this consists of a pair of steel plates, one of which is fixed while the other is operated off of a heavy flywheel. The distance between the plates can be adjusted and solids can be reduced to particles having a diameter of a few millimeters.

Further reduction in particle size is often achieved with a disk grinder. This consists of a pair of vertical metal plates, one of which is rotated by a motor. The sample is fed into the center of the plates and forced outward as the plate rotates; the space between the plates decreases from center to outside so that grinding occurs during the operation. The space between the plates can be varied to produce a fine powder.

Mixing and dividing. After each stage of reduction of particle size, there usually follows a mixing and dividing of the sample. This can be done by coning or rolling and then quartering as described earlier. Division of the sample can also be performed with a riffle. This consists of a series of alternate parallel chutes, half of which lead to the sample pile and half to the reject pile. The sample is fed to the top of the riffle in such a way that it enters all of the chutes and gives a random division of the solid.

TREATMENT OF THE LABORATORY SAMPLE

As it arrives at the laboratory the sample often requires further treatment before it is analyzed, particularly if it is in the form of a solid. One of the objects of this pretreatment is to produce a material so homogeneous that any small portion removed for the analysis will be identical to any other fraction. This usually involves reduction of the size of particles to a few tenths of a millimeter and thorough mechanical mixing. Another object of the pretreatment is to convert the substance to a form in which it is readily attacked by the reagents employed in the analysis; with refractory materials particularly, this involves grinding to a very fine powder. Finally, the sample may have to be dried or its moisture content determined because this is a variable factor that is dependent upon atmospheric conditions as well as physical state of the sample.

Crushing and Grinding of Laboratory Samples

In dealing with solid samples, a certain amount of crushing or grinding is ordinarily required to reduce the particle size. Unfortunately, these operations tend to alter the composition of the sample, and for this reason the particle size should be reduced no more than is required for homogeneity and ready attack by reagents.

Several factors may cause appreciable alteration in the composition of the sample as a result of grinding. Among these is the heat that is inevitably generated. This can cause losses of volatile components in the sample. In addition, grinding increases the surface area of the solid and thus increases its susceptibility to reactions with the atmosphere. For example, it has been observed that the ferrous content of a rock may be altered by as much as 40 percent during grinding—apparently a direct result of atmospheric oxidation of the iron to the ferric state.

The effect of grinding on the gain or loss of water from solids is considered in a later section.

Another potential source of error in the crushing and grinding of mixtures arises from the difference in hardness of the components of a sample. The softer materials are converted to smaller particles more rapidly than the hard; any loss of the sample in the form of dust will thus cause an alteration in composition. Furthermore, loss of the sample in the form of flying fragments must be avoided since these will tend to be made up of the harder components.

Intermittent screening of the material is often employed to increase the efficiency of grinding. In this operation, the ground sample is placed upon a wire or cloth sieve that will pass particles of the desired size. The residual particles are then returned for further grinding; the operation is repeated until the entire sample passes through the screen. This process will certainly result in segregation of the components on the basis of hardness, the toughest materials being last through the screen; it is obvious that grinding must be continued until the last particle has been passed. The need for further mixing after screening is also apparent.

A serious error can arise during grinding and crushing as a consequence of mechanical wear and abrasion of the grinding surfaces. For this reason only the hardest materials such as hardened steel, agate, or boron carbide are employed; even with these, contamination of the sample is sometimes encountered.

Methods of Crushing and Grinding in the Laboratory[5]

The diamond mortar. The so-called *Plattner diamond mortar*, shown in Figure 31.1, is used for crushing very hard materials. It is constructed of hardened tool steel and consists of a base plate, a removable collar, and a pestle.

[5] For further discussion see W. F. Hillebrand, G. E. F. Lundell, H. A. Bright, and J. I. Hoffman, *Applied Inorganic Analysis*, 2d ed., 809-814. New York: John Wiley and Sons, Inc., 1953; A. A. Benedetti-Pichler, *Essentials of Quantitative Analysis*, Chap. 18. New York: The Ronald Press Company, 1956.

The sample to be crushed is placed on the base plate inside the collar. The pestle is then fitted in place and struck several blows with a hammer. This results in reduction of the sample to a fine powder; it is then collected on a glazed paper after disassembling the apparatus and brushing the parts.

The ball mill. A useful device for grinding solids that are not too hard is the ball mill. It consists of a porcelain crock of perhaps 2-liter capacity that can be sealed and rotated mechanically. The container is charged with approximately equal volumes of the sample and flint or porcelain balls having a diameter of 20 to 50 mm. Upon rotation, grinding and crushing occurs as a consequence of the movement of the balls within the container. A finely ground and well-mixed powder can be produced in this way.

Mortar and pestle. The mortar and pestle, the most ancient of man's grinding tools, still finds wide use in the analytical laboratory. These now come in a variety of sizes and shapes. They are commonly constructed of glass, porcelain, agate, mullite, and other hard materials.

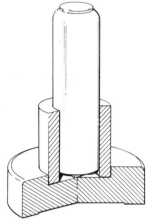

Fig. 31.1 A Diamond Mortar.

Mixing Solid Laboratory Samples

It is essential that solid materials be thoroughly mixed in order to assure random distribution of the components in the analytical sample. Several methods are commonly employed in the laboratory. One of these involves rolling the sample on a sheet of glazed paper. A pile of the substance is placed in the center of the paper and mixed by lifting one corner of the paper enough to roll the particles of the sample to the opposite corner. This operation is repeated many times, the four corners of the sheet being lifted alternately.

Effective mixing of solids is also accomplished by rotating the substance for some time in a ball mill.

MOISTURE IN SAMPLES

The presence of water in the sample represents a common and vexing problem that frequently faces the chemist. This compound may exist as a contaminant from the atmosphere or from the solution in which the substance was formed; or it may be bonded as a chemical compound. Regardless of its origin, however, water plays a part in determining the composition of the sample. Unfortunately, particularly in the case of solids, the water content is a variable quantity that depends upon such things as humidity, temperature, and state of subdivision. Thus, the constitution of a sample may change significantly with environment and method of handling.

In order to cope with the variability in composition owing to the presence of moisture, the chemist may attempt to remove it prior to weighing samples for analysis; if this is out of the question he may try to bring the water content to some reproducible level that he can duplicate at a later date if necessary. As a third alternative he may determine the water content at the time his samples are weighed out for analysis; in this way his results can be corrected to a dry basis. In any event most analyses are preceded by some sort of preliminary treatment designed to take into account the presence of water.

Forms of Water in Solids

It is convenient to distinguish among the several ways in which water can be held by a solid. Although developed primarily with respect to minerals, the classification of Hillebrand[6] and his collaborators may be applied to other solids as well and forms the basis for the discussion that follows.

Essential water. The essential water in a substance is that water which is an integral part of the molecular or crystal structure of one of the components of the solid. It is present in that component in stoichiometric quantities. Thus, the *water of crystallization* in stable solid hydrates (for example, $CaC_2O_4 \cdot 2H_2O$, $BaCl_2 \cdot 2H_2O$) qualifies as a type of essential water.

A second form is called *water of constitution*. Here the water is not present as such in the solid, but rather is formed as a product when the solid undergoes decomposition, usually as a result of heating. This is typified by the processes

$$2KHSO_4 \rightarrow K_2S_2O_7 + H_2O$$

$$Ca(OH)_2 \rightarrow CaO + H_2O$$

Nonessential water. Nonessential water is not necessary for the characterization of the chemical constitution of the sample and therefore does not occur in any sort of stoichiometric proportions. It is retained by the solid as a consequence of physical forces.

Adsorbed water is retained on the surface of solids in contact with a moist environment. The quantity is dependent upon humidity, temperature, and the specific surface area of the solid. Adsorption is a general phenomenon that is encountered in some degree with all finely divided solids.

A second type of nonessential water is called *sorbed water*. This is encountered with many colloidal substances such as starch, protein, charcoal, zeolite minerals, silica gel, etc. The amounts of sorbed water are often large compared with adsorbed moisture amounting, in some instances, to as much as 20 percent or more of the solid. Interestingly enough, solids containing even this much water may appear as perfectly dry powders. Sorbed water is held as a condensed phase in the interstices or capillaries of the colloidal solids. The quantity is greatly dependent upon temperature and humidity.

[6] W. F. Hillebrand, G. E. F. Lundell, H. A. Bright, and J. I. Hoffman, *Applied Inorganic Analysis*, 2d ed., 815. New York: John Wiley and Sons, Inc., 1953.

A third type of nonessential moisture is *occluded water*. Here, liquid water is entrapped in microscopic pockets spaced irregularly throughout the solid crystals. Such cavities often occur naturally in minerals and rocks.

Water may also be dispersed in a solid in the form of a *solid solution*. Here the water molecules are distributed homogeneously throughout the solid. Natural glasses may contain several percent of moisture in this form.

Effect of Temperature and Humidity on Water Content of Solids

In general, the concentration of water contained in a solid tends to become smaller with increases in the temperature and decreases in the humidity of its environment. The magnitude of these effects and the rate at which they manifest themselves differs considerably according to the manner in which water is retained.

Water of crystallization. The relationship between humidity and the water content of a crystalline hydrate is shown by a vapor pressure-composition diagram; Figure 31.2 illustrates such a plot for barium chloride. It was obtained by measuring the equilibrium pressure of water over a mixture of barium chloride and its hydrates in a closed system. The mol percentage of water is shown along the abscissa; the equilibrium vapor pressure is plotted as the ordinate. When anhydrous barium chloride is brought into equilibrium with a dry atmosphere, the pressure of water is zero. When water is added to the salt, however, an amount of the monohydrate is formed and the following equilibrium is established:

$$BaCl_2 \cdot H_2O \text{ (solid)} \rightleftharpoons BaCl_2 \text{ (solid)} + H_2O \text{ (gas)}$$

The vapor pressure of water in the system will be determined by this equilibrium, for which we may write

$$K' = p'_{H_2O}$$

where p'_{H_2O} is the equilibrium pressure of water. As long as both the monohydrate and anhydrous salt are present, their activities are constant and the partial pressure of water is independent of the amounts of these two compounds. This is shown by the horizontal line extending from just above 0 mol percent water to 50 mol percent. At 50 mol percent water, anhydrous barium chloride ceases to exist in the solid, and with its disappearance this equilibrium expression is no longer applicable. Increases in the amount of water result in formation of a new compound, the dihydrate; the vapor pressure over the mixture is now governed by the equilibrium between this compound and the monohydrate

$$BaCl_2 \cdot 2H_2O \text{ (solid)} \rightleftharpoons BaCl_2 \cdot H_2O \text{ (solid)} + H_2O \text{ (gas)}$$

Thus, for this equilibrium

$$K'' = p''_{H_2O}$$

where p''_{H_2O} is larger than p'_{H_2O}, as shown on the graph. Again the equilibrium pressure is constant as long as both the mono- and dihydrate are present.

When the mol percent of water in the system exceeds the molar ratio of water in the dihydrate (that is, 66.7 percent), the monohydrate disappears completely; higher hydrates are not formed, however. Instead, the dihydrate begins to dissolve; the result is a saturated solution that is in equilibrium with the solid dihydrate, that is,

$$BaCl_2 \text{ (sat'd sol'n)} \rightleftharpoons BaCl_2 \cdot 2H_2O \text{ (solid)} + H_2O \text{ (gas)}$$

As long as some dihydrate remains we may write

$$K''' = p'''_{H_2O}$$

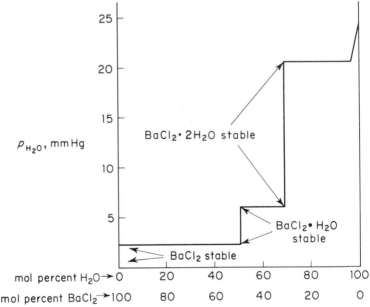

Fig. 31.2 Vapor Pressure-composition Diagram for Barium Chloride and Its Hydrates, 25° C.

Again we have a constant equilibrium pressure of water. This condition is maintained until the solution is no longer saturated (at about 97 mol percent water). Then the solid dihydrate disappears and the vapor pressure of water increases continuously, approaching that of pure water (100 mol percent) at high dilution.

A diagram such as this is useful for it shows clearly the stable forms of a substance at a given temperature as well as the conditions necessary to produce a given form. The behavior of hydrates under various atmospheric conditions can also be predicted. For example, Figure 31.2 indicates that the dihydrate is the stable form of the compound at 25° C when the partial pressure of water in its surroundings lies between about 6 and 21 mm of mercury. This corresponds to a relative humidity range of 25 to 88 percent.[7] The relative humidity in most laboratories will lie well within this range except on very dry or very damp days; thus, the dihydrate will be stable when exposed to typical laboratory conditions. Furthermore, if anhydrous barium chloride were left in contact with the atmosphere with a relative humidity within this range, absorption of moisture would occur until equilibrium had been achieved—that is, until all of the barium chloride had been converted to the dihydrate. Similarly, an aqueous solution of barium chloride would lose water to the atmosphere under these conditions until finally only crystals of the equilibrium species, the dihydrate, remained. The dihydrate would, of course, lose water under some circumstances. For example, if the relative humidity dropped below 25 percent, as might happen during a dry winter day, equilibrium would favor formation of the monohydrate. If the dihydrate were placed in a desiccator with a reagent that kept the partial pressure of water below 2 mm of mercury, quantitative conversion to the anhydrous salt would be the ultimate result. Thus, the composition of a sample containing a hydrate or a compound capable of forming a hydrate is greatly dependent upon the relative humidity of its environment.

As we have pointed out, temperature has a marked effect on equilibrium constants. In general, the equilibrium partial pressure of water over a hydrate increases with temperature so that the horizontal lines in Figure 31.2 are displaced to higher pressures when the temperature rises. Therefore dehydration tends to accompany temperature rises.

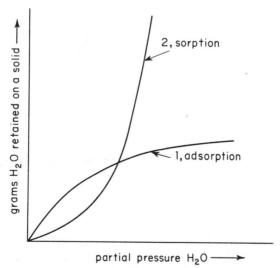

Fig. 31.3 Typical Adsorption and Sorption Isotherms.

[7] Relative humidity is the ratio of the vapor pressure of water in the atmosphere compared with the vapor pressure in air that is saturated with moisture. At 25° C, the partial pressure of water in saturated air is 23.76 mm of mercury. Thus when air contains water at a partial pressure of 6 mm, the relative humidity is

$$6.00/23.76 = 0.253 \text{ or } 25.3 \text{ percent.}$$

Adsorbed water. The amount of moisture adsorbed on the surface of a solid also increases with the amount of water in its environment. This is shown by the adsorption isotherm in Figure 31.3 (curve 1) in which the weight of water adsorbed on a typical solid is plotted against the partial pressure of water in the atmosphere. It is apparent from the diagram that the water content is particularly sensitive to changes in water vapor pressure at low partial pressures.

Quite generally, the amount of adsorbed water decreases with temperature increases and in most cases it approaches zero if the solid is dried at temperatures above 100° C.

Equilibrium, in the case of adsorbed moisture, is achieved rather rapidly, requiring ordinarily only 5 or 10 minutes. This often becomes apparent to the chemist when he weighs finely divided solids that have been rendered anhydrous by drying; a continuous increase in weight is observed unless the solid is contained in a tightly stoppered vessel.

Sorbed water. The quantity of moisture sorbed by a colloidal solid varies tremendously with atmospheric conditions as may be seen from curve 2 of Figure 31.3. In contrast to the behavior of adsorbed water, however, equilibrium may require days or even weeks for attainment, particularly at room temperatures. Furthermore, the amount of water retained by the two processes is often quite different; typically adsorption will involve quantities of water amounting to few tenths of a percent of the solid while sorption may entail 10 or 20 percent.

The amount of water sorbed in a solid also decreases with temperature

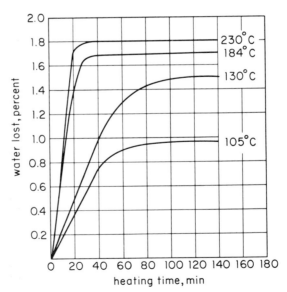

Fig. 31.4 Removal of Water at Constant Temperature. (From data of C. O. Willits, *Anal. Chem.*, **23**, 1058 (1951) by permission.)

increases. Complete removal of this type of moisture at 100° C, however, is by no means a certainty as is indicated by the drying curves for an organic compound shown in Figure 31.4. After drying this material for about 70 minutes at 105°, equilibrium was apparently reached so far as its water content was concerned. It is also clear, however, that additional moisture was removed by elevating the temperature. Even at 230°, dehydration was probably not entirely complete.

Occluded water. This form of moisture is not in equilibrium with the atmosphere and therefore not sensitive to changes in humidity. Heating a solid containing occluded water may cause a gradual diffusion of the moisture to the surface where it can evaporate. Temperatures a good deal greater than 100° C are often required, however, to cause this to occur at an appreciable rate. Frequently, heating is accompanied by *decrepitation*, the crystals of the solid being suddenly shattered by the internal pressure from the steam retained within the internal cavities.

Effect of Grinding on Moisture Content

Often the moisture content, and thus the chemical composition, of a solid is altered to a considerable extent during grinding and crushing. This will result in decreases in some instances, and increases in others.

Decreases in water content are sometimes observed in grinding solids containing essential water in the form of hydrates; thus the water content of gypsum, $CaSO_4 \cdot 2H_2O$, is reduced from 20 to 5 percent by this treatment.[8] Undoubtedly the change is a result of localized heating during the grinding and crushing of the particles.

Losses also occur when samples containing occluded water are reduced in particle size. Here, the grinding process ruptures some of the cavities and exposes the water so that it may evaporate.

More commonly, perhaps, the grinding process is accompanied by an increase in moisture content, due primarily to the increase in surface area exposed to the atmosphere. A corresponding increase in adsorbed water results. The magnitude of the effect is sufficient to alter appreciably the composition of a solid. For example, the water content of a piece of porcelain in the form of coarse particles was zero but upon being ground for some time was found to be 0.62 percent. Grinding a basaltic greenstone for 120 minutes changed its water content from 0.22 to 1.70 percent.[8]

From these remarks we may conclude that water determination should be made upon solids before grinding whenever possible.

Drying the Analytical Sample

The methods for dealing with the moisture content of a sample vary considerably depending upon the physical state of the substance and the in-

[8] W. F. Hillebrand, *J. Am. Chem. Soc.*, **30**, 1120(1908).

formation desired. Often the analytical chemist is called upon to determine the composition of a material as he receives it. Then his concern is that the moisture content of the material remain unchanged during preliminary treatment and storage. Where such changes are unavoidable or probable, it may be advantageous to determine the weight loss upon heating at some suitable temperature—say 105° C—immediately upon receipt of the sample. Then when ready to carry out the analysis, the sample can again be dried at this same temperature so that the data can be corrected back to an "as received" basis.

In some instances, analyses are performed and results reported on an air-dry basis. When this is done, the substance is allowed to equilibrate with its environment before a sample is withdrawn. This usually involves allowing the material to stand in contact with the atmosphere until no changes in weight as a function of time are noted. This method is quite satisfactory for nonhygroscopic materials, such as metals and alloys, which are always analyzed on this basis. Other particulate materials, which do not tend to adsorb moisture strongly, can also be handled conveniently in this way.

We have already noted that the moisture content of some substances is markedly changed by variations in humidity and temperature. Colloidal substances with large amounts of sorbed moisture are particularly susceptible to the effects of these variables. For example, the moisture content of a potato starch has been found to vary from 10 to 21 percent as a consequence of an increase in relative humidity from 20 to 70 percent.[9] With substances of this sort, comparable analytical data between laboratories or even within the same laboratory can be obtained only by carefully specifying a procedure for taking the moisture content into consideration. Often this will involve drying the sample to constant weight at 105° C or at some other specified temperature. Analyses are then performed and results reported on this "dry basis." While such a procedure may not render the solid completely free of water, it will usually reduce the moisture content to a reproducible level.

In many instances, the best procedure for obtaining an analysis on a "dry basis" will require a separate moisture determination on a sample taken at the same time as the samples that are used for the analysis.

DETERMINATION OF WATER[10]

Drying Procedures

Undoubtedly, the most common procedure for the determination of water content of solids involves oven drying of a weighed sample and gravimetric determination of the evolved water either by loss in weight of the sample or

[9] I. M. Kolthoff and E. B. Sandell, *Textbook of Quantitative Inorganic Analysis*, 3d ed., 144. New York: The MacMillan Co., 1952.

[10] For a more detailed discussion of this subject see a series of articles published in the following reference: *Anal. Chem.*, **23**, 1058-1080 (1951).

by the gain in weight of an absorbent for the water. The great virtue of this procedure is its simplicity; unfortunately, however, this simplicity does not necessarily extend to the interpretation of the data obtained by the method, for several processes other than the desired one may occur during the heating. Thus, in addition to evolution of water, one may also encounter volatilization of other components, decomposition of one or more of the constituents to give gaseous products, or perhaps air oxidation of a component in the sample. When the last occurs, increases in weight may result if the products are nonvolatile; decreases may be observed if the oxidation products are gases. Superimposed on these difficulties is the uncertainty with respect to the temperature required to cause complete evolution of water. With adsorbed moisture, and often with essential water, heating at 105° C will accomplish dehydration. On the other hand, removal of sorbed and occluded water is often quite incomplete at this temperature. Many minerals, as well as such substances as alumina and silica require temperatures of 1000° C or more.

Indirect determination. In this method the loss in weight of a solid during oven drying is measured; the assumption is then made that this loss equals the weight of water in the sample. The limitations in the procedure are apparent from the preceding paragraph.

In general, oven drying is carried out at as low a temperature as possible in order to minimize decomposition of the sample. The drying process is continued until the weight of the sample becomes constant at the chosen temperature; arrival at this state cannot be used as a criterion of complete dehydration, however, as is clearly shown by the data in Figure 31.4.

The rate at which drying is completed can often be accelerated by sweeping a stream of dry air over the sample during the heating. This can be done conveniently in a vacuum oven by reducing the internal pressure to a few millimeters of mercury; then while pumping is continued, air dried over a suitable desiccant is allowed to enter the drying chamber slowly and continuously.

Direct determination. In the direct method, the water evolved by heating a sample is collected on an absorbent that is specific for water; the increase in weight of the absorbent is used as the measure of the amount of water present. This procedure circumvents many of the limitations inherent in the indirect drying procedure; accurate results can be expected in most cases provided the sample is heated at a high enough temperature to remove all of the water. Errors arise, however, if air oxidation of the sample produces water; this is often encountered with substances containing organic components.

Figure 31.5 shows a typical arrangement for direct moisture determination. The sample is weighed into a small porcelain boat which is then placed in the Pyrex or Vycor combustion tube. Air, which has been dried by passage through a concentrated sulfuric acid solution and then over a desiccant such as magnesium perchlorate, is forced over the sample. Heating is accomplished by a burner or a tube furnace. The exit gases are led through a U-tube containing magnesium perchlorate or other desiccant; the tube is weighed before and after the analysis. A second guard tube containing desiccant protects the absorbent tube from becoming contaminated by diffusion of water from the atmosphere.

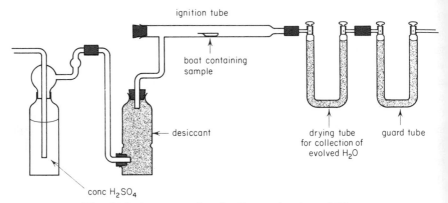

Fig. 31.5 Apparatus for the Determination of Water.

For minerals, rocks, and many other inorganic materials an extremely simple direct method for moisture determination can be employed; this is sometimes called *Penfield's method*.[11] The sample is placed in the bottom of a hard glass tube such as that shown in Figure 31.6. The water, which is driven off by ignition in a Bunsen flame, collects in the upper part of the tube which is kept cool. After the evolution is judged to be complete, the lower end of the tube is softened, drawn off, and discarded leaving the water in the upper end. This is weighed, the water removed by aspiration, and the tube weighed again.

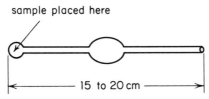

Fig. 31.6 Penfield Tube for Water Determination.

Water by Distillation

A distillation method is useful for the determination of the moisture content of materials that are readily air oxidized. It is widely employed for substances containing organic components such as fats, oils, waxes, cereals, plant materials, and foodstuffs.

The sample to be analyzed is dissolved or suspended in an organic solvent that is immiscible with water; ordinarily a liquid such as toluene or xylene, having a boiling point greater than that of water, is employed. Upon heating, the water in the sample is volatilized and is distilled over with the organic vapors. The distillate is condensed and the volume of the water phase measured to give the water content.

[11] S. L. Penfield, *Am. J. Sci.*, (3), **48**, 31 (1894).

An apparatus such as that shown in Figure 31.7 is often employed. The condensed liquid is caught in a trap so constructed that the heavier water phase collects in the bottom while the organic liquid flows back into the distillation vessel. The trap is calibrated so that the volume of water can be determined directly.

Chemical Methods for Water; The Karl Fischer Reagent

A number of chemical methods for the determination of water have been

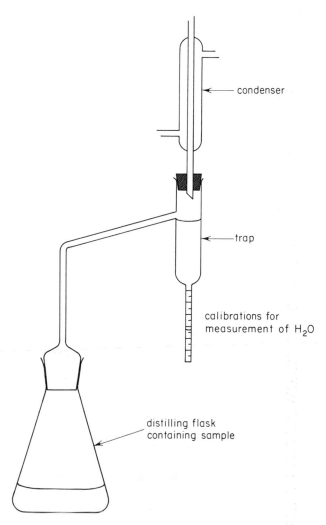

Fig. 31.7 Apparatus for the Determination of Water by Azeotropic Distillation.

devised. Unquestionably the most important of these involves the use of Karl Fischer solution, a relatively specific reagent for water.[12]

Reaction and stoichiometry. Karl Fischer reagent is composed of iodine, sulfur dioxide, pyridine, and methanol. Upon addition of this reagent to water, the following reactions occur:

$$C_5H_5N \cdot I_2 + C_5H_5N \cdot SO_2 + C_5H_5N + H_2O \rightarrow$$
$$2C_5H_5N \cdot HI + C_5H_5N \cdot SO_3 \quad (31\text{-}3)$$

$$C_5H_5N \cdot SO_3 + CH_3OH \rightarrow C_5H_5N(H)SO_4CH_3 \quad (31\text{-}4)$$

Only the first step consumes water, and involves the oxidation of sulfur dioxide by iodine to give sulfur trioxide and hydrogen iodide. In the presence of a large amount of pyridine, C_5H_5N, all of the reactants and products are present as complexes as indicated in the equation.

The second step in the reaction occurs when an excess of methanol is present and is important to the success of the titration, for the pyridine-sulfur trioxide complex is also capable of consuming water

$$C_5H_5N \cdot SO_3 + H_2O \rightarrow C_5H_5NHSO_4H \quad (31\text{-}5)$$

From the standpoint of analysis this reaction is undesirable because it is not as specific for water as is the first; it can be prevented completely by having a goodly excess of methyl alcohol present.

From equation (31-3), it is apparent that the stoichiometry of the Karl Fischer titration involves 1 mol of iodine, 1 mol of sulfur dioxide, and 3 mols of pyridine for each mol of water. In practice, an excess of both sulfur dioxide and pyridine are employed so that the combining capacity of the reagent for water is determined by its iodine content. Methanol is ordinarily used as the solvent for both reagent and the sample, and reaction (31-4) proceeds to the exclusion of (31-5).

End-point detection. The end point in the Karl Fischer titration is signaled by the appearance of a slight excess of the pyridine-iodine complex when all of the water has been consumed. The simplest way of detecting this is by its brown color which is intense enough for a visual end point. The color change observed is from the yellow of the reaction products to the brown of the excess reagent. With some practice, and in the absence of other colored materials, the end point can be established with a reasonable degree of certainty (that is, to perhaps ± 0.2 ml).

Various electrometric end points are also employed for the Karl Fischer titration, the most widely used being the "dead stop" technique described on page 628.

Preparation and stability of the reagent. As mentioned earlier, Karl Fischer reagent is prepared in such a way that its combining capacity for water is determined by the concentration of iodine in the solution. For typical applica-

---

[12] For reviews on this subject see J. Mitchell, *Anal. Chem.*, **23**, 1069 (1951); J. Mitchell and D. M. Smith, *Aquametry*. New York: Interscience Publishers, Inc., 1948.

tion the titer is about 3.5 mg of water per milliliter of reagent; a twofold excess of sulfur dioxide and a three- to fourfold excess of pyridine is provided.

Various methods for preparation of the reagent have been proposed. One involves dissolution of iodine in pyridine and dilution with methanol to give a stable stock solution. Liquid sulfur dioxide is then added to this a day or two before use. A second procedure involves preparation of two stock solutions, one consisting of iodine in methanol and the second of sulfur dioxide in pyridine. Individually, these are both stable and can be combined shortly before use. Stock solutions of the latter sort are available from various chemical supply houses. In addition, a single, stabilized Karl Fischer reagent can now be purchased commercially.

The Karl Fischer reagent decreases in titer with standing. The decomposition is particularly rapid immediately after preparation; it is therefore good practice to prepare the reagent a day or two before it is to be used. Ordinarily its titer should be established at least daily against a standard solution of water in methanol.

It is obvious that a good deal of care must be exercised when employing the Karl Fischer reagent to prevent contamination of the reagent and the sample by atmospheric moisture. All glassware must be carefully dried before use and the standard solution stored out of contact with air. It is also necessary to prevent excessive contact between the atmosphere and the solution during the titration.

Applications. The Karl Fischer reagent has been applied to the determination of water in a variety of substances.[13] The techniques employed vary considerably depending upon the solubility of the material, the form in which the water is retained, and the physical state of the sample. In many instances, particularly where the sample can be dissolved completely in methanol, direct and rapid titration is feasible. For example, this method has been applied to the analysis of water in many organic acids, alcohols, esters, ethers, anhydrides, and halides. Hydrated salts of most organic acids can also be analyzed by direct titration as well as hydrates of a number of inorganic salts that are soluble in methanol.

In those instances where it is impossible to produce a homogeneous solution of the sample and reagent, direct titration ordinarily gives an incomplete reaction. Satisfactory results can often be obtained, however, by addition of an excess of reagent and back titration with a standard solution of water in methanol after allowing time for the reaction to occur. An alternative and often effective procedure is to extract the water from the sample with an anhydrous solution of methanol or other organic solvent; in some instances refluxing the mixture increases the rate of transfer of moisture. The methanol is then titrated directly with the Karl Fischer solution.

Difficulty is also encountered in the analysis of sorbed moisture and tightly bound hydrate water. For these, the preceding extractive techniques are frequently effective.

---

[13] For a complete discussion of the applications of the reagent see J. Mitchell and D. M. Smith, *Aquametry*. New York: Interscience Publishers, Inc., 1948.

Certain substances interfere with moisture determinations by the Fischer procedure. Among these are compounds that react with one of the components of the reagent to produce water. For example, carbonyl compounds react to some extent with methanol to give acetals

$$RCHO + 2CH_3OH \rightarrow R—\underset{\underset{OCH_3}{|}}{\overset{\overset{OCH_3}{|}}{CH}} + H_2O$$

The result is a fading end point in the titration. Many metal oxides will react with the hydrogen iodide formed in the titration to give water.

$$MO + 2HI \rightleftharpoons MI_2 + H_2O$$

Again erroneous data result. In some instances, preliminary treatment of the sample can prevent these interferences.

The presence of good oxidizing or reducing substances is also detrimental to the Karl Fischer water titration inasmuch as these are capable of consuming the iodine in the reagent or reoxidizing the iodide present.

chapter 32. *Dissolving the Sample*

In order to complete most analyses, an aqueous solution of the sample is required; furthermore, the species to be determined must ordinarily be present in that solution in the form of a simple ion or molecule. Unfortunately many of the substances that are of interest to the chemist can only be converted to this form by extensive treatment. Thus, for example, before the chlorine content of an organic compound can be determined, it is usually necessary to convert the element into a form that is more amenable to analysis. Because this will doubtless require rupture of carbon-chlorine bonds, the preliminary treatment of the sample will likely be quite vigorous. Similarly, analysis of the components in a silicate rock requires destruction of the silicate structure by potent reagents; only then can an aqueous solution of the cations be obtained.

Various reagents and techniques exist for decomposing and dissolving analytical samples.[1] Often the proper choice among these is critical to the success

[1] A useful summary of reagents for dissolving or decomposing various inorganic substances is found in the following reference: G. E. F. Lundell and J. I. Hoffman, *Outlines of Methods of Chemical Analysis*, 24-29. New York: John Wiley and Sons, Inc., 1938.

of an analysis, particularly where refractory substances are being dealt with. This chapter is devoted to some of the more common methods for obtaining aqueous solutions of a sample.

Some General Considerations

In most instances, the reagent chosen for attack of a sample should cause complete dissolution of the substance; attempting to leach one or more components from a mixture usually results in an incomplete separation from the unattacked residue.

Considerable care must be exercised in the choice of a solvent in order to avoid one that will interfere in the later steps of the analysis. For example, the use of hydrochloric acid as a solvent for a sample to be analyzed for bromide ion would require the preliminary separation of chloride ion before the analysis could be completed. Furthermore, the solvent should be inspected for impurities that might affect the outcome of an analysis. This is of particular importance in the determination of components that are present in small concentrations. Frequently in trace analysis, the most important consideration in choosing among possible solvents is their purity.

Volatilization of important constituents of a sample may occur during the solution step unless proper precautions are taken. For example, treatment with acids can result in the loss of carbon dioxide, sulfur dioxide, hydrogen sulfide, hydrogen selenide, and hydrogen telluride. In the presence of basic reagents, loss of ammonia is common. Treatment of a sample with hydrofluoric acid will result in vaporization of silicon and boron as their fluorides, while exposure of halogen-containing substances to strong oxidizing reagents may result in the evolution of chlorine, bromine, or iodine. Reducing conditions during the preliminary treatment of a sample can lead to volatilization of such compounds as arsine, phosphine, or stibine.

A number of elements form volatile chlorides that are partially or completely lost from hot hydrochloric acid solutions. Among these are included the trichlorides of arsenic and antimony, tin and germanium tetrachloride, and mercuric chloride. The oxychlorides of selenium and tellurium also volatilize to some extent from this medium. In the presence of chloride ion, certain other elements tend to volatilize from hot concentrated solutions of perchloric or sulfuric acid. These include bismuth, manganese, molybdenum, thallium, vanadium, and chromium.

Boric acid, nitric acid, and the halogen acids are lost from boiling aqueous solutions, while phosphoric acid distills from hot concentrated sulfuric or perchloric acids. Certain volatile oxides can also be lost from hot acid solutions; these include the tetroxides of osmium and ruthenium as well as the heptoxide of rhenium.

Liquid Reagents Used for Dissolving or Decomposing Samples

The most common reagents for attacking analytical samples are the mineral acids or their aqueous solutions. Solutions of sodium or potassium hydroxide also find occasional application.

Hydrochloric acid. Concentrated hydrochloric acid is an excellent solvent for many metal oxides as well as those metals that lie above hydrogen in the electromotive force series; it is often a better solvent for the oxides than the oxidizing acids. Concentrated hydrochloric acid is about 12 F, but upon heating, hydrogen chloride is lost until a constant-boiling 6 F solution remains (boiling point about 110° C).

Nitric acid. Concentrated nitric acid is an oxidizing solvent that finds wide use in the attack of metals. It will dissolve all of the common metallic elements; aluminum and chromium, which become passive to the reagent, are exceptions. Many of the common alloys can also be decomposed by nitric acid. In this connection it should be mentioned that tin, antimony, and tungsten form insoluble acids when treated with concentrated nitric acid; this is sometimes employed to separate these elements from others contained in alloys.

Sulfuric acid. Hot concentrated sulfuric acid is often employed as a solvent. Part of its effectiveness arises from its high boiling point (about 340° C), at which temperature decomposition and solution of substances often proceeds quite rapidly. Most organic compounds are dehydrated and oxidized under these conditions; the reagent thus serves to remove such components from a sample. Most metals and many alloys are attacked by the hot acid.

Perchloric acid. Hot concentrated perchloric acid is a potent oxidizing agent and solvent. It attacks a number of ferrous alloys and stainless steels that are intractable to the other mineral acids; it is frequently the solvent of choice. This acid also dehydrates and rapidly oxidizes organic materials. Violent explosions result when organic substances or easily oxidized inorganic compounds come in contact with the hot concentrated acid; as a consequence, a good deal of care must be employed in the use of this reagent. For example, it should be heated only in hoods in which the ducts are clean and free of organic material and where the possibility of contamination of the solution is absolutely nil.

Perchloric acid is marketed as the 60- or 72-percent acid. Upon heating, a constant-boiling mixture (72.4-percent $HClO_4$) is obtained at a temperature of 203° C. Cold concentrated perchloric acid and hot dilute solutions are quite stable with respect to reducing agents; it is only the hot concentrated acid that constitutes a potential hazard. The reagent is a very valuable solvent and is widely used in analysis. Before employing it, however, the proper precautions for its use must be clearly understood.[2]

Oxidizing mixtures. More rapid solvent action can sometimes be obtained by the use of mixtures of acids or by the addition of oxidizing agents to the mineral acids. *Aqua regia*, a mixture consisting of three volumes of con-

[2] See H. H. Willard and H. Diehl, *Advanced Quantitative Analysis*, 8. New York: D. Van Nostrand Company, Inc., 1943.

centrated hydrochloric acid and one of nitric acid, is well known. Addition of bromine or hydrogen peroxide to mineral acids often increases their solvent action and hastens the oxidation of organic materials in the sample. Mixtures of nitric and perchloric acid are also useful for this purpose.

Hydrofluoric acid. The primary use of this acid is for the decomposition of silicate rocks and minerals where silica is not to be determined; the silicon is, of course, evolved as the tetrafluoride. After decomposition is complete, the excess hydrofluoric acid is driven off by evaporation with sulfuric acid or perchloric acid. Complete removal is often essential to the success of an analysis, because of the extraordinary stability of the fluoride complexes of several metal ions; the properties of some of these differ markedly from those of the parent cation. Thus, for example, precipitation of aluminum with ammonia is quite incomplete in the presence of small quantities of fluoride. Frequently removal of the last traces of fluoride from a sample is so difficult and time consuming as to negate the attractive features of this reagent as a solvent for silicates.[3]

Hydrofluoric acid finds occasional use in conjunction with other acids in the attack of some of the more difficultly soluble steels.

Hydrofluoric acid can cause serious damage and painful injury when brought in contact with the skin; it must be handled with respect.

Decomposition of Samples by Fluxes

Quite a number of common substances—such as silicates, some of the mineral oxides, and a few of the iron alloys— are attacked slowly, if at all, by the usual liquid reagents. Recourse to more potent fused salt media, or *fluxes*, is then called for. Fluxes will decompose most substances by virtue of the high temperature required for their use (300° to 1000° C) and the high concentration of reagent brought in contact with the sample.

Where possible, the employment of a flux is avoided, for several dangers and disadvantages attend its use. In the first place, a relatively large quantity of the flux is required to decompose most substances—often ten times the sample weight; the possibility of significant contamination of the sample by impurities in the reagent thus becomes very real. Furthermore, the aqueous solution from the fusion will have a high salt content, and this may lead to difficulties in the subsequent steps of the analysis. The high temperatures required for a fusion increase the danger of loss of pertinent constituents by volatilization. Finally, the container in which the fusion is performed is almost inevitably attacked to some extent by the flux; this again can result in contamination of the sample.

In those cases where the bulk of the substance to be analyzed is soluble in a liquid reagent and only a small fraction requires decomposition with a flux, it is common practice to employ the liquid reagent first; the undecomposed

---

[3] For methods of removal of fluoride ion see H. H. Willard, L. M. Liggett, and H. Diehl, *Ind. Eng. Chem., Anal. Ed.*, **14**, 234 (1942).

residue is then isolated by filtration and fused with a relatively small quantity of a suitable flux. After cooling, the melt is dissolved and combined with the majority of the sample.

Method of carrying out a fusion. In order to achieve a successful and complete decomposition of a sample with a flux, the solid must ordinarily be ground to a very fine powder; this will produce a high specific surface area. The sample must then be thoroughly mixed with the flux; this operation is often carried out in the crucible in which the fusion is to be done by careful stirring with a glass rod.

In general, the crucible used in a fusion should never be more than half filled at the outset. The temperature is ordinarily raised slowly with a gas flame because the evolution of water and other gases is a common occurrence at this point; unless care is taken there is the danger of loss by spattering. The crucible should be covered as an added precaution. The maximum temperature employed varies considerably depending upon the flux and the sample; it should be no greater than necessary, however, to minimize attack of the crucible and decomposition of the flux. The length of the fusion may range from a few minutes to one or two hours depending upon the nature of the sample. It is frequently difficult to decide when the heating should be discontinued. In some cases, the production of a clear melt serves as a signal for the completion of the decomposition. In others, this condition is not obvious, and the analyst must base the heating time on previous experience with the type of material being analyzed. In any event, the aqueous solution from the fusion should be examined carefully for particles of unattacked sample.

When the fusion is judged complete, the mass is allowed to cool slowly; then just before solidification the crucible is rotated to distribute the solid around the walls of the crucible so that the thin layer can be readily detached.

Types of fluxes. With few exceptions the common fluxes used in analysis are compounds of the alkali metals. Basic fluxes, employed for attack of acidic materials, include the carbonates, hydroxides, peroxides, and borates. The acidic fluxes are the pyrosulfates and the acid fluorides as well as boric oxide. If an oxidizing flux is required, sodium peroxide can be used. As an alternative, small quantities of the alkali nitrates or chlorates are mixed with sodium carbonate. In the paragraphs that follow and in Table 32-1, the properties of the common fluxes are described.

Sodium carbonate. A most common flux for the decomposition of silicates and certain other minerals is sodium carbonate. Heating such materials at 1000° to 1200° C with this substance results in conversion of the metallic constituents to the corresponding carbonates or oxides; these can then be attacked by acids. The nonmetallic constituents are converted to sodium salts which are also amenable to analysis.

The method of treatment of the fused mass depends upon the constituents of the sample and the analysis to be performed. With silicates, for example, the melt is dissolved in dilute acid; this results in decomposition and solution of the metallic carbonates, and partial separation of the silicon as hydrated silica. Further treatment completely removes the silica and leaves a solution that can be analyzed for its metallic constituents.

Table 32-1

THE COMMON FLUXES

Flux	Melting Point, °C	Type of Crucible for Fusion	Type of Substance Decomposed
Na_2CO_3	851	Pt	for silicates and silica-containing samples; alumina-containing sample; insoluble phosphates and sulfates
Na_2CO_3 + an oxidizing agent such as KNO_3, $KClO_3$, or Na_2O_2	—	Pt (not with Na_2O_2) Ni	for samples where an oxidizing agent is needed; that is, samples containing S, As, Sb, Cr, etc.
NaOH or KOH	318 380	Au, Ag, Ni	powerful basic fluxes for silicates, silicon carbide, and certain minerals; main limitation, purity of reagents
Na_2O_2	decomposes	Fe, Ni	powerful basic oxidizing flux for sulfides; acid-insoluble alloys of Fe, Ni, Cr, Mo, W, and Li; platinum alloys; Cr, Sn, Zr minerals
$K_2S_2O_7$	300	Pt, porcelain	acid flux for insoluble oxides and oxide-containing samples
B_2O_3	577	Pt	acid flux for decomposition of silicates and oxides where alkali metals are to be determined
$CaCO_3$ + NH_4Cl	—	Ni	upon heating the flux, a mixture of CaO and $CaCl_2$ produced; used for decomposing silicates for the determination of the alkali metals

In other instances, the fused product is better treated with water than with acid, for this will separate the anions of the sample as the soluble sodium salts from the majority of the cations that remain undissolved as the carbonates. Such treatment would be preferred, for example, in the decomposition of a sample containing insoluble sulfates of barium, calcium, or lead. Treatment of the melt with water would afford a separation of the anions from cations of the sample.

Carbonate fusions are normally carried out in platinum crucibles.

Potassium pyrosulfate. Potassium pyrosulfate provides a potent acidic flux that is particularly useful for the attack of the more intractable metal oxides. Fusions with this reagent are performed at about 400° C; at this temperature the slow evolution of sulfur trioxide, a strongly acid substance, takes place

$$K_2S_2O_7 \rightarrow K_2SO_4 + SO_3$$

Formation of the corresponding metal sulfates results from the treatment of a sample with this flux. Potassium pyrosulfate does not attack the insoluble silicates very rapidly.

The foregoing equation indicates that the flux is converted to potassium sulfate upon prolonged heating. As a consequence, the reagent may loose its ability to attack the sample as the fusion proceeds; it can, however, be regenerated by cooling and adding a few drops of concentrated sulfuric acid. Reheating must be done with care to avoid losses due to the evolution of water vapor.

Potassium pyrosulfate can be prepared by heating potassium acid sulfate.

$$2KHSO_4 \rightarrow K_2S_2O_7 + H_2O$$

Direct use of the acid sulfate as a flux is occasionally recommended; this, however, requires great care at the outset of heating in order to avoid loss of sample by spattering and overflow as the water is evolved. It is better to start with the pyrosulfate.

Pryosulfate fusions can often be carried out in ordinary porcelain crucibles or in vitreous silica containers. Some attack of the porcelain may occur, but this is usually negligible. Platinum ware can also be used although it will result in the introduction of small amounts of the metal into the sample.

Other fluxes. Table 32-1 contains data for several other common fluxes. Noteworthy are boric oxide and the mixture of calcium carbonate and ammonium chloride. Both are employed for decomposing silicates for the analysis of alkali metals. Boric oxide is removed after solution of the melt by evaporation to dryness with methyl alcohol; methyl borate, $B(OCH_3)_3$, distills.

When a mixture of calcium carbonate and ammonium chloride is heated, calcium oxide and calcium chloride are produced; these are sufficiently active reagents to decompose the silicates. Extraction of the fused mass with water results in an aqueous solution of the alkali chlorides ordinarily contaminated only with calcium ions plus any sulfate or borate that may have been present in the original material. These contaminants are readily removed prior to the alkali-metal analysis. Many chemists prefer this procedure, known as the *J. Lawrence Smith method*, to others for the determination of the alkali metals in silicates.

Decomposition of Organic Compounds

Analysis of the elemental composition of an organic substance generally requires drastic treatment of the material in order to convert the elements of interest into a form susceptible to the common analytical techniques. These treatments are usually oxidative in nature, involving conversion of the carbon and hydrogen of the organic material to carbon dioxide and water; in some instances, however, heating the sample with a potent reducing agent is sufficient to rupture the covalent bonds in the compound and free the element to be determined from the carbonaceous residue.

Oxidation procedures are sometimes classified into two categories. *Wet ashing* (or oxidation) makes use of liquid oxidizing agents such as sulfuric or perchloric acids. *Dry ashing* (or oxidation) usually implies ignition of the organic compound in air or a stream of oxygen. In addition, oxidations can be carried out in certain fused-salt media, sodium peroxide being the most common flux for this purpose.

In the sections that follow we shall mention briefly some of the methods for decomposing organic substances prior to the analysis of the more common elements.

Combustion-tube methods.[4] Several of the common and important elemental components of organic substances are converted to gaseous products when the material is oxidized. With suitable apparatus it is possible to trap these volatile compounds quantitatively and use them for the analysis of the element of interest. A common way to do this is to carry out the oxidation in a glass or quartz combustion tube through which is forced a stream of carrier gas. The stream serves to transport the volatile products to a part of the apparatus where they can be separated and retained for measurement; the stream may also serve as the oxidizing agent. The common elements susceptible to this type of treatment are carbon, hydrogen, oxygen, nitrogen, the halogens, and sulfur.

Figure 32.1 shows a typical combustion train for the analysis of carbon and

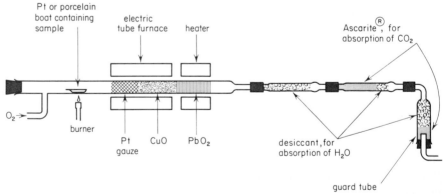

Fig. 32.1 Combustion Apparatus for the Determination of Carbon and Hydrogen.

[4] For a more detailed discussion see C. R. N. Strouts, J. H. Gilfillan, and H. N. Wilson, *Analytical Chemistry*, 1, Chap. 14. London: The Oxford University Press, 1955; J. B. Niederl and V. Niederl, *Micromethods of Quantitative Organic Analysis*, 2d ed. New York: John Wiley and Sons, Inc., 1942.

hydrogen in an organic substance. Here, oxygen or air is forced through the tube to oxidize the sample as well as to carry the products to the absorption part of the train. The sample is contained in a small platinum or porcelain boat that can be pushed into the proper position by means of a rod or wire. Ignition is initiated by slowly raising the temperature of that part of the tube which contains the sample. The sample undergoes partial combustion as well as thermal decomposition at this point, and the products are carried over a platinum gauze packing that is maintained at a temperature of 700° to 800° C; this catalyzes the oxidation. Following the platinum catalyst is a packing of cupric oxide that completes the oxidation of the sample to carbon dioxide and water. Additional packing is often included in the tube to remove compounds that interfere with the determination of the carbon dioxide and water in the exit stream. Lead chromate and silver serve to remove halogen and sulfur compounds, while lead dioxide can be employed to retain the oxides of nitrogen.

The exit gases from the combustion tube are first passed through a weighing tube packed with a desiccant that removes the water from the stream. The increase in weight of this tube gives a measure of the hydrogen content of the sample. The carbon dioxide in the gas stream is removed in the second weighing tube packed with Ascarite (sodium hydroxide held on asbestos). Because the absorption of carbon dioxide is accompanied by the formation of water, additional desiccant is contained in this tube. Finally the gases are passed through a *guard tube* that protects the two weighing tubes from contamination by the atmosphere.

Table 32-2 lists some of the applications of the combustion-tube method to other elements. A substance containing a halogen will yield the free element upon oxidation; this is frequently reduced to the corresponding halide prior to the analytical step. Sulfur finally yields sulfuric acid which can be estimated by precipitation with barium ion or by alkalimetric titration. We have cited the combustion-tube method for nitrogen in an earlier section (see p. 357).

Combustion with oxygen in sealed containers. A relatively straightforward method for the decomposition of many organic substances involves oxidation with gaseous oxygen in a sealed container. The reaction products are absorbed in a suitable solvent before the reaction vessel is opened. Analysis of the solution by ordinary methods follows.

A remarkably simple apparatus for carrying out such oxidations has been suggested by Schöniger (see Fig. 32.2).[5] It consists of a heavy-walled flask of 300- to 1000-ml capacity fitted with a ground-glass stopper. Attached to the stopper is a platinum-gauze basket which holds from 2 to 200 mg of sample. If the substance to be analyzed is a solid, it is wrapped in a piece of low-ash filter paper cut in the shape shown in Figure 32.2. Liquid samples can be weighed into gelatin capsules which are then wrapped in a similar fashion. A tail is left on the paper and serves as an ignition point.

[5] W. Schöniger, *Mikrochim. Acta*, **1955**, 123; **1956**, 869.

Table 32-2

COMBUSTION-TUBE METHODS FOR THE ELEMENTAL ANALYSIS OF ORGANIC SUBSTANCES

Element	Name of Method	Method of Oxidation	Method of Completion of Analysis
Halogens	Pregl	sample combusted in a stream of oxygen gas over a red-hot platinum catalyst; halogens converted primarily to HX and X_2	gas stream passed through a carbonate solution containing SO_3^{2-} (to reduce halogens and oxyhalogens to halides); product, the halide ion X^-, determined by usual procedures
	Grote	sample combusted in a stream of air over a hot silica catalyst; products are HX and X_2	same as above
Sulfur	Pregl	similar to halogen determination; combustion products are SO_2 and SO_3	gas stream passed through aqueous H_2O_2 which converts sulfur oxides to H_2SO_4 which can then be determined
	Grote	similar to halogen determination; products are SO_2 and SO_3	similar to above
Nitrogen	Dumas	sample oxidized by hot CuO to give CO_2, H_2O, and N_2	gas stream passed through concentrated KOH solution leaving only N_2 which is measured volumetrically
Carbon and hydrogen	Pregl	similar to halogen analysis; products are CO_2 and H_2O	H_2O adsorbed on a desiccant and CO_2 on Ascarite; determined gravimetrically
Oxygen	Unterzaucher	sample pyrolized over carbon; oxygen converted to CO; H_2 used as carrier gas	gas stream passed over I_2O_5 $(5CO + I_2O_5 \rightarrow 5CO_2 + I_2)$; liberated I_2 titrated

A small volume of an absorbing solution is placed in the flask and the air in the container is then displaced by allowing tank oxygen to flow into it for a short period. The tail of the paper is ignited and the stopper quickly fitted into the flask; the container is then inverted as shown in Figure 32.2; this will prevent the escape of the volatile oxidation products. Ordinarily the reaction proceeds rapidly, being catalyzed by the platinum gauze surrounding the sample. During the combustion, the flask is shielded to avoid damage in case of explosion.

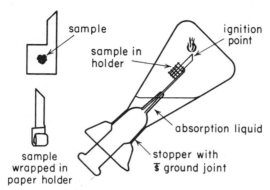

Fig. 32.2 Schöniger Combustion Apparatus. (Courtesy, Arthur H. Thomas Company, Philadelphia, Pennsylvania.)

After cooling, the flask is shaken thoroughly, disassembled, and the inner surfaces rinsed down. The analysis is then performed on the resulting solution. This procedure has been applied to the determination of halogens,[5] sulfur,[6] phosphorus,[7] fluorine,[8] and various metals[9] in organic compounds.

Peroxide fusion.[10] Sodium peroxide is a strong oxidizing reagent which, in the fused state, reacts rapidly and often violently with organic matter converting carbon to the carbonate, sulfur to sulfate, phosphorus to phosphate, chlorine to chloride, and iodine and bromine to iodate and bromate. Under suitable conditions the oxidation is complete, and analysis for the various elements may be performed upon an aqueous solution of the fused mass.

Once started, the reaction between organic matter and sodium peroxide is so vigorous that a peroxide fusion must be carried out in a sealed, heavy-walled, steel bomb. Sufficient heat is evolved in the oxidation to keep the salt in the liquid state until the reaction is completed; ordinarily the oxidation is initiated by passage of current through a wire immersed in the flux or by momen-

[6] W. Schöniger, *Mikrochim. Acta*, **1956**, 869.
[7] R. Belcher and A. M. G. MacDonald, *Talanta*, 1, 185 (1958).
[8] B. Z. Senkowski, E. G. Wollish, and E. G. E. Shafer, *Anal. Chem.*, **31**, 1574 (1959).
[9] R. Belcher, A. M. G. MacDonald, and T. S. West, *Talanta*, 1, 408 (1958).
[10] See C. R. N. Strouts, J. H. Gilfillan, and H. N. Wilson, *Analytical Chemistry*, 1, 301. London: The Oxford University Press, 1955.

tary heating of the bomb with a flame. Bombs for peroxide fusions are available commercially.[11]

One of the main disadvantages of the peroxide-bomb method is the rather large ratio of flux to sample needed for a clean and complete oxidation. Ordinarily an approximate two-hundredfold excess is used. The excess peroxide is subsequently decomposed to sodium hydroxide by heating in water; after neutralization, the solution necessarily has a high salt content. This may limit the accuracy of the method for completion of the analysis.

The maximum sized sample that can be fused is perhaps 100 mg. The method is more suited to semimicro quantities of about 5 mg.

Decomposition with metallic sodium and potassium. Metallic sodium and potassium are powerful reagents for the reductive decomposition of organic compounds. Under proper conditions they will abstract the halogens (including fluorine), sulfur, nitrogen (as cyanide), and other nonmetals from the organic matrix and convert these into water-soluble sodium salts. A preliminary step in the common procedure for qualitative identification of the elemental composition of organic compounds involves this reaction; here the sample is fused for a short period with sodium metal.

Alkali-metal decompositions have proved useful for the quantitative determination of the halogens in general, and fluorine in particular. Compounds of this element are often quite refractory toward the ordinary oxidative reagents; reduction with an alkali metal provides a simple means for obtaining aqueous fluoride solutions. Several variants of the procedure have been proposed. One of these involves heating of the sample with the molten metal at 400° C in a sealed glass vessel for 15 minutes.[12] After cooling, the excess metal is decomposed by addition of ethanol; the entire mass is then extracted with water. After filtration, the analysis is completed on the aqueous solution. Other procedures call for extended refluxing of the organic sample with a solvent containing metallic sodium. The various alcohols, ethanolamine, dioxane, and combinations of these have been proposed. Unfortunately not all substances are completely dehalogenated by this treatment.

A more vigorous reagent than the above solution is metallic sodium dissolved in liquid ammonia.[13] Use of this reagent entails solution of the sample in liquid ammonia or an ether-ammonia mixture to which metallic sodium is added. The mixture is sealed in a glass tube and shaken for several hours at room temperature. The excess metal is then decomposed with alcohol and the residue extracted with water. This procedure appears to be effective for decomposing highly fluorinated materials as well as most other halogenated compounds.

A much simpler and remarkably effective reductive decomposition of substances containing chlorine, bromine, or fluorine makes use of a highly reactive compound of sodium and biphenyl or naphthalene. The reagent is

[11] Parr Instrument Company, Moline, Illinois.

[12] P. J. Elving and W. B. Ligett, *Ind. Eng. Chem., Anal. Ed.*, **14**, 449 (1942).

[13] T. H. Vaughn and J. A. Nieuwland, *Ind. Eng. Chem., Anal. Ed.*, **3**, 274 (1931); J. F. Miller, H. Hunt, and E. T. McBee, *Anal. Chem.*, **19**, 148 (1947).

prepared by heating a mixture of sodium, anhydrous toluene, and the dimethyl ether of ethylene glycol; upon addition of biphenyl, a soluble green compound is produced. The resulting solution is stable if stored in sealed containers and kept cold. It has been found that sodium biphenyl rapidly dehalogenates a wide variety of organic compounds. This is accomplished by shaking a solution of the sample in toluene or other solvent with the reagent for about 30 seconds. Water is then added and the halide extracted. Analysis is performed on the resulting aqueous solution.[14] This is the simplest and most rapid procedure for decomposing organic samples for halogen analysis.

Wet-ashing procedures. Solutions of a variety of strong oxidizing agents will decompose organic samples. The main problem associated with the use of these reagents is the prevention of volatility losses of the elements of interest.

We have already encountered one wet-ashing procedure in the Kjeldahl method for the determination of nitrogen in organic compounds (p. 357). Here concentrated sulfuric acid is the oxidizing agent. This reagent is also frequently employed for decomposition of organic materials where metallic constituents are to be determined. Commonly nitric acid is added periodically to the solution to hasten the rate at which oxidation occurs.[15] A number of elements are volatilized at least partially by this procedure, particularly if the sample contains chlorine; these include arsenic, boron, germanium, mercury, antimony, selenium, tin, the halogens, and phosphorus.

An even more effective reagent than sulfuric-nitric acid mixtures is perchloric acid mixed with nitric acid. A good deal of care must be exercised in using this reagent, however, because of the tendency of hot anhydrous perchloric acid to react explosively with organic material. Explosions can be avoided by starting with a solution in which the perchloric acid is well diluted with nitric acid and not allowing the mixture to become concentrated in perchloric acid until the oxidation is nearly complete. Properly carried out, oxidations with this mixture are rapid and losses of metallic ions negligible.[16] It cannot be too strongly emphasized that proper precautions must be taken in the use of perchloric acid to prevent violent explosions.

Fuming nitric acid is another potent oxidizing reagent that is employed in the analysis of organic compounds. Its most important application is the analysis of the halogens and sulfur by the *Carius method.* The oxidation is carried out by heating the sample for several hours at 250° to 300° C in a heavy-walled sealed glass tube. Where halogens are to be determined, silver nitrate is added before the oxidation begins in order to retain them as the silver halides. Sulfur is converted to sulfate by the oxidation. A critical step in this procedure is that of forming a glass seal strong enough to withstand the rather high pressures that

[14] L. M. Liggett, *Anal. Chem.*, **26**, 748 (1954).

[15] *Methods of Analysis*, 9th ed. Washington, D. C.: Association of Official Agricultural Chemists, 1960.

[16] T. T. Gorsuch, *Analyst*, **84**, 135 (1959); G. F. Smith, *Anal. Chim. Acta*, **8**, 397 (1953).

develop during the oxidation. Occasional explosions are almost inevitable and a special tube furnace is ordinarily employed to minimize the effects of these.

Dry-ashing procedure. The simplest method for decomposing an organic sample is to heat it with a flame in an open dish or crucible until all of the carbonaceous material has been oxidized by the air. A red heat is often required to complete the oxidation. Analysis of the nonvolatile components is then made after solution of the residual solid. Unfortunately a great deal of uncertainty always exists with respect to the recovery of supposedly nonvolatile elements when treated in this manner. Some losses probably arise from the mechanical entrainment of finely divided particulate matter in the hot convection currents around the crucible. In addition, volatile metallic compounds may be formed during the ignition. For example, copper, iron, and vanadium are appreciably volatized when samples containing porphyrin compounds are heated.[17]

In summary, the dry-ashing procedure is the simplest of all methods for decomposing organic compounds but is often unreliable; it should not be employed unless tests have been performed that demonstrate its applicability to a given type of sample.

[17] See T. T. Gorsuch, *Analyst*, **84**, 135(1959); R. E. Thiers in D. Glick, Ed., *Methods of Biochemical Analysis*, **5**. New York: Interscience Publishers, Inc., 1957.

chapter 33. *Analytical*

Separations

A major concern in the application of any given measurement technique to an analytical problem is the presence of extraneous substances that interfere with the determination of the desired quantity. Were it not for such potential interferences, an analysis would be a much more straightforward process. As we have pointed out before, however, the properties and reactions of chemical species do not differ sufficiently among themselves to afford any method that is entirely specific. Thus, the problem of elimination of interferences is more often the rule than the exception in chemical analysis.

There are two general methods for eliminating interferences. The first of these involves the alteration of the system in such a way that the potential interference is immobilized and cannot participate in the measurement operation; clearly this alteration must not appreciably affect the species being determined. This is commonly accomplished by the introduction of a complexing agent that reacts selectively with the interfering substance. Thus, in the iodometric determination of copper, we noted that ferric ions are rendered nonreactive towards iodide by the introduction of a complexing ion such as fluoride or

Table 33-1

SEPARATION PROCEDURES FOR SUBSTANCES X AND Y, BASED ON HETEROGENEOUS EQUILIBRIA

Phase Type	Common Name for Process	Method of Formation of Heterogeneous System	Mechanical Operations Performed on Phases to Complete the Separation
Solid— Liquid	precipitation	$\dfrac{\text{soln}}{X+Y} + \text{soluble reagent} \rightarrow \dfrac{\text{solid}}{X} + \dfrac{\text{soln}}{Y}$	filtration
	electrodeposition	$\dfrac{\text{soln}}{X+Y} + \text{electric current} \rightarrow \dfrac{\text{solid}}{X} + \dfrac{\text{soln}}{Y}$	manual removal of electrode from liquid phase
	leaching	$\dfrac{\text{solid}}{X+Y} + \text{selective solvent} \rightarrow \dfrac{\text{solid}}{X} + \dfrac{\text{soln}}{Y}$	filtration or use of Soxhlet extractor
	ion exchange	$\dfrac{\text{soln}}{X+Y} + \dfrac{\text{solid}}{\text{resin}} \rightarrow \dfrac{\text{solid}}{\text{resin}+X} + \dfrac{\text{soln}}{Y}$	passage through column packed with resin
	adsorption chromatography	$\dfrac{\text{soln}}{X+Y} + \dfrac{\text{solid}}{\text{adsorbent}} \rightarrow \dfrac{\text{solid}}{\text{adsorbent}+X} + \dfrac{\text{soln}}{Y}$	passage through column packed with adsorbent
Solid—Gas	sublimation	$\dfrac{\text{solid}}{X+Y} + \text{heat} \rightarrow \dfrac{\text{gas}}{X} + \dfrac{\text{solid}}{Y}$	condensation or absorption of gas
	selective absorption	$\dfrac{\text{gas}}{X+Y} + \dfrac{\text{solid}}{\text{selective absorbent}} \rightarrow \dfrac{\text{solid}+\text{gas}}{X} \ \dfrac{}{Y}$	gas passed through tower packed with absorbent

Table 33-1 *(continued)*

Phase Type	Common Name for Process	Method of Formation of Heterogeneous System	Mechanical Operations Performed on Phases to Complete the Separation
Liquid—Liquid	extraction	$\dfrac{soln(I)}{X+Y} + \dfrac{soln(II)}{} \rightarrow \dfrac{soln(I)}{X} + \dfrac{soln(II)}{Y}$	separation in separatory funnel or with a continuous extractor
	mercury-cathode deposition	$\dfrac{soln}{X+Y} + \dfrac{liquid\ Hg}{current} \rightarrow \dfrac{soln}{X} + \dfrac{liquid\ Hg}{amalgam\ of\ Y}$	separation in separatory funnel
Liquid—Gas	distillation	$\dfrac{soln}{X+Y} + heat \rightarrow \dfrac{gas}{X} + \dfrac{soln}{Y}$	gas condensed or absorbed
	gas chromatography	$\dfrac{gas}{X+Y} + liquid \rightarrow \dfrac{gas}{X} + \dfrac{soln}{Y}$	passage through column containing liquid held on a solid matrix
	selective absorption	$\dfrac{gas}{X+Y} + selective\ solvent \rightarrow \dfrac{soln}{X} + \dfrac{gas}{Y}$	gas bubbled through liquid

phosphate; neither of these prevents cupric ion from converting iodide to iodine. Similarly, phosphate ion was employed to mask iron in the colorimetric determination of manganese in steel. Numerous other examples have been cited in earlier chapters.

The second method of eliminating an interference involves its physical separation, as a second phase, from the substance to be determined. This general process consists of the following steps: (1) formation of two phases, one of which contains the interference and the other the constituent of interest; (2) mechanical separation of the two phases; and (3) quantitative retention of that phase which contains the substance to be determined. One of the most common ways of accomplishing step (1) is by chemical precipitation with a suitable reagent. Here, either the interference or the substance to be determined is converted to a solid phase by addition of a selective precipitant; then the solid and liquid phases are separated mechanically by filtration. Either the liquid or the solid is retained for the final measurement. Another example would be a distillation in which the more volatile constituent is vaporized and separated from the liquid by means of a distillation column.

Classification of Separation Procedures

In this chapter we shall consider the ways in which the steps mentioned above are carried out. Table 33-1 contains most of the common separation procedures, classified according to the types of phases involved in the mechanical separation. A more complete classification is found in the literature.[1]

Errors Resulting from the Separation Process

None of the methods listed in Table 33-1 will provide an absolute separation, for each is based upon an equilibrium process. At best the separation will reduce the concentration of the interference to a tolerable level. Furthermore, the losses of the component sought during the separation must be smaller than the allowable error in the analysis. Thus the two factors to be considered in any separation are the completeness of the recovery of the species to be determined, and the degree of separation of the unwanted constituent. These factors can be expressed algebraically in terms of recovery ratios or partition ratios R. For example, if we let X be the amount of the wanted constituent recovered in a separation and X_o its amount in the original sample we may write that

$$R_x = \frac{X}{X_0} \tag{33-1}$$

[1] P. J. Elving, *Anal. Chem.*, **23**, 1202 (1951); H. G. Cassidy, *J. Chem. Ed.*, **23**, 427 (1946); L. B. Rogers in I. M. Kolthoff and P. J. Elving, *Treatise on Analytical Chemistry*, **2**, part 1, 920. New York: Interscience Publishers, Inc., 1961.

Clearly, this ratio should be as close to one as possible. We may write a similar partition ratio for the unwanted component Y; that is,

$$R_y = \frac{Y}{Y_0} \qquad (33\text{-}2)$$

The smaller the value of R_y, the better will be the separation process.

In an analytical separation, incomplete recovery of X will result in a negative error. The error arising from the incomplete removal of Y will be positive if this species contributes to the analytically measured quantity, and negative if it lowers the magnitude of this measurement. We must now examine the nature of these errors. We will assume that the analysis is based upon the measurement of some quantity M that is proportional to the concentrations X and Y present in the solution after the separation process (M might be a weight of precipitate, the absorbance of a solution, a diffusion current, etc.). Such being the case, we may write that

$$M_x = k_x X \qquad (33\text{-}3)$$

$$M_y = k_y Y \qquad (33\text{-}4)$$

where k_x and k_y are proportionality constants that measure the sensitivity of the testing procedure for each constituent. (The term k_y will be negative where y interferes by reducing the sensitivity of the test for X).

If both X and Y are present, then the measured value of M will represent the sum of their contributions

$$M = M_x + M_y \qquad (33\text{-}5)$$

In the absence of Y, no separation would be required; if we indicate the measured quantity under these circumstances as M_o, we may write

$$M_o = k_x X_o \qquad (33\text{-}6)$$

The error arising *from the separation process* is then $(M - M_o)$; expressed in relative terms it is

$$\text{relative error due to separation} = \frac{(M - M_o)}{M_o} \qquad (33\text{-}7)$$

Substituting (33-3), (33-4), (35-5), and (33-6) into (33-7) gives

$$\text{relative error} = \frac{k_x X + k_y Y - k_x X_o}{k_x X_o}$$

Substituting (33-1) and (33-2) yields

$$\text{relative error} = \frac{k_x R_x X_o + k_y R_y Y_o - k_x X_o}{k_x X_o}$$

This may be rearranged as follows:

$$\text{relative error} = (R_x - 1) + \frac{k_y Y_o}{k_x X_o} R_y \qquad (33\text{-}8)$$

The first term in equation (33-8) gives the error that is associated with the losses of X during the separation process. Thus if 99 percent of X is recovered during the separation ($\frac{X}{X_o} = R_x = 0.99$), then a relative error of -1 percent results.

The second term in (33-8) takes account of the error resulting from the incomplete removal of Y by the separatory operation. The magnitude of this error is related to the partition ratio R_y. However, it is also dependent upon the ratio of Y to X in the sample. Thus, a highly favorable partition ratio is required for the separation of a wanted minor constituent X from a major component Y. This situation is often encountered in trace analysis where the ratio of Y to X at the outset may be as much as 10^6 or 10^7.

The error incurred in a separation is also seen to be dependent upon the relative sensitivity of the measurement M for the two constituents (k_y/k_x). If a measurement procedure is not greatly affected by Y (that is, if k_y is small), then a quite incomplete separation may be adequate for an analysis. On the other hand, if the measurement system is equally sensitive to both, then a relatively complete separation may be required.

SEPARATION BY PRECIPITATION

Undoubtedly the most thoroughly investigated and widely used methods for separation of inorganic ions are those based on precipitation processes. In some instances the ion to be determined is carried down in the solid phase while the unwanted constituents are left in solution. In others the reverse is the case, and the liquid phase is retained for the analysis.

The fundamental basis for all precipitation separations is the solubility difference between the substance to be analyzed and the undesired components. Solubility product considerations will generally provide guidance as to whether or not a given separation is theoretically feasible and define the conditions required to achieve the separation. Unfortunately, however, other variables also play a part in determining the success or failure of a separation, and some of these are not susceptible to theoretical treatment with our present state of knowledge. Thus, for example, various coprecipitation phenomena may cause extensive contamination of a precipitate by an unwanted component even though the solubility product of the contaminant has not been exceeded. Likewise, the rate of an otherwise feasible precipitation process may be so slow that it becomes useless for a separation. Finally, when the precipitate forms as a colloidal suspension, coagulation may be a difficult and slow process. These latter problems are particularly difficult when the isolation of a small quantity by precipitation is attempted. As a consequence of these considerations, much of the knowledge regarding separations by precipitation is empirical in nature.

Ordinarily a systematic order of separations, such as the classical qualitative scheme, is not employed for quantitative work unless the sample happens to be particularly complex.[2] Rather, special methods are found or improvised to suit the needs for a given type of material.

A variety of precipitating agents are employed for quantitative inorganic separations; we shall limit this discussion to those that have the most general applicability.

Separations Based on Control of Acidity

A wide variety of elements are separated as hydroxides, hydrous oxides, or acids by adjustment of the hydrogen ion concentration of a solution to a suitable level. There are several reasons for the extensive use of pH control for separations in quantitative inorganic analysis, the first and most important being the enormous variation in solubility exhibited by these substances. Then, too, the concentration of the precipitating agent can be varied through a range in excess of 10^{15}. Finally, close control over the pH is readily achieved through the use of suitable buffers.

Separations based upon pH control can be conveniently classified into three categories: (1) those made in relatively concentrated solutions of strong acids; (2) those made in buffered solutions at intermediate pH values; and (3) those made with strong solutions of sodium or potassium hydroxide.

Separations with solutions of strong acids. Several elements form insoluble acidic oxides from strong solutions of the mineral acids. Often these elements are precipitated during the solution of the sample and are thus removed at the outset of the analysis. Included among these are tungsten (VI), tantalum (V), niobium (V), and silicon (IV). All are precipitated as oxides in the presence of concentrated perchloric, sulfuric, hydrochloric, or nitric acids. Tin and antimony form acidic oxides only in the presence of hot concentrated perchloric or nitric acid.

Manganese may be separated as the dioxide by heating a perchloric or nitric acid solution of manganous ion with an oxidizing agent such as potassium chlorate.

Precipitation of basic oxides from buffered solutions. Table 33-2 shows the pH at which precipitation of several hydrous oxides is initiated from aqueous solutions. Somewhat higher values are required for complete precipitation. It is apparent from these data that a number of useful separations should be possible by proper control of pH; this requires the use of buffer mixtures that maintain the hydrogen ion concentration at a suitable predetermined level.

Many of the separations suggested by Table 33-2 are unsatisfactory. The hydrous oxide precipitates are usually gelatinous colloids which are appreciably contaminated by adsorbed impurities. As a consequence several reprecipitations may be required to achieve a clean separation. In addition,

[2] For such a scheme for quantitative analysis of minerals or rocks see G. E. F. Lundell and J. I. Hoffman, *Outlines of Methods of Chemical Analysis*, 33-41. New York: John Wiley and Sons, Inc., 1938.

Table 33-2

pH AT WHICH CERTAIN HYDROUS OXIDES ARE PRECIPITATED[3]

pH	Metal Ion
11	Mg(II)
9	Ag(I), Mn(II), La, Hg(II)
8	Ce(III), Co(II), Ni(II), Cd, Pr, Nd, Y
7	Sm, Fe(II), Pb
6	Zn, Be, Cu, Cr(III)
5	Al
4	U(VI), Th
3	Sn(II), Zr, Fe(III)

precipitation of more soluble constituents of the sample may occur due to local excesses of hydroxyl ion during the addition of base; these frequently do not redissolve readily or completely and thus limit the effectiveness of the separation. Homogeneous precipitation avoids this problem and is often employed in hydrous oxide separations.

Perhaps the most common buffer mixture employed for precipitation of the basic oxides is prepared from ammonia and ammonium chloride. Iron, chromium, aluminum, and titanium are precipitated quantitatively from such a solution while manganous ion and the alkaline earth ions are not. Copper, zinc, nickel, and cobalt also are unprecipitated as a result of formation of soluble ammonia complexes. Quantitative separations of the latter with this buffer are generally not very satisfactory, however, because of their extensive coprecipitation; in addition the precipitates are gelatinous and difficult to handle. Swift has measured the amount of coprecipitation in several instances.[4]

Precipitation of the basic oxides of iron, aluminum, and chromium from a somewhat acidic medium provides a more satisfactory separation from the common dipositive ions. The *basic acetate* method, in which the pH is maintained by an acetic acid-ammonium acetate buffer, is widely used for this purpose. Other acidic buffer mixtures have also been recommended. These include benzoic acid-benzoate, formic acid-formate, and succinic acid-succinate mixtures.

Separations by solutions of strong base. In strongly basic solutions and in the presence of an oxidizing agent such as sodium peroxide, several of the amphoteric elements are quite soluble and can be separated from ions that precipitate under these circumstances. Among the soluble species are aluminum, zinc, chromium, vanadium, and uranium while iron, cobalt, nickel, and the rare earths precipitate.

[3] H. T. S. Britton, *J. Chem. Soc.*, **127**, 2157 (1925).
[4] E. H. Swift, *A System of Chemical Analysis*, 284-285. San Francisco: W. H. Freeman and Company, 1938.

Table 33-3

PRECIPITATION OF SULFIDES UNDER VARIOUS CONDITIONS[5]

Ion	H_2S in 9 N HCl at 100° C	H_2S in 9 N H_2SO_4 at 100° C	H_2S in 0.3 N HCl and 0.3 F NH_4Cl	H_2S in Acetate Buffer pH 6	$(NH_4)_2S$ in NH_3 at pH 9
As(III) or (V)	As_2S_3 or As_2S_5	As_2S_3 or As_2S_5	As_2S_3 or As_2S_5	As_2S_3 or As_2S_5	s
Hg(II)	s*	HgS	HgS	HgS	HgS
Cu(II)	s	CuS	CuS	CuS	CuS
Sb(III) or (V)	s	Sb_2S_3 or Sb_2S_5	Sb_2S_3 or Sb_2S_5	Sb_2S_3 or Sb_2S_5	s
Bi(III)	s	Bi_2S_3	Bi_2S_3	Bi_2S_3	Bi_2S_3
Sn(IV)	s	s	SnS_2	SnS_2	s
Cd(II)	s	s	CdS	CdS	CdS
Pb(II)	s	s	PbS	PbS	PbS
Zn(II)	s	s	s	ZnS	ZnS
Co(II)	s	s	s	CoS	CoS
Ni(II)	s	s	s	NiS	NiS
Fe(II)	s	s	s	s	FeS
Mn(II)	s	s	s	s	MnS

[5] Data taken from Ernest H. Swift, *Introductory Quantitative Analysis*, 422–423. Englewood Cliffs, N. J.: Prentice-Hall, Inc., 1950.

* No precipitate forms.

Sulfide Separations

With the exception of the alkali and alkaline-earth metals, most cations form insoluble sulfides. Their solubilities, however, differ greatly, and since it is a relatively easy matter to control the sulfide ion concentration of an aqueous solution by adjustment of pH, separations based on formation of these compounds have found extensive use. Employment in the classical qualitative scheme is well known. Quantitative applications are also extensive despite the toxic properties of the reagent and the likelihood of difficulties due to coprecipitation.

The theoretical treatment of the ionic equilibria influencing the solubility of sulfide precipitates was considered on page 147. Often, however, such treatment does not lead to realistic conclusions regarding the feasibility of separations because of coprecipitation and the slow rate at which some of the sulfides form. As a consequence, resort must be made to empirical data.

Table 33-3 shows some common separations that can be accomplished with hydrogen sulfide through control of the pH.

Other Inorganic Precipitants

There are no other inorganic ions that are as generally useful for separations as those just discussed. Although phosphate, carbonate, and oxalate ions are often employed as precipitants for cations, their behavior is nonselective so that preliminary separations generally precede their use.

Chloride and sulfate ions are useful because of their relatively specific behavior. The former can be employed to separate silver from a wide variety of other metals while the latter is frequently employed to separate a group of metals that includes lead, barium, strontium, and calcium.

Organic Precipitants

A number of organic reagents are employed for the separation of various inorganic ions; some of these were discussed in Chapter 9. Certain organic precipitants, such as dimethylglyoxime, are useful because of their remarkable selectivity in forming a precipitate with only a very few ions. Others, such as 8-hydroxyquinoline, form insoluble compounds with a host of cations; the solubilities of these differ greatly, however, and by controlling the reagent concentration useful separations can be performed. As with sulfide ion, the concentration of the precipitating reagent is readily controlled by adjustment of pH.

Separation of a Constituent Present in Trace Amounts

A problem often encountered in trace analysis is that of isolating the minor constituent, which may be present in microgram quantities, from the major components of the sample. Sometimes this is done by a precipitation process;

the techniques employed are somewhat different from those used when the desired component is present in more generous amounts.

Even in those instances where the solubility of the precipitate is low enough to allow quantitative isolation of a trace constituent, difficulties are encountered; supersaturation may prevent formation of the solid in a finite length of time, or perhaps coagulation of a tiny quantity of a colloidal suspension is required. There is also the likelihood of loss of an appreciable fraction of the solid during the transfer and filtration operations. In order to minimize these difficulties, a small quantity of some other ion that also forms a precipitate with the reagent is added to the solution. The precipitate thus formed is called a *collector* and serves to carry the desired minor species out of solution. Clearly, the collector must be a substance that will not interfere with the method for completing the analysis for the trace component.

The mechanism by which a collector functions undoubtedly differs in various cases. Sometimes it simply carries down the particles of the trace precipitate by physical entrainment. Other times the process must involve coprecipitation in which the minor component is adsorbed or held in the precipitate by mixed crystal formation.

For example, in isolating manganese as the highly insoluble manganese dioxide, a small amount of ferric iron is often added. The basic ferric oxide precipitate carries down even the smallest amounts of the dioxide. A few micrograms of titanium can be removed from a large volume of solution by addition of aluminum ion and ammonia. Here the aluminum hydroxide serves as the collector. Copper sulfide is often employed as a collector for zinc and lead when traces of these are to be removed as sulfides. Many other examples of the use of collectors are described by Sandell.[6]

Separation by Electrolytic Precipitation

Electrolytic precipitation constitutes a highly useful method for accomplishing separations. Either the wanted or the unwanted components of a mixture can be isolated as a second phase. The method becomes particularly effective when the potential of the working electrode is controlled at a predetermined level (see p. 532).

The mercury cathode (p. 539) has found wide application in removal of many metal ions prior to the analysis of the residual solution. In general, metals below zinc in the electromotive series are conveniently deposited in the mercury, leaving such ions as aluminum, beryllium, the alkaline earths, and the alkali metals.

---

[6] E. B. Sandell, *Colorimetric Determination of Traces of Metals*, 3d ed. New York: Interscience Publishers, Inc., 1959.

EXTRACTION METHODS

When solutes are allowed to distribute themselves between a pair of immiscible solvents, tremendous variations in the equilibrium concentration ratios are often observed. These differences can serve as the basis for quantitative separations. In the usual case, an aqueous solution of the sample is extracted with an organic liquid such as a hydrocarbon, a chlorinated hydrocarbon, an ether, an alcohol, or a ketone. The constituent to be determined may be transferred to the organic layer and leave the unwanted components behind, or the reverse may be the case.

A surprising number of inorganic chlorides, nitrates, and thiocyanates can be transferred quantitatively to various immiscible organic solvents. In addition, organic chelating agents render many inorganic cations readily extractable. Owing to the selective nature of both the chelation and the extraction processes, many useful separations are thus possible.

Theory

The distribution of a solute between two immiscible solvents is governed by the *distribution law*. If we assume that the solute species A is allowed to distribute itself between water and an organic phase, the resulting equilibrium may be written

$$A_w \rightleftharpoons A_o$$

where the subscripts w and o refer to the water and organic phases, respectively. The ratio of activities of A in the two phases will be constant and independent of the total quantity of A; that is, at any given temperature

$$K = \frac{[A_o]}{[A_w]}$$

where the equilibrium constant K is the *partition coefficient*. The terms in brackets are strictly the activities of A in the two solvents; the molar concentrations can often be substituted here without serious error. Generally, K is equal to the ratio of the solubility of A in the two solvents.

In some instances the solute may exist in different states of aggregation in the two solvents; then the equilibrium becomes

$$x(A_y)_w \rightleftharpoons y(A_x)_o$$

and the partition coefficient takes the form

$$K = \frac{[(A_x)_o]^y}{[(A_y)_w]^x}$$

Expressions such as these are useful in describing the behavior of simple systems where the solute species does not associate or dissociate in either solvent phase. Often, however, analytical extractions are complicated by such reactions. For example, in extraction of a metal-organic chelate, one must also take into account the dissociation of the chelate. Thus if A were the extractable chelate, and M and X represented the metal ion and the chelating agent, respectively, the following equilibrium would have to be considered also:

$$A_w \rightleftharpoons M_w + X_w$$

for which

$$K_{inst} = \frac{[M_w][X_w]}{[A_w]}$$

Here, the concentration of A in the organic layer becomes additionally dependent upon the instability constant of the complex in water and the concentration of the chelating agent.

The distribution coefficient is useful because it does allow calculation of the number of extractions necessary to achieve any degree of completeness in the transfer of a solute from one solvent to another. In the treatment that follows, we shall assume that the distribution equilibrium is not complicated by association or dissociation reactions; thus, the equation

$$K = \frac{[A_o]}{[A_w]}$$

is adequate to describe completely the behavior of the solute.[7] We shall now assume an initial volume V_w ml of an aqueous solution containing a millimols of A and that we extract this with V_o ml of an immiscible organic solvent. At equilibrium, x_1 millimols of A will remain in the aqueous layer and we may write

$$[A_w] = \frac{x_1}{V_w}$$

$$[A_o] = \frac{(a - x_1)}{V_o}$$

Substituting this into the distribution coefficient expression and rearranging, we obtain

$$x_1 = \left(\frac{V_w}{V_o K + V_w}\right)a$$

Thus the number of millimols x_1 of the solute remaining in the water after the extraction is a fraction of the original number a. The number of millimols x_2 remaining after a second extraction of the water with an identical volume of solvent will, by the same reasoning, be

$$x_2 = \left(\frac{V_w}{V_o K + V_w}\right)x_1$$

$$= \left(\frac{V_w}{V_o K + V_w}\right)^2 a$$

[7] Suitable modification of this treatment can be made to take into account other equilibria; see H. A. Laitinen, *Chemical Analysis*, 258-269. New York: McGraw-Hill Book Company, Inc., 1960.

After n extractions, the number of millimols remaining is given by the expression

$$x_n = \left(\frac{V_w}{V_o K + V_w}\right)^n a$$

Example. The distribution coefficient of iodine between CCl_4 and water is 85. We shall calculate the number of millimols of I_2 remaining in 100 ml of an aqueous solution that was originally $1.00 \times 10^{-3}\,M$ after extraction with two 50-ml portions of CCl_4.

$$a = 100 \times 1.00 \times 10^{-3} = 0.1 \text{ millimol}$$

$$x_2 = \left(\frac{100}{50 \times 85 + 100}\right)^2 0.1$$

$$= 5.28 \times 10^{-5} \text{ millimol}$$

Thus, the two extractions should reduce the number of millimols of I_2 in the aqueous solution from 0.1 to 5.28×10^{-5}.

The exponential nature of the foregoing relationship shows that a more efficient extraction is achieved by employing several small increments of solvent than by using one large one. We can compare, for example, the results from the following calculation with that just shown.

Example. We wish to calculate the number of millimols of I_2 remaining if the aqueous solution in the previous example had been extracted with one 100-ml portion of CCl_4 rather than two 50-ml portions.

$$x_1 = \frac{100}{100 \times 85 + 100} \times 0.1$$

$$= 1.16 \times 10^{-3} \times 0.1 = 1.16 \times 10^{-4} \text{ millimol}$$

Here the single extraction with the larger volume of CCl_4 leaves 1.16×10^{-4} millimol of I_2 in the aqueous layer compared with 5.28×10^{-5} millimol for the two extractions with half volumes.

Advantages of Extraction Procedures

Extraction procedures are used extensively by the organic chemist to effect separations; only recently, however, has the technique found widespread use in analysis. Separations by extraction have a number of attractive features when compared with the classical precipitation methods. One of these is its freedom from problems of coprecipitation and postprecipitation. Second, extraction procedures are well suited to the separation of microgram quantities whereas real difficulty is encountered in attempting to precipitate and collect traces of a substance. Third, a highly favorable equilibrium constant is not a necessity for a successful extraction procedure inasmuch as multiple extractions can be readily performed. In addition, apparatus that allows continuous extraction of one liquid by another is easily constructed; this usually involves continuous vaporization of the extracting liquid followed by its condensation in such a way that it will pass through the solution containing the sample. Figure 33.1 shows an apparatus that could be used for extracting with a solvent less dense than the aqueous solution. With such an apparatus, quantitative removal of a solute is possible even where the distribution coefficient is relatively unfavorable. In general, a separation by extraction is easier and less time consuming than the equivalent precipitation separation.

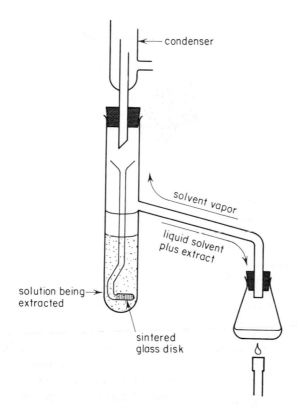

Fig. 33.1 Apparatus for Continuous Extraction of a Sample. Requires a solvent that is less dense than the solution being extracted.

Extraction processes often exhibit a high degree of selectivity. For inorganic cations, a number of organic chelating reagents are available that permit extraction of certain cations but not others. The selectivity of these is often enhanced by control of the pH of the solution being extracted.

Applications of Extraction Procedures

Ether extractions of metal chlorides. The data in Table 33-4 indicate that a wide variety of metal chlorides can be extracted into ether from 6 F hydrochloric acid solution; equally important, a large number of metal ions are unaffected or extracted to only a small extent under these conditions. Thus many useful separations are possible. One of the most important of these is the separation of trivalent iron (99 percent extracted) from a host of other cations. This separation is often employed to remove the greater part of iron from steel or iron ore samples prior to the analysis of traces of elements such as chromium, aluminum, titanium, or nickel.

Table 33-4

ETHYL ETHER EXTRACTIONS OF VARIOUS CHLORIDES FROM 6 F HYDROCHLORIC ACID[8]

Percent Extracted	Elements and Oxidation State
90—100	Fe(III), 99%; Sb(V), 99%*; Ga(III), 97%; Ti(III), 95%*; Au(III), 95%
50—90	Mo(VI), 80-90%; As(III), 80%*†; Ge(IV), 40-60%
1—50	Te(IV), 34%; Sn(II), 15-30%; Sn(IV), 17%; Ir(IV), 5%; Sb(III), 2.5%*
< 1 > 0	As(V)*, Cu(II), In(III), Hg(II), Pt(IV), Se(IV), V(V), V(IV), Zn(II)
0	Al(III), Bi(III), Cd(II), Cr(III), Co(II), Be(II), Fe(II), Pb(II), Mn(II), Ni(II), Os(VIII), Pd(II), Rh(III), Ag(I), Th(IV), Ti(IV), W(VI), Zr(IV)

* Isopropyl ether employed rather than ethyl ether.
† 8 F HCl rather than 6 F.
[8] Data from Ernest H. Swift, *Introductory Quantitative Analysis*, 431. Englewood Cliffs, N. J.: Prentice-Hall, Inc., 1950.

The extraction of iron by ether has been studied in considerable detail. The species extracted appears to be $HFeCl_4$. It has been found that the percentage of iron transferred is dependent upon the hydrochloric acid content of the aqueous phase (below 3 F and above 9 F HCl the percent removed is slight) and to some extent upon the iron content. Unless special precautions are taken, the last traces of iron are not removed by the extraction.[9]

Extraction of nitrates. Certain nitrate salts are soluble in ether as well as other organic solvents; advantage has been taken of this to separate uranium from such elements as lead and thorium. In this application, the aqueous phase is saturated with ammonium nitrate and the nitric acid concentration is adjusted to about 1.5 F; the uranium must be in the + 6 oxidation state. Bismuth and ferric nitrates are also extracted to some extent under these conditions.

[9] See S. E. Q. Ashley and W. M. Murray, *Ind. Eng. Chem., Anal. Ed.*, **10**, 367 (1938).

Extraction of chelate compounds. Many of the organic reagents mentioned in Chapter 9, as well as others, form chelates with various metal ions; often these chelates are readily soluble in such solvents as chloroform, carbon tetrachloride, benzene, and ether. Thus, quantitative transfer of the metallic ions to the organic phase is possible.

A reagent that has found widespread application for extraction separations is 8-hydroxyquinoline (p. 145). Most of the metal chelates with this compound are soluble in chloroform as well as other solvents. Their distribution coefficients are dependent upon the pH of the aqueous phase; as a consequence, control of this variable makes many valuable separations possible, particularly of trace quantities of metals.[10].

Another very useful reagent for separating minute quantities of metal ions is dithizone[11] (diphenylthiocarbazone). Its reaction with a divalent metallic ion can be written as

$$M^{2+} + 2S{=}C \underset{\substack{\diagup N{-}N{-}H \\ \diagdown N{=}N \\ C_6H_5}}{\overset{H \quad C_6H_5}{}} \rightleftharpoons \left(S{=}C \underset{\substack{\diagup N{-}N \\ \diagdown N{=}N \\ C_6H_5}}{\overset{H \quad C_6H_5}{}} \right)_2 M + 2H^+$$

      dithizone

Both dithizone and its metal chelates are soluble in chloroform or carbon tetrachloride; when an aqueous solution of a cation is shaken with an organic solution of the reagent, the following over-all reversible reaction occurs:

$$(M^{2+})_w + 2(HDz)_o \rightleftharpoons (MDz_2)_o + 2(H^+)_w$$

where HDz represents the dithizone molecule and the subscripts indicate the phase. The equilibrium is clearly pH dependent; thus by controlling the pH of the aqueous phase, various separations of metallic ions are possible. The dithizone complexes of many metal ions are intensely colored. Spectrophotometric measurement of the organic extract often serves to complete the analysis after the separation has been made.

A number of other organic chelating compounds are employed for separations by extraction. Further information concerning these is available in the reference works of Laitinen[12] and Sandell.[13]

[10] See E. B. Sandell, *Colorimetric Determination of Traces of Metals*, 3d ed., 179-188. New York: Interscience Publishers, Inc., 1959.

[11] *Ibid*, 151-154.

[12] H. A. Laitinen, *Chemical Analysis*, 269-274. New York: McGraw-Hill Book Company, Inc., 1960.

[13] E. B. Sandell, *Colorimetric Determination of Traces of Metals*, 3d ed. New York: Interscience Publishers, Inc., 1959.

Extraction of Solids with Organic Solvents

The leaching of a soluble component from a solid mixture is not an ideal method for performing a separation. Incomplete removal of the soluble component is the rule because of its occlusion within the residual solid mass. To reduce the errors from this cause, the solid is usually triturated for extended periods with the solvent; even so, complete dissolution of the soluble component is seldom achieved.

Solid extraction is employed for difficult separations that are not readily performed by other means. The best example is within the alkali-metal series, where all of the salts of these metals are too soluble to allow their separation in aqueous solvents. For these the usual procedure is to free the solution of all other metal ions and then evaporate to dryness in the presence of a suitable acid to give the desired salt. The residue is then extracted repeatedly with the organic solvent.

When the perchlorates of the alkali metals are extracted with *n*-butanol or a mixture of this solvent and ethyl acetate, solution of the sodium and lithium salts occurs while those of potassium, rubidium, and cesium remain behind; in this way a group separation can be achieved. Lithium and sodium ions can be separated by extraction of their chlorides with pyridine or dioxane, lithium chloride being the more soluble. The chloroplatinates of the alkali metals also exhibit marked solubility differences in 80-percent ethanol. Thus, the sodium and lithium salts can be leached from those of potassium, rubidium, cesium, and ammonium.

SEPARATIONS BY DISTILLATION

Distillation is widely used for separating organic mixtures into their components. Where the boiling points of compounds differ by only a few degrees, a rather elaborate fractionating column is required to achieve a clean separation. In this device, continuous equilibration between a vapor and a liquid phase is achieved. The liquid and vapor become richer in the more volatile component with each succeeding point up the column; if the column has sufficient length, separation of substances of rather similar volatilities is possible.

Fractional distillation has not been widely employed in the separation of inorganic mixtures; applications have largely been confined to the isolation of a relatively volatile component from a nonvolatile residue. For such a separation a simple batch apparatus can be used; this consists of a distilling flask, a spray trap, and a condenser. Ordinarily the distillate from such a device consists of a mixture of the volatile component and the solvent.

A number of the elements can be converted to volatile compounds by suitable chemical treatment.

Carbon contained in inorganic carbonates is converted to carbon dioxide upon treatment with acid. The carbon dioxide is readily distilled from aqueous solution and is collected by passage of the distillate through a solution containing barium hydroxide or by absorbing the gas on Ascarite (p. 733).

Nitrogen is readily volatilized from basic solution as ammonia. Nitrates and other nitrogen oxides can be reduced to ammonia by Devarda's alloy (p. 356).

Sulfur can be distilled from aqueous solution as hydrogen sulfide or sulfur dioxide. These gases are evolved from sulfides or sulfites upon treatment with a nonvolatile acid. Hydrogen sulfide can be collected by passage of the vapor through a solution of cadmium ion; sulfur dioxide is oxidized to sulfate in a solution of hydrogen peroxide.

Halogens can be distilled from aqueous solution in their elemental form or as the hydrohalides. Separation among the halogens is sometimes accomplished by selective oxidation of one of the halides followed by distillation of the element. Thus, iodide can be separated from chloride and bromide by treatment with nitrous acid. Bromides can be separated from chlorides by selective oxidation with telluric acid or potassium hydrogen iodate.

Fluorine is readily separated from a variety of elements by distillation of a perchloric acid solution of the fluoride in the presence of silica. Fluosilicic acid, H_2SiF_6, is the volatile product.

Silicon can be volatilized as the tetrafluoride by treatment of silicates with hydrofluoric acid.

Boron is often converted to the volatile methyl borate by heating a sulfuric acid solution of boric acid with methyl alcohol.

$$3CH_3OH + H_3BO_3 \rightarrow 3H_2O + (CH_3)_3BO_3$$

The boric ester can be collected by passing the distillate into an alkaline solution.

Arsenic, antimony, and *tin* can be separated from most other elements as well as from one another by suitable distillation procedures. Arsenic, in the trivalent state, can be quantitatively removed as the chloride from an aqueous solution containing sulfuric and hydrochloric acids. Distillation at $110° C$ is required. With the exception of germanium, no other elements distill under these conditions. After removal of the arsenic, antimony trichloride can be separated by raising the boiling point of the mixture to $155°$ to $165° C$. Concentrated hydrochloric acid must be added to the solution during the distillation. Phosphoric acid is added to complex with stannic ion and prevent the partial distillation of its chloride. After arsenic and antimony have been removed, tin can be separated as the tetrabromide at $140° C$ by the addition of hydrobromic acid to the residual solution.

Chromium forms the volatile chromyl chloride, CrO_2Cl_2, when a hot perchloric acid solution of dichromate is treated with hydrochloric acid. This provides a means for separating chromium from a wide variety of other metals.

Osmium and *ruthenium* form volatile tetroxides, and may be isolated as such by distillation from nitric acid solution.

Table 33-5

CLASSIFICATION OF CHROMATOGRAPHIC SEPARATION METHODS

Type of Chromatography	Mobile Phase	Stationary Phase	Type of Equilibrium Involving Species M
Adsorption	solution$_1$	finely divided solid$_1$	$M(soln_1) \rightleftharpoons M(adsorbed\ on\ solid_1)$
Ion exchange	solution$_1$	ionic resin (solid$_1$)	$M^+(soln_1) + X^+R^-_{(solid_1)} \rightleftharpoons M^+R^-_{(solid_1)} + X^+(soln_1)$ $M^-(soln_1) + R^+X^-_{(solid_1)} \rightleftharpoons R^+M^-_{(solid_1)} + X^-(soln_1)$
Liquid—liquid	solution$_1$	solution$_2$ as a film supported by an inert solid	$M(mobile\ soln_1) \rightleftharpoons M(stationary\ soln_2)$
Paper	solution$_1$	solution$_2$ as a film on a coarse paper	$M(mobile\ soln_1) \rightleftharpoons M(stationary\ soln_2)$
Gas—liquid	gas$_1$	solution$_2$ as a film supported by an inert solid	$M(gas_1) \rightleftharpoons M(stationary\ soln_2)$

CHROMATOGRAPHIC SEPARATIONS

Separation methods grouped under the general category of *chromatography* are diverse both with respect to principle and to practice. Many are of considerable importance to the analytical chemist, for they often enable him to separate, isolate, and identify the components of mixtures that are otherwise resolved with difficulty, if they can be resolved at all. Although consideration of this topic must be brief, we shall examine those aspects that are of particular importance to analytical chemistry.

The term *chromatography* is difficult to define rigorously, owing to the variety of systems and techniques to which it has been applied. In its broadest sense, however, chromatography refers to processes that permit the resolution of a mixture as a consequence of differences in rates at which the individual components of that mixture migrate through a stationary medium under the influence of a mobile phase.

The chromatographic separations of analytical chemistry are generally carried out in tubular columns of glass or metal that have been packed with a porous solid. The solid serves as the stationary phase itself or as a support for the stationary phase. The mixture to be separated, after being dissolved in the mobile phase, is introduced at one end of the column. Separation occurs as portions of the mobile phase are forced through the column. The components of the mixture distribute themselves between the two phases; those whose equilibria favor retention by the stationary phase move only slowly with the passage of the mobile phase while those that are less strongly held travel more rapidly through the column. Ideally, the various components of the mixture are ultimately found in bands located in different positions along the length of the column. Separation can then be accomplished by passing enough of the mobile phase through the column to cause these various bands to pass out the end where they can be collected; alternatively, the column packing can be removed and broken up into portions containing the various constituents of the mixture.

Table 33-5 catalogues the various chromatographic techniques according to the physical nature of the stationary and mobile phases, and hence according to the nature of the equilibria upon which separation is based. In those cases where a liquid stationary phase is employed, a finely divided or porous solid is coated with a film of the liquid, thus giving a means by which this phase can be immobilized.

In the sections that follow, we shall discuss briefly the application of the various chromatographic procedures.

Adsorption Chromatography

Historically, adsorption chromatography represents the first type reported. Its discoverer was a Russian botanist, M. Tswett, who, in 1903, employed the technique to separate the various colored components of a plant extract. A quarter of a century was to pass before the significance of his discovery was fully appreciated by investigators engaged in the separation of biological and organic materials.

In all of the early applications, the procedure was limited to the separation of colored substances (thus the name *chromatography*) that could be identified by their appearance. The technique involved packing a glass column with a finely divided adsorbent such as silica, alumina, calcium carbonate, or sucrose; the adsorbent was then wetted with a solvent, and a solution of the sample was introduced at the top of the column. The column was *developed* by washing with further portions of the solvent until colored bands of the solute appeared at various positions along its length. The various colored fractions were recovered by pushing the packing out of the column, cutting it into pieces, and treating each with a solvent that would cause the component to be desorbed.

The method was later simplified by washing the column with sufficient solvent until each of the adsorbed components had been removed in turn and collected. Still later, methods making use of such properties as refractive index and ultraviolet absorbance were employed to indicate when components of the mixture had been washed from the column; this broadened the scope of the procedure to include colorless materials.

Separations by adsorption chromatography rely on the equilibria that govern the distribution of the various solute species between the solvent and the surface of the solid. Wide variations exist in the tendencies of compounds to be adsorbed. For example, a positive correlation exists between the number of hydroxyl groups in a molecule and its adsorption properties; a similar correlation is found with double bonds. Compounds containing certain functional groups are more strongly held than others. The tendency to be adsorbed decreases as follows: acid > alcohol > carboxyl > ester > hydrocarbon. The nature of the adsorbent, however, also plays a part in determining the order of adsorption. Much of the available knowledge in this field is empirical, however, and the choice of adsorbent and solvent for a given separation must often be made on a trial and error basis.

Adsorption chromatography has been used primarily for separation of organic compounds; several monographs on the applications of the technique are available.[14]

Ion-exchange Separations

Ion exchange is a process in which there is an exchange of ions of like sign between a solution and an insoluble solid in contact with the solution. Clearly, the solid must contain ions of its own to exchange; in addition, it must have a permeable structure of large specific surface area so that the solvent and solute

[14] For example, see E. Lederer and M. Lederer, *Chromatography*, 2d ed. Amsterdam: Elsevier Press, 1957; H. G. Cassidy, *Fundamentals of Chromatography*. New York: Interscience Publishers, Inc., 1957; H. H. Strain, *Anal. Chem.*, **21**, 75(1949); **22**, 41 (1950); **23**, 25(1951); **24**, 50(1952); **26**, 90 (1954); **30**, 620 (1958); **32**, 3R(1960).

ions can readily come in contact with the solid surface. Many substances, both natural and synthetic, have these properties. Among the former are clays and zeolites; the ion-exchange properties of these have been recognized and studied for over a century. Synthetic ion-exchange resins were first produced in 1935 and have since found widespread laboratory and industrial application for water softening, water deionization, purification of solutions, and separation of ions.

Synthetic ion-exchange resins are high molecular-weight polymeric materials containing large numbers of ionic functional groups per molecule. For cation exchange, there is a choice between strong-acid type resins containing sulfonic acid groups (RSO_3H), or weak-acid resins containing carboxylic acid ($RCOOH$) groups. The former have wider application. Anion-exchange resins contain basic functional groups attached to the polymer molecule. These are generally amines; strong-base exchangers are obtained with tertiary amines ($RN(CH_3)_3^+OH^-$) and weak-base types with primary and secondary amines.

Ion-exchange equilibria. The equilibrium in a cation-exchange process is illustrated by the reaction

$$x RSO_3^-H^+ + M^{x+} \rightleftharpoons (RSO_3^-)_x M^{x+} + x H^+$$

$$\text{solid} \qquad \text{solution} \qquad \text{solid} \qquad \text{solution}$$

where M^{x+} represents a cation of $+x$ charge and R represents *a part* of a resin molecule. The reaction of a typical anion-exchange resin can be written as follows:

$$x RN(CH_3)_3^+OH^- + A^{x-} \rightleftharpoons (RN(CH_3)_3^+)_x A^{x-} + x OH^-$$

$$\text{solid} \qquad \text{solution} \qquad \text{solid} \qquad \text{solution}$$

where A^{x-} is an anion of charge $-x$.

The composition of an ion-exchange resin can be markedly altered by suitable treatment. For example, a sulfonic acid resin in its *acid form* ($RSO_3^-H^+$) can be converted to the *sodium form* ($RSO_3^-Na^+$) by treatment with a strong solution of sodium chloride followed by washing with water. A basic exchange can be completely converted to its chloride form ($RN(CH_3)_3^+Cl^-$) by a similar treatment.

When a dilute solution of an electrolyte is passed through a column of resin, quantitative exchange of one of the ions may ensue if the number of functional groups of the exchanger is large with respect to the number of ions in solution. Thus if a dilute solution of sodium chloride is passed through a sulfonic acid resin in its acid form, quantitative conversion of the sodium chloride to hydrochloric acid will result.

The attraction by an ion-exchange resin for ions of opposite charge varies considerably with the nature of the ion. In general, the attractive force increases with charge; thus, for example, $Na^+ < Ca^{2+} < Al^{3+}$. For ions of the same charge, the exchange affinity usually increases with increasing atomic number; thus $Li^+ < Na^+ < K^+ < Rb^+$. These differences in affinity sometimes make separations of ions possible, particularly if a column of resin is employed.

Application of ion-exchange resins. The analytical applications of ion exchange have been summarized by Samuelson.[15] A few illustrative examples follow.

Separation of interfering ions. Ion-exchange resins are useful for removal of interfering ions, particularly where those ions are of opposite charge to the species being determined. For example, iron (III), aluminum (III), and other cations cause difficulty in the determination of sulfate by virtue of their tendency to coprecipitate with barium sulfate. Passage of the solution to be analyzed through a column containing a cation-exchange resin results in an exchange of these ions for hydrogen ions. In a similar manner, phosphate ion, which interferes in the analysis of barium or calcium ion, can be removed by passage of the sample through an anion-exchange resin.

Fractional separation of ions. Ion-exchange resins have been used to fractionate ions in a manner similar to that described for adsorption chromatography. Here, advantage is taken of the variation in affinity of an exchange resin for various ions of like charge. The mixture to be analyzed is introduced at the top of a column; the ions are subsequently displaced downward by a suitable solvent. The least strongly held move most rapidly and are thus first to emerge.

Beukenkamp and Rieman[16] employed this technique to separate sodium from potassium. The two ions were introduced upon a column of a sulfonic acid resin in its acid form. The column was then washed with a solution of hydrochloric acid; this caused the sodium ions to move more rapidly than potassium. These were then collected in appropriate fractions of the effluent liquid.

The most important applications of ion exchange have been to the separation of the rare-earth ions, primarily for preparatory purposes. Here, fractionation is enhanced by taking advantage of differences in the formation constants of complexes of the cations involved.[17]

Concentration of traces of an electrolyte. A useful application of ion exchangers is the concentration of traces of an ion from a very dilute solution. Cation-exchange resins, for example, have been employed to collect trace amounts of metallic elements occurring in natural waters by passage of large volumes of sample. The ions are then liberated from the resin by treatment with acid to give a considerably more concentrated solution for analysis.

Conversion of salts to acids or bases. An interesting application of ion-exchange resins is the determination of the total salt content of a sample. This can be accomplished by passage of the sample through a cation-exchange resin; absorption of the cations results in the release of an equivalent quantity of hydrogen ion which can then be collected in the washings from the column and titrated. Similarly, a standard acid solution can be prepared from a salt; for example, a weighed quantity of sodium chloride can be passed through a cation-exchange column in the acid form. The salt is converted to hydrochloric

---

[15] O. Samuelson, *Ion Exchangers in Analytical Chemistry*. New York: John Wiley and Sons, Inc., 1953.

[16] J. Beukenkamp and W. Rieman, *Anal. Chem.*, **22**, 582 (1950).

[17] See a series of papers in *J. Am. Chem. Soc.*, **69**, 2769-2881 (1947).

acid which is then collected in the washings and diluted to known volume.

Treatment of salts with a basic-exchange resin converts the salt to a base by an analogous mechanism.

Liquid-liquid Chromatography

Liquid-liquid chromatography is carried out in much the same way as adsorption chromatography. The fundamental difference between the procedures lies in the types of equilibria involved. In the liquid-liquid form the stationary phase consists of a liquid firmly adsorbed on a solid. The latter simply serves as a support for the liquid film. The sample, in a second immiscible solvent is then passed through the column containing the immobilized liquid phase; this results in the establishment of equilibria between the components of the sample and the two liquids. The rate at which a solute passes through the column is dependent upon these equilibria; separations are possible where the distributions differ.

Liquid-liquid chromatography has important applications in the separation of various organic mixtures. For example, acetylated amino acids can be isolated from one another by employing a mobile phase made up of chloroform, butanol, and water; the stationary phase consists of water adsorbed on silica gel.

Paper Chromatography

Paper chromatography is one of the most important variants of the chromatographic technique. It was first described in 1944,[18] and has become a valued tool of the biochemist for the separation of difficult mixtures such as amino acids.

In this technique, a solution of the sample is placed near one end of a strip of heavy filter paper. The paper is then suspended vertically with the lower edge, containing the sample, immersed in a developing solvent. Capillary action causes the solvent to move up the paper; the components of the sample are carried with it, each at a rate that depends upon the position of the equilibrium involved. After development, the paper may be sprayed with a reagent that forms colored products with the components, thus making their identification possible. The paper may be cut into pieces for recovery of the components, or the analysis may be completed by measuring the size of the colored spots.

Two-dimensional paper chromatography is even more effective. Here, the sample is placed in one corner of a sheet of paper and developed in one direction. The paper is then rotated 90 degrees and developed in the second direction with a different solvent. This technique has made possible some remarkable and useful separations.

There is a good deal of controversy regarding the nature of the equilibria involved in paper chromatography. Some workers are of the opinion that a liquid-liquid distribution is established between water adsorbed on the paper

---

[18] R. Consden, A. H. Gordon, and A. J. P. Martin, *Biochem. J.*, **38**, 224 (1944).

and the developing solvent. Others consider the process to consist of adsorption of the solute upon the paper. Thus the equilibria involved are of the solid-liquid type.

Paper chromatography has found application to both inorganic and organic separations.

Gas-liquid Partition Chromatography

Gas-liquid partition chromatography, sometimes called *gas chromatography*, was first described in 1941; its general use for analysis did not follow until 1952.[19] Since that time there has been a phenomenal growth in the applications of the procedure.

Gas-liquid chromatography employs a stationary phase consisting of a nonvolatile liquid adsorbed on a solid. The sample is carried down a column packed with the solid by the flow of an inert gas such as helium or nitrogen. The rate of movement of the components of the sample is governed by their vapor pressures above the fixed solvent—that is, by the position of the gas-solution equilibria. Where this vapor pressure is very low, essentially no transport of a component occurs; when it is high, rapid movement is to be expected. As in other chromatographic methods, the constituents of the sample are emitted as bands in the inverse order of their affinity for the stationary phase. Clearly, the components of the mixture must have an appreciable vapor pressure (at least 10 mm of mercury) at the operating temperature of the column in order for this method to be applicable.

A variety of materials are employed as the solid support for the liquid phase; Celite and crushed firebrick are most common. The liquids themselves are ordinarily nonvolatile, inert compounds such as polyglycols, Carbowax 1000, silicone oils, or phthalate esters. Frequently the choice of solvent is of prime importance to the success or failure of a given separation and several may have to be tried in order to find the most satisfactory one.

The size of sample employed in an analytical gas chromatography column is remarkably small; it has been demonstrated that the efficiency of separation increases with decreasing sample size. Typically, samples of 0.01 ml or less are used. As a consequence, sensitive detectors are required to indicate the presence of the components in the exit stream. The method of detection most commonly used at the present time depends upon changes in thermal conductivity of the exit gases. Some devices employ thermistors for this measurement and others hot wire filaments. The conductivity of both varies with temperature; this, in turn, depends upon the thermal conductivity of the environment. The measurement, then, is of electrical conductivity; ordinarily this quantity is automatically recorded as a function of time. A typical gas chromatogram is shown in Figure 33.2.

---

[19] A. T. James and A. J. P. Martin, *Analyst*, **77**, 915 (1952); *Biochem. J.*, **50**, 679 (1952).

Gas-liquid chromatography has become one of the most important tools for the separation and identification of the components of organic mixtures. Qualitative analysis is based upon the time required for a peak to appear after introduction of the sample (retention time). The retention times of the various components of the unknown are then compared with the same data for standards. In order for this technique to be effective, close control of experimental conditions is required. Alternatively a chromatogram of the sample is compared with another chromatogram consisting of the sample plus a small amount of a suspected component. An increase in a peak height is taken to confirm the presence of that compound, while the appearance of a new peak indicates the absence of that compound in the original sample. A further technique calls for condensation of the various components as they emerge from the column and identification by infrared or ultraviolet absorption spectroscopy.

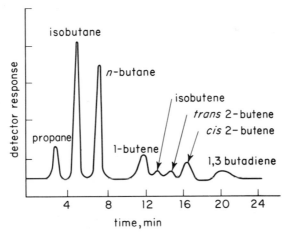

Fig. 33.2 Gas Chromatogram of a C_4 Hydrocarbon Mixture.

Quantitative analysis by gas chromatography is accomplished by measurement of the heights or areas of the peaks in the plots of retention time versus detector response (see Fig. 33.2). Calibration with known mixtures is required. Peak areas generally give more reliable results than the peak heights.

The field of gas-liquid partition chromatography has been reviewed in several monographs which may be consulted for further information.[20]

[20] R. L. Pecsok, *Principles and Practice of Gas Chromatography*. New York: John Wiley and Sons, Inc., 1959; A. I. M. Keulemans, *Gas Chromatography*, 2d ed., New York: Reinhold Publishers, 1959; D. H. Desty, *Gas Chromatography*. New York: Academic Press, 1958.

appendix

Table A-1

Half Reaction	$E°$, volts	Formal Potential, volts
$F_2 + 2H^+ + 2e \rightleftharpoons 2HF$	3.06	
$O_3 + 2H^+ + 2e \rightleftharpoons O_2 + H_2O$	2.07	
$S_2O_8^{2-} + 2e \rightleftharpoons 2SO_4^{2-}$	2.01	
$Co^{3+} + e \rightleftharpoons Co^{2+}$	1.82	
$H_2O_2 + 2H^+ + 2e \rightleftharpoons 2H_2O$	1.77	
$MnO_4^- + 4H^+ + 3e \rightleftharpoons MnO_2 + 2H_2O$	1.695	

[1] The majority of $E°$ values are taken from Wendell M. Latimer, *Oxidation Potentials*, 2d ed., Englewood Cliffs, N. J.: Prentice Hall, Inc., 1952. The formal potentials are from Ernest H. Swift, *Introductory Quantitative Analysis*. Englewood Cliffs, N. J.: Prentice Hall, Inc., 1950. With permission.

Table A-1 (Continued)

Half Reaction	$E°$, volts	Formal Potentials, volts
$Ce^{4+} + e \rightleftharpoons Ce^{3+}$		1.70, 1 F HClO$_4$; 1.61, 1 F HNO$_3$; 1.44, 1 F H$_2$SO$_4$
$H_5IO_6 + H^+ + 2e \rightleftharpoons IO_3^- + 3H_2O$	1.6	
$BrO_3^- + 6H^+ + 5e \rightleftharpoons \frac{1}{2}Br_2 + 3H_2O$	1.52	
$MnO_4^- + 8H^+ + 5e \rightleftharpoons Mn^{2+} + 4H_2O$	1.51	
$Mn^{3+} + e \rightleftharpoons Mn^{2+}$	1.51	
$PbO_2 + 4H^+ + 2e \rightleftharpoons Pb^{2+} + 2H_2O$	1.455	
$Cl_2 + 2e \rightleftharpoons 2Cl^-$	1.359	
$Cr_2O_7^{2-} + 14H^+ + 6e \rightleftharpoons 2Cr^{3+} + 7H_2O$	1.33	
$Tl^{3+} + 2e \rightleftharpoons Tl^+$	1.25	0.77, 1 F HCl
$MnO_2 + 4H^+ + 2e \rightleftharpoons Mn^{2+} + 2H_2O$	1.23	1.24, 1 F HClO$_4$
$O_2 + 4H^+ + 4e \rightleftharpoons 2H_2O$	1.229	
$IO_3^- + 6H^+ + 5e \rightleftharpoons \frac{1}{2}I_2 + 3H_2O$	1.195	
$Br_2 + 2e \rightleftharpoons 2Br^-$	1.065	1.05, 4 F HCl
$ICl_2^- + e \rightleftharpoons \frac{1}{2}I_2 + 2Cl^-$	1.06	
$V(OH)_4^+ + 2H^+ + e \rightleftharpoons VO^{2+} + 3H_2O$	1.00	1.02, 1 F HCl, HClO$_4$
$HNO_2 + H^+ + e \rightleftharpoons NO + H_2O$	1.00	
$NO_3^- + 3H^+ + 2e \rightleftharpoons HNO_2 + H_2O$	0.94	0.92, 1 F HNO$_3$
$2Hg^{2+} + 2e \rightleftharpoons Hg_2^{2+}$	0.920	0.907, 1 F HClO$_4$
$Cu^{2+} + I^- + e \rightleftharpoons CuI$	0.86	
$Ag^+ + e \rightleftharpoons Ag$	0.799	0.228, 1 F HCl; 0.792, 1 F HClO$_4$; 0.77, 1 F H$_2$SO$_4$
$Hg_2^{2+} + 2e \rightleftharpoons 2Hg$	0.789	0.274, 1 F HCl; 0.776, 1 F HClO$_4$; 0.674, 1 F H$_2$SO$_4$
$Fe^{3+} + e \rightleftharpoons Fe^{2+}$	0.771	0.700, 1 F HCl; 0.732, 1 F HClO$_4$; 0.68, 1 F H$_2$SO$_4$
$PtCl_4^{2-} + 2e \rightleftharpoons Pt + 4Cl^-$	0.73	
$C_6H_4O_2$ (quinone) $+ 2H^+ + 2e \rightleftharpoons C_6H_4(OH)_2$	0.699	0.696, 1 F HCl, H$_2$SO$_4$, HClO$_4$
$O_2 + 2H^+ + 2e \rightleftharpoons H_2O_2$	0.682	
$PtCl_6^{2-} + 2e \rightleftharpoons PtCl_4^{2-} + 2Cl^-$	0.68	
$MnO_4^- + e \rightleftharpoons MnO_4^{2-}$	0.564	
$H_3AsO_4 + 2H^+ + 2e \rightleftharpoons H_3AsO_3 + H_2O$	0.559	0.577, 1 F HCl, HClO$_4$
$I_3^- + 2e \rightleftharpoons 3I^-$	0.536	
$I_2 + 2e \rightleftharpoons 2I^-$	0.5355	
$Cu^+ + e \rightleftharpoons Cu$	0.521	
$H_2SO_3 + 4H^+ + 4e \rightleftharpoons S + 3H_2O$	0.45	
$Ag_2CrO_4 + 2e \rightleftharpoons 2Ag + CrO_4^{2-}$	0.446	
$VO^{2+} + 2H^+ + e \rightleftharpoons V^{3+} + H_2O$	0.361	
$Fe(CN)_6^{3-} + e \rightleftharpoons Fe(CN)_6^{4-}$	0.36	0.71, 1 F HCl; 0.72, 1 F HClO$_4$, H$_2$SO$_4$
$Cu^{2+} + 2e \rightleftharpoons Cu$	0.337	

Table A-1 (Continued)

Half Reaction	$E°$, volts	Formal Potentials, volts
$UO_2^{2+} + 4H^+ + 2e \rightleftharpoons U^{4+} + 2H_2O$	0.334	
$Hg_2Cl_2 + 2e \rightleftharpoons 2Hg + 2Cl^-$	0.268	0.242, sat'd. KCl; 0.282, 1 F KCl
$AgCl + e \rightleftharpoons Ag + Cl^-$	0.222	0.228, 1 F KCl
$SO_4^{2-} + 4H^+ + 2e \rightleftharpoons H_2SO_3 + H_2O$	0.17	
$Cu^{2+} + e \rightleftharpoons Cu^+$	0.153	
$Sn^{4+} + 2e \rightleftharpoons Sn^{2+}$	0.15	0.14, 1 F HCl
$S + 2H^+ + 2e \rightleftharpoons H_2S$	0.141	
$TiO^{2+} + 2H^+ + e \rightleftharpoons Ti^{3+} + H_2O$	0.1	0.04, 1 F H_2SO_4
$AgBr + e \rightleftharpoons Ag + Br^-$	0.095	
$S_4O_6^{2-} + 2e \rightleftharpoons 2S_2O_3^{2-}$	0.08	
$Ag(S_2O_3)_2^{3-} + e \rightleftharpoons Ag + 2S_2O_3^{2-}$	0.01	
$2H^+ + 2e \rightleftharpoons H_2$	0.000	$-$ 0.005, 1 F HCl, $HClO_4$
$Pb^{2+} + 2e \rightleftharpoons Pb$	$-$ 0.126	$-$ 0.14, 1 F $HClO_4$; $-$ 0.29, 1 F H_2SO_4
$Sn^{2+} + 2e \rightleftharpoons Sn$	$-$ 0.136	$-$ 0.16, 1 F $HClO_4$
$AgI + e \rightleftharpoons Ag + I^-$	$-$ 0.151	
$CuI + e \rightleftharpoons Cu + I^-$	$-$ 0.185	
$N_2 + 5H^+ + 4e \rightleftharpoons N_2H_5^+$	$-$ 0.23	
$Ni^{2+} + 2e \rightleftharpoons Ni$	$-$ 0.250	
$V^{3+} + e \rightleftharpoons V^{2+}$	$-$ 0.255	$-$ 0.21, 1 F $HClO_4$
$Co^{2+} + 2e \rightleftharpoons Co$	$-$ 0.277	
$Ag(CN)_2^- + e \rightleftharpoons Ag + 2CN^-$	$-$ 0.31	
$Tl^+ + e \rightleftharpoons Tl$	$-$ 0.336	$-$ 0.551, 1 F HCl; $-$ 0.33, 1 F $HClO_4$, H_2SO_4
$Ti^{3+} + e \rightleftharpoons Ti^{2+}$	$-$ 0.37	
$Cd^{2+} + 2e \rightleftharpoons Cd$	$-$ 0.403	
$Cr^{3+} + e \rightleftharpoons Cr^{2+}$	$-$ 0.41	
$Fe^{2+} + 2e \rightleftharpoons Fe$	$-$ 0.440	
$2CO_2(g) + 2H^+ + 2e \rightleftharpoons H_2C_2O_4$	$-$ 0.49	
$Cr^{3+} + 3e \rightleftharpoons Cr$	$-$ 0.74	
$Zn^{2+} + 2e \rightleftharpoons Zn$	$-$ 0.763	
$Mn^{2+} + 2e \rightleftharpoons Mn$	$-$ 1.18	
$Al^{3+} + 3e \rightleftharpoons Al$	$-$ 1.66	
$Mg^{2+} + 2e \rightleftharpoons Mg$	$-$ 2.37	
$Na^+ + e \rightleftharpoons Na$	$-$ 2.71	
$Ca^{2+} + 2e \rightleftharpoons Ca$	$-$ 2.87	
$Ba^{2+} + 2e \rightleftharpoons Ba$	$-$ 2.90	
$K^+ + e \rightleftharpoons K$	$-$ 2.92	
$Li^+ + e \rightleftharpoons Li$	$-$ 3.04	

Table A-2

SOLUBILITY PRODUCT CONSTANTS

Substance	Formula	K_{sp}
Aluminum hydroxide	$Al(OH)_3$	5×10^{-33}
Barium carbonate	$BaCO_3$	4.9×10^{-9}
Barium iodate	$Ba(IO_3)_2$	1.57×10^{-9}
Barium oxalate	BaC_2O_4	1.6×10^{-7}
Barium sulfate	$BaSO_4$	1.0×10^{-10}
Cadmium carbonate	$CdCO_3$	2.5×10^{-14}
Cadmium sulfide	CdS	1×10^{-28}
Calcium carbonate	$CaCO_3$	4.8×10^{-9}
Calcium oxalate	CaC_2O_4	2.3×10^{-9}
Calcium sulfate	$CaSO_4$	6.1×10^{-5}
Cupric hydroxide	$Cu(OH)_2$	1.6×10^{-19}
Cupric sulfide	CuS	8.5×10^{-45}
Cuprous bromide	$CuBr$	5.9×10^{-9}
Cuprous chloride	$CuCl$	3.2×10^{-7}
Cuprous iodide	CuI	1.1×10^{-12}
Cuprous thiocyanate	$CuSCN$	4×10^{-14}
Ferric hydroxide	$Fe(OH)_3$	1.5×10^{-36}
Lanthanum iodate	$La(IO_3)_3$	6×10^{-10}
Lead carbonate	$PbCO_3$	1.6×10^{-13}
Lead chloride	$PbCl_2$	1×10^{-4}
Lead chromate	$PbCrO_4$	1.8×10^{-14}
Lead hydroxide	$Pb(OH)_2$	2.5×10^{-16}
Lead oxalate	PbC_2O_4	3.0×10^{-11}
Lead sulfate	$PbSO_4$	1.9×10^{-8}
Lead sulfide	PbS	7×10^{-28}
Magnesium ammonium phosphate	$MgNH_4PO_4$	2.5×10^{-13}
Magnesium carbonate	$MgCO_3$	1×10^{-5}
Magnesium hydroxide	$Mg(OH)_2$	5.9×10^{-12}
Magnesium oxalate	MgC_2O_4	8.6×10^{-5}
Manganous hydroxide	$Mn(OH)_2$	4×10^{-14}
Manganous sulfide	MnS	1.4×10^{-15}
Silver arsenate	Ag_3AsO_4	1.0×10^{-22}
Silver bromide	$AgBr$	7.7×10^{-13}
Silver carbonate	Ag_2CO_3	8.2×10^{-12}
Silver chloride	$AgCl$	1.82×10^{-10}
Silver chromate	Ag_2CrO_4	1.1×10^{-12}
Silver cyanide	$AgCN$	2×10^{-12}
Silver iodate	$AgIO_3$	3.1×10^{-8}
Silver iodide	AgI	8.3×10^{-17}
Silver oxalate	$Ag_2C_2O_4$	1.1×10^{-11}
Silver sulfide	Ag_2S	1.6×10^{-49}
Silver thiocyanate	$AgSCN$	1.1×10^{-12}
Strontium oxalate	SrC_2O_4	5.6×10^{-8}
Strontium sulfate	$SrSO_4$	2.8×10^{-7}

Table A-2 (Continued)

Substance	Formula	K_{sp}
Thallous chloride	TlCl	2×10^{-4}
Thallous sulfide	Tl_2S	1×10^{-22}
Zinc hydroxide	$Zn(OH)_2$	2×10^{-14}
Zinc oxalate	ZnC_2O_4	7.5×10^{-9}
Zinc sulfide	ZnS	4.5×10^{-24}

Table A-3

DISSOCIATION CONSTANTS FOR ACIDS

Name	Formula	Dissociation Constant, 25° C		
		K_1	K_2	K_3
Acetic	CH_3COOH	1.75×10^{-5}		
Arsenic	H_3AsO_4	6.0×10^{-3}	1.05×10^{-7}	3.0×10^{-12}
Arsenious	H_3AsO_3	6.0×10^{-10}	3.0×10^{-14}	
Benzoic	C_6H_5COOH	6.3×10^{-5}		
Boric	H_3BO_3	5.8×10^{-10}		
l-Butanoic	$CH_3CH_2CH_2COOH$	1.51×10^{-5}		
Carbonic	H_2CO_3	4.6×10^{-7}	4.4×10^{-11}	
Chloroacetic	$ClCH_2COOH$	1.51×10^{-3}		
Citric	$HOOC(OH)C(CH_2COOH)_2$	7.4×10^{-4}	1.74×10^{-5}	4.0×10^{-7}
Formic	$HCOOH$	1.70×10^{-4}		
Fumaric	*trans*-$HOOCCH:CHCOOH$	9.6×10^{-4}	4.1×10^{-5}	
Glycine	H_2NCH_2COOH	4.5×10^{-3}		
Glycolic	$HOCH_2COOH$	1.32×10^{-4}		
Hydrazoic	HN_3	1.9×10^{-5}		
Hydrogen cyanide	HCN	2.1×10^{-9}		
Hydrogen fluoride	H_2F_2	7.2×10^{-4}		
Hydrogen peroxide	H_2O_2	2.7×10^{-12}		
Hydrogen sulfide	H_2S	5.7×10^{-8}	1.2×10^{-15}	
Hypochlorous	$HOCl$	3.0×10^{-8}		
Iodic	HIO_3	1.58×10^{-1}		
Lactic	$CH_3CHOHCOOH$	1.38×10^{-4}		
Maleic	*cis*-$HOOCCH:CHCOOH$	1.5×10^{-2}	2.6×10^{-7}	
Malic	$HOOCCHOHCH_2COOH$	4.0×10^{-4}	8.9×10^{-6}	
Malonic	$HOOCCH_2COOH$	1.58×10^{-3}	8.0×10^{-7}	
Mandelic	$C_6H_5CHOHCOOH$	4.3×10^{-4}		
Nitrous	HNO_2	5.1×10^{-4}		
Oxalic	$HOOCCOOH$	6.5×10^{-2}	6.1×10^{-5}	
Periodic	H_5IO_6	2.4×10^{-2}	5.0×10^{-9}	
Phenol	C_6H_5OH	1.05×10^{-10}		
Phosphoric	H_3PO_4	7.5×10^{-3}	6.2×10^{-8}	4.8×10^{-13}
Phosphorous	H_3PO_3	1.00×10^{-2}	2.6×10^{-7}	
o-Phthalic	$C_6H_4(COOH)_2$	1.3×10^{-3}	3.9×10^{-6}	
Picric	$(NO_2)_3C_6H_2OH$	4.2×10^{-1}		
Propanoic	CH_3CH_2COOH	1.32×10^{-5}		
Salicylic	$C_6H_4(OH)COOH$	1.05×10^{-3}		

Table A-3 (Continued)

Name	Formula	Dissociation Constant, 25° C		
		K_1	K_2	K_3
Sulfuric	H_2SO_4	strong	1.2×10^{-2}	
Sulfurous	H_2SO_3	1.74×10^{-2}	6.2×10^{-8}	
Tartaric	$HOOC(CHOH)_2COOH$	9.4×10^{-4}	2.9×10^{-5}	
Trichloroacetic	Cl_3CCOOH	1.29×10^{-1}		

Taken in part from I. M. Kolthoff and P. J. Elving, *Treatise on Analytical Chemistry*, part 1, 1, 432. New York: Interscience Publishers, Inc., 1959; and in part from I. M. Kolthoff and V. A. Stenger, *Volumetric Analysis*, 1, 282. New York: Interscience Publishers, Inc., 1942.

Table A-4

DISSOCIATION CONSTANTS FOR BASES

Name	Formula	Dissociation Constant, K, 25° C
Ammonia	NH_3	1.86×10^{-5}
Aniline	$C_6H_5NH_2$	3.8×10^{-10}
1-Butylamine	$CH_3(CH_2)_2CH_2NH_2$	4.1×10^{-4}
Ethylamine	$CH_3CH_2NH_2$	5.6×10^{-4}
Glycine	$HOOCCH_2NH_2$	2.3×10^{-12}
Hydrazine	H_2NNH_2	3.0×10^{-6}
Hydroxylamine	$HONH_2$	1.07×10^{-8}
Methylamine	CH_3NH_2	4.4×10^{-4}
Piperidine	$C_5H_{11}N$	1.6×10^{-3}
Pyridine	C_5H_5N	1.4×10^{-9}
Zinc hydroxide	$Zn(OH)_2$	$K_2 = 4.4 \times 10^{-5}$

Taken in part from I. M. Kolthoff and V. A. Stenger, *Volumetric Analysis*, 1, 284. New York: Interscience Publishers, Inc., 1942.

Table A-5

DESIGNATIONS AND POROSITIES FOR FILTERING CRUCIBLES

(Nominal maximum pore diameter in microns is given in parentheses)

Type	Coarse	Medium	Fine	
Sintered glass,[1]				
Pyrex ®	C(40)	M(14)	F(5)	
Porcelain,[2]				
Coors U.S.A. ®		Medium (15)	Fine (5)	Very fine (1.2)
Porcelain,[3]				
Haldenwanger		P3(11-9) P2(9-7)	Pl(7-5)	
Porcelain,[4]				
Royal Berlin ®		A3(8)	A2(7) Al(6)	
Aluminum oxide,[5]				
ALUNDUM ®	RA766(30) RA98(20)		RA360(5)	RA84(0.1)
Fused quartz,[6]				
Vitreosil ®	1(150-90) 2(90-40)	3(40-15)	4(15-5)	

[1] Corning Glass Works, Corning, New York
[2] Coors Porcelain Company, Golden, Colorado
[3] Kern Chemical Corporation, Los Angeles, California, distributors
[4] Fish-Schurman Corporation, New Rochelle, New York, distributors
[5] Norton Company, Worcester Massachusetts
[6] Thermal American Fused Quartz Company, Dover, New Jersey

Table A-6

DESIGNATIONS CARRIED BY ASHLESS FILTER PAPERS[1]

(Tabulated are the manufacturer designations of papers
suitable for filtration of the indicated type of precipitate)

Manufacturer	Fine Crystals	Moderately Fine Crystals	Coarse Crystals		Gelatinous Precipitates	
Schleicher and Schuell	507, 590 589 Blue ribbon	589 White ribbon	589 Green ribbon	589 Black ribbon	589 Black ribbon	589-1H
Munktell[2]	OOH	OK OO	OOR		OOR	
Whatman[3]	42	44, 40	41		41	41H
Eaton-Dikeman	80	70	60		60	

[1] Manufacturers' literature should be consulted for more complete specifications
   Carl Schleicher and Schuell Company, Keene, New Hampshire
[2] E. H. Sargent and Company, Chicago, Illinois, agents
[3] H. Reeve Angel and Company, Inc., Clifton, New Jersey, agents
   Eaton-Dikeman Company, Mount Holly Springs, Pennsylvania

index

Boldface entries refer to specific laboratory instructions.

agent, 254; preparation of standard solutions of, **262**

Potential, asymmetry, 565; decomposition, 526; equivalence point, 401; formal, 394; half-cell, 374; half-wave, 603, *see also* Half-wave potential; liquid-junction, 516; measurement of, 511; oxidation-reduction, 378; standard reduction, 767*t*

Potentiometer, 511-514; accuracy of measurement with, 512, 514; for potentiometric titrations, 550

Potentiometric titration, 549-574; apparatus, 550-552; automatic, 570; of a carbonate-bicarbonate mixture, **573;** determination of end point for, 553; general directions for, **571;** of halide mixtures, 558, **572;** for oxidation-reduction reactions, **573;** of weak acids, **572**

Potentiostat, 543

Power, radiant, 639

Precipitate formation, mechanism of, 157, 162; steps in, 160-161

Precipitates, contamination of, 175-179; colloidal, 164-172; crystalline, 172-174; manipulation of, **115-120;** particle size of, 156-174; rate of formation, 151

Precipitating agents for gravimetric analysis, 193-199, 200*t*

Precipitation, analytical separations employing, 744; gravimetric methods employing, 72; homogeneous, 188, 746

Precision, of data, 35; prediction of, for analytical processes, 54

Primary standard, 212, 213; for acids, 344; for bases, 348; in oxidation-reduction reactions, 423*t*

Prism as a monochromator, 650

Proportional error, effect upon results, 39

Q test, 58

Quinoline yellow as indicator for bromate titrations, 468

Radiant intensity, 643*t*

Radiant power, 639, 643*t*

Radiation, absorption of, 637-639; buffers, 683; detectors, 653-656; electromagnetic, 635

Range of data, 35

Reagent chemicals, purity, 109-110

Real substances, 687-702; analysis of, 687, attainable accuracy in, 700; selection of methods for, 689, 693

Recovery ratio, 742

Reducing agents, 367, 419-430, 485-499; for pretreatment of samples, 425-430; primary standards for, 423*t; see also* specific listings

Reduction potential, 378, 516; standard, 767*t*

Reductors, 425-429

Reference electrode, 506; for polarographic analysis, 601; for potentiometric titration, 550; standard hydrogen, 374; *see also* specific listings

Refraction of light, 650

Rejection criteria, 57-60

Relative half-cell potential, 374

Relative supersaturation, 157

Reprecipitation, minimization of contamination by, 181, 186

Residual current, polarographic, 603, 614

Resins, ion-exchange, 761

Resistance, measurement of, 514

Rest point of balance, 89

Retention time, 765

Reversibility, electrochemical, 371

Richards method, calibration of weights, 89, **96**

Rider of analytical balance, 80

Roll and quarter, method of sampling, 708

Rotating platinum electrode, 627

Rotational level, transitions in, 637

Ruthenium, separation by distillation, 757

Salicylaldoxime as precipitating agent, 201

Salt bridge, 376; effect upon liquid-junction potential, 517

Salts of weak acids and bases, calculation of *p*H, 289-292; conductometric titration of, 583; polybasic, 324-325, 337-338; titration curves for, 315-317

Sample composition, and selection of analytical method, 691

Sampling, 703-708; of liquids, 707; methods for isolation of gross sample, 707; steps in, 704

Saturated calomel electrode, potential of, 507*t*

Schöniger apparatus for decomposition of organic compounds, 733

Schulze-Hardy rule, 170

Semimicro balance, 87

Sensitivity of analytical balance, 82-84

Separations, analytical, 739-765; based upon solubility differences, 146

Short swings, method of, for weighing, **92**

Sign convention, electrode potentials, 377; electrode processes, 373

Significant figures, 60-64

Silica, separation by distillation, 757

Silver as a reductant, 427*t*

Silver chloride, photodecomposition of, 202

Silver coulometer, 516

Silver nitrate, preparation of standard solutions of, **261**

Silver-silver chloride electrode, 508

Single-beam photometer, 658

Single-pan balance, 86; operation of, **92**

Sintered glass as filtering medium, 106, 107

Sodium, decomposition of organic matter with, 736; determination of, in natural waters, 683

Sodium bismuthate, as preoxidizer, 424

Sodium carbonate, as basic flux, 729; as primary standard for acids, 344, **345**

Sodium oxalate, as primary standard, for cerium (IV) solutions, 452, **453;** for permanganate solutions, 436, **437**

Sodium peroxide, as oxidant for organic compounds, 735; as oxidizing flux, 730*t;* as preoxidizer, 424

Sodium tetraphenylboron, 198

785